W9-CVA-035

# PHYSICS
## An Introduction

JAY BOLEMON

*The University of Central Florida*

Illustrated by Ken Snelling

*Prentice-Hall, Inc., Englewood Cliffs, New Jersey 07632*

*Library of Congress Cataloging in Publication Data*

Bolemon, Jay.
Physics: an introduction.

Includes index.
1. Physics. I. Title.
QC21.2.B65 1985 530 84-23720
ISBN 0-13-672221-0

**Editorial/production supervision: Maria McColligan**
**Development editors: Jeannine Ciliotta, Raymond Mullaney**
**Interior and cover design: Christine Gehring-Wolf**
**Cover photo: Ted Cooper**
**Manufacturing buyer: John Hall**
**Credits for photographs begin on page iv.**

© 1985 by Prentice-Hall, Inc., Englewood Cliffs, New Jersey 07632

All rights reserved. No part of this book may be
reproduced, in any form or by any means,
without permission in writing from the publisher.

Printed in the United States of America
10 9 8 7 6 5 4 3 2

ISBN 0-13-672221-0 01

Prentice-Hall International, Inc., *London*
Prentice-Hall of Australia Pty. Limited, *Sydney*
Editora Prentice-Hall do Brasil, Ltda., *Rio de Janeiro*
Prentice-Hall Canada Inc., *Toronto*
Prentice-Hall Hispanoamericana, S.A., *Mexico*
Prentice-Hall of India Private Limited, *New Delhi*
Prentice-Hall of Japan, Inc., *Tokyo*
Prentice-Hall of Southeast Asia Pte. Ltd., *Singapore*
Whitehall Books Limited, *Wellington, New Zealand.*

**to Mildred and Joe and Gertie and Alex,**
*who taught with love.*

# CREDITS

Figs. 1-1, 1-5, 3-25, 5-5, 6-16, 6-19, 6-25, 9-8, 9-12, 10-3, 17-6, 28-9: Deidre M. Bolemon
Figs. 1-9, 1-10, 2-3a,b, 4-27, 10-16, 11-12a, 11-18, 23-1, 23-14, 23-15: Dale Nichols
Figs. 3-4, 3-9, 4-13, 7-16, 12-1, 14-7: NASA
Figs. 7-24, 7-25, 7-26, 16-22: NASA/JPL
Figs. 7-21, 8-26, 15-1a,b, 15-13, 16-19, 16-21, 28-7: NOAA
Fig. 1-6: *Emphasis,* The University of Central Florida, and Tom Netsel
Figs. 1-11, 1-12, 5-7, 5-8, 5-16: H. E. Edgerton
Fig. 2-5: *PSSC Physics,* 3rd ed., 1971, p. 248, Lexington, Mass.: D. C. Heath & Co.
Fig. 3-11: Library of Congress
Fig. 3-16: Thanks to Mike Bolte/Brick courtesy of Bill and Judy Green
Fig. 4-1: *Towayam,* Winter Park High School
Figs. 5-1, 5-15, 10-26: Allen Mathews
Figs. 5-4, 28-6: American Petroleum Institute
Fig. 6-15: Joern Gerdts, Photo Researchers, Inc.
Fig. 7-11: Hughes Aircraft Company
Fig. 8-8: J. P. Hartman, University of Central Florida
Fig. 8-12: Jim and Barbara McConville
Fig. 8-14: Courtesy of Ellen Yates and Paul Doherty
Fig. 10-1: Scala/Art Resource
Fig. 10-2: Bureau of Reclamation, U.S. Department of the Interior
Fig. 10-4: Richard Shepherd, *The Orlando Sentinel*
Figs. 10-11, 23-18: Robert J. Willmann
Fig. 10-13: Innovative Technology International Inc.
Fig. 10-15, 10-22, 11-13: Susan Weinstock
Fig. 10-16: Dale Nichols
Fig. 10-25: General Electric Research Laboratory
Fig. 10-26: Allen Mathews
Fig. 12-11: Philips Electronic Instruments, Inc.
Fig. 13-16: RCGA
Fig. 14-2: Michael J. Bolte
Fig. 14-13: Westinghouse Steam Turbine Generator Div.
Fig. 15-2: Computer calculations courtesy of Rodney Hamilton
Fig. 15-8: M. Timothy O'Keefe
Fig. 15-11: South Carolina Commission of Forestry, courtesy of Charles R. Fernell
Fig. 16-16: *PSSC Physics,* 2nd ed., 1965, D. C. Health & Co. with Education Development Center, Newton, Mass.
Fig. 17-8: Adapted from a cartoon by William C. Oelfke
Fig. 18-10: Adapted from *The Lightning Book,* Peter E. Viemeister, The MIT Press, Cambridge, Mass.
Fig. 20-16: Adapted from *Physics Today,* Vol. 28, No. 8, "Magnetic Fields of the Human Body," 1975, p. 41, Figure 8.
Fig. 20-17: D. Balkwill, D. Maratea
Fig. 21-3: Fermilab
Fig. 21-6: Lockheed Solar Observatory
Fig. 22-3c: *Atlas of Optic Phenomenon,* Springer-Verlag, N.Y., 1962
Fig. 23-12: Computer calculations courtesy of Mark Woodyard
Fig. 23-19b: AT&T Laboratories
Fig. 24-2: Photograph from the Hale Observatories
Fig. 24-12: International Laser Systems, a division of Litton
Fig. 24-13: McDonald Observatory
Fig. 24-14: Professor David C. Auth, University of Washington
Fig. 25-3: Computer graphics courtesy of Paul Doherty, Oakland University
Fig. 26-8: New York Public Library
Page 532: General Electric Research and Development Center
Fig. 27-7: Specimen in the Princeton University Museum of Natural History, Willard Starks
Fig. 27-11: Fermi National Accelerator Laboratory
Fig. 28-3: Florida Power Corporation
Fig. 28-5: Los Alamos National Laboratory
Fig. 28-8: Pacific Gas and Electric Company

# CONTENTS

# TO THE STUDENT

So, you are in a physics course! If you've never taken such a course before, this book is especially for you. It's for students who aren't majoring in a science and who probably won't take another physics course.

Physics is about the world that is right in front of your eyes. It's about matter and light. It's about what things are made of, how they are put together, and how they interact and change. If you've ever wondered why the setting sun is red or why the daytime sky is blue, you will find the answers here. You will learn why there are waves and tides on the oceans and why the front end of your car dips as the car stops. You will find out about the sparks that sometimes leap from your fingers after you've crossed a carpet and about the electricity you use every day. You will read about atoms, the particles they are made of, and the peculiar ways those particles behave. As you use this book, you will probably come upon many things that you didn't know were going on around you. I hope so.

I also hope you will see that physics is a human adventure, one that you can enjoy by understanding its descriptions of your world. Because the physical world around us follows rules and patterns, physics can often make excellent predictions of how things will happen. But in your course you will probably be more concerned with understanding the ideas of physics than with making accurate predictions. Common experience makes you familiar with some of those ideas, and we'll be using your experiences as starting points in your study of physics.

## How to Learn from This Book

Students who had success studying from drafts of this book gave me some tips I'd like to pass along. The teacher in me can't resist.

First of all, the method in the cartoon won't work. Understanding physics can't come that easy for several reasons. Vocabulary, for example—physics has a language all its own. Here's a hint on how to pick it up. To begin, read each assigned chapter quickly. That is, skip any material in boxes, don't fret over any formulas, and stop at the end of the chapter narrative. (Ignore the calculations, review, exercises, and so on.) Then put the book down for an hour or even a day. If you don't understand the material, don't worry. The words, definitions, and examples will be more familiar the second time you read through.

When you reread the chapter, go much more slowly. If there's no one around, read any definitions and equations out loud. Go over the boxed material, too. When you feel comfortable with the chapter (perhaps after a third reading), you are ready for the Review, which follows the tinted Calculations section. You should be able to answer almost all of the review questions without looking back into the chapter. If you can't, put the book down and come back later for another pass. You'll retain more of the material each time you read it.

Once you've done the review, you are ready for some exercises. The exercises let you apply the ideas of the chapter to extra examples. Don't be discouraged if you can't easily answer all of these completely; sometimes they are related to things not in your background, and you'll learn from reading the answers. Exercises with numbers in color have answers at the back of the book.

If you are asked to study the calculations section (or any part of it), save that for last. Everything you'll need for a calculation is on those tinted pages or in the two appendices of the book. Exercises that require calculations are marked with asterisks, and *all* of those have answers at the back. That's about all the advice I have—except that you can't start too early. Last minute studying doesn't work well in physics courses. Begin now, and you'll learn more. Happy studying!

# TO THE INSTRUCTOR

This physics text for nonscience majors follows the traditional order of topics, but with extra features not found in traditional texts. It attempts to build on the habits and experiences of the students and also to build a bridge between the instructor and the students. I've tried to explain the basic ideas well enough so instructors won't need to spend time covering every concept in the lectures, leaving time for them to go over the topics they are most enthusiastic about—especially any that aren't discussed in the book.

Each chapter opens with a vignette on one or more points that will be covered in the chapter. Whether the vignette is historical or a contemporary true story, or an exposition, it is designed to nudge the reader into the physics of the chapter. Science majors learn to read slowly as they study, pondering any tough points as they go along. But nonscience majors are accustomed to reading faster; these chapters were written to permit faster reading. Material that would interrupt the flow of the chapter is enclosed in boxes, and in a quick reading or review, the student can skip those to preserve continuity. Examples involving calculations, normally another stumbling block, are grouped in a "Calculations" section at the end of each chapter. Besides making the chapter easier to read, having the calculations on a few pages means that students don't have to flip back and forth in the chapter looking for help with an example.

Grouping the calculations at the end of the chapter also makes this text easy to adapt to an instructor's course. The chapter discussions are restricted to the concepts and ideas of physics and are independent of the calculations; so instructors who use little or no math in their courses can skip the calculations entirely. Or examples can be assigned that match the math level of a given group of students. (I rarely assign all of the calculational examples in these sections; see the instructor's manual for suggestions.)

To give students immediate feedback, answers to selected exercises are in an answer section at the back of the book. (Those exercises to be answered have a color exercise number.) Some exercises serve as further teaching examples and cannot be answered fully just by reading the chapters. These exercises are always answered, and they serve to point out to the students that physics is found everywhere and not just on a printed page.

One optional feature of this text is the second chapter, which includes independent directions of motion and vectors. If you want, you can skip this chapter since the next chapter briefly summarizes vectors and how they add. Or you could assign Chapter 2 for extra reading, as I've done several times, and have your students answer only the review questions. That worked well, although I prefer to discuss the chapter in class. This text was designed to permit the skipping of other chapters, too, since there is more material here than most instructors would cover in a semester. However, the historical interludes that precede Chapters 3, 7, 24, and 27 should be assigned since they contain background information for following chapters.

There are over 1400 review questions and exercises, an ample number for alternating assignments between consecutive semesters. The student response to the more than 100 demonstrations has been excellent, so be sure to look them over when you are making assignments. In the instructor's manual for this text, I've included some of my favorite digressions, along with numerous tips on effective lecturing, demonstrations, and a large class-tested set of exam questions at three levels of difficulty. If you use this book, don't fail to use the instructor's manual.

## Acknowledgments

This book came about because of daily inspiration provided by students in my introductory physics courses at the University of Central Florida, and it matured under their scrutiny. I am indebted to them both for their able criticisms and their enthusiastic responses. I began writing for them after long encouragement from H. Michael Snell, who, like myself, never imagined how long the project would take. The writing style owes much to Autumn Stanley, whose early constructive criticism made a deep impression. Harriet Bolemon acted as a developmental-editor-in-residence and student advocate extraordinaire for the last 3½ years of writing and rewriting. As a nonscientist, she tried to make sure that her fellow readers would be able to understand each and every sentence. Logan Campbell, former executive editor for science at Prentice-Hall, acquired the project in 1981 and gave excellent guidance and advice. Doug Humphrey ably stepped in after Logan's departure and continued pushing me along, good naturedly perservering despite all the missed deadlines.

Many friends and acquaintances read portions of the evolving manuscript and suggested changes, often contributing examples. Among those were Tim Bandy, Michael Bolte, Doreen Christiani, Diane Froning, Charles Hargraves, Allen Mathews, Donna McKenna, Dale Nichols, and Denise Sellers. George Burgess, Paul Doherty, Allen Mathews, Pat O'Hara, and Dave Trivett shared their real-life adventures that appear as chapter openings. Ron Collins, M.D., and John Popp, M.D., advised on several medical applications of physics. Mark Woodyard did the computer calculations for the sunsets on earth and Venus, and Rodney Hamilton provided the computer scenes of earth to show the seasons.

At various stages a number of people reviewed a part of the manuscript, and their suggestions and criticisms are gratefully acknowledged. Among those reviewers were: Fred Goldberg, West Virginia University; Ruth H. Howes, Ball State University; Philip B. James, University of Missouri–St. Louis; Patrick F. Kenealy, Wayne State University; Peter F. Michelson, Stanford University; William J. Mullin, University of Massachusetts; Jack Prince, Bronx Community College; David M. Riban, Indiana University of Pennsylvania; Jim Watson, Ball State University; and E. W. Winters, San Antonio College. In addition, Paul Doherty, Oakland University, reviewed the entire next-to-last draft of the manuscript.

Four brave reviewers waded through the last two drafts of the manuscript: Jack Brennan, University of Central Florida; Robert Cole, University of Southern California; Lawrence C. Shepley, University of Texas at Austin; and Sheridan A. Simon, Guilford College. The book benefited enormously from their expert and detailed attention. The responsibility for any errors that remain is of course mine, and I'd appreciate hearing from you regarding corrections, comments, and suggestions.

Raymond Mullaney of Prentice-Hall took care of literally hundreds of details, corrected slips of my pen, and handled more than his fair share of worry with this book. Maria McColligan, also of Prentice-Hall, gave new meaning to the word "patience" while taking this book through production. Margaret Parrott of Prentice-Hall gave enthusiastic encouragement at every turn for 3 years. My wife and I at times enjoyed the warm hospitality of Susan Snell, Eileen Campbell, and Judy Snelling during the book's writing. Ken Snelling rendered artwork at all hours, magically transforming crude sketches or sometimes just ideas into effective illustrations. Dale Nichols and Deidre Bolemon contributed the fruit of many hours of their photographic skills to these pages. Thanks to you all.

Ron Edge's *Experiments with String and Sticky Tape* was the source of several of the demonstrations, and readers of *The Physics Teacher* will recognize a few ideas inspired by articles in that journal, which no teacher of physics should be without.

A bouquet of roses to Barbara Anne Bolemon, who had to use earplugs and headphones for the TV and stereo for the duration! And to Deidre and Kevin Bolemon who put up with the typewriter that sat in the middle of their dining room table. I would also like to thank my university for providing a sabbatical leave to get this project underway, and my colleagues in the Physics department who gave me so much encouragement.

Jay Bolemon

# CHAPTER 1
# MOTION

Imagine being alone on the plains of Mars for a day and a night. Astronomers have long known that Mars turns beneath the stars much as the earth does, in just under 25 earth hours. Viking's color photographs from the surface have shown us what you would actually see. Because the thin atmosphere scatters little predawn light, day breaks suddenly. The purple sunrise quickly brightens into the pinkish orange of the daytime sky, colored by the dust from windstorms. As the sun moves across the sky, the shadows of the barren rocks, like shadows on sundials, creep across the sand, tracing the hours of the day. A brief sunset brings on the deep cold of night and brighter starlight than we ever see on earth.

As on earth, the stars and any visible planets move slowly across the night sky, but on Mars you would see two moons, one rising in the east and the other in the west. Phobos, the inner moon, would look about one-third as large as our moon does from earth. Because of its greater distance, Deimos, the outer moon, would look like just a bright point of light. Since their orbits keep them very close to Mars, they often pass through its shadow, so you might very well see an eclipse.

Besides the blowing dust and the heavenly bodies, little else moves on the Martian landscape. This lack of movement might seem to be strangest of all, for we humans are used to motion. Almost from birth, infants follow motion with their eyes, and from then on we are continually aware of things moving about, starting, stopping, turning, bouncing. On earth we see liquids flowing, people moving, and the wind stirring the leaves of trees. Although we cannot see them, we know the very atoms and molecules of matter are constantly in motion. Even mosses and lichens that spend their lives fastened to rocks depend on the movements of gases and liquids to bring them the chemicals essential to life and to carry others away. We take part in motion every day of our lives. When we walk or run, ride on bicycles or in cars, we can describe and compare this motion in terms of *speed* and *acceleration*. In this chapter and the next, we will see how these two quantities describe the motion of things. Chapter 1 deals with how things cover distances, and Chapter 2 discusses directions and turns. ■

## Speed

When you wake in the morning, it's as if your brain is asking, "What's happening?" Your mind digests the signals from your senses and you become aware of your surroundings. Although these mental processes take place without any apparent motion, most things that "happen" around you involve something moving. To live, just to breathe and walk and talk, means motion, and the study of motion, how things move and why, is a large part of physics.

If we just say that something *moves,* someone else will not really know "what's happening." It is one thing to recognize motion but another to *describe* it. To describe motion accurately, we use rates. A **rate** is a quantity divided by a time, so it tells how fast something happens, or how much something changes in a certain amount of time. Suppose a girl runs a course that is 3 miles long. She might sprint at the beginning but tire and slow down along the way, or even stop to tighten a shoelace, so she won't travel at the same rate for the entire 3 miles. But if she finishes in, say 30 minutes, then the ratio

$$\frac{3 \text{ miles}}{30 \text{ minutes}} = 0.10 \, \frac{\text{miles}}{\text{minute}}$$

is the average rate of travel during that time, or her *average speed.*

$$\textbf{average speed} = \frac{\text{total distance covered}}{\text{time used}}$$

The average speed tells little of what actually happened during her run, however. If we are curious about her speed at one certain time or at a specific point along the way, we want to know her *instantaneous speed,* that is, how fast she was moving at one instant, not over a period of time.

**instantaneous speed** = the rate at which something is traveling at a specific instant (or at a specific point)

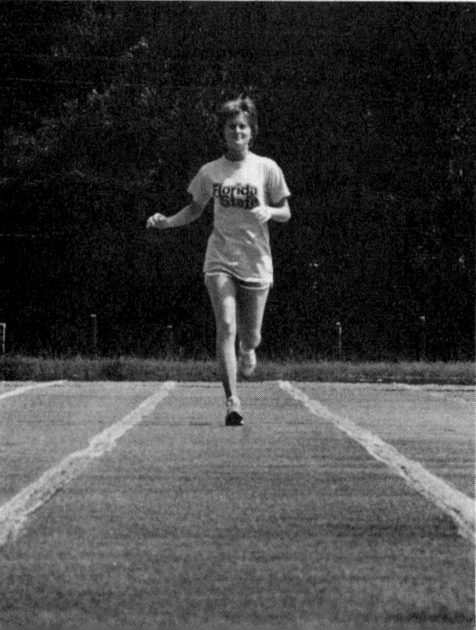

**FIGURE 1–1**
A runner who stays in a straight lane has motion along a single direction, which is the subject of this chapter.

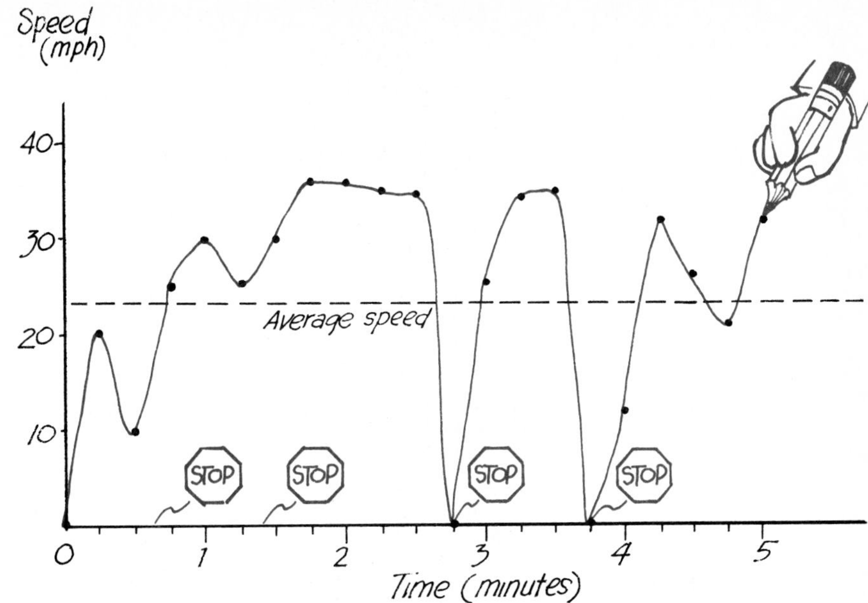


**FIGURE 1–2**
Drawing a graph of speed versus time for a car trip. A smooth line connecting the data points can give a general idea of the car's motion. But the interpolating curve can miss details of the speed in the intervals between data points, as it did in this example.

With just a glance at the speedometer, you can get an estimate of your car's instantaneous speed. If you then watch the road awhile before looking back down, the speed may have changed a bit. However, if someone is jotting down these speeds, say, every 15 seconds, at the end of the ride you could make a graph that shows a lot about the car's motion. Figure 1–2 shows actual data taken at 15-second intervals during a 5-minute trip in a car.

You can get an idea of what the speeds might have been at other instants during the trip if you connect the points on the graph with a smooth line as shown. Where the speed changed quickly, however, such a guess can be misleading. For example, at two of the readings the car was resting at stop signs. But there were two other stops made by the car, at about 40 seconds and 85 seconds into the trip. These appear only as dips in the connecting (or *interpolating*) curve, for the simple reason that they happened entirely within the 15 seconds between two readings.

According to the odometer, the car traveled about 1.95 miles during the 5-minute trip, so its average speed over that time was about

$$\frac{1.95 \text{ miles}}{5 \text{ minutes}} = \frac{1.95 \text{ miles}}{1/12 \text{ hour}} = 23.4 \, \frac{\text{miles}}{\text{hour}}$$

Notice that the graph shows that the car rarely traveled at its average speed. Only if a car's motion is perfectly uniform over some distance (that is, if it moves at a *constant* speed) will its instantaneous speed be the same as its average speed for that time.

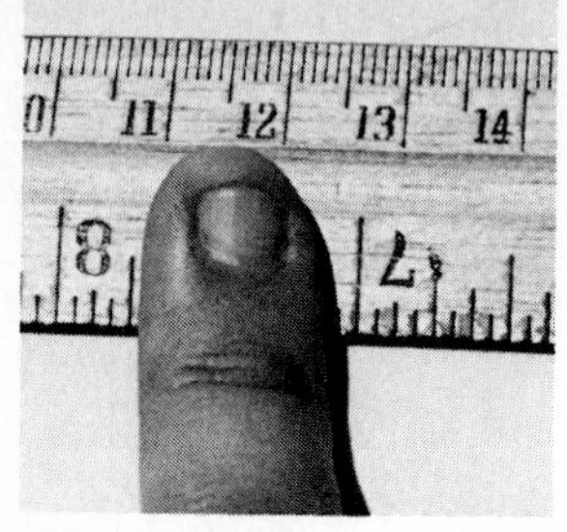


*(a)*

*(b)*

**FIGURE 1–3**
*(a)* A centimeter is about the distance across the fingernail of a person's smallest finger. *(b)* This yardstick and meter stick are aligned at their other ends. This lets you see that a meter is about 9 percent longer than a yard by inspection.

## UNITS OF LENGTH AND TIME

We measure length and time by comparing them to standard units that are picked for their convenience. Two systems of measurement, the English and the metric, are in use today. The familiar English system has the more curious history. An inch was once the distance between the knuckles of the thumb of King Edgar (A.D. 959–975), and the yard was the distance between the outstretched fingers and nose of Henry I (A.D. 1100–1135). The metric system, set up in France soon after the French Revolution, makes no reference to the limbs of nobility. The metric length, the meter, had a curious birth, however. In the 1790s, the scholars in charge carefully estimated the distance between the North Pole and the equator and then proceeded to divide that distance by 10, and then by 10 again, until they came to a length convenient for people to use. After seven divisions by 10, they reached a length equal to 39.37 inches, only a little over a yard in the English system. Voilà—the meter! For convenience, the units of the metric system are related by powers of 10. For example, the meter is equal to 100 of the smaller units called centimeters. For larger distances, we use 1000 meters, the kilometer.

Both the English and the metric systems use the second as the unit of time. The second originally was 1/3600 of an hour, itself 1/24 of an average solar day. But today the standard units are set by the motions of electrons in atoms so they can be accurately reproduced. The modern second flits by in just 9,192,631,770 oscillations of a certain electron in a cesium atom. Precise atomic clocks tick off each second accurately to all those digits. Likewise, the modern definition of the meter is 1,650,763.73 wavelengths of orange light given off by an electron in a krypton atom. It's exotic, but it's more accurate than using someone's arm for a yardstick or measuring time by collecting water flowing from an elevated bucket, as Galileo had to do. For everyday use, of course, less accurate but simpler clocks and meter sticks do just fine.

The following is a short list of the abbreviations of units of measurement:

| | | |
|---|---|---|
| second, s | meter, m | foot, ft |
| hour, h | kilometer (1000 m), km | inch (1/12 foot), in. |
| year, yr | centimeter (1/100 m), cm | yard (3 feet), yd |
| | | mile (5280 feet), mi |

### About Measurements

Speed is a rate that tells how fast something moves, but to *find* a speed requires measurements. Whenever you measure something like a distance or a time, you simply compare two quantities—one of them being a standard, as when you use a tape measure to determine the width of a room or the length of a shelf. No such comparison can be perfectly accurate; there are always uncertainties in the answer when something is measured. For example, the length of a steel tape measure varies with changes in temperature, so an error may be brought in when that particular standard is used for measuring length. But even if the standard is very accurate, human judgment is also a factor. Is the tape straight and taut? Exactly where between the finest marks on the tape is the edge of that shelf? Give the same tape measure to a few of your friends and ask them individually to measure the length of a table as precisely as they can. You'll get a variety of answers, slightly different, even though they use the same equipment.

*Errors are a part of any measurement, but the word "error" doesn't mean the experimenter has made a mistake or a blunder. Because of equipment and human judgment, uncertainties persist even when the most cautious experimenters make the measurements.*

As another example, suppose you are challenged to find the speed of a moving bicycle. You might make two marks on the road and use a stopwatch to see how long the bike takes to travel between them. Dividing the distance between the marks by the time shown on your stopwatch gives the bike's average speed. But this will be only an estimate of the speed, of course. There is a limit to how accurately you can measure the distance, and no stopwatch is perfectly accurate, but it's likely that *you* will be the weakest link in this experiment for yet another reason.

**FIGURE 1–4**
Finding a speed involves measurements of both distance and time.

Even though light from the bicycle reaches your eyes almost instantly, receptors in the eyes need a fraction of a second to turn the information they receive into electrical signals. These signals take another fraction of a second to travel along a network of nerves to your brain. Your brain then interprets what your eyes have seen and makes a decision to act. It sends an impulse to your hands. When this impulse arrives, the muscles contract. Only then does the watch start or stop.

All of this takes time. This eye-to-hand reaction time is normally about 1/7 second. So when you start and stop the watch, there is an uncertainty in your measurement. But that doesn't mean there will necessarily be as much uncertainty as 1/7 second. After a few tries you would be able to judge the speed of the approaching bike and *anticipate* when the bike will hit the marks and react accordingly. (A similar anticipation allows the members of a symphony orchestra or a rock group to perform together smoothly—once established, the beat or rhythm of the music lets them anticipate when to play each note.)

## More about Instantaneous Speed

When a car approaches a street sign, the edge of its bumper will reach the sign at some instant. When it does, that bumper is moving at a definite rate—its instantaneous speed at that point. As simple as this sounds, it can be confusing. An instant is merely a point in time, such as 12 noon. It measures no time interval at all, not even the smallest fraction of a second. At one instant, then, there is no time for something to move, to change its location. If this is true, how can the bumper be traveling at, say, 30 miles per hour?

Of course, the car *does* have a speed at every point during its motion. Don't confuse a solitary instant of time with the time interval you would need to measure its speed. You cannot measure anything's speed at a single instant; there must be a time interval and a distance covered. In other words, you can only *estimate,* not measure, an instantaneous speed.

For example, here's a way you could estimate the instantaneous speed of a bicycle as it moves past a single mark. Draw two parallel lines very close together with that mark in the center, and arrange for improved timing equipment. Then find the average speed over that small distance. If the distance (and the time interval) are small enough, your measurement of the average speed will be a good estimate of the bike's instantaneous speed at the center mark.

### HOW A CAR'S SPEEDOMETER ESTIMATES SPEED

When a car moves, each revolution of its wheels takes it a certain distance down the road. One rotation of the axle carries the car a distance roughly equal to one circumference of a tire. For example, 100 rotations in a minute would move the car about 100 tire circumferences in that time. In other words, *any* rotation rate of the axle corresponds to a certain speed for the car. Basically a car's speedometer mechanically detects the rotation rate of the car's axle. The indicator on the instrument panel is set to show what speed the car will have *if* the car has standard tires inflated to the standard pressure. When the axle turns at a rate that corresponds to 30 mph with standard tires, the speedometer will indicate 30 mph. But suppose a vehicle has oversized tires with a larger circumference (Fig. 1–5). When the axle rotates at the rate that should cause a 30-mph speed, the speedometer will show 30 mph as usual. But each revolution of the larger wheels takes the car farther than a standard tire would, so the car's true speed will be greater than 30 mph. In this case the speedometer *understates* the car's speed. Tires of smaller circumference would move the car a shorter distance for each rotation of the axle and the car would go slower than its speedometer indicates. (See Exercises 28 and 29 at the end of this chapter for more details.)

**FIGURE 1–5**
The nonstandard wheels on this truck will give different odometer/speedometer readings than standard-sized wheels would unless a special odometer/speedometer mechanism is installed.

## Acceleration

Take a ride! Ease your car away from its parking place and reach a steady speed, and when the road is straight and smooth, the ride is very comfortable. As a passenger, you could read a book or pour a cup of tea and drink it; if you were in a van or a large motor home, you could even play a game of darts.

Alas, it is not easy to keep a car's speed steady. Even when the road is straight and without bumps or dips, traffic and the inevitable stop signs and traffic signals make us change speeds. A book you are holding leans forward if the car slows down and then backward if it speeds up. If there is a cup of tea aboard, it sloshes about. Any deviation from a constant speed affects our bodies, too; we shift backward or forward in our car seats, so we *feel* these changes in speed. If the speed

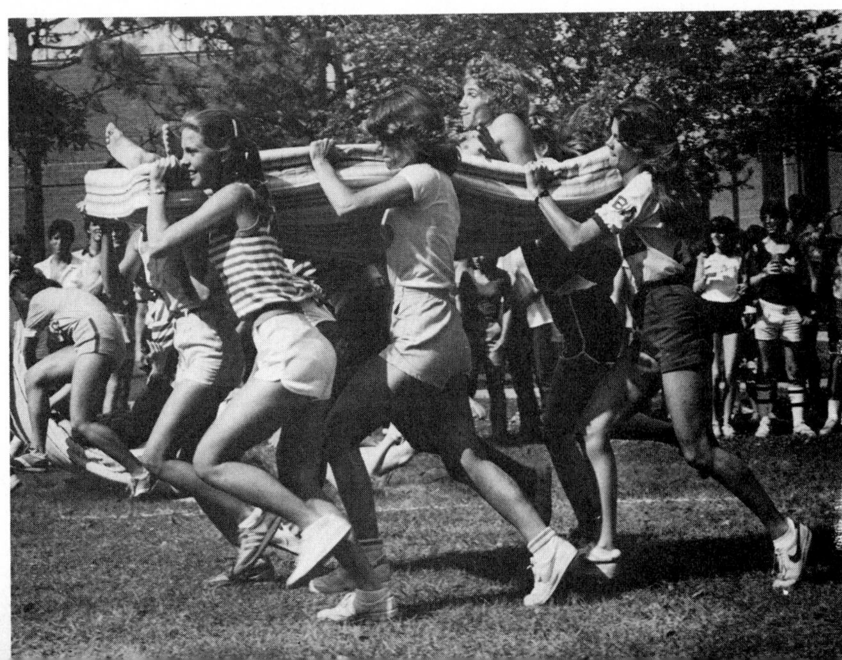

**FIGURE 1–6**
Changing speed makes life interesting. Think about it. Motion with a constant speed is rare in our world. Something's rate of change of speed is called its acceleration, the subject of this section.

**There is, in nature, nothing older than motion, concerning which the books written by philosophers are neither few nor small; nevertheless I have discovered by experiment some properties of it which are worth knowing.**

Galileo Galilei, 1564–1642

changes slowly, we hardly notice it, but any quick change in speed is obvious. It is *how fast* speed changes that matters to us, and that's another rate—the rate of change of speed. We call this rate *acceleration*.

$$\textbf{acceleration}\ \text{(along a straight line)} = \frac{\text{change in speed}}{\text{time required for that change}}$$

Just as for speed, this is the *average* acceleration over a period of time. The instantaneous acceleration tells how fast the speed is changing at any point in time.

You instinctively know a lot about acceleration, for your training starts early and without choice. A new baby picked up or put down too quickly stiffens and throws out its arms and legs, frightened by the sudden unfamiliar acceleration. As the baby begins to play and walk, it copes with the downward acceleration of a falling toy and learns to take a tumble. Later in life this baby will take things like gravity and the accelerations of daily traffic in stride. If you watch a movie filmed from a roller coaster or a race car, you will see how well you've been trained. At all the appropriate places, your mind *expects* accelerations and you feel them in the pit of your stomach even though your stomach isn't actually accelerating.

## Finding an Acceleration

Suppose someone in a dune buggy goes from 22 feet per second to 88 feet per second in exactly 6 seconds. The acceleration is then

$$\frac{\text{change in speed}}{\text{time required}} = \frac{\text{final speed} - \text{initial speed}}{\text{time required}}$$

$$= \frac{\text{88 feet per second} - \text{22 feet per second}}{\text{6 seconds}}$$

$$= \frac{\text{66 feet per second}}{\text{6 seconds}}$$

Notice that the quantity "change in speed" has the units of speed, that is, length over time. These units must then be divided by the units of "time required" to give the units of acceleration. So acceleration always has the units of length over time *over time*. That's easy to remember if you think of it this way: Acceleration is a rate of change of another rate, the speed. Continuing with the calculation, we see that the value of the dune buggy's average acceleration is

$$\frac{\text{66 feet per second}}{\text{6 seconds}} = \text{11 feet/second/second}$$

When two units of time are the same (as they are here), the unit of acceleration can be abbreviated as length/time$^2$. So another way of writing this acceleration is 11 feet/second$^2$.

The word "acceleration" most often brings to mind an increase in speed. But acceleration is a *change* in speed over time, so when anything slows down it is also accelerating. To distinguish slowing down from speeding up, we use the word *deceleration*. The numerical value of an acceleration shows us when this happens. Suppose that dune buggy slows from 40 miles per hour to 25 miles per hour in, say, 5 seconds. Its change in speed is its final speed (25 mph) minus its initial speed (40 mph), or −15 mph. The acceleration is then

$$\frac{-15 \text{ mph}}{5 \text{ seconds}} = -3 \text{ mph/second}$$

So *a negative value of the acceleration means the speed is decreasing,* and we say the dune buggy is **decelerating.**

During an ordinary trip in a car, we've seen that the instantaneous speed changes greatly as time goes by. Normally the instantaneous acceleration changes in a nonuniform way as well. It's very difficult to arrange a steady, or constant, acceleration. For example, suppose you wanted to accelerate a bicycle at a constant rate. On a smooth, straight and level track, you'd have to take exactly the same amount of time to go from 5 mph to 10 mph as you took to go from zero to 5 mph, and so on. Impossible? Perhaps. But in one familiar circumstance, many objects do accelerate uniformly—in the vertical direction.

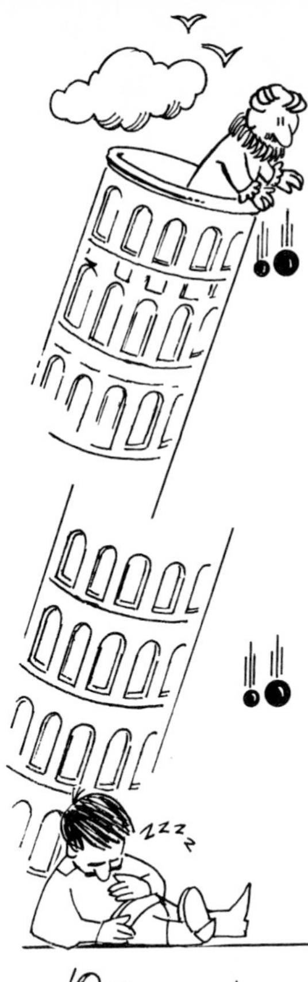

**FIGURE 1–7**
Gee! Legend has it that Galileo dropped two heavy metal balls of different sizes from the Leaning Tower of Pisa to prove they had the same acceleration. If the balls were dropped at the same time they would have reached bottom at almost exactly the same time, having gained speed at the same rate.

## The Acceleration of Gravity

As children we all learned that most things fall to the ground if we let them go in midair, and we use the word "gravity" for this natural acceleration. The value of this acceleration toward the earth is given the symbol ***g*** ("gee") and amounts to about 32 feet per second *per second,* or about 9.8 meters/second/second. In other words, the speed of a freely falling object increases by 32 feet per second for each second that it falls.

But as children we also learned that feathers and small bits of paper accelerate more slowly toward the ground than do toys or shoes. This doesn't mean that gravity's acceleration is different for feathers than for shoes. But if you think so you would be in good company, from Aristotle onward for more than a thousand years.

Galileo Galilei (1564–1642) was the first to understand how earth's gravity affects things near the surface of our planet. From his experiments he argued that if different objects fell "totally devoid of resistance" (without air or anything else to hinder their downward motion), they would fall with the *same* uniform acceleration. A rock and a leaf would reach the same speeds if they fell the same amount of time. Although he didn't have the means to eliminate air resistance to prove that hunch, his conclusions were correct. (See "Beginnings" after Chapter 2 for a discussion.)

Air resists the motion of a falling object and accounts for the various rates of fall we see. Take a balloon, blow it up with a few breaths, and tie the end. Then hold it steady above the ground and let it go. It begins to move downward; it accelerates. But almost at once this acceleration slows until the balloon floats down at a constant speed, its **terminal speed** in air. Its acceleration is then zero.

It's not only small things that act this way. When skydivers leap from an airplane (Fig. 1–8), their fall toward the ground begins at the rate of $g$. But, as with the balloon, air resistance quickly causes the downward acceleration to diminish, and to increase the air drag, they face the ground and extend their arms and legs. During each second, although they are falling faster, the air resistance increases and their downward acceleration becomes smaller. After about 12 seconds, at an air speed of about 120 mph, the air drag brings their downward acceleration to zero. This does not mean that they stop falling, only that they no longer fall faster every second. After reaching their terminal speeds, they fall steadily at those speeds until they open their parachutes. The opened parachutes have more area and run into more air per second than the divers alone do, lowering their terminal speeds for safe landings. We'll study how the air does this in later chapters, but for now we can say that any object falling in a vacuum—that is, without any air resistance—accelerates continuously at the rate of $g$.

Not many years after Galileo's death, Isaac Newton used a vacuum pump to give the first visual proof of Galileo's discovery. After pumping the air out of a tall glass tube, he released a gold coin and a feather at the top of the tube. They landed together. More recently, Apollo astronauts demonstrated the same thing with a hammer and a feather on the moon. There, in the vacuum of space, the hammer and feather accelerated side by side, only at a much lower rate than on earth. The gravitational acceleration at the moon's surface is only 1/6 of the value $g$ at the earth's surface.

**FIGURE 1–8**

## Freezing Motion to See Speed and Acceleration

To prove that your brain is slow to distinguish (or *resolve*) a clear image, just swish your hand in front of your face. It sweeps across your field of view much too fast for your brain to "stop" the motion. But since the first photographs were taken in the early 1800s, cameras have stopped motion. An exposure time of 1/60 of a second might make a sharp picture of someone posing for a portrait. but to catch an unblurred image of your moving hand might take a shutter speed of 1/500 of a second.

The mechanical shutters of many cameras blink open for such short exposures. But suppose you wanted to photograph a midge. This tiny, two-winged fly beats its wings about a thousand times a second, so to stop a midge in flight you need an exposure time much shorter than even 1/1000 of a second.

A clever trick makes very brief exposure times possible. The camera's shutter is left open, but the diaphragm that lets the light in is almost closed. This means that under normal lighting too little light will enter to expose the film. Then, while the shutter is open, an electronic flash floods the scene with a brief, intense blaze of light. A typical

(a)

(b)

**FIGURE 1–9**
*(a)* Taken with a stationary camera, this 1/30 second exposure of a car traveling at 55 mph (88 km/hr) does not freeze the car's image, and neither could your eyes. *(b)* But if the camera or your eyes follow the car's motion, 1/30 second is time enough to resolve an image of it.

**FIGURE 1–10**
An exposure of 1/500 second freezes the car's motion and records a clear image, even with a stationary camera.

**FIGURE 1–11**
Hummingbirds feeding. High-speed film and electronic flash have stopped the birds in mid-flight.

camera flash lasts about 1/50,000 of a second, but custom electronic flashes can deliver bursts of light in a millionth of a second. During that brief period of intense illumination, the film is exposed.

An electronic flash that triggers at regular time intervals is called a **stroboscope.** Directed at someone moving in a darkened room, each intermittent burst of light from the strobe appears to "freeze" the person for a split second. If you watch slow-moving dancers or fan blades by strobe light, you'll see flickering motion like that of early movies. But if you leave a camera shutter open while a strobe flashes, you'll get an interesting record of the motion on a single frame of film (Fig. 1–12). Someone walking by at a constant speed materializes as a row of images, all the same distance apart. If the person changes speed, the spacing between the images varies and you can measure the distances with a ruler to calculate the speeds and accelerations.

## A Familiar Example of Acceleration

Grab your keys and toss them straight up. As you let them go, you give them a certain speed and they are on their own. But because of gravity,

**FIGURE 1–12**
Strobe photograph of a pole vaulter.

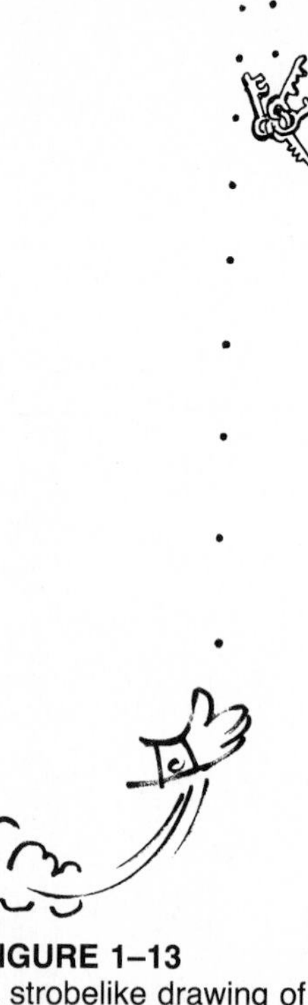

**FIGURE 1–13**
A strobelike drawing of the path of keys in flight. The points show where the key's motion would be frozen by consecutive flashes of a strobe. The points that are farther apart show that the keys are moving faster and covering more distance over the same time interval than when the points are close together.

they slow down while they move up. When the keys come to the highest point in their path, their speed is actually zero (assuming you pitched them straight up). How long is the speed zero? Because the acceleration $g$ never turns off, their speed *changes constantly* and is zero only instantaneously. Then the keys gain speed all the way down at the rate $g$.

This point deserves our attention. The keys do not remain at rest even so long as a thousandth of a second, although it might seem that way as you watch. The keys make a clearer image to us at the top of their path because they are traveling at their slowest there, so we have more time to observe them—to get a clearer image. But the speed of the keys changes constantly while they are in the air, and the value of their instantaneous speed just *passes through* zero at the top of their path. They only *seem* to pause when they change directions.

## Relative Speed

Migrating geese flying overhead in formation on a fall day all have about the same speed. Within the pattern each bird looks motionless to the others (except for its wings) because the distance between them does not change as time passes. That is, there is no relative motion between the birds, and we say their **relative speed** is zero.

Whenever you talk about a speed, you are talking about motion with respect to something else—a car's motion relative to a signpost on the ground or another car, or a sailboat's motion relative to a drawbridge. That's because speed is distance divided by time, and the distance or length must always be measured *from* something *to* something else.

Relative motion is easy to see from a moving car. If you travel at 50 mph over the ground and someone in a car ahead is driving in the same direction at only 30 mph, you approach that car with a relative speed of 20 mph. The relative speed between two cars moving in the same direction is just the *difference* in their speeds. If you ignore the objects along the road (which whiz by your car at 50 mph), you see the same motion beween yourself and the car in front as if it were standing still and you were approaching it at 20 mph. Or, with only a little more concentration, you see the same motion as if *you* were standing still and the car ahead were backing toward you at 20 mph. (That point of view should promote defensive driving.)

**FIGURE 1–14**
Their relative speed at this instant is (60 feet/minute − 6 feet/minute) = 54 feet/minute.

As you drive on a narrow two-lane road, you pay close attention to approaching cars, and well you might. The distance between your car and the oncoming car reduces at a rate that is the *sum* of your individual groundspeeds. In other words, the relative speed between cars traveling in *opposite* directions is the *sum* of their speeds over the ground, not the difference. If you travel at 50 mph and an approaching car travels at 40 mph, the distance between the cars decreases at the rate of 90 mph.

We tend to think that things move only when they aren't at rest on the ground. But of course the earth carries even these resting objects with it around the sun, and the sun itself moves in our galaxy, and so on. So *any* speed you talk about is a relative speed.

As a last example of relative speed, consider the two moons of Mars mentioned in the introduction to this chapter. Astronomers know that both moons circle Mars just as earth's moon circles earth, from west toward east. Because earth's surface turns around beneath the faraway stars much faster than the moon completes its orbit around the

**FIGURE 1–15**
Relative positions of the Martian satellites Deimos and Phobos and an astronaut standing still on the Martian surface at four points in time. At time 1, Phobos is behind the astronaut and Deimos well ahead. By time 4, the speedy Phobos has passed the astronaut and Deimos has fallen behind.

earth, the moon rises on our eastern horizon and sets on our western horizon. But only one of Mars's moons has the same kind of motion. The other rises in the west and sets in the east. The reason the moons of Mars have different motions involves relative speed. The innermost moon, Phobos, circles Mars in less time than a Mars day and so outruns the rotating surface of Mars; while, from a point of view on the planet's speeding surface, the outer moon, Deimos, is left behind, like our own moon (see Fig. 1–15).

Earth's rotation, like that of Mars, makes its surface inconvenient as a reference point to describe the motions of the moons and planets of the solar system. Instead, astronomers refer the locations and speeds of these bodies to the background of faraway stars. Due to their great distances, the stars appear to have no relative speeds unless observed photographically through high-powered telescopes. But for relative motions here on earth, we refer to the landscape without thinking. An automobile moving at 45 miles per hour means "over the ground" to everyone.

## CALCULATIONS

At the end of most chapters you will find a section like this one, inviting you to do some calculations to put the ideas of physics to work. Here you'll find formulas and examples worked step by step. You'll need only to do multiplication, division, addition, or subtraction—the same kind of math you use to balance a checkbook, comparison shop at your grocery, or figure the number of miles per gallon your car gets. Here's an example. If a car travels at 50 mph, how long does it take to go 25 miles? You can probably give the answer, which is half an hour. In these end-of-chapter sections, though, we'll emphasize using formulas to help solve such problems. Although the math is not hard, what is likely to give trouble is the translating of words into formulas so you can do the calculations. For example, suppose that car cruises along at 55 mph and you want to know how long it will take to go 16 miles. This question is no more difficult than the first one, but you have to pause and think about what to do to find the answer. That's when a **formula** (which is an abbreviation of a definition or a statement) can help you keep things straight. Let's work this one through.

**FIGURE 1–16**

---

EXAMPLE 1: To begin, you'll want to use the definition that relates distance, speed, and time:

$$\text{average speed} = \frac{\text{distance}}{\text{time}}$$

To make it a little easier to use, we'll abbreviate the terms, $d$ for distance, $t$ for time, and $v$ for speed (for reasons you'll see in the next chapter). Also, to indicate the average value of any quantity, we'll place a bar above it. Now the definition for average speed becomes the formula

$$\bar{v} = \frac{d}{t} \qquad \textbf{(1-1)}$$

Whenever anything moves from one place to another, this formula gives the relation between its average speed and the distance it goes in the amount of time $t$.

Now let's use this formula to answer the question, **"at 55 mph how long does it take to go 16 miles?"** We need to rearrange the formula so we can find $t$. That is, we need to put it in the form $t$ = "something." Although you might be able to do this without thinking, it is worth working through one step at a time. If you are rusty on working with equations, remember that whatever you do to one side of the equation you must do to the other side to keep the sides equal. First, multiply both sides by $t$ to get rid of the $t$ in the denominator (the bottom term of the ratio).

$$\bar{v} = \frac{d}{t}$$

$$\bar{v} \times t = \frac{d}{t} \times t$$

Now on the right side, $d$ is both multiplied and divided by $t$. Since $t/t = 1$, this is the same as multiplying $d$ by 1. (Anything divided by itself is 1.) At this point you can just draw slashes through the $t$'s on top and bottom and say they cancel. So

$$\bar{v}t = d \quad \text{or} \quad d = \bar{v}t$$

(Read this formula aloud; it says, "distance is equal to average speed times time.") Some more rearranging gives us $t$. Just divide both sides by $\bar{v}$.

$$d = \bar{v}t$$

$$\frac{d}{\bar{v}} = \frac{\cancel{\bar{v}}t}{\cancel{\bar{v}}}$$

$$\frac{d}{\bar{v}} = t \quad \text{or} \quad t = \frac{d}{\bar{v}}$$

Now you can answer the question.

$$t = \frac{16 \text{ miles}}{\dfrac{55 \text{ miles}}{\text{hour}}}$$

Because the unit ''miles'' is in the numerator and in the denominator, ''miles'' cancels. Thus we have

$$t = \frac{16}{\dfrac{55}{\text{hour}}}$$

To eliminate the unit ''hour'' from the denominator and move it to the numerator, we multiply both the numerator and the denominator by ''hour'':

$$t = \frac{16 \times \text{hour}}{\dfrac{55}{\text{hour}} \times \text{hour}}$$

The ''hour'' units in the denominator cancel, since hour/hour = 1. We now have

$$t = \frac{16 \times \text{hour}}{55}$$

$$= \frac{16}{55} \times \text{hour}$$

$$= \mathbf{0.29\ hour}^{*}$$

From what we've just seen, if such a problem involving the average speed (or a constant speed) asks

how fast? use $\bar{v} = \dfrac{d}{t}$

how far? use $d = \bar{v}t$

how long? use $t = \dfrac{d}{\bar{v}}$

## Changing Units

Occasionally you may need to change the units of a quantity. A table relating various units is given in Appendix 1. To illustrate changing units, we'll take a speed of 60 mph and find what this speed is in terms of feet per second. There are 5280 feet in a mile and 3600 seconds in 1 hour (60 minutes/hour times 60 seconds/minute). Therefore,

---

*If you check out this answer using a pocket calculator, you will find that the answer is not exactly 0.29 (the calculator display will read *0.290909*). Rather than use an equals sign, we could use the symbol ≃ to indicate that the answer is approximate, but in this book we'll follow common practice, which is to use the equals sign. Generally, we'll keep only the first two significant figures in an answer, such as 0.**29**, or **9.8**, or **23**00.

$$60\ \frac{\text{miles}}{\text{hour}} = 60\left(\frac{(5280\ \text{feet})}{(3600\ \text{seconds})}\right) = 88\ \frac{\text{feet}}{\text{second}}$$

In other words, when you want to convert a unit, just substitute its equivalent value.

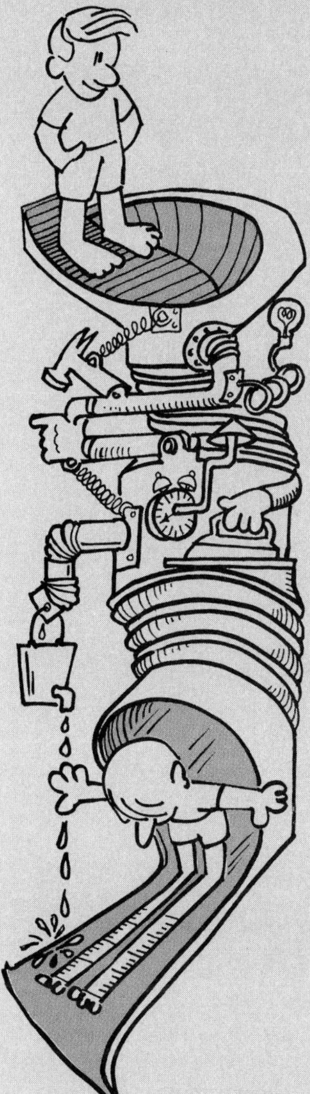

FIGURE 1–17
Changing feet to meters

---

EXAMPLE 2: In units of feet/second/second, the numerical value of $g$ is about 32. Thus each second an acceleration of $g$ increases the speed by 32 feet/second. However, in the United States we are most familiar with miles/hour as units of speed. **Express $g$ in miles per hour per second,** or miles/hour/second:

$$32\ \text{feet/second/second} = 32\left(\frac{\frac{1}{5280}\ \text{miles}}{\frac{1}{3600}\ \text{hour}}\right)/\text{second}$$

$$= 32\left(\frac{1}{5280}\right)(3600)\ \text{miles/hour/second} = \mathbf{22\ mph/s.}$$

Thus, only 3 seconds of free fall could bring a skydiver to a speed of 66 mph if air resistance were eliminated.

## Calculating Accelerations

Average acceleration is defined as

$$\frac{\text{change in speed}}{\text{time required}}$$

If we use $\bar{a}$ for average acceleration, $v_f$ for the final speed, and $v_i$ for the initial speed, this becomes

$$\bar{a} = \frac{v_f - v_i}{t}$$

Multiplying both sides by $t$, we find

$$v_f - v_i = \bar{a}t \qquad (1\text{–}2)$$

---

EXAMPLE 3: A jumbo jet begins at rest and accelerates along the runway. It lifts off some 40 seconds (s) later, traveling at 175 mph. **What was the average acceleration?** Equation (1–2) tells us

$$175\ \text{mph} - 0\ \text{mph} = \bar{a} \times 40\ \text{s}$$

so

$$\bar{a} = \frac{175\ \text{mph}}{40\ \text{s}} = \mathbf{4.4\ mph/s}$$

Compare this to $g$, which is 22 mph/s. Passengers on commercial flights know the feeling of this acceleration (about ⅕ g) well.

Sometimes an acceleration is uniform, as with heavy objects falling short distances. Then it is not hard to relate the distance they fall to the time they accelerate. If an object starts at rest ($v_i = 0$) and has a uniform acceleration $a$ for a time $t$, then

$$d = \frac{1}{2}at^2 \qquad \textbf{(1–3)}$$

Though the derivation of this formula is more difficult than any exercise required in this text, we'll show that this formula comes from $d = \bar{v}t$. When something accelerates uniformly from rest, its *average* speed after a time $t$ is *half* of the final speed, $v_f$. But that final speed is just $v_f = at$. That means $\bar{v} = \frac{1}{2}v_f = \frac{1}{2}at$. Using this for $\bar{v}$ in $d = \bar{v}t$, we get $d = (\frac{1}{2}at)t$ or $d = \frac{1}{2}at^2$.

---

EXAMPLE 4: A construction worker on a towering skyscraper drops a heavy wrench that falls to the ground. **About how far will that wrench have fallen after 1, 2, and 3 seconds?** Since $d = \frac{1}{2}at^2$, and the acceleration for the wrench (neglecting air resistance) is $g$, we have

$$d_1 = \frac{1}{2}(32\ \text{ft/s}^2)(1\ \text{s})^2 = 16\ \text{ft}\,\frac{\text{s}^2}{\text{s}^2} = \mathbf{16\ ft}$$

$$d_2 = \frac{1}{2}(32\ \text{ft/s}^2)(2\ \text{s})^2 = \frac{32}{2}\frac{\text{ft}}{\text{s}^2} \times 4\ \text{s}^2 = \mathbf{64\ ft}$$

$$d_3 = \frac{1}{2}(32\ \text{ft/s}^2)(3\ \text{s})^2 = \mathbf{144\ ft}$$

Each story or floor of a building represents about 10 feet of height, so in 3 seconds the wrench will fall some 14 stories.

## REVIEW

**1.** What two rates do we use to describe motion in one dimension (along a straight line)?

**2.** A rate tells how fast something happens, or how much something changes in a certain amount of time. True or false?

**3.** Define *average* speed and *instantaneous* speed in your own words.

**4.** A measurement is merely a comparison of a quantity to a standard quantity. True or false?

**5.** How long is an instant of time?

**6.** Is acceleration a rate of change of a rate?

**7.** Express acceleration in words.

**8.** The speed of a freely falling object near earth's surface increases by _____ feet per second for each second that it falls.

**9.** Air resists the motion of a falling object, and this accounts for the various rates of falling we see on earth. True or false?

**10.** When a falling object reaches its terminal speed, it is no longer accelerating. True or false?

**11.** An acceleration that has a *negative* value is what is meant by the word *deceleration*. True or false?

**12.** If you throw your keys straight up in the air, what is their speed at the very top of their path? What is their acceleration at that highest point?

**13.** The relative speed between two objects moving in the same direction is the (sum)(difference) in their speeds as measured from the ground.

**14.** Is any speed you talk about a relative speed?

# DEMONSTRATIONS

**1.** Your sense of rhythm gives you an excellent internal clock, and with a little practice you can become an expert timer. Place yourself in front of a clock with a second hand. Pronounce a four-syllable word, such as *Mississippi* or *locomotive,* or four short words, such as "I can do it," and name the second immediately afterward. It's easy to get the cadence.

Start.

I can do it ONE

I can do it TWO

I can do it THREE

I can do it FOUR

Because 5 syllables are spoken during each second, you are splitting each second into 5 parts, each syllable marking the passage of ⅕ second. (⅕ second = 0.2 second) Then, with a friend watching the clock, look away and count to 10. You'll probably determine 10-second intervals with an error of only one- or two-fifths of a second. If you are timing a pencil rolling across a table and it crosses the line as you say "I can . . .," then it hit the mark two-fifths of a second (two syllables) past the last second you named. When you watch a baseball game, time the pitches. Most take about 0.6 seconds to get from the mound to home plate, so you'll be able to say only "I can do. . . ."

**2.** Find several people (perhaps in your class) who are wearing wristwatches with timers or second hands. Have each of them time a pencil or a soft drink can rolling between two marks on a table. The different results you get will show you why in the days before electronic timing the swim meets, foot races, and horse races were all timed by several people to reach an agreement on the time.

**3.** Here is a simple rate demonstration that involves little chance of error. Take your wrist and find your pulse. Then with a clock or watch before you, count the times your heart beats in a minute. This is a rate, of course, an average rate for that minute. It's quite easy to count each heartbeat, so that part of the measurement should have no error at all, unlike a measurement of a distance. And in contrast to speed, it makes little sense to speak of an instantaneous heartbeat rate. The heart is either pumping or resting at any time. (Incidentally, in physics terminology, the heartbeats are **discrete**—they can be counted; there is a whole number of them, whereas a distance is **continuous,** meaning you can have a fraction more or less than any given length. To put it another way, there is no such thing as a fraction of a heartbeat, but you can travel not just whole numbers of miles but any fraction of a mile.)

**4.** Some dark night, find a mercury-vapor or a sodium-vapor street light. Although the lamp seems to send out a steady stream of light, it does not. The brightness of its light varies rapidly, hitting a maximum value each time the power company's electric current peaks. In effect, then, this streetlamp is a strobe. If there is no other lighting nearby, you can see the flashes of light if something beneath the lamp moves fast enough.

Take a stick that is several feet long (a yardstick does nicely) and swish it back and forth at arm's length. You'll see the stick appear and disappear as it moves a noticeable distance between flashes. Or just spread your fingers wide apart and swish your hand back and forth. You'll see 20 or 30 images of your fingers. The 120 separate peaks of light per second coming from that streetlamp appear steady because your brain takes time to form and retain images. For example, a classroom movie projects 24 pictures per second that your brain cannot separate into individual flashes.

**5.** Place two sets of keys in one hand and toss them straight up. Do they seem to stay together in flight? If not, can you explain their variations with air resistance? Does it make a difference if you throw them very high? Why?

**6.** Find your eye-to-hand reaction time. Have someone suspend a ruler (as in Fig. 1–18) and center the lower end between your thumb and forefinger. When your partner drops the ruler without any warning, catch it on its way down and see how many inches slipped through your fingers. Table 1–1 will give you your reaction time. After a

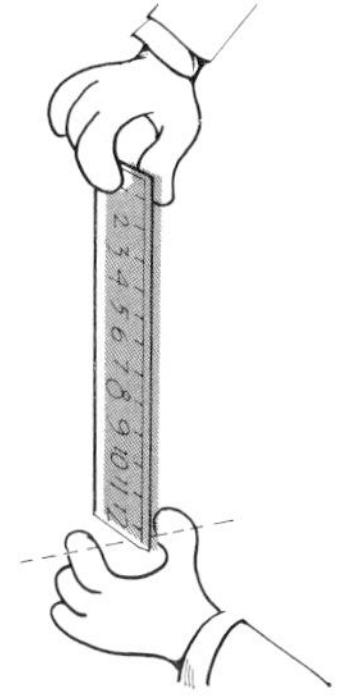

**FIGURE 1–18**

**TABLE 1–1**

| | |
|---|---|
| 3 inches | 0.125 second |
| 4 inches | 0.144 second |
| 5 inches | 0.161 second |
| 6 inches | 0.176 second |
| 7 inches | 0.191 second |
| 8 inches | 0.204 second |

(These times and distances can be checked with the formula $d = \frac{1}{2} gt^2$, which is described in "Calculations.")

few tries, you'll see that 4 or more inches invariably slip past no matter who does the catching. Nerves and brains take at least 1/7 (0.14) second to react. That fact gives rise to a great party trick. Simply use a dollar bill in place of the ruler and have a friend center thumb and forefinger over George Washington's nose. Then tell your friend if he/she catches the bill when you drop it, he/she can keep it! (Explain that moving the hand downward with the bill is illegal.) Your secret? It's only 3½ inches from the edge of the bill to George's nose.

## EXERCISES

**1.** Changes in height/year, gallons/minute, miles/hour, turns/year, and square feet/day are all rates. True or false?

**2.*** Suppose someone handed you a stopwatch and challenged you to walk 1 mile at an average rate of 1 mph. How could you arrange to do this?

**3.** Remember the fable of the tortoise and the hare? Explain their motions in terms of average and instantaneous speeds.

**4.** For some motions, average and instantaneous speeds are the same. Is this true for (a) a swimmer in a race? (b) the hands of an electric wall clock? (c) a car moving uniformly on a straight road? (d) a falling rock? (e) bowling balls moving down a lane?

**5.** When you walk, your average speed is the distance of your stride divided by the time between your footfalls. From this definition, give two independent ways in which you could increase your speed.

**6.** Match the following rates to the actions for which they are most appropriate: inches/second, inches/minute, inches/hour, inches/day, inches/month. (a) reaching for something, (b) a snail crawling, (c) grass growing, (d) a candle burning, (e) a column of water flowing through a wide-open garden hose, (f) sipping a soda through a straw, (g) mercury moving in a thermometer, (h) your fingernails growing.

**7.** A physics major drew the graph in Fig. 1–19, describing an excursion from her home. Her motion was along a single street, that is, in a straight line. (a) In which time interval was her average speed the greatest? What was it? (b) In which time interval was her average speed zero? (c) How far did she get from home that day? (d) What was her average speed for the entire trip? (Careful!)

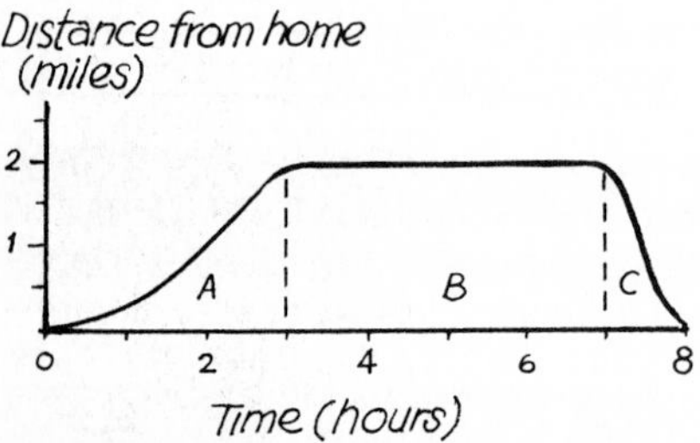


**FIGURE 1–19**

**8.** On a toll parkway, a motorist takes a ticket at one toll booth and travels to another some miles away. At each booth the ticket is stamped by a time clock. Can the authorities at the exit point tell if the motorist exceeded the maximum speed limit for the parkway? Discuss.

**9.** Is it most correct to say that speed and acceleration *predict* motion, that they *describe* motion, or that they *explain* motion? Explain your answer.

**10.** Can speed and acceleration be measured with perfect accuracy? Discuss.

**11.** What practical reasons might skydivers have to increase their air resistance when they are falling?

**12.** The curve in Fig. 1–20 is an estimate of a boy's speed during the first 20 seconds of a 200-meter run. (a) Describe what happens to cause the wrinkles in the curve for the first few seconds. (b) This sprinter's speed peaked at 8 m/s. Will his average speed for these 20 seconds be equal to, less than, or greater than 8 m/s? Why? (c) About what was the runner's average speed over the time $t = 5$ s to $t = 20$ s? (d) About what was the runner's average acceleration for $t = 0$ s to $t = 3$ s?

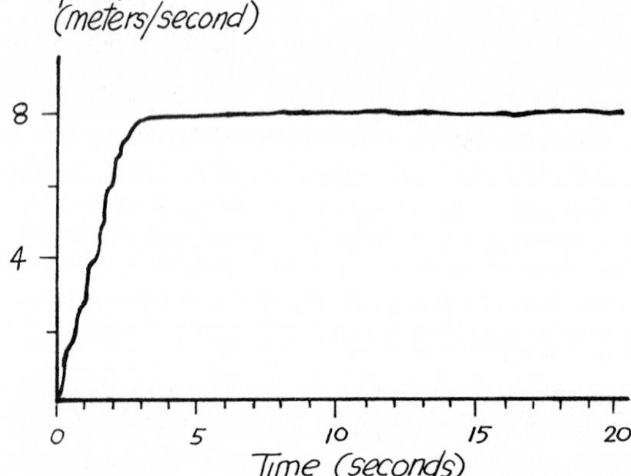


**FIGURE 1–20**

**13.** Have you ever driven a car in a rainstorm so heavy that you couldn't see well between each pass of the wipers across your windshield? What effect did this have on your driving and why? (Notice that this is like a strobe effect.)

**14.** In a physics lab, two students must find the speed of a cart rolling along the floor. Using their new digital

---

*Colored entry numbers indicate exercises that are answered in the back of the book.

watches, they time its passage betwen two lines 1 meter apart. One says to the other, "It took exactly 1.73 seconds." Is the last digit (the 3) significant in this number?

**15.** Are miles per second per second units of acceleration? Is that the same as miles/second/second? Miles/second$^2$?

**16.** What do you think you would see if a strobe light on a dance floor flashed 30 or more times each second? (*Hint:* Read Demonstration 4.)

**17.** If an object has zero acceleration, does that mean its speed is zero? Give an example. If an object has zero speed at some instant, does that mean its acceleration is zero? Give an example.

**18.** Figure 1–21 shows the speed of a toy train. (a) In which time interval(s) was the train's speed constant? (b) In which time interval(s) was the train accelerating? (c) In which time interval(s) was the train decelerating? (d) Which interval has an average speed greater than the average speed for the trip? (e) In which interval does the train decelerate fastest?

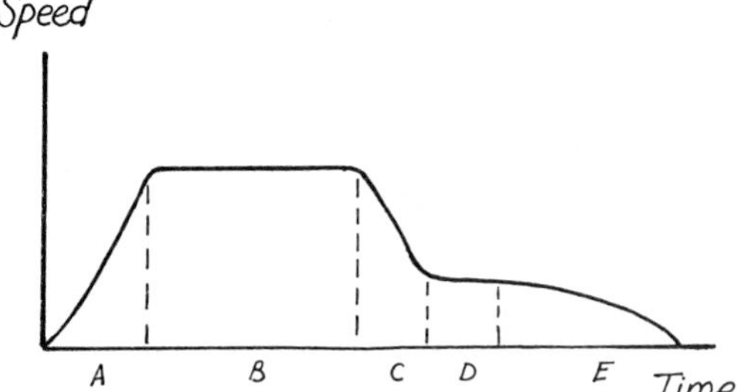


**FIGURE 1–21**

**19.** Explain how a skydiver has more than one terminal speed during a dive.

**20.** In the strobelike illustration of a moving ball in Fig. 1–22, assume the strobe flashed 5 times each second. Explain how you could estimate the *acceleration* of the ball.

**FIGURE 1–22**

**21.** If you took 100 different photos of a bumblebee in flight, you'd see its wings in many different positions. In which positions would the wings appear most distinct (unblurred)?

**22.** Take a flashlight into a darkened room and point the beam of light at the wall (Fig. 1–23). Rapidly flip the flashlight back and forth a short distance, making a straight streak of light on the wall. Where is the streak brightest? Why is this so?

**23.** When you are riding in a car, smooth changes in speed that don't jerk the passengers around are more comfortable than rapidly changing accelerations that do. If the *rate of change* of acceleration is measured, what would be its units in terms of length and time?

**FIGURE 1–23**

**24.** American and Russian spacecraft have docked together while traveling at a speed of some 18,000 mph in earth orbits. Did their high speeds make this maneuver dangerous?

**25.** Some states require bicycle riders to travel on the right-hand side of the road *with* the traffic. Does this make sense in terms of relative motion?

**26.** A car was traveling at 50 mph and collided with the rear of a large truck, but there was very little damage. What can you conclude about their relative speed?

**27.** The needle on a record turntable sits almost still while the record moves beneath it. Does the groove of the record travel under the needle faster when the tone arm is near the record's edge or closer to the center? Why?

**28.** Draw sketches of a tire that is fully inflated and one that is almost deflated. Which tire will go farther along the road for each revolution it makes? So how would deflated tires on a car affect the speedometer reading?

**29.** Professor Dolittle was a physics professor who did very little because he was quite lazy. One thing he liked to do (perhaps because it required so little) was to check the mileage whenever he drove somewhere. After years of checking, he came to the conclusion that it was a little farther from his home to the college in the winter than it was in the summer. Could you suggest why? (Be sure to see the answer to this one.)

**30.** Tell how you could do a simple experiment in a couple of minutes in your living room to estimate how far you could walk in 1 hour.

*31. Top track competitors who run the mile might finish in 4 minutes (or a few seconds less). About how many mph is this?

*32. A marathon runner in the early 1800s ran a distance of 5560 miles in only 59 days. What was his average speed in miles per day? In mph? Discuss your answer.

*33. Show that without air resistance a skydiver could reach the normal terminal velocity of about 120 mph in about 6 seconds.

*34. On the average, human hair grows about ½ inch per month. If an actress bleaches her 1-foot-long hair for a play, she'll have to let her hair grow another foot before it can be back to its normal color and length. How long will that take? If *you* wanted to grow hair down to your waist, how long would it take?

*35. Measurements indicate that the continents of Europe and North America are separating at the rate of about 2 centimeters per year. If Columbus could repeat his famous 1492 voyage in 1992, about how many more feet would he have to go?

*36. A jumbo jet can fly at speeds of 550 mph through calm air. How far can it travel in 8 hours? In 8 minutes? How long would it take to travel 400 miles? How long would it take to go around the world (about 24,800 miles)?

*37. Although you can't stop the dollar bill in Demonstration 6, you can see it fall very clearly. That is because your eyes and brain form an image in less than the eye-to-hand reaction time. Calculate how far something falls from rest in 1/20 second for a better idea of how quickly you can see a change in motion.

*38. From the graph in Fig. 1–2, find the average acceleration the car had between the time it started and 1 minute into its trip. Find the average acceleration from 1 minute into its trip to 4 minutes into its trip.

*39. As you drive along a boulevard at 45 mph, a driver overtakes and passes you at 65 mph. What is your relative speed?

*40. Suppose two trains move in opposite directions. One train moves at 35 mph while the other's speed is 40 mph. What is their relative speed? Does it matter which train has the 40 mph speed? Does the relative speed change after they pass?

*41. A jet flies from New York to San Francisco (about 2800 miles). It meets the headwinds called the *jet stream* that go roughly west to east across the United States. If these winds average 100 mph during the trip, what are the trip times for the plane as it flies westward and again as it returns to New York? Assume the plane would travel at 500 mph if the air were calm.

*42. A girl out for a run passes a boy who is jogging. If she runs at 8 mph while his speed is 5 mph, what is their relative speed? When she is 100 feet ahead of him, he begins to run at 12 mph. How many seconds will pass before he catches up with her? (1 mph is about 1.5 ft/s.)

*43. The catapult on an aircraft carrier sends a jet plane from rest to 132 mph in 2 seconds. The plane's acceleration in terms of $g$ is about (a) 1 $g$, (b) 6.6 $g$, (c) 3 $g$, (d) 6 $g$, (e) ⅔ $g$.

*44. If you drop a golf ball, how far does it fall in ½ second? (a) About 3 inches, (b) 4 inches, (c) 4 feet, (d) 3 feet, (e) 4/3 feet.

*An asterisk preceding an exercise number indicates that the exercise involves calculations.

# CHAPTER 2
# CHANGING DIRECTIONS

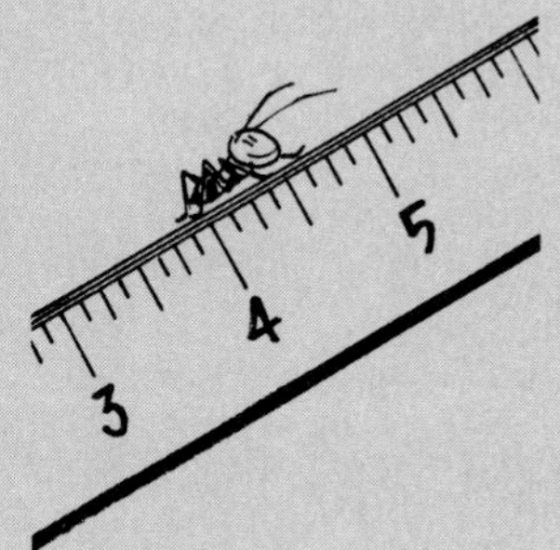


**FIGURE 2–1**
Only one number is needed to locate this ant on the yardstick.

If an ant marches along the edge of a yardstick (Fig. 2–1), you can pinpoint where it is with a single number, the closest mark on the yardstick. Because the ant moves along a straight line and one number (or *coordinate*) locates the ant on the yardstick, we say it moves in **one dimension.** We can describe its stop-and-go motions with speed and acceleration, the subjects of the first chapter.

But a caterpillar, meandering across a red-and-white checkered tablecloth on a picnic table, moves not just along a straight line but on a flat surface called a **plane** (Fig. 2–2). Dodging a pickle jar or a plate in its way, the caterpillar moves in **two dimensions.** The rows and columns of the checkerboard pattern can serve much as lines on a graph do to locate the caterpillar. No matter where it is, we need two coordinates (the row number and column number) to describe its position. In time that caterpillar might become a butterfly fluttering about. Because a butterfly does not move in a single plane but wanders up and down (taking it over and under any number of planes), we have to give a **third** coordinate to fix its position at any moment—the butterfly's elevation above the tabletop, or perhaps the ground. *Since three numbers are the most we need to describe the position of anything, we say the space we live in is* ***three-dimensional.***

**FIGURE 2–2**
Two numbers pinpoint this caterpillar's position on the tablecloth.

(a)

(b)

**FIGURE 2–3**
*(a)* Ducks on a pond move on a two-dimensional surface.
*(b)* Taking off, this duck travels in three-dimensional space.

Just as for the ant, we can use speed and acceleration to define the motions of the caterpillar or butterfly along their paths. But think of how the butterfly flits, soars, and plunges, moving in any direction in three-dimensional space. To describe such actions, we have to include the *changes of direction*. ■

**FIGURE 2–4**
When the keys move on a vertical line, only one of the key's coordinates changes.

## Motion in More Than One Dimension

If you toss your keys straight up over your head, they rise and fall along a straight line, so you need only one number, their elevation, to describe their motion. The keys actually have two more coordinates, namely, your own coordinates on the ground. (A well-known set of ground coordinates is latitude and longitude.) But while the keys were in the air, you stayed in the same place, so those two ground coordinates did not change for you *or* for the keys. In effect, then, the keys moved in only one dimension.

A fishing trawler roams the ocean's curved surface, but to tell where it is, we need only know its precise longitude and latitude. If the ship moves directly north or south, only its latitude changes. Or it might head due east or west without changing its north–south coordinate. There are two independent directions of motion on a plane or a surface, and motion in one direction does not mean any motion has to take place in another, perpendicular direction. There are three independent directions of motion in three-dimensional space. For example, a submarine might dive beneath the ocean's surface without altering its latitude or its longitude. Its depth gauge would give the change in its third coordi-

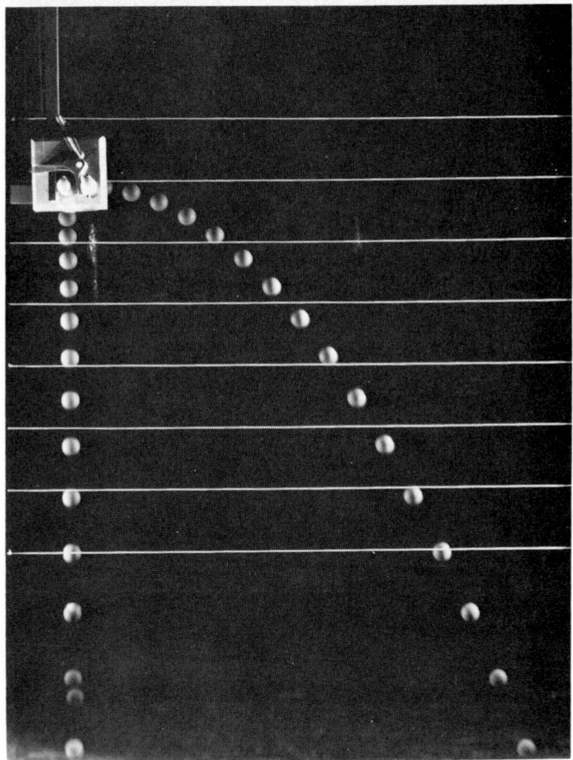

**FIGURE 2–5**
A strobe photograph of two balls that begin to fall at the same instant. One has an initial speed in the horizontal direction.

nate, its depth, which is *independent* of its latitude and longitude.

To see the independence of motion in perpendicular directions, look at the strobe photograph in Fig. 2–5. Two balls fell at precisely the same time from the same elevation. One simply dropped from rest and moved straight down. The other had some speed in the horizontal direction, so it traveled off to the side as it fell. Notice that flash after flash the balls kept pace in the vertical direction. Why? Because gravity accelerated both at the same rate. But notice how the ball with the horizontal motion moved to the side: Its horizontal movements between flashes were equal in length—its speed to the side did not change. Although gravity accelerated the ball downward, it did not affect the speed the ball had in the horizontal (or perpendicular) direction. From experiments like these we can see that *motions in perpendicular directions are independent of one another*.

It's easy to check the beginning and end of motion like this. Using a ruler and two marbles, follow the instructions given in Fig. 2–6.

For another example of independent motions, we can compare the motion of two shots in a hypothetical shotput contest. The shots are thrown in exactly the same way, only at very different places—one on earth, the other on the moon. Figure 2–7 is a strobelike drawing of their paths. (In "Calculations" you can learn to graph these motions for yourself.) The two metal balls have the same horizontal speed and keep pace in that direction. Each also begins with the same speed in the vertical direction, and each accelerates back toward the surface. But the shot thrown on the moon accelerates downward at only ⅙ of the earth's

**FIGURE 2–6**
Place two marbles along the edge of a horizontal tabletop. Align a ruler with the marbles as shown. Holding one end of the ruler steady, pull back on the other end and release it, quickly sending the ruler into the marbles. The marbles are struck at the same time, but the one closer to the end held down has a much smaller horizontal speed than the marble hit by the other end of the ruler. Nevertheless, the marbles accelerate downward together, and you can hear them strike the floor at the same time.

**FIGURE 2–7**
Thrown in the same manner, shots would have very different motions on the moon and earth. Unaffected by gravity, however, their horizontal speeds would be the same (until the shot on earth returned to the ground).

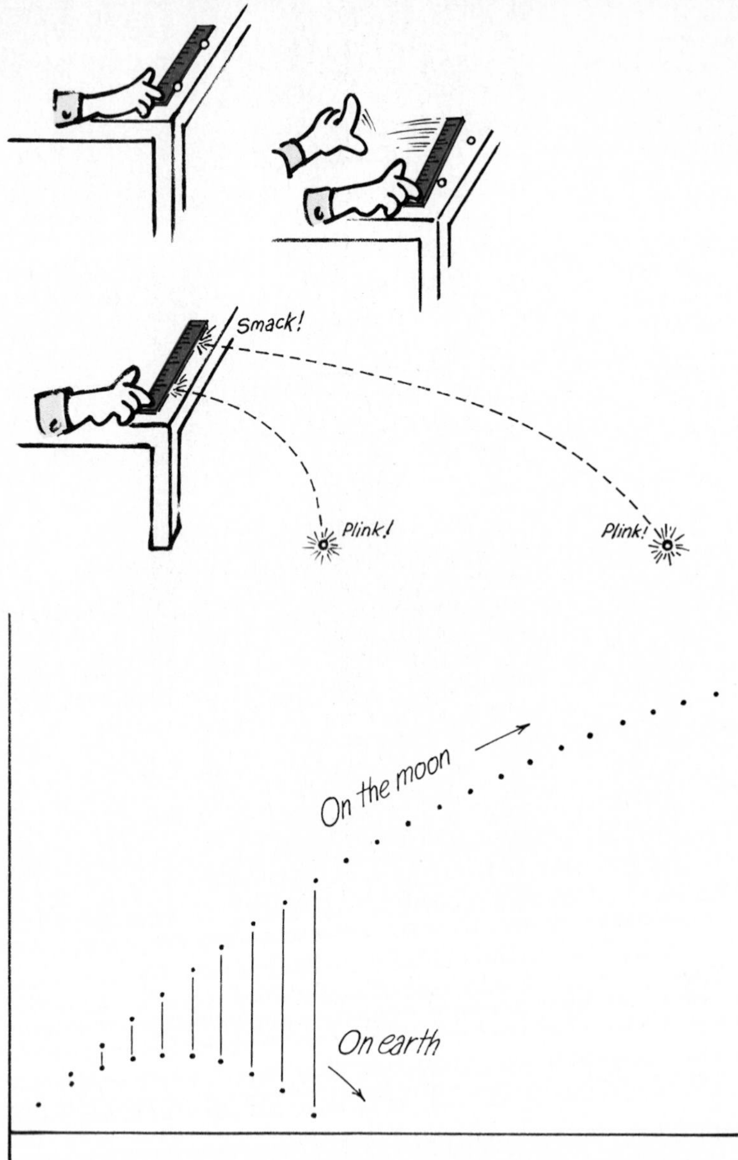


rate of $g$, so it continues to rise even after the shot on earth has reached its greatest height and fallen back to the ground. The moon's weaker gravitational acceleration takes longer to return that ball to the lunar surface, and all the while the shot continues to travel over the lunar landscape with its unchanged horizontal speed. The extra time of flight means the shot on the moon goes much farther over the surface than its earth-bound companion does. (Besides this difference, there are other features of motion on the moon that would seem strange. There's no air resistance there to influence motion. You could probably throw a feather as high and as far as a baseball, and maybe even farther!)

## Showing Directions with Arrows

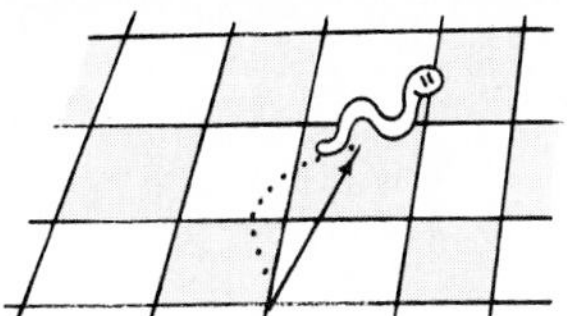

**FIGURE 2–8**
A vector (or an arrow) gives us another way of locating something or describing its motion. Rather than using initial and final coordinate points for this caterpillar, a displacement vector gives the direction and distance traveled from the caterpillar's initial position.

Let's return to the caterpillar crawling on the tabletop. If it has moved from one point to another on the tablecloth, we can describe this action by giving its beginning and ending coordinates, or by stating the direction and the distance it traveled. Even simpler, we can draw an arrow from the caterpillar's first coordinates to its last (Fig. 2–8). This arrow's direction points to the caterpillar's new resting place, and its length is the straight-line distance from where the caterpillar started. The arrow is called a **vector.** This vector represents a displacement from one place to another, so it is known as a **displacement** vector.

*A vector is a device to describe a quantity with direction as well as size*. It can be measured on a graph and gives visual rather than verbal "direction." For example, giving the distance and direction, we could say that a certain town is 10 miles to the northeast. But an arrow properly drawn on a map does the same thing.

## Adding Vectors

In some cities, all the streets run either north–south or east–west. Suppose on a visit you wanted to see a museum that lies to the northeast. If the east–west and north–south blocks are the same length, you could walk 3 blocks north and then 3 blocks east. Figure 2–9 shows these two vector displacements visually. These two consecutive displacements would take you to the northeast. But how far? Since your displacements were not in the same direction, you are closer to where you began than the 6 blocks you actually walked. (If you measure the straight-line distance carefully, you'll see it is between 4.2 and 4.3 blocks.)

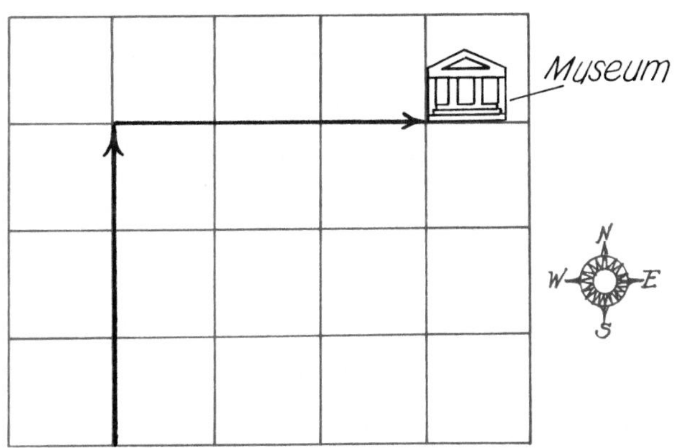


**FIGURE 2–9**
If you walk 3 blocks north and 3 blocks east, your final position is not 6 blocks from your initial position if the distance between these locations is measured along a straight line connecting them.

Because you made these two straight-line jaunts in sequence, the second beginning at the end of the first, we say the two vector displacements were *added*. Furthermore, if you draw another arrow directly from the tail of the first vector to the tip of the second, you get a so-called **net** or **resultant vector** (Fig. 2–10). This is a vector that can do in a single step what the two separate vectors did together. You never necessarily move in the direction of the net displacement vector. It only

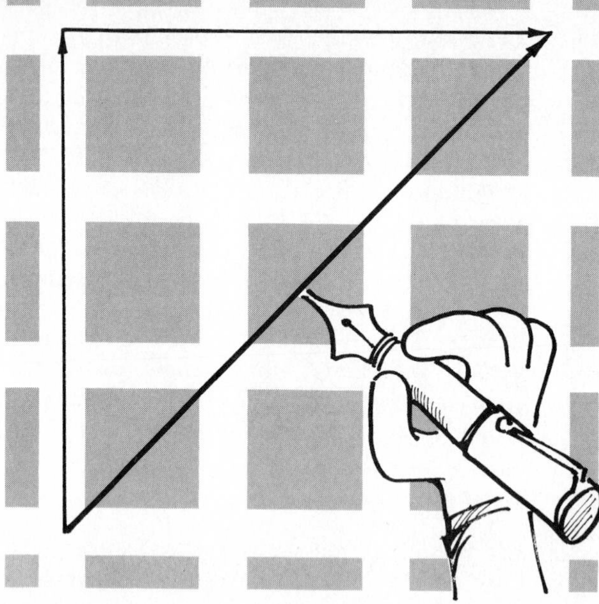

**FIGURE 2–10**
The result of several consecutive displacements can be described with a net, or a resultant, displacement vector. In terms of vectors, the separate vector displacements are added to give a net vector. The net vector always connects the tail of the first vector in the series of displacements to the tip of the last vector.

*summarizes* the result of your several individual displacements. You could say the resultant vector describes your displacement ''as the crow would fly.''

**To add two or more vectors to form a net vector, place the individual vectors tail to tip. A vector drawn from the tail of the first vector to the tip of the last is the resultant vector.**

Notice that you'd get the same net vector if you first walked east, then north. When you add a number of vectors, they may be added in any order, tail to tip, to find their net vector.

Remember how the average speed tells little about a car trip? So it is with a net vector. When a girl skates over a frozen pond, a single vector drawn from her starting position to her stopping place tells nothing about how she arrived there. On the way she may have skated at various speeds and in various directions, back and forth in loops, stars, or circles. To follow her path more clearly, we would need a series of many tiny displacement vectors, placed tail to tip (Fig. 2–11). Her net displacement at any stopping place is a vector that goes from the tail of the first of those tiny vectors to the tip of the last one.

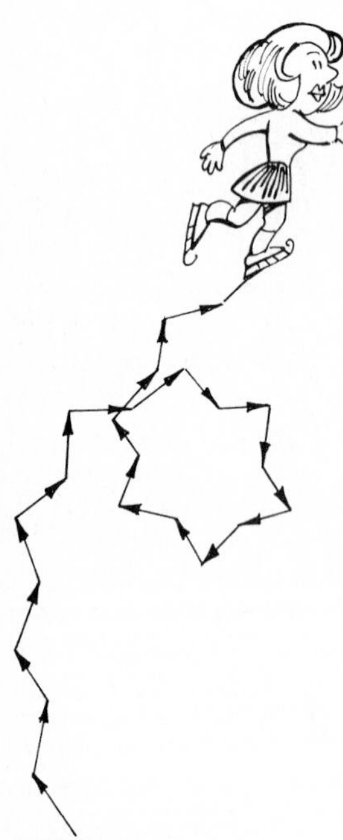

**FIGURE 2–11**
The skater's position at any point in her motion is described by the net displacement vector, which is the sum of the individual displacements to that point.

## Velocity: Speed and Direction

If you are strolling or jogging along a trail, your speed is the distance you cover along the path per unit of time. If the course is winding, you must change directions as you move, taking the path's ''direction'' with each step. To keep track of your motion, you might mark your progress on a trail map, noting your speed at points along the way. At any point where you note your speed, you could draw an arrow to show the direction of your motion, and the length of the arrow could represent your speed. Such an arrow is a special vector called **velocity** (Fig. 2–12). Like other vectors, the velocity vector has a length and a direction. But the length of the velocity vector represents a *speed,* not a distance as for a displacement vector.

The symbol for velocity is $\vec{v}$, where the arrow over the $v$ tells us it is a vector. *The velocity vector of any moving object points in its direction of travel at a given instant, and its length shows how fast the object moves.* If an object is moving along a curved path, its velocity vector is tangent to that path at the object's position (see Fig. 2–12).

**FIGURE 2–12**
As they follow the winding trail, their velocity vectors point along the direction of the trail wherever they are. The lengths of the velocity vectors represent their speeds.

## Components of Velocity

Earlier we saw that some motions in two dimensions become clearer when we ask what happened along perpendicular directions. Perpendicular "pieces," or **components,** of a velocity can also make motions easier to understand. The reason is that a velocity may be either *built up* from several velocity vectors or *separated* **(resolved)** into several other vectors, as the following examples will show.

A girl swims steadily in a river, her smooth strokes giving her a pace of 3 mph through the water. She strokes and kicks directly toward the far side, that is, along a direction perpendicular to the opposite bank. But the water she swims across is also moving. It carries her downstream at 1 mph. It's as if she has two independent velocities. Figure 2–13 shows what happens. If you add these two velocity vectors, the *net* velocity shows her direction of travel as seen from the bank. The length of that net velocity vector represents her speed. She travels across diagonally with a speed that carries her both downstream at 1 mph and crossriver at 3 mph, and those two independent velocities are the *components* of her net velocity. A careful measurement on the graph shows this speed is about 3.15 mph.

Suppose you saw a girl swimming just like this in a river. How could you determine how fast she is moving downstream and how fast she is moving directly across? If you walked along the bank, keeping pace with her as she drifts downstream, your own speed would match the downstream component of her velocity. A friend on a footbridge

**FIGURE 2–13**
*(a)* The swimming girl gives herself a velocity over the surface toward the opposite bank, but the stream itself moves parallel to the bank.
*(b)* The swimmer's net velocity is the sum of the two velocity components, and she moves along the direction of the net velocity.

**FIGURE 2–14**
Two people resolving a swimmer's net velocity into two perpendicular components.

that goes straight across the river could do the same thing to find her cross-current speed. Then you and your friend would have *resolved* her velocity into perpendicular components by tracking her along these independent directions (Fig. 2–14).

## Relative Velocities

The next time you are stuck at home in the rain, try this. Watch the rain outside your window. Large raindrops fall so fast that your eyes see them only as a streak of reflected light, especially if you view them against a dark background. Move briskly to the right or left, peering at the rain as you move. Each time you'll clearly see a change in the direction of the falling drops. Why?

Think of it this way. Your own velocity carries you sideways past the rain's downward path, giving you horizontal speed the raindrops do not have. You see the raindrops being left behind as they fall, giving them a net direction that is diagonal, not vertical, with respect to you (Fig. 2–15). The rain, of course, continues to fall straight down *relative to the ground*. When you move horizontally, the rain moves *relative to you* in the opposite direction; you are leaving it behind with whatever speed you have over the ground. (This book placed in front of you can give the same effect. Move your head to the left and *relative to you* the book moves off to the right with the same speed your head has over the ground.) In terms of vectors, your motion gives the rain an extra component of velocity relative to you that is the reverse (or opposite) of your own velocity. That is, you see the rain as adding a component of velocity equal in length to your own speed but in the opposite direction.

Every motion you see is relative to you, so your everyday experiences include countless relative velocities. You instinctively judge relative velocities, for example, when a class is over and you ease into step with others in a busy hallway or on a sidewalk.

FIGURE 2–15
If rain falls straight down relative to the ground and you are motionless relative to the ground, you see the rain coming down vertically. But if you move sideways as the rain falls vertically, you see it coming down at some other angle. Its velocity with respect to you now has another component that is the opposite of your horizontal velocity.

FIGURE 2–16
An acceleration in the forward direction increases speed, and an acceleration in the backward direction decreases speed. An acceleration perpendicular to something's velocity vector will cause the object to turn.

## Acceleration: A Vector

When you first learn to drive a car, your actions are tentative, to say the least. But before long you get the "feel" of the car, and most of your driving becomes routine, almost automatic. You coordinate the car's position with the scenery before you, estimate the relative velocities to adjust the car's motion, and with the help of the memory bank in your mind, find your way. It's complicated! But Mr. Spock of *Star Trek* fame might say that driving a car is merely controlling its velocity vector.

You control your car's velocity over the ground in two ways. First, you can change its speed, which is the length of the velocity vector. To gain speed (which increases the vector's length), you press the accelerator pedal. Then you hit the brakes to bring the car to a stop, which shortens the velocity vector until it disappears. A second way to change the car's velocity vector is with the steering wheel. When you turn the car, you change the *direction* of its velocity vector (Fig. 2–16).

In one dimension, a change in speed means an acceleration has taken place. But in two or three dimensions, an acceleration is a change in speed *or* a change in direction, or both speed and direction at the same time. That is, *acceleration is a vector,* and it is the *rate of change of the velocity vector.*

To test the fact that you do accelerate as you turn, appeal to your senses! When your car accelerates away from a stop sign, you feel it—your body presses against the back of the seat.* If the car stops quickly, you shift forward. Each time you move opposite to the car's acceleration. Now if you quickly turn the car *without* changing the speed, you

*In Chapter 3 we'll see why you don't sense motion with your body so long as your velocity remains the same, and why you "feel" an acceleration of your body.

will nevertheless shift again, toward the side of the car and in the direction opposite from the direction you are turning in. The sensation is exactly the same as when you changed speed. Acceleration involves a change of either speed *or* direction, or both.

Our reactions in a car tell us even more. Acceleration is a vector, and as such it may be separated into components that are independent of one another. The component of acceleration that changes the speed points forward or backward, and we control it with the brake and accelerator pedals. The component of acceleration that changes the direction points to the right or left, and we control it with the steering wheel. Because we can turn a car without changing its speed and change its speed without turning it, we know these components of acceleration are separate and independent ways to alter the velocity vector. That happens because these different accelerations are perpendicular to one another. In an airplane, however, there is a *third* independent (perpendicular) acceleration to control; it turns the plane's velocity vector up or down.

The simplest motion is movement with constant speed along a straight line. The magnitude of the acceleration vector for that motion is zero, because the velocity vector doesn't change at all. But the complex velocity vector changes of a butterfly are described with a rapidly changing acceleration vector. The butterfly's acceleration can have three independent components at once, changing its horizontal direction (east–west, north–south), its speed and its up-or-down direction, all at the same time.

## Seeing in Three Dimensions

Because we live in a three-dimensional world, our eyes and brains deal with three-dimensional motion all the time. Ballet dancers and gymnasts prove how masterful some people can become at detecting and coordinating their own motion. Even the average person sees and interprets three-dimensional motion very well, or city driving would be much more hazardous than it is. All the techniques we use to see in three dimensions are learned from experience. Unlike us, infants see only a flat world. Their early, inadvertent grasping for fingers or bright toys and those later bumps and bruises play a big part in training them to sense the third dimension, depth.

If you sit absolutely still and close one eye, you will see a two-dimensional scene much like that a camera records.* Everything on a flat photographic print can be located with only two coordinates. But if you move your head at all while one eye is closed, you will sense depth in that scene. Here's how. Objects nearby move across your field of vision when your head moves to the side. If there is a vase on a table in front of you, it shifts more than a chair some distance behind it, but they both move more than the far wall of the room. When you are a

*A single eye can see more than a single photo, however, because tiny muscles within the eye adjust the shape of the lens to bring closer objects into focus. This **accommodation** lets you instinctively estimate distances from a few inches to 20 feet away.

**FIGURE 2–17**
The bobbing and turning motion of an owl's head helps it to use parallax to judge the distance to something.

child, your brain learns to associate these different motions with depth, so the sensing takes place at the subconscious level. Perhaps you've seen owls bob their heads around as they stare at something; they are using this method to help judge the distance to the object (Fig. 2–17).

With both eyes open, you see more than just two-dimensional scenes even if you sit perfectly still. Because your eyes are about 2½ inches apart, they see nearby points from slightly different angles. If you blink one eye and then the other, you'll see the nearest objects shifting positions against the backdrop. This shift is called **parallax.** The tiny shifts of an object from the right-eye view to the left-eye view let your brain estimate its distance even if it is as far as 1000 feet away. This **stereo vision** is also a learned ability.

**FIGURE 2–18**
When you are in motion, close, stationary objects appear to move by faster than more distant objects do.

**FIGURE 2–19**
You may also judge distance by convergence; when focusing on objects that are very close, your eyes cross, or converge, to some extent.

There is another way your brain detects depth. Stare at one finger as you bring it close to your nose. Your eyes rotate inward to focus on close objects, and the brain, sensing how far the eyes turn in, gauges how close the object is (Fig. 2–19). This range-finding mechanism called **convergence** works for distances up to about 100 feet.

Stereo vision aside, we judge our speed on a road or in a crowded room mostly by watching nearby objects, and our motion lets us estimate the distance to them. Doing 55 mph along an interstate highway is comfortable when there are broad, grassy shoulders, but the driver soon tires at the same speed on a narrow two-lane road. Because the roadside scenery really zips past, the driver is aware that it's closer and pays more attention. That's why movies filmed from Grand Prix racecars maneuvering through village streets and alleys strike fear into the hearts of even the most seasoned commuters.

**FIGURE 2–20**
Another way you judge distance is by experience; that is, by knowing the relative size of an object. Then, the larger the angle it takes up in your field of view, the closer you judge the object to be.

## Summing Up

> **To understand motion is to understand nature.**
>
> Leonardo da Vinci

We describe motion in terms of velocity and acceleration, vectors that tell how fast something moves, in which direction, and how rapidly the motion changes. But physics is a science that does not stop with descriptions. A basic goal of physics is prediction, including quite naturally *predicting* why things move the way they do. That's the subject of the next several chapters.

Physics began with scientists like Galileo, who observed motion in experiments and then tried to understand the motion well enough to construct physical laws describing how things behave. These physical laws could be used to know how something will move before it does so. Galileo analyzed the nature of the motion of falling objects and projectiles, and he even discussed how things would move if there were no pushes or pulls to influence their motions. But perhaps his greatest contribution was the method of relying on experiments to show what really takes place. Above all else, experimentation is the basis for modern science.

## CALCULATIONS

In the "Calculations" section for Chapter 1 we expressed words as formulas and did calculations. In this section we will express words as drawings and do estimates. Although vectors and their components can be "calculated" almost exactly using trigonometry, in most cases you can more easily measure a vector's components with a sheet of graph paper and a ruler. Graphs make it easy to add vector components or resolve a vector into components—so long as the grid marks on the graph's perpendicular axes are the same distance apart.

EXAMPLE 1: In Exercise 41 of Chapter 1 we found the effect of a headwind or tailwind on an airplane's velocity. With vectors we can use a graph to find the speed and direction of an airplane experiencing wind from the side (or from any other angle). Suppose a small plane is traveling north at 80 mph, and as it gains elevation, it encounters a 50-mph wind from the west. **Find the plane's velocity over the ground.**

You only need to draw to scale on the grid of the graph these two separate velocities and connect them tail to tip. A vector drawn from the tail of the first vector to the tip of the second is the net velocity over the ground. On the graph in Fig. 2–21 each grid mark represents 20 mph. The net vector shows the direction of the airplane, and its length will give us the speed. To measure that length, place the edge of a piece of paper parallel to the vector and mark its length. Then slide the paper down to the horizontal axis and measure this length in grid units. Using the scale factor of 1 grid unit = 20 mph, you will find the speed of the plane over the ground is **about 94 mph.**

**FIGURE 2–21**

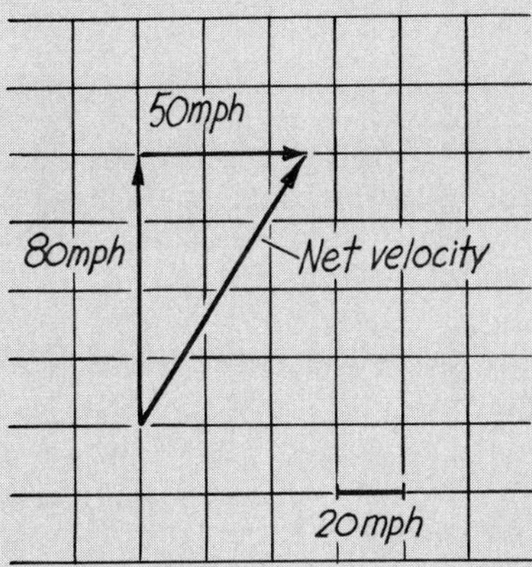


EXAMPLE 2: In a similar fashion, the grid marks on a graph can also help **estimate the acceleration of a girl sliding down the incline of a slippery waterslide.** To begin, draw a vector to represent the acceleration of gravity (Fig. 2–22). (If the slide were not there to support her, the girl would accelerate *straight down* at 9.8 $m/s^2$, or about 10 $m/s^2$.) This vector should point downward and be scaled to 10 $m/s^2$. On this graph 1 grid unit represents 1 $m/s^2$. Next, separate this vector into two perpendicular components, placing one parallel to the slide and one perpendicular to it. The acceleration "down"

FIGURE 2–22

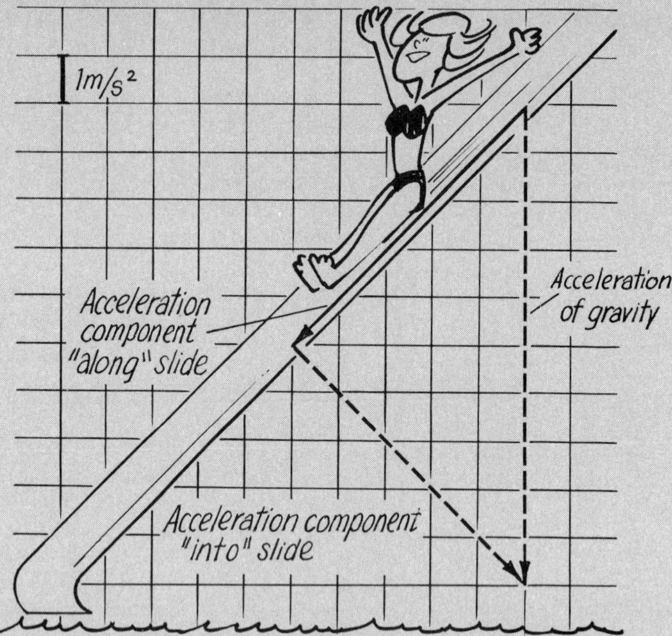


the slide is then independent of the acceleration ''into'' the slide. (The slide's surface does not permit any acceleration into it; we'll study the details of how in the next two chapters. Except for a little resistance from the surface of the waterslide, however, the acceleration along the slide is unchecked.) Using a strip of paper as in Example 1, mark the length of the component representing the acceleration down the plane and measure it against the grid marks. It is slightly over 7 units in length, so the girl gains speed along the slide at **about 7 m/s²**.

Does the tail of the vector have to begin any special place, or as we've drawn it here? *No*. It could be placed *anywhere* on the graph. The slide and girl are shown just to give the correct angle for the slide and to help visualize what is happening. For this problem, the graph and its grid marks represent accelerations, not distances.

---

EXAMPLE 3: Galileo learned that a body in motion tends to move in a straight line with constant speed unless an external influence (such as gravity or air resistance) affects the motion. Using the equation $d = \frac{1}{2} at^2$ and this fact discovered by Galileo, you can **find the path of a heavy projectile,** such as a cannonball.

If there were *no* gravity to slow it, a cannonball shot at an angle of 45° to the horizon would proceed along a straight line with the same speed it has at launch. Therefore, you could find the points where it would be along that line after 1 second, after 2 seconds, 3 seconds, and so on. For example, assume the launch speed is 30 meters/second. This cannonball would be 30 meters along the line after 1 second, 60 meters out after 2 seconds, and so on.

However, since there *is* an influence from gravity, any moving cannonball will also accelerate toward the ground as it travels. *The distance it accelerates relative to the straight-line path it would otherwise take is found using the equation $d = \frac{1}{2} at^2$, where the acceleration, a, is the acceleration of*

*gravity*, $g = 9.8\ m/s^2$, or about $10\ m/s^2$. To find the cannonball's position after 2 s, first find the point along the straight line where it would be after 2 s if there were *no* gravity (Fig. 2–23). Then our equation tells us that after 2 s the true position would be a distance of 20 m *straight down* from the position it would have if there were no gravity.

$$d = \frac{1}{2}(10\ m/s^2)(2\ s)^2 = 20\ m$$

To trace the complete path of the cannonball (Fig. 2–24), you must find the positions at several times, plot them on a graph, and connect them with a smooth curve. The curve becomes more exact if you plot additional points, such as at 0.5 s, 1.5 s, and so on.

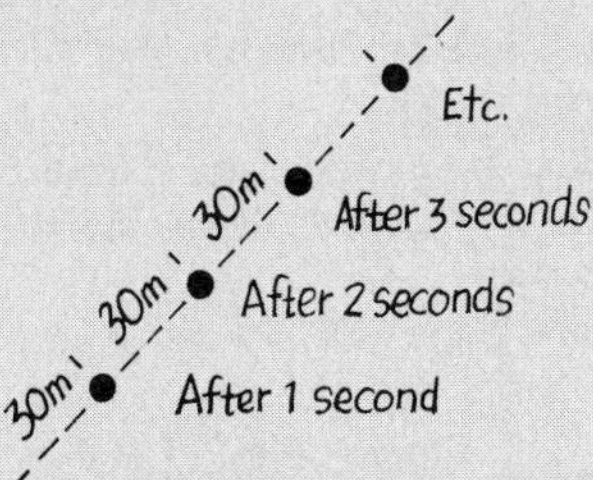


**FIGURE 2–23**
The path a cannonball would take with no gravity or air resistance to affect its motion.

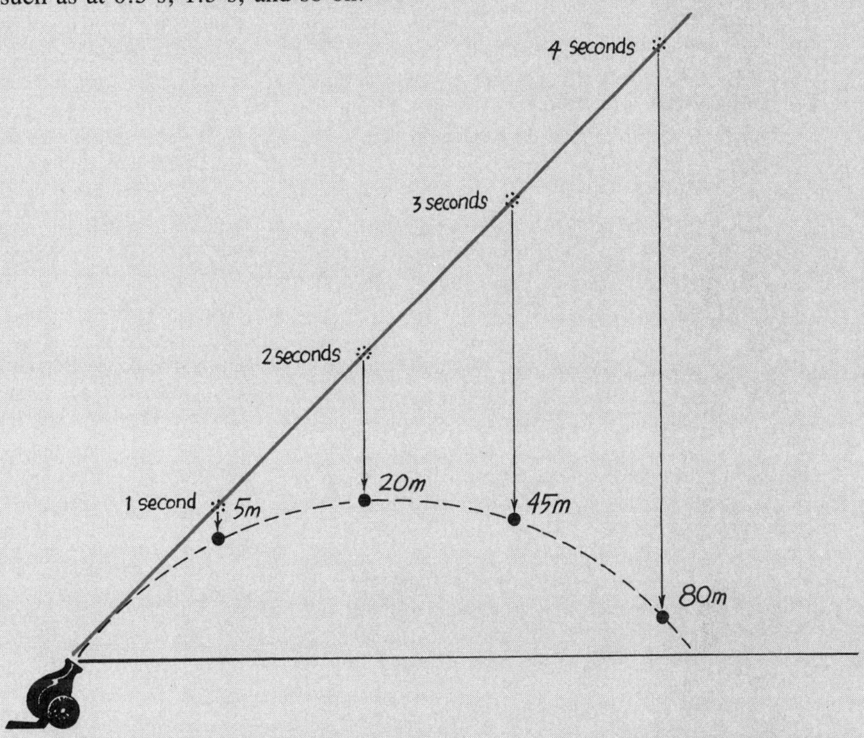


**FIGURE 2–24**
Calculating the cannon ball's path if gravity is present.

EXAMPLE 4: When two perpendicular vectors add to give a net vector, the three connected vectors form a so-called **right** triangle. In the right triangle in Fig. 2–25, $c$ is the length of the net vector and $a$ and $b$ are the lengths of the shorter, perpendicular vectors. You can see by inspection that $c$ will always be a smaller number than $a + b$. This is true whether the vectors are displacements, velocities, or accelerations. But there is a relationship between the lengths of the sides of a right triangle. This relationship, named after Pythagoras, the Greek mathematician who discovered it, is called the **Pythagorean theorem,** $c^2 = a^2 + b^2$.

Very carefully, **measure *a, b,* and *c* in Fig. 2–25 and use those measured numbers in the Pythagorean theorem.** Do they check? Once again, this shows how difficult it is to measure accurately. The formula is exact, but your visual estimates of those lengths (and indeed, the drawing of the triangle figure itself) is not precise.

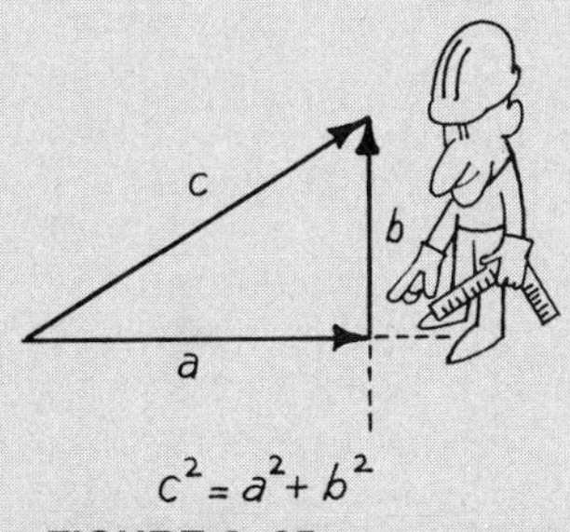


**FIGURE 2–25**

# REVIEW

**1.** To describe where something is in three dimensions, you need three coordinates, in two dimensions, two coordinates, and along a straight line, one coordinate. True or false?

**2.** A train moves on a railroad track between two towns. You'd most likely need ______ coordinate(s) to describe the train's position.

**3.** Does an acceleration along one direction influence an object's speed in a direction perpendicular to that direction?

**4.** A vector is a representation of a quantity with direction as well as magnitude. True or false?

**5.** The vector that represents a change in location is a displacement vector. True or false?

**6.** Velocity is speed and direction taken together. True or false?

**7.** A vector may be built up from several vectors or resolved into several other vectors. True or false?

**8.** Two perpendicular vectors that add to give a third (or net) vector are called components of the net vector. True or false?

**9.** An acceleration changes the direction and/or the speed of an object. Another way of saying this is that acceleration is the rate of change of ______________.

**10.** Everything on this page could be pinpointed with ______ coordinates.

**11.** To add two vectors, you must first, without changing their directions or length, place them tail to tip. True or false?

**12.** Does the net vector of two added vectors join the tail of the first to the tip of the last?

**13.** Because your eyes are about 2½ inches apart, you can use the effect referred to as (a) convergence, (b) parallax, (c) accommodation to judge distance.

**14.** In your car you control the "turn" component of acceleration with the steering wheel and the acceleration "along" (or tangent to) the route with the brake or accelerator pedals. True or false?

# DEMONSTRATIONS

**1.** With only a little effort you can arrange a uniform acceleration of, say, 1 meter/second$^2$. Tilt a board a little so that a marble can roll down the slope. A component of gravity's attraction will accelerate the marble uniformly along this inclined plane. (See Example 2 in "Calculations"; although the marble rolls down rather than sliding, its acceleration is uniform.) The formula $d = \frac{1}{2} at^2$ tells us that an acceleration of 1 m/s$^2$ will carry the marble initially at rest a distance of ½ m in 1 s. Mark this distance off along the board and let the marble start from rest at the top. Check the time it takes the marble to travel to the mark. Then vary the angle of the inclined plane until you find the slope that causes the marble to reach the 0.5-m mark after 1 s of rolling. What you see then is a bona fide, uniform, 1 m/s$^2$ acceleration. Galileo used such inclined planes in his investigations of motion.

**2.** One of the most important devices throughout human history has been the wheel. Its shape is a circle, a curve whose every point is the same distance from a single point, the circle's center. No matter what size, there is a fixed relationship between the distance around the circle (its **circumference**) and the straight-line distance across it and through its center (its **diameter**). The ratio of circumference to diameter is called $\pi$ (Greek letter *pi*, pronounced "pie"). Its value to two decimal places is 3.14. With a ruler and a piece of string, you can check that ancient Greek discovery.

Wrap the string once around a circular shape (use a cup, a can, a plate) and then stretch that length of string along the ruler to estimate the object's circumference, $C$. Next, use the ruler to estimate the diameter, $d$, and form the ratio $C/d$. Do this for a variety of circular objects and compare each of the ratios to the value for $\pi$. (Once again you'll see how difficult it is to measure things accurately.)

**3.** Here's a demonstration to help you visualize what goes on with the tires on a moving car. Using a cardboard tube from a roll of paper towels, tape a pen inside so the point just sticks out beyond the rim, as in Fig. 2–26*a*. Then

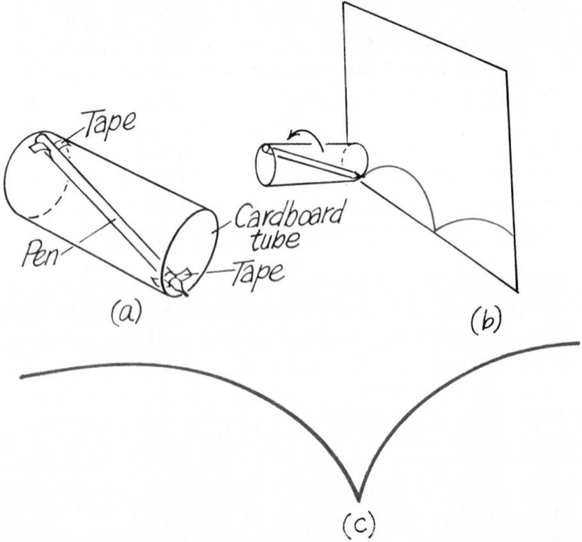


**FIGURE 2–26**

place the tube on a flat surface with the pen's point resting against an upright pad of paper. See Fig. 2–26*b*. Roll the tube along so the pen will trace a path on the paper while the tube rotates. The result is a curve like the one in Fig. 2–26*c*. Any point on the edge of a moving wheel also follows a curve like this. It moves over and down, makes contact with the ground, then abruptly rises up and curves over again. Just when the point touches the ground, it changes directions. At this turning point, as when a set of keys is tossed straight up and turns around, its speed is instantaneously *zero*. Turn the tube slowly and watch the pen as it marks along the paper; you'll see this happen. (For an analysis of this action in terms of relative velocities, see Exercise 35 at the end of this chapter.)

# EXERCISES

**1.** Think about the motion of a shot hurled across a field. It begins its flight in some direction and continually turns downward until it falls to earth. Would you say it moves in one, two, or three dimensions? Discuss. (Also see Exercise 2.)

**2.** Trapeze artists at a circus are dazzling to watch because of their vertical motion and their elevation. Of all performers, they seem to take full advantage of space—or do they? How many dimensions do they normally use during their acts?

**3.** There are (a) 1, (b) 2, (c) 3 independent directions of motion something can have in space, and there are (a) 1, (b) 2, (c) 3 independent directions of motion something can have on a surface.

**4.** Explain how a hiker who stays on a certain trail in the mountains essentially moves on *one* dimension. How could you give the hiker's coordinate?

**5.** On a clear and windless day a plane flies overhead and the villains inside push James Bond out the door without a parachute. Would you say he moves in (a) 1, (b) 2 or, (c) 3 dimensions as he falls?

**6.** Have you ever seen an airplane's shadow zip past over the ground? Is the speed of the shadow the same as the speed of the plane? (See Fig. 2–27.) Discuss.

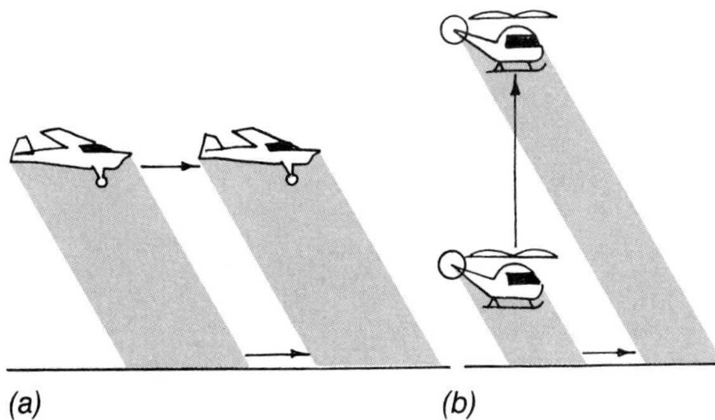


**FIGURE 2–27**

**7.** If the sun were directly overhead and you threw a baseball, describe the motion of its *shadow* on the ground. Should the shadow move with almost constant speed? Discuss.

**8.** Watch a tennis player serve. Ideally, about where is the ball in its path when it is served? Give at least two reasons why that's so.

**9.** Figure 2–28 compares a trajectory on a planet to the path the same object would follow if there were *no* gravitational acceleration. Why are the horizontal positions of the object the same at each instant?

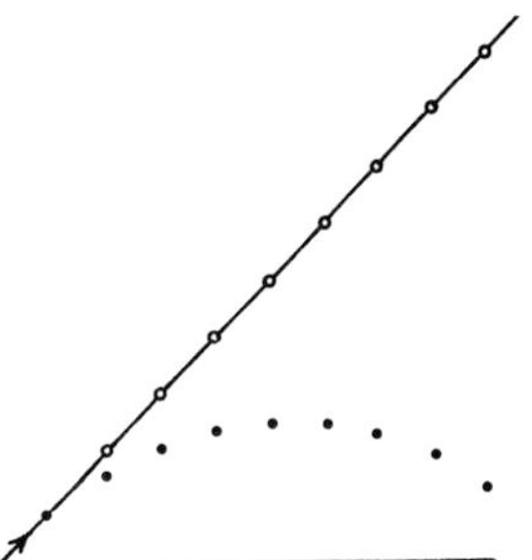

**FIGURE 2–28**

**10.** Two small, heavy objects are dropped simultaneously into a very deep and narrow canyon from opposite sides (Fig. 2–29). One of them is given a shove in the horizontal direction toward the other object at the same instant they are dropped. Explain why, if air resistance is negligible, they will eventually collide.

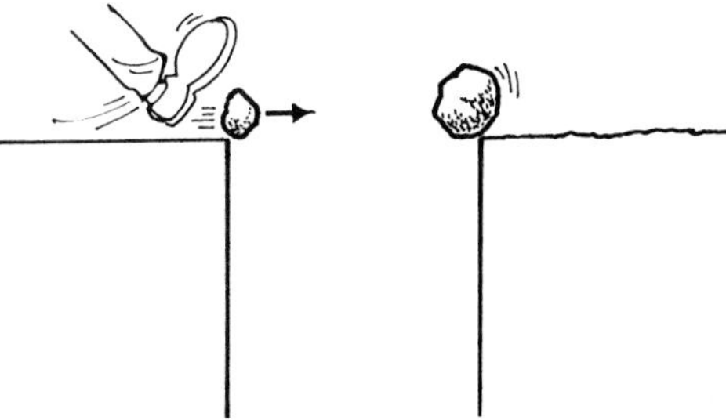

**FIGURE 2–29**

**11.** Which of the following quantities are not vectors? (a) Acceleration, (b) temperature, (c) velocity, (d) a displacement, (e) speed.

**12.** If two snowmobiles have the same speed, does this mean their (a) accelerations, (b) positions, (c) velocities, (d) directions are the same?

13. If an airplane gains elevation or loses elevation at a steady rate, is it necessarily accelerating?

**14.** Can a car travel at a constant velocity while accelerating? Can a car travel at a constant speed while accelerating?

15. Walk around a circle. Your steps never take you any closer to the center. What does this tell you about a tangent line to the circle? (See the answer to this one.)

$\overrightarrow{*16}$. The rowboat shown in Fig. 2–30 has two components of velocity, one from the river's current and another from the efforts of the rower. (a) Draw the net velocity vector. (b) Next, suppose a wind comes along and gives the boat a third component of velocity causing it to stall in midstream. Draw the velocity component that does this.

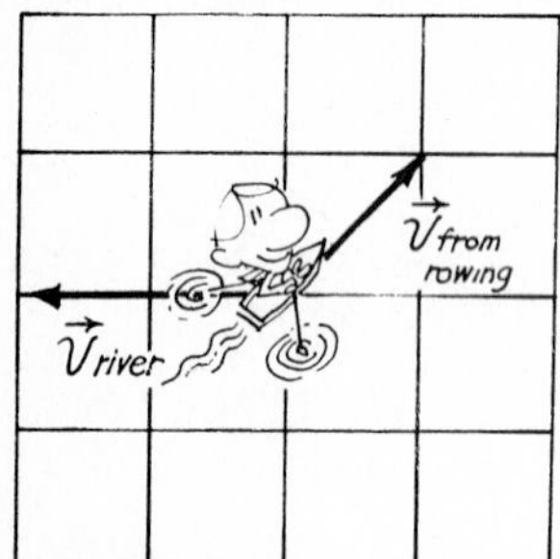


**FIGURE 2–30**

**17.** Can a body have a constant acceleration without changing speed? Explain your answer.

**18.** Three astronauts stand on a level plain on the moon. Acting together, one drops a feather, one throws a baseball horizontally, and one fires an arrow horizontally from a crossbow—each from the same height. Which of these statements is correct? (a) The feather, the baseball, and the arrow hit the surface at the same time. (b) The feather hits first, followed by the baseball and then the arrow. (c) The arrow hits first, followed by the baseball and then the feather.

**19.** When a car moves along the road, does every point on the car's metal body have the same velocity vector? When you walk across the room, does every point on your body have the same velocity vector?

20. A newspaper carrier throws a heavy Sunday paper (negligible wind resistance) straight out from the window of her car onto someone's lawn while moving past at 20 mph. (a) Describe what the carrier sees as the paper is in flight. (b) Describe what someone a block away on the same (straight) road as the car would see. (c) Describe what a squirrel high in a tree in the yard might see.

*The arrow above the exercise number indicates a vector exercise.

$\overrightarrow{21}$. When you walk in the rain with an umbrella and there is no wind, do you carry it straight up or do you tilt it in the direction of your motion? Explain with a drawing.

22. A couple drew the chart in Fig. 2–31, describing their 8-hour Saturday excursion. Assume the motion was *not* along a straight line. (a) When was their speed greatest? (b) When was their speed zero? (Careful.) (c) If they drove for an hour along a circular path centered on their home, when did that take place?

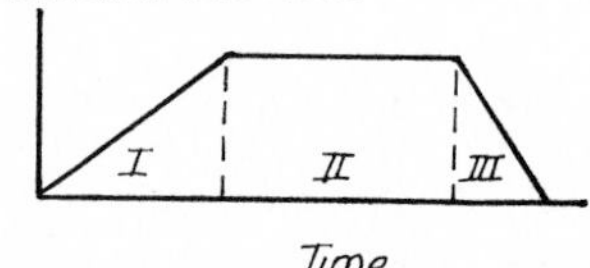


**FIGURE 2–31**

**23.** Figure 2–32 shows a strobelike graph of the motion of a moving object. Identify (a) the regions of constant speed, (b) the regions of deceleration, (c) the regions of acceleration, (d) the regions of constant velocity.

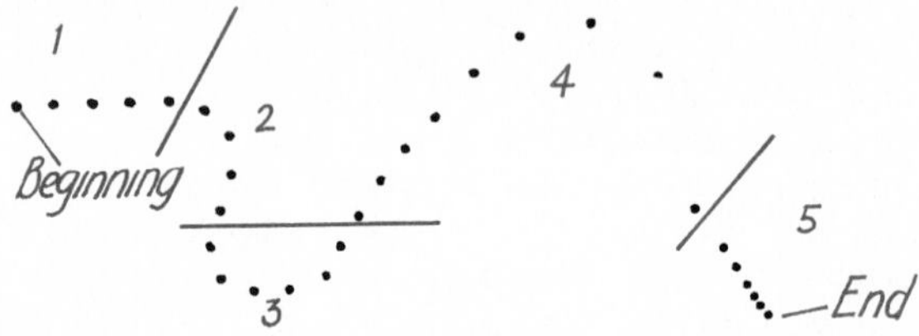


**FIGURE 2–32**

24. If you walk along a trail all day to a campsite and then walk back during the same hours the next day, will it happen at least once that you will be at some point on the trail at the same time on both days? Using the graph in Fig. 2–33, draw several paths and find out.

**FIGURE 2–33**

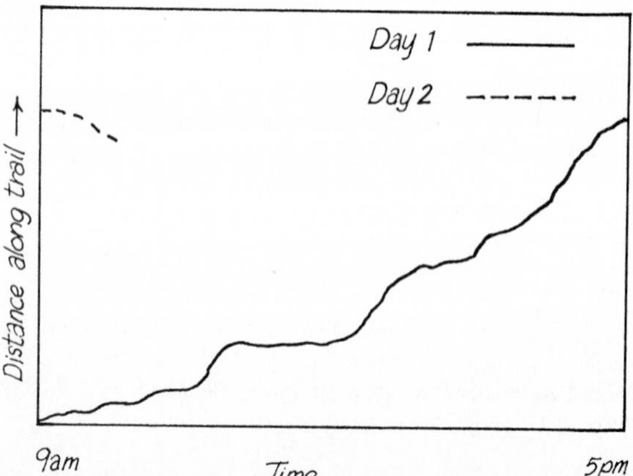

**25.** To get from point (A) on the map in Fig. 2–34 to point (B) in the shortest possible distance, show that there are two different paths you can follow that can be described with only two displacement vectors and that the two displacement vectors are the same ones for either route. Also show that there are two other paths you could follow using three separate displacement vectors that would get you there.

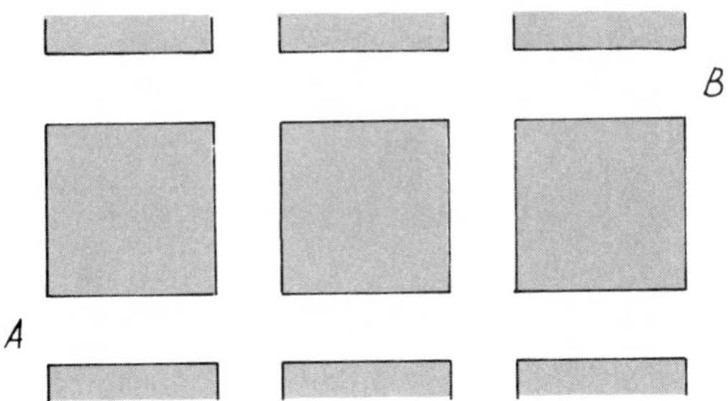


**FIGURE 2–34**

**26.** We see the stars as if they were a painting on a two-dimensional canvas. This is because (a) our eyes are too close together, (b) the earth's orbit around the sun is too small, (c) the stars are too far away, (d) all of these.

**27.** Frogs cannot move their eyes to scan the landscape, and experiments show that they respond only to moving prey. How does this limit the depth perception of the frog as compared to human eyesight?

**28.** Most predators, including humans, have eyes that face forward, although many prey animals such as deer and rabbits have eyes on opposite sides of their heads. Discuss the advantages of each arrangement.

**29.** Draw the resultant of the two vectors shown in Fig. 2–35.

**FIGURE 2–35**

**30.** What is the component in the vertical direction of the vector shown in Fig. 2–36?

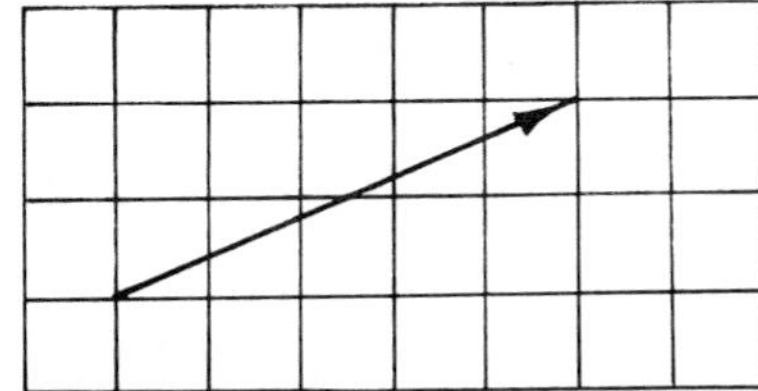

**FIGURE 2–36**

**31.** Add another vector to the sum shown in Fig. 2–37, so the net displacement is zero.

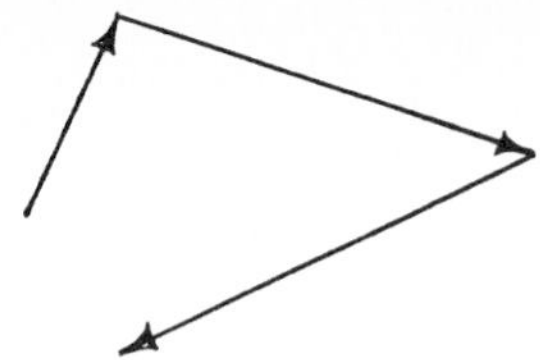

**FIGURE 2–37**

***32.** Taking her time and averaging 2 mph, a girl swims a river that is a mile wide. The river's current is 3 mph. If she swims straight for the other side, how far downstream will she land?

**33.** A pilot flies north with an airspeed of 150 mph. Unknown to the pilot, there is a steady wind from the west at 20 mph. (a) Draw a graph to show the plane's direction of travel. (b) Will its speed over the ground be equal to, greater than, or less than 150 mph? (c) Will its speed over the ground be equal to, greater than, or less than 170 mph?

**34.** Figure 2–38 shows the velocity vector of a band marching diagonally across a football field. How fast are they moving down the field? How fast are they moving across the field? (You will need a ruler to answer this one. First, draw the components of velocity across and down the field that add to give this actual velocity vector. Then measure their lengths and compare them to the net velocity to get the speeds in these two directions.)

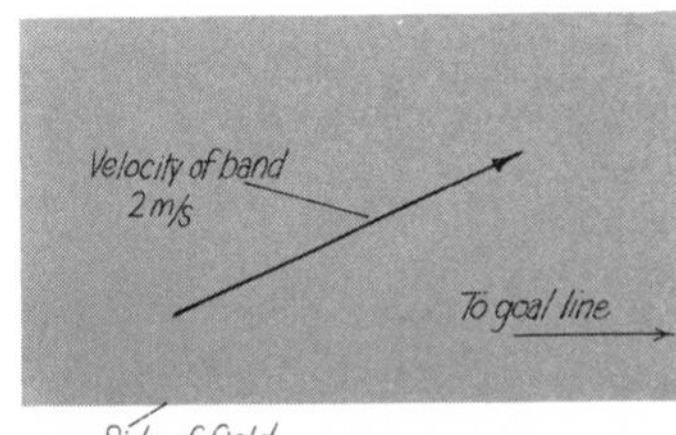


**FIGURE 2–38**

**35.** Figure 2–39 is a diagram of a car's wheel traveling at 55 mph. Exactly what are the *velocities* of the points *a*, *b*, *c*, *d*, and *e* on the wheel with respect to the ground?

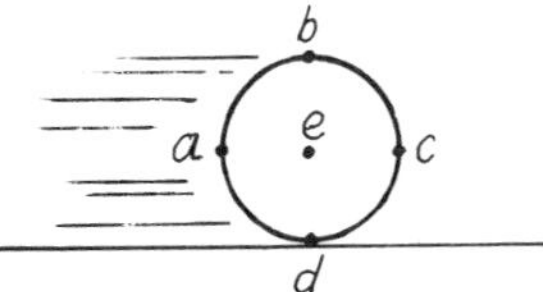


**FIGURE 2–39**

***36.** Using the technique of Example 3 in "Calculations," graph a few points in the motion of two shots thrown at 45° to the horizon, one on earth ($g$) and one on the moon ($g/6$). Compare these to Fig. 2–7.

## BEGINNINGS

When you reach for a glass of water and bring it to your lips, you know what to expect. The glass is at rest, and you accelerate it with your hand—not too fast or you'll spill the water—and you bring it to a halt so you can drink from it. You also know what would happen if it slipped from your grip. More than likely, you would move your feet to avoid the falling glass. Because almost everything you do requires moving something about, whether you're turning a page or merely taking a breath, you know all this ahead of time. That is, you have a feeling that is based on experience for how things move.

The Greek philosopher Aristotle, who took this kind of intuition very seriously, wrote about motion around 350 B.C. Aristotle knew that if he pushed a plate across a table and then took away his hand the motion of the plate would stop. To describe this, he wrote: "All that is moved is moved by something else." He reasoned that when the push from the "something else" stopped, so did the motion; from this he decided that *rest* must be the natural state of any matter.

But this explanation had its problems. For instance, it didn't explain how a spear continues in flight once it leaves the hand, or why an arrow keeps going once it leaves the bow. So Aristotle further decided that the front surface of any object moving through air must compress the air at that surface and cause the air in the space directly behind the object to be rarefied, or thin. He argued that the air from the front must rush to the rear to fill the partial vacuum in the thinner air, and that as the air filled in this space it pushed the projectile along. (To explain why an arrow in flight eventually slows, he said the transfer of air was never complete.) This false premise led to another wrong deduction, namely, that motion must be impossible in the absence of air.

Aristotle deduced his "laws" just from watching things move. Many of the early Greek philosophers like Aristotle who wrote about motion believed that intense mental concentration and pure thought would solve the riddles of nature and that philosophers should never have to perform experiments to gain understanding. Aristotle said, for example, that heavier bodies always fall faster toward the earth than do lighter bodies. (Some do, of course, because of the effect of air resistance.) And since heavier bodies make more noise and larger dents when they strike the ground, that was easy to believe. Furthermore, it is harder to lift a heavier body, so it's certainly attracted more strongly toward the ground.

Although Aristotle's unproved ideas were not universally accepted, they were still taught when the Italian scholar Galileo Galilei (1564–1642) lived and worked. Then Galileo introduced the experimental procedures—*careful observation by measurements*—that made physics a science of accurate predictions. Galileo discovered that all falling objects would move with a uniform acceleration if air were absent. He deduced that force is not necessary to keep things moving, that instead forces of friction bring moving things to a halt. But Galileo fully realized that he had just begun to understand motion. He wrote that he had "opened up to this vast and most excellent science, of which my work is merely the beginning, ways and means by which other minds more acute than mine will explore its remote corners." Isaac Newton made the next steps, and his contributions to physics are so immense that they may be unmatched in greatness in the whole history of science.

Isaac Newton was born on Christmas Day, 1642, in a stone farmhouse in Lincolnshire, England. He was a premature baby, so tiny that his mother said she could have put him in a beer mug. But as a schoolboy he was healthy and very creative in making things, such as waterclocks, sundials, and even a wheeled chair. He boldly carved his name in his desk at school, and one of his notebooks, still preserved, has an article he copied—it tells how to get birds drunk! One of his unique projects, a kite carrying a homemade paper lantern, startled the local populace one night. (This dimly lit spectacle hovering in the dark sky very likely summoned rumors of witches and comets rather than UFOs.) Although Newton's father, who died a few months before Newton was born, had been a farmer, as had his father before him, the local schoolmaster persuaded Newton's mother, who had already decided Isaac would not make a good farmer, to let her 19-year-old son enroll at Trinity College in Cambridge.

Newton came along at an exciting time. Seventy years before, the philosopher–writer Giordano Bruno had visited England and had written that lecturers at the universities were fined if they were critical of Aristotle's ideas. Indeed, only 20 years before Newton's arrival at Cambridge, Galileo had died under house arrest in Italy for writing that the planets revolve around the sun. Besides his experiments in physics, Galileo built and used the first telescope. With it he discovered four large moons orbiting Jupiter, and he saw that the planets were illuminated by the sun, showing "phases" like the moon, depending on where they were in their orbits. Galileo's astronomical discoveries were there for anyone to see through a telescope, and his experiments on motion could be checked anywhere. Progressive scholars* formed groups such as the Royal Society of London for Improving Natural Knowledge. (Today it's known as the Royal Society.) But Newton, who was poor, worked at part-time jobs and graduated without distinction in 1665.

That summer the college closed, for the plague was raging in nearby London, killing over 10 percent of the city's people within 3 months. Newton returned to his family home and in the peace and quiet of the countryside devoted his attention to matters of mathematics and "natural philosophy," as physics was called in those days. During 18

---

*The name *scientist* was not coined until the mid–1800s.

months of intense, uninterrupted study, he accomplished wonders. He discovered how to *predict* motion, he began his investigations of gravity and the colors of light, and he invented the methods of calculus and the famous binomial theorem of algebra. But Newton, being somewhat introverted, kept to himself and was not to make known much of this work for some 20 years.

His studies led him to write down the laws of motion, extending, and in a sense completing, the work begun by Galileo. These three laws together tell us how things move, and today they are known as *Newton's laws*. You'll meet them in the following chapter. (For more on Isaac Newton, see "The Discovery" following Chapter 6.) ■

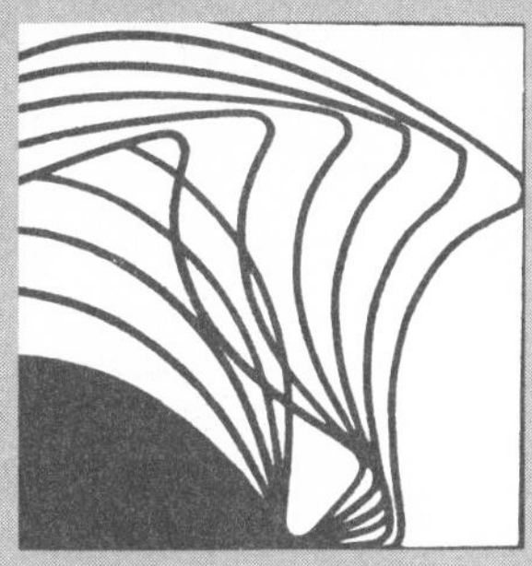

# CHAPTER 3
# THE LAWS OF MOTION

**Open wide the door for us, so that we may look out into the immeasurable starry universe; show us that other worlds like ours occupy the ethereal realms; make clear to us how the motion of all worlds is engendered by forces; teach us to march forward to greater knowledge of nature.**

Giordano Bruno, writer-philosopher; born 1548, burned at the stake for heresy, 1600.

Every morning you wake up to face the same chore; like it or not, you must change your state of motion. Perhaps you stretch and then sit up, swing your feet around, and plant them on the floor. All of these motions involve starting and stopping—that is, accelerations. Then you step away from the bed, and all day long you'll be adjusting your motion. Each time, your brain signals the proper muscles with electrical impulses and these muscles contract, pulling on parts of your body and accelerating them.

A muscle pulls on a bone, and it accelerates. Your hand pulls to open the door of a car, and your shoe pushes to depress the car's brake pedal. Indeed, whenever anything accelerates, a push or a pull causes it. Another word for push or pull is *force*. A walnut that breaks loose from a tree is pulled by earth's gravity and accelerates steadily downward, while many different forces act to guide a taxi through rush-hour traffic in a big city. Because you sense pushes and pulls with your body, you know what a force is. Just walking across a floor, you exert forces and feel other forces acting on you. But you can't *see* a force; you can only see the change it makes in the motion of the thing it acts upon.

Although you can't see forces, you can relate to them. Maybe you've gripped the string of a kite as it sailed along on a stiff breeze, or perhaps you've been drenched by a summer shower when a gust of wind seized your umbrella. If so, you understand what it means to say a bird pushes its wings against the air. Likewise, if a playmate's tricycle ever bumped over your foot when you were a child, you know that a tire pushes against the road. The physics in this chapter makes sense because you live with it all day, every day of your life. ■

## Force: A Push or Pull in a Certain Direction

Finishing your salad at a restaurant, you might shove the plate aside to make way for the main course. The empty plate slides off—not in just any direction, but in the direction you *pushed* it. Every push or pull has

**FIGURE 3–1**
When more than one force acts on you, your body responds to the vector sum of those forces, the *net* force.

a direction, and this property of force is every bit as important as its strength.

Just suppose for a lark two friends grab your arms and before you suspect anything one pulls you northward as the other pulls you toward the east. You know what will happen. If they are pulling at the same time, you won't move in the direction of either force, but at some angle in between (Fig. 3–1). Of course, you'll feel the separate pulls with your arms, but your body responds to both pulls at once. Try it! Equal pulls from your two friends will send you in a northeasterly direction; but if the one to the north tugs harder, there's no doubt you'll move more north than east.

This is what happens whenever several forces act at the same time. Forces, in fact, just like velocity and acceleration, are *vector* quantities. This means you can describe any push or pull on a graph by drawing an arrow to show its direction and using the length of the arrow to represent its strength (Fig. 3–2*a*). If you place the two force "arrows" of your friends tail to tip, you can find the "net" arrow of force (Figs. 3–2*b* and 3–2*c*). And that *net force* describes the effect of the two forces acting together. It's as if only one force were present. If all this seems like common sense to you, don't be surprised. You have a lot of practical experience with pushes and pulls. But that experience really

**FIGURE 3–2**
*(a)* Represent each force with an arrow whose direction is the direction of the pull or push and whose length corresponds to its strength. *(b)* Then slide the arrows together tail to tip in any order to find their sum. *(c)* The net force is a vector drawn from the tail of the first force in the sequence to the tip of the last force.

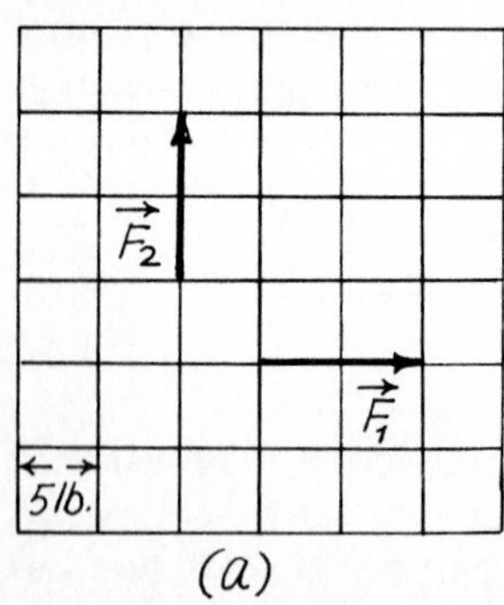


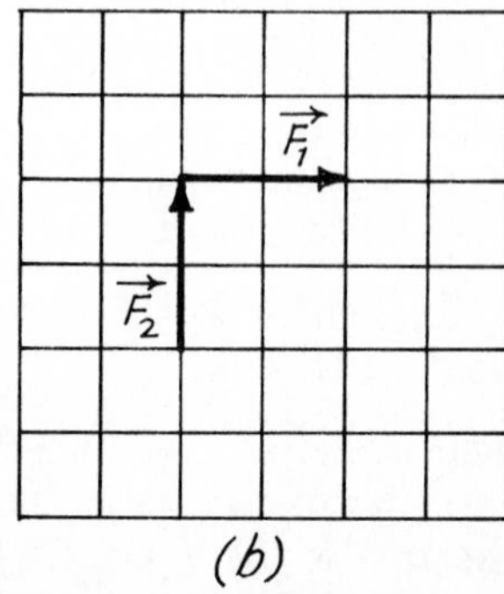


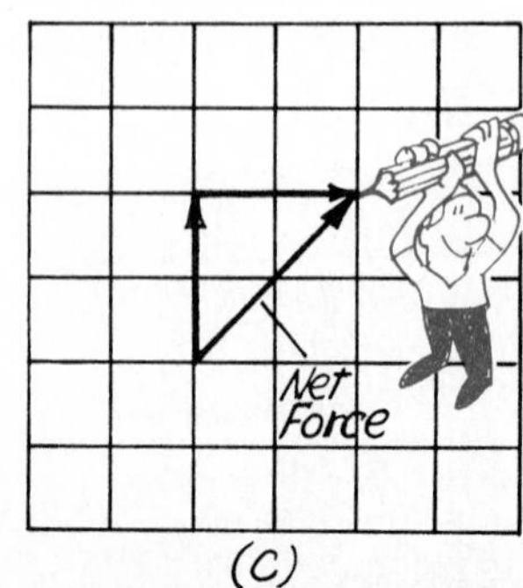

can't tell you *how much* motion is changed when forces act, that is, how fast something accelerates. That's what Newton's laws are about.

**FIGURE 3–3**
This beer mug seems to be following Newton's first law on the slick, horizontal surface (at least for a while!).

## The First Law

Shove this book across your desk. You have to push it to make it move. When you stop pushing, the book stops moving. You could jump to the conclusion that the force causes the motion and that when no force acts, there is no motion. That's what Aristotle thought, but he missed an important point. Forces can *stop* something's motion as well as start it; the frictional force from the desk stops the book. You could say the book did not need more force from your hand to keep moving—it needed less frictional force from the desk.

Galileo got it right. He realized that a rolling ball (or a sliding book) could travel much farther before stopping when there was little frictional resistance. He deduced that with no friction at all, an object should travel without any change in speed or direction. On the other hand, Galileo's written conclusions were vague. He failed to state his results in a way that proved their generality to others. His French contemporary René Descartes did better when he said: "Every body tends to continue motion in a straight line . . . ." However, saying that something "tends" to do something sounds as if it might not *always* do it. It was left for Isaac Newton to say what Galileo really discovered about motion. Newton wrote:

**"Every body continues in its state of rest, or of uniform motion in a right [straight] line, unless compelled to change that state by forces impressed upon it"** (Newton's first law).

This statement seems to be an idealization to us because there are forces all around us. Only astronauts go to an environment where things seem to obey this law. In orbit around the earth they can throw a flashlight and it will move along a straight path until it hits something in its way. Pencils and food containers left in midair can stay in place until small air currents in the spacecraft cause them to drift off. (But even then a force is present, as we'll see in a later section).

This, then, is the essence of the first law: Matter resists any change in its velocity. *Only a push or a pull can change something's speed or direction.* This property of matter, *its resistance to a change in velocity,* is what we call **inertia.** (We'll see how to measure inertia in the next section.)

With a pair of roller skates you can get up to speed and coast a long way. Skateboards, too, take little effort to keep moving. When you enjoy such coasting, tip your hat to the first law! It's your inertia that opposes any change in speed or direction and keeps you moving so straight and steadily.

## The Second Law

Children learn about inertia by pulling and pushing things around. It's easy to run off towing an empty wagon from rest, but just starting one filled with sand is a tough job. Early experiences like these add to what

**FIGURE 3–4**
Alan Bean reading a floating teleprinter tape while orbiting the earth in Skylab in 1973. The tape and the cover cloths behind it had no acceleration relative to Skylab.

we call our common sense. Then, too, it takes practice before our motions become smooth in daily activities. You handle a fork and spoon without thinking because you know what sort of inertia they have before you pick them up. To become good at sports and other games of skill, you must practice enough to get a "feel" for the game. Baseball pitchers, for example, become very good at judging how much force to use to control a ball's motion. But a nonregulation baseball, or one sodden with water, causes even a good player to lose much of the touch that came from so much practice.

The point is this: From experience we all know there is a connection between the *quantity of matter* in a wagon or a spoon or a baseball and how it accelerates when it is pushed or pulled. Experiments can show this connection more precisely than common sense does. For instance, you could pull an empty wagon with a steady pull, say, of 10 pounds. With a strobe photograph of the motion, you could find how the wagon's speed changed. Then you could load an identical wagon on top of the first one and repeat the experiment using the same force. This time, with *twice* as much matter to be moved, you'd get only *half* as much acceleration for the same force. Or, with the wagon empty again, you could double the force, and the acceleration would turn out to be twice as large as before.

Right now, he's busy demonstrating Newton's first law.


**FIGURE 3–5**

Results much like these led Isaac Newton to the second law of motion, the law that *predicts how things will move when forces act*. Newton used the phrase "quantity of matter" to mean the same thing as "amount of inertia," but today we use the word **inertial mass,** or more briefly, just **mass.** The second law gives the relationship between mass, force, and acceleration of any object. The product of the mass $(m)$ of the object times the acceleration $(a)$ it has is equal to the force $(F)$ acting on it. More simply,

$$F = ma$$

Does this law work when *more* than one force acts on something? On a cold night you might sit at a table with your hands wrapped around a mug of hot chocolate to keep them warm. If you are pressing them against the cup, you're exerting forces on it. But the cup doesn't move because the pushes from your hands balance each other and there's no *net* force. When more than one force is acting, it is the net force that causes any acceleration. Experiments show that the net force, $\vec{F}_{\text{net}}$, is equal to the mass of the object times its acceleration, $\vec{a}$, where the object's acceleration is in the same direction as the net force. In words, the net force vector equals the mass times the acceleration vector.

In symbols,

$$\vec{F}_{\text{net}} = m\vec{a} \qquad \text{(Newton's second law)}$$

This law works for baseballs and baby carriages, and for moons and planets. It works for simple motions as well as complex motions. If you know the forces on any object, add them tail to tip to find the net force. The direction of the net force gives you the direction of the acceleration,

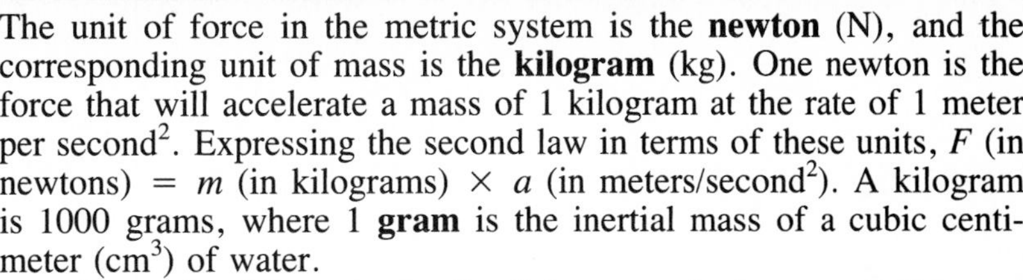

## UNITS OF MASS AND FORCE

The unit of force in the metric system is the **newton** (N), and the corresponding unit of mass is the **kilogram** (kg). One newton is the force that will accelerate a mass of 1 kilogram at the rate of 1 meter per second$^2$. Expressing the second law in terms of these units, $F$ (in newtons) $=$ $m$ (in kilograms) $\times$ $a$ (in meters/second$^2$). A kilogram is 1000 grams, where 1 **gram** is the inertial mass of a cubic centimeter (cm$^3$) of water.

The unit of force in the English system is the **pound** (lb), and the unit of mass is the **slug** (stemming from ''sluggishness''—a large mass responds more sluggishly to a given force). A force of 1 pound accelerates a mass of 1 slug at the rate of 1 foot/second$^2$. So $F$ (in pounds) $=$ $m$ (in slugs) $\times$ $a$ (in feet/second$^2$).

The units of force (N or lb) may also be expressed in terms of the fundamental units of mass and acceleration, since $F = ma$. In this case, force has the units of mass $\times$ length/time$^2$, which can be abbreviated as $ml/t^2$.

About 4.45 newtons equals 1 pound, and 1 slug is about 14.6 kilograms. One kilogram of mass on earth weighs about 2.2 pounds; that is, if $m = 1$ kg, then $mg = (1\text{ kg})(9.8\text{ m/s}^2) = 9.8\text{ N} = 2.2$ lb.

(a)

(b)

**FIGURE 3–6**
*(a)* A mass of one kilogram. *(b)* One kilogram *weighs* 9.8 newtons.

and its magnitude (in pounds or other units) divided by the object's mass predicts the acceleration; this law *predicts* motion.* That is, if you know something's acceleration, you can use the relations between distance, speed, and acceleration that we discussed in the first two chapters to tell where it will be, its direction of motion, and how fast it will be traveling at a later time.

Newton's first law says that matter has inertia; it moves in a straight line with constant speed (or lies at rest) unless there is a net force on it. In other words, when there's no net force on an object, its velocity is constant. *Newton's second law states how a net force changes something's velocity*. Suppose at an airport you need to pull a freely rolling hand truck loaded with several suitcases. If the combined mass of the truck and suitcases is 3 slugs, a 6 pound pull will accelerate them at a rate of 2 feet/second$^2$. ($F = ma$, so 6 pounds $=$ 3 slugs $\times$ 2 feet/second$^2$.) But if three more suitcases with a total mass of 2 slugs are piled on this handtruck, the force needed to accelerate it at the same rate is now 10 pounds. If a mass is small, little net force is needed to accelerate it, while if the mass is huge, a great net force is needed to change its velocity at the same rate. The larger the mass, the greater the object's inertia, that is, the greater its resistance to a change in its velocity.

*Exactly how accurate are the predictions of this law? See the last section of Chapter 4.

Anytime the velocity of matter changes either in magnitude or direction, there is a net force on it. Because you are reading these lines, your eyes must turn back and forth. If you look up, the lens must adjust to focus the light from a different distance. Extremely small muscles exert extremely small pulls to cause these accelerations. But whenever matter gets a push, Newton's second law predicts what will happen. There is still more to know about changes in motion. Each time something gets a push, something else *gives* it. The book slides across the desk because your hand pushes it. You can *feel* the book resist this change in motion; it pushes back against your hand. Newton's third law deals with this feature of motion.

## The Third Law

When two children collide on a playground, both get pushed, not just one. When a cue ball meets an eight ball on a billiard table, the motions of both change. Newton studied the actions of colliding objects in many experiments and came to this conclusion:

**For every force (or action) there is an equal but opposite force or reaction** (Newton's third law).

But won't equal but opposite forces balance each other, giving a net force of zero? The important thing to realize about this law is that the action is on one object and the reaction is on the other. These forces can't oppose each other because *they act on different things*. It is only when equal and opposite forces act on the same object that they neutralize one another. In a playground collision, the force on one kid won't cancel the force on the other.

FIGURE 3–7
Action: The skater pushes on the wall.
Reaction: The wall pushes on the skater.

Let's think about this with a few examples. Lean against a wall, and the third law says *the wall pushes back*. If that sounds strange, imagine what would happen if the wall suddenly vanished while you were leaning on it—you would take a spill. The wall really does push to hold you up. Why, then, don't you move in response? Friction between your feet and the floor. If you want to test this, try standing on roller skates which are nearly frictionless while you push against a wall (Fig. 3–7). (You might prefer to say the wall *resists* as you push against it, but that resistance is a force.)

Place a lamp on a table. Gravity pulls on the lamp and presses it against the tabletop. The solid table supports the lamp; it reacts to the lamp's downward push with an upward push. In fact, the reaction force from the table exactly cancels the gravitational pull on the lamp. How can we tell? The lamp doesn't accelerate into the table and the table doesn't throw the lamp upward; the lamp is motionless, so the net force on it must be zero. Action equals reaction, and the lamp sits quietly on the tabletop.

Throw a football. As you push it with your hand, it accelerates; at the same time, you feel a resistance from the ball. It presses against your hand as you push it—with the same amount of force. The football's push would cause you to accelerate backward except for the friction between the ground and your feet. If you stand on a frozen pond and throw that ball, you *will* drift backward (Fig. 3–8). Fortunately,

**FIGURE 3–8**
For every force there is an equal but opposite force. But if the masses those two forces act on are not equal, the accelerations will not be equal either: $M \times a = m \times A$.

you won't accelerate as fast as the ball does. That's because your *greater* mass gets a *smaller* acceleration from the same amount of force that hurls the football. When we work or play, most of the things we move around are *less* massive than ourselves, and the third law is reason enough!

Have you ever stepped ashore from a small boat and felt the boat slip backward beneath you? That's action and reaction. A motorboat pushes through the water by propelling water in the opposite direction, just as an aircraft pushes through the air by forcing large volumes of air backward. *Every time a force acts on something, there is a reaction force on something else.*

Let's pause for a moment and state the key ideas in this chapter:

**Matter has inertia; it resists changes in motion.** *(First law)*

**An object's mass is "how much" inertia it has. Force changes matter's motion; it accelerates matter. The magnitude of the net force divided by the magnitude of the acceleration it causes is the measure of its mass.** *(Second law)*

**A push on any object must come from something else, and the something else gets an equal and opposite push.** *(Third law)*

That's all: three laws. So to see how things move, all you need to do is add all the forces acting and use these laws. But first you have to know which forces are acting. What follows is a brief description of some you use and feel every day. Remarkably, most of these forces are linked to two fundamental forces of nature called the *gravitational* force and the *electric* (or electromagnetic) force. These forces, which influence and control the motion in our world, are familiar to us all.

## Weight

Perhaps nothing is so ingrained in our senses as the perpetual pulling of the earth on our surroundings. It's always there, never changing. It's been hugging solids and liquids and gases to the earth's surface for over 4 billion years. Earth's gravity is built into our descriptions of our world with words like *up, down,* and *weight.*

Exactly what is weight? A weight is a *force,* nothing more. Your **weight** is the *pull of earth's gravity on your body*. Likewise, the weight of your car is the force of the earth's attraction for it. The greater the mass, the larger the attraction. Two identical pickup trucks weigh exactly twice as much as one. And, of course, there's twice as much matter, or mass, in two trucks as in one. But mass and weight are not the same; they are measures of two different things, inertia and force.

For example, consider the rocks brought from the moon's surface by astronauts. Because of the earth's stronger gravitational attraction, these rocks *weigh* more on earth, about six times as much as they weighed on the moon. But their mass, their resistance to a change in velocity, is still the same; they have the same quantity of matter on earth as they did while on the moon.

**FIGURE 3–9**
Astronaut Alan B. Shepard inspects a large rock on the moon in February, 1971. Rocks and soil samples the astronauts collected on the moon weighed six times as much when brought back to earth.

Even though weight and mass are not the same, most of us do not make a distinction between them. Suppose someone hands you two books and asks which is the more *massive*. Almost certainly you would "weigh" one in each hand and choose the *heavier* book. That's okay, because the heavier one *does* have more mass. But if the two books were on a smooth table, you could just push each book back and forth to see which has the larger inertia. (Their weights don't come into play, being balanced by upward pushes from the table.) Even then, pointing to the one that's harder to accelerate, you might from habit still say "That one is heavier." The point here is "that one" is harder to accelerate only because it has a greater mass. An astronaut could pick up a large rock on the moon with much less force than required on earth (Fig. 3–9). But if the astronaut shoved the rock in a horizontal direction, it would take just as much of a push to accelerate it at, say, 5 feet/second$^2$, as it would take on earth. There *is* a difference between weight and mass.

To measure your weight you can use a bathroom scale, which is a glorified spring that stretches if it is pulled (or compresses if it is pushed). As you step onto the scale, the spring's pointer registers a larger and larger force until you are at rest, supported entirely by the scale. The scale then shows you how much force (from the spring) balances gravity's pull on your mass, and this force is equal to your weight. If you step down and drink two cups of coffee and then step back up on the scale, you'll weigh about 1 pound more.

**FIGURE 3–10**

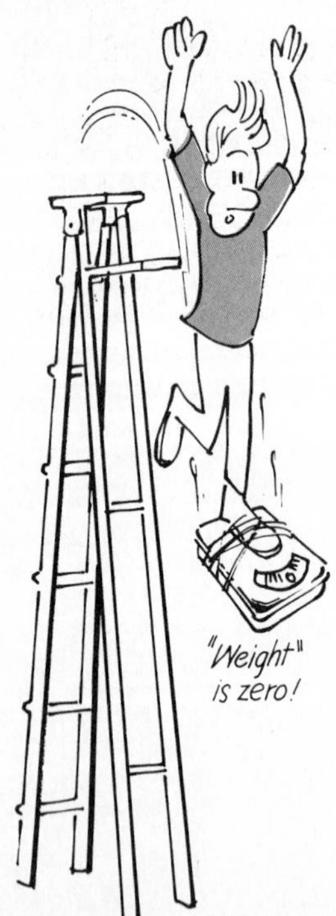


But suppose some fellow strapped a small scale to his feet and jumped from the top of a stepladder (Fig. 3–10). You can imagine what would happen, although you should not actually try it. While he was falling, the scale would fall with him—it wouldn't support him, and he couldn't press against it. In this situation, the scale would show a reading of zero. Gravity's pull would still be there, of course, pulling on him as he fell. He would still have weight, the pull of gravity on his body. It's just that nothing would stop that fall, there would be no supporting force opposing the gravitational pull, so he would *feel* weightless.

To jump with a scale would be awkward (and dangerous). But if you strap on a small backpack stuffed with books and hop down from a chair, you can feel the pack's weight vanish from the shoulder straps while you are falling. Perhaps you've jumped piggyback with a friend into a swimming pool. If your friend is on your back and you jump, your friend's weight disappears from your back while the two of you are in midair. Nevertheless, the weight of your friend doesn't disappear; it causes your friend to accelerate right along with you, at the rate of $g$, toward the water. In the same way, if you are on the back of someone who jumps, you'll hover over that person's back all the way down to the water. This is why news reporters often say astronauts are "weightless" when they are in orbit. But a better way to describe their condition is to say they are in *free fall*.* Since everything in a spaceship falls

*Gravity constantly pulls them, and they constantly "fall"—along an orbit that takes them around the earth. We'll talk more about orbits in Chapter 7.

FIGURE 3–11
During a jump, the jumpers experience "weightlessness"; that is, no force counteracts their weights, and they have no weight relative to each other.

*together* around the earth, nothing inside supports anything else. It's true that the astronauts hover and float within their spacecraft as if they were weightless, but gravity still pulls on their bodies, so they do have weight. The term *weightlessness* is a misnomer, but it gets the idea across. While in free fall, things seem to have no weight *relative to each other*.

Provided there's no air resistance, everything near the earth's surface falls with acceleration $g$. We can use this fact and the formula $F = ma$ to find the weight of an object. If something is falling freely (in a vacuum), its weight is the only force acting, so its weight is the net force. The acceleration $a$ is simply $g$, and substituting in the formula, we find $weight = mg$. (When anything is at rest, the acceleration is zero, of course, because the force from the ground balances the weight.) We measure weight in pounds or newtons, the usual units of force.

## Forces Between the Particles of Matter

If you jump feet first from a diving board, the air resists your motion only a bit. There isn't much matter in a few cubic feet of air, and down you plummet, pushing the particles of air aside. Then your feet smack the water's surface, and the water resists your passage more than the air did. But you plunge on through, moving aside the particles of water. Later, you might stand on the edge of the pool. The particles there do not move to let you pass; the solid material supports you. That is because the solid's tiny particles are bonded tightly to each other with attractive forces called **bonding** forces. The bonding forces in solids such as metals and cement are extremely strong, but they are usually relatively weak in liquids and gases.* (These bonds are electrical in

*Though relatively weak, the bonding forces in liquids and gases (except helium) are strong enough to hold even these particles together as solids if their temperature is low enough.

nature, and they have their origins in atoms and molecules, the particles that make up matter and are discussed in Chapters 9, 24, and 25.)

In addition to the bonding force, another force between the particles of matter helps to hold you up when you stand on firm ground. When atoms of any kind touch and are pressed together, a strong repulsive force arises between them and prevents them from getting closer. It pushes them straight apart, *normal* (or perpendicular) to their point of contact. It is called the **contact force,** or the **normal force.** The atoms on the solid ground push straight out on the atoms of your feet with this normal force, perpendicular to the ground. When this normal force supports your weight, it is equal to $mg$.

This is how the bonding force and the contact force work together as the ground holds you up: Your feet exert a force on the particles at the ground's surface, which in turn pushes back with a force normal to the surface. Because the particles in the solid are locked in place with bonding forces, they cannot move to let you pass. Furthermore, the ground's particles repel one another with the contact force when they are pushed together under your weight, so they cannot compress and let you sink in. These two forces are everywhere. As you are standing by the pool, the bones in your skeleton support one another with normal forces, and they remain rigid because of bonding forces. There are, however, solids with bonds too weak to support you. Try standing on jello.

The strength of the bonds in any solid has limits. Although a kitchen chair keeps its shape under normal use, it can be broken. If a solid splinters or cracks, it means some of the bonds within the solid have broken. The contact forces between the particles in a solid are much stronger than the bonding forces, however. For example, a hydraulic press at a junkyard can squeeze a car into a small cube, breaking molecular bonds of steel and glass. But when there is no more air space between the crumpled parts, the press can be turned off. The crushed solid material compresses no more, because of the contact force at work between the atoms and molecules. You cannot push two quantities of solid or liquid into the same space at the same time with any ordinary means, and it's as true in a grocery bag as it is in that junkyard press.*

**FIGURE 3–12**
The directions of the spring force. (Notice the arrow over the symbol for force doesn't necessarily point in the direction of the force; that arrow only symbolizes that the force is a vector, that it has *some* direction.)

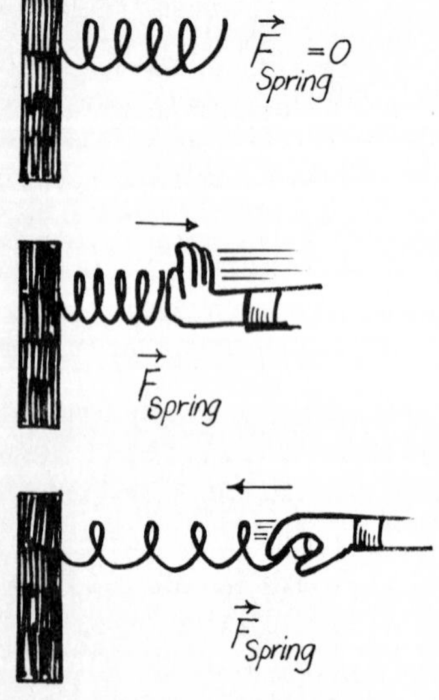


## Elasticity

Mash a spring, and it compresses and pushes back on your hands. Pull on it, and it stretches, pulling back on you (Fig. 3–12). We say it is *elastic. An elastic solid is one that can recover its original size and shape after being deformed.* When you squeeze a solid, the atoms and molecules crowd together and the contact force between them grows and pushes back. If you stretch the solid, those small units ease apart; but the intermolecular bonds pull them back toward each other. The combination of these two separate actions gives the solid **elasticity.** In effect, this causes the atoms to behave as if they were connected by small invisible springs.

---

*Though much stronger than the bonds between atoms, the contact force has limits. Under very extreme circumstances, atoms can be pressed very close together, or as we'll see much later, even broken and compressed.

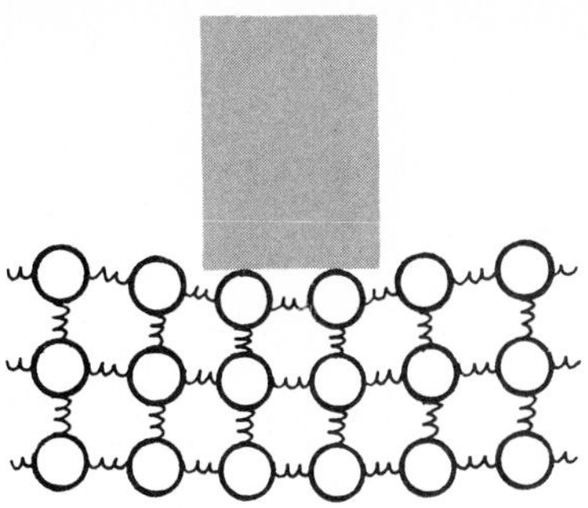

**FIGURE 3–13**
The molecules in a solid respond to a force as if the molecules were all connected by springs.

No solid is perfectly rigid; all show some degree of springiness, or elasticity. You can bend in the side of an aluminum soft drink can, and it can pop out when you let go. (If you squeeze too hard, however, you'll break some bonds, making a permanent dent.) Even your fingertips have elasticity. Press a pencil point against one. Take away the pencil, and the fingertip returns to its normal shape.

Place your hand on a countertop and press down. Even though you can't see it happen, the surface's molecules under your hand squeeze a little closer together as they push back, supporting your hand. Below them, countless others share the weight, shifting a tiny bit and passing the load along to the floor below. When you lift your hand, all the molecules return to their former positions. However slight their shifting, that's elasticity.

## Friction

Slide a large dictionary across a desk top. If you push gently, the dictionary won't budge. All solid surfaces, even the smoothest, have microscopic bumps and dips and notches that catch when the surfaces are pressed together. When you push the dictionary gently, these irregularities push against each other. As you push harder, the growing force from your hand will finally break their holds and the dictionary will slide along. Then there is usually less resistance (or friction) because the surfaces are skimming over each other and fewer of the "ups and downs" catch and grab.

The force of friction between the table and book while the book is stationary is called **static friction** (from the Greek word *statikos,* "causing to stand"). Notice that the force of static friction grows in response to your push and matches it ounce for ounce up to some maximum force. When the frictional grip of the two surfaces on one another has been exceeded, the surfaces slip. The frictional force that remains between them as the dictionary moves is called **kinetic friction** (from the Greek word *kinetikos,* "of motion").

Frictional forces, like all forces, occur in pairs. When the desk resists the dictionary's motion, the desk gets an *equal but opposite* tug from that sliding dictionary (see Fig. 3–14). The direction of a frictional force is easy to find: The frictional forces on objects always oppose the motion or the "pending" motion.

**FIGURE 3–14**
The frictional force the table exerts on the book opposes the force that the hand exerts on the book. In return the table gets an equal but opposite frictional force from the book. (Not shown are *mg* and the contact force from the table on the book.)

Frictional forces between solids also depend on *how hard the two surfaces press together*. Put another dictionary on top of the first one and push on that bottom dictionary again. You'll find it is much harder to set into motion; the maximum value of the static friction is greater than before. (See Example 4 in "Calculations" for more details.)

Liquids and gases, as we've said, resist the passage of a solid object. That's friction, too. Air drags on anything that moves through it; air molecules collide with the oncoming surface, and the contact forces retard its motion. Just swish your hand back and forth rapidly and you can *feel* the air with your hand. If you swish your hand through water, there's much more resistance.

Our bodies are well adapted to using friction in everyday activities. The fingerprint ridges on the elastic skin of our fingers and palms increase friction when we grasp things. (One reason a dish can slip from wet hands so easily is because water has filled in the ridges, making a smoother surface.) The soles of our feet have similar patterns that aid in gripping. You use friction when you turn a doorknob, tie your shoe-

**FIGURE 3–15**
The frictional force is greater when the force of contact between the book and the table is larger.

**FIGURE 3–16**
Using friction from his hands, Mike holds a brick. Pressing directly into the brick provides the contact force that gives rise to friction. The upward frictional forces on the brick balance the brick's weight. The corresponding frictional forces on Mike's fingers point downward.

**FIGURE 3–17**
The bricks are holding Mike up with friction. The frictional forces on his feet and hands point upward. If he doesn't press against the bricks hard enough, the contact forces will be too small, and there will not be enough frictional force to balance his weight.

lace, or scratch an itch. Be sure to inspect Figs. 3–16 and 3–17 to make certain you understand the directions of frictional forces.

**FIGURE 3–18**

Close your eyes and imagine waking up in the morning and going to your kitchen for breakfast. That part isn't hard to do. Now fantasize that as you walk in, all frictional forces cease to exist! You'd really have some problems. To stop your motion over the frictionless floor, you'd have to grab a cabinet knob or something else that is bolted down. The table and stove would have to be perfectly horizontal, or the frictionless plates and pans would slide downhill and fall off. Even then you'd have to set them down with zero speed or they'd drift away, courtesy of the first law. As you stirred a recipe of muffin batter, the reaction force would stir you in return. Your feet would follow a miniature path over the floor like the larger one your hand and spoon made in the bowl. Don't despair; it probably wouldn't take you long to adjust. Newton's laws, our familiar friends, would still govern the performance. Astronauts get along without "weight," so surely you could find a way to function without friction. Just as long as you didn't try scrambling eggs . . .

# CALCULATIONS

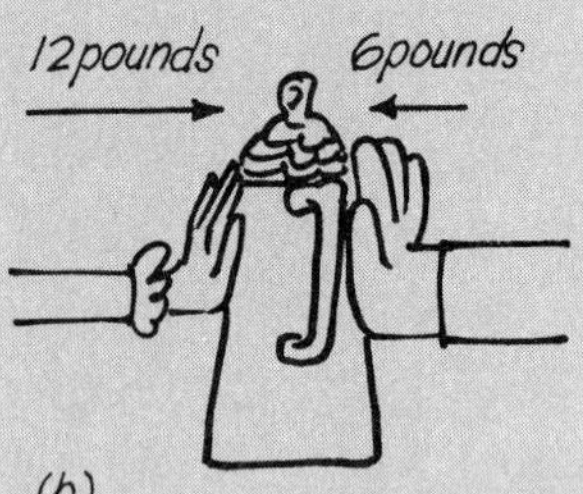


To explain something's motion, start with a drawing. For every force that acts, sketch an arrow to show its direction and magnitude. (A 10-pound force should have an arrow that is twice as long as the one for a 5-pound force, etc.) These arrows, or vectors, only *represent* the forces, so it doesn't matter where you place them on the drawing so long as their directions and lengths are correct. In fact, to find the *net* or *resultant* force, you must rearrange (or add) the arrows as follows: One by one, in any order, unite the arrows tail to tip. A new vector, drawn from the tail of the first vector to the tip of the last vector, represents the *net* force.

FIGURE 3–19

EXAMPLE 1: A two-handled beer stein rests on a very slick bar that offers no friction. Two revelers engage in a tug of war to see who gets the goods (Fig. 3–19). The one on the right pulls with a force of 12 pounds, while the other manages only a 6-pound pull. **What happens?**

Add the force vectors on the drawing or on a graph. Just before you add them, the arrows' lengths and directions should look like this: ⟶ ⇄ Now slide them together so that the tail of the 6-pound force starts at the tip of the 12-pound force. The net force vector goes from the tail of the 12-pound vector to the tip of the 6-pound one, so it is 6 pounds "long" and points to the right. The beer mug will slide in that direction with an acceleration equal to the net force divided by the mass of the mug, or $\boldsymbol{a_{mug} = 6\ \mathrm{lb}/m_{mug}}$. (If the mug isn't a massive one, it will really scoot.)

Suppose toward the end of their merrymaking, the tiplers push instead of pull. If the drinker on the left pushes with 12 pounds while the other shoves with 6 pounds of force, what happens? The same thing happens as before. Pushes or pulls, it doesn't matter; both are forces.

EXAMPLE 2: Suppose two friends grab your arms and pull as shown in Fig. 3–20. **Use a graph to find the net force from these two, and then find what force the friend pulling from behind would have to exert to keep you at rest.**

Add vector 1 to vector 2 (or vice versa) and draw the net vector. If you are to stay at rest, the net force must be zero. So if friend 3 pulls with a force that balances (or cancels) this force from the other two, the net force from all three will be zero (Fig. 3–21).

FIGURE 3–20

FIGURE 3–21

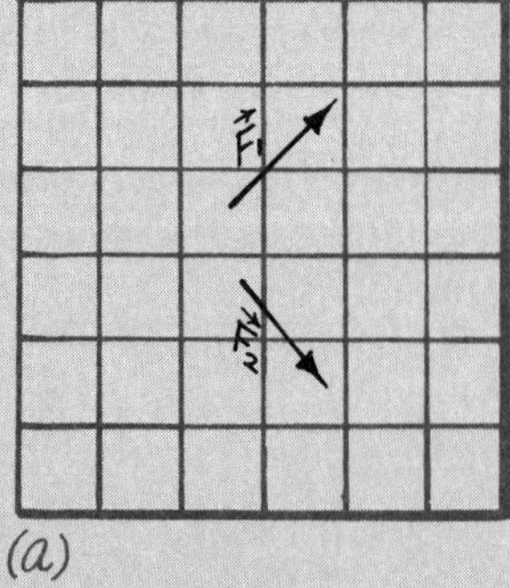


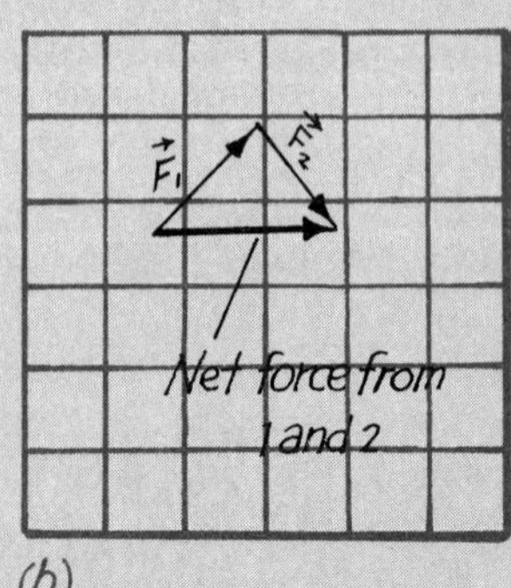


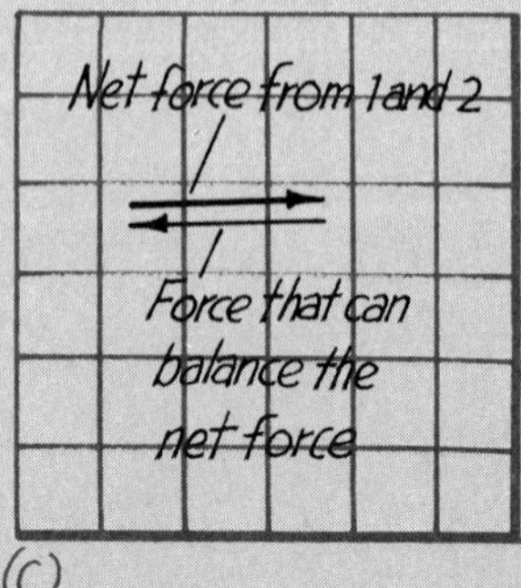

**EXAMPLE 3:** Pull gently on a spring; as it stretches, the spring pulls back on you (Fig. 3–22). Then stretch it until the end moves twice the distance as before; the spring will pull on your hand twice as hard as before. Triple the initial displacement of the end of the spring, and the spring pulls back with 3 times the initial force. This is the same as saying the force the spring exerts is *proportional to its displacement*. (The displacement is the spring's new length minus its unstretched length.)

Expressing the spring's behavior as a formula, we write

$$F_{\text{spring}} = k \times d \qquad \textbf{(3–1)}$$

where $d$ is the displacement of the end of the spring and $k$ is called the **spring constant.** (It's also called the *proportionality constant* between $F$ and $d$.) From

**FIGURE 3–22**

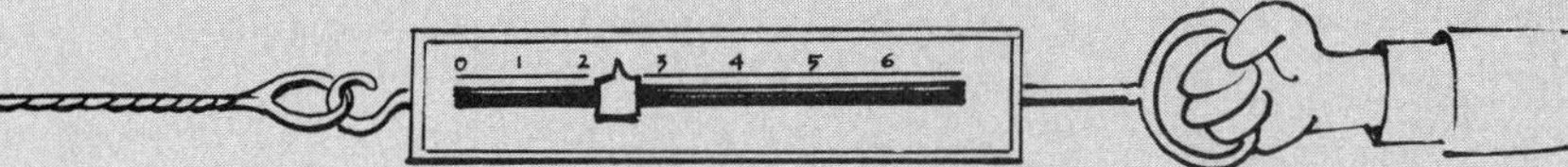


the equation we see that $k$ must have units, namely, the units of force divided by the units of distance. Let's use this equation to see exactly what it means.

When a weight of 0.4 N hangs from a certain spring, it stretches 0.01 meters (or 1 centimeter). Using these quantities in the formula for the spring force, we see that $k$ = 0.4 N/0.01 m = 40 N/m. Now that is the $k$ for this spring. Once you know the $k$ for any spring, Eq. (3–1) predicts that spring's force for *any* displacement $d$ (so long as the spring is not deformed or broken). For this spring, then,

$$F_{\text{spring}} = (40 \text{ N/m})d$$

If $d$ is twice as large as before (0.02 m), then $F_{\text{spring}}$ = (40 N/m) × 0.02 m = 0.8 N, a force that's also twice as large as before.

The spring force always opposes the displacement even if the spring is loosely wound so that it can be compressed as well as stretched. When a spring is stiff and strong, $k$ will be very large compared to the $k$ for a weaker spring.

As you saw earlier, elasticity (including that of springs) comes about because the molecules in a solid behave as if they are connected by small springs. The countertop that yields a bit under the pressure of a hand or a typewriter resists with the spring force, and its $k$ is very large.

**EXAMPLE 4:** Pull a magazine across a table; the resistance from the friction between the tabletop and the magazine's cover is easy to overcome. Now place something heavy (a telephone or a potted plant) on the magazine and pull again. The frictional resistance between the surfaces of the magazine and tabletop increases dramatically. *The frictional force depends on how hard the magazine cover and tabletop are pressed together.*

In very many cases the size of the frictional force between two surfaces is *proportional to* the size of the normal (or contact) force $F_n$ between them. Double (or triple) $F_n$ and the frictional force doubles (or triples). In the form of an equation,

$$F_{\text{friction}} = \mu F_n \qquad \textbf{(3–2)}$$

where $\mu$ (Greek letter *mu,* pronounced "mew") is a number (without units) that shows the constant value of the ratio of the frictional force to the normal force, $F_{\text{friction}}/F_n$. The constant $\mu$ is also called the **coefficient of friction.** Because the resistance of friction usually decreases once the surfaces begin to slide, there is a different value of $\mu$ for each case; the coefficient in the equation for a static friction force is normally larger than the coefficient for a kinetic frictional force. If the coefficient is large, the friction can be great when the surfaces are pressed together. If $\mu$ is small, the friction is small, unless the normal force is really large. So a small coefficient of friction usually represents slippery surfaces, whereas rough, scratchy surfaces have large $\mu$'s. It's a fact that the frictional force rarely exceeds the normal force, even between rough surfaces. That is, $\mu_{static}$ for a static frictional force rarely has a value as great or greater than 1.

The resistance the magazine gets from the tabletop opposes the motion of the magazine. The tabletop gets an equal but opposite frictional force from the magazine. Notice that both of these forces are perpendicular to the normal force that appears in the equation for friction. But while the normal force is necessary for there to be friction, the frictional force itself is caused by your pull, which acts to slide the surfaces over each other.

## REVIEW

**1.** All forces are vector quantities; that is, they have size and direction. True or false?

**2.** Does a net force describe the effect of two or more forces acting on an object at the same time?

**3.** State the first law of motion in your own words.

**4.** The property of matter that resists a change in its velocity is called ________ .

**5.** Can mass be described as the quantity of matter something has? Can mass be described as a measure of something's inertia?

**6.** What is the formula that expresses the second law of motion?

**7.** The acceleration of an object is always in the direction of the *net* force. True or false?

**8.** State the third law of motion in your own words.

**9.** Is the pull of earth's gravity on your body the force you call your weight?

**10.** Mass and weight are measures of two different things. What are they?

**11.** Which force locks atoms or molecules together in solids?

**12.** Does the contact (or normal) force always push things straight apart when they are pressed together?

**13.** When you are in free fall, as when you jump into a swimming pool, are you without weight? Do you experience weightlessness?

**14.** Name two types of friction between solid surfaces. Which type is usually the stronger?

## DEMONSTRATIONS

**1.** To demonstrate action–reaction, try this: Stand on a bathroom scale, arms stretched out horizontally. Twist your body so that your arms move to the front and back in a horizontal plane; notice your weight on the scale. Next move your arms up and down with about the same degree of motion. What happens to your weight? Explain this in terms of Newton's third law.

**2.** Don't believe in weightlessness? Maybe the backpack routine in this chapter didn't convince you. Well, if feeling isn't believing, try seeing! Rest a styrofoam cup in your hand while you fill it with water. It gets heavier in your hand because the water presses against the cup's bottom with its own weight. Punch a small hole near the bottom and a stream of water will squirt straight out (Fig. 3–23). Holding the cup by the rim, release it. Immediately the stream vanishes as cup *and* water fall freely toward the ground. The water in the cup is "weightless" so it doesn't press against the bottom or the sides. Hence, it doesn't squirt out as it falls. (Just wait until it lands!)

**3.** Hang some small weights from a spring. Test the spring force to see that if you double the weight (force), the distance the end of the spring moves will also double. (See Example 3 in "Calculations.")

**4.** A well-known magician's trick is to pull a tablecloth from underneath plates, dishes, and glasses without disturbing them. If the cloth is jerked away rapidly, the fric-

FIGURE 3–23

5. As trees grow, they sometimes exhibit an interesting ''action–reaction'' effect. Look for it in tree trunks or logs sawed for a fireplace. Very often a young tree is tilted by a fallen limb or branch from another tree; even a strong wind can leave it leaning. As that sapling grows, it builds extra wood on the underside of its trunk to strengthen the tree against falling. This ''reaction'' wood makes the growth rings much thicker on the lower side of the tree than on the upper side (Fig. 3–24). A few years of this lopsided growth may straighten the tree back up, and later growth rings will have an even width all the way around.

tional force between it and the objects acts for such a short time that there can be little actual change in speed for the tableware. Use a sheet of notebook paper and a dry glass to do this trick for yourself. (If the paper extends a few centimeters over the edge of the tabletop or desk, it is easier to grasp.) Then be sure to pull the paper straight to the side. Of course, there is always some friction. So in which direction does the glass move? How does its motion depend on *where* the glass sits on the paper? Do it and see.

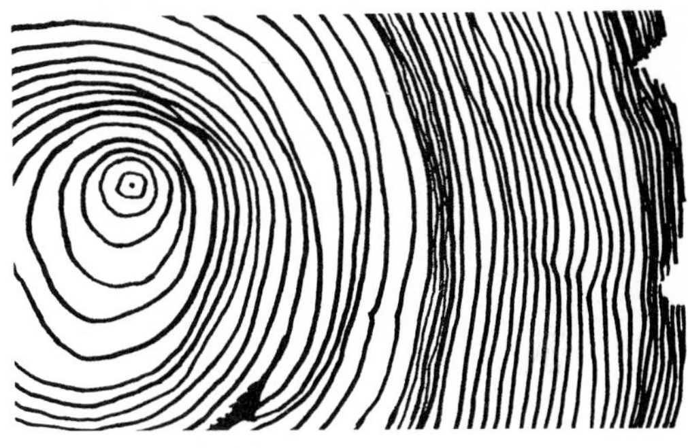

FIGURE 3–24

# EXERCISES

**1.** When you watch something move, can you see its acceleration? Its velocity? Can you see its inertia? Can you see the force that acts on it?

**2.** Is it correct to say that you can *feel* this book's inertia? All humor aside, discuss.

**3.** Two ruffians have a shoving match on the playground, and one is finally pushed over. The victor shouts, ''I pushed the hardest!'' Is he right?

**4.** Should astronauts choose pencils of hard lead or soft lead for taking notes in orbit? Explain your answer.

**5.** Why does the head of someone who is pushed from behind seem to snap backward? How does Newton's first law explain this whiplash motion? Exactly how do headrests in cars help guard against this type of injury?

**6.** If action equals reaction, why don't you see a ping-pong paddle rebound when a ping-pong ball strikes it during a game?

**7.** Two wagons, one empty and one fully loaded, are parked side by side. If you grab the handles and begin to run, pulling the wagons behind you and keeping them perfectly even with each other, do both have the same (a) inertia? (b) force on them? (c) acceleration? (d) velocity? (e) speed?

**8.** When a large car and a small car collide, the forces they receive are equal and opposite. But what about their accelerations and the accelerations of the passengers?

**9.** On the moon an astronaut can pick up a rock with less effort than on earth. But when the astronaut throws the rock horizontally, it's just as hard to do as on earth. True or false? If the rock is dropped on the astronaut's toe, should it hurt more or less than on earth?

**10.** You can't see a push or a pull, so how do you sense it with your body?

**11.** Place a ball in the front of a wagon. Then pull sharply on the handle. As the wagon accelerates, what does the ball do? Now *stop* the wagon suddenly. What does the ball do? Explain the motions with the first law and your knowledge of frictional force.

**12.** There is only one force acting on a freely falling object. True or false? Are skydivers really in free fall after they jump?

**13.** When you carry a heavy book in one hand, your hand is more likely to get bruised or scratched should it bump into something. What property of matter does this illustrate best?

**14.** If you've ever helped push a stalled car, you've probably noticed that it is much easier to keep it rolling than it is to accelerate it from rest. Why is that true?

**15.** Would it be more exact to say that a person diets to lose *mass* rather than *weight?*

**16.** Have you ever noticed that it is easier to pull a paper towel from a full roll if you jerk the towel quickly? Explain why.

**17.** Bathroom scales often show kilograms as well as pounds. The reading indicates your mass, not your weight. Suppose the astronauts had taken such a scale to the moon to weigh themselves. Would the pound reading be correct? Would the kilogram reading be correct?

**18.** Cheetahs are bigger and faster than smaller gazelles, but more often than not these gazelles escape a pursuing cheetah by zigzagging. Exactly why does this give the cheetah a hard time?

**19.** Suppose you are in a spaceship in free fall between the earth and the moon. How could you distinguish between a lead brick and an ordinary brick if you had on a blindfold and gloves?

**20.** Would it be easier for you to accelerate while running on the moon or on the earth? (Ignore air resistance on earth.) Be sure to see the answer to this question.

**21.** Can a 1-kilogram mass ever weigh more than a 2-kilogram mass? Explain your answer.

**22.** If something is traveling at a constant velocity, does this mean there are no forces acting on it? If two things accelerate at the same rate, must they have the same mass? Are the forces acting on them the same size?

**23.** Suppose you know the mass of an object and you experimentally determine its acceleration. Using Newton's second law, can you find all of the forces acting on it? If not, what can you find?

**24.** A car is compressed until it is the size of 1 cubic yard. Has its mass changed? Its weight?

**25.** Looking at the equation $F = ma$, do you think that a constant force will *always* result in a constant acceleration of an object? Explain using an example.

**26.** A rocket's engine causes a constant thrust on a rocket during its flight. So in the equation $F = ma$, the force $F$ is constant. But as the rocket travels, it uses fuel that initially accounts for most of its mass. What does the second law predict will happen during the flight?

**27.** As the space shuttle orbits the earth, are the motions of the astronauts inside due to Newton's first law?

**28.** What force explains why paint clings to a wall? What force makes adhesive tape sticky? What force makes wax stick to a car?

**29.** When you clap your hands, what force stops them when they collide?

**30.** When a book lies at rest on a table, which of Newton's laws of motion tells us the book has a net force of zero?

**31.** When a book lies at rest on a level table, does the normal force from the table arise only because the book presses against it? Does the normal force have the same magnitude as the force of gravity on the book?

$\overrightarrow{\textbf{32}}$**.** When you are writing, how many independent components of force do your fingers apply to the pencil?

**33.** Newton's second law says the acceleration of an object is always in the direction of the net force. But does an object always move only in the direction of a net force? Give an example to explain your answer.

**34.** Four people stand side by side at a party with their arms stretched out as far as possible and holding hands. Without warning, a person at one end pulls hard, and the line of people begins to move. Which pair of hands feels the most force? Why? Which pair feels the least force? (Do this as a demonstration, exchange places, and do it again!)

**35.** In a park in Peking, China, there lies a huge boulder that was transported from 100 km away. At some time in antiquity, people moved that boulder, which was much too heavy to drag because the friction would be so great. It is also too heavy to roll or lift. How did they accomplish that feat? Archaeologists have discovered the answer. The clue lies in a trail of ancient water wells, about a kilometer apart. Can you solve the mystery?

**36.** If the tires of your car slip, skid, or spin over the road, you won't be able to accelerate or decelerate as fast as you could otherwise. True or false? Why?

***37.** A 1-pound box of crackers generally has the extra notation "454 grams" (Fig. 3–25). The pound is its weight, while the 454 grams is its mass. Use $F_{weight} = mg$ to prove that a mass of 454 grams weighs 1 pound on earth. Which of these measurements would also be accurate on the moon?

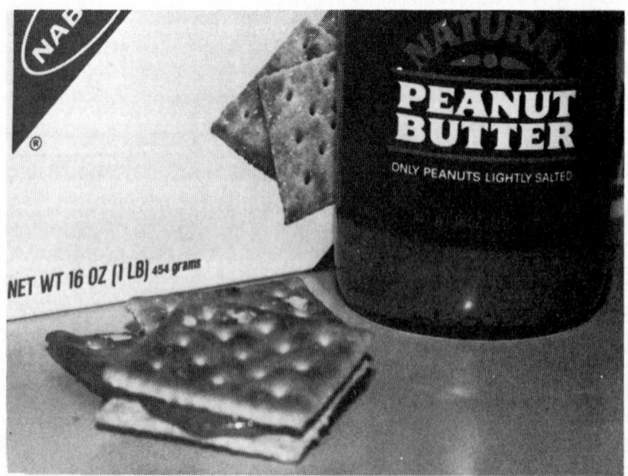


**FIGURE 3–25**

***38.** If you use a kitchen measuring cup to measure a cup of water, how much does that cup of water weigh? (Use 1 liter = 1000 $cm^3$ = 1.057 quarts. There are 4 cups to 1 quart, and 1 cubic centimeter of water has a mass of 1 gram.)

***39.** Twelve people at a party decide they want to lift a VW Beetle off the ground. It weighs about 1800 pounds. What force will each person have to exert on the average?

***40.** The weight of a woman wearing a parachute is 128 pounds. She jumps from an airplane, and after a few seconds the air drag on her at one instant is 50 pounds. What is her acceleration at that instant?

***41.** If your mass were 50 kg, how many pounds would you weigh on earth? (Use your own weight to find your mass in kilograms.)

***42.** An average-sized apple weighs about 1 newton. There are 16 ounces in a pound, so about how many ounces does that average apple weigh?

***43.** A car on a level street accelerates at 4 $m/s^2$. If its mass is 900 kg, what is the net force on the car?

***44.** How much horizontal force does a car seat exert on a person with a mass of 65 kg to accelerate that person at 2 $m/s^2$? Do you think you would feel such a force on your back? Why or why not?

***45.** A certain bullet of mass 10 grams accelerates for 0.002 seconds and leaves the barrel of a rifle with a speed of 380 meters per second. Calculate the average force on the bullet and compare it to a weight that you might lift, say a 10-lb bag of potatoes.

***46.** At a roller rink a boy and girl push straight apart. The girl's mass is 45 kg, and during the push she accelerates at 1 $m/s^2$. If the boy's mass is 60 kg, what is his acceleration while they push off?

***47.** A rock placed on a scale registers 100 pounds. A boy passing by tries to lift the rock unsuccessfully; but while he is exerting a steady pull, the scale registers only 40 pounds. How much force is he applying? If he stands on another scale while trying to lift the rock (Fig. 3–26), what will that scale register? If both of these scales are on another large scale, what will its reading show?

***48.** A frugal fisherman uses a spring and a ruler to measure the weight of his catch. When he hangs a 10-pound weight from the spring, it stretches 2 inches. What is the spring constant (in pounds/foot) of this spring? How far would a 1.8-pound fish stretch this spring?

***49.** At many self-service cafeterias the plates are stacked on a movable platform that sinks into the countertop under their weight. Remarkably, the top plate always shows itself at countertop level. When a customer takes that plate, the stack moves up just enough to reveal the next plate! That's the property of a spring force. For each extra centimeter a spring is depressed or extended, the change in force is the same. If a cafeteria plate weighs 1.5 pounds and its depth is 1 inch, what is the spring constant in pounds/foot for the spring device?

**FIGURE 3–26**

***50.** A rubber tire on a dry concrete road has a coefficient of static friction of 1.0. When the same road is wet, the coefficient of static friction is only 0.7. If a car can stop from 40 mph in as little as 2.5 seconds on the dry road without skidding, how long will it take to stop on the wet road without slipping? How *far* does the car take to stop in each case?

***51.** The joints between the bones in our bodies are so well lubricated that the coefficient of static friction is a mere 0.003. Repeat the calculation in Exercise 50 to see how long it would take that car to stop on a surface with this coefficient of static friction and how far it would go during that time.

***52.** When the Apollo astronauts accelerated from earth orbit into the trajectory that took them to the moon, their spacecraft weighed about 260,000 lb. A single rocket propelled them with an initial thrust of 178,000 lb. What was their initial acceleration?

# CHAPTER 4
# UNDERSTANDING FORCES

If you understand the laws of motion, you can understand the everyday motions you see around you. These laws also apply to any motion you take part in, although you probably would never stop to think, *''Aha, that's physics!''* A basketball player never calculates the ball's path before taking a shot, nor does a racecar driver use an equation to take a sharp curve. Practice gives them instincts for what they do, and you have the same instincts for the things you do. In this chapter you can test these instincts, because we'll apply the laws of motion to some of the things most of us do almost every day. Even when your educated guess is correct, you'll find that you can understand these ordinary events more clearly in terms of Newton's laws. To visualize what happens in the following examples, you should draw your own sketch of each situation. Newton's laws explain the motion *only* if you get the forces and their directions right, and a simple drawing helps every time. ■

**FIGURE 4–1**
Lots of small forces balancing a much larger force.

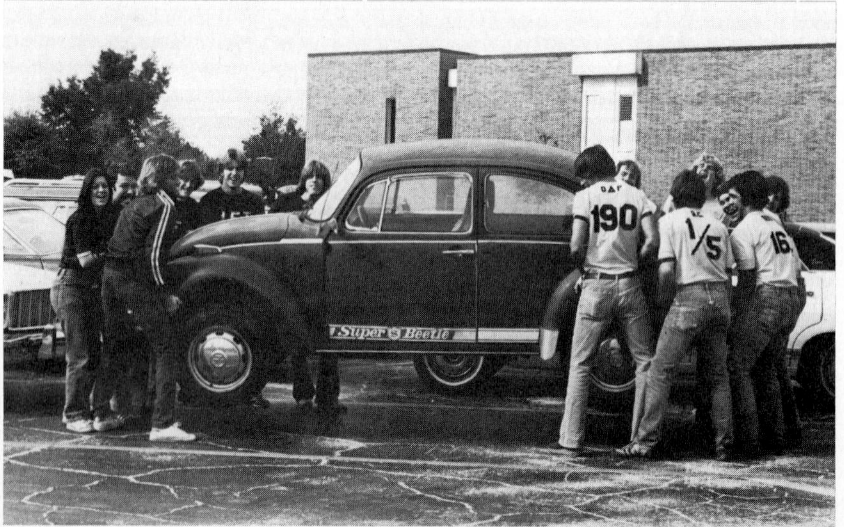

## Standing and Walking: Normal Forces and Friction

After a long day your feet can ache just from holding up your weight. When you stand at rest on a level floor, each foot supports about half your weight as the normal force from the floor balances the pull of gravity on your body (Fig. 4–2*a*). You don't accelerate because the net force on you is zero. However, if you then take a step, you must accelerate horizontally. To do that takes another force because neither gravity nor the normal force from a level floor points horizontally at all. Because your feet and the floor are firmly pressed together, there is friction if your foot pushes off against the floor (Fig. 4–2*b*). The floor pushes back on your foot—action and reaction. The frictional force *from the floor* acts *on your mass* and *you* accelerate horizontally: $\vec{F} = m\vec{a}$.

Don't be confused by all the body movement involved in the act of walking. Certainly lots of muscles pull, tendons stretch, and bones swivel and rotate. But these forces and motions are internal to your body, only changing orientations of your limbs and so on. For your entire body to accelerate, there must be a force on you from *outside* your body; and when you walk, that force is friction. Inside actions help you push against the ground, but unless the ground pushes back, you go nowhere. Imagine trying to walk or run in a puddle of grease: No

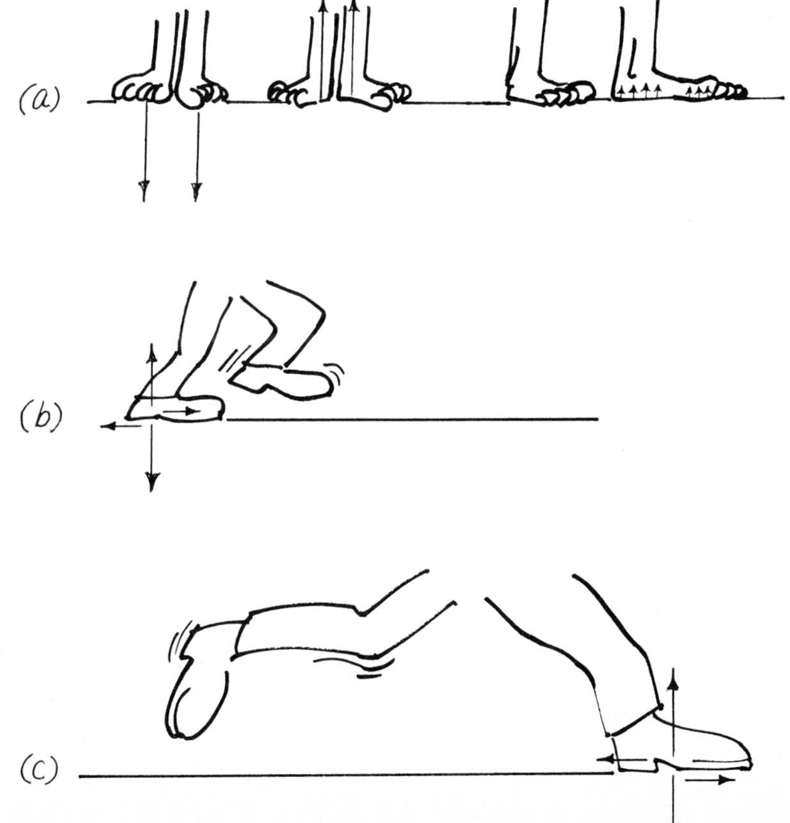


**FIGURE 4–2**
*(a)* You press against the ground with a force equal to your weight, about half of it on each foot. The ground presses back with the same amount of force, so the net force on you is zero, and you don't accelerate. Actually, each square centimeter of your feet that touches the ground supports some of your weight. *(b)* As you take a step, you push off against the ground. In reality, the frictional force from the ground, which points in the direction opposite to your push, pushes on your feet and you move forward.
*(c)* When you take a giant step, you depend on friction with the ground to oppose your motion and stop your foot.

amount of body motion will propel you horizontally then. *Only an external force can make you accelerate.*

Once you've taken the first step, you have overcome your inertia and acquired a walking speed. Then walking step by step is like a series of short, controlled falls. As you step forward, your body drops a tiny distance. You extend a foot to stop your fall, and that foot makes use of friction again—only in the opposite direction—to keep it from slipping forward while your body catches up (Fig. 4–2c).

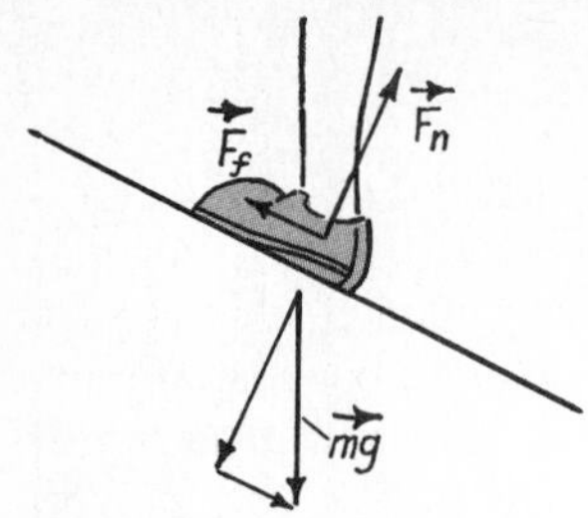


**FIGURE 4–3**
On a slope, not all of your weight presses against the surface. A component acts to pull you down the slope. The contact force $\vec{F}_n$, which is the reaction force to the component of your weight pressing against the surface, gives rise to a frictional force that opposes the component of your weight pointing down the slope.

When you stand or walk on a slope or incline, the situation is only a little different. Your weight pulls you straight down, but the normal force pushes your feet straight away from (perpendicular to) the surface. The normal force, remember, is the surface's response to a force against it; the normal force is a *reaction* force. In this case the action force is *not* the full weight of your body, but only the component of your weight that presses directly "into" (or perpendicular to) the surface. The other component of your weight pulls your body down the incline, parallel to the slope. (See Fig. 4–3.) You can stand still on a slope only if the static friction between your feet and the surface balances the component of your weight that is parallel to the slope. Otherwise, there will be a net force downward along the slope, and you'll slip.

## Riding Up and Down with Newton's Second Law

Let's imagine that you take your bathroom scale into an elevator, place it on the floor, and stand on it. If the elevator starts upward, the sensations in your stomach and from your feet tell you something is happening. A glance at the scale shows a force greater than your usual weight, and if you have an armful of books, they feel heavier. What's going on?

The first step in understanding this is to identify the forces acting on you. Your weight *(mg)* pulls you downward, and that pull is constant as long as you stay near the earth's surface, *no matter what you are doing*. Then there's the normal force from the scale that pushes upward on you, preventing you from falling through the floor. That's it, two forces. Newton's second law tells us the rest.

Because of the upward acceleration of the elevator floor (Fig. 4–4), the scale pushes you upward with a normal force that is *greater* than your weight, causing you to accelerate upward at the same rate as the scale and the elevator. The net force on you is equal to that normal force minus your weight, *mg*. The excess force (over your weight) that you see on the scale is the net force on you, and your upward acceleration is equal to that net force divided by your mass. If you are holding books, your arm supplies the extra force to accelerate them, which is why they, in turn, feel heavier to you. Again, action and reaction.

After about a second of acceleration, an elevator's speed usually levels off. Then you and the books and the scale and the elevator all travel up *at a constant speed* (Fig. 4–5). Whenever anything moves at a constant speed, the net force on it is *zero*. Now the normal force from the scale exactly balances your weight (including the weight of any

**FIGURE 4–4**
There are two forces on the woman in the elevator, the contact force from the floor and her own weight, *mg*.

**FIGURE 4–5**
When the contact force balances the weight, the acceleration is zero and the passenger feels normal, even if moving up or down.

**FIGURE 4–6**
While the passenger is in free fall, only the pull of gravity acts on the passenger. With no contact force from the floor, the passenger seems to be weightless inside the elevator.

books you hold), as shown on the scale's indicator, and you no longer feel heavier than usual.

Many older elevators are raised and lowered with overhead cables. Suppose you step into such an elevator and the cable suddenly slips, its tension vanishing for a second. For that second, the elevator floor, the scale, and you accelerate downward at the rate of $g$ (Fig. 4–6). Only one force, the pull of gravity, acts on you (and the scale and the elevator), and you are in free fall. In free fall, remember, nothing supports you against the action of gravity. It feels as if the supporting floor has literally dropped out from under you, and the scale beneath you shows a force of zero. Then, when the cable takes hold and suddenly stops the elevator's motion, the scale exerts a normal force again. While the scale decelerates you, stopping your fall, it registers a force greater than your weight. (A force equal to your weight would only balance *mg*, and you'd continue to fall at the same speed.) But as you come to rest, the normal force becomes equal to your weight, because the net force on you is zero once more.

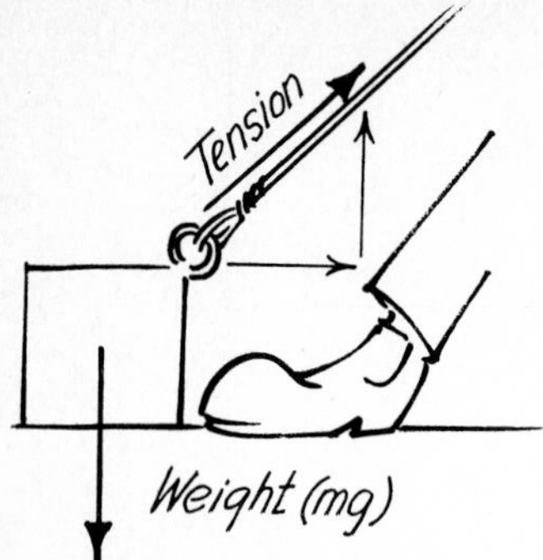


**FIGURE 4–7**
One component of the applied force subtracts from the weight, and the other tends to accelerate the object horizontally. (Not shown are the contact force and the frictional force.)

## Using Components of Force

Suppose you need to drag a heavy footlocker across a floor. You might lean over and pull it along, or if it were especially heavy you might tow it with a rope over your shoulder to avoid straining your back (Fig. 4–7). Either way, you can better understand what happens if you split your pull into a horizontal force and a vertical force. (These two force components must, of course, add together as vectors to equal your pull.)

The upward component of your force opposes the weight of the trunk and lessens the force the trunk exerts against the floor. But unless this vertical part of your pull exceeds the trunk's weight, the trunk remains on the floor. Even so, because the normal force between trunk and floor is now less, the friction between trunk and floor is less and the trunk will be easier to accelerate in the horizontal direction.

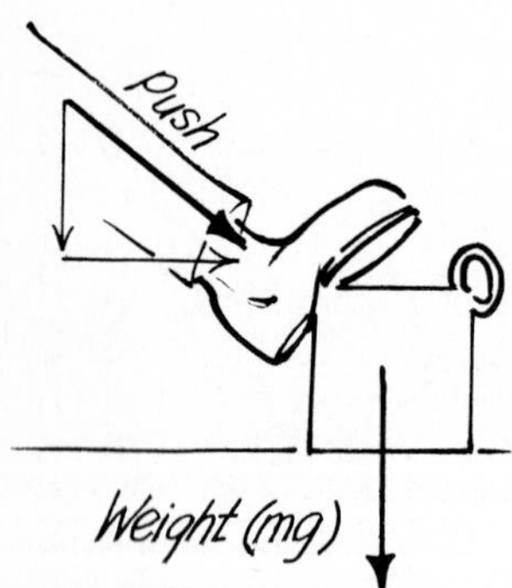


**FIGURE 4–8**
Here the downward component of the applied force adds to the weight of the box. This increases the contact force between the box and the floor and hence the frictional force between them.

For the trunk to accelerate toward your feet, there must be a net force in that direction. The trunk won't move at all unless the horizontal component of your pull exceeds the static friction between the trunk and the floor. If you pull hard enough to break static friction's hold, the sliding (kinetic) friction will be less. Then the component of your pull that is parallel to the floor *minus* the force of kinetic friction from the floor gives the net force. You might then choose to pull with *less* force and let the trunk slide with a constant speed. The horizontal component of your pull would then balance the kinetic friction on the trunk. (If the trunk were really heavy, you could use oil or grease on the floor to reduce the friction. Such lubricants provide layers of molecular "beads" or "logs" for the footlocker to roll over.)

Instead of pulling, you could push this trunk with one foot (as in Fig. 4–8). Then your push has a downward component of force in the same direction as the trunk's weight. This extra vertical force increases the normal force from the floor, which will increase the frictional force between trunk and floor. So the trunk is easier to pull along than it is to push along, as you might have guessed. From this we can see why it's easier to pull a lawnmower through high grass than it is to push it, and why you would pull a wagon through soft sand rather than push it. As the grass got higher (or the sand softer) you would most likely direct your pull even more toward the vertical direction. That reduces the downward force on the ground, lessens the horizontal resistance from the grass (or sand), and makes the towing easier.

## Air Resistance: What a Drag

Even on a day when the air is still, a runner feels a breeze. A slow jog brings only a breath of air, and a quicker pace brings a whistle of wind past the ears and a flutter to any loose fold of clothing. That breeze may feel good on a warm day, but it means the air is resisting the runner's motion through it. A wind against the competitors at a track meet has a slowing effect. A breeze that is with them, however, gives them less resistance than they would get from still air, letting them run faster for the same effort. In fact, a mere 4-mph breeze in the runners' direction so reduces the usual air resistance that record times set under such conditions are not officially recognized.

Air resistance, or air **drag,** is the force that opposes a solid object that moves through the air. This force comes from collisions with the molecules of the air. *The faster an object moves, the greater the drag on the object becomes.* (See Example 6 in "Calculations.")

Air drag affects the motion of an object *falling* through air, too. Here's how. If a squirrel breaks loose a pine cone in a tall tree, gravity's pull accelerates the cone downward. If gravity were unopposed, the cone would accelerate at the rate of $g$. But as soon as it begins to move, air drag also begins. As the pine cone gains speed, the air drag grows, and this resistive force opposes the pull of gravity and decreases the pine cone's net acceleration. Soon the air drag gets to be as large as the pull of gravity on the cone, so the net force on it becomes zero and the falling pine cone does not gain any more speed. As we've seen, its speed of fall when this happens is called its **terminal speed** (or *terminal velocity*). It continues to fall at this terminal speed until it hits the ground.

Skydivers can vary their terminal speeds in air by changing body positions during their falls. If a skydiver spreads arms and legs wide and falls face down, his or her body catches a lot of air. The terminal speed for a skydiver in this position is usually about 120 mph. But if a skydiver dives nose first, he or she may plunge groundward at speeds approaching 200 mph. Of course, even 120 mph is much too fast for a safe landing, so the skydiver must use a parachute to lower the terminal speed further. When the parachute opens, it presents much more surface to the air than the skydiver's body does and collides with a far greater number of air molecules per second. Those extra molecular collisions increase the air resistance dramatically and immediately cause the skydiver to decelerate. The skydiver's speed drops rapidly to a new terminal velocity, perhaps 10 to 12 mph, a speed much safer for landing.

When an object moves through the air, both the *area* meeting the air and its *shape* help determine the actual force of air resistance at any given speed. However, this fact about air resistance is not always very obvious. If a smooth wooden ball and a steel ball with identical diameters and smooth surfaces are dropped together from a window several stories high, the steel ball will arrive at the ground a noticeable distance ahead of the wooden ball. Yet, at the same speed the force of air resistance on the two balls is equal. But the wooden ball has less mass, so the *same* air drag decelerates it *more* than the steel ball. That's why lighter objects of a given surface shape are affected more by air resistance.

**FIGURE 4–9**
*(a)* The skydiver accelerates until the air drag balances the skydiver's weight, no matter which position is taken. *(b)* When less surface area is presented to the air, a greater speed is reached before the drag equals the weight.

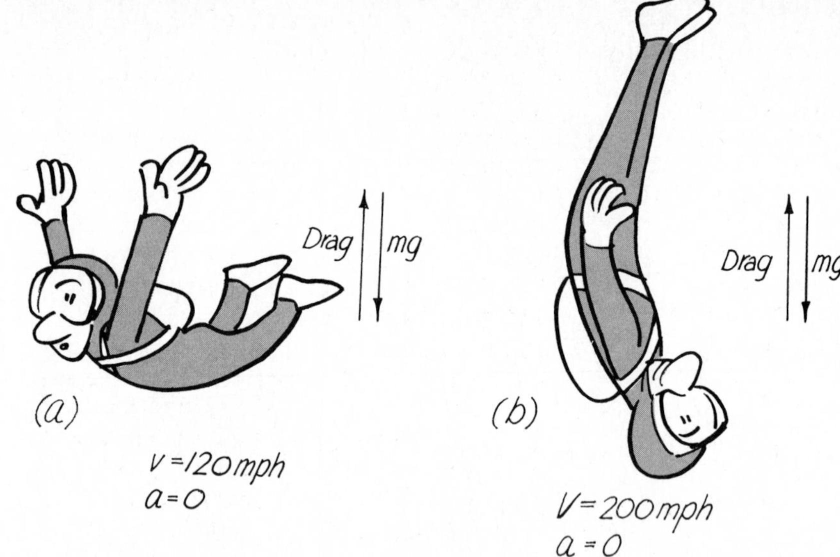


Autumn leaves float down lazily when they fall. Without drag, they would drop as if made of lead. But they have little mass, and the air drag on their large surface areas really slows them down. A bedsheet billows and gently sinks when you make up your bed. Without air to resist its motion, it, too, would descend more like a rock. Films of astronauts cavorting in the Lunar Rover show moondust kicked up by the wheels. Unresisted on that airless surface, those fine particles sprayed outward for great distances, like miniature projectiles. In the air of earth, such tiny particles would slow to a stop in only a centimeter or two.

## Using Air Resistance

Stick your hand out of the window of a moving car. What you feel is air resistance. Flatten your hand and hold it horizontally, and then slightly tilt the forward edge upward. Your hand is pushed upward and backward as the air hitting your palm is pushed downward and forward. The upward component of air resistance on your hand is called **lift,** and since the Wright brothers' triumph at Kitty Hawk, airplanes have made use of this force to take people aloft. The backward component of air resistance, the one that opposes the hand's motion through the air, is called **drag.**

If an airplane flies horizontally and at a constant speed, the forces on it are easy to understand. The undersides of its wings get lift and drag from the air they push through, just as the palm of your upward-tilted hand does. The body of the plane gets drag from the air, but almost no lift. The net drag points straight back, parallel to the ground. There is a forward-pointing force called **thrust** that is supplied by the plane's engine(s) (see Figure 4–11). When the plane's horizontal speed

**FIGURE 4–10**
Ngaio feeling the drag and lift on her physics book, while Annette avoids running over the physics instructor.

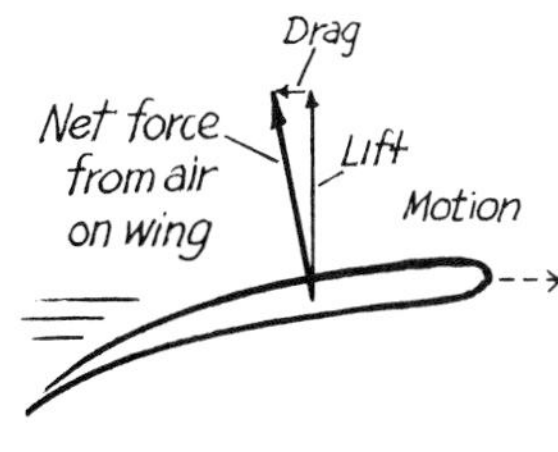


(a)

Thrust
Total drag
Total lift
Weight

(b)

**FIGURE 4–11**
*(a)* This diagram, which shows the forces on the wing of an airplane, could apply to the book in Fig. 4–10. (Not shown is the force on the air, pushing the air downward and to the right.) *(b)* The forces on an airplane as it flies horizontally with constant speed.

is steady, we know there can be no net force in the horizontal direction, so the thrust and the drag cancel. These two forces are equal in magnitude and in opposite directions. Similarly, since there is no vertical motion when a plane flies horizontally, there can be no net force in that direction. The lift on the wings, then, balances the airplane's weight. Just knowing that the net force is zero tells us a great deal.

What forces act on the passengers inside a cruising jetliner? Inside the cabin they are shielded from the onrushing air. Only the pull of earth's gravity, their weight, has to be supported. The chairs and floor of the level-flying aircraft serve exactly as they would on earth. But if the plane flies through an updraft or a downdraft, those aboard feel momentarily heavier or lighter as the plane (and the floor and chairs beneath them) accelerates up or down. It's exactly the same effect we feel in elevators. In a severe downdraft, however, a plane can be accelerated downward at a rate *greater* than $g$. In the rare instances when this has happened, passengers who were not buckled into their seats have actually hit the ceiling as the plane fell from beneath them. Those passengers buckled in at the time were pulled down with the plane, and they had the memorable experience of watching their drinks and ice cubes, which accelerated downward at only $g$, fly right out of the cups in their hands and up to the ceiling.

A pilot can guide an airplane upward and then pull the nose sharply downward. Done properly, this act causes the upward lift on the wings to vanish, and for a short while the airplane falls toward earth with an acceleration of $g$. That is, the plane travels along an arc just like the path a stone takes when it is thrown from the ground (Fig. 4–12). At that moment the passengers are in free fall. Any loose magazines or plastic cups will hover in the air of the cabin as they accelerate downward at the same rate as the plane. Then the pilot must pull out of

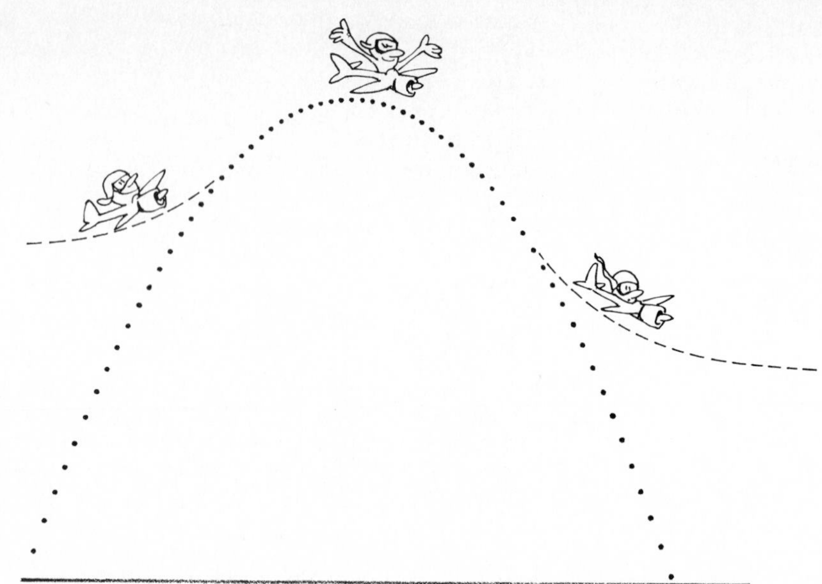

**FIGURE 4–12**
When the plane is in free fall, it follows a path that would be taken by a heavy stone thrown upward from the ground. While on that path, everything inside the plane experiences weightlessness.

the free fall path before the plane returns to earth just as a stone would. This is the way the National Aeronautics and Space Administration (NASA) gives its new astronauts a few moments of free-fall experience before their first orbital flights.

**FIGURE 4–13**
Anna Fisher, astronaut candidate, during a zero-*g* training session in a NASA aircraft.

## Applying Force: Pressure

Most likely you are sitting as you read this page. The normal force that supports your weight acts on the areas of your body in contact with the chair and the floor. Each square centimeter that touches the chair or floor receives a certain amount of force, and the (vector) sums of those forces balance your weight. When you stand, the bottoms of your feet get all that force. They have less supporting area, so each square centimeter there gets a bigger share, and if you stand on one foot for a moment, you'll really feel the difference. It probably hurts you just to imagine what happens to a ballerina's big toe as she dances on point.

*The force per unit of area* is called *pressure*. More precisely,

$$\textbf{pressure} = \frac{\text{force normal to the area of application}}{\text{area of application}}$$

or

$$P = \frac{F}{A}$$

where $F$ is perpendicular to the surface area. The units of pressure are typically pounds per square inch or newtons per square meter.

You can calculate the pressure your feet exert on the ground when you are standing. First you need to know the area of contact between

**FIGURE 4–14**
The smaller the area that supports a given force, the greater the pressure on that surface.

## UNITS OF PRESSURE

The common units of pressure in the English system are **pounds per square inch** (lb/in.$^2$), abbreviated as **psi,** and **atmosphere** (atm), where 1 atmosphere = 14.7 lb/in.$^2$, the accepted standard pressure of the earth's atmosphere at sea level.

In the metric system the pressure of 1 newton/meter$^2$ is called a **pascal** (Pa). One atmosphere of pressure in these units is about 101,000 N/m$^2$, or 101 kilopascals (kPa). The pascal is a relatively new standard unit, adopted in 1960.

Barometers, which measure atmospheric pressure, typically indicate a value in *inches* or *centimeters*. This measurement refers not to pressure but to the height of a column of mercury that the air's pressure will support wherever the barometer is. In Chapter 12 we'll have more to say about barometers and other units of pressure.

your feet and the ground. Here's one way to measure it: Moisten the bottom of your bare foot with water and step on a clean sheet of graph paper marked with squares. Count the number of squares contained in the damp footprint. Divide your weight *mg* by *twice* this area to account for two feet, and that's the average pressure you exert on the ground while you are standing still. (Of course that means the ground exerts the same pressure on your feet. Action and reaction.)

Now, $F/A$ is an *average pressure;* when you stand in sand, the balls and heels of your feet press in deeper than the instep portion because the pressure is higher on these regions. If you stand barefoot on a hard floor, even less area supports your weight, since the floor does not conform at all to your feet. Soft shoes and sand and plush carpets feel good because they do conform to your feet, distributing your weight more evenly and over more area. That's why you subconsciously think about pressure each time you choose a chair to sit in. You weigh the same in any chair, but a soft, cushiony one conforms more to your body than a hard, straight chair. The different pressures are easy to feel.

The definition of pressure shows us how sharp knives and razors cut so well. The slight force you apply with the tiny cutting area of the thin blade edge produces a large pressure. Naturally, the sharper the blade, the thinner the region of contact. This is why it's easy to nick your finger on the thin edge of a piece of paper. If the area is very small, the applied force on it doesn't have to be large.

When you take a walk, the rolling motion calls for many small adjustments of pressure on the bottoms of your feet. Only your hands require more fine tuning of pressure during the day. So it's no wonder that half the bones in your body are in your feet and hands (over 100). You might break a bone if you trip and fall on a small area such as a wrist. Therefore, it is important sometimes to maximize the force-bearing area in order to lessen the pressure. By taking care to land on large areas of their bodies, professional wrestlers, motorcycle riders, skiers, and acrobats can take spectacular falls and walk away unharmed.

## Round and Round

When a car accelerates along a straightaway, its tires get a push from the road, just as your feet get a push from friction with the ground when you take a step. Then when the car stops, friction from the road decelerates it as well. Whenever a car's speed changes, an outside (or *external*) force has to change it.

Velocity, however, is speed *and* direction. When a car turns, then, a force causes that, too, whether or not the car's speed changes at the same time. A force component parallel to the direction of travel changes an object's speed, while *a force component perpendicular to its direction of travel turns it*. In this section we'll explore the properties of the "turning" component of force. To keep things simple, we'll look at objects moving in circular paths. (But the results apply to *any* curve, not just circles; see Exercise 40 at the end of the chapter.)

Whenever anything moves along a circular path, the turning force is always perpendicular to the object's path and points in the direction of the center of the circle; it is called the **centripetal** (center-seeking) **force.** Attach a string to a small stone and whirl it around (Fig. 4–15). The tension in the string pulls the moving stone toward the center of its circular path. The string supplies the centripetal force. All kinds of forces can act as centripetal forces. For instance, the moon perpetually travels around the earth because the pull of earth's gravity tugs on it, turning it and keeping it in its orbit. In this case, gravity supplies the centripetal force. Centripetal force is only the name we give to the turning component of a net force, no matter what force or forces provide that component.

The actual value of the centripetal force depends on how fast an object travels and on the size of the circular path. You know this from your experience in cars. When you take a curve, a centripetal force turns you. It comes from friction with the seat, or contact with the seatbelt, or even contact with a door. At the same speed a sharp turn (a small circle) gets you a bigger tug than a gentle turn (a large circle)—

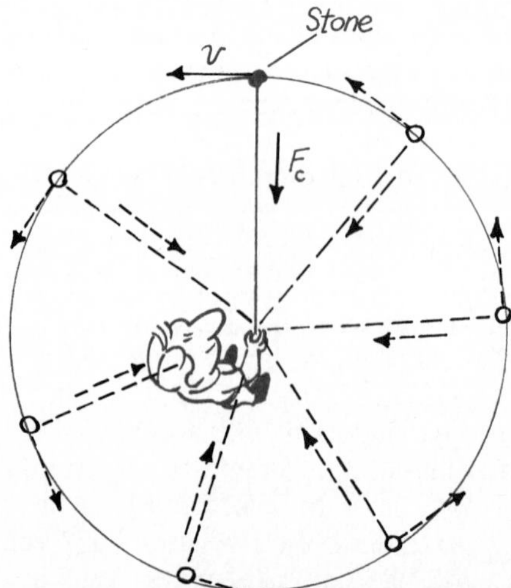


**FIGURE 4–15**
The center-seeking force acts continually to turn a moving object in a circular path.

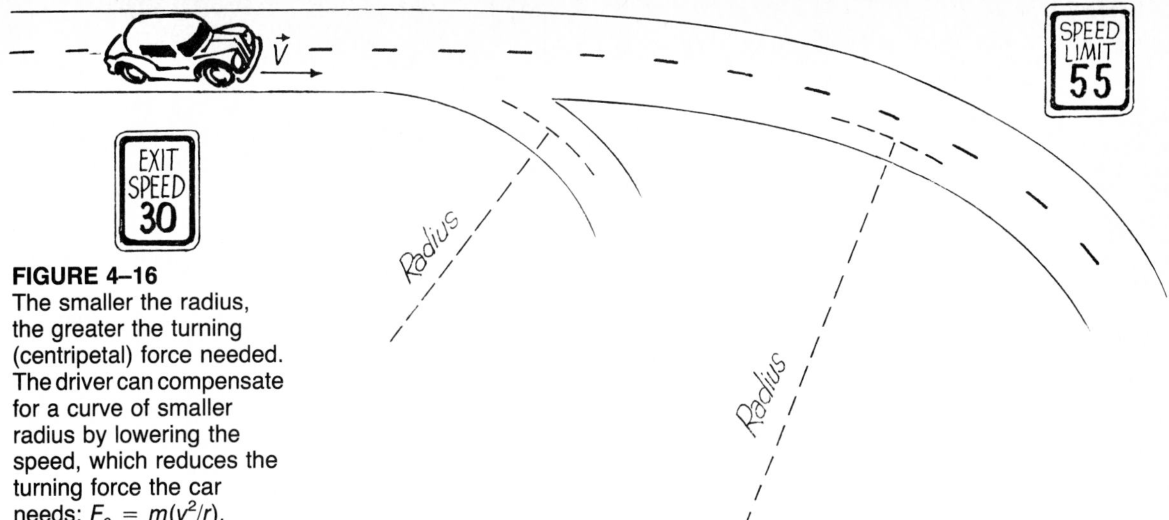


**FIGURE 4–16**
The smaller the radius, the greater the turning (centripetal) force needed. The driver can compensate for a curve of smaller radius by lowering the speed, which reduces the turning force the car needs: $F_c = m(v^2/r)$.

Fig. 4–16. On the gentle turn your *direction changes more slowly,* so the sideways acceleration (and hence the force) is less. So a bigger radius means a smaller centripetal force. Likewise, on any given curve you can move slow or fast. Although a low speed might be comfortable, a high speed might require a stomach-shifting force to turn you along the same circular path. At higher speeds your *direction changes faster,* meaning you accelerate faster; in other words, the centripetal force is larger.

The relationship* between the centripetal force, the speed on the circle, and the radius of the circle is

$$\textbf{centripetal force} = \text{mass} \times \frac{\text{speed} \times \text{speed}}{\text{radius}}$$

or

$$F_c = \frac{mv^2}{r}$$

Notice what the two factors of speed, $v$ times $v$, mean to the value of the centripetal force. If you double the speed along a circular path (say, from 20 mph to 40 mph), the force is *4* times as much. That is, $(2v_i)^2$ is always 4 times as large as $(v_i)^2$, where $v_i$ stands for the *initial* speed.

Perhaps as a child you played on a merry-go-round at a park—the kind you push yourself. When it was turning about as fast as a small child could run and push it, hanging onto the edge was easy. But the ride got much more exciting if an adult came over to spin the platform faster. Only a *little* extra speed made a *great deal* of difference in how firmly you needed to hang on. That was $v^2$ at work. For more discussion of the centripetal force formula, see Example 5 in "Calculations."

*We won't derive this formula, but the interested reader could find a geometrical derivation in almost any advanced physics text.

## Taking a Curve

When you drive a car along a circular curve, centripetal force turns your car. What pushes it toward the center of the circle? The friction between the tires and the road. That friction is *static* friction, because the tires are rolling and not sliding.* On a curve along a level road, then, a car's tires must grip the road with a sideways force equal to $mv^2/r$. If the tires slip, the frictional force is less, and the car goes off the curve along a straighter path. (What if the road is "banked"? See Exercise 45.)

On the moon the car couldn't take a level curve nearly as fast as it can on earth. Whatever the mass of the car, it would weigh only ⅙ of its earth weight, so the tires would press against the surface only ⅙ as hard. Static friction, remember, depends directly on the normal force, so the most sideways frictional force the tires could muster on the moon would be about ⅙ of the maximum value on earth. The car's mass is the same on either the moon or the earth, of course, so at the same speed the car would need just as much centripetal force to turn on either surface.

Driving safely on the moon would take some practice. A level curve that your car could take without skidding at up to 50 mph on earth would be impossible to stay on at more than 20 mph on the moon. To stop from 40 mph might require 2.5 seconds on earth, but on the moon it would take 15 seconds. (See Exercise 54.)

### HOW ACCURATE ARE NEWTON'S LAWS

Isaac Newton's discoveries in physics explained and predicted for the first time some of the grandest actions of nature. With his laws of motion and his law of gravity (Chapter 7) he explained the paths of the planets around the sun. He explained the ocean tides caused by the moon and the sun, and he showed that the earth should bulge somewhat at its equator because of its rotation. This effect was measured later; the earth is some 27 miles thicker through the equator than it is through the poles. Newton's discoveries, published in 1687, gave scholars of natural philosophy powerful predictive laws, and the newborn science of physics, begun by Galileo, began to take an important, permanent place in human affairs.

For the most part Newton's contributions stood unaltered for more than 200 years. But Newton himself knew that his findings could not be the final, absolute word on how things behave. When he published these laws he wrote, ". . . I hope the principles here laid down will afford some light either to this or some truer method of philosophy." They did indeed.

It was not until early in this century that revolutionary discoveries in physics revealed limitations to the accuracy of Newton's laws

*See Demonstration 4 in Chapter 3; when a wheel rolls, the part of the wheel that touches the ground has zero speed.

of motion. Albert Einstein found that time, distance, and mass are not the same for moving objects as for those same objects at rest (Chapter 26). Since these quantities appear in the second law, $F = ma$, Newton's famous result was shown to be in error. But differences between the predictions with $F = ma$ and Einstein's laws aren't normally measurable unless the speed of an object is a large fraction of the speed of light (186,000 miles per *second*). For ordinary everyday events that most of us deal with, speeds are never so great. In fact, when NASA scientists calculated the paths and motions that took the astronauts to the moon, they used Newton's laws, not Einstein's.

Another discovery in the early decades of this century concerns the behavior of the subatomic particles, electrons and protons, and even individual whole atoms. Even though these tiny particles make up the matter around us, they do not behave like the matter we all deal with everyday. Physicists discovered that with such small particles predictions of precise motions are actually impossible. A new theory quite unlike anything that Newton or Einstein discovered was developed to account for their behavior. However, even the new theory predicted that when large numbers of atoms bond together to make the everyday objects around us, Newton's laws do predict their actions with great accuracy.

All of the laws and theories you read about in this text share this possibility of being disproved. Physical laws are only our best descriptions of how nature works, and the physics that explains the things we know today won't provide explanations for everything that physicists may discover in the future. Physics evolves through discovery today as it has since Newton's time, offering a frontier that can never disappear with progress: Progress in physics only opens the door to new questions, ones that will need (to borrow from Newton's words) "some truer method of philosophy" for their answers.

## CALCULATIONS

EXAMPLE 1: A typical elevator accelerates upward only briefly before it reaches a steady speed. Suppose an elevator's acceleration is a steady 1 $m/s^2$ (about 3 $ft/s^2$). If a student who weighs 534 newtons (about 120 pounds) stands on a scale in this elevator, **how much force would the scale show** during the brief upward acceleration?

Even with this simple problem, you should first sketch the forces. But which forces? Only those acting on the student are needed because it is the upward force on the student that the scale shows. In your sketch let $F_n$ represent the normal (or contact) force from the scale. $F_n$ points upward, and the student's weight $mg$ points downward. The difference between these opposing forces gives the net force on the student, and that difference is equal to $ma$, the student's mass times the student's acceleration. In symbols, $ma = F_n - mg$. We know that $mg$ is 534 newtons and $a$ is 1 m/s$^2$ (the elevator's acceleration is also the student's acceleration), but what is $m$? To find $m$, divide $mg$ by $g$: $m$ = 534 N divided by 10 m/s$^2$ = about 53 kg. After rearranging terms, the normal force on the student from the scale while the elevator is accelerating is found to be

$$F_n = mg + ma = 534 \text{ N} + 53 \text{ N} = \mathbf{587\ N\ (132\ lb)}$$

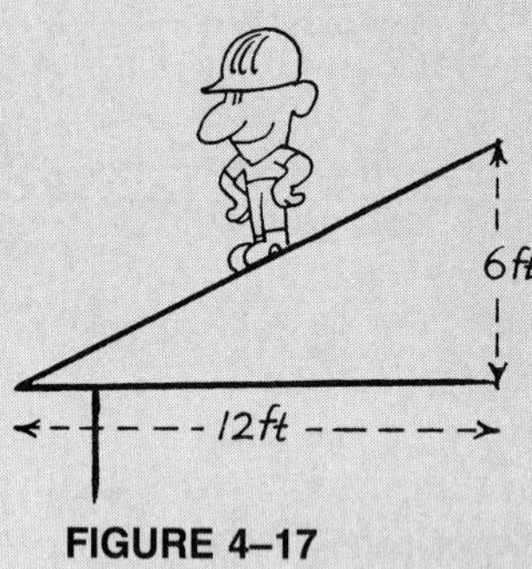

FIGURE 4–17

EXAMPLE 2: A 150-pound roofer stands still on a roof, as shown in Fig. 4–17. **How many pounds of frictional force keep him from slipping?** Sketch the forces on graph paper so you can measure the components. First, draw the weight vector, using some convenient scale. Then resolve the weight into component forces along (parallel to) the roof and against (perpendicular to) the roof. Just use a right angle (90°) such as the corner of this page to help you sketch the line for the perpendicular component's direction.

Measure the lengths of these component forces and use the scale to find their values. You'll find the 150-pound roofer presses against the roof with a force of 135 pounds, while a force of about 66 pounds acts to drag the roofer downward along the roof. If this component were unopposed, he would accelerate along the slope at quite a clip, as though he were on a water slide (see Example 2 in "Calculations" in Chapter 2). Therefore, **the roof exerts 66 pounds of static friction on his feet** or else he could not stand there. To check the values of the components, use the Pythagorean theorem. $(150)^2$ should be nearly equal to $(66 \text{ lb})^2 + (135 \text{ lb})^2$.

EXAMPLE 3: If a 120-pound gymnast stands with one foot each on two scales placed side by side, how much weight is each scale supporting? Each scale reads 60 pounds. If the gymnast hangs from two vertical ropes suspended side by side, how much force is each hand supporting? Just as for the scales, if the vertical acceleration is zero, the net vertical force on the gymnast is zero. So each hand supports half the gymnast's weight. This means each rope has a **tension** (the upward force the rope supports) of 60 pounds also. See Fig. 4–18.

Suppose the gymnast hangs from ropes that are not vertical but at an angle halfway between vertical and horizontal (45° from either of those directions). **What would the tensions in these ropes be?** In other words, how

FIGURE 4–18

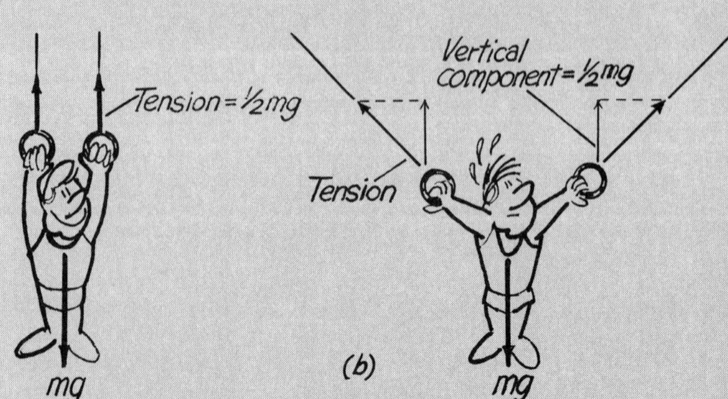

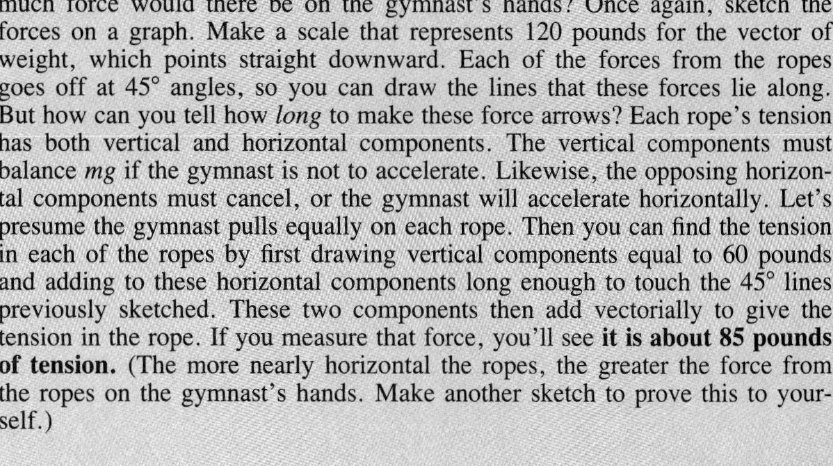

much force would there be on the gymnast's hands? Once again, sketch the forces on a graph. Make a scale that represents 120 pounds for the vector of weight, which points straight downward. Each of the forces from the ropes goes off at 45° angles, so you can draw the lines that these forces lie along. But how can you tell how *long* to make these force arrows? Each rope's tension has both vertical and horizontal components. The vertical components must balance $mg$ if the gymnast is not to accelerate. Likewise, the opposing horizontal components must cancel, or the gymnast will accelerate horizontally. Let's presume the gymnast pulls equally on each rope. Then you can find the tension in each of the ropes by first drawing vertical components equal to 60 pounds and adding to these horizontal components long enough to touch the 45° lines previously sketched. These two components then add vectorially to give the tension in the rope. If you measure that force, you'll see **it is about 85 pounds of tension.** (The more nearly horizontal the ropes, the greater the force from the ropes on the gymnast's hands. Make another sketch to prove this to yourself.)

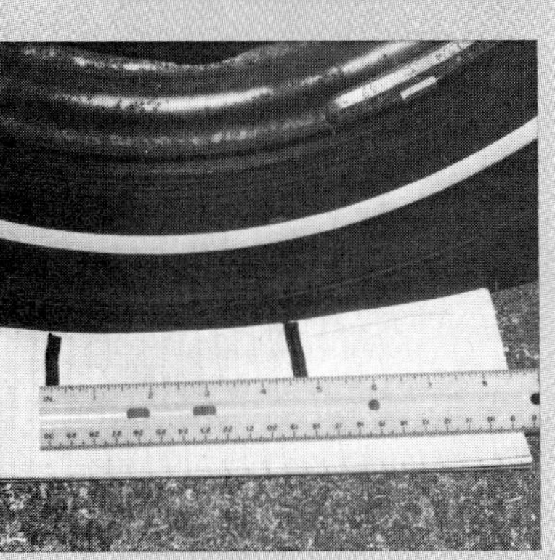

FIGURE 4–19

EXAMPLE 4: **Estimate the average pressure the tires of a Volkswagen Beetle exert on the ground.** A Beetle with two passengers weighs about 2000 lb, so the average force on each of the four tires is 500 lb. However, pressure is force divided by the area of application, so you need to estimate the area of the tire that is in contact with the ground. The standard tire on a Beetle is relatively narrow, only a palm width or so wide. (See Fig. 4–19.) The width of the area that touches the ground is about 4 in., the length a shade longer, say 5 in. The area of a rectangle is equal to length times width, so the estimated area of contact (ignoring the open areas of the tire's tread) is 20 square inches. From the formula for pressure, then,

$$P = \frac{F}{A} = \frac{500\ \text{lb}}{20\ \text{in}^2} = \mathbf{25\ lb/in.^2}$$

for each tire. (This is, in fact, exactly the manufacturer's recommended average air pressure for a Beetle's tires—20 lb in the front tires and 30 lb in the rear.)

EXAMPLE 5: When a car rounds a corner at an intersection, it is temporarily on a circular path (see Fig. 4–20). The radius of that circle is typically about 15 feet. Let's **calculate how fast the VW Beetle of Example 4 can make that turn without skidding.**

In good driving conditions, the maximum static friction from the road (which keeps the car turning on its circular path) is about equal to the car's weight. (See Example 4 in Chapter 3.) So the maximum centripetal force we can expect the Beetle's tires to supply is about 2000 lb. That means $F_c = mv^2/r = 2000$ lb. We know $r$, but we need to find $m$, the car's mass, before we can solve for $v^2$, the square of its speed. To find $m$, we divide the weight $mg$ by the value of $g$: 2000 lb divided by 32 ft/s$^2$ = 62.5 slugs = $m$. Now using the formula for centripetal force, we find $v^2 = 480$ ft$^2$/s$^2$. Trial and error with a pocket calculator tells you $v$ is about 22 ft/s, which is 15 mph. (For example, $20^2 = 400$ and $25^2 = 625$, so try squaring numbers between 20 and 25 to approximate 480. When your answer is less than 480, increase the next try. When it is more than 480, try a smaller number.) To keep from skidding on the curve, then, the Beetle should take the corner at a speed **no greater than 15 mph.**

FIGURE 4–20

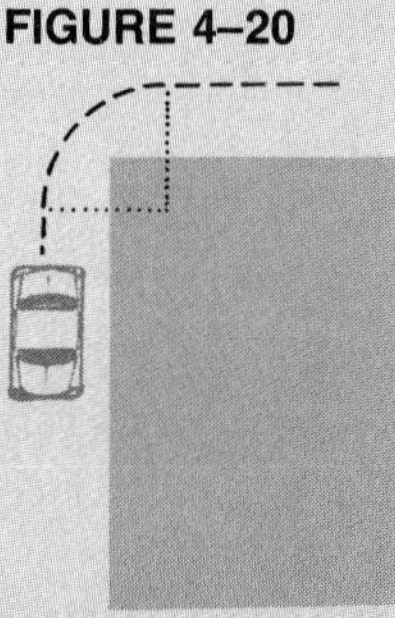

EXAMPLE 6: Air resistance, or resistance from any other fluid, is a complex phenomenon. Nevertheless, a simple picture gives a good account of air drag on ordinary things at everyday speeds. A solid object pushing through air collides with the air's molecules in its path. As a result, the molecules are pushed away, and the object gets a push that opposes its motion. The greater the object's speed, the more collisions it has per second. For instance, if the speed doubles, about twice as many collisions take place per second. For that reason, the drag on an object depends on its speed $v$. But when the speed $v$ doubles, the forces between the air molecules and the object's surface double as well. Accordingly, a doubling of speed *quadruples* the drag because there are *two* factors of $v$ contained in the drag formula. The expression for the air drag on something is

$$F_{\text{air drag}} = bv^2$$

where $b$ is a constant number that depends on an object's shape and the amount of surface area that collides with the air. The units of $b$ are force/speed$^2$.

If we know the air resistance on an object for any one certain speed, we can find the value of $b$. Then the formula $F_{\text{drag}} = bv^2$ lets us determine the air resistance at any other speed. For example, if a 150-lb skydiver's terminal speed is 120 mph (176 ft/s), the air drag at that speed is equal to $mg$, or 150 lb. Rearranging $F_{\text{drag}} = bv^2$ to find $b$, we find

$$b = \frac{150 \text{ lb}}{(176 \text{ ft/s})^2} = 0.0048 \frac{\text{lb}}{\text{ft}^2/\text{s}^2}$$

With this value for $b$, **what is the air drag on this skydiver at 60 mph (88 ft/s)?** We find

$$F_{\text{drag}} = bv^2 = \left(0.005 \frac{\text{lb}}{\text{ft}^2/\text{s}^2}\right) \times (88 \text{ ft/s})^2 = \mathbf{39\ lb}$$

(You could as well find $b$ in terms of lb/mph$^2$, of course, and express $v$ in mph when calculating $v^2$.)

# REVIEW

**1.** Only an *external* force can cause your body to accelerate. True or false?

**2.** As you walk, do you make use of friction with the ground in both the forward and the backward directions?

**3.** If you stand on an incline, does the full force of your weight press into the incline?

**4.** Sliding a heavy object over the floor is usually easier if you direct your force (a) somewhat upward, (b) horizontally, (c) somewhat downward, (d) vertically.

**5.** Define the terminal speed for a falling object.

**6.** When a solid object moves through air, does its speed affect the amount of drag on it? Does its shape affect the drag? Does its size? Its mass?

**7.** Which statement is correct when a plane travels horizontally with a constant speed? (a) Lift balances drag; (b) drag balances thrust; (c) thrust balances lift; (d) drag balances weight.

**8.** Define pressure, and give its units in the metric and English systems.

**9.** When something moves in a circular path, the centripetal force is the force that turns it. True or false?

**10.** Is the centripetal force always perpendicular to a moving object's direction at any instant?

**11.** Does the centripetal force on an object depend on its speed? On the radius of the circular path it moves on? On the mass of the object?

**12.** Either friction, gravity, or a contact force can supply the centripetal force on an object if the circumstances are right. True or false?

## DEMONSTRATIONS

**1.** Pick up a book with one hand and a small sheet of paper with the other. Hold them horizontally and drop them side by side. The sheet of paper descends slowly while the book zips down, accelerating at the rate of $g$. Next, hold the book horizontally and then lay the sheet of paper on the book's cover. When you drop the book again, there'll be no air drag on the paper. How does this affect the paper's acceleration? Now crumple the paper into a small wad and drop the paper and the book side by side again. Although the mass of the paper is the same as before, the change in the size and shape of its surface area obviously affects the paper's acceleration.

**2.** Fill a cup or glass half full with water. Holding it at arm's length, twirl it in a circular path by turning your body around (Fig. 4–21). To keep from spilling the water, tilt the container toward you. Spin faster and hold the container horizontally. Why doesn't the water spill?

**FIGURE 4–21**

**3.** Use a sheet of graph paper and estimate the pressure on your feet when you stand normally (as explained in this chapter). Convert that pressure to atmospheres. One atmosphere (atm) is about 15 lb/in.$^2$. (The average in the author's classes is about ⅓ atm, or 5 lb/in.$^2$, but individuals range between ¼ atm and a bit over ½ atm.)

**4.** Have two of your strongest friends tug at the opposite ends of a short rope, keeping its length parallel to the ground. Then hang a 10-lb sack of potatoes from the rope midway between them and challenge them to keep the rope straight. They can't, of course. To support the 10-lb downward force, each end of the rope must bend upward from that point; otherwise, the tension in the rope will have no upward component to counter the downward pull from the sack of potatoes.

**5.** Stand still on a bathroom scale. Then lift one foot slowly as you watch the indicator. Put the foot back down, then lift that foot abruptly. Why does the greatest acceleration cause the largest force change?

**6.** The next time you are in an elevator, try this. Before it begins to move, hold one arm straight out (to the side or in front). No matter how strong you are, you won't be able to keep your arm level when the elevator accelerates. Also try this with a book in your hand. The force your muscles exert to hold the arm up is suddenly too little if the elevator starts upward or too much if the elevator starts downward. You must adjust the force to accelerate the arm and book at the same rate as your body accelerates. *That involves your reaction time*. In most elevators the acceleration time is small, and it may be over before you can make the adjustment. Be sure to try this—that book really does feel heavier or lighter.

## EXERCISES

**1.** If a wagon is pulled at a constant velocity, the net force on it is zero. (a) True, (b) false, (c) you can't tell from this information.

**2.** A skydiver momentarily accelerates upward when the parachute opens. True or false?

3. How much force (in terms of its weight) must you exert on a bag of groceries (a) to hold it while talking to someone in the parking lot? (b) to carry it at a steady speed while walking? (c) to lift it from the car seat? (d) to put it down into the trunk of a car?

**4.** If you find it takes a push of 3 lb to slide your dictionary at a constant speed across a table, then what is the value of the kinetic frictional force retarding its motion?

5. If a speeding car suddenly goes over a crest in the road, do the passengers feel heavier or lighter? If the speeding car is going down a hill and then the road suddenly rises, do they feel heavier or lighter? Explain why.

6. On your next visit to the kitchen, pick up a small can of food. Do you think you could throw it farther if it were empty? Why? (The answer tells you why bullets are made of lead rather than an aluminum alloy.)

**7.** When a weight lifter cleans and jerks a 250-lb barbell (lifts it from the ground and pushes it overhead), the force exerted on the barbell must be (a) less than, (b) equal to, (c) greater than 250 lb. Why?

8. If there is a wind blowing, an eagle or an osprey will land on its nest *into* (or against) the wind. Why does it do this? (It's the same reason airplanes take off and land into the wind.)

**9.** What reason can you think of for the bottom of an elephant's foot to be so flat? Why should the small foot (hoof) of a deer be so hard?

**10.** When you get up from a sitting position on the floor, each foot pushes down with (a) half, (b) more than half, (c) less than half of your weight.

**11.** With a force diagram, show the net force a football player feels when he's double teamed, as in Fig. 4–22.

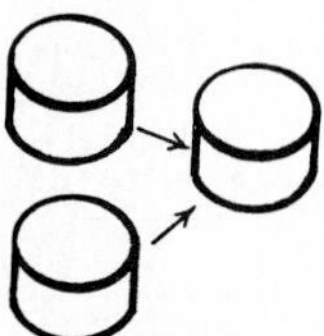

**FIGURE 4–22**

**12.** Faced with getting a shopping cart over a curb in a parking lot, would you push it or pull it? Explain why.

**13.** A defending lineman on a football team often tries to block by getting his body under his opponent's and pushing upward. What does this do to the frictional force on the opposing lineman's feet? How does that affect the opposing lineman's acceleration?

**14.** Which of the forces shown in Fig. 4–23 can push the book sideways? Why?

**FIGURE 4–23**

**15.** To stop faster, a driver pushes harder against the brake pedal. But that doesn't always work. If the driver pushes too hard, the wheels will lock and the tires skid along the ground. Tell why the car decelerates *less* at that point than it would if the brakes weren't applied quite hard enough to lock the wheels.

**16.** Analyze the statement (made in Chapter 2) that on the moon you could throw a feather as far as a baseball. Can you give an argument that you should be able to throw the feather even farther than the baseball if you were on the moon?

**17.** If you drop a nickel coin and a wooden nickel the same size together, is it likely they will strike the ground at the same time? Why?

**18.** Before a jump, could you pick out which of several parachutists might have the greatest terminal speed *after* the chutes are opened, provided all the parachutes were the same size?

**19.** Why does the coffee in a cup you are holding in a car slosh forward when the car stops?

**20.** A long-distance swimmer must pass through shark-infested waters. The swimmer stays inside a shark-proof cage that is towed at the swimmer's speed by a motorboat preceding the swimmer. Think of a *second* way the shark-proof cage could make the swim easier on the swimmer.

**21.** If you pulled a low but heavy footlocker with a rope and greased its bottom to reduce friction, would it help more to use a short rope or a long one? (Do a sketch to see the answer.)

**22** Sketch the forces on the sailplane shown in Fig. 4–24.

**FIGURE 4–24**

**23.** A student starts his car to leave a parking area on the beach, but the wheels slip. Fearing they will continue to slip and dig deeper, he lets some air out of the tires. How will this help get the car out of the sand? (If the wheels are already deep into the sand, this technique can't help.)

**24.** Why is a faulty trailer hitch more likely to break if a car is going uphill at a constant speed than on a level road at that same constant speed?

**25.** A car accelerates quickly from rest. Its tires are more likely to spin if it is going uphill than if it is on a level road. Why? Are its tires more likely to spin if it's going downhill rather than on a level surface?

**26.** In late June, 1893, the Norwegian vessel *Fram* sailed to the coast of Siberia. The crew's goal was to prove there is no solid continent in the Arctic Ocean, no land mass at the North Pole. In September their ship became frozen in the ice, as expected. But whenever the ice pressed in on its sides, the specially designed angular hull of the *Fram* directed the pressure *upward*. Instead of being crushed, the *Fram* popped upward in the ice. *Show with a diagram how this could work.* The solid ice pack surrounding the ship moved, thanks to the wind and ocean currents; 3 years and

1000 miles later, the *Fram* broke from the ice into the open sea between Spitsbergen and Greenland, having passed within 250 miles of the North Pole. The frozen Arctic *is* only ice that forms on the ocean's surface.

**27.** The bones in your fingers are flat on the gripping side. What's the advantage of flattened finger bones over round finger bones when you grip something firmly or lift something heavy?

**28.** You've been traveling on an interstate highway just after a rain, and you see a curve ahead. You want to maintain your speed, minimizing the risk of skidding. Which lane should you take and why?

***29.** For purposes of comparison a large acceleration is often expressed in terms of $g$ rather than in m/s$^2$ or ft/s$^2$. For example, a modern fighter plane can bank and make a circular turn so fast that it accelerates toward the center of its circular path at a rate as great as $5g$. That means the pilot receives a (centripetal) force from the contoured chair in the cockpit equal to 5 times the pilot's weight. How much force does the 5-lb flight helmet exert on the pilot's head during such a turn?

***30.** If you step barefoot on the point of a single nail, it can penetrate your foot. But you can safely lie on a bed of closely spaced nails. If the bed of nails supports your body with 300 nails, there is ______________ times as much area in contact with you as there would be with one nail, and each nail will support ______________ times as much weight as one nail alone.

**31.** What happens when the string in Fig. 4–25 is cut? Discuss.

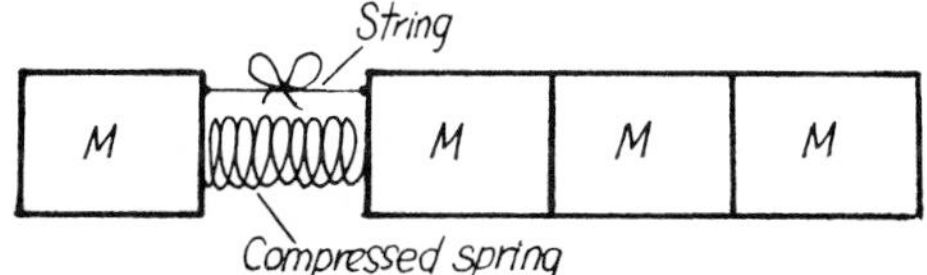


**FIGURE 4–25**

**32.** Professor Quick wanted to save time for his students who spent extra minutes every day in the slow elevators on campus. He suggested to the administrators that the elevators should accelerate uniformly half the way up and decelerate uniformly at the same rate for the remainder of the trip. Would you prefer a slow elevator or a "Quick" elevator? Why?

**$\overrightarrow{33}$.** A plane climbs at a steady rate with its nose pointed somewhat upward. Draw the forces acting on the plane. Does thrust balance drag? Does lift balance weight?

**34.** Give the effects on the force needed to turn a car if (a) the car's mass is doubled; (b) the car's speed is doubled; (c) the radius of the car's path is doubled. (d) $m, v$, and $r$ are all doubled.

**35.** A rock is lodged in the tread of a tire. When it works loose, what direction does the stone take as it leaves? Is it more likely to come loose at low speeds or high speeds? Why?

**36.** If you walk on your heels at the beach, why do you sink deeper into the sand? To avoid sinking in snow, people wear snowshoes or skis. Suppose you were faced with retrieving an errant Frisbee from the thin ice over a just-frozen pond. Fearing a breakthrough, what could you do to help get out and back again without crashing through?

**37.** When your car travels with constant speed on level ground, the forward push your tires get from the road exactly balances the air's drag on the car. True or false? Why?

**38.** Draw a picture to show how a single layer of oil molecules between two flat solid surfaces can reduce kinetic friction. Could there be *static* friction between the molecular "logs" or "beads" and both moving surfaces?

**39.** What is your motion like if you let go from the edge of a child's merry-go-round that is spinning fast?

**40.** On a sheet of paper make a sketch of a smooth but winding road. Take a compass (one that's used for drawing circles) and show that on any small segment of that path you can find by trial and error a circle whose arc matches the road's curve over that segment. What you've found is the radius $r$ of the road's curvature there; to figure out the centripetal force on a car at that point, use that measured $r$ in the formula $mv^2/r$.

**41.** You and your friend have a friendly tug of war with a rope, but it's a standoff. Make a drawing showing the forces that act on you, then a separate drawing of the forces on your friend.

**42.** Every sailor knows that wrapping a rope once around a sturdy rail allows a tiny force on one side to support a *huge* force on the other side (Fig. 4–26). Analyze why. (The reason is the same one that keeps the knots in shoelaces from coming apart.)

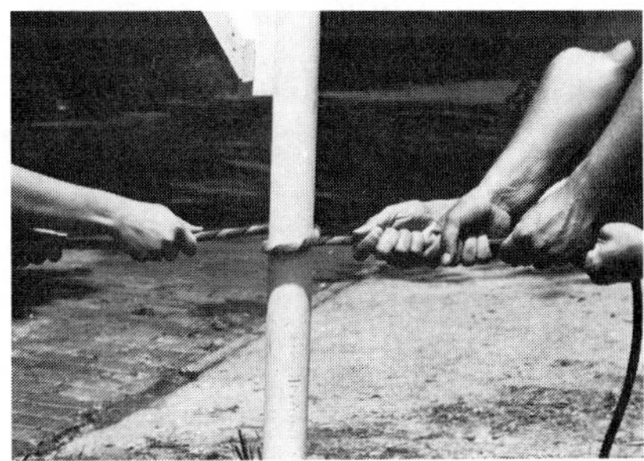

**FIGURE 4–26**

**43.** As the earth rotates around its axis, you move in a circular path around that axis about once every 24 hours. What supplies the centripetal force to keep you turning? What would happen to you if that force suddenly vanished for you?

**44.** Connie flies a model airplane attached to a string. To make her sport easier on her hands, should she prefer to replace her model with one that weighs the same but goes twice as fast, or one that goes just as fast but weighs twice as much?

45. Curves on roads are "banked" for safety (Fig. 4–27). Make a sketch to show that the normal force from the banked road helps to turn a car, decreasing (or even eliminating) the need for a frictional force from the road for that purpose.

**FIGURE 4–27**

**46.** After he jumps, a skydiver reaches terminal speed after 12 seconds. Does he gain more speed during the first second of fall or the eleventh second of fall? Does he fall farther during the first second or the eleventh second?

47. Have you ever wondered why seats on 10-speed bicycles are so narrow? Guess why.

**48.** Would it take a longer distance to stop your car on the earth or on the moon? Discuss.

49. You might think that such a streamlined bird as a peregrine falcon could fall with its wings tucked in through the air much faster than a falling baseball. Yet the greatest measured terminal speed for that bird is 82 mph compared to a 140 mph terminal speed for a baseball. Discuss.

**50.** Analyze how you hold a book when you walk between classes. Do you use friction or the contact force more?

51. If you are driving at 50 mph on a level road, your car will need some minimum distance to stop without skidding. If you are driving downhill at this same speed, could you stop in that same distance? Explain why or why not.

*52. After a skydiver's parachute balloons open, it takes only about 2 seconds to bring the skydiver to 12 mph from 120 mph. (a) Calculate the deceleration of the skydiver in $ft/s^2$. (b) Express this in terms of $g$.

*53. A certain car can accelerate along a road at a maximum rate of 2 $meters/second^2$ without spinning its tires. If it is hooked up to a trailer with twice its own mass, how fast could it accelerate without burning rubber?

*54. Show that at any speed it would take about 6 times as long (in seconds) to stop your car on the moon than it does on earth.

*55. When a skydiver makes a jump, it takes only about 12 seconds to reach terminal speed. What uniform rate of change of acceleration in the vertical direction $(a_f - a_i)/t$ would bring this about?

*56. A fully loaded 747 airplane weighs 600,000 pounds. How much lift must this plane have from its wings when it is cruising with zero vertical motion? Its wings have about 7000 square feet of area contributing to the lift, so what is the average pressure above the atmospheric pressure on the wings? Compare this pressure to the atmospheric pressure at sea level.

*57. Calculate the pressure beneath the solid base of a waterbed whose dimensions are 5 feet by 6.5 feet if it holds 180 gallons of water. (One gallon of water weighs 8.3 pounds.) Many apartment complexes won't allow waterbeds in second-story rooms. Do you think the pressure they exert is the reason, or the total weight? Or just the chance that they might spring a leak? Discuss.

*58. The VW Beetle in Example 4 in "Calculations" travels at 55 mph. What is the minimum safe radius of a level curve that it could take without slipping?

# CHAPTER 5
# MOMENTUM

FIGURE 5–1

Sunlight slipped through a break in the clouds and struck the glacier looming a mile above the four climbers. Created from centuries of snowfalls, the huge mass crept under the force of its own weight at an imperceptible pace across the mountaintop. The wind and sun had carved and melted fantastic shapes into its profile, and many deep cracks (called crevasses) split the vertical face of ice. Below the glacier the mountain's snowy slopes bore vertical grooves left by huge chunks of ice that had cracked loose and fallen from the glacier's edge. Day by day the climbers, mere dots beneath the blue ice and gray shadows, inched along a knife-edge ridge that led upward to the glacier's icefall.

Stirring in the middle of the night, one member of the team, Allen, heard a sharp cracking sound. Unseen, a block of ice the size of a house had broken from the glacier and was plummeting downward, gathering speed along its path. The avalanche it caused passed several thousand feet to the side of the tiny tents on the ridge and gradually came to rest at the foot of the mountain more than a mile below. The rumbling subsided, leaving Allen once again with only the noise of the wind.

As we have seen in the last two chapters, forces describe the motion of such an avalanche. But falling ice will crack, shift, and tumble, and the frictional forces along its path change from one instant to the next. It's a situation too complicated to describe in any detail with $\vec{F} = m\vec{a}$. However, we can still understand much about even such complicated events, as you will see in this chapter. We know an avalanche plunges down a mountain because of the pull of earth's gravity. And Newton's third law tells us there is an equal but opposite force on the earth from the gravitational attraction of the falling ice and snow. Action equals reaction. This causes the great mass that is our planet to move toward a moving avalanche a tiny, tiny bit! Using the properties of the quantity called *momentum,* we can actually calculate the distance the earth moved toward the avalanche that disturbed the climber's sleep. (See Example 5 in "Calculations.") ■

## Impulse and Momentum in One Dimension

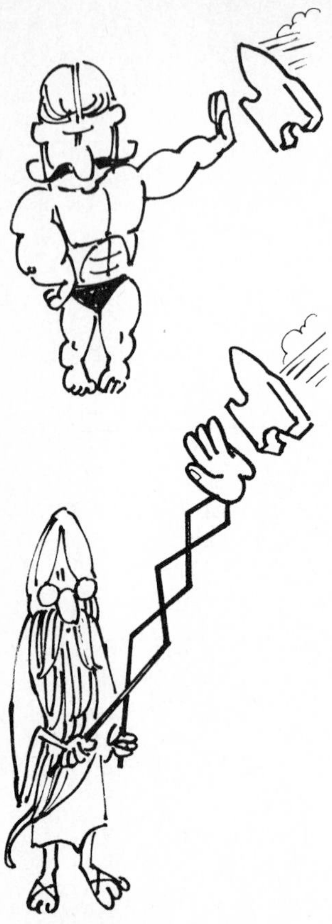

Anytime you drive a car, you eventually need to come to a stop. To do this, you can slam on the brakes to stop quickly, or you can apply the brakes gently for a longer time. In other words, a large force for a short time brings about the same change in speed as a small force for a long time. With Newton's second law we can see exactly how time and force bring about a change in speed. For motion along a straight line, *average force = mass × average acceleration,* or

$$\text{force}_{\text{average}} = \text{mass} \times \frac{\text{change in speed}}{\text{time}}$$

Multiplying both sides by time so that force and time appear together, and dropping the subscript "average," we find

$$\text{force} \times \text{time} = \text{mass} \times \text{change in speed}$$

The quantity (force × time) is called **impulse.** This formula tells us it is the impulse $(F \times t)$ that brings about a change in speed for your car. If 200 pounds of force can bring it to rest in a time of 3 seconds, then 50 pounds of force can do the same job in 12 seconds. That is, (200 pounds × 3 seconds = 600 pound-seconds) or (50 pounds × 12 seconds = 600 pound-seconds) causes the same change in speed. *Any* product of force and time equal to this value would bring your car to rest.

**FIGURE 5–2**
Two ways of stopping a flying object: Captain Forceful applies a large force for a short time. The anvil stops suddenly. Father Time applies a small force for a long time. The anvil stops gradually as the spring collapses.

Notice what the right-hand side of the formula for impulse tells us. A car that's twice as massive as another needs twice as much impulse to stop from the same speed. That makes sense. If you've ever pulled a trailer or a boat or driven a carload of people you know extra mass makes your car more difficult to stop. Also, you're aware that the faster you go, the harder it is to stop. If your car goes twice as fast, the right-hand side of the impulse formula tells you that you need twice as much impulse to stop it. So both increased speed and increased mass have equal effects on how much impulse is needed (Fig. 5–3). To show the relationship between impulse, mass, and speed more clearly, we can rewrite the right-hand side of the formula. Let $v_i$ be the initial speed and $v_f$ be the final speed. Then

$$\text{impulse} = m(v_f - v_i) = mv_f - mv_i$$

The product of mass and speed, $mv$, is called **momentum.** This formula says

$$\textbf{impulse} = \text{change in momentum}$$

**FIGURE 5–3**
Because momentum = mass × speed, a loaded school bus moving slowly can have the same momentum as a speeding sportscar.

The units of impulse and momentum have no special names. Newton-seconds (N·s) and pound-seconds (lb·s) are typical units for impulse, while kilogram-meter/second (kg·m/s) is most often used for momentum. For numerical examples of impulse and momentum, see "Calculations."

## Using Impulse and Momentum

**FIGURE 5–4**
Because of its great mass, this cruising VLCC (very large crude carrier) takes a long time to stop, traveling well over a mile before it can be brought to a halt.

If you jump down from a small stool or step ladder, you probably don't think about how you do it, but you invariably flex your legs at the knees. Why? Because that gives the upper part of your body more time to stop. If you hit the ground stiff-legged, you stop abruptly and the force on your feet which is passed on by the bones in your body is enough to jar your teeth. However, if you bend your legs, they act as shock absorbers, reducing the *force* your body feels by increasing the *time* of impact. (Notice that the final *change* in momentum is the same no matter how you land.) You use the same technique when you walk down stairs. Bending your knee spreads the impact of each step over a longer time to ease the shock of decelerating your body. We often manipulate impulse to make life easier and safer.

If a moving car suffers a front-end collision, its bumper and fenders crumple, and their "giving" action brings the rest of it to a slower stop. If the passengers are wearing shoulder harnesses and seatbelts, they take as long to decelerate as the car does. If they are not fastened down tightly, they leave the car seat as it slows beneath them (Newton's first law) and continue moving until the dashboard or steering wheel stops their motion. There they decelerate very quickly and feel a much larger average force than if they remained with the car seat to take advantage of the delaying action caused by the bending of the bumpers and fenders. A sudden change in momentum such as this requires a large force on the body compared to the force that causes a gradual change in momentum. FORCE × time or force × TIME, the result is the same.

Momentum itself is a familiar concept. We all learn early just how hard it is to stop a moving object and that it's not just speed that matters. Most of us would try to catch a softball moving at 20 mph (a slow pitch) without a glove, but we would hesitate to get in front of a mass as large as a firetruck moving at 5 mph and try to stop it with our bare hands. Because of its huge mass, the firetruck's momentum is much greater than the momentum of a softball even though the ball moves much faster. MASS × speed versus mass × SPEED. So if you pushed as hard as you could ($F$), it would take a much longer time ($t$) to bring the firetruck to rest. $F \times t =$ change in $mv$.

Sometimes you want a large change in momentum. Baseball players swing hard at the ball, and the high speed of the bat helps make the force large. But they also follow through when they swing, increasing the time the bat acts on the ball. Golfers on tee shots, football players on kickoffs, and billiard players on break shots all follow through to give the ball the largest possible velocity. They maximize both the force *and* the time it's applied to boost the impulse and get a larger change in momentum.

High speed *or* great mass can make momentum large. That's why the linemen of professional football teams should be massive. Opposing linemen are positioned close together in a game and can't gather much speed before making contact. Nevertheless, their large masses give each substantial momentum. And yet momentum is why lighter-weight running backs traveling at full speed can be so hard to stop. This leads us to a question: What happens to the momentum of objects that collide?

## Simple Collisions

Much of what we know in physics today comes from studies of how things behave during collisions. For a personal experience with collisions, you might go to an amusement park and take a ride in the bumper cars, those slow-moving, electric-powered cars with bumpers all around (Fig. 5–5). Collisions and rebounds are the objects of the ride. Now if you're flying along and bump into the rear of a car in front of you that's moving at a slightly slower speed, you'll get the same bump as if that car were at rest and your car were barely moving. In other words, it is the *relative speed* of the moving objects that makes a collision soft or hard, no matter what the individual speeds are. The hardest knocks will come when two bumper cars meet head-on at full speed, when their relative speed is greatest.

Let's recall what relative speeds mean for cars on highways. If your car is traveling at 30 mph and strikes the rear end of a parked car, the relative speed just before the collision is 30 mph. If your car is doing 45 mph and collides with the rear of that same car while it is traveling in your lane at 15 mph, the relative speed is 30 mph as before (45 mph − 15 mph) and the potential for damage during the collision would be exactly the same. Finally, if that car is backing toward you at 10 mph and you are traveling at 20 mph at the time of the impact, the cars once again come together at 30 mph (20 mph + 10 mph), and the likelihood is that the damage would be the same.

Different things take place in simple, straight-line collisions, but

**FIGURE 5–5**

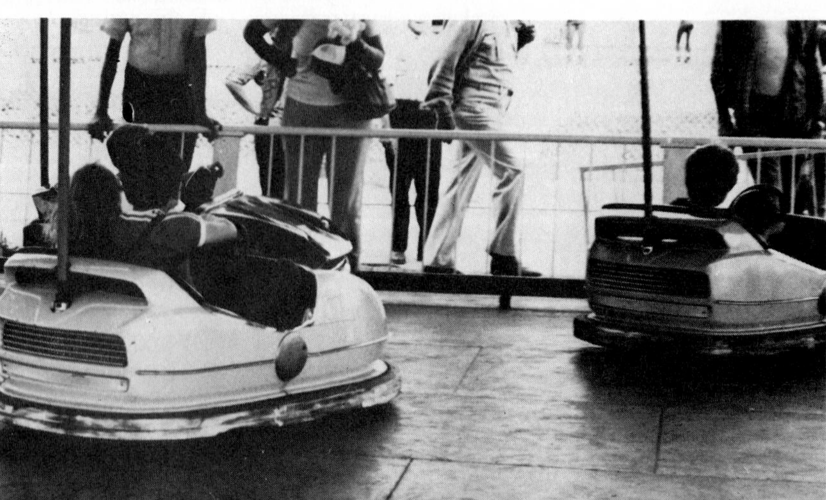

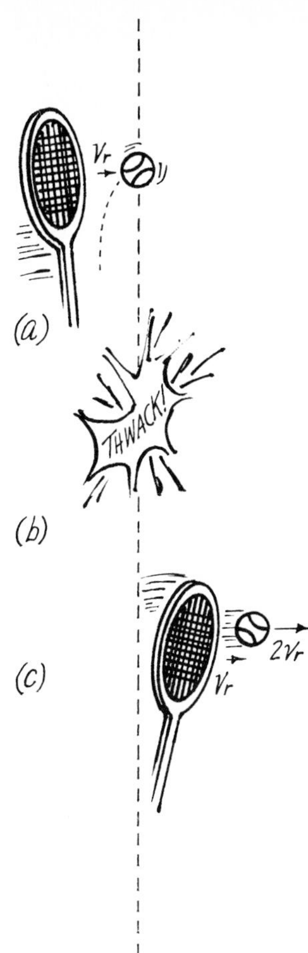


**FIGURE 5–6**
The relative speed after a collision cannot exceed the relative speed before.

there are limits to what can happen. For instance, if you throw two balls of soft clay at each other, they can stick together and their relative speed after the collision is zero. Such a collision (where the relative speed afterward is zero) is called a **perfectly inelastic** collision. A different kind of collision can be seen on a pool table. An incoming ball can slam straight into a target ball at rest. The moving ball stops dead, and the target ball glides over the table with (almost) the same speed the incoming ball had. That is, even though the individual speeds for the balls have changed, the relative speed after the collision is almost the same as the relative speed before. If *no* relative speed is lost, the collision is called **perfectly elastic.** Most collisions fall somewhere between these two limits, so that some relative speed is lost, but not all. Bending or breaking or friction during a collision generally reduces the relative speed after the collision.*

If you drop a steel ball or a toy superball onto a thick steel plate, the ball might bounce upward with 90 percent of the relative speed it had just before the collision. That's a very elastic collision; the metal surfaces suffer little damage. But throw a grapefruit onto that same steel plate, and the bruised grapefruit bounces back with less than 10 percent of its initial relative speed. That's a very inelastic collision. In sports springiness is important. A regulation baseball is designed to rebound from a hard surface with a speed equal to 54.6 percent of its incoming speed. The strings of a tennis racket are stretched and tied (usually with a tension of 58 pounds). Because of the high tension there is little deformation of the strings when the tennis ball is served or received, making the collisions very elastic.

Let's use the facts about collisions to find out what we can about the speed of a serve in tennis. A tennis ball is gently tossed upward to be hit near the top of its path by an onrushing racket. Because the tennis ball's speed is essentially zero, the relative speed of the ball and racket is just the speed of the oncoming racket, $v_{\text{racket}}$ (see Fig. 5–6a). After the collision the very most the relative speed can be is the same as the relative speed before the collision. (In a simple collision, remember, the perfectly elastic collision is the most springy, where the relative speed is the same before and after.) Now the racket is very massive compared to the ball and loses almost no speed when they collide. Therefore, the relative speed after impact will be $(v_{\text{ball}} - v_{\text{racket}})$. See Figure 5–6c. Why the minus sign? The ball and racket will be moving in the same direction after they make contact. If we set this equal to the relative speed before the collision, $v_{\text{racket}}$, we find

$$v_{\text{ball}} - v_{\text{racket}} = v_{\text{racket}}$$

Or, adding $v_{\text{racket}}$ to each side of this equation, we find

$$v_{\text{ball}} = 2v_{\text{racket}}$$

*To a degree, the elasticity of solid matter (as in Chapter 3) determines how elastic its collisions might be. If, during a collision, an object is deformed but then springs back, the collision is more elastic than if the deformation is permanent. But even an elastic deformation causes heating, and in the next chapter we'll see that the appearance of heat means some relative speed must be lost.

FIGURE 5–7

So a 100-mph serve, for example, can come about only if the racket moves at 50 mph or better. To confirm this calculation, look at the strobe photo of a golf ball being struck in Fig. 5–7. As the ball leaves the tee, its speed is clearly about twice the speed of the incoming golf club, meaning the collision was very elastic.

## Momentum and Impulse as Vectors

Forces are vectors, and a force can change the direction of the object's motion, not just its speed. Consider the baseball that was tossed to the batter in Fig. 5–8. The baseball's direction changed as well as its speed. The momentum of the ball and the impulse it gets from the bat are *vector* quantities.

FIGURE 5–8

Average acceleration is the change in the velocity vector per unit of time. In vector symbols the average force is

$$\vec{F}_{av} = m \frac{(\vec{v}_f - \vec{v}_i)}{t}$$

or

$$\vec{F}_{av} \times t = m\vec{v}_f - m\vec{v}_i$$

or

$$\textbf{vector impulse} = \text{change in vector momentum}$$

The diagram in Fig. 5–9 shows the vector momentum of the baseball in Fig. 5–8 and the vector impulse it got from the bat.

**FIGURE 5–9**

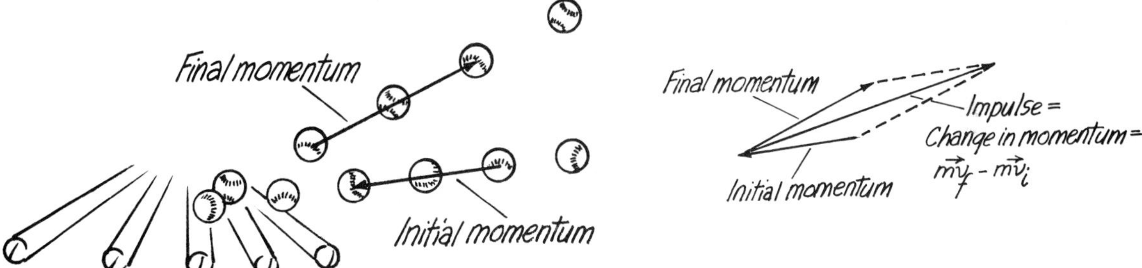


## Conservation of Momentum

Isaac Newton called mass the "quantity of matter," and momentum he called the "quantity of motion." It was from his studies of collisions that Newton discovered his third law of motion—for every action there is an equal but opposite reaction. In fact, Newton even wrote his second law, $\vec{F} = m\vec{a}$, in the form

$$\text{force} = \frac{\text{change in momentum}}{\text{time}}$$

where the change in momentum is in the same direction as the applied force.

Newton learned an interesting fact in his study of collisions. The "quantity of matter," or mass, was presumed in his day (and for centuries before) to be constant, no matter what took place, collision, or chemical reaction, or whatever. What Newton found experimentally was that the total "quantity of motion," or momentum, of colliding bodies is also constant, no matter what sort of interaction took place. To be specific, in any collision of two bodies, the vector momentum of each will change, but whatever momentum is lost by one is gained by the other so that **the sum of the vector momenta of the colliding objects is constant.** The total momentum before the collision is the same as the total momentum after the collision.

If you think about the impulses in a collision, you can see why the net momentum won't change. If a railroad engine collides with a boxcar, each gets an impulse that is equal to the change in their momenta. That is, a force from one acts for a certain time on the other. But the other gets an equal force in the other direction *for exactly the same amount of time*—the time of contact during the collision. So the boxcar's *change of momentum is equal but opposite* to the engine's. The sum of the momentum changes, then, is zero. Total $mv$ before the collision = total $mv$ after the collision (see Figs. 5–10 and 5–11).

**In any interaction where momentum is exchanged, the net vector momentum before and after is the same.** This property of motion

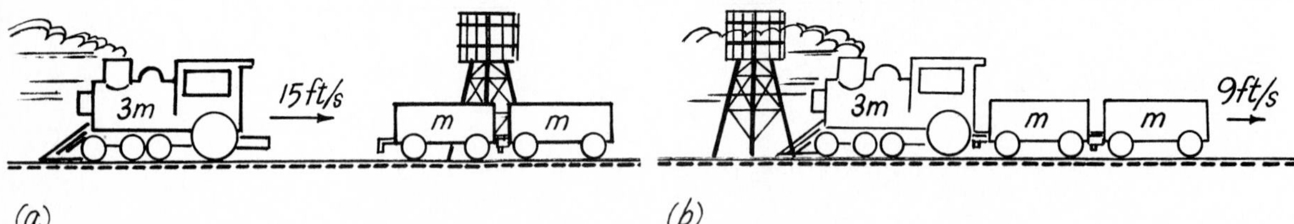


**FIGURE 5–10**
A perfectly inelastic collision. The momentum (= mass × speed) before the collision equals the momentum afterward. (Check it out!)

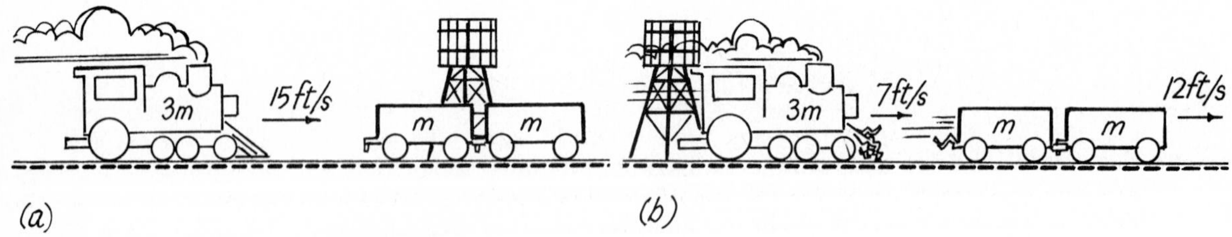


**FIGURE 5–11**
An inelastic collision. The momentum before is equal to the momentum afterward, once again.

**FIGURE 5–12**
Momentum is conserved in every interaction.

is called **conservation of momentum.** It's been true in any reaction that has ever been checked, so it has the status of a physical law. The net momentum never changes during a collision. Even if interacting objects have hidden internal springs that release to push them apart faster than they came together, or explosions take place to hurl the matter in every direction, no matter what sort of interaction, the net vector momentum of all the participants stays the same. So it is during a complex situation such as a gigantic avalanche. As the mass of ice and snow gains momentum on the way down, the earth gains an equal amount of momentum in the upward direction. Then when the avalanche comes to rest due to forces at the base of the mountain, the earth gets an equal but opposite impulse from the avalanche and stops as well.

**FIGURE 5–13**
The player and the ball begin with zero momentum, so they have zero momentum after the ball is thrown. Their momenta are equal but opposite.

Does this explanation sound too easy? There's only one catch. When more than one interaction takes place at once, you might forget to take an interaction into account. Throw a baseball. You've changed its momentum, but you took part; you're part of the interaction. You held the ball to begin with, and your momentum and the ball's momentum were both zero. When you gave the ball momentum, the law of conservation of momentum tells us you should have an opposite amount of momentum so that the sum would still be zero, right? *Wrong*. You interact with the earth at the same time. Friction stops you from moving backward over the surface, so the earth and you together move backward (an undetectable amount, to be sure) while the ball is traveling through the air. Then, when the baseball hits the ground, friction brings it to rest. The ball's tiny forward momentum is transferred to the earth, including you as you stand there, stopping the ever-so-tiny backward motion of the earth and you.

**FIGURE 5–14**
Once again, equal but opposite momenta add to give zero.

## The Momentum of the Solar System

People born 5000 or 10,000 years in the future might have some really spectacular vacations. No road maps of the countryside for them; they could plan their trips on three-dimensional star maps. On such a map of nearby stars, only a small dot would represent our sun. But we know what lies at that spot on the map, a system of interacting particles. They're large particles by our standards: a sun and 9 planets, 50 or more moons, and a host of asteroids and comets. This is the sun's *system,* our solar system, and it's the sun's gravity that keeps all the lesser bodies from sailing their own ways, never to return. Yet all the particles in this system attract one another. That means the momentum of each planet and moon constantly changes because there are always forces tugging on them. For example, Jupiter tugs on earth with a force that's strongest when the two planets are closest, and earth tugs back on Jupiter. Each and every second the momentum change of Jupiter because of earth's attraction is equal and opposite to the momentum change of earth due to Jupiter's attraction. And so it goes, moon by moon and planet by planet, between all the bodies in the solar system. Momentum is swapped back and forth all the while, but the total momentum of the system is constant. This is as true for the solar system as for a collection of billiard balls on a billiard table, and that is the real significance of a physical law. Conservation of momentum is a universal truth, holding for particles as small as atoms and molecules as well as for the stars.

Of course, the previous warning still holds. If we miss any interactions, there will be a discrepancy. To be sure, the net momentum vector of the solar system does slowly change, principally because the sun and planets orbit the distant, massive nucleus of our galaxy. The pull from the galactic nucleus, which lies far behind the constellation *Sagittarius,* slowly turns our solar system around it, once every 250 million years. But that's such a slow rate of turn that on that futuristic vacation travelers from earth won't have to worry about the path of the solar system. The force is small so the rate of change of momentum is small. "Straight and steady, as she goes!"

## Getting Back to Earth

A moving freight train, a drifting iceberg, or a large river carry large amounts of momentum. Some glaciers have moved at rates of several hundred feet per day, and because of their great mass those slow-moving juggernauts have large momenta as well. As thin as air is, the winds in a storm can represent considerable momentum. In Chapter 8 you'll read about the climatic condition called El Niño that slowed the earth's rate of rotation by a fifth of a millisecond (0.0002 s) per day in 1983. The eastward-moving winds fanned by El Niño gave the earth's atmosphere over the Pacific Ocean a net momentum toward the east. In this interaction between the earth and the atmosphere, the earth received the other part of the action-reaction; its eastward momentum dropped ever so slightly, lengthening the day.

Of the large-scale changes in momentum on earth, few compare to the abrupt actions of volcanoes and earthquakes, which set vast areas of

**FIGURE 5–15**
The huge chunk of Mt. Huascarán that is missing at the top left of this 1983 photo came off in the May 31, 1970, earthquake. The avalanche path is still visible in the foreground.

the earth's surface into motion and sometimes cause avalanches. Unlike the ice and snow avalanches described and pictured at the beginning of this chapter, some avalanches involve great volumes of rock. On May 31, 1970, an earthquake occurred in Peru, leveling many coastal towns and leaving a million people homeless. Farther inland in the Peruvian mountains, the quake triggered a massive rock slide on Mt. Huascarán. A huge part of its north face collapsed and 10 to 15 million cubic yards of granite fell downward for a vertical distance of 4000 meters (see Fig. 5–15). The speed of the avalanche reached 300 km/hr as it rode a cushion of air. Because of its great momentum, a part of the moving rock, snow, and ice jumped a ridge some 200 meters high and buried the town of Yungay and its 18,000 inhabitants. When friction from the ground finally reduced the great momentum of the avalanche to zero, some of the rock had traveled 65 kilometers (40 miles) from its starting point.

## CALCULATIONS

EXAMPLE 1: (a) A small car traveling at 18 m/s (about 40 mph) approaches a stop sign. The driver applies the brakes and brings the car smoothly to rest in 5 s. If the car's mass is 1000 kg, **find the average frictional force from the road that decelerates the car.***

*The car's brakes aren't the only decelerating forces on the car. Both air friction and the friction between wheels and axles will bring any coasting car to a halt in time. But we can neglect those forces when compared to the force from the brakes that stops the car in only a few seconds, as in this problem.

$$\text{Impulse} = \text{change in momentum}$$

$$\overline{F} \times t = mv_f - mv_i$$

so

$$\overline{F} = \frac{0 - (1000 \text{ kg} \times 18 \text{ m/s})}{5 \text{ s}} = \mathbf{-3600 \text{ N}}$$

This large braking force of about 800 lb (about 200 lb from each tire) points out how large the momentum is for even small cars at moderate speeds.

(b) Suppose this car had stopped more suddenly by colliding with a telephone pole. In this case the car might lose its momentum in about a third of a second. **Find the average force the telephone pole exerted on the car.**

$$\overline{F} = \frac{0 - (1000 \text{ kg} \times 18 \text{ m/s})}{1/3 \text{ s}} = \mathbf{-54{,}000 \text{ N}}$$

or about 12,150 lb, or 6 tons. That's a force over five times the car's weight.

(c) The change in the driver's momentum is even more sudden than is the change in the car's momentum if the driver isn't held by a seatbelt. Suppose the steering wheel stops the driver in about half the time that the telephone pole caused the car to stop. **Find the average force on a driver** who weighs 120 lb. The driver's mass is about 55 kg (1 kg weighs about 2.2 lb). Using

$$\text{impulse} = \text{change in momentum}$$

we find

$$\overline{F} = \frac{0 - (55 \text{ kg} \times 18 \text{ m/s})}{1/6 \text{ s}} = \mathbf{-5940 \text{ N}}$$

or about 1300 lb, *over 10 times the driver's weight.*

---

**EXAMPLE 2:** Two large booster rockets propelled the Apollo spacecraft into an earth orbit. Then a large engine burned for a few minutes to boost the craft and its astronaut passengers into a trajectory that would carry them to the moon. After that brief burn the spacecraft coasted with engines off, influenced only by the gravitational pulls of earth and moon. But during this coasting period its velocity had to be fine-tuned, and this was done by a small rocket engine with a thrust of 70 lb. **How long would this engine burn** in order to change the 150,000-lb spacecraft's speed by 2 ft/s? (Ignore the tiny loss of mass as the rocket used fuel during the burn.)

$$\text{impulse} = F \times t = \text{change in momentum} = m_{\text{craft}} \times 2 \text{ ft/s}$$

so

$$t = \frac{\left(\frac{150{,}000 \text{ lb}}{32 \text{ ft/s}^2}\right) \times 2 \text{ ft/s}}{70 \text{ lb}} = \mathbf{133.9 \text{ s}}$$

or a little more than 2 min.

---

**EXAMPLE 3:** Skylab had a roomy interior, and astronauts who spent time in orbit there became accustomed to moving about in free fall. Suppose an astronaut floating in such a spaceship pushes a cargo box away from him. We know he will drift backward, and conservation of momentum can tell us about the speeds the box and the astronaut will have as they move apart.

total initial momentum (of astronaut + box)
= total final momentum (after they part)

To begin, the astronaut (A) and the box (B) he holds are both at rest. $v = 0$, and $mv$ is zero, which means the total initial momentum is zero as well. This means that whatever momentum the box gets from the shove, $m_B v_B$, the astronaut's momentum will be *equal* in value but *opposite* in direction. Only then will these final individual vector momenta add to give zero. In symbols,

$$m_B v_B = - m_A v_A$$

Suppose the astronaut's earth weight is 160 lb; his mass would be about 75 kg. If the mass of the box is, say, 40 kg, this formula becomes

$$-v_A = v_B \times \frac{m_B}{m_A} = v_B \times \left(\frac{40 \text{ kg}}{75 \text{ kg}}\right)$$

so

$$v_A = - 0.53 \times v_B$$

Then whatever the speed of the box, whether the astronaut shoves gently or very hard, the astronaut will go backward at a speed that's a little more than half the box's speed. For example, a moderate push sends the box off at 2 m/s. Conservation of momentum tells us the astronaut drifts backward at the speed

$$v_A = -0.53 \times 2 \text{ m/s} = -1.06 \text{ m/s}$$

## Powers of 10

Example 4 and many other examples and calculation problems from this point on involve handling large (or small) numbers that are most conveniently expressed in powers of 10 notation. If you aren't familiar with this method of writing numbers and the rules that govern their multiplication and division, please read Appendix 2 at the back of the book.

FIGURE 5–16

EXAMPLE 4: Golf balls are tough; if you stand on one, its shape doesn't change. But, in fact, when a golf club strikes a golf ball, the ball is severely squashed by the force it receives (see Fig. 5–16.) A strobe photograph reveals the time such a collision takes and the speed with which the golf ball leaves. A typical time of collision is 0.0005 s, causing the ball to travel at approximately 150 mph, or about 70 m/s, for a regulation golf ball with a mass of 46 grams. **Calculate the average force on the ball and its acceleration during the impact.**

$$\overline{F} \times t = \text{change in } mv$$

In powers of 10, we have 0.0005 s = $5 \times 10^{-4}$ s, and 46 g = 0.046 kg = $4.6 \times 10^{-2}$ kg. So

$$\overline{F} = \frac{4.6 \times 10^{-2} \text{ kg} \times 70 \text{ m/s}}{5 \times 10^{-4} \text{ s}} = \left(\frac{4.6 \times 70}{5}\right) \times 10^{-2} \times 10^{4} \text{ kg·m/s}^2$$

$$= \mathbf{64.4 \times 10^2 \ N}$$

This force that squashes a golf ball is about the same as the weight of seven 200-lb men. To find the golf ball's acceleration, we use $F = ma$. So

$$a = \frac{F}{m} = \frac{6440 \text{ N}}{4.6 \times 10^{-2} \text{ kg}} = 1400 \times 10^2 \text{ m/s}^2$$

$$= \mathbf{140{,}000 \text{ m/s}^2} \text{ (about 14,000 } g\text{!)}$$

---

**EXAMPLE 5:** A block of ice the size of a house cracks loose from a mountaintop glacier and comes to rest a mile below. Conservation of momentum lets us estimate how far the earth moves upward while the avalanche falls. As in Example 3, the initial momentum of earth and the block of ice is zero, so throughout the motion that follows, the total momentum remains zero. That means the momentum gain of the avalanche (A) is equal and opposite to the momentum gain of the earth (E) at all times, so that their *sum* is zero. In symbols,

$$m_A v_A = -m_E v_E$$

If we know the masses, this formula gives the relative rates of travel of earth and avalanche at any time during the motion. It tells us the relative distance each moves, too, because $v_A = d_A/t$ for whatever time $t$ the avalanche might move at the speed $v_A$, and for the earth $v_E = d_E/t$ during that same time period. Substituting, we find that at any point throughout the motion

$$m_A\left(\frac{d_A}{t}\right) = -m_E\left(\frac{d_E}{t}\right)$$

so that

$$m_A d_A = -m_E d_E$$

The earth's mass is about $6 \times 10^{24}$ kg. Let's estimate the mass of a house-size avalanche. The volume of a house is *length* × *width* × *height*, so let's use 70 ft × 30 ft × 12 ft = about 25,000 cubic feet. A cubic foot of ice weighs about 56 lb. So $m_A$ is roughly

$$25{,}000 \text{ ft}^3 \times 56 \frac{\text{lb}}{\text{ft}^3} = 1{,}400{,}000 \text{ lb}$$

Since 1 kg weighs about 2.2 lb, this avalanche has a mass of about 636,000 kg, or $6.4 \times 10^5$ kg. The avalanche falls a mile, so $d_A = 1 \text{ mi} = 1.609 \text{ km}$, or about $1.6 \times 10^3$ m. Using these values in $m_A d_A = -m_E d_E$, we find

$$d_E = -\frac{m_A d_A}{m_E} = -\frac{(6.4 \times 10^5 \text{ kg}) \times (1.6 \times 10^3 \text{ m})}{6 \times 10^{24} \text{ kg}} = 1.7 \times 10^{-16} \text{ m}$$

Now $10^{-16}$ is truly a small number, so the distance the earth moves is quite invisible. Indeed, there isn't anything you can see that compares to a distance this tiny. Nonetheless, in terms of atomic dimensions, as you will see in Chapter 9, this distance is only about one-tenth of the diameter of an atomic nucleus. But such a distance is *not* so tiny that it can't be detected. Physicists who are today searching for gravitational disturbances from nearby collapsing stars hope to use instruments to detect motions in huge bars of metal that are only one hundredth as large as this—$10^{-18}$ m.

# REVIEW

**1.** Define impulse in words and symbols.

**2.** Define momentum in words and symbols.

**3.** Give the formula that relates an impulse to the change in momentum it causes.

**4.** What are typical units for impulse and momentum?

**5.** If the relative speed of two objects immediately after a collision is zero, was the collision perfectly inelastic? What is a perfectly elastic collision?

**6.** Isaac Newton referred to a "quantity of motion." What name do we give to this quantity today?

**7.** Is momentum a vector? Is impulse a vector?

**8.** In any interaction where momentum is exchanged, the total vector momentum before and after is the same. This property of matter in motion is called ______________.

**9.** Does conservation of momentum apply to any system of interacting objects? Give examples.

# DEMONSTRATIONS

**1.** A children's game, "Catch the Egg," demonstrates the effect of impulse on momentum (Fig. 5–17). In this outdoor game the players have to give special attention to the way an egg's momentum is brought to zero, *or else!* Here's how it's played: A pair of players face each other and toss an egg back and forth. Each time they catch the egg without breaking it, they both take a step backward, adding to the distance that the egg must be thrown. As they throw harder so that the egg doesn't fall short, they give it more momentum—a fact they realize with each contact between the egg and the hand. The trick is to increase the time you use to stop the egg. That is, the extra impulse needed should come from increased time, *not* increased force. To do this, you must let your arm and hand swing backward (in the egg's direction) as the egg arrives, yielding with the egg as it decelerates. If you don't do this properly, you'll be the first to know.

**FIGURE 5–17**

For a variation of this game, just before washing your bedsheets, have two friends hold a sheet up by two corners. Throw the egg at the sheet. If they hold it loosely so the sheet can give, you can throw the egg very hard without breaking it. If the sheet is inclined away from you, the egg can roll down to the floor without breaking.

**2.** With a skateboard and two bricks anyone can demonstrate conservation of momentum for you. Standing on the skateboard with a brick in each hand, a friend can swing the bricks back and forth at arm's length. As the bricks and arms move one way, the skateboarder moves in the opposite direction. If the total momentum is zero before the motion, it will be zero throughout the motion; the skateboarder's momentum always cancels the brick's momentum. Consequently, your friend gets nowhere. Next, ask your friend to sling the bricks away in the direction that gives forward momentum to the skateboard. Then tell your friend that's exactly how a rocket operates. The momentum the rocket gains is precisely equal but opposite to the momentum of the exhaust gases as they leave the engine.

**3.** Place three or four marbles of the same size in a straight row so they touch. From several inches away but in the direction of that line, shoot another marble at the end marble. Only one marble, the one on the *far* end, will pop out of its position. It will move outward from the others with about the same speed as the incoming marble; the series of bumps that take place between the touching marbles in the row are very elastic, so the relative speed is almost the same between each collision. This demonstration works just as well with balls on a billiard table.

# EXERCISES

**1.** An alley cat, being chased by a full-grown St. Bernard, runs just fast enough to keep a few feet ahead of its pursuer. Which has more momentum? Which would be more difficult to stop if you stepped in front of them?

**2.** If a moving object has a constant momentum, what do we know about the forces acting on it?

**3.** Which has the greater momentum, a freight train at rest or a drop of water escaping from a leaky faucet?

**4.** Dumbo the elephant shoots a peanut from his snout. The same force is applied to the peanut and the elephant for the same fraction of a second. Which has the greater momentum change?

**5.** To ease the pain, should you catch a fast-moving softball while your hand is moving toward the ball, while it is at rest, or while it is moving in the incoming ball's direction? Why?

**6.** A person rides a bicycle at full speed. Which likely has more momentum, the person or the bike? What is the total momentum of their system? Does this help explain why the rider goes over the handlebars if the bike hits a curb head-on?

**7.** A leap from a 50-ft height to land on concrete is usually fatal. But a swimmer can dive from that height into water without injury. (a) In each case the jumper receives the same impulse on landing. True or false? (b) One important difference is the time over which the impulse is delivered. True or false? (c) Another important difference is how great the average force is. True or false? (d) The change in the jumper's momentum is the same in either case. True or false?

**8.** When a bat hits a baseball, which suffers the larger change in momentum? (a) The bat. (b) The baseball. (c) They both have the same change in momentum.

**9.** How can an airbag that expands from under the dashboard during a crash help the driver? (Explain in terms of impulse, momentum, and average force.)

**10.** Inspect a wrecked car. Is it safe to say the crumpled parts of the car's body stopped quicker than the unaffected portions? Discuss.

**11.** How fast can a golf ball leave the scene, compared to the speed of the incoming club? Does the club's mass matter? Why, or why not?

**12.** Explain why it's easier for a batter to hit a home run from an incoming fastball than from one that the batter tosses up in practice.

**13.** To increase a moving wagon's momentum by a factor of 2, you can double its mass by adding sand while it moves with a constant speed *or* double its speed without changing its mass. True or false?

**14.** If momentum is conserved during a tennis serve, the momentum gained by the ball was lost by the racket. So how could we claim (as we did in this chapter) that the racket's speed doesn't change very much?

**15.** If you either double the force *or* double the time it is applied, you double the impulse. True or false?

**16.** Observe motorcross racers as they go over the bumpiest parts of the course. Exactly why do they stand on the foot pegs rather than sit on the motorcycle seats?

**17.** Traveling in opposite directions, two soccer players in pursuit of the ball leap into the air and barrel into each other. The collision brings them to rest in midair. What does the conservation of momentum tell you about their individual momenta *before* their collision?

**18.** If an inventor wanted a golf ball to gain more speed when struck and hence go farther than the regulation ball in Fig. 5–7, should the inventor design the ball to deform more or less than a regulation ball?

**19.** Why do drag racers remove the seats and spare tires from their cars?

**20.** Farmers and outdoor adventurers who must sometimes tow mired vehicles from muddy fields prefer ropes to towchains. Why?

**21.** Why does a prizefighter move toward his opponent while throwing a punch rather than just stand still?

**22.** Imagine that Congress passed a law requiring cars to be perfectly elastic in order to lower the damage done in collisions. What effect would that have on the cars and the drivers?

**23.** As an avalanche progresses down a mountainside, it may gather more snow and ice. What does this do to its momentum?

**24.** Exactly why does it help to have tennis shoes on if you jump down from a stepladder?

**25.** Before it lands but while still high in the air, the space shuttle turns in giant S-curves to the left and right. How does this increase the impulse delivered to the shuttle to lower its speed?

**26.** Give *two* reasons why a parachutist doesn't land stiff-legged but instead rolls and turns his or her body to one side so one leg, hip, and the side of the chest strike the ground.

**27.** Baseballs have a small cork center surrounded by layers of rubber, but the thick outer layers are made of top-quality wool yarn. During World War II the Allied Forces used all the best wool fleece for the war effort, and baseballs were made from inferior-quality wool. A 1943 National Bureau of Standards research paper showed that the balls rebounded with only 41 percent of an 85-ft/s incoming speed as opposed to 54.6 percent with today's baseballs. What effect could this have had on the game?

**28.** If you stand in the front of a small rowboat and walk to the rear, what happens to the boat? What does conservation of momentum tell you about the boat's momentum?

**29.** Which of these situations likely requires momentum changes (or impulses) in two or three dimensions? (a) A tugboat captain guides a passenger ship into its berth at a wharf. (b) An astronaut ‘‘burns’’ one or more rockets of known thrust for a precise number of seconds. (c) A freight train leaving a terminal gains speed along a straight stretch of rail. (d) A bus travels an S-curve in a road.

**30.** Fuel is the largest part of any rocket's mass. As a rocket burns, the thrust (the force from the operating engine) can be constant but its mass is not. During the first second of thrust (as compared to the last second), does the rocket's momentum change (a) more, (b) less, (c) the same? What does this say about the rocket's acceleration?

**31.** In multistage rockets (such as the space shuttle boosters or the Saturn/Apollo launch vehicles) a ‘‘spent’’ stage is jettisoned before the next stage ignites. To do this, small rockets fire to separate the stages a few feet before the ignition of the next stage. To maximize the efficiency of the impulse, should those rockets be located on the spent stage or on the stage that will continue?

**32.** See Fig. 5–18. Which person has the greater impulse exerted on his shield? Which likely receives the greater force? (You can answer this by considering the momentum changes of the arrows.)

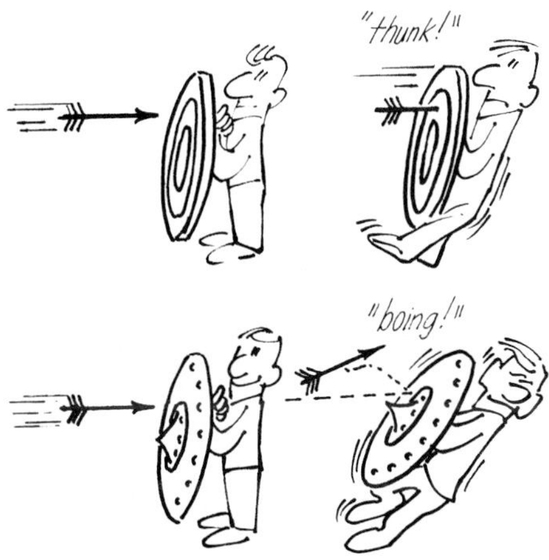


**FIGURE 5–18**

**33.** Someone throws a medicine ball to you while you are standing on a skateboard. You catch it and roll backward with the skateboard. Is your momentum change greater if (a) you catch the ball and hold it; (b) you catch the ball then throw it back; (c) your outstretched hands don't yield to the ball; they stop it and it falls straight down to the ground as you roll backward?

***34.** Compare the momentum of a softball moving at 40 mph with a firetruck moving at 5 mph. ($m_{\text{ball}}$ = 0.014 slug, $m_{\text{typical ``pumper''}}$ = 1000 slugs.)

***35.** Two ice skaters talk on an ice rink. One weighs 200 lb, the other half as much. After an argument they shove each other. Using the conservation of momentum, discuss their final speeds.

***36.** Calculate the force needed to stop a fastball with a catcher's mitt in 1/10 of a second. ($m_{\text{ball}}$ = 149 g; $v_{\text{fastball}}$ = 90 mph, or about 40 m/s.)

****37.** A person jumping from a 3-foot-high stepladder and landing stiff-legged stops in about 0.0045 seconds. Use $F = ma$ and the impulse–momentum relation to show that the force on the feet (and transmitted through the legbones) is about 100 times the person's weight, because the acceleration is about 100 $g$. Bending knees during the impact is really helpful, spreading the impulse over a longer time.

****38.** A Saturn V rocket had a dry (or empty) weight of 288,000 lb and a weight of 5,031,000 lb when loaded with fuel. Assume the thrust remained constant during its flight. Use the impulse–momentum relation to compare its change in speed during the last second of its flight to the change in speed the rocket has during the first second of liftoff.

***39.** During a single beat of a girl's heart, about 50 g of blood is pushed upward in about 0.1 seconds, moving at a speed of 1 m/s. Use the impulse–momentum relationship to find the average upward force on the blood during this time. (The body, of course, receives an equal and opposite force. If you could stand on very sensitive scales, you could watch the effect of your heartbeats on your body.)

***40.** A diesel engine weighs 3 times as much as a flatcar. If the diesel coasts into the car at 5 mph and they couple and coast off together, what is their final speed?

****41.** If rather than coupling together the engine and flatcar in Exercise 40 collide and then drift apart with a relative speed of 25 percent of the initial speed, what are the final speeds of engine and car? (Assume both travel in the same direction after the collision.)

***42.** Use conservation of momentum to show that the small mass in Exercise 31 in Chapter 4 will travel 3 times as far as the large mass. (*Hint:* See Example 5 in this chapter's ‘‘Calculations.’’)

***43.** A fastball comes across the plate at 90 mph. The bat travels forward at 70 mph. (a) What is their relative speed before they collide? (b) Suppose the batter knocks a line drive back to the pitcher. If the relative speed after the ball is struck is 50 percent of the relative speed before, what is the speed of the ball as it comes at the pitcher?

**44.** Because of conservation of momentum, spacecraft being planned to use telescopes for deep-space photography won't have human beings aboard. Why would people interfere with the operation of the telescope?

---

**A double asterisk means the calculation involves more than one step, or is otherwise more difficult.

# CHAPTER 6
# ENERGY

*You can't get something for nothing*. That bit of wisdom has more than a bit of truth in it. Take this morning, for example, when you got up, got dressed, and got busy doing whatever you did. Forces did all that, of course. Muscular actions got you going. But what gives your muscles the ability to contract? Where do those pushes and pulls really come from? Your exertions aren't free; you really can't get something for nothing.

This morning's actions came from yesterday's breakfast, lunch, and supper. The food you eat is fuel that your body burns to run your muscles. Eat a hot fudge sundae, and in short order you'll probably feel hyperactive—the extra fuel turns you on. Perhaps you would want to do something active to "work off" that dessert. If you exercise enough to get tired, however, you might even use another familiar expression and say, "I'm out of energy." We say we get energy from food, and energy is what your muscles use when they contract. It takes some doing to change that ice-cold sundae into pushes and pulls and heat, and it is food's energy that keeps you warm, keeps you alive, and lets you perform. This chapter is about what energy is and what it means, and it is about those energy-related words, work and power. Although the common usage of these words doesn't always agree with their technical meanings, in physics they have precise definitions and values that can be measured and even predicted. The idea of energy, you will see, applies not just to us but to every physical process in nature. Energy has special properties that make it one of the most important concepts in all of science.

**FIGURE 6–1**
You might be surprised to hear neither one of these fellows is doing any work. The first section of this chapter explains why.

To begin our study of energy, let's look closely at a word that's familiar to every student: *work*. ■

## Work

The word work means the opposite of play to almost everyone, and we also equate work with mental or physical effort. In physics the word *work* is related to physical action only, and in a very specific way.

**FIGURE 6–2**
When a force acts along the line of a displacement, work $W$ is done.

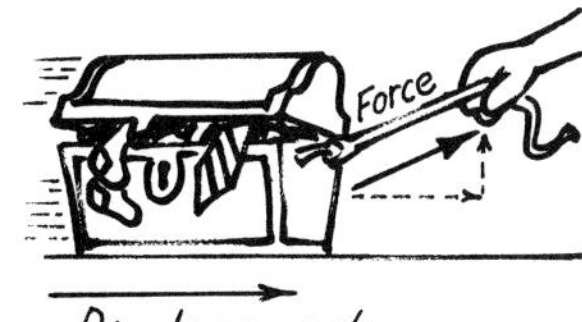


**FIGURE 6–3**
The component of an applied force that is parallel to a displacement does work.

**FIGURE 6–4**
Force that you use to support the suitcase does no work as you walk along.

Push as hard as you please for as long as you wish on the wall of a building, and the wall won't budge. Use the same force on a stalled Toyota, and it will roll. There is a difference in the way those forces perform. Sometimes a physical effort is productive, and sometimes it goes for nothing. You might say the push on the car did something, while the one on the wall did not.

*Work* is a measure of how productive an applied force is. When it pushes or pulls something *through a distance* (Fig. 6–2), we say the force does work. If an applied force isn't entirely in the object's direction of motion (as shown in Fig. 6–3), the result is the same as if two *separate* forces were acting. The component of force parallel (||) to the object's displacement does work while the component perpendicular ($\perp$) to the displacement does none. The work of a force is the *average force parallel to the displacement multiplied by the distance it acts through*, or

$$W = \overline{F}_{\parallel} \times d$$

In the metric system a newton-meter is the unit of work; it is also called a joule (J, pronounced "jool"). In the British system the unit of work is a foot-pound (ft·lb). One joule is about 0.74 ft·lb. On a smooth table a force of ½ pound slides this book along easily. If you use that push to move the book a distance of 2 feet, you've done a foot-pound of work, or about 1.35 joules of work.

Sometimes when you're working hard, you're not doing any physical work, $W$. Suppose you lift a heavy suitcase. To do this, you use a force that raises it some distance. That is work, $\overline{F}_{\parallel} \times d$. But then if you carry it by your side as you walk (Fig. 6–4), the force used to hold up that suitcase does no work at all; that force is perpendicular to the suitcase's displacement. It's true you are exerting yourself to support the luggage; perhaps you're even "working up a sweat" as you walk. But that doesn't mean you are doing physical work $W$ on the suitcase. We'll see later that you should say it takes *energy* to support the suitcase as you carry it.

FIGURE 6–5
All works are not equal.

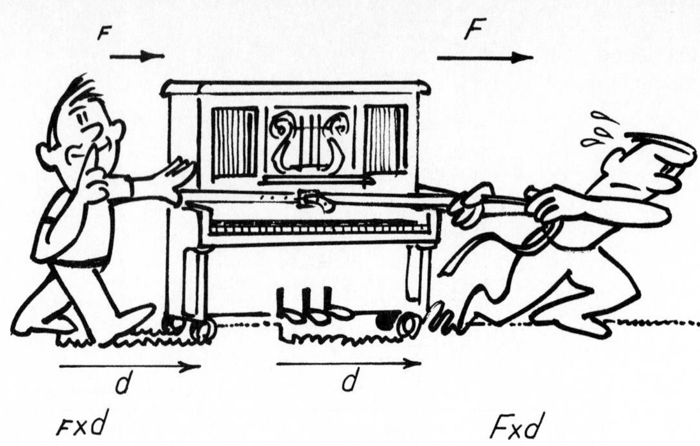


## Making Work Easier: Simple Machines

It won't surprise you to hear there are easy ways and hard ways to do the same amount of work. The secret to making work easy lies in applying a bit of ingenuity. The key is in the work formula, in which $F_{\parallel}$ and $d$ are multiplied.

Think about the job of building a stone wall. Lifting a 100-pound stone straight up from the ground to the top of a 5-foot-high wall isn't very easy. $W = 100 \text{ lb} \times 5 \text{ ft} = 500 \text{ ft·lb}$. But there is a less strenuous way: Rest one end of a strong plank on the top of the wall; place the stone on rollers and push it up the ramp. To get it to the top, you would push with a force equal to only a *fraction* of the stone's weight. However, there's a catch. You would have to push the stone farther to get it to the top than if you merely lifted it straight up from the ground. If the plank's length is such that you use *half the force* (50 lb), you would have to push that stone *twice as far* (10 ft). That is, the work you would do is $W = 50 \text{ lb} \times 10 \text{ ft} = 500 \text{ ft·lb}$, the same amount of work you'd use to lift the stone directly up from the ground. Figure 6–6 shows why this is true for any length of plank (or any angle the inclined path makes with the ground). What you do with such a ramp, called an *inclined plane,* is to trade extra distance for a smaller force. But the product of this new force and distance, the work *W,* remains the same.

FIGURE 6–6
*(a)* With an inclined plane, a little force through a longer distance does the same amount of work as a larger force through a shorter distance. *(b)* The ratios of the force components to the force of gravity are the same as the ratios between the sides of the inclined plane and its length. That's why $f \times D = F \times d$ as in part *a*.

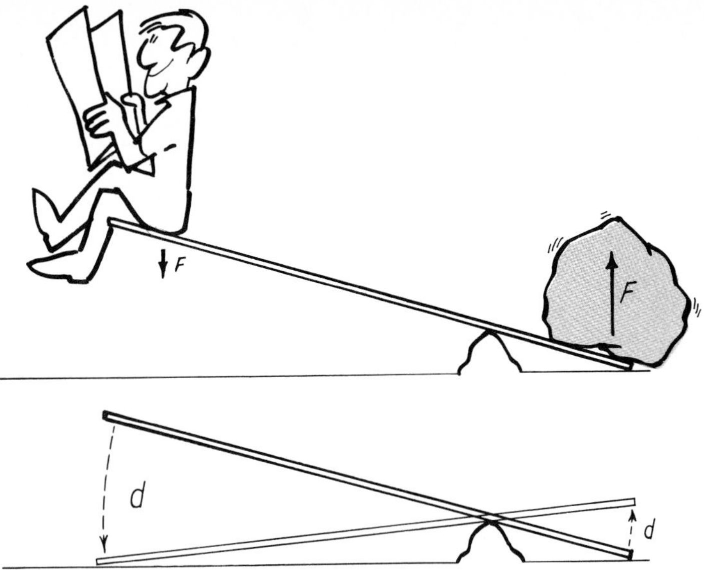


**FIGURE 6–7**
The smaller weight descending through a larger distance lifts the larger weight through a smaller distance.

**FIGURE 6–8**
Can you identify the directions of the forces and the distances through which they move?

When used this way, the inclined plane becomes a simple machine, a device to let you do work with less force. But *it doesn't save you any work, W*. It only adjusts the product of distance and force to help you do the work with less strength.

Levers also trade distance for force. With a sturdy lever you could move a heavy boulder in a field that might be impossible for you to dislodge otherwise. Figure 6–7 shows how. You use the same technique on a smaller scale to open a bottle of pop with a bottle opener or a paint can with a screwdriver. Pliers, scissors, and wrenches have long handles for the same reason, to trade distance for force (Fig. 6–8). Even wood screws with their wrap-around inclined planes work on this principle. A simple machine never changes the amount of work you have to do, it just lets you do it more easily.

**FIGURE 6–9**
The work done divided by the time it takes to do it is called the power.

## Power

A very strong person can lift one end of a small car. A child can do the same thing using a jack for a lever. A weight lifter lifts the car more quickly, but the child eventually performs the same amount of work. In both cases, the car's front end is raised through the same distance, and $W = \overline{F}_{\|} \times d$. The only difference is *how fast* the car reaches that position, or the rates at which the weight lifter and the child do the work. The rate of doing work is called *power*. **Power is the work done divided by the time it takes to do the work.**

$$P = \frac{W}{t}$$

The child that takes a longer time to lift the car with a jack uses less power than the weight lifter. The units of power are foot-pound/second and joule/second. One joule/second is called a **watt** in honor of James Watt (1736–1819), the inventor who improved and commercialized the steam engine. Another common power unit is the **kilowatt** (1000 watts).

It takes more power for you to run up a flight of stairs than to walk up. Either way, you do the same work—lifting yourself through a distance. The stairs are really just an inclined plane with steps, so you lift your weight *(mg)* through a vertical distance equal to the height *(h)* of the stairs. If you weigh 120 pounds and dash up a 10-foot height in 3 seconds, your power output is

$$\frac{mg \times h}{t} = \frac{1200 \text{ ft·lb}}{3 \text{ s}} = 400 \text{ ft·lb/s}$$

Not bad, considering 1 **horsepower** is defined as 550 ft·lb/s.

## What Can Work Do?

Now that we have a notion of what work is, it's time to ask what work can do. From what we've learned about forces, we know that work can change something's speed and certainly its position. And if you've ever used a piece of sandpaper on wood, you know that work can rearrange material and create heat. As we deal with these actions in the next few sections, you'll meet the concept of *energy*. Energy is a physical quantity that can take on different forms. One type (*kinetic* energy) has to do with motion, another (*potential* energy) deals with position, and a third (*thermal* energy) relates to the temperature of things. Energy is an abstract concept, so don't be surprised if you need to read some of these sections more than once.

## Net Work and Kinetic Energy

Almost any moving thing has more than one force acting to influence its motion. Whenever any such force has a component parallel to the object's displacement, there is work done. We can find the *net,* or total, work done on something by adding the work each acting force contributes. The net work measures the productiveness of all the forces acting.

**FIGURE 6–10**
If the force opposes the displacement, negative work is done, which decreases the kinetic energy.

We can show that *any time net work is done on an object, its speed changes according to the formula**

$$W_{\text{net}} = \left(\frac{1}{2}mv^2\right)_{\text{final}} - \left(\frac{1}{2}mv^2\right)_{\text{initial}} \qquad \textbf{(6–1)}$$

The net work done by the forces acting on something brings about a change in the quantity $\frac{1}{2}mv^2$, where $m$ is the object's mass and $v$ is its speed. The expression $\frac{1}{2}mv^2$ is called the **kinetic energy** (KE) of the moving object, or its **energy of motion.** In words, the formula in Eq. (6–1) says

**net work done on object = change in object's kinetic energy**

The kinetic energy $\frac{1}{2}mv^2$ has the units of work, joules or foot-pounds. Notice that the kinetic energy is *not* a vector quantity. In any direction a bird flies its kinetic energy is just ½ its mass times its speed squared. Also notice what happens if the work *brakes* an object, decreasing its speed (Fig. 6–10). Then (final kinetic energy) − (initial kinetic energy) is a negative number, so work that slows is *negative* work. (The force opposes the displacement.)

## Examples with Work and Kinetic Energy

Imagine pushing a friend who's wearing new roller skates along a smooth and level floor. There's no frictional force to speak of, and gravity does no work on the skater. So the net work on your friend is the work you do, your push (or average force) times the distance you push through. The final kinetic energy your friend has after the push-off will be equal to this work, provided that your friend was at rest initially (Fig. 6–11). Your friend's energy of motion is $\frac{1}{2}mv^2$ and will remain at that value until another force, such as an opposing push from someone else, does work to change it. Only an equal amount of nega-

*Example 6 in ''Calculations'' shows where this formula comes from.

**FIGURE 6–11**
*(a)* This fellow is doing work on the skater. The work is equal to the average force he applies times the displacement. The work done on the skater goes into changing her kinetic energy, $\frac{1}{2} mv^2$. *(b)* $W_{net} = \frac{1}{2} mv_f^2 - \frac{1}{2} mv_i^2$. Since the skater's initial kinetic energy was zero (her initial speed $v$ was zero), the work $W$ done on her by the fellow is equal to her final kinetic energy, $W = \frac{1}{2} mv^2$.

**FIGURE 6–12**
*(a)* The skater sees the end of her ride ahead. The fellow's hand will do negative work on the skater, decreasing her kinetic energy. *(b)* The average force the fellow's hand applies is opposite in direction to the skater's displacement. Since all of the skater's kinetic energy disappears, the fellow knows how much work he's done, that is, $W$ equals $-\frac{1}{2} mv^2$, the kinetic energy the skater had.

tive work can bring the skater back to rest, as in Fig. 6–12. Though the works come at different times, their total would be zero. And if the net work is zero, so is the change in kinetic energy.

If you gently slide this book several feet across a table, you'll do about 1 foot-pound of work on the book. But a frictional force from the table opposes the book's motion. The frictional force does negative work on the book at the same time you are doing positive work. As soon as you stop pushing, friction brings the book to rest. *So the change in the book's kinetic energy from start to finish is zero.* That means the net work from the two forces is also zero, so the work done on the book by the friction is the negative of the work you did.

$$W_{\text{net}} = 0 = W_{\text{friction}} + W_{\text{your hand}}$$

If the work you did was exactly 1 foot-pound, then the work of friction is $-1$ foot-pound.

Much the same thing happens if you lift a bowl from a countertop to a shelf. Your push moves the bowl up through a distance, doing work on it (see Fig. 6–13). There's also an opposing force that acts on the bowl during its displacement, the force of gravity. Your force is greater than $mg$ to begin with, so the bowl gains speed (and kinetic energy) and moves up. But as the bowl nears the shelf, you push less hard, and gravity slows the bowl. It comes to rest on the shelf with a final kinetic energy of zero. The change in kinetic energy from one position to the other is zero, so the net work done on the bowl is zero. Overall, gravity's (negative) work canceled your (positive) work.

$$W_{\text{net}}\text{ (on bowl)} = (\overline{F}_{\text{hand}} \times d) + (-mg \times d) = 0$$

So your average force is equal to $mg$. You pushed just enough to bring the bowl to rest on the shelf. If you had pushed too hard, the bowl's speed would have carried it past the shelf, and the change in kinetic energy at the shelf's position would have been greater than zero. (See the examples in "Calculations" and the exercises for more examples of work and kinetic energy.)

**FIGURE 6–13**
*(a)* The bowl is initially at rest. *(b)* While your hand does work on the bowl, so does gravity. *(c)* Afterwards the bowl is still at rest. Since there was no change in kinetic energy, no net work was done. The negative work done by gravity cancelled the positive work done by your hand.

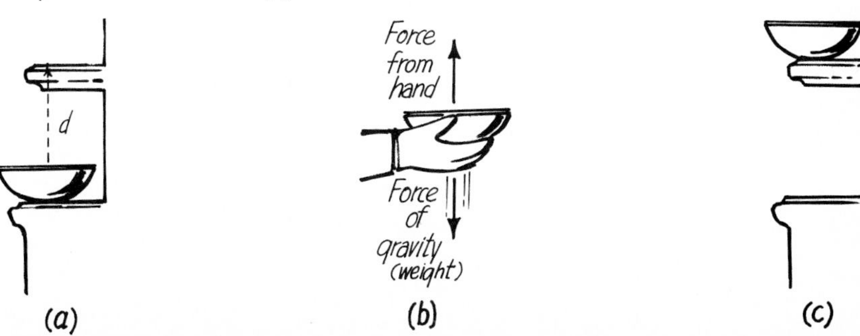

## What Is Kinetic Energy?

The skater in Fig. 6–12 moves along with her kinetic energy, $\frac{1}{2}mv^2$. The guy does work *on her* to stop her. If her kinetic energy was 100 joules, he did work equal to $-100$ joules to bring her energy of motion to zero. That's all the formula *net work on girl = change in girl's kinetic energy* tells us. But let's see what else happens. The push the guy gave the girl to do work on her and stop her wasn't the only force in action. There was an equal but opposite force on the guy's hands, and his arms moved backward through the same displacement that the girl moved through as she stopped. So the girl did 100 joules of work on the guy, which is just equal to the kinetic energy she had before she ran into him. Her kinetic energy was a measure of the work she could do on him. That's what energy of motion means: *Kinetic energy is the measure of a moving object's ability to do work.*

Of course, her force on him was *not* the net force he felt. Because he wasn't on skates, friction with the ground kept him from gaining kinetic energy. So what happened to the work she did? We'll see later in this chapter that when something's kinetic energy decreases, it doesn't always go into work that results in a net displacement of matter.

### ABOUT THE $v^2$ IN KINETIC ENERGY

Brakes must do work to stop a moving car: They must bring the car's kinetic energy from $\frac{1}{2}mv^2$ to zero. The $v^2$ term in the kinetic energy plays the part of a "heavy" in this operation. Because of it, when a car goes *twice* as fast, the brakes must do *four* times the work to stop the car. To put this another way, doubling the speed increases the energy of motion 4 times. This can make even small increases in speed deceptively dangerous.

Compare $\frac{1}{2}m(v_i)^2$ to $\frac{1}{2}m(2v_i)^2$. In the second term the speed is twice that of the first term. Rearranging the second term gives

$$\frac{1}{2}m(2v_i)^2 = \frac{1}{2}m(2v_i)(2v_i) = \frac{1}{2}m(4)(v_i^2) = 4 \times \frac{1}{2}mv_i^2$$

or 4 times the kinetic energy of the first term.

Here's what $v^2$ does in a different situation. Suppose a car is speeding at 70 mph. The brakes have to do about the same amount of work just to slow it to 50 mph (that is, to reduce its speed by 20 mph) as they will have to do to stop it completely from 50 mph! (See Exercise 44.) The dramatic increase in kinetic energy at high speeds makes high-speed collisions especially dangerous. Any extra speed ensures that the car will do a *lot* of extra work on whatever stops it, and the car itself will receive a lot of work in return.

## Potential Energy

Throw your keys straight up over your head and catch them when they come down (Fig. 6–14). When the keys leave your hand, they have kinetic energy, $\frac{1}{2}mv^2$. As they rise, the force of gravity pulls on them, doing negative work that reduces their kinetic energy. Over every centimeter along that upward path, the force of gravity does the same amount of negative work, $-mg \times 1$ centimeter. So for each centimeter they rise, they lose the same amount of kinetic energy. Finally, at the top of the path where their speed is zero instantaneously, the keys have *no* kinetic energy. Gravity's work has brought their kinetic energy to zero. There is, however, more work to be done.

From the top of the path, they fall downward, and the force of gravity does work once again. For each centimeter they fall, gravity does the same amount of work as it did when the keys were moving upward. Only this time the work is positive because the pull is in the direction of the displacement. As they fall, the keys gain kinetic energy step-for-step in the identical amounts they lost earlier. Just as they reach your hand, the keys have the same amount of kinetic energy as when they left your hand. Overall, nothing is lost. The pull of gravity first takes away the kinetic energy and then gives it back. It's as if the energy of motion is momentarily traded for elevation, and then vice versa. We can say the key's energy of motion is stored by the action of gravity as the keys rise.

**FIGURE 6–14**
As the keys rise, gravity opposes their motion, doing negative work and decreasing the key's kinetic energy, $\frac{1}{2}mv^2$. When their kinetic energy is finally depleted, the keys begin to fall and gravity does positive work on them, increasing their kinetic energy again.

When the keys are a distance $d$ above your hand, gravity has done an amount of work, $-mg \times d$, on them. But as they return from that elevation to your hand, gravity will do an amount of work, $+mg \times d$, on them. That is, just because of their height, or position, they have the potential gain in work (or kinetic energy) of $mg \times d$. That is what we call the *potential energy* of the keys due to gravity. Now as the keys fall, their potential energy is traded for kinetic energy, which gives them the ability to do work. **Potential energy,** then, is *the measure of an object's potential to do work* (Fig. 6–15). The potential energy due to gravity of a mass $m$ that's elevated a vertical distance $d$ is

$$\text{potential energy} = mgd$$

A squirrel of mass 0.5 kilograms that climbs 4 meters up a tree has a potential energy (relative to the ground) of

$$mgd = (0.5 \text{ kg})\,(9.8 \text{ m/s}^2)\,(4 \text{ m}) = 19.6 \text{ J}$$

Should it slip, 19.6 joules is the amount of work that gravity's pull would do as the squirrel fell to the ground. (Because of air resistance, squirrels can fall from even greater heights without harm.)

For there to be potential energy, there must be a *force ready to do work.* Were it not for gravity, neither keys nor squirrel could fall and gain kinetic energy. But gravity isn't the only force that can store potential energy. An arrow held in an archer's hand has a small amount of potential energy due to gravity. If dropped, it falls the few feet to the ground. Placed in the archer's drawn bow, however, the arrow has much more potential energy. It's in a position to receive a lot of work as a result of the tension in that taut bow. A flick of the fingers, and the bow and string unbend, releasing stored energy to the arrow by doing work on it. The arrow leaves with a lot of kinetic energy.

As another example, a metal spring can be compressed under a jack-in-the-box. So long as the top of the box is closed, the coiled spring remains compressed, ready to do work the instant the lid is released. That spring, then, stores potential energy. And that brings us to an interesting point. Remember that atoms and molecules act as if they are connected by springs. That is, the interatomic forces are springlike when the connected atoms are pushed together or stretched apart, giving matter the properties of resistance to compression and elasticity. During chemical reactions, molecules are put together or taken apart, and as they are rearranged interatomic bonds are formed or broken. To put it another way, the bonds that hold atoms together resist as the atoms are torn apart. The atoms move, and the bonding forces act through distances, and they can either store potential energy or give it up, much as springs can. *There is, therefore, potential energy stored in the bonds of molecules*. This is the kind of potential energy (or *chemical* energy) that comes from gasoline molecules as they burn in car engines. It's also the kind of energy we get from food molecules to run our bodies. In each

**FIGURE 6–15**
Few people would picnic beneath this boulder on a windy day. Because of its elevation it has the potential to do a lot of work on anything below. That is, the boulder has a large potential energy.

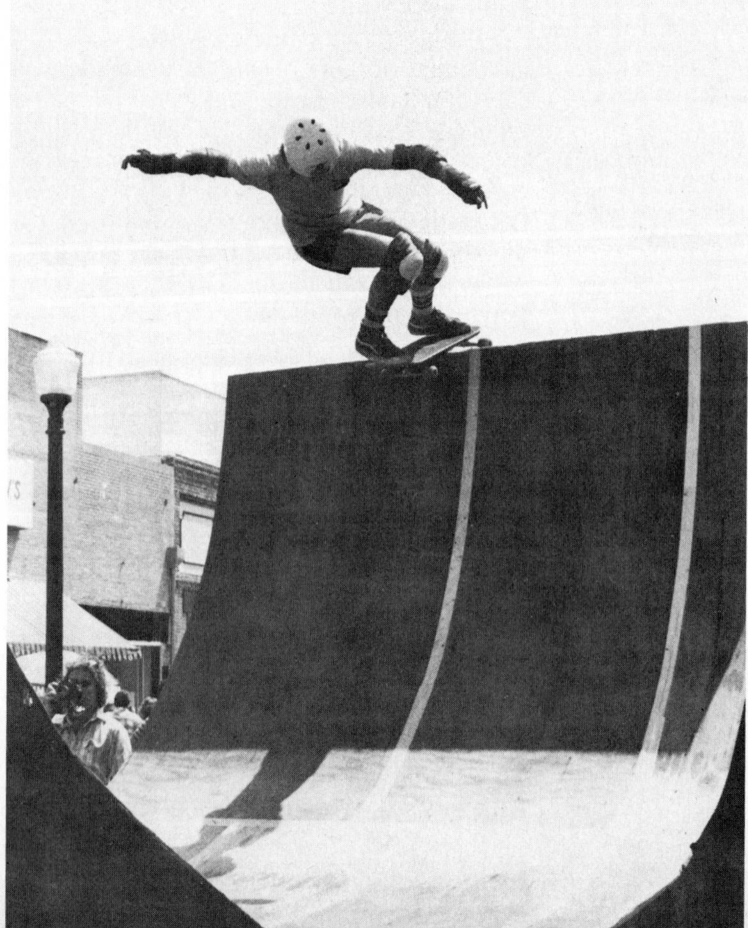

**FIGURE 6–16**
Getting ready to trade potential energy for kinetic energy.

## MOVING ALONG WITH POTENTIAL ENERGY

Most animals on the run store some of their energy of motion between strides. Elastic tendons connect muscles to the feet and leg bones, and with each step and foot placement, these tendons are stretched. While under tension, they retain some of the energy it took to stretch them, and during the next stride, they snap back, helping the foot to push off the ground. Like springs, the tendons store some kinetic energy as potential energy and give it back again.

Kangaroos, hopping as if on pogo sticks, have a perfected running technique that sets them apart from all other animals. Treadmill tests show that when a red kangaroo (the largest of them all) *triples* its speed, for example from 8 mph to 24 mph, it uses up to 10% *less* energy. (A cheetah or a horse must use over 100% *more* energy for a similar gain in speed.) The kangaroo actually uses less energy if it goes faster! At cruising speeds (up to 40 mph), the hopping kangaroo bends far over when it lands and straightens out as it pushes off for the next hop. *It uses its entire body like a spring*.

Scientists speculate that this process has survival value for kangaroos in the Australian desert. Thunderstorms are rare and miles apart, but they bring green grass out very quickly. To get to this food before the desert sun parches it (often in just a few hours), the kangaroos must move fast. But they must move efficiently, too, or else they might expend more energy just getting to the faraway table than they can consume once they arrive.

process potential energy is changed into work, but all the work doesn't go into kinetic energy, as we'll see in the next section.

### Friction and Heat: Thermal Energy

Throw your keys once more, but this time toss them along the floor. They'll skitter along for a bit as friction does negative work on them, reducing their kinetic energy as gravity did earlier (Fig. 6–17). Quickly the keys come to rest, and the kinetic energy you gave them is gone. But in this case they can't turn around (as they did when you pitched them up), gather speed from the same frictional force that acted to slow them, and jump back into your hand with the same kinetic energy. Once the motion stops, the friction stops, and it cannot restore the kinetic energy of the keys as gravity was able to do. *Frictional forces don't store energy in the form of potential energy*. But all the kinetic energy that you gave the keys when you tossed them doesn't just disappear.

The keys skid across the surface of the floor, scraping and catching, causing friction. The affected surface molecules of the keys and the floor are pushed through some tiny distance, giving them extra kinetic energy. Molecules throughout any solid object such as your keys bounce

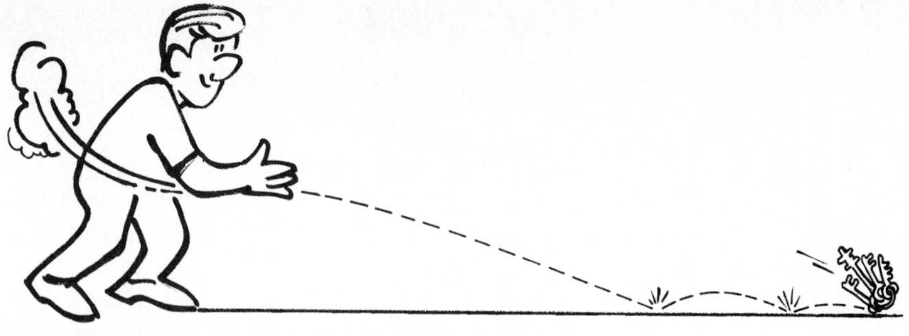

**FIGURE 6–17**
The work done by friction slows the keys. The keys cannot recover the lost kinetic energy this time, as they could when gravity acted to take them back over a similar displacement. The friction force does not give the keys potential energy.

**FIGURE 6–18**
The energy released in a chemical reaction spreads quickly, so a hand held high above a lighted match is safe.

around in all directions even though they are held in place by bonds. When molecules on the surface are struck, however, they bounce around even faster. They slam into their nearest neighbors (in all directions), and these, too, move a bit faster. In this way the kinetic energy scatters aimlessly among the molecules. All of the organized kinetic energy of the keys disappears, and most of it goes into this disorganized kinetic energy of the molecules. We call this chaotic energy of motion *thermal* energy, or *heat* energy. In Chapter 13 we'll see the exact relation between the thermal energy in matter and the average speed of the matter's molecules. The kinetic energy of the keys is largely transformed into heat. (We'll see shortly that a portion of the kinetic energy goes into sound and into work that deforms, or scratches, the keys and the floor.)

*Because of its chaotic nature, heat energy is much less effective at producing work than the more organized kinetic energy or potential energy of a solid object*. If you hold a match to a spoon and heat one spot for a moment, you know what happens. The heat travels outward, even into the spoon's handle. The energy in the hot spot spreads and is shared by more molecules. Molecules of the surrounding air that are hit by the spoon's hotter molecules carry away extra kinetic energy, leaving the spoon with less heat. As the heat energy disperses, there is less energy per unit of volume (or per molecule). The heat energy will never be in such a concentrated form in the spoon again. Although heat energy doesn't disappear, it spreads; it becomes more dilute—and less available to do work.

In a car, for example, the potential energy stored in the gasoline becomes random kinetic energy as the gasoline molecules burn. The heated gases push pistons downward, but *all* the heat energy can't be converted to work on the pistons. The energy those pistons get from the hot gas is only about 25% of the energy released from the chemical bonds. The rest of the heat escapes, leaving with the hot gases through the exhaust system or moving through the cylinder walls.

## COLLISIONS

In every collision, the total energy before equals the total energy afterward. This places limits on what can happen in a collision. Using both conservation of energy and conservation of momentum (and a lot of algebra), we can obtain the following facts:

***Perfectly inelastic collision*** (objects stick together, their relative speed after impact is zero): There is a maximum loss of kinetic energy when this happens. The lost kinetic energy produces heat or does work deforming the bodies, as when a bullet penetrates a block of wood. (But "maximum loss" doesn't mean all the kinetic energy can disappear. Conservation of momentum tells us that if two masses have a net momentum *not equal* to zero before the collision, they must have the same momentum afterward. If they have any momentum $\overrightarrow{mv}$, the kinetic energy, $\frac{1}{2}mv^2$, cannot be zero.)

***Perfectly elastic collision*** (the relative speeds before and after the collision are the same): In this situation *no* kinetic energy is lost. The kinetic energy of the approaching objects changes into potential energy during the collision, and *all* that potential energy turns into kinetic energy again as they part. Each object might take away a different amount of kinetic energy than it brought in, but their *sum* remains the same.

*Other **"in-between" collisions*** (the relative speed after the collision is less than before, but not zero): This is the kind of collision that usually happens in the everyday world. Some relative speed is always lost, which means that some energy of motion disappears. For example, if two cars with a small relative speed run together in a parking lot, the bumpers may bend like springs, stop the cars, and push them gently apart. But even if there is no visible damage, there is always some molecular friction when solids stretch or bend. The energy that the friction turns into heat disappears, and the cars will separate with less relative speed than when they came together. Of course, when the impact is hard and bumpers and fenders crumple, even more kinetic energy disappears as work is done to change the shapes. *The more work done to deform material, the less elastic the collision*. Should the cars stick together during the collision, they will lose as much kinetic energy as is possible; the collision is *perfectly* inelastic, and a maximum amount of heat and/or deformation occurs.

***Collisions where objects gain kinetic energy*** (the relative speed afterward is greater than the relative speed before): Imagine that a toy truck with a compressed spring attached to its front end runs into another toy truck. When they make contact, the spring releases and pushes on both trucks until they part. The potential energy the compressed spring had adds to the initial kinetic energy of the trucks, and they fly off with more kinetic energy than they had before. *Only when hidden* (potential) *energy is released can the total kinetic energy increase during a collision*. If no extra potential energy appears the relative speed afterward cannot be greater than the relative speed before.

Heat energy is ordinarily measured in **calories,** where one calorie is equivalent to 4.18 joules. (One calorie will raise the temperature of a gram of water 1° Celsius.) That means 4.18 joules of kinetic energy or potential energy can be transformed into 1 calorie of disorganized kinetic energy, or heat.

Almost every time anything moves through some distance, friction from some source does work and transforms some kinetic energy into heat energy. The connection between work and heat, however, was not fully understood until the 1840s. Once that connection was made, it led quickly to an important insight into nature, which is the subject of the next section.

## Conservation of Energy: The First Law of Thermodynamics

More than just heat is produced when you throw your keys along the floor. You can *hear* the keys hit and jingle. Sound travels across the room and vibrates your eardrum, which moves a collection of three little bones, the lightest bones in your body. To do this takes a force through a distance, so sound carries energy away from those sliding keys. Furthermore, any scratches on the floor or on the keys mean that molecules have moved, knocked away from their neighbors, their bonds broken. Any such deformation takes work and uses energy.

The kinetic energy you gave to the keys disappears, but not without a trace. It changes into heat and sound, and so on, but *all of it is accounted for in these other forms*. Careful measurements in a host of different experiments involving every sort of reaction have proved that energy may change forms but its total value is *constant,* or conserved.* This fact is expressed as a law called the **conservation of energy: In every interaction of any kind, the total energy afterward is always the same as the total energy to begin with.** Because heat is a form of energy, this law is also known as the **first law of thermodynamics.** Thermodynamics is the area of physics that deals with heat and matter.

Watch a child using a playground swing, or better yet, give a push and stand back. The child becomes a human pendulum (Fig. 6–19). The kinetic energy is a maximum at the bottom of the path, and the energy becomes entirely gravitational potential energy at each end where the motion reverses. Energy of motion turns into energy of position, and the cycle repeats. Were there no air to resist the motion, were there no frictional forces at the swing's points of support, the motion along the arc would continue for a long time. Conservation of energy tells us that *the sum of the potential energy and the kinetic energy would stay the same*. But those tiny frictional forces do interfere, changing kinetic energy into heat, and the swing's motion finally slows and stops. The energy didn't vanish, however. The air around the swing is a little

**Conserved* in the physics sense means "never disappears"; when you hear of energy conservation on the news, however, it means "don't waste" or "use less."

**FIGURE 6–19**
For just an instant, at the top of her path, Carey is in free fall. Notice her hair–it's weightless! At that point her energy is entirely potential energy. She can swing only so long as she provides extra energy to make up for the energy lost through friction.

warmer, and so are the connections between the swing and the supporting bar above it. The first law of thermodynamics guarantees it!

Energy and energy changes are used to describe the exercise of joggers, the production and output of thunderstorms, the destructive power of earthquakes, the changes in population of prey and predators, the output and input in chemical reactions, and the generation of light and heat from stars. It is the conservation of energy that makes energy so valuable to describe changes in nature. If only part of the energy can actually be measured in a situation, the rest can be deduced from the application of this law. Energy has been called the "common denominator" of the natural sciences because its conservation law makes it so useful in understanding any physical process.

## Another Law for Energy: The Second Law of Thermodynamics

A cup of hot coffee left on the breakfast table soon loses its extra heat and *cools* to room temperature. The heat energy disperses into the room. As it does, the energy is spread over more volume and over more matter. It becomes more dilute (less concentrated) in the sense that there is less heat energy per unit of volume. Similarly, a glass of iced tea left on the lunch table undergoes an energy change. It, however, *gains* heat energy from the matter in the room and *warms* to room temperature. Once again, heat energy (from the room) flows to include more matter (the glass and the tea) and volume. Heat energy always flows from hot to cold. In spreading out into more matter, it becomes less organized. This situation also happens every time something with kinetic energy encounters friction. Heat energy appears as its kinetic energy decreases, and that heat disperses. This tendency toward randomness from organization, disorder from order, leads to a principle called the **second law of thermodynamics.** Though clear when stated in mathematical language, this law sounds almost mystical when only words are used to describe it. Nonetheless, here it is in words: **Whenever energy changes forms, it tends to go from a more ordered state to a less ordered state.**

Energy changes form any time matter interacts, any time there are forces. When a basketball is dribbled, it hits the gym floor, compressing slightly (changing the ball's kinetic energy into potential energy), and then rebounds upward (changing potential energy into kinetic energy). But it won't regain all of the kinetic energy it had as it hit the floor. During the compression and rebound some of the ball's kinetic energy goes into "wear and tear" of the ball's material, and some goes into sound, but even more goes into heat as the basketball flattens and then springs back. The spot on the court where the ball bounced is a little warmer, as is the point of impact on the basketball. Left alone, the ball loses much of its kinetic energy during each successive bounce and quickly comes to rest. The organized kinetic energy the basketball had winds up, for the most part, as heat, just as the second law of thermodynamics predicts.

So despite the fact that energy is conserved, it can "run down" into forms less useful for doing work. This principle is even connected to the anticipated worldwide energy crisis. The depletion of fossil fuels means there is less chemical potential energy concentrated in a form we can easily use. The coal, oil, and natural gas we use today came from plant and animal life eons ago, which brings us to another interesting point. Plants and animals are more concentrated forms of energy than the simple ingredients they come from. Does this imply that living organisms violate the principle of the second law?

## Living with the First and Second Laws of Thermodynamics

The first and second laws of thermodynamics apply to energy processes of all kinds. Because all life depends on energy, these laws govern the energy exchange of living organisms, too. Conservation of energy, for example, guarantees that you can't use more energy than you consume. The second law guarantees that you won't be able to put all of the energy you do consume to work for you; a large quantity will be wasted as heat.

The human body very efficiently extracts potential energy from food, retaining about 95 percent of the available energy of the food that passes through it. By comparison, the chemical processes used to run vital organs, to keep the brain functioning, to provide warmth, and to do any physical work are much less efficient. On the average humans (and other animals) use only 10 percent of the potential energy obtained from food. The other 90% of food's energy generates only heat. Yet even this dismal percentage makes us among the most energy-efficient organisms on earth. Growing plants manage to use only 1 percent or less of the energy they absorb from sunlight, storing it as potential energy in the chemical bonds of molecules.

The effects of the first and second laws of thermodynamics can be seen in the comparative numbers of organisms that live on earth. For example, every higher organism feeds on certain lower forms that in turn feed on lower organisms in a progression that eventually ends with plant life. This progression of dependence is called a *food chain*. The inefficiency of each organism in using its available energy has a dramatic, cumulative result on a food chain. Suppose you use a pound of tuna in a recipe. To grow that pound of meat, the tuna ate about 10 pounds of herring. (The tuna, like us, uses only about 10 percent of the energy it consumes to build molecules of protein and fat.) In turn, the 10 pounds of herring required some 100 pounds of smaller fish and other prey that grew from consuming about 1000 pounds of algae. *The inefficient use of energy at each level means the lowest levels of a food chain must necessarily be the most abundant*. Although plants and animals use potential energy from food to construct complex molecules from simpler ones, *overall* the energy flow is toward disorganization; more potential energy is degraded into heat than is used productively.

Your personal energy consumption probably amounts to about 2500 Calories per day (about average for the United States). One food Calorie (capital C) is 1000 calories, or 1 kilocalorie. Normally, about this much energy goes into heat and life processes in your body. However, if you take in more Calories than you use during the day (if you overeat), your body stores the energy you don't use in the form of fat. Fat cells store a lot of potential energy. One pound of fat contains about 4200 Calories, over twice the energy content for a similar quantity of carbohydrate or protein. Running a marathon, a conditioned runner goes those 42 kilometers without using up as much energy as is stored in only 1 pound of fat. Exercise alone, then, won't shed extra pounds quickly or easily. Those who need to lose weight generally diet instead, consuming less food energy per day than the total energy requirement of the body. The energy it takes just to live must come from somewhere (so says the first law), and the body automatically turns to its fat reserves to make up the difference. Successful dieters, in effect, burn their fat, since about 90 percent of its energy turns into heat.

Even as you rest, your life processes (or energy exchange processes) go on. Just lying down and breathing quietly, the average person uses about 90 Calories (or kilocalories) per hour, an energy consumption rate of about 100 watts. The rate of energy usage while at rest is called the *basal metabolic rate*. The metabolism, the name given to the collective chemical processes of the body, depends on how fast the chemical processes of life proceed. These rates in turn depend strongly on the body's internal temperature. For every 2°F the body's temperature rises above normal, the basal metabolic rate increases about 10 percent. This same relation works in reverse: The metabolic rate drops about 10 percent for each temperature drop of 2°F. Animals that hibernate in winter have very low body temperatures during that period, so they burn their fat stores at a relatively slow rate. In the course of a good night's sleep, your own body temperature might drop 1° or 2°F, so even you "hibernate" on occasion. But an abnormal temperature drop can be dangerous. In cold weather, or in cold water, a person can lose heat too rapidly. To prevent the internal temperature from dropping to a dangerous level, that person shivers as muscles vibrate (contracting and relaxing) at a fast rate. Those muscles go to work automatically, generating heat to maintain the normal internal temperature. During strenuous exercise, the body's larger muscles release up to 20 times more heat than resting muscles do.

## Converting Energy to Force with Muscles

About 40% of your body weight is muscle. Skeletal muscles are bundles of long (several centimeters or more), thin muscle cells, or fibers. These fibers shorten when stimulated by nerve cells, and the muscle contracts. The energy to do this comes from special high-energy molecules that give up a large amount of potential energy when they are split. Called *adenosine triphosphates,* or ATP, these tiny molecules are the source of instant energy for life processes in all cells, not just muscles.

ATP molecules act something like shuttle buses for energy. Each day your body breaks down and rebuilds *more than your entire body weight* in ATP molecules. This is done by means of three chemical pathways that convert the energy of either glucose (sugar) molecules or fatty acids (from fat cells to build ATP molecules). During normal activities, two of these methods break down glucose to supply energy. The most efficient uses oxygen from the bloodstream (this method is called *aerobic*). From the energy of each glucose molecule, the aerobic method produces 34 ATP molecules ready to give up energy to muscles or to run other chemical activities in the cells of your body. During strenuous exercise such as running, this process alone cannot supply enough energy, and the second process must break down more glucose. Called *anaerobic* because it doesn't use oxygen, this method is less efficient than the first and gives only 2 molecules of ATP for each converted glucose molecule. One by-product of this energy pathway is lactic acid, thought to be the cause of muscular cramps when it accumulates in muscles during a hard workout.

If you run far (or fast) enough, you literally run out of glucose. Your body then switches to the third chemical pathway to keep you going. Adrenalin pumps into the bloodstream to cause fat cells to release fatty acids that are broken up by another aerobic reaction. This process cuts in when you get your second wind. (Fat contains much more energy than glucose, which is why it takes a lot of physical activity to lose a few pounds of fat.)

Incidentally, maybe you're one of the millions who wake up by drinking a cup or two of coffee. Besides stimulating the nervous system, it also stimulates fat cells to release fatty acids, so caffeine really does give you extra energy. Treadmill tests show those who've had 2 cups of coffee an hour before a workout can do about 20% more work before they're exhausted. (One exception: people of 50 kg mass (110 lb weight) or less. Two cups of coffee may give them an overdose of caffeine, and their bodies may respond *less* efficiently rather than more efficiently.)

Earlier in this chapter we saw that you can exert yourself without doing any physical work. You don't do any work just supporting a suitcase as you walk or pushing steadily against a piece of furniture that doesn't budge. Nevertheless, your muscles do contract to support that suitcase or exert force against the furniture, and they use energy at a high rate. In this case, the energy that comes from those high-energy molecules all winds up as heat, so while you support a heavy suitcase you really do work up a sweat!

## Other Forms of Energy

Conservation of energy is such a general principle that it covers many situations we've yet to discuss. Energy changes, too, and the principle of the second law of thermodynamics applies to every interaction in nature. So from this chapter on, you'll meet the word energy time and time again. Before going on, however, we'll mention some other forms of energy just for perspective, leaving the details to later chapters.

### Solar Energy:

Sunlight carries energy that warms the ground, the oceans, and the air. The warm environment sunlight creates at earth's surface is the home of the only life environment we know of. Light energy from the sun is used by growing plants much as chemical energy from food is used by our bodies.

### Heat Rays:

Heat radiation, those invisible rays that warm you by the campfire, are also a type of light called infrared radiation. Your face and hands detect the heat energy these rays carry, even if your eyes can't see them.

### Electrical Energy:

The electricity (or electrical energy) we use daily is generated from other sources of energy in commercial power plants. Potential energy of some variety produces heat energy, which is used to operate turbines to run the electrical generators. (This energy-producing chain is very much like a food chain in its inefficiency; 60 to 70 percent of available energy disappears as waste heat.) The potential energy stored in fossil fuels (oil, coal, and natural gas) is used to generate most of the electricity in the United States. Nuclear power plants and hydroelectric plants (where potential energy of elevated water drives turbines) account for less than 15 percent of the electrical power in the United States.

### Nuclear Energy—Energy From Mass:

In 1905, even before the nucleus of atoms had been discovered, Albert Einstein showed that mass itself is a form of energy, more concentrated than any other form of energy in our world. (His equation relating energy and mass, $E = mc^2$, is quite famous.) And on the earth, the nuclei of atoms is where 99.9 percent of mass is located. Many thousands of nuclear reactions are known today that exchange some of the potential energy in the form of mass into kinetic energy.

The nuclear energy produced by present-day nuclear power plants comes from the breaking up of certain uranium nuclei. Mass actually disappears in the reaction, the initial nucleus weighed more than the products of the reaction. That missing mass becomes kinetic energy, which is carried away by the pieces of the "exploding" nucleus.

The sun generates nuclear energy at its center (though with a different process from the one used in nuclear reactors). That energy slowly filters out to the sun's outer regions to the heat that keeps its outer edges hot and glowing, filling the space around it with the sunlight that sustains life on earth.

## CALCULATIONS

EXAMPLE 1: Paul is an avid rock climber who in the winter months climbs a rope to stay in shape.

a) When Paul, whose mass is 60 kg, pulls himself up the rope a distance of 4 m, **how much work has he done?** Paul's arms pull him upward while gravity pulls downward. If Paul starts at rest and ends at rest, the net work on him is zero. Then $W_{arms} + W_{gravity} = 0$, or $(F_{arms} \times d) + (-mg \times d) = 0$. The only way for these two terms to be equal to zero is for the average force from his arms to be equal to $mg$. So

$$W_{arms} = mg \times d = (60 \text{ kg} \times 10 \text{ m/s}^2) \times 4 \text{ m} = \mathbf{2400 \text{ J}}$$

b) If Paul walks up stairs until he is 4 m above the ground, **how much work has he done?** Stairs resemble a ''notched'' inclined plane, and the work needed to get to the top of any inclined plane is the same as the work needed to go straight up.

$$W_{Paul} = mg \times d = mg \times 4 \text{ m} = \mathbf{2400 \text{ J}}$$

c) Once Paul has lifted himself 4 m by either method, **can he get his energy investment back? Yes;** whenever he is 4 m above the ground, he has gravitational potential energy relative to the ground. In this case, the potential energy is

$$\text{PE} = mgd = (60 \text{ kg})(10 \text{ m/s}^2)(4 \text{ m}) = \mathbf{2400 \text{ J}}$$

EXAMPLE 2: Ellen accelerates down a frictionless waterslide at an amusement park. **Calculate her kinetic energy when she has lost 2 m of elevation.** Her mass is 50 kg. (Does the angle of the slide matter? No. Inclined planes only trade force for distance, and the work the force of gravity does on her between any two elevations is the same as if she had jumped straight down from the top of the slide to fall that distance.)

$$W_{gravity} = mg \times d = (50 \text{ kg} \times 10 \text{ m/s}^2) \times 2 \text{ m} = 1000 \text{ J}$$

This must be equal to the change in her kinetic energy, $(\frac{1}{2}mv^2)_f - (\frac{1}{2}mv^2)_i$. If she starts with no kinetic energy, then $(\frac{1}{2}mv^2)_i = 0$, and her kinetic energy after dropping 2 m is **1000 J.**

EXAMPLE 3: A car moving at 10 m/s (about 22 mph) hits a telephone pole and stops in a distance of 0.4 m. Use the formula *net work = change in kinetic energy* to **find the car's deceleration.**

$$W_{on\ car} = \left(\frac{1}{2}mv^2\right)_f - \left(\frac{1}{2}mv^2\right)_i$$

or

$$F_{on\ car} \times d = -\left(\frac{1}{2}mv^2\right)_i$$

since $v_f = 0$. But $F = ma$, so

$$ma \times d = -\left(\frac{1}{2}mv^2\right)_i$$

Dividing each side of the equation first by $m$, then by $d$, we find

$$a = \frac{-\frac{1}{2}(v^2)_i}{d} = \frac{-\frac{1}{2}(100 \text{ m}^2/\text{s}^2)}{0.4 \text{ m}} = \mathbf{-125\ m/s^2}$$

Now $g$ is about 10 $\text{m/s}^2$, so the car's deceleration in terms of $g$ is $a = -12.5\ g$.

---

EXAMPLE 4: A barefoot person jumps from a 1-meter stepladder and lands virtually stiff-legged. Use the formula *net work = change in kinetic energy* to estimate the jumper's deceleration. If $F$ is the average force from the floor on the jumper and $d$ is the distance the foot compresses,

$$\overline{F} \times d = \left(\frac{1}{2}mv^2\right)_f - \left(\frac{1}{2}mv^2\right)_i$$

If you press on the sole of your foot, you'll see that it could compress about 1 cm, so $d$ is about 0.01 m. Using $\overline{F} = m\bar{a}$, and $v_f = 0$, and realizing that the kinetic energy the jumper has when arriving at the floor is $mg \times h$, where $h$ is the height the jumper falls from, we find

$$m\bar{a} \times d = -mgh$$

Dividing both sides by $m$ and $d$ gives us

$$\bar{a} = -g \times \frac{h}{d}$$

or

$$\bar{a} = -g \times \frac{1 \text{ m}}{0.01 \text{ m}} = \mathbf{-100\ g}$$

*The jumper experiences a brief deceleration of 100 times the acceleration of gravity*. Bending the legs upon impact lessens the deceleration enormously.

---

EXAMPLE 5: A large car typically uses a force of about 150 lb just to overcome air resistance at a speed of 60 mph. **How much power does this require from the car?** Since the car exerts a force of 150 lb through a distance of 88 ft in 1 s, we have

$$P = \frac{W}{t} = \frac{150 \text{ lb} \times 88 \text{ ft}}{1 \text{ s}} = 13{,}200 \text{ ft-lb/s}$$

$$= \textbf{24 horsepower} \text{ (almost 18,000 watts)}$$

Notice that $P = W/t = (F \times d)/t = F \times d/t = F \times v$. If you know the force and the speed, $P = Fv$ is the simplest way to calculate the power.

EXAMPLE 6: When more than one force acts while something moves, you can sum the works of the individual forces to get the net work. Or you could just as well sum the vector forces to find the resultant net force and multiply it by the distance $d$ the object moves. The net work is then $W_{net} = F_{net} \times d = ma \times d$. **Can we show that $W_{net} = (\frac{1}{2}\, mv^2)_f - (\frac{1}{2}\, mv^2)_i$ from this formula for $W_{net}$?**

We need to express both $a$ and $d$ in terms of $v_i$ and $v_f$ if we are to show that the right-hand sides of the two equations above are the same. For $a$ we use $(v_f - v_i)/t$, and for $d$ we use $d = \bar{v}t$, where $\bar{v}$ is the average speed during this average acceleration. Because $v_f$ is the speed at the end of the acceleration and $v_i$ is the speed at the beginning, the *average* speed is $v = (v_f + v_i)/2$. This gives

$$W_{net} = ma \times \bar{v}t$$

or

$$W_{net} = m\left(\frac{v_f - v_i}{t}\right) \times \left(\frac{v_f + v_i}{2}\right) \times t$$

Multiplying the terms yields

$$W_{net} = \frac{1}{2}m(v_f^2 - v_i^2) = \left(\frac{1}{2}mv^2\right)_f - \left(\frac{1}{2}mv^2\right)_i = \mathbf{KE_f - KE_i}$$

# REVIEW

**1.** What is work? Give the formula and its units.

**2.** A simple machine allows you to do the same amount of work with less force. Name three household items that are simple machines.

**3.** Define power. What are the units of power?

**4.** Work can cause a change in something's speed, its position, or its temperature. True or false?

**5.** State the formula that relates net work and change in kinetic energy. What are the units of kinetic energy?

**6.** Which form of energy is the measure of a moving object's ability to do work?

**7.** Which form of energy is the measure of an object's potential to do work because of the object's position?

**8.** What is the formula for the gravitational potential energy of a squirrel in a tree?

**9.** For there to be potential energy there must be a force ready to do work. True or false?

**10.** What's another word for the potential energy stored in the bonds of molecules?

**11.** What are the common units of heat energy?

**12.** Heat energy is a less organized form of which other form of energy?

**13.** State the law of conservation of energy. Discuss its meaning. State the first law of thermodynamics, and discuss what it means.

**14.** Energy changes form any time matter interacts, any time there are forces. True or false?

**15.** What law expresses the fact that heat energy always spreads outward and never concentrates of its own accord?

**16.** Name several forms of energy besides kinetic, potential, and thermal energy.

**17.** Give an example of a nearly elastic collision and a nearly inelastic collision.

## DEMONSTRATIONS

**1.** Try cracking a walnut with your hands; then use a simple machine, the nutcracker, to help you do the work. After you crack the nut, measure the distance between the ends of the open handles and the distance between them at the walnut's position. Using these measurements, form the ratio $d_{hand}$ divided by $d_{nutcracker}$. *This shows you about how much your force was multiplied by the nutcracker.* (Why? $F_{hand} \times d_{hand} = F_{cracker} \times d_{cracker}$, so $F_{cracker}/F_{hand} = d_{hand}/d_{cracker}$.)

**2.** Find a long board and prop one end to make an inclined plane. Then attach one or more rubber bands to the front end of a child's toy, perhaps a heavy truck or tractor. Use the rubber bands to tow the toy up the incline to the top. Next, place the toy on the floor (or ground) beside the board and lift it by the rubber bands straight to the top of the incline. Notice how much farther the rubber bands stretch and how much more force you have to use when you lift the toy directly up. It's hard to believe the amount of work done by each method is the same, but they are. Simple machines help that much.

**3.** If you don't believe a screw is a wrap-around inclined plane, try this. Cut a triangle of any shape from a piece of paper—this is the inclined plane. Holding a pencil vertically at the high end of the incline, wrap the inclined plane around the length of the pencil to form the threads on a "screw." The more *threads per inch* the inclined plane makes on the screw, the more distance will be traded for force when the screw does work.

**4.** Prove to yourself there is potential energy in food. Skewer a shelled pecan on a straightened paperclip and use a match to set the nut on fire. Then hold a small aluminum cup of water over the burning pecan and watch the water boil from the heat produced by the pecan's potential energy. (This is the method used by laboratories to determine the number of Calories in food, except that the *calorimeters* they use are carefully insulated against heat loss. A quantity of beans, or carrots, or broccoli, or some other edible item is completely burned, and the heat energy given up by the food is measured in Calories.)

**5.** At a sink, fill a small bowl with water and let it stand until a thermometer shows that the water's temperature is not changing. Then it is at "room temperature." (Temperature is one measure of how much heat energy something has, as you'll see in Chapter 13.) Next put the beaters of an electric mixer into the water and let the mixer run. After a few minutes, turn the mixer off and measure the water's temperature again. The beaters do work on the water, pushing its molecules through a distance as they move. The extra kinetic energy the water molecules gain is random kinetic energy, or heat energy, and the water temperature increases. (For a quicker demonstration to prove work can generate heat, rub a piece of sandpaper quickly and firmly over a board. After several strokes, you'll *feel* the heat energy.)

**6.** Stand a meterstick (or yardstick) vertically on a hard floor. Drop a tennis ball or a racquetball from the top of the meterstick and measure how far the ball returns after it bounces. At the top of its rebound, its energy is entirely potential energy, just as it was before you dropped the ball. Your measurement can tell you what fraction of its initial energy it retains after the bounce.

$$\frac{\text{potential energy after bounce}}{\text{potential energy before bounce}} = \frac{mgh_{after}}{mgh_{before}} = \frac{h_{after}}{h_{before}}$$

The fraction of energy retained after the bounce is just $h_{after}/h_{before}$. For example, if a tennis ball is dropped from a height of 1 m and rebounds to a height of 40 cm, or 0.4 m, it retains only $0.4/1 = 0.4$ of its energy after the bounce, or 40 percent. The other 60 percent of its energy disperses as heat, sound, and so forth. (Very little energy is lost to air resistance if the ball doesn't fall any farther than a meter.)

Drop the ball again and let it bounce *twice*. Investigate whether it loses the same percentage of energy during its second bounce as it does during the first bounce. (*Note:* A concrete or tile floor gives the most elastic collisions; a wooden or linoleum-covered floor "gives" and absorbs more of the ball's energy.)

## EXERCISES

**1.** Does gravity do work on you while you are lying in bed?

**2.** Lazy in the morning? Discuss exactly how high your bed should be so you can get up and about with the least effort (with the least work).

**3.** When a pool ball rolls across a level billiard table, does gravity do work on the ball?

**4.** The Incas, though quite advanced in building techniques, never used the wheel. With respect to work, how is the wheel important?

**5.** Does carrying a heavy backpack along a level trail require work on the backpack?

**6.** Simple machines can do which of the following? (a) Save you work, (b) do work for you, (c) make work easier, (d) multiply energy.

**7.** When you push a lawnmower across the lawn, do you do work? Does gravity do work on the lawnmower (when the lawn is level)?

**8.** Which person in Fig. 6–20 will be moving faster when he hits the water? Which will get to the water first?

FIGURE 6–20

**9.** When you walk down a flight of stairs, does gravity do work on you? When you walk up the stairs? Discuss.

**10.** An applied force on a moving object always does work on it. True or false?

**11.** Walking along on a level surface and carrying a heavy suitcase, you do no work $W$ on your suitcase. But you do work lifting and lowering your feet as you walk with that suitcase across the floor. Do you do more, less, or the same amount of work when you walk with the suitcase than you would do without it?

**12.** Can work be done by more than one force at a time?

**13.** Anytime something's velocity has changed, net work has been done. True or false?

**14.** When you carry a heavy suitcase up a flight of stairs, do you do more work on it if you run rather than walk?

**15.** Is the fellow shown in Fig. 6–21 doing any work? Discuss.

FIGURE 6–21

**16.** Does the tension in the rope or chain of a swing, as in Fig. 6–22, do any work on a child who is swinging?

FIGURE 6–22

**17.** Only a force (or force component) in the direction something moves can give the object kinetic energy or take it away. True or false?

**18.** Can something have potential energy and kinetic energy at the same time?

**19.** The farmer in Fig. 6–23 must throw all of this dirt over the wall. About how much work could he save himself by beginning to shovel on top of this pile?

FIGURE 6–23

**20.** Since all forces can do work, can all forces store potential energy?

**21.** After waiting 3 years, Professor Quick got a reply to his suggestion for "Quick" elevators at his school (see Exercise 32 in Chapter 4). The administrators politely turned down his request, saying the power requirement to propel his elevators would be too high and cost too much. What should Quick say to this?

**22.** Which of the methods of jogging shown in Fig. 6–24 uses more energy? (Presume the two joggers have the same speed, and give an explanation in terms of work.)

**FIGURE 6–24**

23. It is thought that porpoises "porpoise" (arc out of the water in a smooth motion) in order to save energy when they move at high speeds. Discuss.

**24.** A single stick of dynamite is small in size and weight, yet it contains all of the energy released when it explodes. Where does this energy come from?

25. A cat climbs a tree. Describe its ascent in terms of potential energy, the work the cat does, and the work gravity does.

**26.** Analyze this slogan: "Get more for your energy—do work!"

**27.** What form of energy enables your body to produce forces?

**28.** When an avalanche comes to rest at the base of a mountain, where has its potential energy gone?

**29.** Does work against the force of friction always generate heat energy?

30. What happens to a car's kinetic energy when it comes to a stop?

**31.** Books on diets and jogging often have tables that show the approximate energy consumption from various activities. Would it surprise you to read that walking down a flight of stairs requires more energy than walking at the same speed on level ground? Explain why it's true!

32. Bicycle manufacturers go to great lengths to make bike frames rigid. Explain how this helps a biker with regard to energy.

33. After exercise, you can probably see your pulse in the veins of your arms and feet. Each heartbeat forces blood through your system of arteries and veins, raising the pressure, and the elastic blood vessels expand. But what about the time *between* heartbeats—does the blood flow stop? *No*, and the answer has to do with potential energy. Can you guess why the blood keeps flowing?

**34.** A child throws a ball of clay against a wall and it sticks. Has energy been conserved?

**35.** Explain briefly in terms of energy why the lowest levels of a food chain are the most abundant.

**36.** When you are at rest, the work of breathing consumes only about 2 percent of the energy you use. But if you exercise strenuously, your breathing can take as much as 25 percent of your total energy consumption. Explain.

**37.** Can someone save energy while walking by shuffling along rather than stepping?

38. In theater productions, scenery backdrops are suspended from the ceiling and lowered when needed. They are balanced with counterweights, a principle employed by elevators as well. How does this help with regard to work? (Also see Fig. 6–25.)

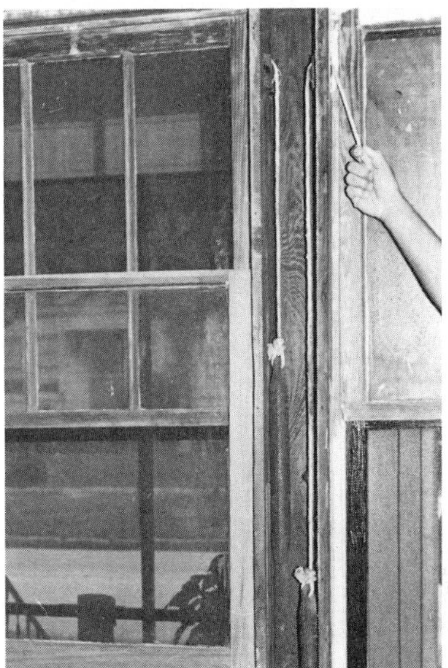

**FIGURE 6–25**
Old-style windows had counterweights in the walls that made them easier to lift or close.

**39.** If energy is conserved, why don't you gain a pound for every pound of food you eat? Discuss.

40. A person who runs 10 miles in 60 minutes uses approximately 1200 Calories of food energy. But a walk for this distance uses one-sixth of this energy. Why should there be such a big difference in the amounts of energy used? (*Hint:* Muscles used in running give off up to 20 times the heat they do at rest.)

**41.** If a child pulls an old wagon up a steep incline on a sidewalk, what does the child's work do? Against what forces must the child's pull act?

**42.** The **efficiency** of a machine or an animal (or any other energy processor) is defined as **work** ($W$) **done/total energy used**. The actual efficiency of a machine or an animal can never be equal to 1 (or 100 percent), however. Why?

***43.** (a) Calculate your kinetic energy (in joules) when you stroll along at 1 m/s. (1 kg weighs 2.2 lb.) (b) Calculate the energy you use (the work you do) to lift yourself off the floor. If you're lying on the floor and stand up, you've done work $mg\bar{h}$ where $\bar{h}$ is the *average* distance each kilogram of your body ascends. Since your feet don't move upward at all and your head gains an elevation almost equal to your height, what might the average gain in distance be for each kilogram? The height of your navel, at least for most people. Estimate the potential energy you use to pick yourself up off the floor, $mg \times h_{navel}$, and compare that to your kinetic energy of walking, found in part (a). (c) When you lifted yourself from the floor in one second [see part (b)], how much power did you use?

***44.** Show that a car's brakes must do about the same amount of work to slow a car from 70 mph to 50 mph as they must do to stop that car from a speed of 50 mph.

***45.** A roller coaster starts from rest at the top of a steep incline. As it falls, should its velocity depend on how many riders are aboard? Use *net work = change in kinetic energy* to prove your answer.

***46.** If your body could convert food energy to physical work with 100% efficiency, you'd have to eat very little. Calculate the distance the 2500 Calories you might consume today could lift you against the pull of gravity. (Remember that 1 Calorie = $4.18 \times 10^3$ joules.) Convert the height you calculated to stories of stairs at 3 meters/flight and ask yourself if you could walk up that many stories on a day's rations. (For comparison, Mt. Everest is about 8850 meters above sea level.)

***47.** Because kinetic energy is 1/2 $mv^2$ and any speed is *relative*, kinetic energy is also a relative quantity. A summertime dragonfly hovering over a road has nearly zero kinetic energy. But a T-shirted motorcyclist who at 55 mph (about 25 m/s) sees the dragonfly *approaching* (relative to him) thinks otherwise. When the 5-gram insect smacks the motorcyclist on the chest, how many joules of energy does the dragonfly deliver? (After the collision, the dragonfly has the motorcycle's speed over the ground.)

***48.** A steel ball drops onto a thick steel plate. Its speed just before it hits is 10 m/s, and it rebounds with a speed of 9 m/s. What percentage of its kinetic energy did it lose during the collision? What would you expect its speed to be after the second rebound? (See Demonstration 6.)

***49.** Calculate how long you could live on just your body's supply of fat. (Ideally, 15 percent of an adult male's weight and 22 percent of an adult female's weight is in the form of fat.) If you use 90 Calories/hour even while resting, how many days would it take to get rid of your fat if you were on a starvation diet? (1 pound of fat = about 4200 Calories.)

***50.** Suppose you need 2400 Calories per day just to stay at your present weight. If you went on a strict diet of 2000 Calories/day, how long would it take you to lose 5 pounds? (1 pound of fat = about 4200 Calories.)

***51.** Using the example quoted in this chapter, estimate how many joules of sunlight began the process that put a 1 pound can of tuna on the shelf at the grocery store. (1 ounce of tuna contains about 30 Calories.)

***52.** A rockclimber tires and slips from a vertical wall in Yosemite Valley. He falls 100 ft before his rope straightens out and begins to brake his fall. What is the average deceleration if the rope stretches only 10 ft before bringing him to a stop? If the rope stretches 25 ft? (This shows why nylon ropes, which stretch 20 percent or more of their length under tension, are so important in the sport of rockclimbing.)

***53.** Show that 90 Calories/hour is equivalent to about 100 watts. So even at rest your body releases heat energy at about the same rate as a 100-watt light bulb.

# THE DISCOVERY
## ACT 1

1665 was a plague year, and at the end of the term Trinity College in Cambridge shut down for the duration of the scourge. Students packed their belongings and carted them off in carriages or wagons. So it was that 23-year-old Isaac Newton made his way back to his mother's farm in the English countryside. There he concentrated on natural philosophy for the next year and a half. He later remembered it as a time when he thought of little else.

One of the mysteries young Newton turned his attention to was gravity. The attraction from the earth seemed to give everything in Newton's world (just as in ours today) a virtuous trait—specifically, that all masses share exactly the same acceleration, *g,* in the absence of air resistance. Why, he must have wondered, should earth pull on all objects, no matter what their mass, with precisely the right amount of force to accelerate them all the same? With a great leap of imagination, he solved its puzzle. He later told a biographer he was inspired by a falling apple, so that legend is apparently true.

Perhaps Isaac Newton was the first to see the moon as an object "falling" through space, just like that apple that fell on his mother's farm. A force must act to keep the moon from gliding out of sight along a path tangent to its orbit. What else but earth's gravity could keep the moon traveling in an orbit centered on earth? (See Fig. 1.) We can't know Newton's precise questions to himself, but we do know his answers. He could estimate the acceleration of the moon. Moving in a nearly circular path, it has an acceleration toward earth of $v^2/r$ (see Chapter 4). The distance $r$ was estimated by the early Greeks, and its value was even more refined in Newton's day. The moon's speed $v$ was the distance around its circle, $2\pi r$, divided by the time the moon needed to circle the earth, a

**FIGURE 1**
The earth's gravity keeps the moon in an orbit centered on earth.

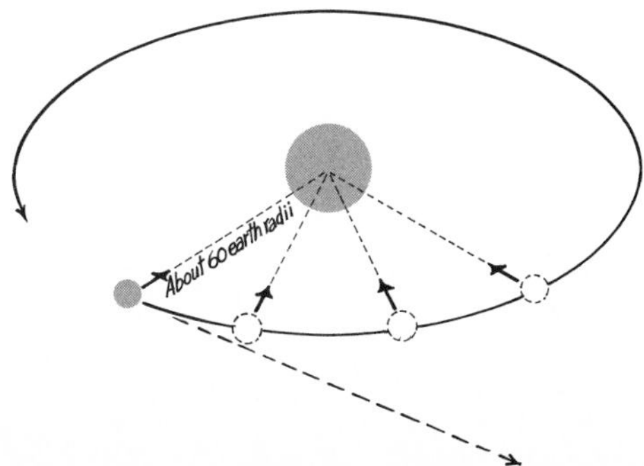

number known from simple observations. The moon's acceleration works out to be 0.009 feet per second squared, as compared to 32 feet per second squared for the falling apple. The earth's gravity was obviously much *weaker* at the moon's distance. Newton wanted to find a way to calculate gravity's strength at the moon's distance and *predict* the moon's acceleration. This calculation was to provide the key to calculating the force of gravity between any two objects at any given distance $d$ away from each other.

The objects near earth's surface that accelerate at the rate $g$ are all the same distance from earth's center, namely $R_{earth}$, the radius of the earth. Newton knew the moon is about $60R_{earth}$ away. Comparing the acceleration for each, he could see the acceleration of the moon toward the earth was about g/3600, or about $g/(60)^2$. The acceleration apparently decreases with distance according to the factor $1/d^2$. The force that causes the acceleration should do the same. That force must depend on the mass of the attracted object, too, since something twice as massive on earth weighs twice as much. Likewise, because action equals reaction, the mass of the attracting object must figure in as well. Putting the facts together, Newton deduced the formula for the force of gravity,

$$F_{gravity} = G\frac{m_1 m_2}{d^2}$$

where $G$ is a constant known only by experiments, $m_1$ and $m_2$ are the masses of the attracting objects, and $d$ is the distance between them.* It worked for the earth and the moon, but would it hold for other distances and even other objects? Newton needed a test for his idea.

The scholars of Newton's time had a great legacy from earlier astronomers. The model of the solar system created by Copernicus (1473–1543) had set the five moving planets visible to the naked eye and the earth in circular orbits around the sun. However, the circles of Copernicus did not predict future positions for the planets quite as accurately as they could be observed. Johannes Kepler (1571–1630), a contemporary of Galileo, discovered why. In 1604 Kepler found that the planets actually orbit the sun along elliptical paths (curves that resemble flattened circles). He went on to find very precise mathematical relations between their average distances from the sun and how fast they move (1609) and how long it takes them to complete their orbits (1619). This remarkable achievement showed *how* the planets move, but no one could say *why* they moved that way.

---

*Details are given in Chapter 7. Newton was somewhat astonished by one feature he discovered: The gravity of earth's huge bulk attracts the things around us, and the moon as well, *just as if all its mass is located at its center*. He eventually proved why it should, using mathematics he began to develop during the two plague years.

Newton considered the planets and the sun. In his day the distances of the planets to the sun were not accurately known, so Newton couldn't compute the accelerations of the planets as he could for the moon. What he could do was to see if his law of attraction would provide just the right force at each point to cause the planets to move according to Kepler's laws. He was able to work it out; his law was correct. His law of gravity showed that the attraction from the sun reached out and pulled on the planets, causing them to move in elliptical orbits. It was a magnificent discovery. Then Newton did an unthinkable thing—at least to almost anyone but Isaac Newton. He put his calculations away and left them, perhaps because he had satisfied himself. The discovery of the law of gravity was put to rest, to lie in dark and silence for almost 20 years.

## ACT 2 (17 YEARS LATER)

In 1683 Edmund Halley (Halley's comet is named for him) and Christopher Wren (famous for his architecture) put a question to the English scientist Robert Hooke (who was the first to describe the spring force). They asked what force from the sun might cause the planets to move in elliptical paths.

Hooke, by all accounts a man unwilling to hide his candle under a basket, made a guess. He answered that it must decrease with distance according to the factor $1/d^2$. Perhaps Hooke drew an analogy between the sun's gravity and the sun's light; scientists knew that a candle's light grows more feeble at points farther away from the source according to $1/d^2$. At any rate, when they asked him to prove it, he never could. Hooke went to the tops of buildings in London to do experiments to see if earth's gravity was less there, and he went down into wells to see if earth's gravity increased, but all to no avail. Nor could Hooke prove that an attraction proportional to $1/d^2$ would cause the elliptical orbits of the planets. Hooke's remarkable guess went unproved.

The next year Halley journeyed from London to Cambridge to put the still unanswered question to Newton, by then a renowned professor in mathematics and physics. Halley asked what path a planet would follow should the force from the sun be proportional to $1/d^2$, to which Newton replied, "Ellipsis." When asked how he knew, Newton said, "Why, I have calculated it." With great joy and enthusiasm, Halley insisted that Newton publish his discoveries. Newton couldn't find his old calculations that day, somewhat to his embarrassment, but he promised Halley he would submit a written account later.



Newton began to write, and in 1687, with Halley's financial backing, he published (in Latin) his *Principia (The Mathematical Principles of Natural Philosophy).* In this book were his three laws of motion, the law of gravity, and a number of applications of his laws, explaining things that had never before been understood. He was able to *derive* Kepler's three laws of planetary motion, explain why there are tides in the oceans, show why earth and Jupiter bulge at their equators, and much more.

After the publication of *Principia,* Newton continued to teach at Cambridge. In 1696 he was appointed warden of the English mint, then in 1699, master of the mint, government positions that brought security and a better salary. Newton quickly put his originality to use in his new job. Coins in those days were cut from thin sheets of precious metal and handstamped with a design. Some of the enterprising populace shaved the edges from all the coins that passed through their hands and hoarded the bits of gold and silver.* To eliminate this practice, Newton began scoring grooves into the edges of new coins. That made clipped coins easy to notice, and they weren't legal tender, so no one would accept them. (Examine the change in your pocket; each dime and quarter in use today still carries Isaac Newton's marks, even though the metals used aren't nearly so precious as gold or silver.) Before his death in 1727, the output of the mint had increased by tenfold and British currency was on a firm basis; and for the first time, Britain became a major merchant power in Europe.

In 1704 Newton published a book based on his experiments with light. Called *Opticks,* it is a milestone in *experimental* philosophy, as opposed to the *Principia,* a product of mental processes and inventiveness, or *theoretical* philosophy. This second great book was written in English and appealed to a much wider audience. In 1705, Isaac Newton was knighted by Queen Anne at Trinity Lodge while he stood beneath a portrait of Galileo.

Isaac Newton put physics on solid ground. Experiments proved the predictions of Newton's laws to be true time and time again. Great truths are rare in human knowledge, and these rare gems ultimately affected thinking in such diverse areas as philosophy, psychology, politics, and law. Newton's accomplishments were a catalyst for the Age of Reason, when science first emerged as an important influence in human culture. As the English poet Alexander Pope (1688–1744) said of Newton: "Nature and Nature's laws lay hid in night: God said, Let Newton be! and all was light." ■

*The poet John Dryden once received money so badly clipped that he couldn't spend it—and it was from his publisher!

# CHAPTER 7 GRAVITY AND TIDES

Long ago in the high desert plains of the American West, Papago Indians explained their surroundings in legends. Then, as now, there was little to hide the sky, neither forest, clouds, smog, nor city lights, and soon after sunset the open sky was crowded with stars. Because year after year the stars rose in the same patterns, the Papagos believed they were somehow connected to earth. They told their children that in the beginning spider people had used their webs to sew the heavens to the desert. Spider webs, so nearly invisible and yet remarkably strong, seemed to be the perfect threads to whirl the glittering stars around.

Today we know the Papagos were not far wrong. There is a connection between our planet and the heavens. Now we call this web *gravity,* and its threads really are invisible, and they really do stretch all the way to the stars. Like the Papagos, we too pass our explanations along, only in lectures and books, and as science rather than legend. ■

## The Law of Gravity

There is a force in nature that's everywhere; it pervades everything, all of space and every particle that exists. It's a mutual attraction between all particles in the universe, an attraction that acts over long distances, through empty space, and directly through any other particles in between. Every particle with mass exerts this pull on every other particle and in return gets an equal and opposite pull from each of those particles. This remarkable property of matter is called gravity. This force is the same one you must overcome each morning when you get out of bed. The earth has mass, and it attracts you because you, too, have mass. Your right arm has mass and so attracts your left arm, and so it goes throughout the universe.

The law of gravity tells us that any two particles, with masses $m_1$ and $m_2$, pull on each other, *directly toward each other*. These pulls are equal and opposite, and the magnitude of the force is given by

$$F_{\text{gravity}} = G\frac{m_1 m_2}{d^2}$$

where $d$ is the distance between their centers and $G$ is a constant that we'll discuss shortly. The farther apart the particles are, the smaller the attraction they have for each other. As $d$ becomes larger, $(Gm_1m_2)/d^2$ becomes smaller at an ever greater rate. Nevertheless, the distance must become infinitely large before $F_{gravity}$ vanishes completely. Here's a force that acts over a distance as great as you can imagine, straight through anything that's in its way! The quantity called $G$ is a constant of proportionality, a number found by experimentation. It tells how strong the gravitational force is between two masses a distance $d$ apart. Called the universal constant of gravitation, $G$ is a very small number in ordinary units because the force of gravity between two people-sized masses a few meters apart, for instance, is very tiny. You'd certainly never *feel* your left arm's attraction for your right arm. In metric units, $G = 6.67 \times 10^{-11}\ \text{N·m}^2/\text{kg}^2$. All the indications are that $G$ has the same value throughout the universe and throughout time.

**FIGURE 7–1**
People who were forced to use gravitational attraction alone to get them together would have to have patience.

If you use this law to calculate the gravitational force between a man (whose earth weight is 160 pounds) and a woman (whose earth weight is 120 pounds) standing 10 feet apart, you'll find their mutual attraction is only 0.0000000064 ($6.4 \times 10^{-9}$) pound, a force far too small to be felt or noticed under ordinary circumstances. Were they in space, however, 10 feet apart and at rest, their gravitational attractions would cause them to drift together. As they crept toward each other, the force would increase slightly (because $d$ gets smaller), but some 17 hours would pass before they would touch. (The woman would accelerate a bit faster than the man, who gets the same force but has a larger mass.) At home on earth, however, both of them are attracted far more to the great mass of the earth than to each other. Compare their leisurely rate of acceleration toward each other in space to a swimmer who in less than 1 second falls 10 feet from a diving board into a swimming pool.

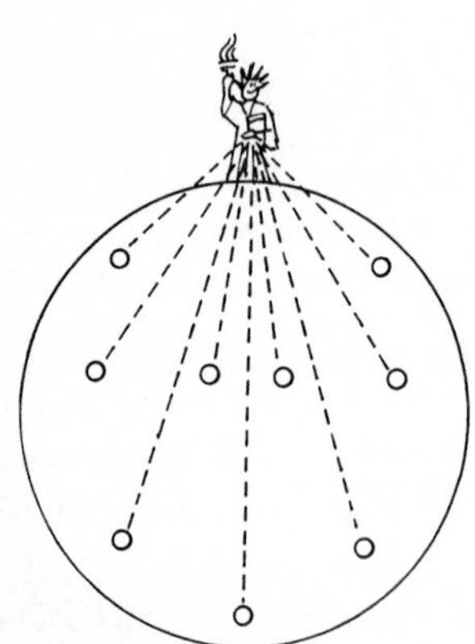

**FIGURE 7–2**
An object on the surface of earth is attracted by the gravity of all the mass within the earth.

## How Earth's Gravity Attracts You

Think of how the earth attracts you. Every particle in the earth pulls on every atom of your body according to $F_{gravity} = (Gm_1m_2)/d^2$. But things like atoms and even grains of sand have incredibly small masses, and their quantities are staggering. What a problem it would be to calculate your weight using those individual attractions! Especially because each tiny pull is a vector—and all those little vector forces have to be added. Figure 7–2 diagrams the problem.

A particle-by-particle accounting is impossible to do. But as we've seen, Isaac Newton discovered (somewhat to his amazement) that the earth attracts things on or above its surface (even the moon) just *as if all the mass of the earth were concentrated at its center*. That seems strange because all of the earth's mass isn't at its center.

Despite the fact there are almost innumerable attracting particles to consider, there's only one *net* force. Now it's not too hard to see that the "average" direction of all the particles in the earth is the direction toward its center. (See Fig. 7–3.) That the value of the net force should be the same as if all the particles are at the earth's center is *not* obvious.

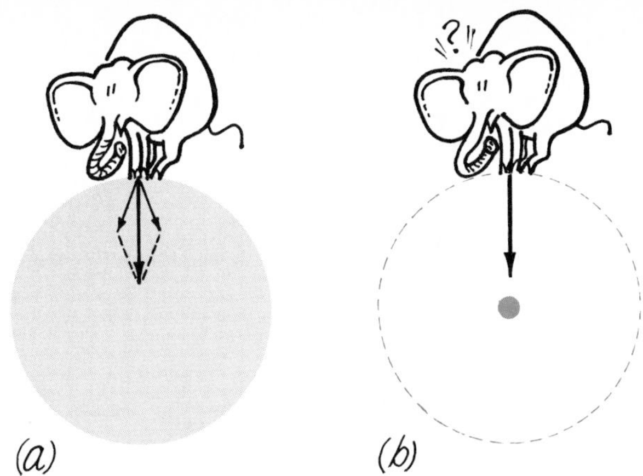


**FIGURE 7–3**
*(a)* Because the earth is very symmetrical, the net attraction of all its parts is toward the center of the earth. *(b)* The result of the attractions of all the mass of the earth, both near and far from the object, is the same as if all the mass were at the earth's center.

Newton was eventually able to prove this result with a branch of mathematics called calculus, which he invented himself. Though we can't show this here, if the downward components of all those tiny forces were summed, the answer for the net force toward the center is the same as if all the particles *were* together at the center of the earth. This means you can find its attraction for objects at its surface (or even out in space) by assuming all the earth's mass is at its geographical center.

Every object has such a point, called a **center of gravity**. For a human standing on earth, his or her center of gravity lies somewhere in the lower abdomen. The equal and opposite attractions between that person and the earth are given by the law of gravity, where $d$ is the distance between their centers of gravity.

## *g* Explained

Drop a shoe and a coin together from shoulder height. Notice that they accelerate toward the ground side by side, despite their great difference in weights. They have the same acceleration, even though the force of gravity on them is very different. Newton's discovery of the law of gravity explained this puzzling feature of earth's gravitational attraction.

The net force on both the shoe and the coin is essentially the pull of gravity. On their way to the ground, they never move fast enough for air drag to become a factor. As always, the net force is equal to $ma$. Therefore, for either shoe or coin, we can set

$$ma_{\text{gravity}} = G\,\frac{mM_{\text{earth}}}{R^2_{\text{earth}}}$$

where $R_{\text{earth}}$ is the earth's radius, or the distance from the earth's center to where these things fall on the surface. In this equation, the mass $m$ of the shoe (or the coin) appears on both sides. If each side is divided by $m$, we get an equation for the acceleration caused by gravity—*and the mass of the object does not appear!*

$$a_{\text{gravity}} = G\,\frac{M_{\text{earth}}}{R^2_{\text{earth}}}$$

This is the acceleration we call $g$, and its value is about 9.8 m/s$^2$. Anything on earth's surface, then, has this acceleration unless air resistance plays a part in its fall. Since everything on earth's surface is about the same distance from its center, we earthbound people can use $mg$ to calculate weights, rather than $(GmM_{\text{earth}})/d^2$. Weight, remember, is the force of earth's gravity on an object.

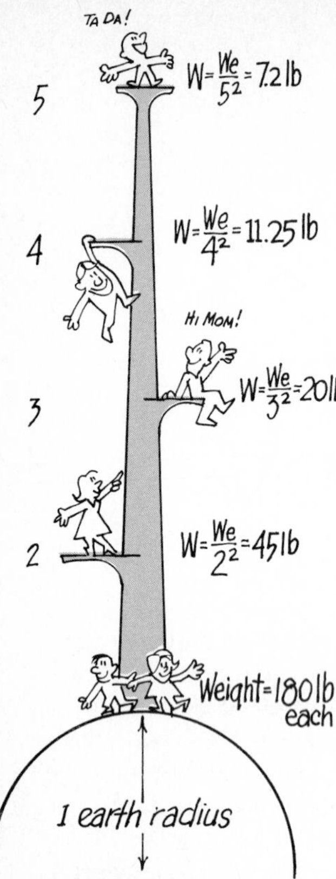


**FIGURE 7–4**
If people move away from the center of the earth, their weights diminish. The force of gravity decreases with distance; $F_g = G\frac{m \times M_{\text{earth}}}{r^2}$

## Visualizing Gravity: The Force Field

Put a particle or any bit of matter anywhere in the neighborhood of another, and it gets a tug straight toward that second piece of matter, and vice versa. The closer they are, the greater the pull. To describe this property of matter, think of mass as modifying the space around it, putting some sort of stress on space and changing its properties to cause any nearby matter to accelerate toward it. The mass exerts a relatively strong influence on nearby space, but the influence weakens with increasing distance. To visualize this modification of the surrounding space, or the field of influence of a mass, sketch lines at any equal intervals straight outward from a bit of matter. These lines, called **field lines**, show the direction of the attractive force pulling on matter within the **force field**. Where the field lines are dense, or close together, the value of the force is larger. Farther from the mass, where the field lines are spread out, the force is proportionally smaller. Figure 7–5 shows gravitational field lines in the space around earth. Like the spider webs of the Papagos, all field lines are imaginary, of course, but they give us a helpful picture of the force of gravity throughout space.

**FIGURE 7–5**
Imaginary "field" lines drawn toward the earth's center indicate the direction and strength of gravity around the earth.

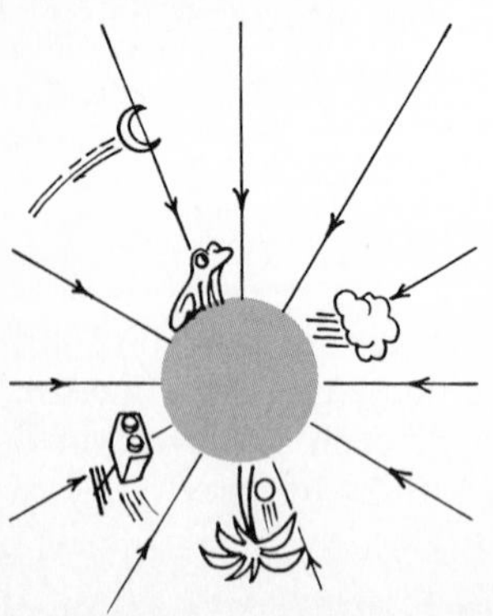

## More on Earth's Gravity

Let's try a famous old imaginary experiment to investigate the force of gravity inside the earth. First we'll need a bit of preparation. Starting at the North Pole, drill a hole straight through to the earth's center and on to the South Pole. While we do, let's think away the atmosphere to eliminate air drag and forget about the high temperatures within the earth. Now we're ready. *You* step into the tunnel and see what happens. Instantly you accelerate toward the center because of earth's gravity. At the surface you accelerate at the rate of $g$, but what about inside the earth?

As soon as you move down in the tunnel, there will be some matter *above* you. This matter pulls backward on you in the opposite direction from your motion, so you should expect the acceleration due to gravity to diminish. Indeed, when you reach the earth's center, the net pull on you is *zero* (Fig. 7–6). Even though all those tons of earth are still attracting you, there is just as much mass in one direction as any other, and the *net* force is *zero*. At the center, you will have a very high

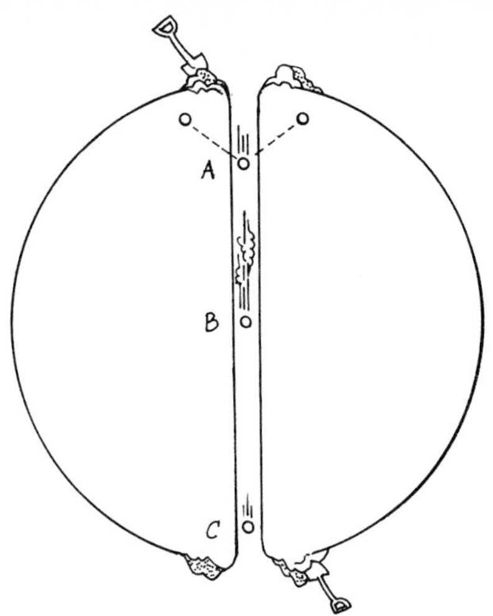


**FIGURE 7–6**
At point *A,* the ball's acceleration begins to diminish as the mass it leaves behind retards it. At point *B,* the center, the net force of gravity from the earth is zero since the ball is tugged equally in all directions. There the ball has its greatest kinetic energy, having been pulled toward the center during its entire plunge. At point *C,* the ball has slowed continually since leaving the center, since most of earth's mass was then "behind" it.

speed, so you'll zip right past into the other half of the tunnel. Eventually you slow and come to a stop at the surface on the far side. That's because the force at every depth on the far side is everywhere equal and opposite to the force at that depth on the entry side. Therefore, at any point you decelerate at the same rate as you accelerated previously. Then, unless someone there helped you step out of the tunnel, you'd be in for a round trip, like it or not. Probably one time through would satisfy most people—a one-way crossing would take about 42 minutes.

Roughly, that's what would happen. But if you had been wearing a device called an accelerometer to show changes in speed, you might be surprised. For quite a few miles into the hole, you'd find that your speed changed at a rate *greater* than 32 feet/second$^2$ even though the earth above you pulled backward on you. Gravity's pull increases below the earth's surface initially because the materials in the core have been squeezed until they are about 7 times as dense as surface rocks (Fig. 7–7). This condensed mass exerts a disproportionally large pull as you approach it, and the pull temporarily increases faster than the backward pull of the matter you leave behind, accelerating you at a rate a little greater than $g$. How far this goes on depends on exactly how the density of the earth varies with depth, information that is unknown today. Nev-

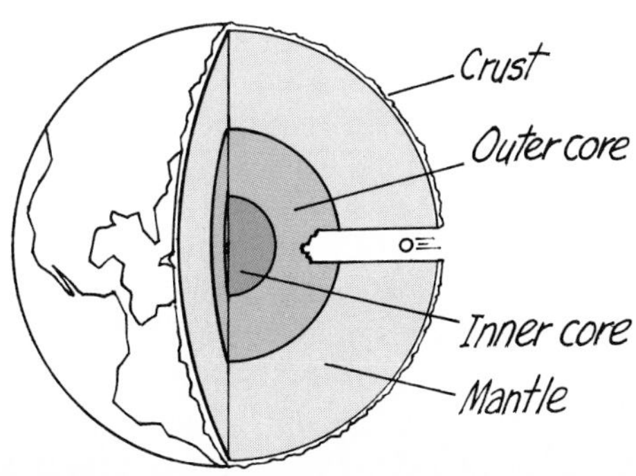


**FIGURE 7–7**
The regions of the earth beneath our feet.

ertheless, after some distance the pulls of the matter surrounding you begin to cancel, and the acceleration due to gravity begins to diminish. At the earth's center, the net pull of gravity from all the earth's mass is zero.

## Orbits and Other Paths Caused by Gravity

An arrow leaves a bow extremely fast and follows a curve high above the ground. At the circus a human cannonball explodes from the mouth of the cannon to arc over three rings before finding the net. A mighty swing and a whack, and a baseball climbs fast from home plate, levels out somewhere in center field, and falls into the stands. Each of these is a **projectile,** something that is given a velocity and released to the influence of gravity and air drag. The path of a projectile is called a **trajectory.**

In the *Principia* Isaac Newton took trajectories a step beyond human experience of his day. In an illustration he placed a cannon on an imaginary mountaintop far above earth's atmosphere. Using the law of gravity and his second law of motion, he derived the paths a cannonball could follow in the absence of air, *paths caused by gravity alone*. Figure 7–8 is a strobelike version of paths like those that Newton drew 300 years ago.

As you can see from this modern rendition, he projected the cannonball horizontally from a mountaintop with various speeds. The low-speed projectiles, such as paths *A* and *B,* arc over and fall to earth, as baseballs and arrows do. You can also see the effect of the earth's curvature on where the projectile lands. The greater the projectile's speed, the farther it goes around the earth. Finally, at some very great horizontal speed, a cannonball might perpetually fall toward earth but never reach the ground. Such a trajectory, path *C,* is called an **orbit.** If the orbit is circular, the projectile falls at just the right rate so the distance between it and the curved horizon of earth does not change.

Newton proved that the gravitational orbits of one body around another are **ellipses,** curves that resemble flattened circles.* The paths Newton drew for cannonballs that did not have enough speed to orbit the earth are also elliptical as far as they go, that is, until the cannonballs strike the ground. Even the path of a bullfrog that jumps from a lily pad into a pond is part of an ellipse that starts and ends at the pond's surface.

To orbit the earth requires a very great speed if the projectile is not far away. At the relatively close distance of 100 to 200 miles above earth's surface, for example, the necessary speed is about 17,000 mph. That sort of orbit was achieved by Sputnik 1 in October 1957, and since then thousands of artificial satellites have been placed in orbits. A rocket lifts a satellite, gives it the necessary speed to orbit, and releases it.

*See Demonstrations 1 and 2 for the details of ellipses.

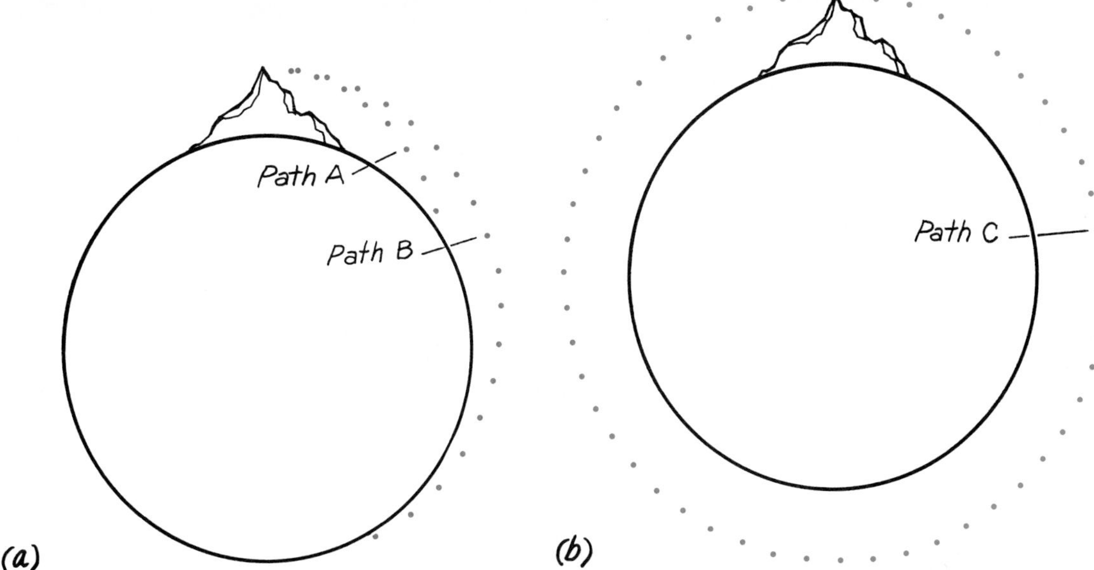


**FIGURE 7–8**
*(a)* Fired from an imaginary mountaintop far above the atmosphere, a cannonball would fall far "around" the earth before hitting the surface. Path *B* would be taken by a cannonball with a greater initial speed than one that would follow path *A*. *(b)* At one certain launch speed the rate of travel to the side (the horizontal speed of the ball) would be great enough so that as the ball fell its distance to the earth's surface would remain the same. The ball would then follow path *C*, a circular orbit.

Then the satellite becomes a projectile, gliding in an elliptical path dictated by earth's gravity, affected only a little by drag, from the farthest remnants of earth's atmosphere, the pull of the moon, and other factors. The moon, too, follows an elliptical path around earth. It, like artificial satellites, continually accelerates toward us, moving to the side as it does. Its *average* distance from earth, about 240,000 miles, changes very little from year to year. At the moon's great distance earth's gravity accelerates it more slowly, and the orbital speed is only 2300 mph.

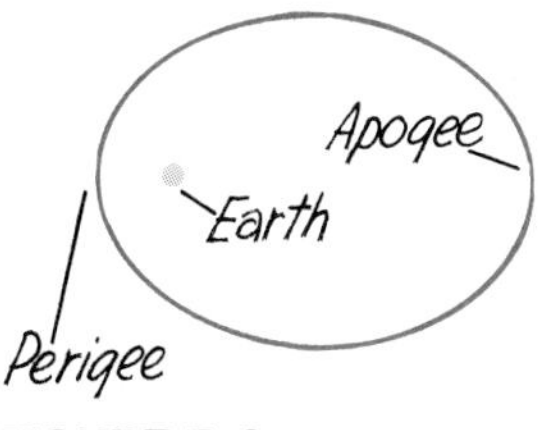


**FIGURE 7–9**

## MORE ABOUT ORBITS

When the space shuttle or an artificial satellite orbits our planet in an elliptical orbit, earth's center isn't at the center of the ellipse. It is at a point toward one end called a **focus.** (See Demonstration 1.) An elliptical orbit brings an orbiting body sometimes closer and sometimes farther away from its parent mass (Fig. 7–9). The closest point in an earth orbit is called the **perigee;** the farthest, the **apogee.** Corresponding points for an elliptical orbit around the sun are **perihelion** and **aphelion,** and for such an orbit around the moon, *perilune* and *apolune*.

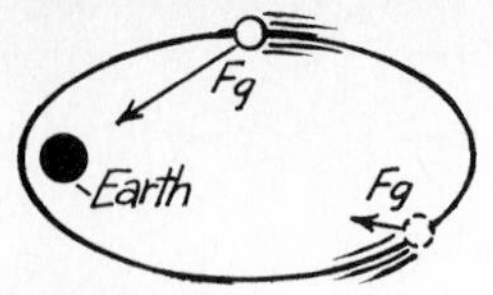


**FIGURE 7–10**
The directions and relative strengths of the earth's gravitational force at two points in the path of an orbiting body.

A body leaving apogee descends toward the earth, losing elevation as it falls to the side, until it eventually reaches perigee. (See Fig. 7–10.) As it falls, a component of the earth's gravity pulls it along its path, doing *positive* work on it and increasing its speed. (In terms of energy, some of its potential energy due to gravity is converted into kinetic energy.) At perigee it has its maximum speed because of that work and scoots around this part of the orbit to rise in the earth's gravitational field. On the second half of the satellite's orbit, there is a component of attraction pulling backward, doing *negative* work and slowing the satellite as it travels. The satellite is moving slowest at apogee. Then, the process repeats. At apogee the orbiting body has its maximum potential energy and its minimum kinetic energy. At perigee it has its maximum kinetic energy and minimum potential energy.

When a smaller mass orbits a larger one, both masses get the same gravitational tug. The earth gets the same pull as an artificial satellite does, but because of its huge mass, the earth's acceleration is negligible. Even as the moon orbits the earth, we don't think about the earth's small motion. Nonetheless, it traces out a miniature of the moon's orbit. Figure 7–12 shows a pair of orbiting bodies; one is 10 times as massive as the other. In comparison, the earth is 81 times as massive as the moon, so earth's ellipse is *very* small.

**FIGURE 7–11**
The space shuttle and artificial satellites orbit the earth in elliptical orbits.

Let's return to Isaac Newton's mountaintop above the atmosphere. The paths of projectiles shot with even greater speeds than those drawn by Newton are shown in Fig. 7–13. A speed somewhat greater than 17,000 mph flings the projectile along a large elliptical orbit. A launch speed of about 25,000 mph catapults the cannonball into a path called a **parabola**. This type of curve doesn't orbit the earth. It takes the cannonball to infinity, or at least it would if there were no other nearby bodies like the moon to perturb its parabolic path. That speed, 25,000 mph, is called the **escape velocity** from earth. (A projectile can never get "beyond" earth's gravity, but it can move fast enough so that earth's gravity cannot pull it back.) Should the cannonball start with a speed greater than the escape velocity, it moves along a path called a **hyperbola,** a curve whose shape is even more open than a parabola.

Astronauts first achieved escape velocity from earth on December 21, 1968, when Apollo 8 left an earth orbit and traveled to the moon. Upon arrival in the moon's vicinity, a rocket fired forward slowed the craft, leaving it in a lunar orbit. The crew made 10 revolutions in an elliptical orbit there, and then a rocket was fired to increase their speed to more than the escape velocity from the moon. Since the moon is less massive than earth, the escape speed from the lunar *surface* is smaller, only 5100 mph. (A very small rocket could do that job, as it did when the lunar landers from the later Apollo spacecraft lifted from the moon's surface.) Their trajectory brought them back toward earth, where other burns corrected their speed and direction for a descent into the atmosphere.

**FIGURE 7–12**
Two bodies in slightly elliptical orbits, one with a mass that is ten times greater than the other. The smaller mass has the greater speed and acceleration, as the strobelike points reveal.

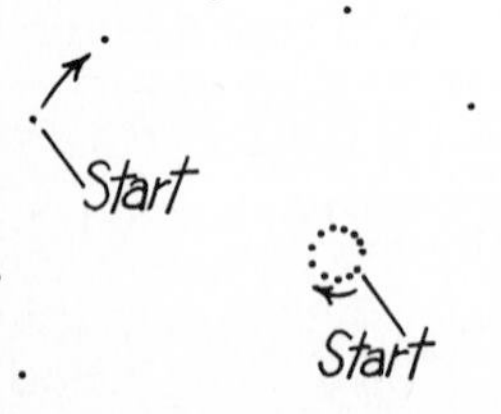


The United States has launched several spacecraft that are travel-

FIGURE 7–13

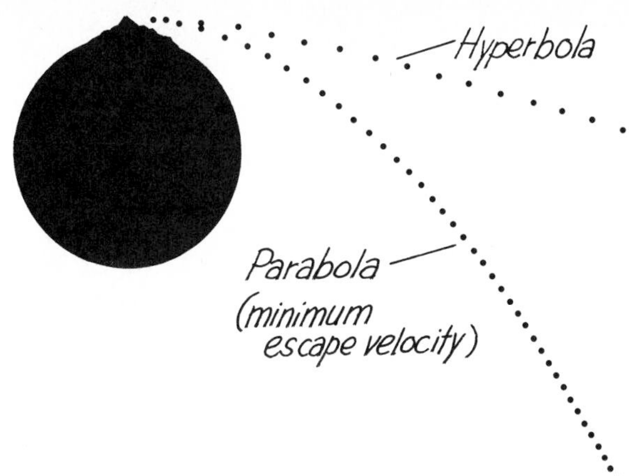


**There will certainly be no lack of human pioneers when we have mastered the art of flight. . . . Let us create vessels and sails adjusted to the heavenly ether, and there will be plenty of people unafraid of the empty wastes. In the meantime, we shall prepare, for the brave sky-travelers, maps of the celestial bodies—I shall do it for the moon, you Galileo, for Jupiter.**

Johannes Kepler to Galileo, April 1610, upon hearing of Galileo's discovery of Jupiter's four large satellites.

ing fast enough to escape not only earth's gravitational pull but that of the sun and other planets as well. One of these, Pioneer 10, passed the orbit of Pluto in mid-1983. As it eventually travels past other stars in our galaxy, it will nevertheless still be in the firm grip of a gravitational field. It will, in fact, orbit the same center of attraction our sun orbits, the massive nucleus of the center of our galaxy, the Milky Way.

The escape velocity from a planet has everything to do with determining how much atmosphere the planet can have. The earth, for instance, continually loses fast-traveling molecules that break loose from the upper edges of its atmosphere. A molecule heading outward at 25,000 mph won't return. Mars's gravity is weaker than earth's, and it can't hold onto any outward-bound molecules that travel at speeds greater than 11,200 mph. Because of its low escape velocity, the moon can't hold gas molecules at all. The sunlight that heats the moon's surface soil to 100 degrees Celsius would quickly heat any gases there, increasing the kinetic energy of the molecules enough to let them escape. Even the exhaust gases left by the liftoffs of the lunar landers are long gone.

## THE GRAVITY OF OTHER PLANETS

The law of gravity can tell you how much you would weigh on another planet, such as Mars. You'd need to know the mass of Mars and the distance from its center to its surface. Mars's radius has long been known from telescopic measurements (see Table 7–1). The mass of Mars can be estimated by observing one of its moons. Deimos and Phobos speed around the red planet in nearly circular orbits, and Mars's gravity turns them, providing the centripetal force. Therefore, $F_c = F_g$ for each moon.

$$\frac{mv^2}{r} = G\,\frac{mM_{\text{Mars}}}{r^2}$$

where $r$ is the distance of the moon from the center of Mars, and $r$ is measured using a telescope. Dividing the $m$ term from both sides of the equation and multiplying both sides by $r^2/G$, we find

$$\frac{v^2 r}{G} = M_{\text{Mars}}$$

Knowing $r$ makes the moon's speed easy to find. All that is needed is the time $T$ the moon needs to circle the planet once. The distance around its orbit is $2\pi r$, so $v = d/t = 2\pi r/T$. Using these values in the formula gives the mass of Mars.

The same sort of method works for any planet with a satellite; an elliptical orbit makes the calculation only a bit more complicated. (See Table 7–2 and Fig. 7–14.) A planet orbiting the sun reveals the sun's mass in the same way. Planets without moons, such as Venus, require a bit more detective work. When Mariner 5 flew close to Venus in 1967, scientists carefully watched how Venus's gravity affected the trajectory of the passing spacecraft. Once they knew the deflection, they used computers and the law of gravity to find how massive Venus had to be to cause Mariner 5 to respond as it did. Before that time Venus's mass could only be inferred by carefully tracking the orbits of Mercury, Venus, and earth. Their mutual attractions perturb their paths, causing slight deviations in their otherwise elliptical orbits around the sun. These small effects are called **gravitational perturbations.**

Once the radius of a planet and its mass are known, its surface gravitational acceleration can be calculated using

$$a_g = G\frac{M_{\text{planet}}}{R^2_{\text{planet}}}$$

See Table 7–3 for a comparison.

**TABLE 7–1**

| RADIUS IN TERMS OF EARTH'S RADIUS (EARTH'S RADIUS = 1.0) | |
|---|---|
| Sun | 109 |
| Mercury | 0.38 |
| Venus | 0.95 |
| Moon | 0.27 |
| Mars | 0.53 |
| Jupiter | 11.19 |
| Saturn | 9.47 |
| Uranus | 3.69 |
| Neptune | 3.5 |
| Pluto | 0.47? |

**TABLE 7–2**

| MASS IN TERMS OF EARTH'S MASS (EARTH'S MASS = 1.0) | |
|---|---|
| Sun | 330,000 |
| Mercury | 0.06 |
| Venus | 0.81 |
| Moon | 0.012 |
| Mars | 0.11 |
| Jupiter | 318 |
| Saturn | 95.2 |
| Uranus | 14.6 |
| Neptune | 17.2 |
| Pluto | 0.1? |

**TABLE 7–3**

| SURFACE GRAVITATIONAL ACCELERATION ($g$ = 1) | |
|---|---|
| Mercury | 0.4 |
| Venus | 0.9 |
| Moon | 0.16 |
| Mars | 0.39 |
| Jupiter | 2.6 |
| Saturn | 1.1 |
| Uranus | 1.1 |
| Neptune | 1.4 |
| Pluto | ? |

**FIGURE 7–14**
How much would a 150-lb earthling weigh elsewhere in our solar system?

**FIGURE 7–15**
A cattle guard in New Mexico. Cattle won't cross through the opening in the fence for fear that their hooves will slip between the rails and trip them up. Because of their mass (and hence weight) a fall is a serious matter.

## Gravity and You

Single-cell organisms that float about in water probably don't know up from down; the closest scrutiny of biologists has found no trace of physiological characteristics in those microscopic creatures indicating that they respond to gravity in any way. As large land dwellers, however, we are built to perform in the pull of the earth. Each time someone stands up, pressures and angles readjust, and muscles do work against the force of gravity. Unlike the one-cell floaters, we can *fall,* too, and at considerable peril. The danger from falling increases with size and mass. American bison shy away even from crossing a railroad track because of the chance they might trip and fall. The first transcontinental railroad laid down in the 1800s permanently divided the buffalo into a northern herd and a southern herd because the buffalo refused to step over the rails and crossties. Elephants in India are captured by enticing them into an area laced with sugar cane and nearly surrounded by a shallow ditch. The ditch need be only deep enough so the elephants' trunks won't touch the bottom. At a disadvantage because of poor eyesight, an 11,000-lb elephant won't risk a fall to get out of the enclosure. The bigger they are, the harder they fall.

Orbiting the earth in spacecraft, astronauts and cosmonauts demonstrate how much gravity has to do with the physical condition of our bodies. The first Skylab crew was weightless for 29 days (in 1973). Functions like eating and excreting are normally aided by gravity, and like previous astronauts, Skylab crews found they couldn't even summon belches without help from gravity. Other modifications were more serious. In spite of a vigorous exercise program, the members flunked their self-administered cardiovascular exams after two weeks in orbit. In free fall even the distribution of blood in the body changes. Blood is not pulled downward into the lower limbs as it normally is, and the astronauts' faces puff up with unaccustomed body fluids. Moreover, during the first few days of weightlessness, each astronaut lost up to 25% of his red blood cells and 10% of his blood's plasma and, for a few weeks, calcium from the large bones of his legs. After touchdown, the group had trouble walking properly, and each had very erratic blood pressures. (They also complained of things *falling* when they were released in midair.) When the second crew stayed for 59 days, however, their on-board tests revealed that these changes level off in about a month as their bones and muscles adjust to life without weight.

The Soyuz-26 crew, who spent 96 days in earth orbit, took 3 weeks to adapt to normal earth gravity when they returned. Soviet cosmonauts now wear suits with built-in elastic cords that cause them to strain whenever they move about, providing extra exercise. They have also worn other special suits for up to 3 hours a day to put negative pressure on their lower bodies to stimulate blood flow there. After an even longer stay in space, for over 200 days, cosmonauts were taken in wheelchairs from their spacecraft and still had problems with walking a month later.

Doctors find similar types of degenerative changes in patients bedridden for long periods. Although such a patient is not weightless and in fact still feels his or her normal weight, the person in bed does little

**FIGURE 7–16**
Charles Conrad, Jr., gives Paul Weitz a haircut while in earth orbit aboard Skylab. Notice the vacuum cleaner to collect the floating hair. The puffiness around astronaut Weitz' eyes is due to fluid redistribution in the zero-*g* environment.

work against gravity. The pressures on the body are at a minimum, too, because the patient rests on larger surface areas. The heart does less work pumping blood through a horizontal circulatory system, and skeletal bones experience much less stress and strain. When a person is lying quietly in bed, about the only work the body does against gravity is breathing. To that degree, then, the effects of such inactivity compare to those of weightlessness. Each time you climb out of bed in the morning, you begin a battle vital to the performance of a healthy body.

## Tides

On a pleasant summer day over 100 years ago, a survey ship was anchored at the mouth of a fiord about 60 miles southeast of Juneau, Alaska. An adventurous young seaman named Ford decided to row in with a small boat. The fiord was (and is) T-shaped, created by the action of two glaciers that came together head-on, twisted around, and pushed out to the sea. These giant bulldozers of ice dug deep at the back of the fiord, but they left the narrow exit to the ocean very shallow, only 10 to 20 feet deep. As luck would have it, the tide was neither rising nor falling when Ford made his entrance. But luck, so the story goes, wasn't with him for long. Young Ford was inside for only a short while before the tide began to fall, and the extra water in the inlet began to withdraw. The waters from the fiord backed up at the shallow exit, causing a roaring, 7-foot-high cataract there. Water splashed high, and whirlpools and rapids formed around Ford's wooden dinghy. Icebergs, broken loose from the glaciers, drifted toward the cleft and collided with the sheer rock walls and with each other; deep crunching sounds came through the water. Ford managed to avoid the current, the whirlpools, and the icebergs, and in 6 hours or so the cataract subsided and he returned to the ship. Ford, it seems, was quite shaken, because the surveyors gave the place the name by which it's known today. Rather than Ford's fiord, its called *Ford's Terror*.

TABLE 7–4

**TIDES AT THE BATTERY, NEW YORK, 1982**

| | | |
|---|---|---|
| October 26 | high | 3:11 A.M. |
| | low | 9:36 A.M. |
| | high | 3:22 P.M. |
| | low | 10:12 P.M. |
| October 27 | high | 4:10 A.M. |
| | low | 10:28 A.M. |
| | high | 4:23 P.M. |
| | low | 10:57 P.M. |
| October 28 | high | 5:02 A.M. |
| | low | 11:16 A.M. |

## Tides in the Oceans

Since the dawn of navigation, fishermen and mariners have marked deep channels in ports and rivers to avoid sandbars and mud flats at low tide. For centuries large ports have kept tidal records, both the times and levels of the high and the low tides. Table 7–4 lists the times of the high and low marks at the Battery in New York City for several days in 1982. It reveals a general feature about tides: There are usually two high tides and two low tides each day; the high tides are about 12 hours and 25 minutes apart, as are the low tides. This worldwide regularity was the key to discovering the cause of the tides. While 24 hours is the period of the earth's rotation with respect to the sun (sunrise to sunrise, say), 24 hours and 50 minutes is earth's rotation period with respect to the moon (or moonrise to moonrise). The moon pulls on all of earth's matter with the pull of gravity. But the response of earth's fluid oceans to this pull is somewhat different from that of the solid earth, which (for now) we'll assume is rigid and moves with an *average* acceleration toward the moon. As we'll see in the next section, *the change in strength and direction of the moon's gravitational force from one place to another over the earth causes the tides.*

## How the Moon's Pull Causes the Tides

If a straight line passed through the centers of the earth and the moon, it would intersect the earth's surface at two points. Any matter on the surface of earth near the point that's closest to the moon gets a *larger* pull from the moon than the earth does on the average. If earth's gravity didn't tug on the people, sticks and stones, and ocean waters in that area, they would out-accelerate the earth toward the moon. Earth's pull is much larger than the moon's, of course. The ocean waters and the people in that region only press a tiny fraction less against earth's surface because of the moon's opposing pull. Yet, as we'll see shortly, that small difference acts to create a small bulge in the ocean waters

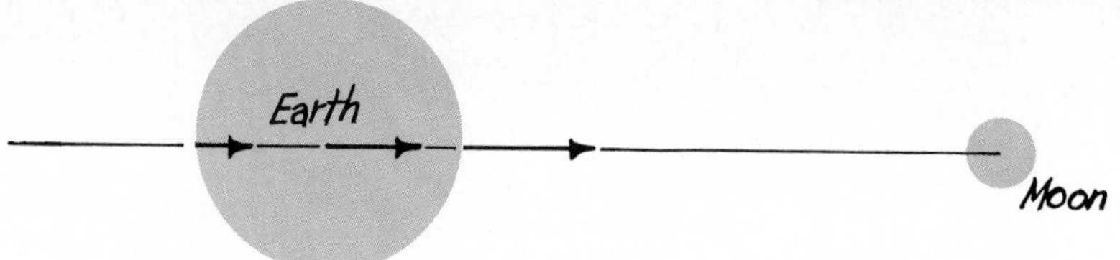


**FIGURE 7–17**
A comparison of the relative strengths of the moon's gravitational pull at three points, the closest and farthest points to the moon from earth's surface and at the center of the earth. (The moon is actually much farther away from the earth than shown here, and the forces are more nearly the same at the three points.)

that pass through that area as the earth rotates. *That bulge is the cause of one of the daily high tides on the world's beaches*.

Another bulge in the ocean occurs at the far side of the earth from the moon. *That causes a second high tide during the daily rotation of a beach or coastline about the earth's axis*. Things near that point get a *smaller* pull from the moon's gravity than the earth does. If the earth's pull on them vanished, the earth would accelerate toward the moon faster than the pencils, people, and ocean water there. You are a tiny fraction lighter against earth's surface when the moon is on the other side of the earth, just as you are noticeably lighter if an elevator floor you are standing on suddenly accelerates downward. You don't leave the floor of the elevator because earth's gravity is strong enough to keep you snug against it even as it accelerates downward. Likewise, neither you nor the ocean is left behind by the earth because earth's own tug on you is far too great. But the oceans do rise in this region as a result of being a trifle lighter there.

Finally, imagine a line on the earth's surface that circles both the center of the earth *and* the earth–moon line. Figure 7–18 shows the two points on that line that appear in a cross-section view. On that line, and in the region off to each side, the moon's pull actually makes things a tad *heavier* against earth's surface. As the figure shows, the moon's force on anything in this beltlike region has a vector component pointing down toward the center of the earth. (The other component of force is parallel to the earth–moon line and helps keep those objects moving

**FIGURE 7–18**
At points on earth's surface that are perpendicular to the earth-moon line, the moon's gravitational pull has a small component that presses matter toward earth's center, making things slightly heavier there.

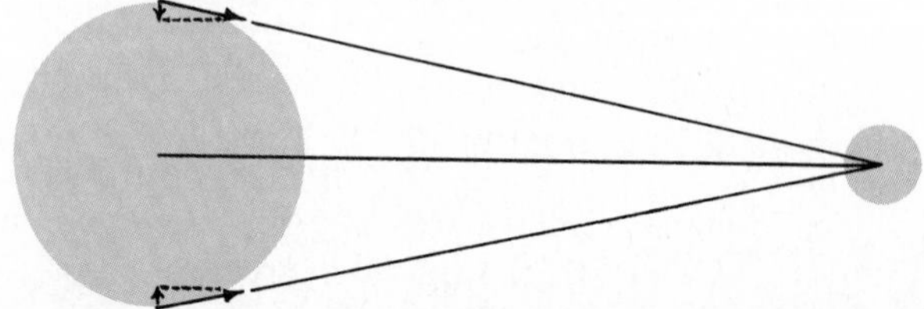

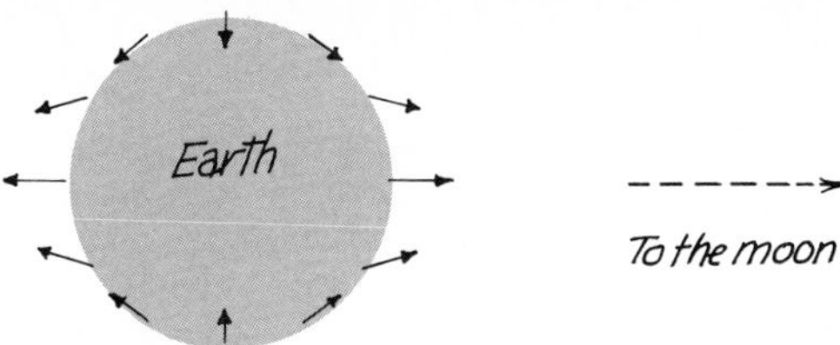


**FIGURE 7–19**
If the vector forces at earth's surface due to the moon's gravity are drawn and the pull of the moon's gravity at earth's *center* is subtracted from each, the result is the action of the moon's gravity with respect to the earth itself. That force is called the tidal force. It tends to cause a bulge (high tide) in the oceans under the moon and on the opposite side from the moon, and low tide regions in between.

toward the moon at the same rate as the earth.) *This squeezing action depresses the oceans by a small amount as they pass through that belt twice each day, causing low tides.* Once again, when you are there you don't feel heavier, but the ocean level goes down. In the tidal bulge regions you don't feel lighter, yet the oceans rise. Why?

All the oceans over the entire surface of the earth are connected. As a consequence, if the surface level of water goes down in the slightest degree somewhere, it has to rise somewhere else. Even as the tidal force squeezes the global water in one region, it lightens it in another. At any point, the tidal force is tiny, but it acts over all the enormous ocean (Fig. 7–19). The ocean waters move together, much as a gelatin salad does when a spoon touches it.

We can see why there's no obvious tidal effect in a pond and especially in a person's body. Top, bottom, center, and sides of a pond or a person get almost exactly the same gravitational pull from the moon. There isn't much of a difference in the force from point to point, and hence, little of that squeezing or stretching effect. But whenever a gravitational field changes appreciably in size and/or direction across the dimensions of a body, there will be a tidal effect. Tidal forces are *everywhere,* but for the most part they are too tiny to matter.

Far from any continent, the tidal rise and fall of the oceans because of the moon's pull amounts to a net change of a little more than 1½ feet. As we'll see in the next section, the sun also exerts tides on the oceans, and when the high tides from the sun and the moon are together, the midocean level rises and falls through a distance of about 2½ feet.

## Tides from the Sun

The sun's gravity also raises tides on earth's oceans, but they are about half the size of the moon tides. Yet if you use the law of gravity to calculate the attractions the moon and the sun have for the earth, you'll find that the sun's gravity pulls on earth with a strength almost 200 times greater than the moon's does. Why, then, does the moon have more tidal influence on the oceans than the sun? As we've seen, tides come from the *difference in gravitational force* from point to point over the whole body. While the moon's pull differs by some 6.7% across the earth, the sun's difference in pull is only about 0.017%. It is only because the sun's tug is almost 200 times larger than the moon's that the sun tides are even half as high ($200 \times 0.017\% = 3.4$ percent, or about half of $1 \times 6.7$ percent).

Because the earth turns beneath the sun once every 24 hours, the sun-caused tidal bulges and depressions show up on the beaches within a period of 24 hours. About twice a month, when the moon is in its *new* and *full* phases in the sky, the moon and sun align so their tide-producing forces add, giving higher than average high tides and lower than average low tides. These are the **spring tides.** (See Fig. 7–20.) When the moon and the sun are about 90° apart in our sky, the moon's tidal bulge area coincides with the sun's depression area, and vice versa. Then the forces subtract, giving lower than usual high tides and higher than usual low tides. These twice monthly tides are the **neap tides.**

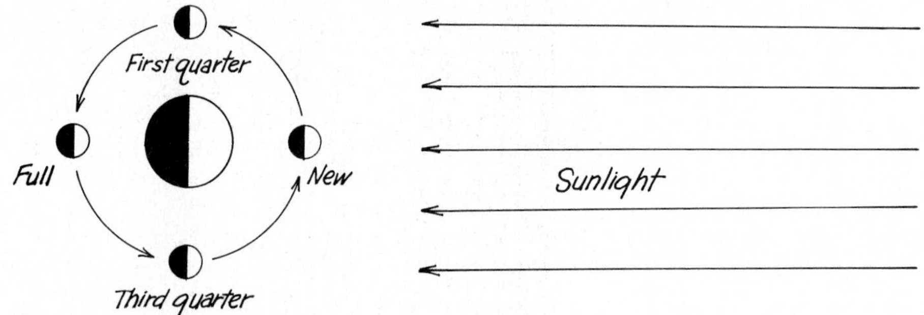


**FIGURE 7–20**
The tides from the moon and sun add when the moon is full or new, and they subtract when the moon is in the first quarter or third quarter phase.

## Tidal Bores

Coastline indentations, such as bays and rivers, can funnel the rising tidal waters and bring about a great horizontal current and a higher rise in the local water level. A noteworthy example are the tides in the Bay of Fundy in Nova Scotia, where the water sometimes rises 50 feet! Occasionally the tidal waters sweep into this bay (as they do in certain other estuaries, like the mouth of the Amazon river) with such swiftness that a wall of water builds and moves upstream *over the surface of the water coming downstream*. These walls of water are called **tidal bores,** or **bore waves.**

In G. H. Darwin's book *The Tides* (1898), he recounted an incident with bore waves recorded by the British Admiralty. Late one morning in 1888, three British survey ships were pushing up the Tsien-Tang-Kiang, a river that empties into the Sea of China. With no warning, a bore wave came upstream and caught the ships. They turned about to try to hold their positions, but even with their anchors out and their steam engines chugging full speed, they were swept upstream. Aware that the next tide could bring another bore wave, they secured the ships against it. That night was calm and quiet, but the tides were at work. Just before midnight the crews heard the bore wave coming up the river "with a roar." Despite their precautions, the steamships were soon overpowered by the current; one was grounded and damaged, the other two (once again running their engines on full and dragging their anchors) were helplessly pushed *3 miles* upriver.

**FIGURE 7–21**
A bore wave is a wave that rushes *over* standing or moving water. This spectacular bore wave occurred not because of the tides but because of an earthquake that shifted a huge volume of water.

## Tidal Strengths: Roche's Limit

Ocean tides are well known to beachgoers, especially when a neglected float or towel washes away in the rising surf. For all their visibility, however, the ocean tides are a minor result of the moon's attraction for earth. In the next section we'll see that the moon's gravity also causes tides in the solid earth, raising and depressing its entire surface twice a day, though this effect is unnoticed by those beachgoers and almost everyone else. It's important to realize that the tidal action of a gravitational field acts on *anything* in its vicinity, whether solid, liquid, or gas. Every satellite that orbits a planet receives a tidal force from that planet's gravity, just as the planet receives a tidal force from the orbiting body. With an eye toward exploring these less familiar tides, we need to know more about the strength of the tidal force.

*For any massive object, the strongest tides it can exert are found in the closest regions around it.* Figure 7–22 plots gravity's strength versus distance, and you can see that the closer something is to the attracting body, the greater the *difference* in gravity's pull across its surface. That difference in force causes tidal effects.

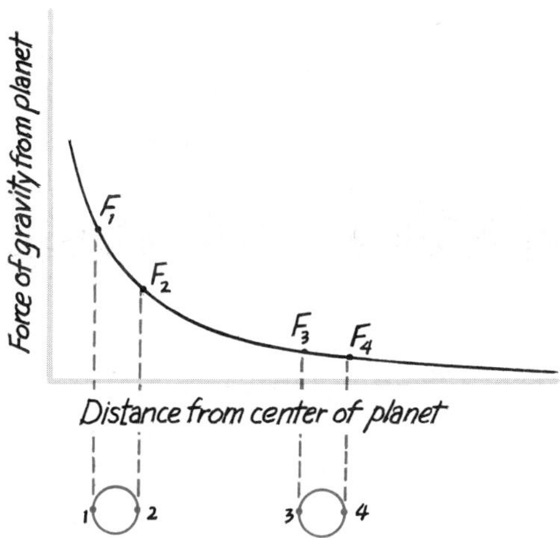


**FIGURE 7–22**
The farther an object is from an attracting mass, the smaller the percentage difference in the gravitational force from one side of the body to the other. So the tidal force diminishes if the body is farther away and increases if the body comes closer.

Also, the larger the mass, the larger its tidal force at any given distance. Figure 7–23 compares the gravitational field strengths of earth and Mars. When the mass of the attracting body is larger, the force of gravity changes more rapidly at a given distance. A satellite 10,000 miles from Mars's center would receive a smaller tidal force than if it were 10,000 miles from earth's center.

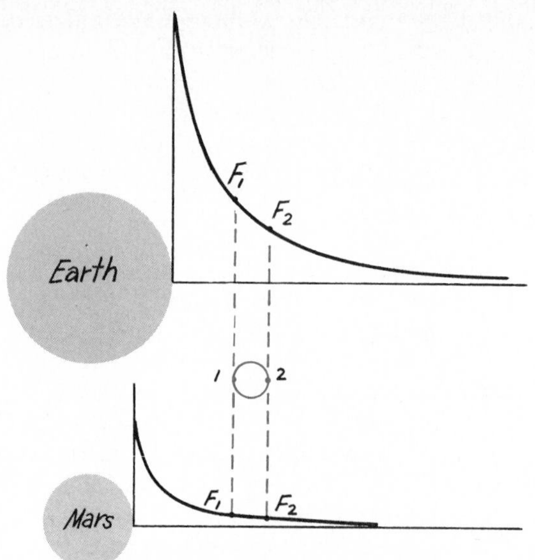


**FIGURE 7–23**
A comparison of the tidal forces due to earth and Mars at an object that is the same distance from each. The curved lines indicate the force of gravity from each planet. As you can see here, the force on the near side, $F_1$, differs from the force on the far side, $F_2$, for both planets. So both planets would exert a tidal force on the object. But $F_1$ is greater than $F_2$ *by a larger percentage* if the body is near the earth, meaning earth exerts a stronger tidal force at the same distance than Mars does.

The beautiful rings of Saturn exist only because of the strong tidal force close to that planet. Myriad small clumps of matter orbit Saturn in flat, disklike regions above its equator (Fig. 7–24). Ordinarily gravity would slowly pull them together to form one or more large satellites, like the 17 (or more) moons in equatorial orbits farther out. But when two particles in Saturn's rings drift together, *the stretching effect of Saturn's tidal force is stronger than their mutual gravitational attraction,* and they promptly separate. So it is for *any* planet in our solar system. Only at distances greater than about 2.5 radii of a planet can matter accumulate under the action of gravity. That distance is called **Roche's limit.** Saturn's rings of matter are all within 2.3 radii of Saturn's center, whereas its nearest *large* satellite (Janus, discovered in

**FIGURE 7–24**
The beautiful rings of Saturn are due to Saturn's tidal force.

1966) is 2.6 radii from the center of the planet. In 1977 astronomers discovered several rings around Uranus, and Voyager 1 in 1979 discovered a ring around Jupiter (Fig. 7–25). These ring systems, too, are within Roche's limit for those planets. But these beautiful and delicate examples of the tidal force pale beside the strength the tides show as they act across the large, solid bodies of the moons and the planets.

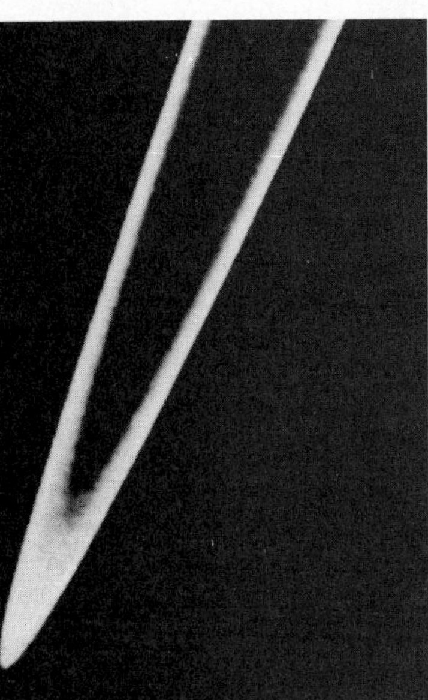

**FIGURE 7–25**
Jupiter's rings were discovered only when a passing Voyager spacecraft photographed them.

## Tides in Solid Bodies

Though the earth is rigid compared to water, the moon's tidal action stretches and squeezes the solid earth throughout its volume, and it yields. Even solid rock is slightly elastic. There is a tide in the solid earth that at the surface amounts to about one foot, high to low. Its strength is impressive: Twice a day the earth tide lifts Mt. Everest up a foot and lets it down again. We don't see the up and down motion of the ground, however, since we all rise and fall with the buildings and lakes and everything else around us. On the other hand, petroleum engineers who monitor pressure in large underground reservoirs of petroleum can watch an effect of these earth tides caused by the moon. The liquid-filled cavity in the rock below them is stretched and squeezed as the tides deform the solid earth, and the pressure rises and falls on their gauges twice each day.

The moon itself is squeezed and stretched by the earth's gravitational field. But there are no daily highs and lows at points on the moon's surface, because the moon doesn't rotate with respect to earth. It keeps the same face toward earth all the time. Nevertheless, it still suffers consequences from earth's strong tidal grip. The moon orbits earth along an ellipse that brings it as near as 226,000 miles and sends it as far away as 252,000 miles. The earth's tidal force on it is strongest at perigee, its closest point of approach. Seismographs left on the moon by astronauts detected swarms of small tremors each time the moon passed through perigee. At apogee, the farthest point in its orbit, those moonquakes are more rare.

Suppose the moon did turn around with respect to earth. (Perhaps it once did!) Calculations show that the tides in the moon's surface would be 25 feet or more from high to low. The tremendous tidal stretching and compressing would surely cause great internal grinding and friction. Heat so generated in that rocky body would eventually escape into space. As this happened, *the moon's rotation would inevitably slow*. The work done in the lifting and squeezing would come from the kinetic energy of the matter's motion around the axis of rotation. Conservation of energy dictates that, as the moon heated, the rate of spin would decrease until that smaller body faced the earth as it does today. Only then would great tides no longer sweep across its surface, changing kinetic energy into heat.

Strong tidal forces from Jupiter and Saturn grip their major satellites, too, none of which rotate with respect to their parent planets. When the Voyager spacecraft (1979) got close looks at Io, Jupiter's nearest large satellite, scientists discovered another tidal effect. Io (pro-

**FIGURE 7–26**
A volcanic plume blossoms in Io's weak gravitational field.

nounced either *eye-oh* or *ee-oh*) is about the size of our moon, and it is very close to Jupiter. Scientists had presumed that Io, like the moon, would be solid throughout, having long ago lost any internal heat left over from its formation. Yet the pictures beamed to earth showed *volcanic plumes*. Matter was spewing from fissures on Io's surface even as the Voyagers passed by (Fig. 7–26). Researchers now think Jupiter's next closet moons, Europa and Ganymede, sometimes disturb Io's almost circular orbit, *making it temporarily more elliptical*. During these periods, Jupiter's tidal force on Io changes, reaching the maximum at the satellite's nearest point of approach and decreasing to the minimum when it's farthest from Jupiter. This is the same kind of action that sets up moonquakes at the moon's perigee. But on Io the internal grinding from these fluctuations produces sufficient heat to melt rock. Hence the volcanic displays.

## CALCULATIONS

To measure the value of $G$ in Newton's law of gravity required measuring the force of gravity between two known masses, a fairly difficult thing to do since the gravitational attraction between everyday objects is so weak. It wasn't accomplished until 1798, when Henry Cavendish succeeded in measuring the force between two sets of lead balls with a delicate balance arrangement. Newton never knew the value of $G$. He derived his many results with this law by comparing forces directly with the surface gravity of earth; that is, he made ratios to eliminate $G$ by division from the comparison formulas.

The currently accepted value for $G$ is $6.67 \times 10^{-11}$ Nm$^2$/kg$^2$.

EXAMPLE 1: **Calculate the value of $g$ at the earth's surface to two significant figures.** The earth's mass is $5.98 \times 10^{24}$ kg, and the earth's average radius $R_{earth}$ is 6370 km.

The acceleration of a mass $m$ due to earth's gravitational force at its surface is found by setting $ma$ (the net force *always* equals $ma$) equal to the force of gravity:

$$ma = G\frac{mM_{earth}}{R_{earth}^2}$$

or

$$a = g = G\frac{M_{earth}}{R_{earth}^2}$$

Thus,

$$g = (6.67 \times 10^{-11}\ \text{N}\cdot\text{m}^2/\text{kg}^2)\frac{5.98 \times 10^{24}\ \text{kg}}{(6370 \times 10^3\ \text{m})^2} = \mathbf{9.8\ m/s^2}$$

(Cavendish used the formula $g = G(M_{\text{earth}}/R^2_{\text{earth}})$ to find the earth's mass. He presented his findings in a scientific paper entitled "Weighing the Earth.")

---

EXAMPLE 2: Deimos completes its nearly circular orbit around Mars in 1.262 earth days, and its distance from Mars's center, $r$, is $23.5 \times 10^6$ m. (These are numbers astronomers can measure by watching the moon with telescopes.) **Use these facts to calculate Mars's mass.**

In a circular orbit the centripetal force on the moon (mass $m$) is equal to the gravitational pull of the planet. So

$$\frac{mv^2}{r} = G\frac{mM_{\text{Mars}}}{r^2}, \quad \text{or} \quad M_{\text{Mars}} = \frac{v^2r}{G}$$

The satellite's speed is $2\pi r/T$, where $T$ is the period of its orbit. Now $G = 6.67 \times 10^{-11}\ \text{N}\cdot\text{m}^2/\text{kg}^2$, and remembering that a newton is $1\ \text{kg}\cdot\text{m/s}^2$, we see $T$ must be expressed in *seconds* (and $r$ in meters) for the units to agree in the equation above. Using $\pi = 3.14$ and doing the multiplication and division, we find

$$M_{\text{Mars}} = \mathbf{6.4 \times 10^{23}\ kg}$$

---

EXAMPLE 3: Show that **the planets closer to the sun must move faster in their orbits than the farther planets.** (Assume the planets move in circular orbits.)

To do this, set the centripetal force equal to the force of gravity from the sun.

$$\frac{mv^2}{r} = G\frac{mM_{\text{sun}}}{r^2}$$

where $r$ is the radius of the planet's orbit and $m$ is the mass of the planet. By solving for $v^2$, you can see how the speed $v$ of a planet is related to its distance $r$ from the sun.

$$v^2 = G\frac{M_{\text{sun}}}{r}$$

Now both $G$ and the mass of the sun are the same for any planet. The distance $r$ can be a very large number, as it is for Pluto, or a much smaller number, as it is for Mercury. This formula tells us $v^2$ will be much *smaller* for Pluto (large $r$) than it is for Mercury (small $r$), and hence Mercury zips around in its orbit by comparison to Pluto. (Mercury moves at 47.8 km/s and Pluto pokes along at 4.7 km/s.) **The nearer a planet is to the sun, the faster it travels in its orbit.**

## REVIEW

**1.** Are these statements true? (a) Gravity is the attraction of mass for all other mass. (b) Gravity acts over long distances, through empty space, and through matter. (c) The force of gravity pulls any two masses directly toward each other.

**2.** What equation gives the *strength* of the gravitational force between two objects.

**3.** Anything on earth's surface is attracted toward its center just as if all earth's mass were concentrated there. For any object, such a point is called the ______________.

**4.** How does the strength of the gravitational field around a mass weaken with distance?

**5.** The gravitational orbit of one body around another has the shape of an ______________.

**6.** The path of a projectile launched from a point near earth's surface at 25,000 mph (the escape velocity from earth) is a parabola. At greater than this escape velocity, the path is a hyperbola. True or false?

**7.** Which property of the moon's gravitational force create tides in earth's oceans?

**8.** Daily there are usually two high tides about ______ hours and ______ minutes apart, and ______ low tides about 12 hours and 25 minutes apart.

**9.** The sun's gravitational force on the earth is about 200 times larger than the moon's, yet it produces tides only half as high. Why?

**10.** What are the spring tides and the neap tides? How often do each occur?

**11.** The strength of the tidal force depends on the distance and mass of the attracting body. True or false?

**12.** Particles orbiting between a planet's surface and Roche's limit for that particle cannot coalesce by their mutual gravitational attractions to form a larger body. True or false?

# DEMONSTRATIONS

**1.** An ellipse is a curve such that every point on the curve is at *the same total distance from two points*. Each of these points is called a *focus* of the ellipse (see Fig. 7–27*a*). An ellipse is easy to draw with a pencil, a piece of string, and two thumbtacks. Try it! When a satellite follows an elliptical orbit caused by the earth's gravity, *the center of the earth is at one focus of the ellipse*.

One measure of the flatness of an ellipse is the **ellipticity,** $(a - b)/a$, where $a$ is the **semimajor axis** and $b$ is the **semiminor axis** (see Fig. 7–27*b*). For a circle the ellipticity is zero, and for a completely flattened ellipse, it is 1.

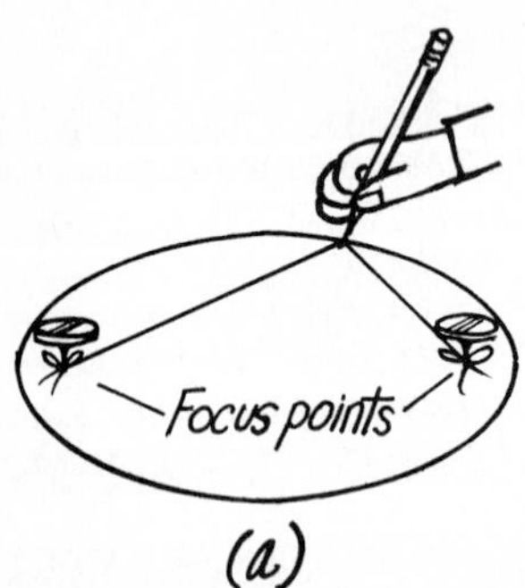


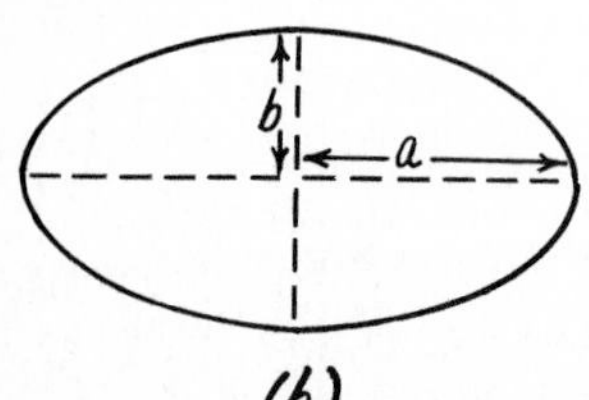


FIGURE 7–27

**2.** Turn on a single bright light (preferably without a shade) in an otherwise dark room, step to the far wall, and hold a phonograph record so that its shadow is projected there. The outline of the shadow of the record disk is an ellipse. If the light comes in perpendicular to the wall and the record is held vertically, the shadow's edge is circular. A circle is an ellipse with no ellipticity, or *zero* ellipticity. (For the definition of ellipticity, see Demonstration 1.) Turn the record at an angle, and the shadow becomes flattened, displaying more ellipticity. When the record is turned edge-on toward the light, its shadow (except for the record's thickness) is a straight line. This is an ellipse of ellipticity equal to 1. If viewed from an angle, any two-dimensional circular shape, such as a basketball hoop, has the shape of an ellipse.

**3.** Most of us are sensitive to rhythmic beats, or sounds repeated at regular intervals of time. It's fun to use this sense to demonstrate that $g$ is a constant for things falling near earth. You'll only need 6 marbles, 10 feet of string, and some tape.

The object is to tape the marbles to the string, hold the string vertically, and then drop it. You'll hear the marbles strike the floor in rapid succession. If the marbles are spaced correctly along the string, they'll hit and click at equal time intervals and make a constant beat that's obvious to your ear. Since $g$ is constant, we can use the formula $d = \frac{1}{2}gt^2$ to tell how far, $d$, a marble will fall in a given time $t$.

For example, if you wish to hear the marbles ⅛ of a second apart, use the following spacings: 3 in. (⅛ s); 12 in. (2/8 s); 27in. (⅜ s); 48 in. (4/8 s); 75 in. (⅝ s); 108 in. (6/8 s). That is, the first marble should be taped 3 in. above one end of the string, the second marble 12 in. above that end of the string, etc. Stand on a chair and hold this string so the bottom of it just touches the floor. Let the marbles fall into an empty metal trash can for emphasis. Then try it another time to see if your ear can tell that the beat is off. To do that, move the top marble along the string so it is in a new position. If your ceiling is too low and you can't try this outside, space the marbles at 6, 24, 54, and 96 in. In a manner of speaking, this experiment lets you listen to the constancy of *g!*

**4.** With a ballpoint pen (the kind with a tiny hole in the barrel), write a line or two on a piece of paper. Next, turn the pen upside down, hold the paper over it, and try to write again. The pen won't write! The ink needs gravity's assistance to flow. Such pens won't work in orbit either, while in weightlessness. The pens used by the Apollo astronauts on their way to the moon and back had ink cartridges pressurized to 3 atmospheres with nitrogen gas to force the ink to flow without the help of gravity.

# EXERCISES

**1.** Discuss this verse in light of the law of gravity: ". . . thou canst not stir a flower without troubling of a star." (Francis Thompson)

**2.** If the sun were twice as massive, would its pull on earth be twice as large? *Would the earth's pull on it then be twice as large?*

**3.** Do you think you would weigh more or less on the top of Mt. Everest than you do at home? Why?

**4.** Is there some point between the moon and earth where the *net* gravitational force on a spacecraft drifting there would be zero? Discuss.

**5.** All mass attracts all other mass. So why aren't you round like the sun and the planets? (Take a peek at Exercise 14.)

**6.** What properties of earth does *g* depend on?

**7.** If its mass remained constant but the earth began to shrink, what would happen to the value of *g*? ($g = GM_{earth}/R^2{}_{earth}$.)

**8.** Earth and the moon exert equal but opposite gravitational pulls on each other. True or false?

**9.** If you know the acceleration of gravity on another planet, you'd only need to multiply that acceleration by your mass to find your weight on that planet. True or false?

**10.** How much gravitational force does your body exert on the earth?

**11.** When a bullet is fired horizontally from a rifle, does it begin to fall as soon as it leaves the barrel? Or does air friction have to slow it so it will fall?

**12.** An object suspended by a cord doesn't usually point *exactly* toward the center of the earth. For example, near the base of the Himalayas, a suspended object is measurably deflected a small amount *toward the mountains.* Why?

**13.** A rock lying on earth has a certain weight. If it were lying on the moon's surface instead, its weight would be different. What *two* factors would change to cause the difference in weight?

**14.** Why are the planets, the sun, and the moon all more or less spherical in shape?

**15.** Calculations show that if you went deep into a mine shaft on the moon or on Mars your weight would *decrease* rather than increase as it would on earth. What does that tell you about these bodies?

**16.** A barrel of oil weighs less than half as much as a barrel of ordinary rocks. What effect might a huge underground oil reservoir have on the value of *g* at the surface above it?

**17.** A 747 airplane travels almost horizontally at 35,000 feet, while above it the space shuttle travels almost horizontally at an elevation of 125 miles. Why do the passengers on the airplane feel their normal weights, yet the shuttle's passengers are "weightless"?

**18.** When a satellite is in a circular orbit about a larger mass, does gravity do any work on the satellite?

**19.** Explain what would happen to the earth if something took away its orbital speed.

**20.** With respect to work done by gravity, discuss the speeds an earth satellite should have at apogee and perigee.

**21.** A simple first aid procedure to use on someone who has fainted is to lower the person's head to the level of their heart. Why should this help?

**22.** The weather pictures you can see daily on the TV news come from two GOES spacecraft (Geostationary Operational Environment Satellites) that are in stationary positions 22,300 miles above earth's equator. Exactly how can they stay in position over the equator day after day? What would happen if they were closer in? Farther out? (It helps to draw a sketch of this.)

**23.** Do the speeds of the planets around the sun depend on their masses? Their distances from the sun? The sun's mass?

**24.** A large asteroid collides with the moon, adding its mass to the moon's. Does earth's force of gravity on the moon increase?

**25.** Why couldn't a satellite have an orbit only 6 miles above our planet? (It could over the moon.)

**26.** Figure 7–28 is a graph of altitude, for the first 14 hours, of the Soyuz and Apollo spacecrafts that docked together in orbit on July 17, 1975. (a) About how many revolutions of the earth did the Soyuz make before the Apollo launch? (b) Describe the orbits these crafts had during the times shown on the graph. (c) At the arrow a rocket changed the Soyuz speed by 3.1 m/s. Was the speed increased or decreased? Was its speed in its new orbit increased or decreased (compared to the old orbit)?

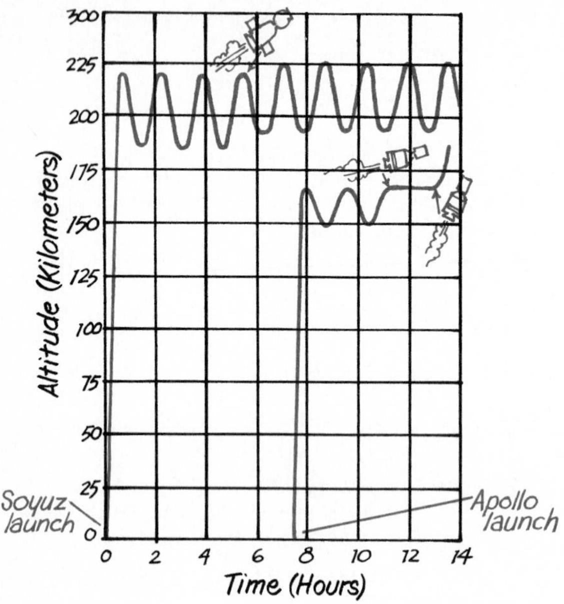


**FIGURE 7–28**

**27.** The earth is closest to the sun in January and farthest from the sun in July. In which month is the earth traveling fastest along its orbit?

**28.** Could an astronaut drop something to earth from an orbiting space shuttle? Discuss.

**29.** Suppose Professor Phate developed an "antigravity" machine that rendered him unaffected by gravitational fields. Consider his fate if he were sitting on the equator when he turned on the machine. Figure 7–29 shows his path and the position of his launching pad at several subsequent times. Describe how his path would look to an accomplice on the ground. (His upward acceleration would initially be only 0.1 ft/s$^2$, so they would have ample time to say goodbye. Farewell, Phate!)

**FIGURE 7–29**

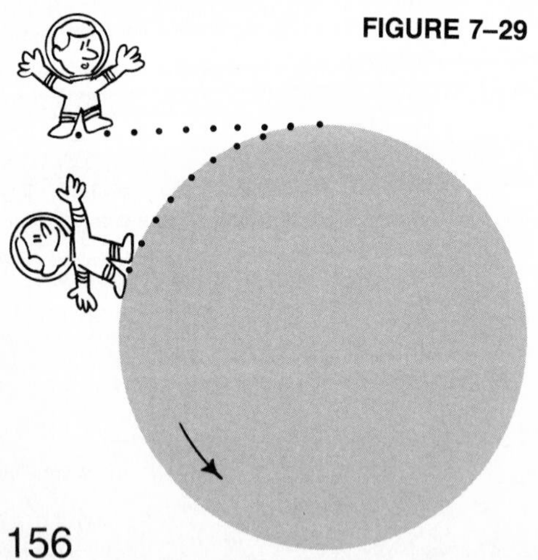

**30.** After his initial mistake, Professor Phate decided that rather than get into space, he wanted to get onto TV! He turned his machine on *partially* so that the gravitational field on him was merely reduced to ⅕ of its normal value. Would this make him as strong as a "bionic" person? Explain.

**31.** Use a ruler to find the ellipticity $(a\text{-}b)/a$ of the ellipse in Fig. 7–27*b*.

**32.** (a) Find the *average* time between the consecutive high tides in Table 7.4. (b) Find the *average* time between consecutive low tides in that table. Do these correspond to an average time of 12 hours and 25 minutes?

**33.** Which positions of the moon relative to earth and sun bring the highest ocean tides? (Refer to Fig 7–20.)

**34.** If a lifeguard at the beach tells you high tide is at 9 A.M., when should you expect the *next* high tide? The next low tide? When will high tide be the next morning?

**35.** Does the strongest tidal force on our bodies come from (a) the moon, (b) the sun, or (c) earth?

**36.** If the moon had its present mass but was somewhat larger or smaller in diameter, would the tidal force on earth change?

**37.** If the solid earth had its present mass but was larger in diameter, would the ocean tides be larger or smaller?

**38.** The earth's atmosphere, like the earth's oceans, responds to the lunar tidal force. Would you expect these tides to be detectable?

**39.** Describe what the ocean tides would be like if the moon did not exist. (They would *not* vanish!)

**40.** Convince yourself that if the ocean waters followed the moon, the ocean would be racing around the earth relative to the continents at hundreds of miles per hour. (*Hint:* Consider the speed due to earth's rotation of a point on the equator.)

**41.** When scientists discovered that moonquakes occurred most frequently at the moon's perigee, they decided to check the dates of recorded earthquakes to see if the moon's tidal force on earth could be involved. They found no correlation between the times of earthquakes and times when the moon was at perigee. Discuss.

**42.** Most of the artificial satellites placed in orbit are much closer than Roche's limit for earth. Why aren't they torn apart by the tidal force?

**43.** The sun is closest to earth in January, farthest in July. When are the sun's tides on earth's oceans largest?

**44.** Would you expect the sun to be at rest in the center of our solar system? Why or why not?

*45. The moon is 384,400 km from earth on the average, and it takes 27.3 days to orbit the earth. Calculate its average speed in its orbit.

*46. If you were in orbit one earth radius above earth's surface, what would be the force of earth's gravity on you? If you were 2 earth radii up?

*47. The acceleration due to gravity at the surface of Mars is $a = GM_{Mars}/R^2_{Mars}$. Use Tables 7.1 and 7.2 to get values for Mars's radius and mass in terms of earth's radius and mass. Use $M_{earth} = 5.98 \times 10^{24}$ $g$ and $R_{earth} = 6.4 \times 10^6$ m to derive $a_{Mars}$. Check your answer with Table 7.3.

*48. In "Calculations" we found the mass of a planet with a satellite in a circular orbit to be $M_{planet} = (v^2r)/G$, where $v$ is the satellite's speed and $r$ is its radius. Use the numbers in Exercise 45 to find the mass of the earth.

*49. Jupiter's moon Europa (one of four visible with binoculars) orbits in essentially a perfect circle. It takes 3.55 days to go around Jupiter once, and it is $671 \times 10^3$ km from Jupiter's center. Use $M_{Jupiter} = (v^2r)/G$ to find Jupiter's mass.

*50. Calculate the force of gravity between two people who weigh 160 lb (72.6 kg) and 120 lb (54.5 kg) if they are standing 10 ft (3.05 m) apart.

*51. Calculate the gravitational force the sun exerts on your body and compare it to your weight. $M_{sun} = 2 \times 10^{30}$ kg; $\bar{d}_{sun\ to\ earth}$ = 93,000,000 mi, or about $1.49 \times 10^8$ km.

*52. If the earth's mass were twice as great and its radius half as large, what would the value of $g$ be on its surface?

*53. Show how fast the earth speeds along as it travels around the sun. (The distance to the sun is $1.49 \times 10^8$ km.)

*54. Show that the sun pulls on the earth with almost 200 times the force of the moon on the earth. $M_{moon} = 7.35 \times 10^{22}$ kg, $d_{moon} = 384 \times 10^3$ km; $M_{sun} = 2 \times 10^{30}$ kg, $d_{sun} = 1.49 \times 10^8$ km.

**55. Suppose on earth you can jump 2 feet high from a flat-footed stance. How high would the same amount of energy carry you on Mars? (*Hint:* The kinetic energy you can give yourself stays about the same, equal to your maximum muscular force for that activity through the same distance. It becomes, at the top of your trajectory, equal to $mg_{Mars}h$.)

# CHAPTER 8 ROTATIONAL MOTION

The planet we live on rotates, a fact everyone learns in grade school. The earth's rotation whirls us around, and for about half of each rotation, we pass "under" the sun. From our self-centered view, we say the sun rises and sets each day. At sunset we enter the earth's own shadow, and many of the nighttime stars rise and set as the rotation continues. These motions in the sky are the most apparent effect of our daily round trips. Earth takes its time; you can stand on one foot and spin completely around in less than a single second, but your planet takes about 24 hours for one rotation.

Yet turn it does, and because of earth's size, the points on its surface farthest from the axis of rotation move very fast to get around in 24 hours. We think of the 200 plus miles per hour qualifying speeds at the Indianapolis 500 as fast, and they are. But if you are in Boston or Detroit, your speed as you whirl around earth's axis is 770 mph; in New York City, Pittsburgh, or Salt Lake City, your speed is about 789 mph; in Washington, D.C., St. Louis, or Colorado Springs, you're moving at about 810 mph. Seattle travels a little slower, about 700 mph, and Anchorage only 500 mph. Farther south, Los Angeles and Atlanta move at 862 mph, while Dallas and San Diego move at 874 mph. New Orleans and Houston roll on at 900 mph, while Orlando

**FIGURE 8–1**
Rotation is a common form of motion.

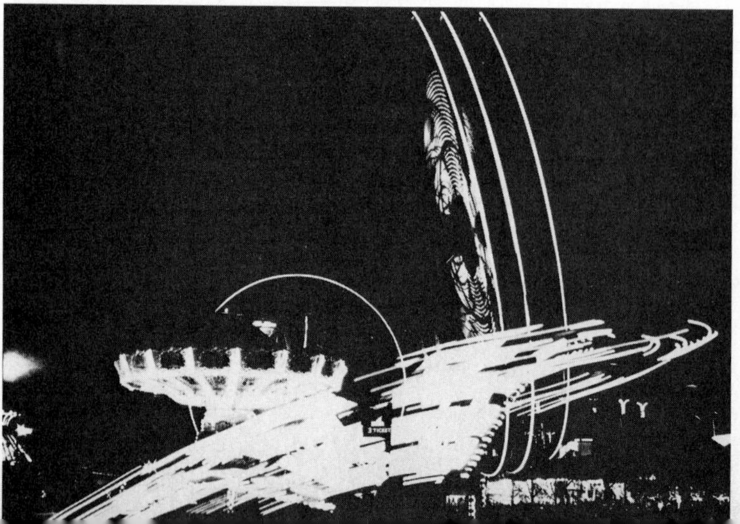

circles at 913 mph and Miami whips round at 936 mph. Luckily for us (wherever we are), the earth's crust is solid, and everything on its surface is more or less anchored by the force of gravity. All the relative velocities would be tough to handle otherwise.

Rotation is a common form of motion. Doors rotate about their hinges, and jar lids rotate to go on or off. And where would we be without the rotating wheels of bicycles and cars? This chapter is about the physics of rotations, how they can change, how to cause them, and how to prevent them. ■

## Rotations and Revolutions

A wheel spinning on its axle and an ice skater doing a pirouette share the same type of rotational motion. Each turns around some **axis** (a straight line about which a body rotates) within its body. This kind of motion is called a **rotation.** We also say a rotating body **spins.** When an object turns about an *external* axis, that's a different type of rotational motion called a **revolution.** A couple on a dance floor who lock arms and swing around revolve about an axis somewhere between the two of them. The earth has both types of rotational motion: it revolves around the sun while it rotates around an axis that passes through its geographical poles.

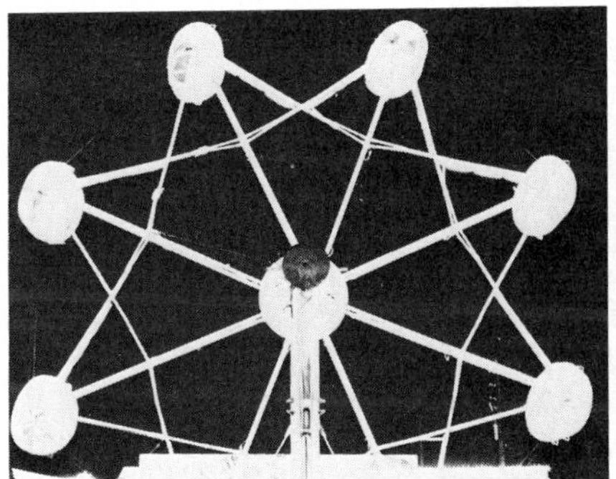

**FIGURE 8–2**
Fair rides have well-defined axes of rotation and (sometimes) revolution.

Any object with rotational motion has a rate of rotation or a rate of revolution. The earth, for example, revolves around the sun *once per year,* which is its rate of revolution. At the same time it rotates around its polar axis at the rate of about *once per day,* its rate of rotation. As a rigid solid object rotates, all its individual particles circle the axis of rotation *in the same amount of time*. The parts farthest from the axis travel faster than the nearer parts, because they move on larger circles. (One look at a record on a spinning turntable will convince you of this.) However, every point shares the same rate of rotation. It is the *number of turns per unit of time* that all parts of a rotating solid have in common. It's true for a wheel or a stereo record, and it's true as well for all the locations on the solid continents of the rotating earth. "Rotations per second'' or ''revolutions per minute'' or any other such units apply to all particles in a rigid solid object.

Another way to describe a rate of spin is to tell what *angle* (or how much of the circular path) all the particles in a rotating solid turn through in a unit of time. This particular rate of rotation is called the **angular speed,** and its symbol is $\boldsymbol{\omega}$ (the Greek letter *omega*).

## The "First Law" of Rotational Motion

Newton's first law of motion (Chapter 3) stated that in the absence of external forces (or a net force) an object's velocity is unchanged. Rotational motion is just motion of another type, where particles are constrained by forces to move in circles. Rotating particles have speed and direction, they have kinetic energy, and they can have potential energy. And they are subject to Newton's laws of motion.

Toss a pen into the air with a twist to set it spinning. While it is in the air, its spin rate stays the same until you catch it. *Once rotating, any rigid object continues to turn at the same rate if it is left alone.* This is in effect the ''first law'' of rotational motion. Only an *external* force, one applied in a definite manner, can change a rigid object's rate of rotation.

**FIGURE 8–3**
If you spin a book with your finger, you'll see that you have a feeling for where the force should be applied to give the book the greatest increase in spin rate.

## Torque: The "Force" of Rotational Motion

Some things begin to spin or change their rates of spin with little effort. A flick of a wrist sends dice skittering along a gameboard. A coin dropped at a vending machine almost always prefers to roll away (usually under the machine). But some rotations you can predict and alter to suit yourself.

A book at rest on a table can show you a lot about simple rotational motion. Press your forefinger on the book's center to hold it and spin it with the other hand. Your finger becomes the axis of rotation. Notice how you get the book to start spinning. Intuitively you push at its corner, a point that's as far as possible from the axis of rotation. It's easier to spin something if the force is applied farther from the axis. Try other points on the book and compare. The direction of your push is just as important as the place it is applied. If you push inward toward the book's center, the book won't turn. Probably you automatically pointed your push perpendicular to a line between your hand and the axis of rotation. (See Fig. 8–3.) Experience has taught you that this is the direction that gives the greatest amount of rotation for your effort. Think of how you open a door, for example.

Anyone who opens a hinged door uses these two principles. Since a door rotates about its hinges, you'd never grab the doorknob and push or pull it *sideways* to get the door to turn. You push perpendicular to a right-angle line between the doorknob and its axis, which passes through the hinges. You can watch the other principle being applied to doors at grocery stores and libraries. With arms full of shopping bags or books, a person might lean against one edge of the door, and

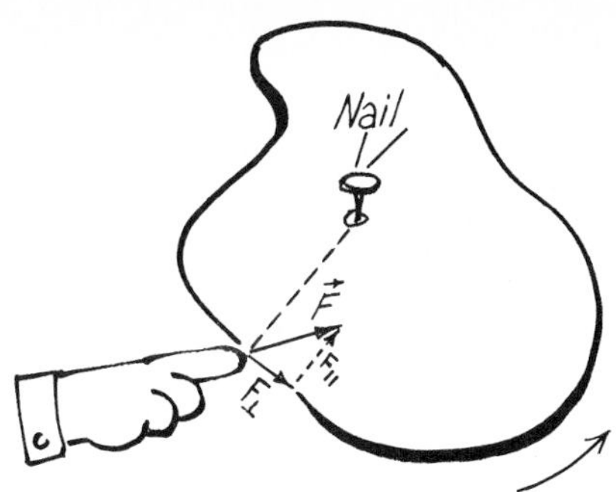


**FIGURE 8–4**
When a force acts to rotate something, only the component that is perpendicular to the line from the axis of rotation to the point of application acts to rotate the object. The product of the distance to the axis along that line and $F_{\perp}$ is called the torque of the force.

then when nothing happens, slide over to the other edge to push the door open. To describe this action, you might say the person needed more leverage to get the door to open. The distance from the axis to the point of contact is called the **lever arm** of the person's perpendicular force. To get the greatest change of spin rate, you want (1) to use the longest lever arm possible, and (2) to apply the largest possible force at a right angle to the lever arm.

If the force is not at a right angle to the lever arm, only the perpendicular component of the force, $F_{\perp}$, will act to change the rate of rotation. Simple experiments show that an increase in the lever arm has the *same* effect on the rotation rate as an increase in the perpendicular force. These two quantities multiplied together are what we call *torque*.

$$\textbf{torque} = \text{force}_{\perp} \times \text{lever arm}$$

The units of torque are foot-pounds and newton-meters, but don't confuse torque with work, which has the same sort of units. With torque, the force doesn't move through the distance of the lever arm; the force is perpendicular to it. (See Fig. 8–4.)

Here's what a torque can do. A torque can set something spinning, as when you push along the top edge of a tire to roll it on the ground. It can also stop that tire's rotation if the push from your hand is in the other direction. That, too, is a change in the rate of rotation. Two such torques can neutralize each other if they have the same magnitude. (Torques can also act at an angle that tends to turn the axis of a rotating body. (See Fig. 8–13). We'll discuss the motion that results from such

## ANOTHER DEFINITION FOR TORQUE

To find the torque, it is sometimes easier to use the entire value of the force, not just its perpendicular component, and a different lever arm in a different location. Some fiddling around with vectors sets up the following alternative definition of torque: *Torque is the product of the applied force (all of it) and a lever arm that is the shortest perpendicular distance from the axis of rotation to the line along which the force acts.* Figure 8–5 illustrates the whole force and a lever arm as described above.

**FIGURE 8–5**
Another useful definition of torque is the product of the entire applied force with the *lever arm,* the closest distance of the line of the force itself to the axis of rotation.

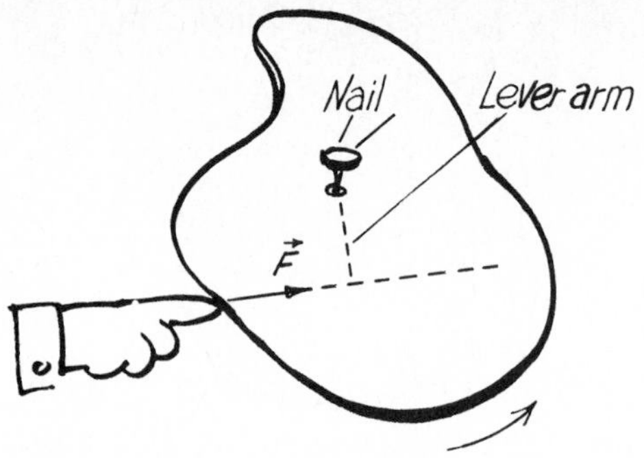


a torque in the section called "Gyroscopes," but for the rest of the chapter we'll only work with torques that tend to change something's rate of rotation.)

Just as the net force determines how fast something's velocity changes, the net *torque* determines how fast something's angular *speed* changes. Let's see how fast a given torque changes a rate of rotation.

## The "Second Law" of Rotational Motion

Every rigid object—book, wheel, or planet—resists a change in its rate of rotation. Only an applied torque can cause a rigid body's spin rate to change. How fast that rate changes for a given amount of torque depends on the object's *rotational inertia*. Inertia, as you'll recall from Chapter 3, is a body's resistance to a change in speed (in straight-line motion). **Rotational inertia** is a body's resistance to a change in *angular* speed, or rate of rotation. The mass $m$ of an object is the measure of its (straight-line) inertia, and it's equal to the ratio of the net force to the rate of change of speed. For rotational motion, the measure of something's rotational inertia is called its **moment of inertia,** and it is equal to the ratio of net torque to the rate of change of spin. So the "second law" of rotational motion is

**net torque = moment of inertia × rate of change of spin**

A spinning planet, we all know, would be most difficult to stop, so it must have a large moment of inertia, no doubt due to its large mass and

### TORQUE, A TWIST

The everyday word for a force is a push or a pull. There's also an everyday word that means the same as "torque." When you loosen or tighten a jar lid, one that rotates on or off, you say you give it a *twist*. That is, you exert a frictional force along the edge, perpendicular to the axis of rotation of the lid. A twist such as this is a torque. The word *twist* is usually applied only to small objects, probably because we associate the word with wrist action, such as "a twist of the wrist." One doesn't give a twist to a door, or a planet.

great size. It's easier to change the spin rates of smaller, less massive things, so their moments of inertia must be correspondingly small. We need to know more about the moment of inertia, and the "second law" will help us explore its properties.

## Two Facts About Rotational Inertia

(a)

(b)

**FIGURE 8–6**
When the cans are close to the axis of rotation, as in *(a)*, it's fairly easy to twist the yardstick. When the cans are moved away from the axis of rotation, as in *(b)*, twisting the yardstick becomes more difficult.

A yardstick (or meter stick), two cans of soup from your kitchen, and some household tape are all you need for a moment of inertia experiment. Leaving room for your hand at the center of the yardstick, tape the cans close to each side of the center point. Then holding the stick at its center, apply some torque by twisting it back and forth. Next, tape the cans to each end of the yardstick and twirl it back and forth again with your hand at the center as before. The mass of the cans and yardstick is the same as before, but most of the mass is farther from the axis of rotation, and the increase in rotational resistance, the moment of inertia, is impressive. Simple experiments show the relationship between the distance $d$ of a mass $m$ from its axis of rotation and its moment of inertia, called $I$:

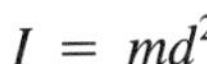

$$I = md^2$$

Each can of soup contributes $md^2$ to the rotational inertia of your homemade dumbbell, where $m$ is the can's mass and $d$ is its distance to the center of the stick. If you neglect the rotational inertia of the yardstick itself, the dumbbell's moment of inertia is $md^2 + md^2 = 2md^2$. If you move those cans, say, 3 times as far from the center of the stick as they were initially, their rotational inertia becomes *9* times as great, whereas if you use cans that are twice as massive, $I$ increases only by a factor of 2 (see Fig. 8–6).

The moment of inertia of a solid object is the sum of $md^2$ for all its particles, where $d$ is taken from the axis of rotation. To find the moment of inertia $I$ for a solid object like a planet or a book, then, you'd need to calculate $md^2$ for each particle in it. This fact leads to another important observation about the moment of inertia.

Unlike mass, rotational inertia can take on different values for the same object. *The moment of inertia of a solid body depends on the location of its axis of rotation*. Another experiment with a yardstick or meterstick will show you that it is easier to change the rate of rotation of a symmetrical object if it spins about its center than if you hold it near one end to turn it. Hold the stick at its center (with or without the cans taped on) and twist, first one way, then the other. Your hand applies torque to set it turning. Move your hand down to one end and do it again. The difference in rotational inertia is easy to feel. Try the same thing with this book. After pressing your finger to the book's center and spinning it, move your finger to one corner and twirl the book about that new axis. It's not as easy to give it a good spin. It takes much more torque to make it spin at the same rate. *The moment of inertia of a uniform geometrical shape spinning about a central axis is less than for any other axis*. That is the axis where the average of $md^2$ for all the particles is smallest.

## A Point of Balance: The Center of Mass

You can easily balance a book horizontally on your finger by finding a point near its middle where it won't rotate around your finger and fall off. This balance point lies under a point in the book called the **center of mass.** For symmetrical shapes, the center of mass is at the geometrical center. Use a piece of tape to hold that same book closed and toss it in the air, giving it a twirl as you release it. Watch carefully and you'll see that the book spins naturally about this balance point. *Any freely rotating object* always spins about its center of mass.

**FIGURE 8–7**
The center of mass of a horseshoe is in the space surrounded by the arch of the rim. The horseshoe rotates about the center of mass as it travels along its trajectory.

We saw in the last section that symmetrical objects like a book rotate most easily about axes through their geometrical centers, where the center of mass lies. That's true even for nonsymmetrical objects. Hold a baseball bat horizontally and find its center of mass. Take the bat there with one hand and twist it back and forth. Next, slide your hand in either direction a few inches and twist again. The bat rotates most easily about its center of mass because its smallest moment of inertia is about that axis. Then put a piece of tape around the bat at its center of mass, take it outside, and have a friend throw it spinning through the air like a baton. You can watch the bat rotate about its center of mass.

The softball player who chokes up on the bat can accelerate it around faster because the point of rotation (roughly the hands) is closer to the center of mass, and this minimizes the rotational inertia.

## Causing and Preventing Rotations: Stability

Take a horizontal ruler and balance it on one finger. Gravity pulls on all its particles. On one side of your finger, torques try to rotate the ruler clockwise, while torques on the other side try to rotate it counter-

**FIGURE 8–8**
This bridge span rests on a greased platform that has a gear on its rim (barely visible in the photo). To turn the bridge 90° (1/4 turn) the operator must walk around a vertical gear axis about 6 times, rotating the axis with the help of the 6-foot bar for leverage. The linked gears act like inclined planes in that they trade distance for force, letting a single person rotate a huge mass.

clockwise. The result is no rotation. See Figure 8–9. We can describe this, however, by saying that for the rigid ruler gravity pulls as if all the ruler's mass were at the center of gravity.* That point is right above your finger, so as the weight of the ruler presses down on your finger, it has no lever arm and doesn't tend to rotate the ruler. But move your finger even 1 centimeter to the side, and the weight, acting at the center of gravity, has a lever arm about your finger and exerts a torque. Your finger becomes an axis of rotation until the ruler falls off.

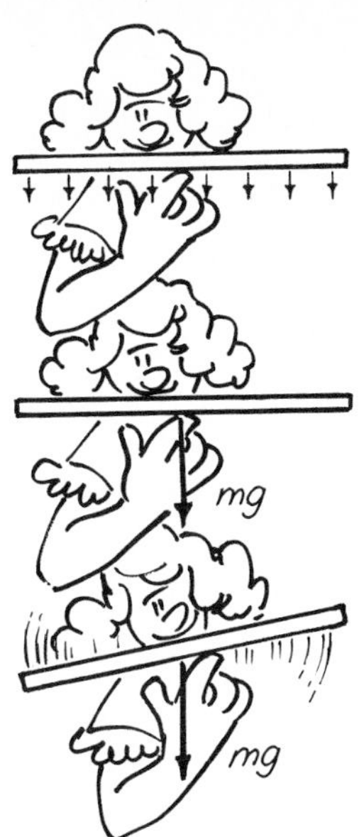


**FIGURE 8–9**
The net effect of gravity on the board is as if the net force acted directly on the center of mass. If the person moves her finger off center, the rod will twist.

Try balancing a pencil vertically with the point on the tip of your forefinger. It's not easy; the smallest motion to the side will put its center of gravity to the side of the pencil's point, and gravity's torque will make it topple. The vertical pencil, we say, is unstable. Its center of gravity easily moves to the side of its base of support, letting gravity exert an unopposed torque. *A solid object is said to be more stable if it is more difficult to get its center of gravity to the side of its base of support where gravity's torque will cause it to rotate.*

Think of tipping over two cement blocks (see Fig. 8–10), one on its side and the other resting on one end. Grabbing the block that's on its side, you begin to lift, and the block pivots about its far edge where it meets the ground. That's its axis of rotation, and you're exerting a torque. But your torque is opposed by another torque which the weight exerts at the center of gravity. Unless your torque is greater, the block won't rotate. As Fig. 8–11 shows, it's easier to tip over the block that's on its end because gravity's lever arm is so much shorter.† The block lying down is more difficult to turn, so it is more stable than the block on its end. For the same reason, a double-decker bus is less stable against tipping over than a low-slung sports car is.

A person who is standing is, like a vertical ruler, not very stable. If you stand with your feet together, your base of support is small, and a slight push can move your center of gravity to one side of your feet. Then if you don't shuffle your feet or grab something to steady yourself, down you go! Subway riders know enough to stand with feet planted

**FIGURE 8–10**
Which is harder to rotate clockwise?

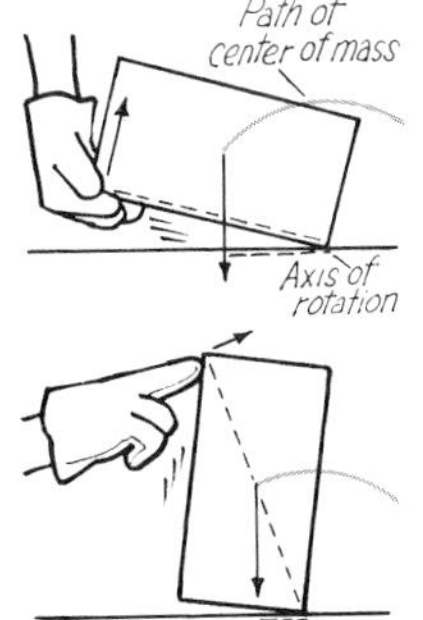


**FIGURE 8–11**
It is easier to rotate the block that is standing, because the force of gravity, which acts at the center of mass, has a smaller lever arm and so exerts a smaller torque to oppose the rotation. Also, as you can see from the paths of the center of mass, the block on its side must have its center of mass raised more directly in order to rotate, meaning more work must be done.

*The center of gravity, as defined in Chapter 7, is for our purposes the same point as the center of mass.

†You can also see it would take more work $W$ to tip over the block that is lying on its side. To raise that block would require raising its center of gravity (or its center of mass) a good deal farther, and the work you do equals $mg \times \bar{d}$, where $\bar{d}$ is the distance you raise the center of mass. That distance is the *average* distance each particle in the block is raised.

**FIGURE 8–12**
Double decker buses cannot turn fast without risking turning over because of their relatively high center of mass.

## AN EVERYDAY EXAMPLE: STOPPING CARS AND BICYCLES

In brisk traffic, it's not uncommon to see a car come to a screeching halt. During such a quick stop, the car takes a nose dive. That is, its front end dips toward the road while the rear actually rises. The car begins a rear-to-front rotation. Torques show what causes this to happen.

When a car stops, the frictional forces on the tires point backward along the ground where the tires touch the road. These forces have lever arms with respect to the car's center of mass. (See Fig. 8–13c.) The torques from the forces on the tires begin to rotate the car. The direction of rotation is the same as you would get by lifting the rear bumper. This lightens the load on the rear tires and increases the weight on the front tires. That very action soon stops the car's rotation. When the front end dips, the car pushes its front tires harder against the ground, and the ground pushes back. The ground pushes back just as hard, and the extra push up on the front wheels gives the car a torque that counters its rotation from the friction's torque on the tires. (See Fig. 8–13d.)

This rotation makes the front brakes do more of the work required to stop a car. The rear tires don't press against the road as much as the front tires and so can't supply as much frictional force. Normally the brake shoes or brake pads on the front wheels wear out sooner than the rear shoes do.

A bicyclist feels this rotational action. Flying along at top speed, a cyclist who brakes hard with the front brake risks somersaulting over the front wheel. That's because as the rotation begins, the front brake can apply even more friction to stop the bike. Applying only the rear brake stops the bike more slowly, but the risk of rotation is less.

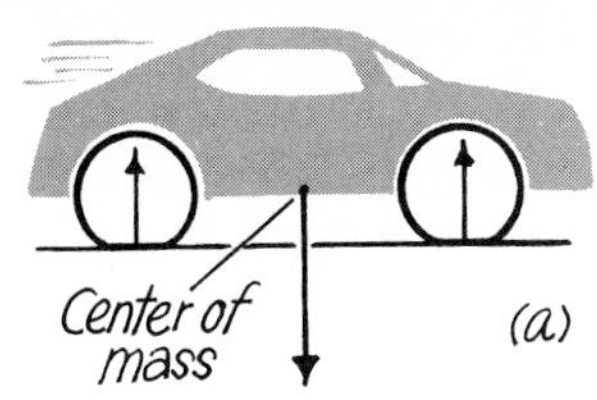


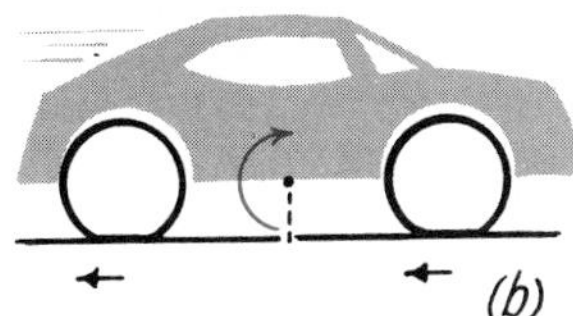


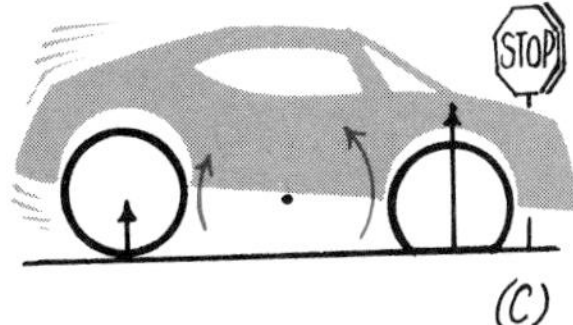


**FIGURE 8–13**
*(a)* The forces on a (symmetrical) car during motion at a constant velocity. *(b)* The frictional forces from the road that stop the car when the brakes are applied tend to rotate the car clockwise about its center of mass. *(c)* The clockwise rotation presses the front wheels harder to the ground while lessening the pressure of the rear tires on the ground. The ground pushes back more on the front tires and less on the rear tires, and the net torques from those forces opposes the clockwise rotation and brings it to a halt.

far apart for stability as the cars sway. Toddlers trying out inexperienced legs wobble as they learn to summon the small torques from back and leg muscles, and feet and toes, that keep pedestrians upright. A person who's had too much alcohol has slower than normal reaction times, and the delayed responses to correct his or her balance lead to "tipsiness." When you hold a heavy bag of groceries in front of you, or a basket of wet clothes, you invariably lean backward to place the new center of gravity (of you plus bag or basket) over your feet. A backpacker, on the other hand, must lean forward to shift the new center of gravity (including the backpack) over the feet. Otherwise, the backpacker will topple over backward.

**FIGURE 8–14**
What happens if Ellen drops her arms? Why? Would Paul then play a part in what happens?

## Angular Momentum

Picture an Olympic ice skater spinning with her arms outstretched. As she brings them in close to her body, her spin rate soars until she is just a blur. This is a new wrinkle—her spin rate changes without torques from external forces. You can get the same effect with a playground merry-go-round. Pack some friends around the edge, start the rotation, and then have all step to the center. The rotational speed of merry-go-round plus friends picks up greatly. These situations are different from any others we've discussed.

Only an external torque can change the spin rate of a rigid body, but the skater's spin rate changed when she changed her shape by bringing her arms in closer. That happened because the mass of her arms moved closer to the axis of rotation. This caused her moment of inertia to decrease, which made her rate of rotation increase. The same thing took place when the people on the merry-go-round moved to the center. The moment of inertia of the system (of merry-go-round and riders) became less as the rate of rotation rose. Measurements of such motion where there are no external torques show the quantities of the moment of inertia and the angular speed change in such a way that their product remains constant. This product is called the *angular momentum.*

$$\textbf{angular momentum} = I \times \omega$$

In the absence of external torques, *the angular momentum of a system is a constant.** This fact is called **the law of conservation of angular momentum.**

The law of conservation of angular momentum is broader than the examples of the merry-go-round and the ice skater show. There are two types of rotational motion, revolutions as well as rotations, and the conservation of angular momentum covers both. For some systems, the total angular momentum is the sum of a *spin* angular momentum and a revolutional (or *orbital*) angular momentum. The earth, for example, has both types. The moment of inertia of the earth about its polar axis times its spin rate is its *spin* angular momentum, while earth's moment of inertia about the sun, $md^2$ (where $d$ is the distance to the sun), times its angular rate of revolution is its *orbital* angular momentum. Later we'll see that sometimes angular momentum of one type may be traded for the other.

## Gyroscopes

Only a torque can change the angular momentum of a rigid spinning object. That's true not just for its *rate* of spin but also for its *direction* of spin—that is, for the direction of the spin's axis. (The force that acts to produce a torque is a vector, and technically speaking, so is a torque. Then, too, as momentum $m\vec{v}$ is a vector, so is angular momentum. However, we won't analyze the vector natures of torques and angular momentum here.)

A child's top demonstrates what happens when a torque acts to change the direction of an axis of spin. Figure 8–15 shows such a top. Gravity's pull on that top acts to topple it over, changing the direction of the top's central axis. Figure 8–16 shows what happens *when the top is spinning*. Gravity's torque doesn't topple the top, it causes it only to swing around, to "circle" the vertical direction, keeping the same angle it begins with. This fascinating motion is called *precession*. If you've never seen this happen, borrow a top and demonstrate it for yourself.

**FIGURE 8–15**
A top that isn't spinning will fall (rotate) to its side because of the torque of gravity.

*Here's why the angular momentum within a system of particles remains constant. Only *internal* forces act, and these occur in equal and opposite pairs at the same points; their lever arms are the same, so their torques cancel. That is a "third law" for rotational motion. Remember that the third law of motion leads to the conservation of the total momentum $m\vec{v}$ for a system where there are no external forces. Likewise, this "third law" of rotational motion leads to the conservation of angular momentum.

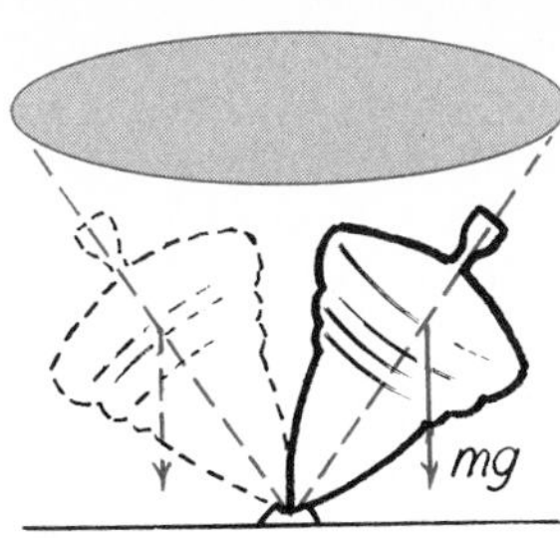


**FIGURE 8–16**
A top that is spinning won't fall over, even though gravity exerts a torque that would ordinarily topple it. Instead, the torque turns the direction of the axis of spin.

This action is why quarterbacks throw spiral passes with footballs. The spin axis of the football keeps the ball traveling through the air in its most streamlined position, making its airborne motion more predictable. Should the axis not be perfectly aligned when the ball is launched by the quarterback, a slight torque from the air will cause only precession of its spin axis. That is, it will wobble, but won't turn end over end and increase its air resistance as a nonrotating football might. This is the same reason a gun barrel contains rifling grooves. The grooves set the bullet spinning, and a spinning bullet travels more truly than one that doesn't spin.

The greater the angular momentum of a spinning object, the harder it is to swing its axis, and the smaller the rate of precession is for a given torque. The angular momentum ($I\omega$) can be large because of a large moment of inertia ($I$), as with the earth, or because of a large angular speed ($\omega$), as is the case with a *gyroscope*. A gyroscope (Fig. 8–17) is a very rapidly spinning toplike object that is mounted on nearly frictionless swivels and bearings. The mount is built to turn in any direction without exerting a torque on the gyroscope. As its mount changes directions, the whirling gyroscope stays pointed almost perfectly in the same direction. A gyroscope is used to keep a predetermined reference direction in spacecraft, guided missiles, airplanes, and submarines, even through the most complicated motions. (The most sensitive gyros are very delicate. When starting their rotations from rest, their spin rates must be increased very slowly. Some missile gyroscopes need half an hour to bring their spin rates to operating levels.)

**FIGURE 8–17**
The mass of the gyroscope is mounted on bearings in an axis that is anchored in a ring. The ring is also movable, rotating on an axis that is perpendicular to the axis of rotation of the gyro mass, meaning the gyro is free to move in virtually any direction.

## Clockwork Earth

Earth's rotation ranks with earth's gravity as an effect we take for granted—the sun always comes up in the morning. With its great rotational inertia, earth turns beneath the sun once every 24 hours in the frictionless vacuum of space. No torques there, so it would seem. But let's not be hasty.

The earth's spin rate *is* changing, in fact; it is slowing down. Studies of ancient fossils reveal that a day was once only 17 hours long. By the time the dinosaurs reigned, it had lengthened to about 23 hours. With telescopes to magnify their view of the sky, astronomers keep track of earth's spin rate by accurately clocking the passage of a given star as the earth spins beneath it. Earth is slowing today at a rate that will cause this century to last about 28 seconds longer (in earth days) than the previous century. But it slows only *on the average*. Sometimes earth's rotation rate quickens for a few months or even longer before it levels off and begins to slow again.

The changing rotation rate of the earth makes keeping time difficult; we want our clocks to read "12 noon" when the sun is highest in the sky. As earth's rotation rate changes, the International Time Bureau in Paris must periodically add or subtract a second to the world's standard times to match the passage of the sun.

The decrease is for the most part caused by a torque that comes indirectly from the moon's gravity. The moon tides on earth's oceans are slowing the earth's rate of rotation. This is what is happening: When the earth's rotation carries a continent into a tidal bulge in the ocean, that land mass gets a backward push from the water. That frictional push has a lever arm equal to the earth's radius, so the tidal torque (from the moon) slows the earth. This means the earth is losing its spin angular momentum. Yet for every action, there is an equal but opposite reaction, even with rotational motion. *The total angular momentum* of the two interacting bodies is *constant*. Therefore, if the moon is slowing the earth's spin and causing it to lose angular momentum, then the *moon must be gaining angular momentum in a like amount*. Here's how that happens.

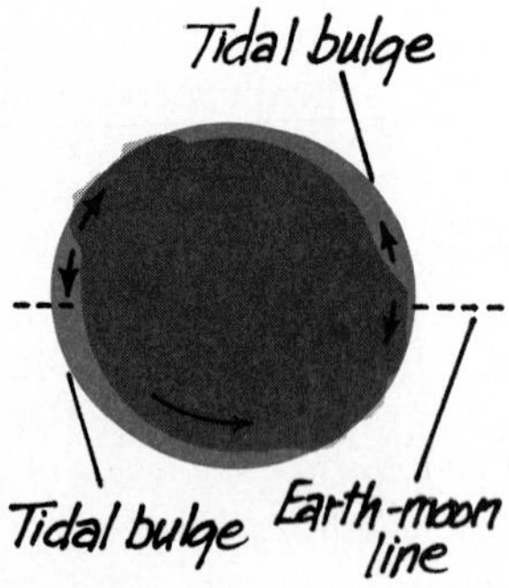

Tidal bulge

Tidal bulge

Earth-moon line

**FIGURE 8–18**
Friction with the spinning earth sweeps the tidal bulges away from the earth-moon line. These bulges exert gravitational pulls on the moon that aren't in line and don't cancel, resulting in a tiny acceleration of the moon along its orbit, which in turn means the moon moves farther out from the earth.

The tidal bulges push against the land masses as they rotate toward the earth–moon line, and *the continents push back on those bulges*. As a result of friction, the bulges are pushed away from the earth–moon line. Specifically, the bulge in the region directly under the moon is pushed *ahead* of the earth–moon line (see Fig. 8–18). That bulge pulls back on the moon. (So does the far bulge, but its pull is less.) Most important, its tug has a component that *speeds up* the moon in its orbit, and the result is that the faster-moving moon drifts slowly outward to orbit earth from farther away. The moon is *gaining* angular momentum (of the revolution variety) as the earth is *losing* angular momentum (of the spin variety).

But why should the earth sometimes speed up? Geophysicists think it is the liquid regions of earth's core that do it. As earth slows, the inner core may not slow as fast, causing friction at the boundary between the liquid and the solid. This friction would tend to speed up the outer solid regions while slowing the faster-rotating liquid a bit. You can see this effect with a fresh egg. Spin it, really get it going, then stop it dead with your fingers, and quickly let the egg go. The yolk inside will still be spinning. It rubs against the shell, and the entire egg is soon spinning again.

A recent study has shown that even unusual weather conditions can temporarily affect the earth's rotation rate. In 1976 the U.S. National Meteorological Center in Washington, DC, began keeping records of the wind speeds world wide. These records let researchers estimate the total angular momentum of the earth's atmosphere. In January 1983, when El Niño (a huge warming of the Pacific Ocean) was at its peak, winds moving from west to east raised the angular momentum of the atmosphere by nearly 10 percent. Scientists involved in measuring the earth's rotation rate noticed a drop in the earth's angular speed during that same period, and the loss of angular momentum by the solid earth matched the gain of angular momentum of the atmosphere. Action-reaction works even with rotational motion.

## Rotating Frames of Reference

From a viewpoint on the ground, motion seems relatively simple. Things at rest remain at rest (if they are on the ground), and a bowling ball can roll straight down the lane with almost constant speed. So ex-

## THE MOST ACCURATE CLOCKS IN THE UNIVERSE

It's out there right now as you are reading this. No matter that it is so far away. The signals it sends are loud and clear to the big-ear radio telescopes, even if the signals *are* 8000 years old when they arrive. Made of matter unlike anything on earth, the source of these signals is one of nature's most awesome creations. It has mass, perhaps more than twice the total mass of our sun and all its planets. But its matter is not the kind of matter that's in the sun. This thing's matter is, well, "compressed." Anyway, that is as close a word as the English language has. An average cubic centimeter of its mass would weigh a billion tons on the surface of earth. And yet this heavyweight monster is only 10 miles or a little more in diameter. There should be an orbiting signpost that reads DO NOT APPROACH, and that signpost's orbit should not be too close. At about 2000 miles from the edge of this thing, the tidal force would discomfort a sign painter. The orbiting sign painter's downward-pointed feet would feel a force equal to their earth weight toward this thing while the head would feel its earth weight in the opposite direction. Much closer and the tides would rip the sign painter apart.

Astronomers call this body a *pulsar,* one of more than 340 known today, and refer to it as PSR 1937+21. The signals it sends are radio waves, a pair of "bursts" for each time it turns around. Most pulsars emit several radio pulses each second, which means they make several rotations during that time. But as of 1983, PSR 1937+21 held the record for pulsar spin rate. A normal household fan on the high setting might spin around 15 times in a second; the propeller of a light airplane rotates about 40 times a second; and here's an object a dozen or so miles in diameter with twice the mass of an entire solar system turning around 642 times in a second!

Astrophysicists think pulsars are neutron stars, collapsed stars whose death is brought about by depletion of nuclear fuels. These giant stars are so massive that their own gravitational attraction causes them to collapse. As they do, the mass rushes in close to the axis of rotation. Very close—it shrinks from a diameter of more than a million miles to a diameter of slightly more than 10 miles. The star matter that was rotating perhaps once in several months crashes to the center; the moment of inertia of the star shrinks, and its rotation rate soars because angular momentum is conserved.

Because of its great mass, the moment of inertia of a neutron star and its angular momentum are still monumental. The signals received from pulsars show that successive spins take almost exactly the same time intervals. Though the pulsars lose energy by sending out radio waves, heat, and light, their rotational energy is so great that the spin rate slows only infinitesimally. The radio waves from PSR 1937+21 show that its signals, and hence its rate of rotation, are slowing down at the rate of about $3.8 \times 10^{-12}$ seconds per year, the lowest rate of change of all pulsars. It puts the rotating earth to shame as a clock; in fact, this pulsar is the most accurate clock in the known universe, more accurate over long periods of time than any present-day atomic clock. At this rate of slowing down, it would lose only 4 seconds after a trillion years of timekeeping.

cept for gravity, which accelerates vertically everything that isn't resting on the ground, objects here seem to obey Newton's first law. If an observer sees Newton's first law of motion hold (in *every* direction), the observer is in an inertial frame of reference.

From the platform of a moving carousel at the fair, however, the world at large looks a lot different. The horizon just beyond the riders seems to whirl around. Children hold on, convinced by the feelings in the pits of their stomachs that if they let go they'll be flung outward into that crazy, spinning world. Motion taking place on a rotating platform such as this doesn't look as simple either. Should someone at the middle of the carousel drop a coin or roll something in the direction of someone on the outer edge, it won't behave on that moving platform according to Newton's first law. That is because the person on a moving carousel is *not* in an inertial frame of reference and does not see the world around from the same perspective. That person sees the surroundings from a *rotating* frame of reference. In other words, as the riders circle, they themselves are accelerated *over the ground*. Therefore, to the observer in a rotating frame of reference, something that moves in a straight line *as seen from the ground* seems to accelerate, changing speed and direction, all because of the observer's own acceleration. To put this another way, a rotating frame of reference is not an inertial frame of reference but an *accelerated* frame of reference.

We'll investigate the curious motions seen and felt in rotating frames of reference because the ground, indeed all the surface of earth, is actually a rotating frame of reference. Earth's rotation is slow, so the effects on local motions (such as for bowling balls) aren't great, but a few are striking.

## Motion in a Rotating Frame of Reference

A phonograph turntable is a rotating platform, a miniature carousel. That's all you need to investigate motion in a rotating frame of reference. First, put an old record on the turntable and gently place a marble anywhere on the record. Switch the turntable on. The marble will roll, scooting across the record until it falls off the edge (Fig. 8–19). As you see this happen, you might say, "Friction sets the marble into motion, and once moving it followed a nearly straight line until it left the record." Try it and watch the motion.

On the other hand, if you could have a viewpoint from somewhere *on* the rotating record (Fig. 8–20), you'd see the marble's motion differently. Suppose the marble was at rest on the spinning platform, perhaps taped down there, and you released it. It would immediately accelerate toward the edge in a line straight away from the axis of

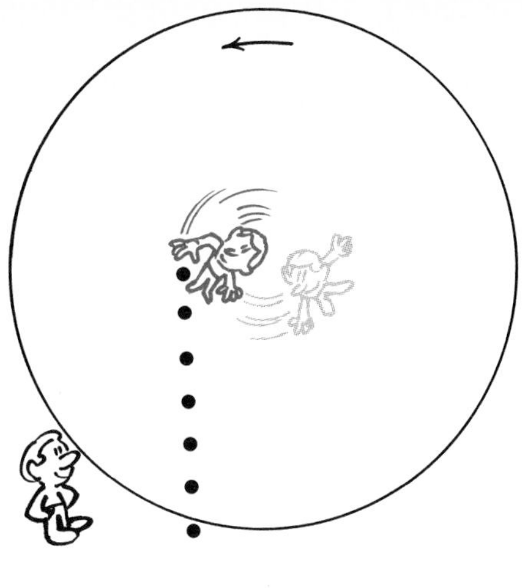

**FIGURE 8–19**
A marble released on a rotating platform by someone on that platform moves in a straight line with constant speed, as shown in the strobelike illustration. Its speed is the same as the speed it had in "orbit" before the passenger aboard the platform released it.

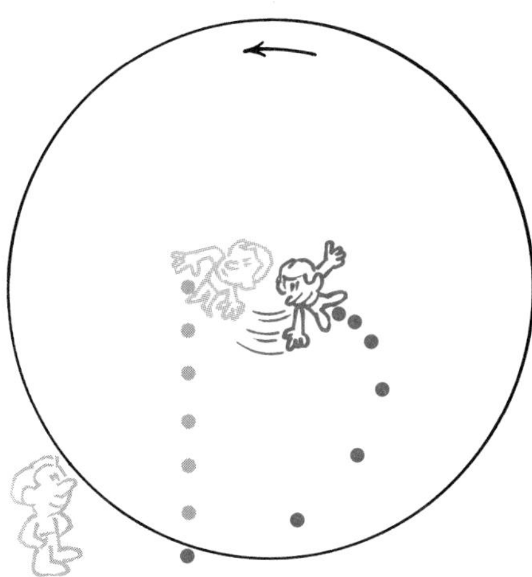

**FIGURE 8–20**
From the point of view of the rotating passenger, however, the marble he released accelerates outward to the edge of the platform, curving as it goes. To him it seems as if there are forces present that act on the moving marble. These forces are the subject of the next two sections.

rotation. Then, as it gained speed, you would see it curve off to the right, accelerating all the way until it finally left the turntable.* From your point of view within the rotating frame of reference, we'll see that you would need two forces to explain the marble's acceleration. The forces that influence the marble's motion are called the *centrifugal* force and the *Coriolis* force. The centrifugal force acts on anything that is within a rotating frame of reference (unless it is located directly on the axis). The Coriolis force acts only when something moves in the rotating frame of reference where it is being viewed.

As we explore the properties of these two special forces in the next few sections, keep in mind that these forces are only seen by observers who are *in* the rotating frame of reference, *because of their own accelerations due to their own rotations*. Someone in an inertial frame of

*If you'd like to see the marble's track, use a piece of cardboard rather than a record, and dip the marble into water with a little food coloring added before you put it on the cardboard. It will leave a trail that's easy to follow.

reference explains the same motions without the use of these forces. From that point of view, these two forces are fictitious. Yet they are real in a rotating frame of reference where an observer there can *measure* them and watch the accelerations they cause.

**FIGURE 8–21**

## Centrifugal Force

In a rotating frame of reference such as a carousel, there is a centrifugal (center-fleeing) force present on everything there. It points straight away from the axis of rotation, and its strength is equal to $mv^2/r$, where $r$ is the distance of the mass $m$ from the central axis. Figure 8–21 shows how this force is viewed both from the rotating frame and from a nearby inertial frame.

If you visit a child's merry-go-round at a playground, you can *feel* this force. Hold onto the safety rail at the edge and have someone spin the platform *fast*. You are definitely thrown outward, straight away from the center. For a given rate of rotation, the centrifugal force is strongest when you are near the edge because your speed is greatest there.

Figure 8–22 pictures a simple demonstration to show how the centrifugal force affects things at different distances from the axis. It's as if a force field points outward from the central axis and grows *stronger* farther from the axis. Someone on the ground would tell you there isn't really a force field—everything on the rotating platform is merely trying to move in a straight line. They see the force you use to stay on the merry-go-round as a centripetal (center-seeking) force that carries you around in a circular path.

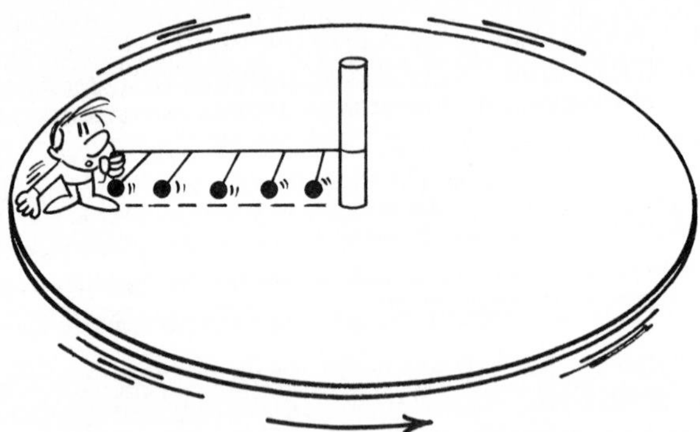

**FIGURE 8–22**
The farther an object is from the center of rotation, the greater the centrifugal force.

## The Coriolis Force

If the centrifugal force were the only force acting in a rotating frame of reference, the marble you release on a rotating record would travel straight out from the axis of rotation. But as seen from the record's rotating frame of reference, the marble veers off to the right once it begins to move. And the faster it moves, the greater its sideways acceleration. Detailed observations would reveal the properties of this second force, the *Coriolis* force.

First of all, as seen by an observer, this force acts only when something moves on the record (Fig. 8–24). Then it acts only to *turn* the moving object. That is, its direction is always perpendicular to the object's velocity vector. This means it cannot do any work ($F_{\parallel} \times$ d) and so cannot change the object's kinetic energy. Furthermore, it acts to turn the object in the opposite sense from the rotation of the frame of reference from which it is viewed. A marble dipped in water leaves a trail that veers to the right, or clockwise, as the record (as seen from above) rotates counterclockwise.

When the marble began to roll on the turning record, the Coriolis force began to act. However, the centrifugal force didn't cease then; it acts whether or not the marble is moving in the rotating frame of reference. (A small frictional force continues to act as well.) As the marble travels faster over the record, the Coriolis force increases in proportion to its speed, and the curve of the marble's path becomes more pronounced. (See Demonstration 7 for an easy way to trace some paths as seen from a rotating frame of reference.)

On a playground's merry-go-round you can feel the Coriolis force. Stand on the edge of the moving platform, bend over, and swing an arm toward and away from the axis of rotation. It's best if you plant your feet apart and concentrate on watching the floor, which is the rotating frame of reference, while you swing your arm back and forth like a pendulum. You'll feel the Coriolis force push your arm first to one side, then to the other. It's even more obvious if you hold a book in that hand.

**FIGURE 8–23**
The centrifugal force keeps these riders pinned to the wall of their ride; they feel as if they are pressed to the wall. Someone outside, however, observes the riders being pushed in a circular path by the wall, which is centripetal force.

**FIGURE 8–24**
The Coriolis force is always perpendicular to the velocity vector.

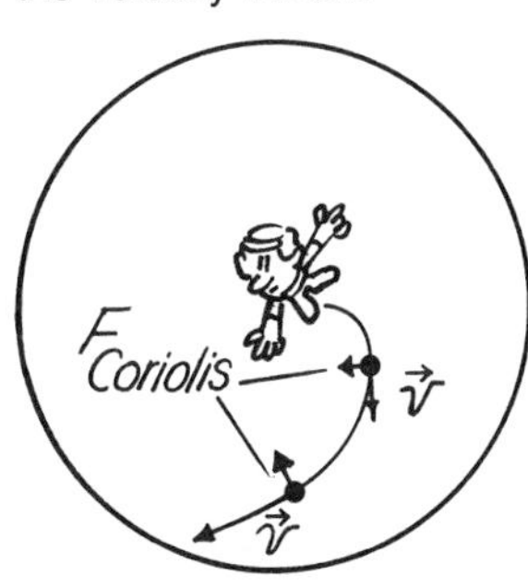


## The Centrifugal and Coriolis Forces on Earth

Because the earth rotates, both centrifugal and Coriolis forces come into play on its surface. The centrifugal force acts on the matter throughout the rigid earth as well. It has its greatest value at points farthest from the axis of rotation, and its direction is always straight out from the earth's axis, perpendicular to it. Consider a mountain on the equator, say in Ecuador, or on the other side, in Sumatra. It is attracted to earth by the force of gravity, *mg*. But because it is in a rotating frame of reference, it is subject to the centrifugal force, $mv^2/r$, and presses down a bit less on the earth's surface because of that force. The mountain (and anything else at the equator) exerts about 0.35 percent less force than its weight against the surface. A person who weighs 140 pounds at the poles, where there's no centrifugal force, will weigh about ½ pound less at the equator because of this force. This effect, acting throughout the earth, causes the earth to bulge at the equator. Because of the centrifugal force, matter is "lighter" at equatorial latitudes and "heavier" at the poles, so the would-be spherical earth is squeezed a little flatter at the poles. The distance through the earth's center at the equator is about 27 miles greater than the straight-line distance between the poles.

The Coriolis force acts only when something moves in the rotating frame of reference, and so no mountains feel its push. Because earth takes about 24 hours for one rotation and because the Coriolis force is small for small speeds, its effect on things we do is usually insignifi-

**FIGURE 8–25**
In moving from New Orleans to St. Louis, an object travels around the earth on a path of smaller radius.

cant. But when motion takes place over many miles or at high speeds, it can make a moving object curve a good deal. Consider the following thought experiment to see why it is so. St. Louis, Missouri, is almost due north of New Orleans, Louisiana. Suppose someone wanted to catapult a message capsule from New Orleans to St. Louis. (Ignore air friction.) Shooting the capsule directly north would *not* work. The projectile would leave New Orleans with a northern component of velocity, but it would also have a certain easterly component of velocity because of New Orleans's rotation about earth's axis. As noted at the beginning of this chapter, this eastward speed is some 900 miles per hour. St. Louis, on the other hand, has an eastward speed of 810 miles per hour. Therefore, as the capsule travels north, *it goes faster toward the east than St. Louis does*. So as it travels, it curves eastward as seen from the slower-moving ground beneath it. This is the Coriolis effect (Fig. 8–25). A capsule similarly launched but in a direction south from St. Louis would curve westward. In either case the object curves to the *right,* or clockwise as seen from above. The Coriolis force affects the paths of missiles, long-distance airplanes, and artillery shells in relation to the surface of the earth rotating beneath them.

In the Southern Hemisphere the Coriolis force is *opposite* in direction to the Northern Hemisphere's. In the middle latitudes there, an object traveling northward in the direction of the equator moves over ground traveling to the *east* faster than it is. As seen from earth's surface, the object is deflected westward, or counterclockwise as seen from above. That is, it curves to the left as it moves, as seen from the ground. In World War I the British ''rediscovered'' that property of the Coriolis force the hard way. Naval ships whose guns were checked for accuracy in the seas near England fought a battle with German ships around the Falkland Islands, far south of the equator. To the dismay of the British, their shells fell consistently to the left of their targets. Their gunsights and gunnery tables were coordinated for a Coriolis force that

**FIGURE 8–26**
A hurricane in August 1980 centered between Cuba and the Yucatan peninsula. Notice the counterclockwise circulation pattern of the clouds.

would deflect their shells to the right, as it does in the Northern Hemisphere. Adjustments were the order of the day!

The direction of the winds in hurricanes and tornadoes shows the influence of the Coriolis force. Winds go from areas of high atmospheric pressure into areas of low pressure. In the middle latitudes of the Northern Hemisphere, winds that come into low pressure areas from every side *swerve to the right*. This causes a counterclockwise circulation round the center of the low-pressure area, as you can see from Fig. 8–26. In a hurricane (or typhoon) the wind circulation is counterclockwise in the Northern Hemisphere and clockwise in the Southern Hemisphere, all because of the Coriolis force.

## CALCULATIONS

Dating back (at least) to the ancient Greeks, a simple balance is one of the oldest devices for comparing weights. A uniform beam suspended (or supported) at its middle has two weight pans, one on each end. With no weights in the pans the beam balances horizontally. If a weight to be measured is placed in one pan, the beam rotates as the weight exerts a torque about the beam's point of suspension. Known weights placed in the other pan can bring the beam back to horizontal again. At that point the torques exerted by the total weight in each pan cancel. Since the lever arms of the weights (their distances to the pivot point) are equal, their weights must be as well because torque = weight × lever arm when they are perpendicular. The symbol for torque is $\tau$ (the Greek letter *tau,* pronounced to rhyme with "now").

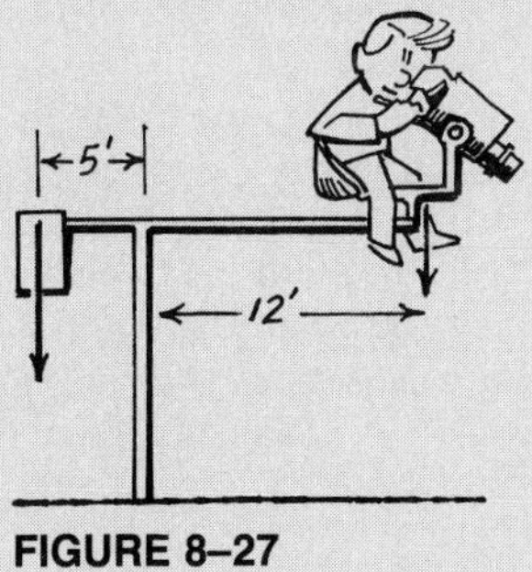

FIGURE 8–27

EXAMPLE 1: The camera operator and camera on the lightweight boom in Fig. 8-27 weigh 500 lb. **How heavy must the counterweight be?** (Ignore the weight of the boom.) The camera plus operator exert a torque that would turn the boom clockwise, while the counterweight's torque would rotate the boom counterclockwise. The torques must balance:

$$\tau(\text{counterweight}) = \tau(\text{camera} + \text{operator})$$

$$mg \times 5\text{ ft} = 500\text{ lb} \times 12\text{ ft} = 6000\text{ ft·lb}$$

so

$$mg = \frac{6000\text{ ft·lb}}{5\text{ ft}} = \mathbf{1200\ lb}$$

A ruler and a table's edge make a simpler balance for lightweight objects. Lay the ruler flat on the table and slowly slide one end off. The ruler will pivot downward just as the halfway point leaves the table's edge. The ruler's center of mass is there, and when it is no longer supported, gravity's force exerts a torque to rotate the ruler. This torque is unopposed, and the ruler then turns about the edge of the table, which becomes its axis of rotation.

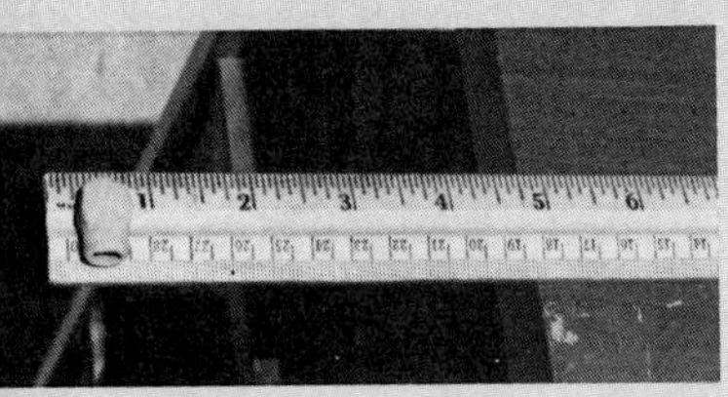

FIGURE 8–28
Measuring the mass of a small object with a ruler.

With the ruler back on the table, put a small object on one end of the ruler and slowly slide that end off the table again (Fig. 8-28). When the ruler begins to teeter-totter on the table's edge, you've found the new balance point. That is where the torque from the object's weight on the end, which acts to rotate the ruler off the table, balanced the torque from the ruler's own weight, acting at the center of the ruler. Measure *both* the distance from the object on the end to the new balance point ($d_{obj}$) and the distance from the ruler's center to the edge of the table ($d_{ruler}$). Since the torques are equal, you know

$$m_{obj} \times g \times d_{obj} = m_{ruler} \times g \times d_{ruler}$$

So if you know either the weight or mass of the ruler, you can find the weight or mass of the object, and vice versa.

---

EXAMPLE 2: A certain ruler is perfectly balanced when its center is 2¼ inches from the table's edge and the center of two quarters stacked at the overhanging end are 3¼ inches from the table's edge. The mass of an ordinary quarter is 5.6 grams (Bicentennial quarters are 5.8 g). **What is the mass of this ruler?**

$$11.2 \text{ g} \times g \times 3\frac{1}{4} \text{ in.} = m_{ruler} \times g \times 2\frac{1}{4} \text{ in.}$$

$$m_{ruler} = \mathbf{16.2\ g}$$

---

We've seen that the moment of inertia is less if an object rotates about its center of mass. For example, tape two cans of food of equal mass $(m)$ on the ends of a meterstick or yardstick. With one hand twist the stick about its center. Then pivot the stick by holding it at one end near one of the cans. Although only half the mass (ignore the mass of the meterstick) is revolving about the new axis, it is still harder to twist the meterstick.

---

EXAMPLE 3: Calculate the moment of inertia $I$ for two cans of the same mass $m$ taped at the ends of a stick (of negligible mass) of length $L$. First **calculate $I$ for a rotation about the *center* of the stick and then calculate $I$ for a rotation about one *end*.**

When a mass $m$ is a distance $d$ away from an axis of rotation, its moment of inertia is $md^2$. The total $I$ for the stick is the sum of the rotational inertias of the two cans. Let $L$ be the length of the stick. Then

$$I_{about\ center} = m\left(\frac{1}{2}L\right)^2 + m\left(\frac{1}{2}L\right)^2 = \frac{1}{2}mL^2$$

When you turn this solid about one end, however, the can at the axis of rotation only spins, contributing almost nothing to the rotational inertia (its distance $d$ from the axis is zero, so it has no "revolutional" angular momentum). So the rotational inertia $I$ for the rigid body about one end (neglecting the mass of the stick again) comes from the can at the other end. That is,

$$I_{about\ end} = md^2 = mL^2$$

or *twice* the rotational inertia that it has about its center of mass.

EXAMPLE 4: Two skaters at a skating rink hold hands at arms' length and revolve around each other. Each has the same mass *(m)*, they are 3 feet apart, and they whirl around once each second. Then they pull together until they are only 2 feet apart (center to center). **What is their new rate of spin?**

Conservation of angular momentum tells us that

$$I_i \omega_i = I_f \omega_f$$

Their initial rotational inertia about their axis of rotation (the point midway between them, the center of mass of their system) is $m(3/2 \text{ ft})^2 + m(3/2 \text{ ft})^2$. Their initial spin rate $\omega_i$ is 1 turn/second.* Their final moment of inertia is $m(2/2 \text{ ft})^2 + m(2/2 \text{ ft})^2$. So

$$\left(\frac{9}{2} m\right) \text{ft}^2 \times 1 \text{ turn/second} = (2m) \text{ ft}^2 \times \omega_f$$

or $\omega_f = 9/4$ turns/second = **2.25 turns/second.**

# REVIEW

**1.** As a solid object rotates, do all of its individual particles circle the axis of rotation in the same amount of time? Do they all have the same speed?

**2.** Define (a) axis of rotation, (b) rotation and revolution, (c) angular speed.

**3.** The longest possible lever arm and the largest possible force produce the fastest change in rate of rotation. True or false?

**4.** The product of the perpendicular component of force and a lever arm is ________. Its units are ________.

**5.** A body's resistance to a change in angular speed is its rotational inertia. True or false?

**6.** The *measure* of rotational inertia is called the ____________.

**7.** The moment of inertia of a rotating object depends on how its mass is distributed about the axis of rotation. True or false?

**8.** Is an object considered to be unstable if its center of gravity moves easily to the side of its base of support?

**9.** The angular momentum of a rotating object is a product of what two quantities?

**10.** State the law of conservation of angular momentum. Give an example of its application.

**11.** Compare an inertial frame of reference to a rotating frame of reference in terms of Newton's first law.

**12.** Identify and define the two special forces that act only in a rotating frame of reference.

# DEMONSTRATIONS

**1.** *Watch* the earth rotate! (Indirectly, of course.) On a bright, sunny day notice the tip of the shadow of a tall flagpole. Or you could watch a slit of sunlight that comes in a window at an angle to strike the floor or a wall. Pay close attention, and you'll see the shadow or sunlight move slowly as the earth turns. (A diagram you can draw should convince you that the tip of a flagpole's shadow moves faster in the early morning or late afternoon than it does at lunchtime, although you can see it move even then.)

**2.** Here's an old barroom trick, but it works almost anywhere. Hold a penny or a dime as shown in Fig. 8–29. Slowly decrease your fingers' pressure until the coin falls. It's almost impossible to keep that coin from rotating a

*Since $\omega$ appears on both sides of the equation, it may be measured in any units we want, turns/second or degrees/second, and so on. In some other equations of rotational motion, however, $\omega$'s units aren't arbitrary. (While you won't meet these equations in this text, you will if you take a higher-level physics course.)

little. Why? The chances of both fingers releasing it simultaneously are almost zero, which sets up this scene. The last finger touching the coin acts for a brief interval as an axis of rotation for the coin. Gravity acting at the coin's center of mass exerts a torque that makes it slowly rotate about this axis as it falls. Once free, the rotating coin continues to rotate about its center of mass. Try it when you are with some friends. Then when you have their attention, do a sequel to this trick. Twist the coin around as you drop it, keeping it horizontal as you do. Then it is relatively easy because of its angular momentum to get the coin to remain horizontal as it falls.

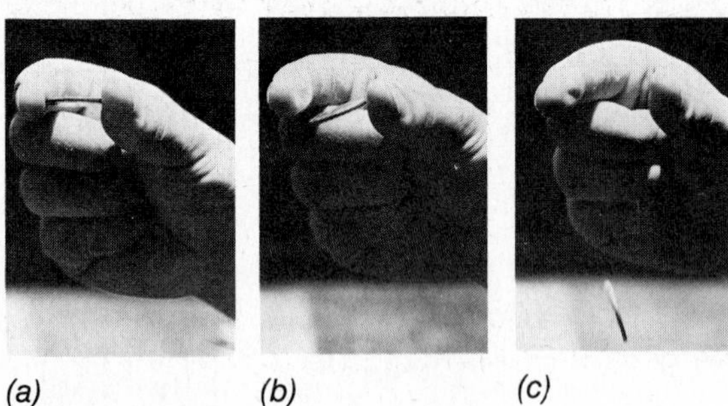

*(a)* *(b)* *(c)*

**FIGURE 8–29**
*(a)* Holding the coin horizontally and . . .
*(b)* . . . releasing it. *(c)* A typical result.

**3.** You may have seen a circus performer who balances rotating dishes on a very long, flexible pole. This demonstration is similar. About one-fourth of the way from one end of a meterstick or yardstick, tape a can of beans (or whatever). Balance the meterstick vertically on one finger, first on one end, then on the other. You'll find it's much easier to do when the can is farther from your hand. Its rotational inertia about your finger is much greater, and so the torque from gravity can't change its spin rate very fast. It is slower to swivel around your finger. Since the rotation takes longer, you have more time to move your finger around and keep the stick vertical.

**4.** Pin a sheet of paper against the top of your desk with a pencil eraser, pressing lightly at a right angle to the desktop. No matter where you place the pencil's end, you can spin this sheet, and the pencil shows you the axis of rotation of the paper; all the molecules in the sheet travel in circles whose planes are perpendicular to this axis. A flick of the finger applies torque and sets it whirling.

Move the pencil point to different locations to make the paper rotate about different axes. A moment's experimenting will show you that an applied torque sets the paper spinning most rapidly when the pencil is at its geometrical center. There is less resistance there to a change in rate of spin; in other words, there is less rotational inertia.

**5.** You can find the center of mass of a solid by hanging it from a string. Cut any shape from a piece of heavy cardboard and tape one end of a string to any point along its perimeter. Then hang it by the string and let it come to rest. Its center of mass must lie directly under the string, or else gravity will exert a torque to cause it to swing. Draw a line straight down from where the string is attached. Hang the shape from another point and draw another vertical line. The point where the two lines intersect is the center of mass. Make a dark mark there and toss the cardboard shape upward. You'll see that it turns about the center of mass you found.

**6.** Add yourself to the list of thousands who have tried this ancient and amusing demonstration. Stand with your back to a wall, heels pressed against the baseboard. Bend over slowly as if to touch your toes. As your center of mass moves in front of your feet, you will rotate forward (probably more than you bargained for). If you don't believe your center of mass moves, even to points outside your body, cut out a cardboard shape similar to someone bending over and try Demonstration 5 with it.

Next stand with one side against the wall, and press the side of your foot against the baseboard. Lift the other foot up an inch or two; your rotation is inevitable.

**7.** This vivid demonstration gives you superb pictures of how uniform straight-line motion in an inertial frame of reference appears when it is seen from a rotating frame of reference. Punch a small hole in the center of a sheet of paper and fit the paper over a record on a turntable. Prop a ruler horizontally about an inch above the rotating paper. Rest a felt-tipped pen against the ruler at a point near the edge of the paper, and let it lightly touch the paper for one rotation. The pen will trace a circle, which is the shape of the path of a stationary object as seen from the point of view of the *rotating* frame of reference. (For example, a child on a merry-go-round sees its parents, waiting at the edge of the ride, going around in a circle at the edge of the platform.)

Next, slide the pen along the ruler's edge at a constant speed. The curve it traces on the paper is how an observer rotating in that frame of reference would see the tip of the pen moving. *The observer sees the pen move under the influence of the centrifugal and Coriolis forces.* But

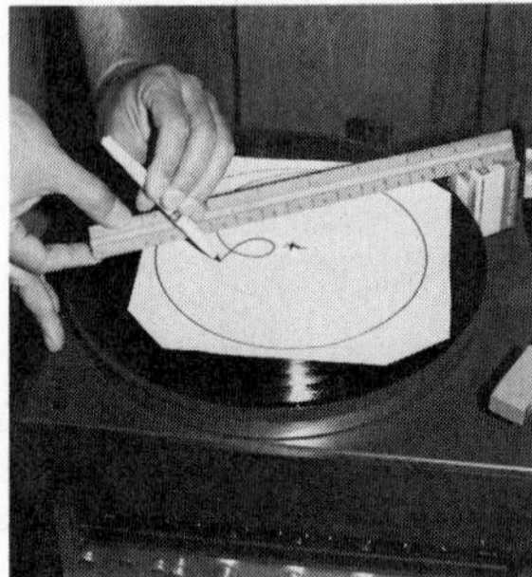

**FIGURE 8–30**
Move the pen at a constant speed along the ruler.

you know the pen only had a straight-line motion at constant speed because you controlled it yourself (Fig. 8–30).

Use different speeds, each steady, to trace different paths. Then move the pen in the opposite direction at a steady speed with the ruler in the same place. Move the ruler inward so the pen passes close to the center of the turntable, and make some shorter runs across the edge of your first circle. You'll find some beautiful curves by experimenting (Fig. 8–31).

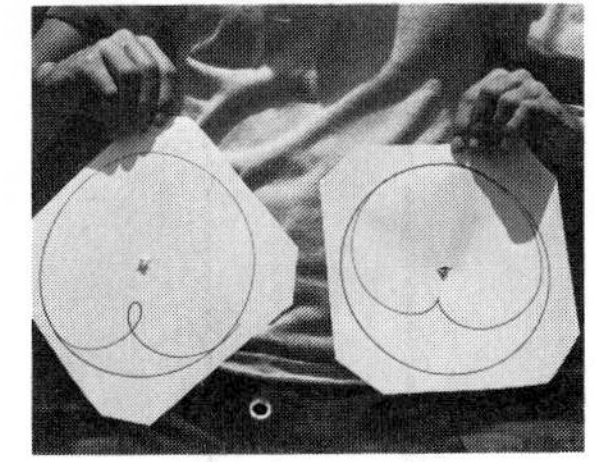

**FIGURE 8–31**
Typical results showing the result of centrifugal and Coriolis effects from the point of view of the rotating platform.

# EXERCISES

**1.** What is the rotation rate (in revolutions per minute) of the second hand on a clock? The minute hand? The hour hand?

**2.** Why is it harder to get a big wheel spinning than a small one even if they both have the same mass?

**3.** Have you ever noticed that large fan blades take longer to stop rotating than small ones do after the current has been turned off? Why should that be?

**4.** Which is the hardest way to do sit-up exercises (Fig. 8–32)? Should taller people have a harder time doing sit-ups?

**FIGURE 8–32**

**5.** Draw the moon in orbit around the earth and throw in a few background stars. Remember the same lunar plains and mountain ranges face earth all the time. Then convince yourself that the moon's rate of rotation about its center has the same value as its rate of revolution.

***6.** When you push open a heavy door, how much harder would you have to push at the center than you would at the edge far from the hinges just to get the door moving?

**7.** Which statements are correct? (a) Only an external force can change a body's rate of rotation. (b) Only an external torque can change a rigid body's rate of rotation.

**8.** As a tennis player brings her racket around in a smooth stroke, it revolves about her. If it slips from her grasp, it flies off, rotating about its center of mass. Where does that rotation come from? (*Hint:* See Exercise 5.)

**9.** A bowling ball and a volleyball are about the same size. Why is it so much easier to set the volleyball spinning?

**10.** A large can of corn ($m$) placed on a turntable adds $md^2$ to the turntable's moment of inertia, where $d$ is the distance from the can's center to the turntable's axis of rotation. How does this contribution to the moment of inertia change if the distance $d$ is doubled? What effect does the can have on the turntable when it starts up?

**11.** Should it be easier to turn a corner while carrying a long and narrow suitcase or a short and stubby one of the same mass? While carrying two 8-foot boards or four 4-foot boards (of the same cross section)?

**12.** Describe how your *arm* can become the lever arm of a torque when you open a new jar of peanut butter.

**13.** A bowling ball that is released with no spin slides along the lane only for a second or so and then begins to roll. Why?

**14.** Discuss the moment of inertia of a frisbee, the most popular gyroscope available today. Besides serving as a place to grip it, what does the curved rim do?

**15.** The Hale telescope and its mount weigh 530 tons, and yet an electric motor with only one-twelfth horsepower is sufficient to turn them as they follow the stars across the sky. What does this tell you about this 200-inch telescope on Mt. Palomar?

**16.** Can the center of mass of a solid object lie outside its body? Give an example.

**17.** Does the moment of inertia of a rigid, solid body depend on the angular speed the body has?

**18.** A fresh egg that is rotating will resume a slow rotation when stopped quickly and then released again. Why? Would this work if the egg were hard-boiled?

**19.** Racing bicycles have large-diameter wheels. Give two reasons why it is so important that the tires be lightweight.

**20.** Tell why the seesaw in Fig. 8–33 is balanced.

**21.** Trees are stable only because their root systems can summon torques from the ground to counter the torques exerted by wind (Fig. 8–34). Otherwise, should a tree lean until its center of mass is beyond its base of support, it would topple. The problem is worse for a tall tree than a short one, because its center of mass is higher above the ground. Even a slight lean places its center of mass over

FIGURE 8–33

nothing but air. That's why redwoods grow so vertically. Another factor, first pointed out by Galileo (who didn't know about redwoods), also limits how tall trees can grow. He estimated that 300 feet would be the tallest a tree would ever grow, which is about right. Can you guess that second reason?

FIGURE 8–34

**22.** Why does a washing machine bump and jump if the clothes are not evenly distributed in the tub?

**23.** Some women have backaches in the last months of pregnancy and then again after delivery. Analyze how a woman must compensate for the change in her center of mass. (Men aren't exempt from similar troubles! As seen in X rays of men who have long-established paunches, their backbones "curve" to compensate for the added torque.)

**24.** If there is a strong wind, you have to lean into it to remain on your feet (other holds barred, of course). Explain why.

**25.** If you bend over a sink to wash a lot of dishes, your back will feel the strain. Discuss why.

**26.** Is this statement true? You need not apply a force equal to the weight of a cement block in order to tip it over.

**27.** As a pendulum swings to and fro about its pivot point, its angular momentum obviously isn't constant. What torque is at work? (It helps to draw a picture.)

**28.** Why does the front end of your car rise a little when you accelerate rapidly?

**29.** Aristotle believed that the earth must not rotate, or else he'd be flung off! Discuss his conjecture.

**30.** Explain why a cyclist on a 10-speed bike will slide as far back as the seat will allow when forced to make an emergency stop.

**31.** If you ever see a penguin in a zoo, take a look at its feet. Discuss the penguin's mobility, stability, and flexibility on land.

**32.** Why are hiking boots such a good idea for backpackers?

**33.** Modern running shoes often have very wide heels. This design reduces pressure, but how else can it help?

**34.** You are temporarily in a rotating frame of reference in a car going around a curve. Suppose the car turns left. Describe what happens to a roll of candy mints lying on the seat beside you.

**35.** It isn't too hard to break off an aluminum key in a lock. Analyze why.

**36.** It is easier (in terms of work and energy) to put a spacecraft into orbit if it is launched in a direction toward the east rather than the west. Why?

**37.** If you are in a fast-flowing stream or river, you have to lean into the current just to stand still. Why? (*Hint:* See Exercise 24.)

**38.** Why do motorcycle riders have to lean as they take curves?

**39.** An orange and a grapefruit roll down an incline. Which will reach the bottom first? Why?

**40.** (a) Which of the bricks in Fig. 8–35 is most stable against a rotation when pushed from the left side? (b) Which is least stable? Why?

FIGURE 8–35

**41.** Why is it better to set bullets spinning before they leave the barrels of firearms? Why are most passes in football thrown with a spiraling motion?

**42.** How would an observer in a spacecraft hovering near the earth explain earth's equatorial bulge from the *inertial* frame of reference of the craft?

**43.** The next time you are at a large party, try to get the partygoers to go outside and run toward the east. What might this do to the length of earth's day? (If it's a *big* party?)

**44.** Perhaps you (or one of your friends) can balance a basketball on the tip of a finger by spinning it with the other hand. What would happen if when you tried to balance it the basketball was not spinning? (Try it and see!)

**45.** Why is a soft drink bottle more stable when balanced on its base than on its mouth?

**46.** A swimmer pauses on the 10-meter diving board, then launches herself with a very slow rate of rotation. But then she tucks her knees up close to her body and does two quick rotations before breaking into the water. What principle of physics is she using?

**47.** When someone you are riding with takes a curve too fast and you slide up against the car door, what force should you say moved you over?

48. Tom the Tinker had a terrific idea: "I'll make handles for bathroom faucets twice as long. They'll be easier to turn on and off, since they'll have more leverage." Discuss his idea.

**49.** In the earth's core there are liquidlike regions. Some geophysicists think iron (with relatively high mass) is sinking in this region even today. If true, what effect would this have on earth's rotation rate?

50. Suppose the ice on the continents near the polar regions all melted. The earth's ocean level would rise substantially (perhaps a hundred feet), even at the equator. What effect would this have on the earth's spin rate?

**51.** Why do tightrope walkers carry those long, long poles? Discuss.

52. Why is it easier for you to walk over a log across a creek (or on one of the iron rails of a railroad track) if you hold your hands out to the side? Why does it help even more if you hold one of your shoes out in each hand as you walk?

*53. Jimmy Physicsmajor wanted to time his hiccups, but there was no clock in the dorm room. So he turned on his turntable (33⅓ rpm) and used the rotations to find his rate of hiccups. Prove that each revolution of a turntable takes about 0.55 seconds, or about half a second.

*54. Levers are just beams that rotate about a point (called a *fulcrum*) somewhere along the beam. So when Melanie and Ashley had a flat tire and the car jack was missing, Melanie used a concrete block for a fulcrum and a strong board as a lever. She lifted that corner of the car by its bumper so Ashley could prop it up and change the tire. Melanie weighs 110 pounds and sat sideways on the board 4½ feet from its axis of rotation. The bumper was 1 foot from that fulcrum. What was the force applied to lift that corner of the car?

*55. Susie, who weighs 60 lb, sits 6 ft from the axle of a seesaw. Johnny, who weighs 80 lb, hops aboard on the other side. Oops, he's heavier than Susie, so to balance the seesaw, he slides forward. How far is Johnny from the axle when the seesaw becomes balanced? (Draw a sketch first.)

*56. An old turntable takes 3 seconds to reach 33⅓ rpm from rest. Calculate its rate of change of angular speed, $(\omega_f - \omega_i)/t$, which is its angular acceleration, in rotations/min/s.

**57. Calculate how fast the earth would have to spin before something at the equator would float because of the centrifugal force. (Presume the earth could keep its present shape even at that high rate of speed, although it really couldn't.)

*58. A swimmer who stands stiffly on the edge of a high-diving board and topples over toward the pool has an initial spin rate of 1 revolution per second. Then on the way down, the diver curls up into the tuck position and spins 3 times each second. What is the diver's moment of inertia in the tuck position as compared to the standing position?

*59. When you see the sun at sunset, it is about ½° wide, or ½° in diameter. Suppose you see it when it just touches the horizon on its way down. How long does it take it to disappear? (*Hint:* the earth rotates 360 degrees in 24 hours.)

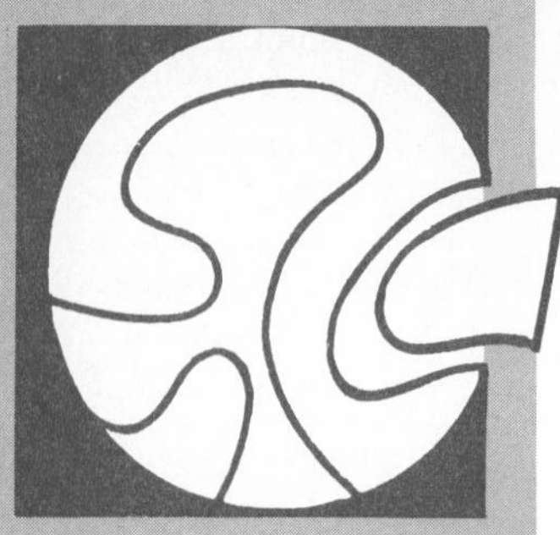

# CHAPTER 9 ATOMS AND MOLECULES

FIGURE 9–1

*"Why is the moon following our car?" "What holds the clouds up?"* Children can ask awfully tough questions about the world. Some are possible to answer, but others aren't. One of the simplest childlike questions about the world is, *"What's this made of?"* That's a question that was around for a long time before it was answered. Over 2000 years ago in Greece, philosophers including Democritus and Leucippus (about 400 B.C.) tried to get the answer to this question by guessing at what they couldn't see. These thinkers believed every substance was made of invisible units that could not be broken or divided, and they called these tiny units **atoms.*** (The Greek word *atomos* means "indivisible" or "uncuttable.") But the philosophers of those times never tried to prove their ideas with investigation. Many centuries passed before scholars realized that only experiments could prove or disprove what they might guess; where once there was only speculation, science had come forth.

Eventually the science of physics grew to the point where investigators could probe matter to solve the ancient question. Yet it was only at the close of the last century that experiments began to show the nature of atoms and the structures they build, which we call **molecules.** ■

## The Size of the Smallest Structures

When astronauts look at the sunlit earth from less than 200 miles above its surface, they can see little sign of life. In just that relatively short distance the naked eye cannot reveal the teeming cities and crowded highways. We stand over the world of the atom from a distant overlook, too, and it takes only a moment to appreciate how difficult it was to

*The Roman poet and philosopher Lucretius (96?–55 B.C.) reported on these speculations in his work *De Rerum Natura* ("On the Nature of Things"):

"We see that wine flows through a strainer as fast as it is poured in; but sluggish oil loiters. This, no doubt, is either because oil consists of larger atoms, or because these are more hooked and intertangled and, therefore, cannot separate as rapidly, so as to trickle through the holes one by one."

**FIGURE 9–2**
The incompressibility of atoms and molecules is what limits the number of people that can squeeze into a car or a telephone booth.

discover the nature of objects too small to be seen even with a microscope. What can you tell about the *structure* of water from looking at a single drop? Left alone, it evaporates in quantities too small for you to see. However, water does have one striking property that gives us a clue about the smallest units. If you squeeze a handful of water, it runs out through any opening between your fingers. But if you put water into a steel cylinder, insert a snug-fitting piston, and push really hard, the volume of water doesn't seem to get any smaller. Water may flow and evaporate, but in the subvisual world it has a tough structure. This near *incompressibility* is common to other liquids and solids as well (Fig. 9–2). But no matter how hard you try, you can't directly observe the tiny units that make matter hard. Merely to estimate their sizes requires experiments.

An everyday example can give us a crude idea about the size of atoms. A tiny piece of soot (carbon) from a campfire can be rubbed between your fingers until it disappears in a "layer" too thin to be seen. Likewise, in normal driving a car tire wears away, leaving as it rolls along an irregular layer of material too thin to see. Radial tires on small cars typically wear off a 1-centimeter thickness of tread or less in 40,000 miles. The answer to Exercise 38 shows the *average* thickness of the layer of material (mostly carbon) that such a tire leaves on the road is about $3 \times 10^{-8}$ centimeters. This extremely thin layer of material should be at least an upper limit to the size of the smallest units of rubber, whether atoms or molecules.

There are other experiments that don't use sophisticated equipment and yet measure the diameters of molecules with some accuracy.* But more sophisticated experiments with modern apparatus have given better values for the sizes of atoms and molecules. Carbon atoms are a little more than $1 \times 10^{-8}$ cm across, and the diameters of even the largest atoms are only three times larger. So each rotation of the tire wears off a layer that is a few *atoms* thick on the average.

Let's see what this small diameter means when compared with an everyday object such as a pencil. A single carbon atom is about $1 \times 10^{-8}$ centimeter across. So if you could count the atoms that lie side by side along a centimeter of a pencil lead, how many would there be? That number is 1 centimeter divided by (about) $1 \times 10^{-8}$ centimeter/atom, or roughly $10^8$ carbon atoms. That's one hundred million, and it's not easy to relate to a number that large. Let's draw on imagination: Suppose you were of atomic dimensions and you stood at one end of that centimeter of pencil lead (Fig. 9–3). Without guessing at their real appearance, imagine you were the right size to stroll along from one atom to the next, using them as stepping stones. If each stepping stone were 3 feet across, you'd find a path of them stretching for about 60,000 miles. All that just to get across 1 centimeter of a pencil lead—or across 1 centimeter of this page, or of any solid material, since all atoms are roughly the same size.

**Where the telescope ends, the microscope begins. Which of the two has the grander view?**
Victor Hugo, *Les Miserables*

*See, for example, Eric Rodgers, *Physics for the Inquiring Mind,* Princeton University Press, 1960, pp. 101–104. But be forewarned; the very nature of a single atom or molecule is such that it is impossible to assign an *exact* diameter to it.

**FIGURE 9–3**
If a person could be shrunk until the atoms along this centimeter of exposed pencil lead appeared to be 3 feet across, the person would have to walk 60,000 miles to cross from one end of the lead to the other. That's about the number of miles most people walk in an entire lifetime.

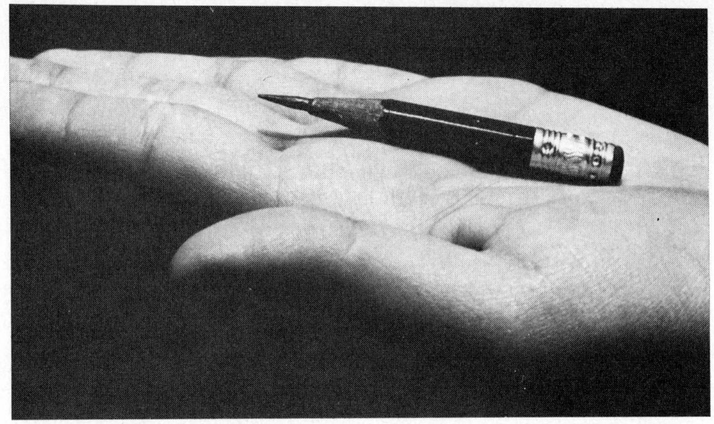

## The Chemical Elements and Their Combinations

Even before experimenters could measure the size of atoms, chemists were studying what happens when various substances combine or break apart, causing changes in their physical properties. To do this, they mixed precise quantities of materials, sometimes heating them to bring about the changes, and then carefully weighed the different products that appeared. With this simple procedure they gained insight into the unseen world of the atom.

**FIGURE 9–4**
Early chemists studied chemical reactions by weighing the reactants and the end products, gathering evidence of atoms and molecules long before any direct proof of their existence.

What they found comes down to this. Underlying all material things—the air you breathe, the foods you eat, the fragrances you smell—is a marvelous simplicity. The fact is that *there are only 90 species of atoms found on earth,* and every pebble and twig is built up from one or more of these varieties of atoms. These 90 species are called the **chemical elements.** Only about a dozen of the elements were known in antiquity because most of the 90 types of atoms stick together in all sorts of groups and combinations, making them hard to separate and identify. The most common elements are listed in Fig. 9–5.

To illustrate chemical combinations, we can consider the two elements copper (chemical symbol Cu) and phosphorus (P). In a piece of ordinary copper wire, the copper atoms stay together because they have a natural, if weak, affinity for each other. Likewise, the atoms in a very pure sample of phosphorus (a colorless waxy solid) attract each other. But if either of these elements is left open to the air, the phosphorus (or copper atoms) will unite naturally with oxygen (chemical symbol O) to make another substance. When this happens, we say there is a **chemical reaction** between the two elements.

The changes are easy to see. Phosphorus atoms react quickly in air to combine with oxygen, giving off a lot of light and heat in the process. This chemical reaction takes place so fast we call it *burning,* and it continues until all the phosphorus atoms have combined with oxygen. (When phosphorus and oxygen burn, they unite in the ratio of 5 oxygen

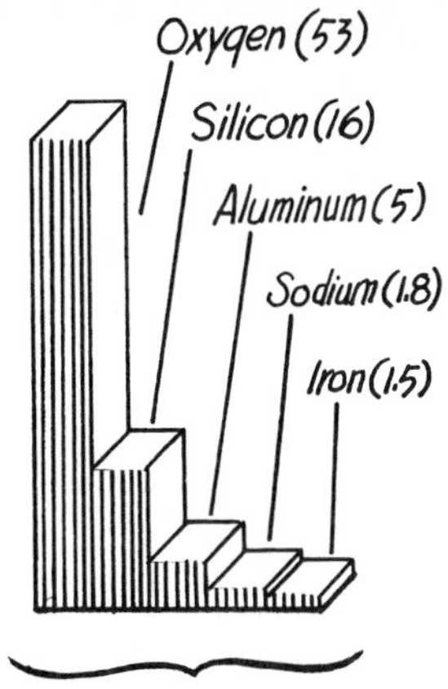

**FIGURE 9–5**
The most abundant elements in the crust of earth, its oceans, and its atmosphere. The numbers shown are the approximate numbers of atoms per one hundred atoms; of every 100 atoms in our environment, 53 are of the chemical element oxygen.

atoms for every 2 phosphorus atoms, so the *chemical formula* for the end product is $P_2O_5$.) In comparison, copper reacts very slowly. A neglected copper teapot develops a characteristic green-black coating after a few months, showing that the copper atoms on the surface have combined with oxygen. Once this layer forms, however, the air doesn't penetrate well, and the copper beneath it is preserved. (For black copper oxide, the formula is CuO because 1 oxygen atom unites with each copper atom.)

Such combinations of different chemical elements are called **compounds.** All animal, vegetable, and mineral matter in our world comes about only because the atoms of the various elements can stick together in various ways by means of *chemical bonds*. These bonds are the subject of later sections in this chapter.

## Molecules

Very often atoms join into bundles called **molecules** that have their own chemical identities. That is, the identical bundles of atoms (or identical molecules) often react with various atoms or groups of atoms as if the molecules were individual particles themselves. Water is an example of matter that is composed of molecules. A water molecule consists of two hydrogen atoms bound to an oxygen atom with such strength that they will separate only under extreme conditions. Of course, some molecules are not held together so strongly as water's molecules are; nitroglycerine is a famous example. A molecule of nitroglycerine bursts apart into smaller molecules—water ($H_2O$) and carbon dioxide ($CO_2$) among others—with the slightest of shocks. Sometimes atoms of a single species form molecules. Nitrogen and oxygen, the major ingredients of earth's atmosphere, are mostly paired up in the molecules $N_2$ and $O_2$. These are examples of *diatomic* molecules.

## Subatomic Particles

Until the present century, atoms seemed truly indivisible; no one knew how to divide an atom into smaller parts, to uncover its inner structure. But today physicists can explore atoms in a variety of ways. They can scatter radiation (X rays, visible light, or heat rays, for example) from matter to see how much is reflected or absorbed or if the matter changes the radiation in some way. They can break atoms up and use their pieces as projectiles to pierce other atoms, see what comes out, and work backward to find out what must have happened. These investigations have shown us what atoms are like.

An atom is made up of three kinds of particles so tiny we cannot see them by any means. We call these particles **electrons, protons,** and **neutrons.** An atom that these three particles build occupies a roughly spherical volume, but its outer boundary isn't well-defined as the surface of a billiard ball is. That's because the sphere is not a solid ball of matter but instead is almost entirely empty space! That sphere represents a volume where the lightweight electrons move. The electrons' behavior in (and out of) atoms is actually rather bizarre, and we'll study it in detail in Chapter 25. For the present you might think of electrons and their motion as physicists did early in this century before the whole picture was revealed. Imagine electrons to be planetlike particles whirling at superhigh speeds about a speck at the atom's center called the **nucleus.** That speck is the home of the protons and neutrons, particles much more massive than electrons. In this picture a typical electron, with motions like those of a planet, revolves around its heavyweight nucleus, which is nearly stationary like the sun. But the electron's orbiting motion is severe: It circles the nucleus $10^{15}$ times every second, so the comparison to planets and stars is only poetic. A visualization of any model with such properties cannot be realistic; nevertheless, the whirlwind orbiting leaves an impression that one might see a uniform *blur* from even a slow-motion model of an atom. Remind yourself that this model is only a way of thinking about atoms and not the way they are. When a football coach draws OOO and XXX to illustrate a play, it's a convenient representation of the play. That's what this atomic model is to the real atoms. Just as a football player isn't really a little X with an arrow and line for motion, neither are electrons little planets orbiting a miniature sun.

## The Electric Force: How Atoms Stay Together

Atoms stay together because of a powerful force of attraction between the protons in the nucleus and the electrons in the electron cloud. Except for this attraction, the orbiting electrons would go their own ways very quickly—the atom would break apart. This **electric force** acts between any proton and electron, and it is some $\mathbf{10^{40}}$ times as strong as the force of gravity between them. Just like gravity, however, it loses strength with increased distance between the particles. But its tremendous strength is not the only difference between the electric force and the force of gravity. The electric force works in two ways: Protons and electrons attract, but a proton repels any other proton and any electron repels any other electron, even within the atom. (You might like to sneak a look at the first two sections of Chapter 18.) To describe these interactions, we say the proton has a *positive charge* and the electron has a *negative charge* of the same size. That gives us a simple rule: *Like* charges *repel* each other; *unlike* charges *attract* (Fig. 9–6). Neutrons don't participate in this; they have no charge.*

*Another force, the strongest yet detected in nature, keeps the protons together in the nucleus. The **nuclear force,** or **hadronic** force, a hundred times stronger than the electric force, overcomes the protons' repulsion and keeps the protons (and neutrons) in that small, dense ball at the center of the atom.

**FIGURE 9–6**
Any two protons repel each other and any two electrons repel each other. If the distance between them is the same, the repulsions are the same. If a proton and an electron are the same distance apart as the pair of protons or the pair of electrons were, they attract each other with the same force as the repulsive force between the others. If the distance between two electrons (or between two protons) is halved, the repulsion increases by a factor of 4. (The electric force is discussed in detail in Chapter 18.)

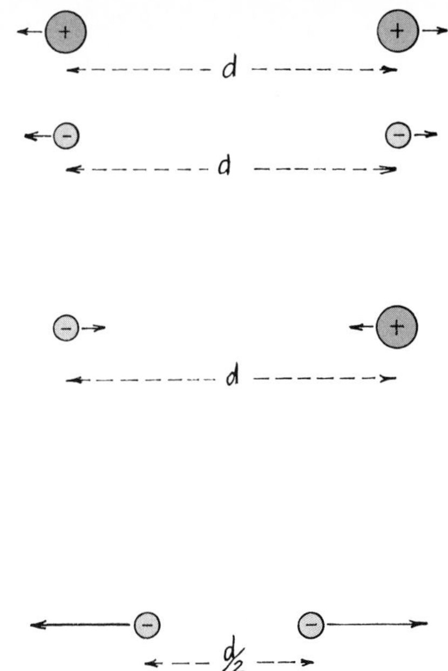


## Most Atoms and Molecules are Neutral

Since the electric force is so much more powerful than the gravitational force, you might wonder why you don't see more evidence of it around you. The reason is that most atoms and molecules are *neutral*. They don't exert a *net* force on anything else because they have the same amount of positive and negative charges. Consequently, another charged particle, say an electron, that happens to pass by a neutral molecule is repelled just as much by the electrons as it is attracted to the protons in the nuclei. Those forces cancel just as if there were no charge there at all, and that's what the word *neutral* means (Fig. 9–7).

There seem to be just as many electrons as protons throughout the universe. Because the electric force is so strong, these charges tend to pull themselves close together, forming the units we call atoms, neutralizing themselves to the regions outside. Even in stars, where the temperatures are so high that the collisions between atoms knock off some or all of their electrons, there cannot be a large-scale imbalance of electric charge. In a fistful of star-stuff, there is always the same amount of positive and negative charge; were it otherwise, the powerful electric force would quickly pull these opposite charges to each other, even if it is too hot for them to form whole atoms.

## The Reason Atoms Are So Hard

Despite their neutrality, any two atoms that get close enough will interact through the electric force. In fact, when you clap your hands or chomp down on a piece of celery, the electric force helps keep the colliding atoms and molecules from going right through each other. When two atoms touch, the edges of their electron-regions come together. Suddenly, electrons from two atoms are trying to share the same region of space, and if those atoms move closer together, the electrons have a chance to find themselves on top of one another. But this won't happen. First of all, the electric force becomes enormous as the distance between them decreases, so the electrons push off with great strength. Second, it is known that by their very nature more than two electrons *cannot* be at the same place with the same speed at the same time. We'll see the details in Chapter 25, but at this point we'll just say the electrons in groups are kept somewhat apart because they are a special type of particle that follow a rule called the **exclusion principle.*** When these two atoms separate, their electrons withdraw from contact, the repulsive electric force disappears, and the atoms seem neutral to each other again.

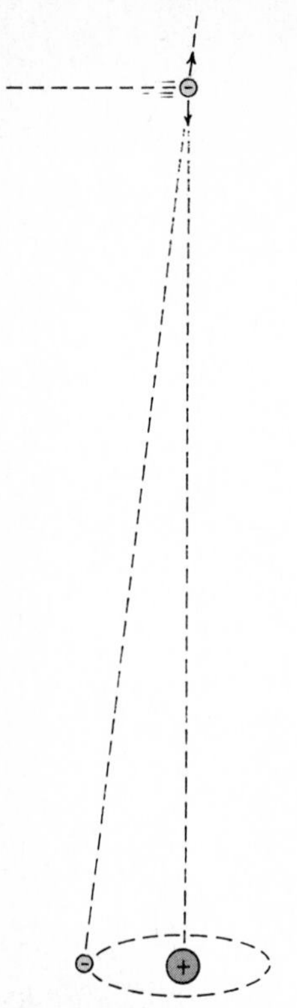

**FIGURE 9–7**
An atom or molecule that has equal numbers of electrons and protons appears neutral to a passing charge, because the net electric force on that passing charged particle is approximately zero. That is, no matter whether the passing particle is positive or negative, it receives (about) equal and opposite electrical forces from the protons and electrons (see the vectors on the charge at the top).

This mighty repulsion between the electrons of atoms is why atoms, molecules, and the material things they make are so hard, so nearly incompressible. They cannot collapse much when they are squeezed together with ordinary pressures. It is this property of atoms that is at the root of the contact, or normal, forces we studied in Chapter 3, which together with gravity are the most obvious forces in our surroundings.

Another obvious property of matter is its mass. The mass of any object is very nearly the sum of the masses of its atoms (or molecules). Though earlier chemists could measure the relative weights of reacting substances accurately, it was well into the twentieth century before the masses of the individual atoms of each element were known.

## The Masses of Atoms and Molecules

An ordinary hydrogen atom, lightest of all, has a mass of only $1.7 \times 10^{-24}$ grams. The masses of other atoms show a surprising feature: *Every single atom, whatever the element, has a mass that is very nearly equal to that of some whole number of hydrogen atoms.* The mass of a carbon atom is typically about 12 times as much as a hydrogen atom, and an oxygen atom is about 16 times more massive than a hydrogen atom. On a scale, then, 16 hydrogen atoms balance just 1 oxygen atom. Thus the mass of **$1.7 \times 10^{-24}$ grams** is a convenient unit to use for comparing the masses of atoms (and molecules). It is called the **atomic mass unit** (amu).†

---

*When electrons are squeezed together into spaces much smaller than atoms in the hearts of collapsing stars, the exclusion principle keeps them apart nevertheless. There is no known force in nature strong enough to cause electrons to merge.

†Oxygen, rather than hydrogen, was long used by chemists as a standard for comparing atomic masses. Oxygen combines with most of the elements, making comparisons of other atomic masses by experiment easy. It was *assigned* the mass of 16.00000 amu. Both hydrogen and oxygen are gases at ordinary temperatures (those found on earth) and this makes them somewhat inconvenient for everyday laboratory references. Carbon is the standard today, being assigned the atomic mass of 12.00000. That gives hydrogen a mass of 1.0079 amu and oxygen a mass of 15.9994 amu. Such small differences will not concern us here, however.

**FIGURE 9–8**
This VW and its driver together weigh about 1850 pounds, while the cheese the driver is holding weighs 1 pound. That is the relative weight of a proton and an electron. The electron is the real lightweight of the atom.

The collection of electrons in any atom weighs almost nothing compared to its nucleus. A proton's mass is more than 1836 times as great as an electron's (Fig. 9–8). A neutron is slightly heavier than a proton, but only by 0.1 percent, so protons and neutrons have about the same mass. To find the mass of an atom, then, first *count the protons and neutrons in its nucleus*. That number is the **atomic mass number** of the atom. Multiplying that number by $1.7 \times 10^{-24}$ grams gives a good estimate of the mass of the atom. (Chemists sometimes use the word *atomic weight* instead of atomic mass for this number, no doubt because of the way in which mass is found in most experiments—by the method of weighing.)

The hydrogen atom is not just the lightest atom; it is the simplest as well. Ordinary hydrogen has a nucleus that is a single proton, and only one electron circles that proton. Its atomic mass number is 1. A carbon nucleus normally contains 6 protons and 6 neutrons, so its atomic mass number is 12. The carbon nucleus rests at the center of 6 orbiting electrons.

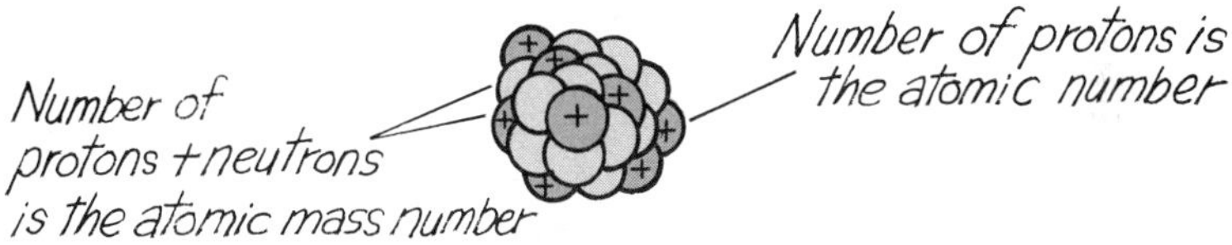

Number of protons + neutrons is the atomic mass number

Number of protons is the atomic number

**FIGURE 9–9**
The atomic number is the number of protons in an atom's nucleus. In a neutral atom it is also the number of electrons circling the nucleus. The atomic mass number is the number of protons and neutrons found in the nucleus of the atom. The actual mass of the atom is about $1.7 \times 10^{-24}$ grams times the atomic mass number.

**FIGURE 9–10**

Proton

Hydrogen nucleus

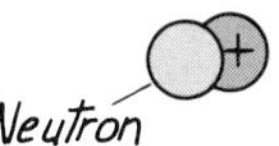

Neutron

Deuterium nucleus

Tritium nucleus

There is, however, a form of hydrogen atom that acts normal, entering into chemical reactions just as every other hydrogen atom does, except that it is *twice as heavy* as the majority of hydrogen atoms. Its atomic mass number is 2. About 15 of these turn up in every 10,000 hydrogen atoms on our world. The reason for the extra mass is a neutron that clings to the hydrogen's proton. It sometimes happens that a hydrogen atom has two neutrons in its nucleus in addition to the proton, making it three times as massive as ordinary hydrogen, but that is even more infrequent.

There are, then, three varieties, or **isotopes,** of the chemical element hydrogen. The double-mass atom is called *deuterium,* and the triple-mass hydrogen is called *tritium* (Fig. 9–10). The other elements have isotopic forms, too. For example, the most common carbon atom has an atomic mass number of 12, but about 1 percent of all carbon atoms have one more neutron in their nuclei, giving them a mass of 13. Both forms of carbon have 6 electrons that balance the charge of the 6 protons, and those electrons act the same when chemical bonds are formed. So in a practical sense, *it is the number of protons in a nucleus that determines the chemical element to which it belongs*. The number of protons is what we call the **atomic number.** Carbon has the atomic number 6 and hydrogen 1, regardless of how many neutrons are in the nucleus.

The mass of a molecule is (as closely as chemists can measure) just the sum of the mass of its atoms. Two oxygen atoms unite with a single carbon atom to form a molecule of carbon dioxide, so its chemical formula is $CO_2$. If the carbon atom has the atomic mass number 12 and each oxygen has 16, the mass of the molecule is $12 + 16 + 16 = 44$ atomic mass units.

## The Organization of the Electrons in Atoms

In the 1800s, chemists discovered some remarkable similarities in how some of the chemical elements combine with others. For example, the element sodium (Na) combines with nitrogen (N) to make a compound called sodium nitride. In this compound three sodium atoms unite with one nitrogen atom, so its chemical formula is $Na_3N$. Potassium (K), another soft, silvery-white metal like sodium, combines with nitrogen *in the same ratio* to make potassium nitride, $K_3N$. Whenever sodium or potassium combines with other atoms or groups of atoms, they combine with them in the same ratios. Though sodium's atomic mass is 23 and potassium's is 39, they have the same tendencies in chemical reactions; that is, they have similar chemical behavior. In 1869 a Russian chemist, Dmitri Mendeleev (1834–1907), arranged the elements known at the time in a chart that's now called the **periodic table.** The columns in the chart contain groups of elements with similar chemical behavior. They are arranged by their weights, which increase from left to right and from one row to the next. The modern periodic table is shown in Table 9–1.

The reason why the elements have these patterns of behavior was discovered by physicists in the early 1900s. Electrons move rapidly in the space around the nucleus, but they don't move at random. Each electron is found in a three-dimensional pattern, traveling only in certain areas. In a way, the electrons keep to onionlike layers, or spherical shells, around the nucleus, and that picture helped to explain a great deal.

Physicists found that each shell-like region could accommodate only a specific number of electrons. Many atoms have their innermost shells filled to capacity. If the outermost shell of an atom is filled as well, the atom tends to be inert, that is, it doesn't combine easily with other atoms to make molecules or compounds. However, if the outer shell in some atom is one or two electrons shy of being filled to capacity, it is as if that atom had a hole in its outer shell, an empty slot or two that could be filled by electrons from some other atom. Another atom may have only one or two electrons in an outer shell that could accommodate 8 electrons, for instance. These two types of atoms can unite chemically if the atoms with only a few electrons in their outer shells can share or give up those electrons to atoms that have holes. This leaves both types of atoms with filled (or more nearly filled) outer

**TABLE 9–1**

## THE PERIODIC TABLE OF THE ELEMENTS

| PERIODS | IA | IIA | IIIA | IVB | VB | VIB | VIIB | GROUPS<br>VIII | | | IB | IIB | IIIA | IVA | VA | VIA | VIIA | 0 |
|---|---|---|---|---|---|---|---|---|---|---|---|---|---|---|---|---|---|---|
| 1 | 1.008<br>H<br>1<br>hydrogen | | | | | | | | | | | | | | | | | 4.003<br>He<br>2<br>helium |
| 2 | 6.941<br>Li<br>3<br>lithium | 9.012<br>Be<br>4<br>beryllium | | | | | | | | | | | 10.811<br>B<br>5<br>boron | 12.011<br>C<br>6<br>carbon | 14.007<br>N<br>7<br>nitrogen | 15.999<br>0<br>8<br>oxygen | 18.998<br>F<br>9<br>fluorine | 20.179<br>Ne<br>10<br>neon |
| 3 | 22.990<br>Na<br>11<br>sodium | 24.305<br>Mg<br>12<br>magnesium | | | | TRANSITION | | ELEMENTS | | | | | 26.982<br>A1<br>13<br>aluminum | 28.0855<br>Si<br>14<br>silicon | 30.9738<br>P<br>15<br>phosphor-us | 32.06<br>S<br>16<br>sulfur | 35.453<br>Cl<br>17<br>chlorine | 39.948<br>Ar<br>18<br>argon |
| 4 | 39.0983<br>K<br>19<br>potassium | 40.08<br>Ca<br>20<br>calcium | 44.956<br>Sc<br>21<br>scandium | 47.90<br>Ti<br>22<br>titanium | 50.9415<br>V<br>23<br>vandium | 51.996<br>Cr<br>24<br>chromium | 54.938<br>Mn<br>25<br>manganese | 55.847<br>Fe<br>26<br>iron | 58.933<br>Co<br>27<br>cobalt | 58.71<br>Ni<br>28<br>nickel | 63.546<br>Cu<br>29<br>copper | 65.37<br>Zn<br>30<br>zinc | 69.72<br>Ga<br>31<br>gallium | 72.59<br>Ge<br>32<br>german-ium | 74.922<br>As<br>33<br>arsenic | 78.96<br>Se<br>34<br>selenium | 79.904<br>Br<br>35<br>bromine | 83.80<br>Kr<br>36<br>krypton |
| 5 | 85.468<br>Rb<br>37<br>rubidium | 87.62<br>Sr<br>38<br>strontium | 88.906<br>Y<br>39<br>yttrium | 91.22<br>Zr<br>40<br>zirconium | 92.9064<br>Nb<br>41<br>niobium | 95.94<br>Mo<br>42<br>molyb-denum | 98.906<br>Tc<br>43<br>technetium | 101.07<br>Ru<br>44<br>ruthen-ium | 102.906<br>Rh<br>45<br>rhodium | 106.4<br>Pd<br>46<br>palladium | 107.868<br>Ag<br>47<br>silver | 112.41<br>Cd<br>48<br>cadmium | 114.82<br>In<br>49<br>indium | 118.69<br>Sn<br>50<br>tin | 121.75<br>Sb<br>51<br>antimony | 127.60<br>Te<br>52<br>tellurium | 126.90<br>I<br>53<br>iodine | 131.30<br>Xe<br>54<br>xenon |
| 6 | 132.906<br>Cs<br>55<br>cesium | 137.33<br>Ba<br>56<br>barium | 138.906<br>*La<br>57<br>lanthanum | 178.49<br>Hr<br>72<br>hafnium | 180.948<br>Ta<br>73<br>tantalum | 183.85<br>W<br>74<br>tungsten | 186.2<br>Re<br>75<br>rhenium | 190.2<br>Os<br>76<br>osmium | 192.22<br>Ir<br>77<br>iridium | 195.09<br>Pt<br>78<br>platinum | 196.967<br>Au<br>79<br>gold | 200.59<br>Hg<br>80<br>mercury | 204.37<br>Tl<br>81<br>thallium | 207.2<br>Pb<br>82<br>lead | 208.981<br>Bi<br>83<br>bismuth | (209)<br>Po<br>84<br>polonium | (210)<br>At<br>85<br>astatine | (222)<br>Rn<br>86<br>radon |
| 7 | (223)<br>Fr<br>87<br>francium | 226.025<br>Ra<br>88<br>radium | (227)<br>**Ac<br>89<br>actinum | (261)<br>[Rf]<br>104<br>ruther-fordium | (262)<br>[Ha]<br>105<br>hahnium | (263)<br>[ ]<br>106 | (262)<br>[ ]<br>107 | | (266)<br>[ ]<br>109 | | | | | | | | | |

| | | | | | | | | | | | | | | |
|---|---|---|---|---|---|---|---|---|---|---|---|---|---|---|
| *Lanthanides | 140.12<br>Ce<br>58<br>cerium | 140.908<br>Pr<br>59<br>praseo-dymium | 144.24<br>Nd<br>60<br>neodym-ium | (145)<br>Pm<br>61<br>prometh-ium | 150.4<br>Sm<br>62<br>samar-ium | 151.96<br>Eu<br>63<br>europ-ium | 157.25<br>Gd<br>64<br>gadol-inium | 158.925<br>Tb<br>65<br>terbium | 162.50<br>Dy<br>66<br>dyspro-sium | 164.930<br>Ho<br>67<br>holmium | 167.26<br>Er<br>68<br>erbium | 168.934<br>Tm<br>69<br>thulium | 173.04<br>Yb<br>70<br>yttrium | 174.967<br>Lu<br>71<br>lutetium |
| **Actinides | 232.038<br>Th<br>90<br>thorium | 231.031<br>Pa<br>91<br>proact-inium | 238.029<br>U<br>92<br>uranium | 237.048<br>Np<br>93<br>neptun-ium | (244)<br>Pu<br>94<br>pluton-ium | (243)<br>Am<br>95<br>amer-icium | (247)<br>Cm<br>96<br>curium | (247)<br>Bk<br>97<br>berkel-ium | (251)<br>Cf<br>98<br>cali-fornium | (254)<br>Es<br>99<br>einst-einium | (253)<br>Fm<br>100<br>fermium | (256)<br>Md<br>101<br>mendel-evium | (253)<br>No<br>102<br>nobel-ium | (257)<br>Lr<br>103<br>lawren-cium |

metals; metalloids; nonmetals; noble gases

1.008 ← atomic mass
H ← symbol
1 ← atomic number
hydrogen ← name

Numbers below the symbol of the element indicate the atomic numbers. Atomic masses, above the symbol of the element, are based on the assigned relative atomic mass of $^{12}C$ = exactly 12; ( ) indicates the mass number of the isotope with the longest half-life. [ ] indicates not officially approved or named.

shells. As long as they are bound together, they are relatively inert. *The transfer or sharing of one or more electrons between different chemical elements is the mechanism that provides the glue we call* ***chemical bonds*** *in molecules and compounds.*

The swapping of electrons when atoms combine to form compounds makes matter more stable. If chemical bonding takes place naturally, the atoms lose energy as the electrons rearrange themselves. The combination of atoms then has a lower total energy and is more stable. That is, energy would have to be *added* to break the combination apart. The great majority of the elements on earth are normally found only in compounds for this reason, explaining why only a handful of chemical elements were known at the beginning of the science of chemistry. You could even say the stability of compounds is why, for example, your teeth don't stick to a piece of celery when you eat it.

As you look along the rows of the periodic chart from left to right, each element has more protons in the nucleus of its atom than the preceding one and so is heavier. Since any neutral atom has one electron for every proton, heavier elements also have more electrons. It follows that the progression of atoms fills up one electron shell after another across the periodic table. The reason sodium (Na) and potassium (K) form many compounds in the same way is that they both have similar outermost shells. Each has only one electron in its outermost shell. It doesn't matter much in the chemical reaction that potassium has an extra shell full of electrons beneath its chemically reactive shell. Chlorine (Cl), fluorine (F), and bromine (Br) each *lack* only one electron to fill their outermost shells, so they too form similar compounds.

## Types of Chemical Bonds

The chemical bonds between the outer layers of the electrons of atoms come in several varieties, with shades in between. Sometimes an electron from one atom transfers almost entirely to an atom of another type. This is called an **ionic bond,** from the word **ion,** which means charged particle. (Any atom with an imbalance of charge is an ion.) Sodium and chlorine bind together this way in ordinary table salt. The sodium atom gives up an electron and becomes a positive ion, while the chlorine atom takes that electron into its outer shell and becomes a negative ion. The ionic bond is very strong, since the positive and negative ions attract each other directly with the electric force.

Sometimes outer electrons are shared in the regions between two atoms. Called **covalent bonds,** they are the glue for the diatomic molecules like nitrogen ($N_2$) and oxygen ($O_2$), the major components of air. In a covalent bond two electrons share the space between the two atoms. When atoms form covalent bonds in a molecule, the bonds regulate the positions of the atoms because the electron concentrations in each bond affect each other; covalent bonds are *directional* bonds. Carbon is notable for covalent bonding. A single atom of carbon may form covalent bonds with up to four other atoms at once, including other carbon atoms. That helps to account for the many geometrically complex organic (carbon) compounds found on earth. In strength the covalent bonds average slightly less than ionic bonds.

In metallic elements such as copper, silver, gold, and aluminum, the atoms bond in a unique manner. Atoms of metallic elements (and combinations thereof) stay together by sharing outer electrons in the most general way possible. In **metallic bonding,** the outermost electrons of the atoms are able to wander from one atom to another, shared not just between two atoms but by the entire metal structure. It's this mobility that makes many metals good conductors of electric current and heat. Metallic bonds are generally much weaker than covalent or ionic bonds.

Another type of bond happens between *any* two neutral atoms or molecules that get close together. These **van der Waals bonds** are the weakest bonds of all. (If they weren't weak, the world would be very different indeed.) Because of the motions of electrons in atoms or molecules, at any instant there might be more on one side than on the other. Then we say the atom (or molecule) is **polarized.** Suppose a polarized atom has its negatively charged side facing a neutral atom. The electrons on that neighboring atom are repelled and move somewhat to the far side of their nucleus, leaving the side toward the initially polarized atom positive. The two atoms then have oppositely charged regions facing each other, giving rise to an attraction. That is the van der Waals bond; it comes from the natural fluctuations of the charge throughout the atom due to the electron's motion.

**If, in some cataclysm, all of scientific knowledge were to be destroyed, and only one sentence passed on to the next generation of creatures, what statement would contain the most information in the fewest words? I believe it is the *atomic hypothesis* (or the atomic fact or whatever you wish to call it) that *all things are made of atoms—little particles that move around in perpetual motions, attracting each other when they are a little distance apart, but repelling upon being squeezed into one another.* In that one sentence, you will see, there is an enormous amount of information about the world, if just a little thinking and imagination are applied.**

R. P. Feynman, *The Feynman Lectures on Physics*

## The Energy of Chemical Reactions

The strength of chemical bonds is not measured in terms of force, but in terms of the energy the bond represents. Ionic and covalent bonds take more energy to break than the others, and they give up more energy when they are formed. *Every* chemical reaction, however, *releases* or *absorbs* energy. If energy is released, the chemical reaction is **exothermic;** and if energy is absorbed, the reaction is **endothermic.** The outer electron shells of atoms have a great deal of potential energy (as well as kinetic energy), and any rearrangement results in a greater or smaller total potential energy for the atoms. That is where energy comes from (or goes to) during chemical reactions.

As we've noted, if a chemical reaction proceeds *spontaneously,* it is always exothermic. But many important chemical reactions that are exothermic do *not* occur spontaneously and need an energy nudge before they proceed. Gasoline, vaporized and mingling with air, is a good example. The molecules of $O_2$ and gasoline collide and bounce off without reacting at ordinary temperatures. A simple spark, however, provides enough extra energy to a few of these molecules to make them react, or burn, and the energy they release provides the **activation energy** to the surrounding molecules of gasoline and oxygen. An analogy to such a reaction is a car that is parked next to a curb at the top of a cliff. If someone pushes it up over the curb (adds the activation energy), gravity takes over, and the change of potential energy into kinetic energy (the exothermic reaction) is automatic.

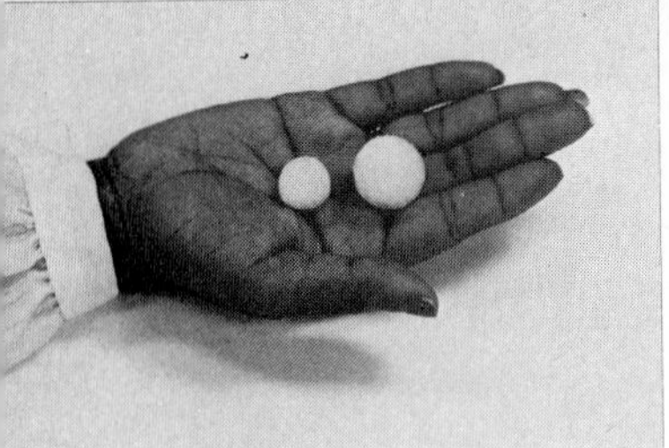

**FIGURE 9–11**
A comparison of the relative sizes of a radium atom and a lithium atom. Though the lithium atom has only 3 electrons while the radium atom has 88, the radium atom is only a little larger than the lithium atom. Their masses, however, differ greatly. Radium atoms have masses of 226 atomic mass units, while the lithium atoms have masses of 7 atomic mass units.

## Another Look at the Size of Atoms and Molecules

You might suspect that an atom with more occupied shells should be much bigger, but the very largest atoms have diameters that are only about three times larger than the smallest. For example, a radium atom (atomic number 88) is not much larger than a lithium atom (atomic number 3); see Fig. 9–11. The 88 protons in a radium nucleus exert a huge electrical attraction on their innermost electrons—in fact, almost 30 times greater than the attraction from the three protons of a lithium nucleus. Because of the stronger attraction, the inner shells of electrons in a radium atom are more tightly constructed, and overall this leads to the fact that atoms aren't very different in size.

We should ask next what a molecule might look like. Because the outermost electrons are the seat of the chemical bond, the atoms just touch at precise points and angles, their size adjusting only slightly as the chemical bonds form. The inner electron shells are essentially unaffected. So the behavior Mendeleev discovered, nearly all of the science of chemistry, and most of the physical properties of matter around us have to do only with the electrons in the farthest shell from the nucleus. Those electrons make mercury flow, make lead soft, and make diamond hard. It is the outermost electrons' behavior that governs the chemical reactions of all the substances in living things.

## On Describing Matter

To describe the world of the atom, we must summon fairyland visions. Only three types of particles, the tiny electrons, protons, and neutrons in perpetual and frantic motion, build 90 chemically different units. Incompressible and spherelike, atoms are nearly the same size despite their differences in weight. But there is magic in their actions. They stay in motion on our world, and, depending on the species and the speeds of their collisions, atoms may stick together to form the larger structures called molecules. The kaleidoscopic forms and functions of the matter around us come from the properties of these molecules.

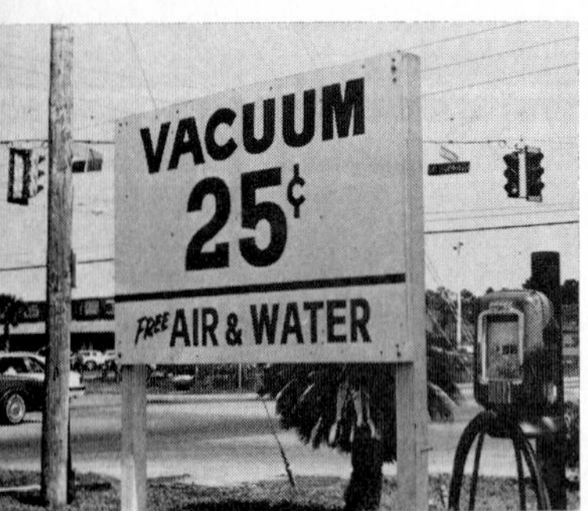


**FIGURE 9–12**
Nothing for something and something for nothing.

As the artist uses design, colors, and textures, and the poet uses words, certain groups of carbon-based molecules weave a vision much greater than the individual parts could ever show. Life is the greatest abstraction of matter, an organized symphony of reactions whose highest level is capable of the subtle actions we call independent thought and creativity. The philosophers of old responded to that side of human nature that causes us to look beyond the surface of things, to see deeper into our surroundings and our lives. There are philosophers still, in today's academic disciplines. Like their predecessors, they have understanding as their goal. Psychologists try to understand mental activities in terms of the underlying social and physical environment. Sociologists study the collective behavior of human groups in terms of less visible interactions and patterns. Scientists study the behavior of nature, and within that group certain physicists study the most fundamental and basic levels of matter. To get around their clumsiness and blindness (if only one finger is raised, it disturbs trillions of unseen molecules of air), some investigate matter by experimenting. Others explore with the tool of mathematics, trying to predict matter's properties.

The next three chapters explore the **states** of matter, the three ways that atoms and molecules are most often found on earth. Solids, liquids, and gases have properties that are familiar to us all, and others more subtle but just as important. The atomic picture of matter helps us understand these forms of matter at a level beyond our ordinary perceptions.

## CALCULATIONS

EXAMPLE 1: The element chlorine (Cl), found on earth in compounds and used in bleach, treated water, and swimming pools, has two isotopes. One, of atomic mass 35, is the most abundant, accounting for about 75.5% of all chlorine atoms. The other. representing 24.5% of chlorine atoms, has an atomic mass of 37. **Show that the *average* atomic mass of chlorine is about 35.5, in agreement with the value seen on the periodic table on page 193.**

An easy way to solve this is to work with 1000 representative chlorine atoms. Then 755 have an amu of 35 and 245 have an amu of 37. The average amu for chlorine atoms is then

$$\frac{(755 \times 35 \text{ amu}) + (245 \times 37 \text{ amu})}{1000} = \mathbf{35.49\ amu}$$

(The difference in this number and the number in the table, 35.45, is because the atomic masses are not *quite* equal to 35 and 37. The chlorine-35 atom's mass is actually 34.968851, for example.)

## REVIEW

**1.** Atoms (and the combinations of atoms called molecules) are the smallest structures of ordinary matter. True or false?

**2.** The 90 species of atoms found on earth are called the ______________.

**3.** A compound is a substance that's formed when two or more species of atoms unite by means of chemical bonds. True or false?

**4.** Two or more atoms can unite to become a unit known as a molecule. True or false?

**5.** Name three subatomic particles, and describe an atom in your own words.

**6.** State the facts of attraction and repulsion between particles that are charged. "Like charges ______________," and so on.

**7.** The atomic mass unit (amu) is a convenient unit for comparing masses of atoms. Its value is about $1.7 \times 10^{-24}$ grams. True or false?

**8.** The atomic mass unit multiplied by the atomic mass number gives a good estimate of the mass of an atom. True or false? The atomic mass number is the sum of the numbers of protons and electrons in the atom. True or false?

**9.** The different isotopes of an element contain different numbers of neutrons in their nuclei. True or false?

**10.** What special number distinguishes one chemical element from another?

**11.** The periodic table groups the elements in columns according to their chemical behavior. True or false?

**12.** Name four types of chemical bonds.

**13.** Define exothermic and endothermic chemical reactions.

## DEMONSTRATIONS

**1.** One very familiar chemical reaction is rapid oxidation, or *burning,* which we associate with flames. Most fuels (oil, gasoline, alcohol, natural gas, coal, kerosene, candle wax) are *organic hydrocarbons,* containing carbon and hydrogen and derived from once-living organisms. When they burn completely (in exothermic chemical reactions), the carbon combines with oxygen from the air to form carbon dioxide ($CO_2$) and the hydrogen combines with oxygen to form water, $H_2O$. These two compounds are always among the end products of burning hydrocarbons.

You can test what's going on in the flame of a candle or a match or a gas stove. The flame has a central cone that is bright blue. That cone of rising gases is the vaporized fuel. Oxygen from the air can't easily get into the fast-rising stream of gases, so the burning to form $CO_2$ and $H_2O$ takes place mainly at the edges of the flame. At the very top of the blue cone, the last of the rising fuel is burned, and the flame is hottest at that point. If the flame comes from a burning match or a candle, there's a yellow area in the flame around the central blue part. The yellow light is given off when carbon burns incompletely, and there are many particles of unburned carbon in the yellow portion. If you hold the bottom of a cool dish in this yellow part, it blackens quickly with particles of carbon, a fact known to every camper who's cooked over a campfire. You'll also notice that the bottom of the dish gets damp. That water comes from the $H_2O$ formed during the burning. Mixed with enough air, alcohol and natural gas burn the carbon with almost 100 percent efficiency and have only blue or colorless flames. A gas stove that has a yellow flame will blacken pots, and the air-fuel mixture should be adjusted.

If your instructor can provide you with a glass tube (about 1 foot long) that has been drawn into a fine opening at one end, you can show that the blue cone contains unburned fuel. With a potholder, hold the tube at a 45° angle with the large end placed directly in the blue cone of the flame. After a few seconds you can light the unburned fuel coming out of the small end of the tube with a match.

An amusing trick can be performed with unburned fuel. Let a candle burn for a moment or two and blow it out. A stream of smoke will rise from the snuffed candlewick. Immediately hold a lighted match an inch or more above the candle in that stream of smoke. The flame from the match will race *downward* to relight the candle.

## EXERCISES

**1.** In the deepest mine in the crust of the earth, the rocks must hold up over 16,000 pounds per square inch, yet the rocks aren't different from hard rocks found at the surface. What property of matter does this show?

**2.** What simple observation about our world shows you that as atoms unite to form molecules, their distinct physical and chemical properties change?

**3.** The small compressibility of matter comes more from (a) the electrons, (b) the protons, (c) the nuclei, (d) the neutrons.

**4.** Water molecules are nearly incompressible. This is one reason an advancing glacier moves faster in the summertime than in winter, when it accumulates extra layers of snow and ice. Can you speculate why?

**5.** Which property of the elements has the greater range of variation, atomic size or atomic mass?

**6.** The fact that some compounds are stable while others are unstable has to do with the strength of ________________.

**7.** The contact forces of matter come more from (a) the electrons, (b) the protons, (c) the nuclei, (d) the neutrons.

**8.** Ten of the chemical elements were known in antiquity. Do you think they would be the *most* reactive elements or the *least* reactive? (Guess a few of these ten.)

**9.** What force keeps atoms together? What force keeps nuclei together?

**10.** If electrons are so light compared to nuclei, why don't atoms just slide through each other when they collide (unless their nuclei run together)?

**11.** The weight of matter comes more from (a) the electrons, (b) the protons, (c) the nuclei, (d) the neutrons.

**12.** The volume of matter comes more from (a) the electrons, (b) the protons, (c) the nuclei, (d) the neutrons.

**13.** When nitroglycerine is made in a laboratory, would you expect the chemical reaction to be exothermic or endothermic? Why?

**14.** By 1970 chemical publications listed about 3 million pure substances. Over 2.5 million of those were organic, containing carbon atoms. Comment on this fact.

**15.** A **mixture** is a combination of substances that are not chemically connected, so the ingredients can be in any percentage whatsoever. Classify each of the following as (a) mixture, (b) compound, (c) element: ink, milk, ice, apple pie, copper, bread, air, silver, sugar, a soft drink.

**16.** In days of yore, alchemists spent a lot of time trying to change mercury (Hg) into gold (Au). Why mercury? What would they have needed to do to mercury to turn it into gold?

**17.** The chemical properties of matter come more from (a) the electrons, (b) the protons, (c) the nuclei, (d) the neutrons.

**18.** What is the mass of 1 atom of gold?

**19.** The nucleus of nitrogen has 7 protons and 7 neutrons. What is its atomic weight? Its atomic number?

**20.** Some 0.2 percent of all oxygen atoms weigh 18 amu rather than 16 amu as most do. What is the difference in their structures?

**21.** The periodic table tells you that a gold (Au) nucleus has 1 more proton than a platinum (Pt) nucleus. Yet the gold nucleus on the average weighs 3 amu more than platinum. Explain why.

**22.** Copper (Cu) and silver (Ag) are both excellent electrical conductors. Would you expect gold (Au) to be one? Why?

**23.** Consult the periodic table to see if calcium (Ca) should form compounds more as potassium (K) does or more as magnesium (Mg) does.

**24.** Find the molecular weight in amu of the following compounds: (a) $Fe_2O_3$, (b) $H_2O$, (c) NaCl, (d) $CO_2$, (e) $N_2$, (f) $O_2$.

**25.** Which has the most protons in its nucleus? (a) gold (Au), (b) silver (Ag), (c) copper (Cu).

**26.** Explain why a radium atom with an atomic mass number of 226 isn't much larger than a lithium atom with an atomic mass number of 7.

**27.** The most accurate balances detect differences in mass of about one-millionth of a gram ($10^{-6}$ g). What does this imply about finding an atom's mass?

**28.** Think of how flexible the insulated copper wires on the extension cords of electrical appliances are. Metals are often malleable (easy to shape) and ductile (easy to draw into wires). What does this say about the strength of the metallic bonds?

**29.** The planets have potential energy because they are in the force field of the sun's gravity, which can do work on them. What force field gives electrons in atoms potential energy?

***30.** When you buy a pound of table salt (NaCl), what percentage of that pound is sodium?

***31.** About how many atoms of carbon are in a piece of soot that weighs $1 \times 10^{-6}$ gram?

***32.** Is it possible to have two samples of hydrogen gas, $H_2$, with the same number of molecules but one weighing (a) 2 times the other? (b) 3 times the other? (c) 3/2 times the other? (d) 5/2 times the other? Explain.

***33.** There are twice as many hydrogen atoms as oxygen atoms in water ($H_2O$). Show that water normally contains 11⅑ percent hydrogen by mass *and* weight. Since "a pint's a pound" as far as water (or beer) is concerned, how many ounces in a pint of water (or beer) are due to hydrogen? (16 oz = 1 lb.)

***34.** Show that a single cubic centimeter of graphite should contain roughly $10^{24}$ carbon atoms. (*Hint:* The volume of the cube is height × width × length, and each side is roughly $10^8$ carbon atoms long. Actually, because carbon atoms average somewhat more than $1 \times 10^{-8}$ centimeter across in graphite, there are only about $10^{23}$ atoms in a cubic centimeter of graphite.)

***35.** Refer to Exercise 34. Do you think you will ever see $10^{24}$ separate objects in your lifetime? Assume 80 years and show that in each and every second of your life you'd have to inspect between $10^{14}$ and $10^{15}$ objects.

***36.** Refer to Exercises 34 and 35. If you had a fast electronic counting device to help you *count* to $10^{24}$, show that you'd need more than 32 million years to count to $10^{24}$ even at the rate of a billion counts a second.

***37.** A red blood cell (erythrocyte) in your body is about $7 \times 10^{-6}$ meter in diameter. If you could count the atoms in a line across that cell, about how many would you find?

****38.** The radius of a new tire for a VW Beetle is about 13 inches, and its tread is 1 centimeter thick. If it goes 40,000 miles before its tread is used, what is the average "thickness" of tire lost during each revolution?

# CHAPTER 10
# SOLIDS

Early in history, building stones were shaped or carved to fit together. But no matter how stones are stacked to make a wall, there are limits to the wall's stability and strength. The Romans were the first to perfect a stonelike cement to anchor building stones together (Fig. 10–1). Their natural cement was a lime paste added to volcanic ash that would set (harden) in the presence of water. This cement was used in the construction of the Colosseum, the Pantheon, the Appian Way, and the aqueducts. After 2000 years the remains of these structures prove how well the Romans mastered that art. With the fall of the Roman Empire, however, the secret of cement making was lost for many centuries.

In 1796 in England, James Parker burned impure limestone, added water, and watched the mixture slowly harden into a stonelike solid. He named his discovery *Roman* cement. It was used to build the Erie Canal in the 1820s. Then in 1824 Joseph Aspdin, an English bricklayer, patented a cement of superior strength. He burned limestone and clay together; the clay melted and fused with the lime, and he cooled the resulting substance and ground it into a powder. When water was added, this powder slowly combined chemically with the water in an exothermic reaction. The paste hardened into a stonelike material resembling limestone quarried from the region called Portland on the coast of England. Today all over the world, this *Portland cement* is in those thousands of cement-mixer trucks that roll out of cement plants to building sites.

Concrete is a mixture of cement and an aggregate such as sand or gravel. As the cement hardens, it tightly locks the aggregate particles into place. The largest concrete structure in the world is the Grand Coulee Dam on the Columbia River in the state of Washington (Fig. 10–2). The total amount of heat released as the poured concrete cured in this dam was enormous. Left alone to cool, the concrete dam would have taken decades to come to the temperature of its surroundings. To speed up the cooling process, several thousand miles of water pipes were laid throughout the concrete as it was poured, and cold water was pumped through the pipes. This great manufactured stone, precision-fit to a river channel, backs up water in a lake that stretches 151 miles upriver and provides much-needed flood control and cheap electricity. ■

**FIGURE 10–1**
Cement was used in parts of the Colosseum in Rome and in the waterways of the aqueducts.

**FIGURE 10–2**
The Grand Coulee Dam.

## The Solid State of Matter

What's a solid? Sometimes familiar things are hard to put into words. "Rock solid." "Frozen solid." "Solid gold." A good definition of a **solid** is that it tends to keep its shape when it is left alone. But that doesn't mean a solid is necessarily rigid. Rubber bands, books, the clothes you wear—these flexible materials maintain their shapes to some degree. They aren't rigid, but they are solid.

**FIGURE 10–3**
Every solid has its melting point, whether it is a stick of butter or a gold ring. While this butter melts at room temperature, 14-karat gold melts at a much higher temperature (about 1600°F).

At the atomic level, the atoms or molecules bonded together in a solid *stay in place* with respect to their nearby neighbors. They jiggle about, but it is safe to assume they don't normally change places. If they did, they would flow, eventually turning the solid into a shapeless puddle. A solid's strength and rigidity depend to a large degree on how strong the bonds are between its molecules or atoms.

Of course, no solid's molecules or atoms can stick together when exposed to extreme temperatures; they simply jiggle faster and faster until their bonds break and they *melt* (go into the liquid state) as an ice cube does, or *sublime* (go directly into the gaseous state) as moth balls do. So a solid is only one form matter can take, and it is the state that needs the coolest environment (Fig. 10–3). In a sense every piece of solid matter is frozen into the solid state because the temperature is low enough for it to exist. The 14-karat gold ring worn on someone's finger is frozen solid only so long as it remains colder than about 900° Celsius (roughly 1600° Fahrenheit).

Solids are not the most common form of matter in the universe. The stars are large, dense collections of matter, but none of it is in the solid state. They are immense furnaces of gases and **plasma** (a gas of ions) far too hot for the particles to hold each other in one place. Plasma is sometimes called the fourth state of matter. The temperatures here on earth are cool enough for many elements and molecular compounds to rest as solids, yet not so cold that all of them freeze. This important factor lets earth have a gaseous atmosphere and its accompanying liquid oceans. On the other hand, we stand and walk on terra *firma,* the 29 percent of planetary crust that lies above sea level. Moreover, almost everything we do is tied to our solid environment: living in solid houses, creating and marketing solid goods, even reading all those solid books.

**FIGURE 10–4**
When these houses were built in Belle Glade, Florida, in the 1920s, they were built on pilings on level ground. Soil erosion in the area has left them up in the air. The ground we live on isn't as unchanging as it might seem.

## The Structure of Solids

**FIGURE 10–5**
Obsidian is melted rock (magma) that cooled so fast that there was no time for atoms to jostle around into a crystalline structure. Glasslike, obsidian was once prized by Indians for the making of arrowheads.

In any solid the atoms or molecules are in fixed positions. When there is an order, that is, a pattern in the placement of the molecules or atoms that repeats throughout the solid, it is called **crystalline.** Examples of crystalline solids are table salt, diamonds, quartz, and ice. If the molecules or atoms in a solid have no particular arrangement, fitting together in a seemingly random way, the solid is called **amorphous.** Plastics, glass, and the cement in concrete are examples of amorphous solids. However, many solids have mixed structures. Rocks such as sandstone and granite are amorphous composites of small crystals of different chemical compositions.

Whether a solid is crystalline or amorphous depends on how it is formed. For example, suppose melted rock (called **magma**) cools very fast, as when magma vents from a volcano at earth's surface. The molecules have no time to find places in a crystalline pattern; besides, there's little incentive for the cooling atoms to get together in an orderly arrangement unless they're under pressure. That magma hardens into an amorphous solid; sometimes it even looks like glass. When magma cools while underground, it cools more slowly and under pressure. The resulting rock has grains of mineral crystals in it, giving it a rough texture. (A **mineral** is a naturally occurring inorganic compound, and over 2000 are known. *Inorganic* means "containing no carbon atoms.") Especially slow cooling can result in very large crystals on occasion. This same process affects the quality of ice cream. To get the smooth consistency prized in top-quality ice cream, commercial producers control the crystallization process. They must take the new ice cream mixture to $-40°$ Fahrenheit as quickly as possible. Ice cream that is frozen too slowly is very grainy in texture because of the large crystals; rapid freezing of the mixture produces only microscopic crystals.

**FIGURE 10–6**
Granite forms underground when mixtures of molten minerals slowly cool, and the crystalline grains are the result.

Even if in trace amounts, impurities in a crystalline solid often affect its physical properties such as color and even hardness. Ordinarily a natural diamond (a crystal of carbon atoms) has a faint blue color due to the presence of one boron atom for every million carbon atoms. If a diamond has one atom of nitrogen interspersed among 100,000 carbon atoms, it is no longer clear and blue, but yellow instead. Clear, colorless aluminum oxide, $Al_2O_3$ (the mineral corundum), becomes pink sapphire if a small percentage of chromium atoms are interspersed throughout the corundum crystal. A slightly larger percentage of chromium turns the corundum into the deep red mineral called ruby.*

## The Strength of Solids

The bonds between the atoms or molecules aren't the only factor that determine the strength of a solid. Diamond, the hardest mineral known, and graphite, which is so soft and slippery that it's used to lubricate door locks, are both pure forms of carbon atoms, held together with

*Impurities even affect ice cream, though in a different manner. If the workers who make the mix wear nail polish or perfume, the ice cream picks up the "flavor."

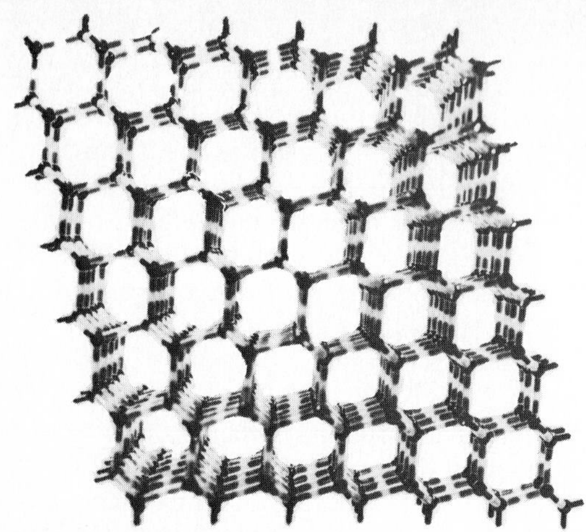

**FIGURE 10–7**
In the diamond crystal, the carbon atoms are bound by strong bonds in various directions.

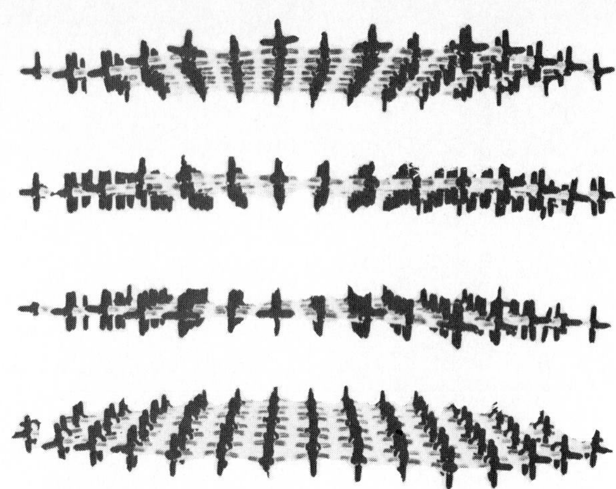

**FIGURE 10–8**
In graphite, atoms of carbon are bound with strong bonds only in a single plane.

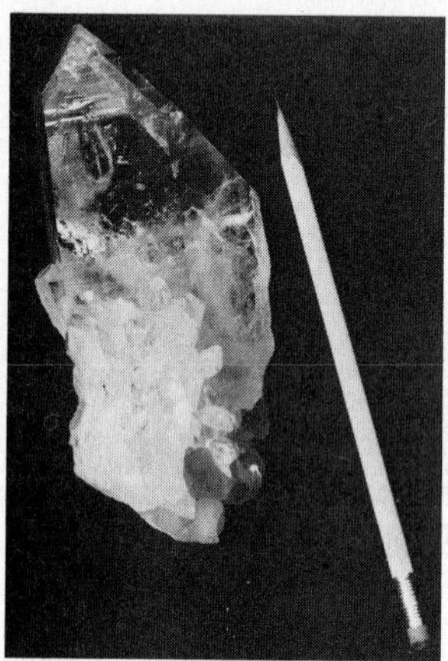

**FIGURE 10–9**
Faces in a natural crystal of quartz.

covalent bonds. In this case, the difference that makes one hard and the other soft are the arrangements of the atoms in their particular crystalline structures. The atoms in the diamond structure are in a closely packed pattern, held rigid by bonds pointing at various angles in all three dimensions (Fig. 10–7). In graphite, strong covalent bonds bind the atoms together in planelike sheets. These sheets are held in place by relatively weaker attractions between the neighboring parallel layers. Thus, the individual sheets of atoms can easily slip past each other, allowing the graphite in a door lock to act almost like a liquid when a key slides in to turn it (Fig. 10–8).

Large crystals often have surfaces that consist of intersecting planes. These planes are called **faces** of the crystal. A crystal's faces are planes of symmetry in the crystal structure itself. (See Fig. 10–9.) Along planes parallel to these crystalline faces, the atomic attractions are usually much weaker than along other directions. As a result, even the strongest crystals can be **cleaved** (or cleanly broken) with relatively little force along planes parallel to crystal faces (Fig. 10–10). Even diamond, the world's hardest mineral, cleaves easily along certain planes with a single directed tap of a gemcutter's mallet.

Often a crystal can yield, or slip, along crystal planes without cleaving into two pieces. Such slippage is called a **dislocation.** We'll see shortly that dislocations can affect the strength of materials.

Since there are no planes of symmetry in an amorphous substance such as glass, it breaks or shatters with fewer limitations due to its structure. A break like this is called a **fracture** (Fig. 10–11). Crystalline solids can fracture too if a large force acts in a direction not parallel to a crystal face. The great care a gem cutter takes in cleaving a diamond is partly to ensure that the gem won't fracture, or crack along lines not controlled by its internal symmetry.

It is possible to make a crystalline solid stronger by blocking the easy directions of yield in the crystal structure. Sometimes dislocations

FIGURE 10–10
Though the atoms themselves are extremely hard, crystals can be cleaved between planes of atoms relatively easily. A macroscopic model of a crystal that shows this behavior can be made if billiard balls are stuck together with glue. Struck with a sharp object between the planes of the balls, the "crystal" yields, whereas striking one of the balls itself might not crack the structure.

in a crystal block themselves. This can happen if enough dislocations occur throughout the crystal so that they become twisted and tangled and stop each other's motion (Fig. 10–12). A perfect crystal of copper several inches long easily bends to one side with only finger pressure. Once it is bent, however, the many microscopic dislocations that resulted crisscross and interrupt the easy planes of slippage, and the crystal cannot be bent back to its original shape with all the force a person can muster.

Two other methods serve to harden crystalline materials by making dislocations more difficult. One is simply to make the solid grainy, composed of smaller crystals locked together in all directions. This can sometimes be done by heating the solid to its melting point and cooling it quickly. Then the scrambled grains, if bonded tightly, make dislocations more difficult and cleavages impossible. The second method is to mix an impurity atom or molecule or a larger particle into the material in a specific amount. Particles interspersed in the crystal disrupt the perfect repetition of symmetry and bond strength. Such impurities can serve to block dislocations and give greater strength to the material.

When metallic elements are mixed with other metals (or even nonmetals) for the purpose of altering the physical properties such as strength, the new material is called an **alloy.** Certain alloys played a large role in the early development of civilization, and others are indispensable in our current technology.

## Alloys

Like gold and silver, the metal copper (Cu) is sometimes found in the pure state in nature, so it's not surprising that archaeologists have found bits of worked copper that date back to about 8000 B.C. Durable, yet easy to hammer into various shapes, no doubt it was highly prized. There is evidence that about 6000 B.C. copper had been melted in open fires, and around 5000 B.C. copper artifacts were placed in graves in

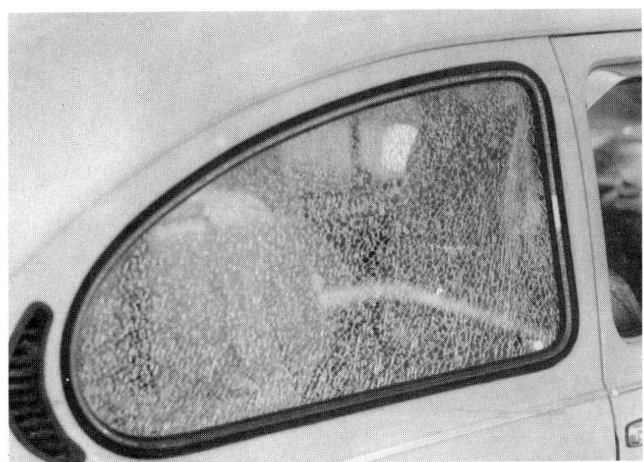

FIGURE 10–11
Glass, without a crystalline structure, fractures in random directions.

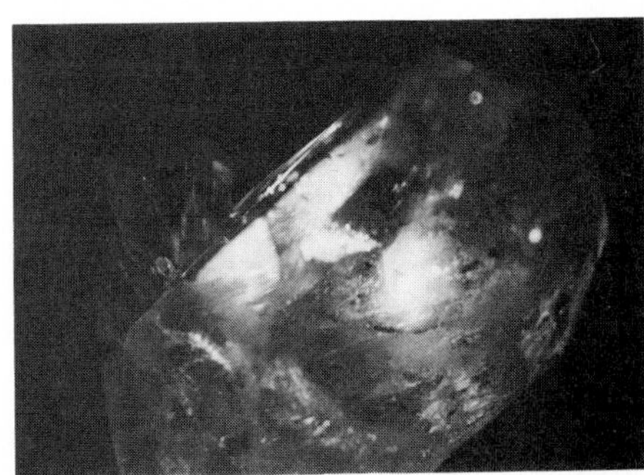

FIGURE 10–12
Light from a laser beam scatters from the fractures and dislocations in the crystal of Figure 10–9.

## THE STRENGTH OF NATURAL AND FABRICATED MATERIALS

A blade of grass or a piece of bamboo splits easily along its length, but either stubbornly resists tearing across its width. Each has long, cellular fibers aligned along its principal direction of growth. In terms of structure, it's relatively easy to cleave the fibers but not to cut across them. These materials can be woven into containers like baskets because the weave interlocks the fibers so that in any direction at least one component gives strength. Wood, too, is easiest to split with the grain. Some kinds of wood, however, have fibers that twist around each other, and it's actually easier to saw those kinds in a direction perpendicular to their fibers rather than along the grain. A sheet of paper is strong because the wood fibers (and sometimes even cotton fibers) twist and overlap in every direction, forming a flexible mat.

A woven rope or even a steel cable gains strength because the component fibers or wires act together to resist tension. The weave does more than just keep them neatly together. The bending and twisting around each other brings the mighty contact force into play between the fibers, and friction between the strands brings them all into action so that none of the single strands gets overloaded. Even so, a woven rope is very flexible in the direction perpendicular to its length and may be coiled for storage or transport.

Cloth is woven to be stretchable or flexible even though an individual thread of its material is not. In a simple vertical and horizontal *weave,* such as in handkerchiefs and tablecloths, the material will not stretch much in either the horizontal or the vertical direction. When pulled at a diagonal angle, however, it stretches in that direction. (People who sew call that diagonal angle the *bias* of the material.) Other materials called *knits* allow varying degrees of flexibility in all directions; inspect the fabric of a T-shirt and a synthetic garment to see how this is done.

You can easily turn or crease the pages of this book, and you can fold your clothes into a small space. It's the *macroscopic* structure of the mat or the weave that lets you do these things. The individual fibers are flexible because the materials they are made of have weak spots at the molecular level (the bonds are either weak or nonexistent). The strength of these fabricated materials comes more from their macroscopic structures than from their molecular order (as in crystalline solids).

Egypt.* Copper ores were melted in kilns as a coloring agent on pottery, so copper paint probably took humans eventually from the Stone Age into the Age of Copper.



*One copper mine in southeastern Europe had tunnels over 5 miles long, and one million tons of copper ore were removed over a 1000-year period.

Another element occasionally found as a free metal in nature is tin (chemical symbol Sn), a soft, bluish-silver-white element. At some time, perhaps around 4000 B.C., someone melted copper and tin together and mixed them. That metalworker gave birth to **bronze,** a copper–tin alloy (with 10–12 percent tin), and civilization was soon in the Bronze Age. (A bronze rod was found in a pyramid that dates to 3700 B.C.) Bronze was far stronger than copper and held a cutting edge. Razors, sickles, chisels, even safety pins appeared. Bronze rims made wooden wheels durable for the first time. The Bronze Age was to last until after 1200–1000 B.C., when iron ores were melted in furnaces. By 600 B.C. the alloys of iron had replaced bronze, and civilization entered the Age of Iron.

Carbon mixed with molten iron (Fe) in small amounts provides alloys with a variety of properties. When iron has 1 percent carbon or less, it is called *steel*. Structural beams are made of steel with less than 0.2 percent carbon. One-percent carbon steel is good for low-speed cutting tools. If even more carbon is added to molten iron, it begins to flow as freely as water. Iron with up to 4 percent carbon is *cast iron* that is poured into molds and cast into shapes such as frying pans and engine blocks.

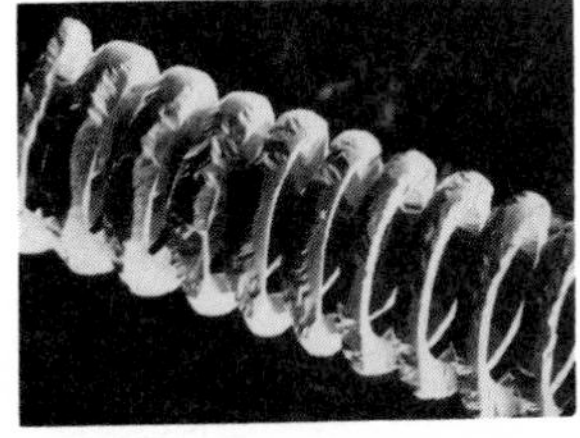

**FIGURE 10–13**
Highly-magnified photo of the tungsten filament of a 40-watt lightbulb after many hours of burning. Tungsten, which has a high melting point, is used to harden steel.

To increase its durability as coinage, gold was eventually alloyed with copper (90 percent Au, 10 percent Cu). In jewelry, 16-karat gold is two-thirds gold and the rest copper and silver, while the 14-karat gold of wedding rings is 58 percent gold alloyed with various percentages of copper and silver. Platinum gold, one form of white gold, is 60 percent gold and 40 percent platinum (Pt).

Many hundreds of alloys are used today for all sorts of special purposes; some, such as tungsten steel and stainless steel, for strength or durability; others, such as spring steel and Bell metal, for special elastic properties. Aircraft alloys of aluminum, magnesium, and titanium provide strength with low density. There are dozens of kinds of brass (copper and zinc); there is pewter (tin and copper with traces of other metals), and German silver (which contains only copper, zinc, and nickel), and solders that are useful for low melting points (and sometimes good electrical conductivity). Even the metals used for typewriter keys and typing elements are special lightweight alloys developed just to take beatings. The metal filling material used for teeth is an alloy of 70 percent mercury and 30 percent copper that expands and contracts at about the same rate as teeth do when the temperature changes.

A recently discovered alloy of nickel and titanium, called Nitinol, reacts with great strength when heated and then cooled. A temperature difference as small as 18 degrees Fahrenheit will send a wire of this alloy into its magic act. As the wire expands during heating, it retains an exact memory of its previous cooler state. When it cools and contracts, its memory proves to be a mighty one. If the wire is held by the ends, it exerts a pressure of 60,000 lb/in.$^2$ as it shrinks. That is a force equal to the weight of 30 Toyotas on 1 square inch! This material could have valuable practical applications. Since very small amounts of heat produce huge forces, machinery made with Nitinol could be used to drive irrigation pumps with just sun-warmed parts. Motors made with

## YOUR BONES: A DIFFERENT KIND OF ALLOY

Your bones are made of two principal materials that are strong only in combination. One is a very elastic organic substance called *collagen* and amounts to a little more than half of a bone's volume. The other is a brittle mineral of calcium, phosphorus, oxygen, and hydrogen that is interspersed with the collagen. A bone whose calcium–phosphorus mineral has been dissolved can be twisted into a circle without breaking. Without the mineral component your skeleton would be too flexible to support you. Dissolve the collagen instead of the mineral, and the calcium–phosphorus compound, though rigid, collapses under only a little pressure. Without the elastic collagen to fill out its porous structures, your bones would disintegrate, fracturing immediately under your weight.

Nitinol wire would run on small amounts of heat and could be used to generate electricity (see Fig. 10–14). The memory of this alloy is astonishing. Tests have taken wire of Nitinol through many millions of cycles with no sign of memory fade or breakage.

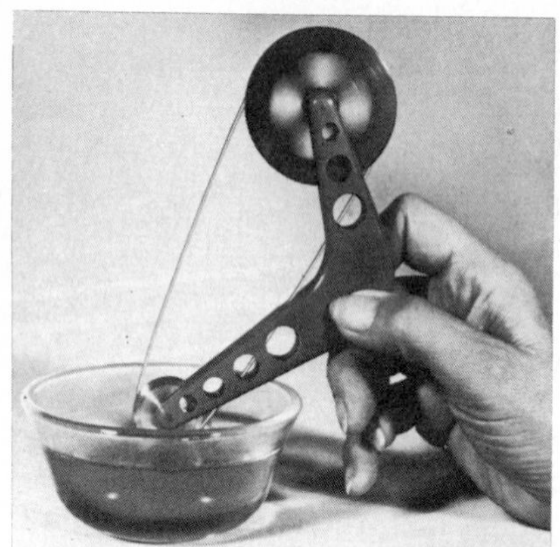

**FIGURE 10–14**
The rapid expansion upon heating and contraction upon cooling of a Nitinol wire drives this machine. The liquid has to be only a few tens of degrees Fahrenheit warmer than the air for the machine to work.

## The Elasticity of Solids

Stretch *or* compress a steel spring, and it will snap right back as soon as you let it go. That's because the bonds between the atoms don't break; instead they act like springs themselves. If the atoms strain directly apart or to the side, the bonds simply attract a bit more to oppose the stretching action. Push the atoms closer together, and their electrons resist; the atoms won't compress much. If the stretching or compressing isn't too great, the atoms will return to their original positions as soon as you stop pushing or pulling. Even the most rigid solid objects show some degree of such elasticity. But every substance also has a point where bonds break and the deformation becomes permanent.

Materials that deform very easily and permanently, like lead and modeling clay, are called **inelastic.** Substances like steel and rubber spring back even after considerable extension; these are **elastic.** Sometimes solids that don't seem to be elastic are. You may think of a building as a perfectly rigid structure, but in a strong wind, even tall buildings sway. Watch the wingtips of an airplane move up and down when you take a flight. Your body, too, is elastic, perhaps more than you know. As you walk around during the day, the weight of your body gradually compresses the elastic connections between the vertebrae in your spinal column. By nighttime, you may be as much as half an inch shorter. Then during sleep, you literally stretch out again. When the astronauts who stayed a month or more in the weightlessness of Skylab returned to earth, they were taller than when they left. Without the daily compressions of their spinal columns, the elastic tissues supported their vertebrae farther apart than normal.

**FIGURE 10–15**
People, too, are elastic, though some are more elastic than others.

## The Great Holdup: Why You Don't Sink into the Ground

When you take a step, your foot puts pressure on the ground. The ground pushes back to stop your downward acceleration, but what happens to the solid ground under your foot? Visualize the molecules directly beneath your foot transferring force to the other molecules below and toward the sides. As the influence spreads among more and more molecules farther and farther away, each molecule is displaced less and contributes less to holding you up.

*It is the strength of the solid's bonds that supports your weight.* If the chemical bonds did not hold those molecules and atoms together, it really wouldn't matter that the atoms and molecules are incompressible. Should the bonds fail, you would sink into the ground as it became fluid.

Even if you stand on sand, the contact forces between the grains cause friction, and the grains you stand on pass your weight among more grains below and to the sides. Once again, the molecules in those grains don't yield because of their molecular bonds, and they bring your descent to a halt.

If you stand on snow or ice, the result is different. To be sure, you can stand there, and the crystals of frozen water will seem to support you just fine. But the bonds that hold water molecules together in ice are not nearly as strong as the bonds in ordinary rock, so the molecules will move under pressure. If you stood in one place on ice, even in well-insulated boots, you would gradually sink as time went on. Storage buildings erected by expeditions on the Antarctic ice cap sink noticeably in a period of months because the ice gives under their pressure. Ice skaters make use of this fact. The skates' thin blades support the skater's weight on a small area, so the pressure on the ice is large. That pressure melts the ice, and the skater glides over a microscopic layer of water with little resistance.

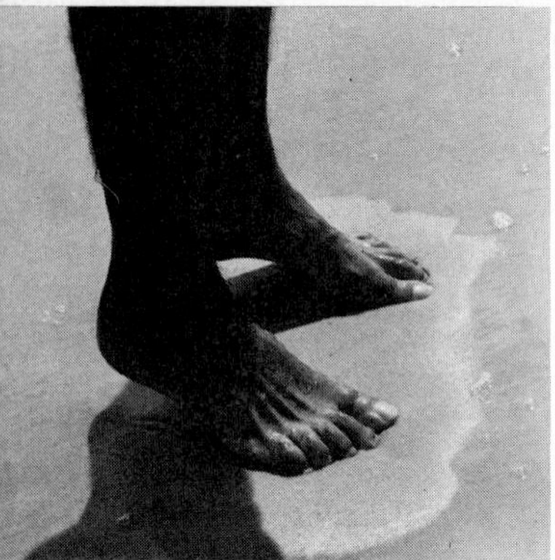

**FIGURE 10–16**
Try this at the beach. Standing on tip-toes just after a wave recedes raises the pressure under your feet. The sand responds by rising around your feet for quite a distance, leaving the water below.

## Describing Solid Objects: Volume, Surface Area, and Density

Aside from the *microscopic* (invisible to the naked eye) view of solids, the *macroscopic* (large enough to be seen with the naked eye) properties of solids are evident everywhere around us. The first step in understanding many interesting properties of the solids in our world is to describe them accurately.

Ask a child to describe something, some object such as a cookie box. Besides giving color, the child will probably compare the size and weight to some other object the two of you know in common. Of course, when we measure things like dimensions or weights, we're doing exactly the same thing—comparing. When we compare objects, one way to talk about physical size is to specify the volume.

All material things occupy space—they have *volume*. Children often have trouble understanding what this means. Pour water from a slender glass into a fat one, and they may say the slender glass had "more" water in it. A small ball of clay seems "bigger" to them if you flatten it. This is a confusion between the height and volume of a glass and between the surface area and volume of a clay ball, each of which has *size*. But the **volume** of anything, regardless of its shape, is always just *the number of cubic centimeters (or other units) enclosed by its surface*.

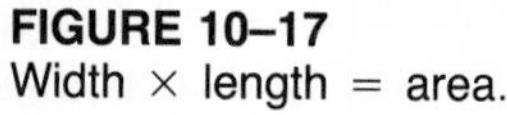

**FIGURE 10–17**
Width × length = area.

FIGURE 10–18
Volume = area × height.

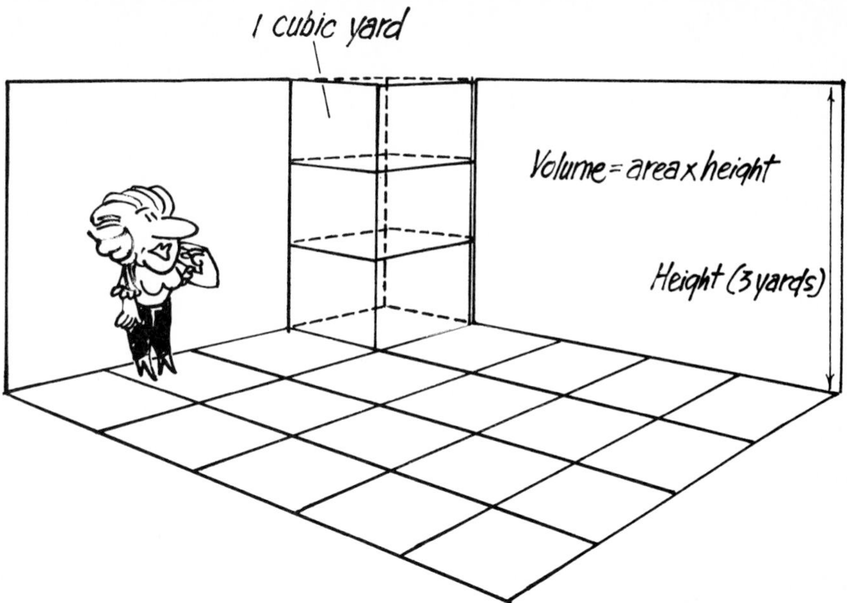


Several ounces of gold may be pressed into a thin leaf or foil to cover the dome of a statehouse; although its surface area becomes enormous, its volume remains the same.

Well, then, how do we measure volume? If something doesn't have a solid rectangular shape, it's not easy to measure its volume by counting cubic centimeters. But you can use an indirect way of measuring. For example, if you totally immerse a nonabsorbent solid of any shape into water, it pushes aside, or *displaces,* a volume of water equal to its own volume (Fig. 10–19). A solid submerged in a graduated cylinder raises the water level, and the extra cubic centimeters tell you exactly what the object's volume is.

(a)

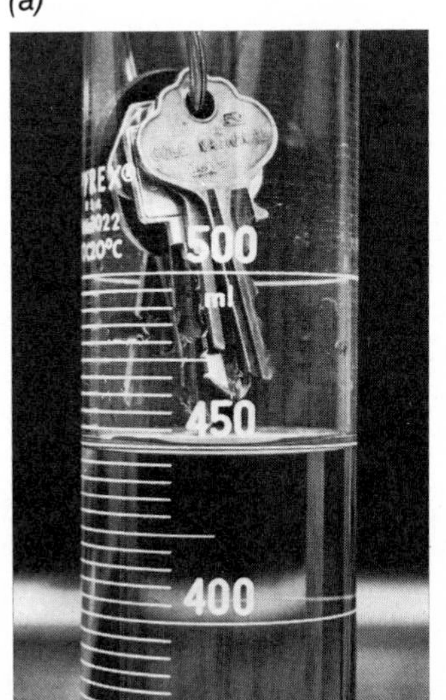


(b)

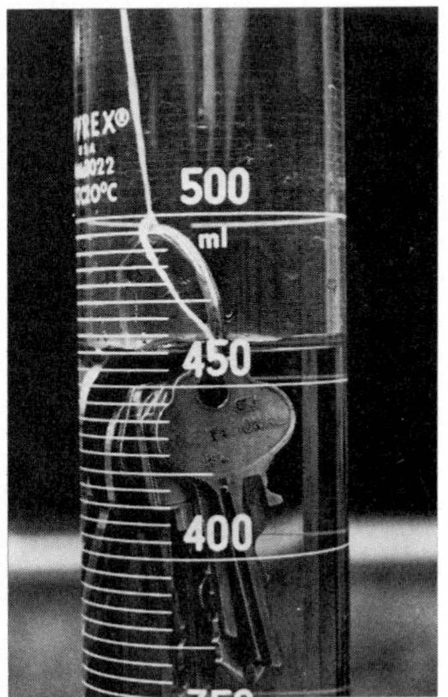


FIGURE 10–19
The number of cubic centimeters the water level rises is equal to the volume displaced by the submerged object.

**FIGURE 10–20**
The density depends on both the mass per object (apple, orange, etc.) and the number of objects per unit of volume.

Sometimes the same amount of mass will occupy different volumes (Fig. 10–20). A quantity we use to show the relationship between the mass and the space it fills is *density,* the ratio of mass to volume.

$$\textbf{density} = \frac{\textbf{mass}}{\textbf{volume}}$$

Whenever matter is spread evenly throughout a volume, its density is *uniform*. Density is usually measured in grams per cubic centimeter. For instance, granite rock averages about 2.7 grams per cubic centimeter; gold's density is about 19.3 grams per cubic centimeter (Fig. 10–21). Compare these with the "standard" of water, which has a density of 1 gram per cubic centimeter. (See Tables 10–1 to 10–4.)

## Pressure and Solids

Here's a good question for you. Which of the two cubes shown in Fig. 10–22 (of the same material) exerts more pressure on its base? One cube is twice as long, twice as wide, and twice as tall as the other. The larger cube has more mass and weighs more. Think of the large cube as a collection of 8 cubes, each equal in size to the small cube. The larger cube, then, weighs 8 times as much as the smaller. Yet that doesn't mean the pressure on its base is 8 times more than the pressure on the base of the smaller cube. Because pressure is force/area, you have to take into account the larger area the big cube rests on. Each face of the

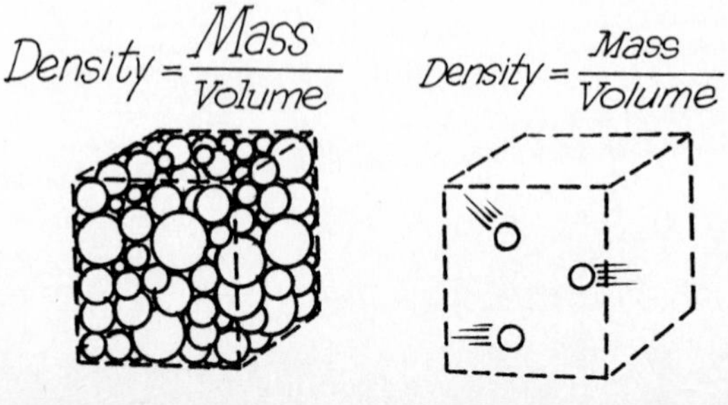

Density = Mass/Volume    Density = Mass/Volume

**FIGURE 10–21**
Though the volume of these two cubes is the same, the number of atoms they contain is quite different. Even if the atoms are of the same type, the density of matter in the cubes is very different.

**TABLE 10–1**

**APPROXIMATE DENSITIES (IN GRAMS PER CUBIC CENTIMETER)* OF SOME ELEMENTS**

| | |
|---|---|
| Calcium | 1.5 |
| Carbon: graphite | 2.25 |
| diamond | 3.5 |
| Aluminum | 2.7 |
| Iron | 7.8 |
| Copper | 8.9 |
| Silver | 10.5 |
| Lead | 11 |
| Gold | 19 |

*For example, the mass of one cubic centimeter of gold is about 19 grams. For comparison, the density of water is 1.0 gram per cubic centimeter at room temperature.

**TABLE 10–2**

**APPROXIMATE DENSITIES (GRAMS PER CUBIC CENTIMETER) OF SOME COMMON OBJECTS AND SUBSTANCES**

| | |
|---|---|
| Cork | 0.25 |
| Paper | 0.7–1.15 |
| Butter | 0.86 |
| Fat | 0.9 |
| Ice | 0.917 |
| Beeswax | 0.96 |
| Human beings | 1.0 |
| Muscle | 1.1 |
| Nylon | 1.09–1.14 |
| Sugar | 1.6 |
| Bone | 1.7–2.0 |
| Brick | 2 |
| Clay | 2 |
| Granite | 2.7 |
| Cement (set) | 3 |

**TABLE 10–3**

**APPROXIMATE AVERAGE DENSITY (GRAMS PER CUBIC CENTIMETER) OF THE SUN AND PLANETS**

| | | | |
|---|---|---|---|
| Sun | 1.4 | Jupiter | 1.3 |
| Mercury | 5.4 | Saturn | 0.7 |
| Venus | 5.2 | Uranus | 1.2 |
| Earth | 5.5 | Neptune | 2 |
| Moon | 3.3 | Pluto | Unknown |
| Mars | 3.9 | | |
| Phobos and Deimos | 2 | | |

**TABLE 10–4**

**APPROXIMATE DENSITIES (GRAMS PER CUBIC CENTIMETER) OF SOME SEASONED (CURED) WOOD**

| | |
|---|---|
| Pine | 0.35–0.6 |
| Cedar | 0.3–0.4 |
| Spruce | 0.5–0.7 |
| Hickory, maple, oak | 0.6–0.9 |
| Walnut | 0.7 |
| Ebony | 1.2 |

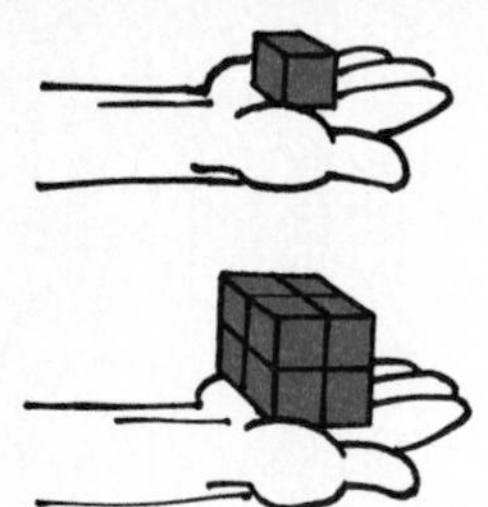

(a)

(b)

**FIGURE 10–22**
*(a)* If one cube is twice as long as the other along any side, it has 4 times as much surface area and 8 times as much volume as the smaller one. *(b)* These two elephants have much the same shapes, but the larger one has a much greater volume. Consequently, *its surface area per unit of volume* is less than that of the smaller elephant.

large cube has 4 times the surface area of a single face of the small cube, so the large cube rests on 4 times as much surface area as the small cube does. With 8 times the weight ($F$) supported by 4 times the area ($A$), the large cube exerts only *twice* as much pressure on its base as the small cube does.

This example points out a very general fact about the relation between volume and surface area. The volume of the large cube is 8 times the smaller cube's volume, but the surface area of the large cube is only 4 times the surface area of the small cube. *The larger cube, then, has more volume per unit of surface area than the smaller cube*. That is, the ratio $V/A$ is greater for the larger cube than for the smaller cube. This general result holds for any shape. **The larger size of any shape has more volume per unit of surface area. Likewise, the smaller size of any shape has more surface area per unit of volume.**

To put this another way, if a fixed shape expands (or is scaled up in size), its volume increases more rapidly than its surface area. You can see the changing relation between volume and surface area as you blow up a balloon. When you first begin, the surface area increases rapidly with each breath. But as the balloon grows, equal breaths of air change the size of its surface less and less. Each breath adds about the same amount of volume* but brings a smaller percentage of change in surface area. Larger balloons, then, have less surface area for each unit of volume, just as for those cubes.

The relation between volume and surface area has important consequences. Imagine this highly improbable but perilous scene: You are traveling in a boxcar with a mouse, a cat, and a hippopotamus when the train suddenly stops. Innocently following Newton's first law, all four of you will slam against the front wall of the boxcar (Fig. 10–23). The pressure from the wall is less on the smallest creatures because they have less mass per unit of surface area of contact. If all of you hit the wall in the same manner, the mouse will be better off than the cat, and the cat will be better off than you. The hippo will have the worst time (unless you are caught between the hippo and the wall!). This is why a small pet might well endure a car crash better than its owner, and why an insect along for the ride is in even less danger.

## The Great Squeeze Play: Pressure and the Solid Earth

At Carleton, South Africa, miners descend each day to work about 2.4 miles beneath the earth's surface. At such depths the pressure on the rock is so great that there can be no such thing as a cave-in; a collapse there means the rock literally *bursts* in. Each 12½-foot depth of rock adds the pressure of about 1 atmosphere, about 15 pounds per square inch. (See Example 3 in "Calculations.") The bonds that keep the rock together around that shaft are marvelously strong.

*It is true that you force the air into the balloon under pressure, which has an effect on the volume of air, as we'll see in Chapter 12. But after the first puff or two, the pressure in a balloon stays nearly constant as the rubber expands.

FIGURE 10–23
The smallest creature, with a greater surface area per unit of volume, receives less pressure when it collides and hence has an easier time of it.

Some rock isn't that strong, of course, especially the porous rock where oil and natural gas are found. When a prospective oil well is drilled deep into the earth, the cylindrical hole made by the drill is subject to almost certain cave-ins. To prevent this from happening, the drilling crews start with a large-diameter drill. At a certain level the drill is pulled out, and a smaller diameter hollow pipe is inserted into the drill hole. The area between the sides of the hole and that casing is then pumped full of—what else—cement! This artificial rock, stronger than the porous rock around it, preserves the expensive well. Drilling to greater depths continues with smaller diameter drills inside the casing, and the process is repeated. The special cements used at great depths can cure rapidly even under high temperatures and pressures. Finally, when the well is as deep as geologists think necessary, small

## THE SHAPE THAT MINIMIZES SURFACE AREA

Work a large lump of modeling clay with your hands. No matter how you shape it, you don't change the density of the clay or its volume. Each solid cubic centimeter of clay still contains the same number of molecules (so its mass is the same), and the lump still displaces the same volume if totally submerged in water. You only change its surface area.

To make the lump as compact as possible so it has the least surface area for its volume, you want to round it into a ball. *A sphere has less surface area per unit of volume than any other shape*. Or from another viewpoint, *a sphere encloses more volume per unit of surface area than any other shape*.

Use your fingers to press a large elephant ear out from each side of the ball of clay and pull a nose out. You've obviously increased the surface area significantly without changing the volume. When you smooth them back into a round ball, the surface area for the same volume decreases.

explosive charges are lowered to the proper depth and set off. That *perforates* the cement-lined well to let any oil or gas through to be pumped to the surface.

Deep in any well or mine, the temperature rises, the result of heat released from the natural radioactivity of some of the minerals in the earth's interior.* At the lowest level of that mine in South Africa, the temperature is 131°F. One natural-gas well drilled in Oklahoma reached a depth of 5.95 miles. The temperature there was 475°F.

Even deeper into the earth the pressure grows relentlessly, squeezing the atoms tighter to make the rock more dense. At depths from about 5 to 25 miles, at the boundary between the earth's crust and mantle, the very molecular structure of the minerals changes (Fig. 10–24). Because of the extreme pressures and temperatures, the rock there can flow where there is an imbalance of pressure. This contributes to the movement of the earth's continents above at rates of a few centimeters per year. Another unseen happening takes place at depths of about 100 miles and temperatures of about 2700°F; crystals called diamonds slowly form in the molten rock.

**FIGURE 10–24**
These rocks were subjected to pressure after they had formed, wrinkling the layers.

**FIGURE 10–25**
Artificial diamonds are manufactured from grains of carbon (in the background), the same material a pencil lead is made of.

Understandably, the pressure is greatest at the very center of the earth. The atoms there must support all the overlying weight of the earth and the atmosphere. They squeeze close together, increasing the density to nearly 20 grams per cubic centimeter for combinations of atoms that might have densities of only 7 to 9 grams per cubic centimeter if they were at the surface.

Geophysicists study the behavior of matter under high pressures in a diamond anvil cell that forces two cut diamonds together with a mechanical device. Since diamonds are transparent, lasers can reveal what happens to the specimen of matter in the cell as the pressure rises. New solid structures form from ordinary minerals, and chemical processes occur that explain how iron could separate and sink toward the earth's center. But this type of experiment has its limits. When the pressure gets as high as the pressure at the boundary between the mantle and the core of earth, some 25 millions pounds per square inch, the molecular bonds of the hardest mineral in nature finally give way: The diamonds in the cell flow as freely as melted plastic.

---

*Geologists think some of that heat also comes from friction when plastic-liquid matter moves slowly past more solid matter.

## CALCULATIONS

Density, mass/volume, has the symbol $\rho$ (Greek letter *rho,* pronounced "row"). Thus,

$$\rho = \frac{m}{V}$$

If a rock has a volume of 41 cubic centimeters and a mass of 117 grams, its *average* density, $\rho$, is 117 g/ 41 $cm^3$ = 2.85 grams per cubic centimeter. When you know the density, you can find the volume if you know the mass, or the mass if you know the volume.

The surface areas and volumes of solids enter into many practical situations, as you will see in the exercises. Here we'll inspect their values for two of the simplest solid shapes, rectangular solids and spheres.

From Fig. 10–17 you can see that the surface area $A$ of a rectangle is $A = l \times w$ where $l$ and $w$ are the lengths of its two perpendicular edges. From Fig. 10–18 you can see that the volume $V$ of a rectangular solid is $V = l \times w \times h$, where $l$, $w$, and $h$ are the lengths of its three perpendicular edges. The surface area $A$ enclosed by a circle is $A = \pi r^2$, where $r$ is the circle's radius. For a sphere, its surface area is $A = 4\pi r^2$, while the volume it encloses is $V = \frac{4}{3}\pi r^3$.

---

EXAMPLE 1: A dedicated student taking this course decides to wallpaper a wall of his room with the pages of this book to make it easier to study. The pages are about 7 inches by 9 inches, and his wall is 8 feet high and 12 feet long. **How many pages are needed to cover a wall?**

In terms of square inches, one page is about 7 in. × 9 in. = 63 $in.^2$. One square foot is 12 in. × 12 in. = 144 $in.^2$. The area of the wall is 8 ft × 12 ft = 96 $ft^2$. So the area of the wall divided by the area of one page gives

$$\frac{96\ \text{ft}^2 \times 144\ \text{in.}^2/\text{ft}^2}{63\ \text{in.}^2/\text{page}} = \textbf{219.4 pages}$$

(But unless he used pages from 2 copies of the book, he'd be able to study only every other page.)

---

EXAMPLE 2: An architect is told to design an economical house with 2500 square feet of living area. **To save on the costs of materials, should he design the house in the shape of a square or an elongated rectangle?** Calculate the length of wall material needed to enclose a floor area 50 ft by 50 ft versus a floor area 100 ft by 25 ft.

Even though both houses would have the same amount of floor area, they would have different total lengths of wall to enclose those areas. **The square house** would have **200 feet of walls** (4 × 50 ft) as measured along the baseboards, and **the rectangular house** would have **250 feet of walls** (100 ft + 100 ft + 25 ft + 25 ft). A circular house would save even more material.

---

EXAMPLE 3: Suppose you were given many identical cubes of granite rock, 1 inch along each edge. **How many would you have to stack on edge for the pressure beneath the bottom block to be equal to 1 atmosphere** (14.7 lb/$in.^2$)?

From Table 10.1, the density of granite is 2.7 g/$cm^3$. Since 2.54 cm = 1 in., $(2.54\ \text{cm})^3 = (1\ \text{in.})^3$, and 1 $in.^3$ = 16.4 $cm^3$, so 1 $in.^3$ of granite contains 44.3 g of granite. Its weight is *mg*, or 0.0443 kg × 9.8 m/$s^2$ = 0.43 N = 0.098 lb. To have a weight of 14.7 lb/$in.^2$ on the base of a stack of such cubes, you'd need 14.7 lb/0.098 lb = **150 cubes,** which would be 150 in./(12 in./ft), or about 12½ feet tall.

You should convince yourself that a column of granite 12½ feet high of *any* uniform cross section will exert one atmosphere of pressure on its base.

## REVIEW

**1.** Define a solid in your own words.

**2.** Name and define the two general kinds of structure in solid inorganic matter. Give some examples of each.

**3.** Both the arrangement of the atoms or molecules and the bonds between them help determine how hard a solid is. True or false?

**4.** A cleavage in crystalline solid occurs between planes parallel to its faces. A fracture does not. True or false?

**5.** What effect can impurities or grains have in crystalline solids?

**6.** Name several properties of solids that may be obtained by combining elements into alloys.

**7.** Is it true even the most rigid solids have some degree of elasticity?

**8.** The strength of molecular bonds keeps us from sinking into the ground. True or false?

**9.** Give the formula for density, which is the relation between mass and the space it fills. Give some typical units of density.

**10.** The volume of space enclosed by the surface of a solid is described by units of length cubed. True or false?

**11.** Far beneath the earth's surface, the weight of the overlying material produces great pressure. (a) Does this pressure *increase* the *density* of matter there? (b) Does this pressure *decrease* the volume of matter there? (c) Does this pressure *change* the amount of mass there?

## DEMONSTRATIONS

**1.** Adding a little at a time, stir salt (NaCl) or sugar ($C_{12}H_{22}O_{11}$) into a cup of warm water until no more will dissolve. Warm the solution on the stove—don't boil it—and add more salt (or sugar) while stirring. Pour the solution (without any undissolved salt or sugar) into a clean glass and suspend a thread. Make certain the thread almost touches the bottom of the glass. Let the solution and string stand undisturbed for a few days.Crystals will form on the string. If you grow a really big crystal of salt, use a single-edge razor blade and a small hammer to cleave it. If you grow crystals of sugar instead, that's rock candy. Help yourself!

To make an *amorphous* solid from sugar, heat a few spoonfuls in a small pan and *slowly* melt it. (If you heat it too fast, you'll get the broken sugar molecules called caramel.) Place the hot pan into a larger container of cold water. The liquid sugar cools quickly into the clear, glossy candy familiar in cough drops, candy mints, and lollipops.

**2.** A few plastic drinking straws and some string let you demonstrate the strength of various geometric structures in two dimensions. Feed some string through three straws; bring the ends together and tie a knot, tightening the string to make sure the ends of the straws touch and form a triangle. Do the same with four straws (a square), then six straws (a hexagon). Pull and push on the straw shapes. The triangular pattern is the only *rigid* framework. It is used in the construction of homes, bridges, and automobile frames.

**3.** Cut some scrap paper into three or four identical regular pentagons, five-sided figures with equal sides and equal angles between the sides. Try to fit the sides of the individual pentagons together to make a pattern that repeats. They won't fit without leaving holes or without overlapping. That's why no crystals in nature have pentagonal arrangements of molecules or atoms; that pattern can't be repeated symmetrically. The same is true for a regular seven-sided figure; try it and see.

**4.** Using a pair of pliers, hold one end of an iron nail in a candle flame. The nail may get red hot but not much else happens. Then use the pliers to hold a piece of steel wool (which is essentially the same stuff) in the flame. It will burn brightly. The increased surface area is responsible for the fireworks. Since the steel wool has much more surface area per unit of mass than the nail does, as oxidation proceeds over the surface of the fine strands of steel, their temperature is raised enough to increase the rate of oxidation to the point we know as burning.

**5.** This demonstration requires caution and a pair of thick work gloves. Find two plates of window glass, clean and *dry them thoroughly,* and press their faces together firmly. Because the faces are very flat, a large number of molecules across the two faces will touch. The surfaces will stick to each other because of the molecular attractions. Often plates like this cannot be pulled directly apart without fracturing. The two sheets of glass can be separated, however, by *cautiously* sliding the faces over each other. This is analogous to "cleaving" them along their planes of weakest attraction. (P.S. It's cheating to use a drop of water between the panes of glass. That involves adhesion and surface tension, subjects of the next chapter.)

# EXERCISES

**1.** How do we know without seeing them that the atoms or molecules in a solid generally keep their places with respect to each other?

**2.** The Incas did not use cement to build the walls shown in Fig. 10–26, yet they've withstood the tests of time. Comment on the reasons for their stability.

FIGURE 10–26

**3.** Both diamond and graphite are pure crystals of carbon atoms. How can you explain their different densities (as seen in Table 10–1)?

**4.** What do you think would happen if the force of gravity suddenly ceased to exist throughout the earth?

**5.** The top of the Jefferson Memorial Arch in St. Louis, Missouri, sways back and forth over a distance of 3 feet in high winds. What property of solid matter does this demonstrate?

**6.** If you break a cube of ice with your teeth, are you most likely breaking *atomic* bonds or *molecular* bonds? Why?

**7.** In terms of the structural arrangements, why are the layers of bricks and concrete blocks in home construction staggered? (See Fig. 10–27.)

**8.** Why should the bricks on the brick road shown in Fig. 10–28 be staggered as they are? That is, why shouldn't this pattern be rotated by 90°?

**9.** How does it help to have sugar and salt in such fine grains for use at the table?

**10.** A heavy metal object resting on a metal surface at only a few points can stick to that surface, a phenomenon know as "cold welding." Explain what could cause that to happen.

FIGURE 10–27

FIGURE 10–28

**11.** Why might a truck loaded with sand or salt travel up an icy incline when a small car cannot? Why does it help to put tire chains on the smaller car?

**12.** To cool a drink quickly, should you use shaved ice or ice cubes? Why?

**13.** Give an argument for keeping coffee beans whole and grinding them just before making coffee.

**14.** Pure uranium metal reacts with oxygen in the air as many metals do. In fact, if a small amount of pure uranium is machined, any small particles of uranium burst into flame, whereas a larger chunk will not. Discuss.

**15.** Exactly why does it help to use small twigs rather than large limbs and branches to start a campfire?

**16.** During a great earthquake in Alaska in 1964, half a mountain collapsed onto the Sherman glacier in the Chugachs range, covering about 10 square miles of its surface What effect might this have had on the glacier's movement?

**17.** Why does it help your digestive system if you chew food well before swallowing?

**18.** What is the advantage of using a phonograph needle made from diamond, the world's hardest mineral? (If you can't guess, be sure to see the answer at the back.)

**19.** Explain why snowflakes fall more slowly than hailstones.

***20.** A single hydrogen atom has a mass of 1 amu (atomic mass unit) and a diameter of about $1 \times 10^{-8}$ cm. One of the largest molecules known has a molecular weight of about 10 million amu, yet its diameter is only about $1 \times 10^{-5}$ cm. Explain how something that's only a thousand times bigger than something else might weigh 10 million times more.

**21.** When the winds have been right, extremely tiny particles of sand from the Sahara Desert have drifted thousands of miles through the air, even passing over Florida, where they were visible as a white haze in the sky. Discuss in terms of surface area and mass (or weight) how this is possible.

**22.** When a milk shake is mixed, it froths with bubbles as air is beaten into the mixture. Does the density of the milk shake change? Does its surface area per unit of volume increase, decrease, or remain the same?

**23.** The next time you see a rotary-type lawn sprinkler, watch it carefully. Why doesn't it just wet a circle at a certain distance rather than the entire area within the circle?

**24.** Why does the government demand that a box of crackers display the weight rather than the volume of the box?

**25.** Bush pilots use this trick to deliver fresh eggs to expeditions in wilderness areas in Alaska. They wrap each egg in a crumpled sheet of newspaper and put them in a cardboard box, which is dropped from the plane as they fly over the adventurers. Discuss the factors that contribute to the soft landing of the eggs.

**26.** Why will an adult skydiver whose weight is twice that of a child have a higher terminal speed than the child will despite having a greater surface area?

**27.** Drops of mist fall more slowly through the air than raindrops because they have more surface area per unit of mass. True or false?

**28.** A child puts a penny on a railroad track and picks it up after a train passes over it. (a) Has its mass or weight increased, decreased, or remained the same? (b) Has its surface area increased, decreased, or remained the same? (c) Has the ratio of surface area to volume increased, decreased, or remained the same? (d) Has its density increased, decreased, or remained the same?

**29.** The rate at which a lollipop dissolves in your mouth should depend on how much surface area there is. True or false?

**30.** Two lollipops weigh the same, but one is spherical and the other is a "squashed" sphere. Which would dissolve at the faster rate initially?

**31.** There is more surface area per unit of volume for the last cubic centimeter of a lollipop to dissolve in your mouth than there is for the new lollipop. True or false?

**32.** Refer to Exercises 29, 30, and 31. (a) Will you dissolve more lollipop per second from a new lollipop or one that's almost disappeared? (b) Will the last cubic centimeter of a lollipop that dissolves in your mouth disappear faster, slower, or at the same rate as the first cubic centimeter you get from a new lollipop? (Careful!)

**33.** Which factors influence the rigidity of a solid? (a) Temperature, (b) pressure, (c) crystalline structure, (d) chemical bonding.

***34.** Suppose you have a cube of modeling clay. If you slice it in a direction parallel to any side, you will add two extra faces of surface area without changing the total volume of the clay. How many such slices does it take to double the surface area of the original cube?

***35.** If 16-karat gold is ⅔ gold and 14-karat gold is 58% gold, what should 24-karat gold be?

***36.** A solid gold nugget weighs 1.4 pounds (so its mass is 0.63 kilograms). What is its volume in cubic centimeters? (Use Table 10-1.)

***37.** Twenty-nine percent of the earth's surface is *not* covered by water. Using 4000 miles for the earth's radius, find the number of square miles of solid land (or ice) on earth.

***38.** A lead brick in a physics lab is 25 cm long, 10 cm wide, and 6 cm thick. What are the mass and weight of such a brick? (Use Table 10-1.)

***39.** A 1-lb rock of basalt displaces 758 $cm^3$ of water. What is the rock's average density? (2.205 kg *weighs* 1 lb.)

***40.** A solid silver coin has a mass of 70 grams. What is its volume? (See Table 10-1.)

***41.** A gallon of a certain paint will cover 400 square feet, according to the manufacturer. About how many gallons will cover a 9-ft-high exterior wall on a rectangular house 50 ft long and 35 ft wide?

***42.** A 50-pound bag of fertilizer should cover 2000 square feet according to the information on the package. How many bags will cover a rectangular lot that is 150 feet by 90 feet if an 1800-square-foot house is on the lot?

***43.** A call to a local pizza place revealed the following prices: cheese pizza, 9 in. diameter, $4.73; cheese pizza, 13 in. diameter, $7.77; cheese pizza, 15 in. diameter, $9.98. Find the costs in cents per square inch.

***44.** A certain telephone pole (treated with creosote) has a density of 1.3 grams per cubic centimeter and is 12 meters high. What pressure does it exert on its base when it is placed erect in the ground? Find the answer in newtons/meter$^2$ and convert that to atmospheres.

***45.** Suppose the telephone pole in Exercise 44 were made of solid aluminum. What pressure would it exert on its base? (Use Table 10-1.)

***46.** A survey of a university bookstore reveals that the average book has the dimensions 8 in. by 10 in. by 1.5 in. If you had five courses a semester, each of which used a different book, and kept your books through four years of school, how many cubic feet of storage would your books occupy?

**47.** The deposits of diamonds found on earth usually occur in ancient volcanic "pipes" or vents. Why should that be?

***48.** Do you think you might be able to support a cubic meter of ice? Find its weight using Table 10-2.

# CHAPTER 11
# LIQUIDS

Ka-thump, ka-thump, ka-thump. About once a second the muscle that is your heart squeezes down, putting pressure on the blood inside and pushing it into the aorta. Like water, blood is nearly incompressible, so when your heart muscle pushes, the blood throughout your body's circulatory system moves. It flows through the major channels called *arteries* and *veins* and into the tinier vessels called *capillaries*. Four to five quarts of this liquid carry molecular fuel and the oxygen needed to "burn" it to all the cells in your body. The oxygen is transported by the hemoglobin molecules in the red cells, which typically account for about 45 percent of the blood's volume. The blood also carries away the molecular waste products from the cells and the waste heat that's generated as the cells burn their "food." The molecular waste is filtered from the blood by your liver and kidneys, and the heat escapes when the blood circulates in the many small vessels close to your skin.

When you are standing, the pressure of the blood in your feet is greater than in your brain because the blood in your feet supports the weight of the blood above it. Lie on the floor for a minute and then stand up quickly. You might feel a little lightheaded until the heart increases the pressure to pump blood up to your head. You can get the same feeling at the fair if a ride accelerates you upward. Along with everything else in your body, your blood becomes "heavier" with an upward acceleration, and it's harder for the heart to pump it up to your head. In fact, if such an acceleration exceeds three times the acceleration of gravity (3 *g*'s) while you are standing, your heart isn't strong enough to pump *any* blood to your head, and you will have a blackout. Your cardiovascular system is built for 1 *g*, and your heart works all the time to pump the blood against the force of gravity from your feet up to your scalp. To keep the rising blood from draining back down between heartbeats, small bands of muscles (called shunts, or cuffs) around the ends of the smaller vessels clamp down, ensuring that blood flows in only one direction. When one of these one-way valves fails for some reason, the blood pools, and the result is a varicose vein.

Without liquid, there could not be life as we know it. A **liquid** has distinct properties: Its molecules are close enough to touch, as those of a solid are, but the molecules of a liquid attract each other with weaker forces than do a solid's molecules. That is why a quantity of liquid will flow, assuming the shape of its container under the action of gravity and contact forces. As it flows, a liquid keeps a constant volume unless it is exposed to air and some of its molecules are lost to evaporation. This chapter is about properties of the *liquid state of matter*. ■

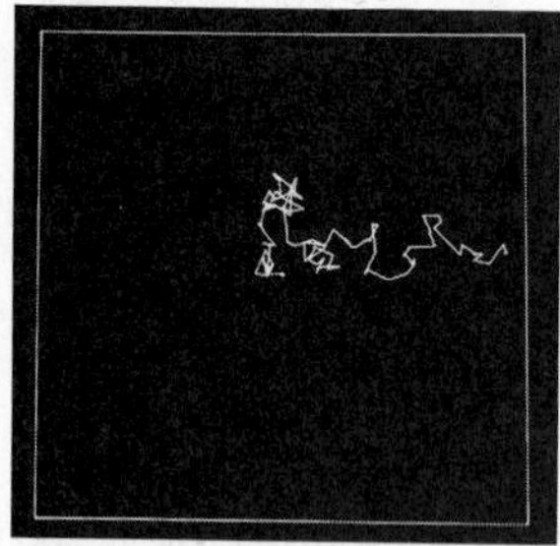

*(a)*

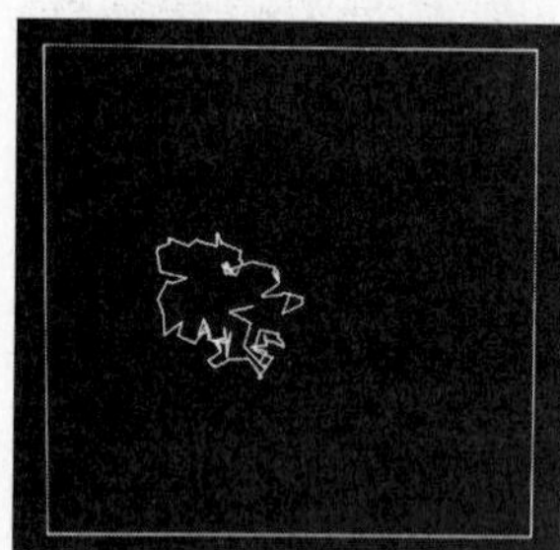

*(b)*

**FIGURE 11–1**
Two instances of particles undergoing Brownian motion (each began at the center of the illustration). Because of the random nature of the collisions and the resulting changes in direction, such particles in a liquid (or even in the air) often get nowhere fast.

## Liquids and Pressure

If you look through a microscope at a sliver of ice, chances are you'll find a tiny particle of dust or lint trapped inside. Within the crystalline array of the frozen water, it seems to have no motion—at least none that you can see. But when the ice melts, the same dust particle becomes animated, jerking frantically like a puppet on a string. In the liquid state, the unseen water molecules continually move past each other and, in doing so, strike this larger, visible particle from all directions. Sometimes the speck gets smacked more frequently on one side than another, and sometimes really fast-moving molecules give it an extra-hard shot. Of course, you can't see the water molecules through the microscope, but you can see the dust particle recoil, dancing in the sea of invisible torpedoes. This three-dimensional lurching of the small particle in water is called **Brownian motion** (Fig. 11–1).

Brownian motion, then, gives us a clue to what is happening in water. But the details of the molecules' motion are even more astonishing. In a glass of water from your kitchen, the invisible molecules have speeds of about a thousand miles per hour on the average. But they get nowhere fast because in that water they are packed tightly enough to touch; as a result, each molecule averages several billion collisions in a single second, and like that jiggling dust particle, moves in a "drunkard's walk," changing directions continually.

The nonstop activity of the molecules causes a constant beating on the sides of the glass. The net force of so many molecules hitting on each square centimeter every second produces a steady pressure on the container. What's more, the water molecules themselves feel pressure because of the pounding from their neighbors. The collisions and rebounds occur at all angles and in all directions, up, down, and sideways. This aimless, random motion of liquid molecules gives the pressure exerted by a liquid a special property: *At any point in a liquid, pressure is transmitted equally in all directions*. This is called **Pascal's principle,** after Blaise Pascal (1623–1662).

Suppose you apply some pressure to a liquid in a closed container, such as a capped tube of toothpaste (Fig. 11–2). There is no extra space between the molecules in a liquid; liquids are nearly incompressible. So the pressure you apply shoves one molecule into the next all throughout the tube in a chain reaction. The pressure you apply is transmitted to every nook and cranny in the confined liquid, and since it is transmitted in every direction (Pascal's principle), all the wrinkles in the tube straighten out if you push hard enough.

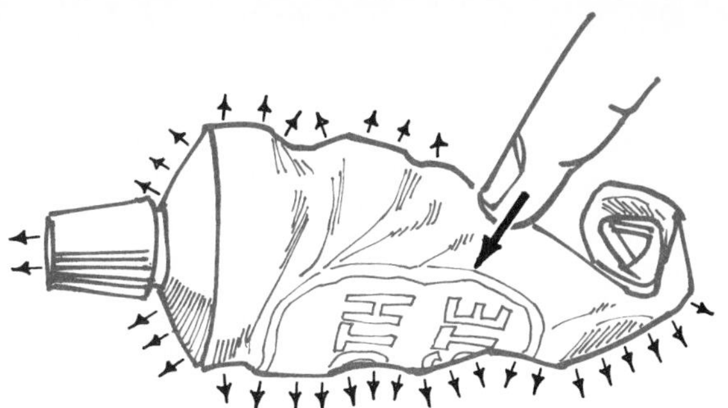

**FIGURE 11–2**
Each time you squeeze a tube of toothpaste, you make use of Pascal's principle: Pressure in a fluid is transmitted in *all* directions.

If you uncap the tube of toothpaste and squeeze on the sides, the thick liquid oozes out of the open end even though the force you apply is not in that direction. This is Pascal's principle again. The molecules, no longer completely confined, respond to the unbalanced pressure and flow. Because the intermolecular forces in a liquid are weak, *a liquid will flow from a region of greater pressure toward any region of lesser pressure*.

## The Pressure from the Weight of Liquids

Each tiny atom or molecule in a solid has weight, so the atoms or molecules at the base of a solid must support the combined weight of all the connected atoms and molecules above. They, in turn, transmit that weight to the surface they lie on.

### USING PRESSURE IN LIQUIDS

When you step on the brake pedal to slow a car, you make use of two characteristics of an enclosed liquid: It is nearly incompressible, and it transmits pressure in every direction (Fig. 11–3). A push on the brake pedal sends a plunger into a closed reservoir of brake fluid. The plunger raises the pressure of the liquid in the reservoir. That pressure passes through the fluid in the metal tubes, called brake lines, that lead to a cylinder on each wheel. When the pressure rises throughout the system, movable pistons in the cylinders are pushed outward, pressing brake pads or brake shoes onto drums or disks that rotate with the wheels. Friction between them slows the wheels. Push harder, and they press together harder, increasing the frictional force and slowing the car faster.

The same properties of an enclosed liquid help to *multiply force* in certain construction equipment, hydraulic lifts, and hydraulic presses. This occurs in a strong chamber that has two cylindrical holes in its walls. The holes have different diameters, and each is plugged with a tight-fitting but movable piston (Fig. 11–4). Liquid (usually a lightweight oil) fills the chamber. For a simple example, suppose one piston has an area of 1 square inch and the other has an area of 5 square inches. If someone pushes on the small piston from the outside with a force of 10 pounds, the piston presses on the liquid in the chamber. The piston exerts a pressure *(P = F/A)* of 10 pounds per square inch. This pressure is transmitted throughout the reservoir; the pressure everywhere in the liquid goes up by 10 pounds per square inch. That pressure is felt by the large piston. Since it has 5 square inches of area, the force on that piston is 50 pounds *(F = P × A)*. It presses outward with a force that is 5 times as great as the force the person applied to the smaller piston.

Though this simple machine multiplies force, you still can't get more out of it than you put into it—in terms of work or energy, that is. Remember that levers and pliars multiply force, too, and in each case there is a trade-off between force and distance (since work = force × distance). There's no principle of physics that says you can't multiply force, but the law of conservation of energy says you can't multiply energy. Example 1 in "Calculations" shows what this means in the case of a hydraulic device. In the example above, it means the large piston would move ⅕ as far as the small piston does.

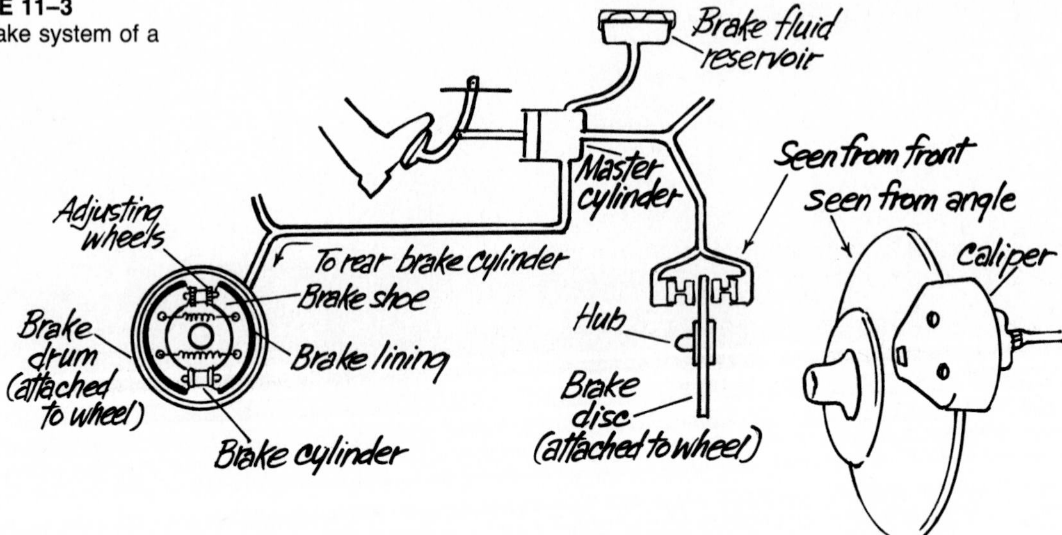


**FIGURE 11–3**
The brake system of a car.

**FIGURE 11–4**
The pressure exerted by the small piston is transmitted everywhere throughout the fluid. Since the large piston has more surface area than the small piston, the total force it receives from the fluid is greater: $F = P \times A$.

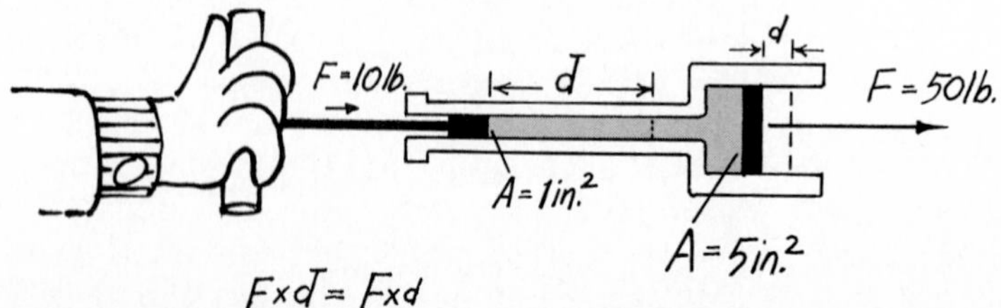

**FIGURE 11–5**
The weight of the water, *mg*, is supported by the area of the bottom of the glass.

**FIGURE 11–6**
The volume of water in this glass is equal to the water's depth times the area of a cross-section of the glass: $V = A \times d$.

**FIGURE 11–7**
The water pressure presses perpendicularly on each point on the surface of the finger. Points at equal depths receive equal pressures.

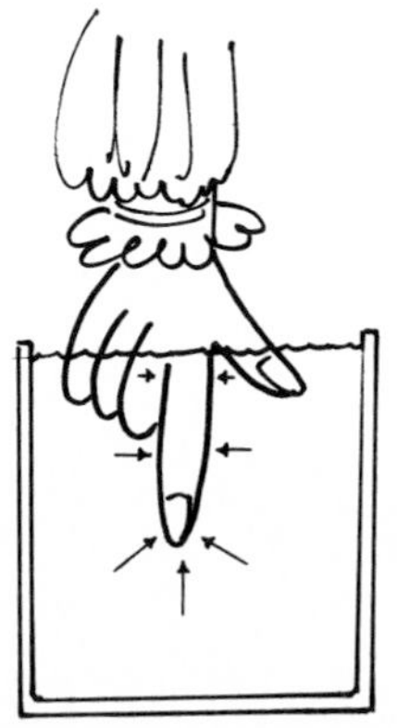

Liquid molecules don't stay still and rigidly support the molecules above them. But at any level the molecules in a glass of water, or in a water puddle, or in the ocean must bear the weight of the water above. And the pressure from the weight of the liquid, just as with solids, is passed on to the solid bottom of the liquid's container. The liquid molecules at various depths don't necessarily move faster or have more collisions per second than those at the liquid's surface; the pressure is merely passed along by the ceaseless molecular bombardment of the nearly incompressible molecules.

We can find a formula for this gravity-caused pressure. Fill a glass as in Fig. 11–5 with water. The bottom of the glass supports the force of the water's weight, *mg,* and the pressure on the bottom of the glass (whose area is $A$) is $P = F/A = mg/A$. We can change this formula for pressure around so it will be useful for any liquid, not just water. For liquids as well as solids, *density* = *mass/volume,* so we can replace the mass $m$ with the quantity (*density* × *volume*). Further, because the glass has straight, vertical sides, we can substitute (*surface area* × *depth*) for *volume*. The pressure due to the liquid's weight is then

$$P = \frac{mg}{A} = \frac{(\text{density} \times A \times \text{depth}) \times g}{A} = \text{density} \times g \times \text{depth}$$

Notice that nothing pertaining to the glass itself appears in the final formula for the pressure. This equation, although derived here for a glass of water, is perfectly general. The pressure 3 feet deep anywhere in a swimming pool is three times as great as the pressure 1 foot down. The pressure halfway down into a glass full of water is half the pressure at the bottom, *regardless of the shape of the glass*. Depth in this formula is *depth beneath the surface of the liquid*. Remember, however, that this is only the pressure due to the liquid's weight. The atmosphere above the surface of a swimming pool or a glass of water exerts pressure on the water's surface, and that pressure too is transmitted throughout the liquid.

You can actually feel water pressure change with depth. If you plunge your arm into a sink filled with water, the pressure is so uniform (even around the hairs on your arm) that you don't really feel the change. But wrap your arm in a waterproof plastic bag and slowly submerge your arm up to your elbow. As the pressure grows, the wrinkles in the bag become taut and pull unevenly on your skin, making you aware of the change in pressure.

Many cities use elevated water towers to supply gravity-fed water to homes and businesses (Fig. 11–8). Because pressure = density × $g$ × depth, no matter what the shape of the container, the water pressure at a home faucet depends on the vertical distance between the faucet and the surface of the water in the tower. (Most large systems use pumps at some point along the water lines to supply extra pressure. This extra pressure passes throughout the connected lines, just as it does in the brake systems of cars.)

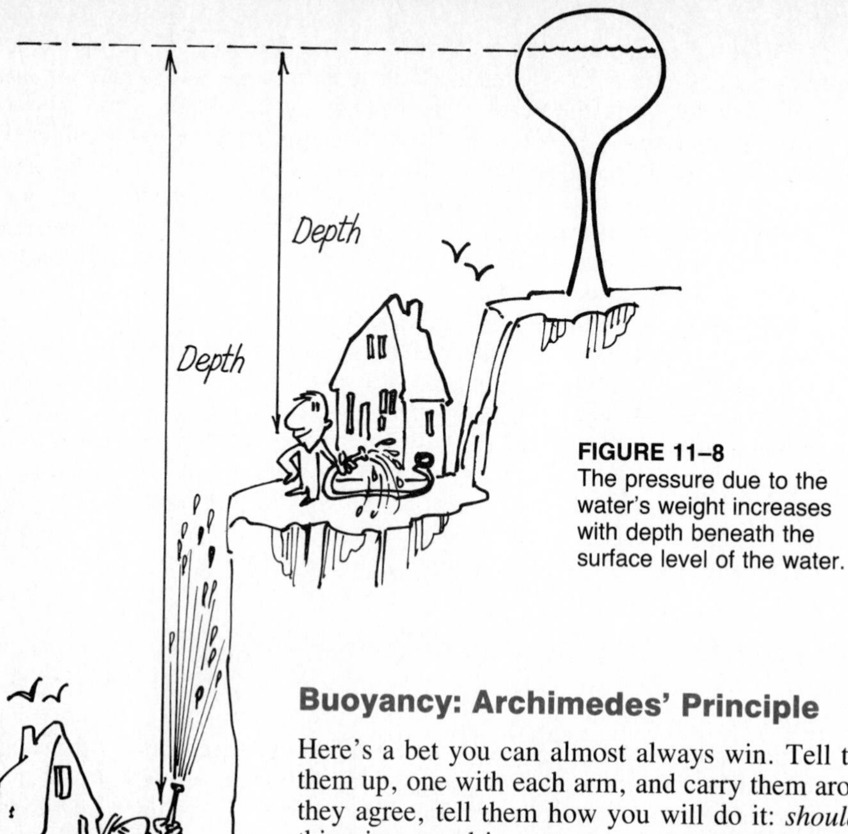


**FIGURE 11–8**
The pressure due to the water's weight increases with depth beneath the surface level of the water.

## Buoyancy: Archimedes' Principle

Here's a bet you can almost always win. Tell two friends you can pick them up, one with each arm, and carry them around for 5 minutes. Once they agree, tell them how you will do it: *shoulder-deep in water*. Anything immersed in water seems to weigh less, and water pressure is the reason.

If you lower your hand, palm down, into water, pressure from the water on the top of your hand pushes it downward, while pressure from beneath it pushes upward (Fig. 11–9). The water under your hand, however, is farther below the surface than the water that's just over your

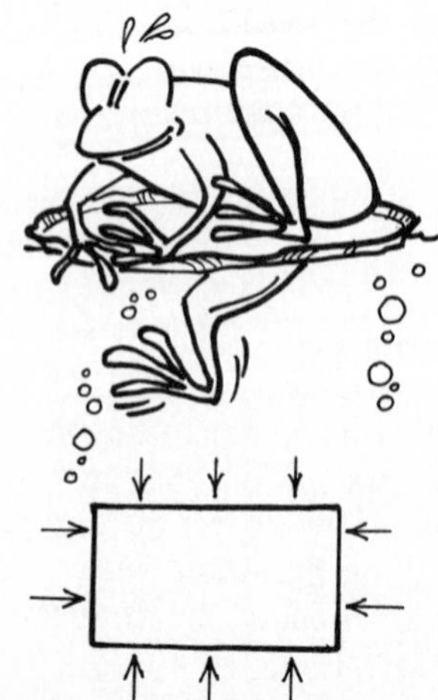

**FIGURE 11–9**
The pressure exerted upward on the bottom of a submerged object is greater than the pressure exerted on its top. The pressures on each side are equal at equal depths. The result is a net force that pushes upward. It is called the buoyant force.

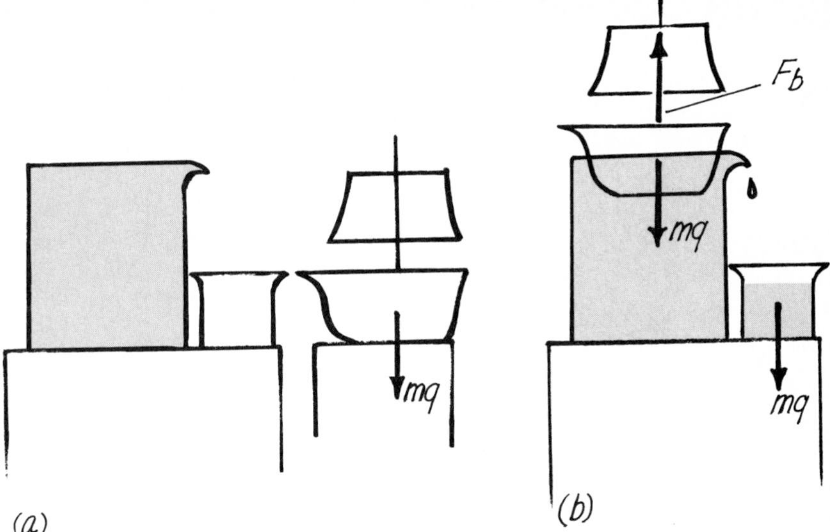


**FIGURE 11–10**
This experiment demonstrates that the weight of the water displaced by the floating boat is equal to the buoyant force on the boat.

hand. That means the upward pressure on your palm is higher than the downward pressure on the back of your hand. $P = density \times g \times depth$. Consequently, your hand gets a net push in the upward direction. We call that net push from the water's pressure the **buoyant force,** $F_b$. The buoyant force is why your friends will seem to weigh less in water. Their weights won't change, of course, but each gets a buoyant force from the water that pushes upward, counteracting (to some extent) their weights. The buoyant force affects you, too. Get in up to your shoulders and you can stand on the very tips of your toes with no trouble at all. On dry land, it hurts to do that—if you can manage it at all.

Simple experiments show exactly how large the buoyant force can be. First, fill a container to the brim with water. Carefully place a toy boat on the surface, catching the overflow with another container (Fig. 11–10). In this case you can tell immediately what buoyant force is acting on the boat. Since the boat neither rises nor falls, the net force on the boat is zero. The buoyant force, then, is equal and opposite to the boat's weight. Next, weigh the overflow, and you'll find *the weight of the water displaced by the floating boat is equal to the boat's own weight*. **The buoyant force on a floating object is equal to the weight of the displaced water.**

Another experiment takes this result further. Suspend a rock from a spring scale with string (Fig. 11–11). When you completely immerse this rock in a container filled with water, its pull on the scale is less than its weight, $mg$. Notice how much force is missing—that's the buoyant force on the rock. Next, weigh the overflow as before. You'll find the weight of the displaced water (this time equal in volume to the submerged rock's own volume) is once again equal to the buoyant force. Just as for the floating boat, *the buoyant force equals the weight of the displaced water*. This is called **Archimedes' principle.**

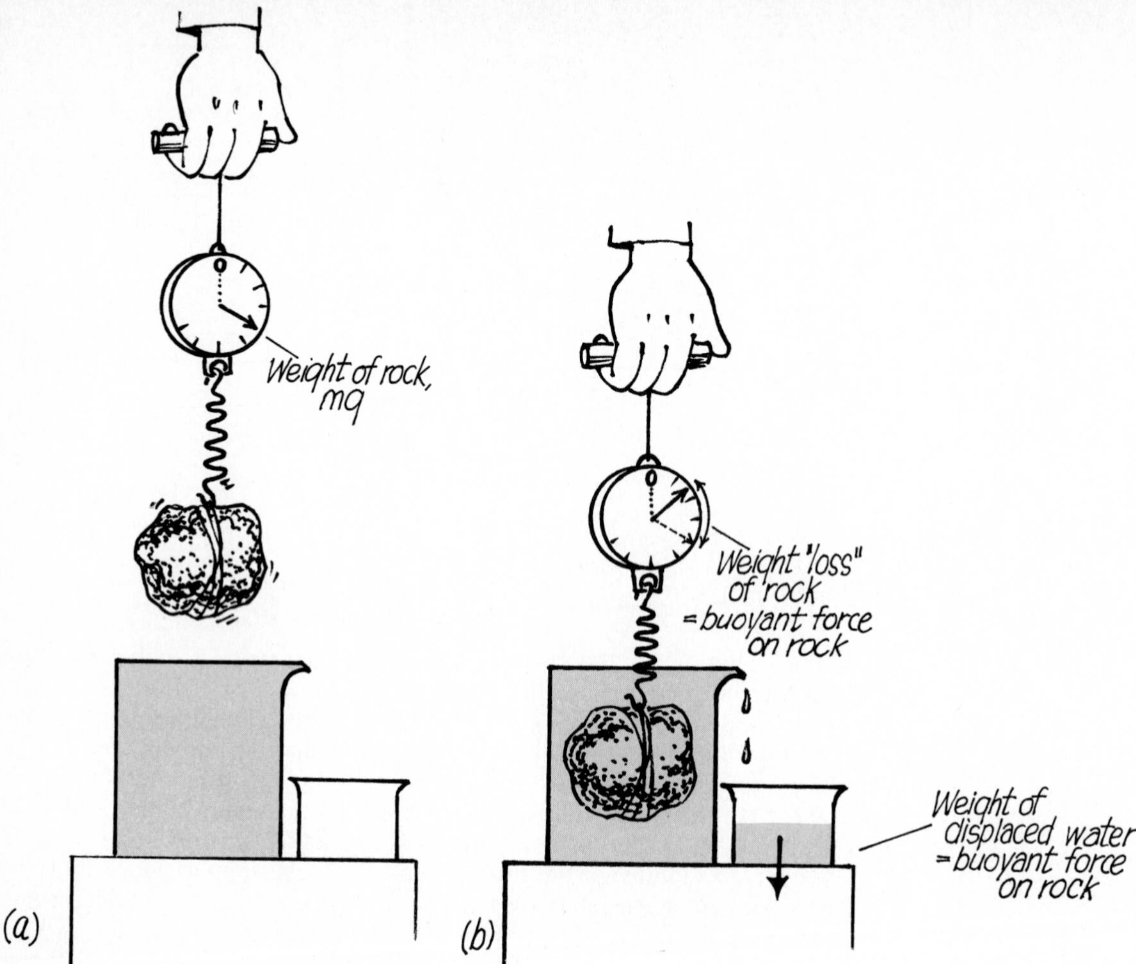


**FIGURE 11–11**
The upward force that buoys the rock is equal to the weight of the water that the rock displaces.

## Floating

If you hold a bar of soap under water and let it go, two forces act on it, its weight and the buoyant force. They point in opposite directions, so their difference is equal to the net force on the soap. Should the buoyant force be exactly equal to the soap's weight, it will hover wherever you leave it under the surface. Should the buoyant force be less than the soap's weight, it will sink. If the buoyant force is greater than the weight of the bar of soap, the soap will rise to float at the surface.

The submerged bar of soap displaces a volume of water equal to the bar's own volume. This leads to an easy way to compare the weight of the bar to its buoyant force, which is the weight of the water it displaces. If the density of the soap is greater than the density of the water, the bar weighs more than the water it displaces, and the bar sinks. If the soap's density is less than the water's, the soap rises. A hovering bar of soap has a density equal to water's density.

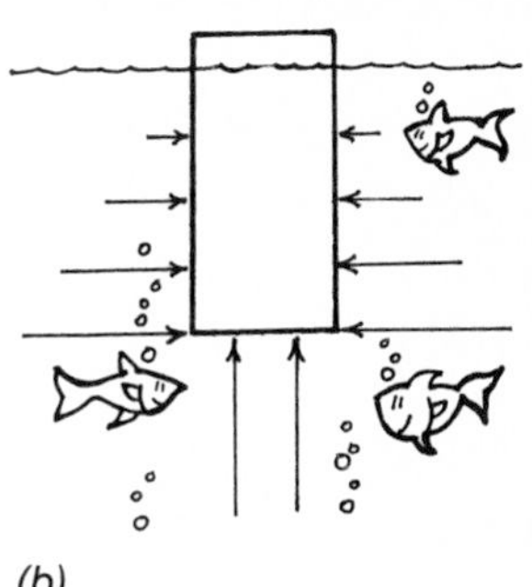

**FIGURE 11–12**
*(a)* The average densities of these sea otters are slightly less than the density of water. Hence they float. *(b)* Details of the water's action on a floating object. The upward water pressure against the bottom gives a buoyant force that balances the object's weight. (The weight isn't shown in this drawing.)

Suppose the bar of soap rises to the surface. As it pops up through the surface, the buoyant force on the bar immediately changes because the portion of the bar that's above the water's level displaces no water. The bar of soap will settle to float at a level where its submerged portion displaces a volume of water that weighs exactly as much as the soap weighs.

Most people have a density about equal to water's density, 1 gram per cubic centimeter. But unlike the bar of soap, you can change your average density a bit. Try this in a bathtub almost full of water. Breathe in deeply and you will float higher in the water. Then exhale as far as possible and you'll sink. Expanded lungs add volume to your body, and since a breath of air has little mass, your average density, $m/V$, decreases. Likewise, deflated lungs mean your body's volume is less, so your average density increases. Other animals vary their buoyancy regularly. Hippopotamuses eat underwater vegetation. They exhale violently so they can sink and walk along the river bottom to feed. Then they must push up to the surface or walk to shallow water when they need to breathe. Loons, too, expel the air from their lungs to let them fish underwater. Armadillos do exactly the opposite to swim across a wide lake. They swallow air and inflate their stomachs and intestines to make them more buoyant for their long swim.

The next time you swim in a pool, try this. Go under water and exhale air until you can walk across the bottom. The experience is much

**FIGURE 11–13**
Even a massive hippo is buoyant—unless it exhales a large volume of air.

like walking on a tiny planet or a large asteroid where your weight would be next to nothing. Muscles used to working against gravity have an easier time; lifting an arm above your head takes perhaps 5 percent of the usual tension. Back muscles that are always tense when you stand on dry land relax completely when you stand immersed to your neck in water.

## Diving

As a scuba diver descends, the pressure on the body increases. A descent of about 34 feet in fresh water brings on an extra atmosphere of pressure, almost 15 pounds per square inch. And yet scuba divers easily go to depths of 150 feet or more, adding 4 or 5 atmospheres of pressure to their bodies in the process.* That's a lot! Most car tires carry about 2 atmospheres of air pressure (30 pounds per square inch or so) over and above the air's normal pressure, and if a tire has a blowout, the sound is loud and sharp. When a truck tire inflated to 5 atmospheres of pressure has a blowout, it's more like an explosion. Though people sometimes feel fragile the morning after a party, our bodies are tough. They are almost incompressible, except for our lungs. To keep the water's great pressure from compressing the lungs, a scuba diver must breathe compressed air from scuba tanks. As the air comes from the tank, it is automatically regulated to match the water pressure at the diver's depth. The internal pressure in the diver's lungs keeps pace that way with the external water pressure that changes with the depth during the dive. Otherwise, the water's pressure, which is transmitted through the closely packed molecules of the diver's body, would crush the lungs.

Before a common diving beetle goes under, it traps a bubble of air under its wings. It uses this air to breathe while underwater, exchanging carbon dioxide for oxygen in the air pocket. The trapped air also serves another purpose. The beetle squeezes the air in the bubble, compressing it so the beetle can sink. Then when it's time to rise, it releases the tension and the air expands, making the beetle buoyant.

## Liquids in Motion: What Happens at a Boundary

A stream or river flows more slowly near its edges than at its center. Along the banks, flowing water meets solid earth, and the result is friction. The molecules at the surfaces of the solid don't move, and any liquid molecules pressed firmly against them won't move along either, any more than your hand can move sideways if you press it hard against a sheet of sandpaper. *The net sideways speed of the layer of liquid molecules next to a solid surface is zero*. The immobile liquid molecules

*The record for an unassisted descent with standard compressed-air scuba equipment is 390 feet. Scuba, by the way, is an acronym for *self-contained underwater breathing apparatus*.

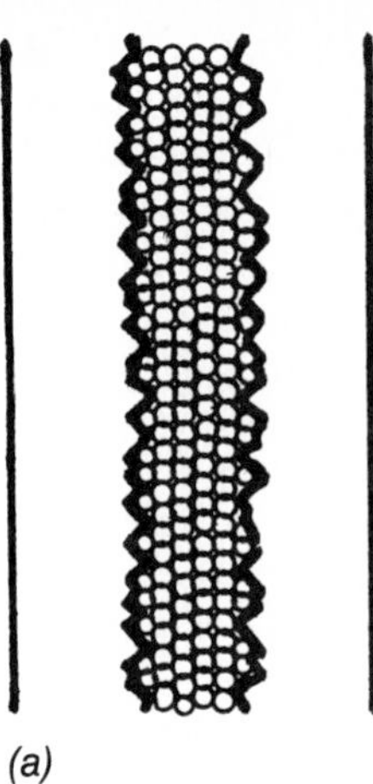

(a)

(b)

**FIGURE 11–14**
*(a)* Because of boundary effects, several layers of oil molecules are necessary to provide proper lubrication between moving parts in an engine. *(b)* In addition to providing more cushion for a straight-on collision, two layers of tires against this barrier can better deflect a racing car that sideswipes them than only one layer could. This is much like the boundary effect molecules experience against a solid surface.

at the solid surface partially check the movement of the molecules next to them, which in turn slow their neighbors, and so on. Farther from the boundary, this interference diminishes, and the water flows freely.

The boundary layer (or film) of any liquid next to a solid is immobilized. When a film of oil is used to lubricate the moving parts in a car engine (or electric motors in sewing machines or fans), it must be a few molecular layers thick to work effectively. The oil molecules can't move freely unless they are several layers away from the two passing metal surfaces. (See Fig. 11–14.)

**FIGURE 11–15**
Salt water taffy is a very viscous liquid at 80°F. Fresh from the refrigerator, however, it can be rock hard.

## Viscosity and Turbulence

Friction with boundaries isn't the only thing that slows a liquid. The passing molecules attract and even collide, interfering with the flow. This *internal* friction is called **viscosity.** It is greater for liquids like honey and syrup and less for those like water and gasoline. Viscosity keeps the thick liquid toothpaste lying on the bristles of a toothbrush.

Viscosity depends greatly on temperature. When a liquid becomes warmer, the faster molecular motion makes it easier for the molecules to slide past one another. Honey that's very slow to move flows freely if the honey jar sits in warm water for awhile. Thicker motor oils are sometimes recommended for summer months when a car's engine runs warmer.

Internal friction in liquids comes about in yet another way, because of the different speeds of neighboring portions of a moving stream. If these differences are great enough, a smooth flow will become turbulent. In **turbulent** flow, small whirlpools or eddies occur. These interfere with the stream's passage, retarding the flow, sending some of the liquid sideways and even backwards.

Even when a liquid flows smoothly through a pipe, the boundary effect and viscosity influence the rate of flow. If the diameter is large, the flow at the center isn't influenced much by the friction at the sides, and the flow is much faster. If the diameter is small, the viscosity of the fluid ensures that the stream will be slowed even at the center by friction from the pipe enclosing it. The onset of turbulence is easier in a pipe of smaller diameter because there's less distance between the edge, where the stream's speed is zero, and the center, where the flow is fastest. Once turbulence starts, extra pressure on the liquid at one end of the pipe won't cause an increase in the rate of flow. Any extra work (or energy) given to the stream just goes into making more (and bigger) eddies.

The flow of blood through your arteries and veins is ordinarily smooth. But a constriction in a major artery can cause turbulent flow as the blood squeezes through. That fact is behind the pressure cuff used by nurses and physicians to check blood pressure. The pressure cuff, complete with an air pressure gauge, circles the arm. It is pumped up until the flow of blood through the artery in the arm is cut off. Then a valve is turned to reduce the pressure slowly. At some point the blood's pressure is great enough to squeeze through the compressed artery, but the flow is turbulent and *audible through a stethoscope*. The pressure at the point at which the turbulence is first heard is called the *systolic* pressure. It is the larger number (in millimeters of mercury) that the listener records as part of the blood pressure reading. It is the maximum pressure in this artery caused by the beat of the heart. As the pressure continues to decrease in the cuff, the artery gradually assumes its normal diameter, whereupon the flow becomes smooth once again, and the turbulence and noise vanish. The pressure the cuff registers on the gauge at this point is the *diastolic* pressure, or the blood's pressure in your arm between heartbeats. This is the lower number in the reading. Similarly, a physician can hear a murmur in a patient's heart if there is a too-narrow valve or other constriction that causes turbulent flow during a heartbeat.

**FIGURE 11–16**
Although a molecule inside a body of water feels no net attraction to the surrounding molecules, a molecule at the surface does. The net attractions for the molecules at the surface cause a volume of water to act as if it had a weak "skin." The tension exerted by this skin of attracted surface molecules is called surface tension.

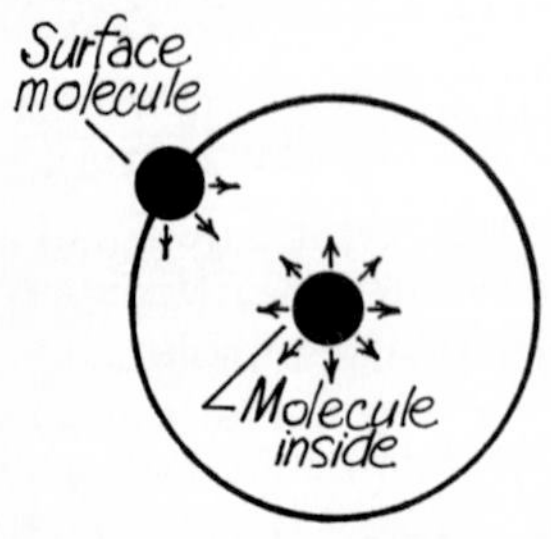


## Surface Tension of a Liquid

Any water molecule inside a drop of dew bounces around furiously. However, the *average* force on a molecule over a period of time is actually *zero*. Because there are just as many neighbors in any direction, the tugs and pushes from its surrounding neighbors balance out.

The story is different at the surface of a dewdrop. The molecular attractions for a water molecule on the surface are all on one side—the inward side, into the dewdrop. When a molecule at the surface gets pushed outward by a collision with a molecule from inside, its outward motion is slowed by the attractions from the molecules behind it. (See Fig. 11–16.) In effect, then, it's as if there is an invisible skin at the

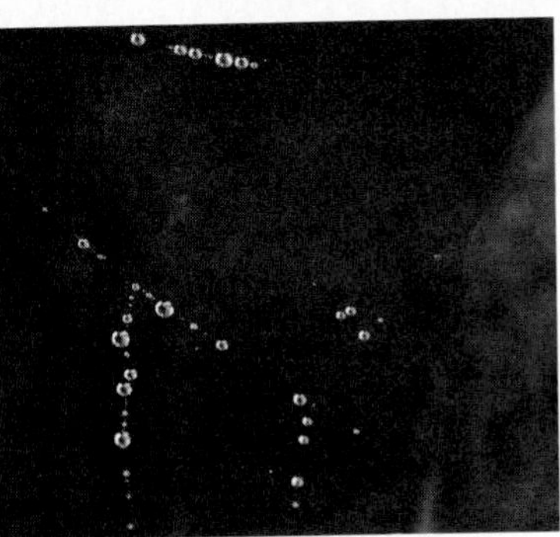

**FIGURE 11–17**
In a droplet of water, such as these drops of dew on a spider's web, the surface tension pulls the collection of molecules into spherical shapes.

surface due to the intermolecular attraction that makes it difficult for molecules in the surface layer to escape. Likewise, if a jostled molecule from below tries to squeeze its way into the surface layer, that net inward force resists the intrusion. In other words, *the surface of the liquid resists any change in shape that would add surface area.* That is, the surface of a liquid is under tension. It's called **surface tension.**

A *small* dewdrop on a leaf or on a spider's web is almost perfectly spherical in shape except for its points of contact; its surface tension tends to pull it into that shape because a sphere has less surface area per unit of volume than any other shape. A larger drop lying on a leaf is much flatter in shape, however, because the drop's own weight presses it downward. This not only does work against the surface's tense skin but increases the surface area of the drop. In a volume as large as a glass of water, the surface of the water is almost flat. There is still tension on its surface, but it is not nearly strong enough to round such a large mass or surface area into a sphere against the force of gravity. In Skylab, however, with no weight to contend with, very large drops of water pulled themselves into spheres as they floated in the air of the cabin.

The strength of surface tension varies greatly from liquid to liquid. Water has two to four times as much surface tension as gasoline, ether, or alcohol. If you've ever had the misfortune to break a mercury thermometer, no doubt you noticed how persistently the small drops of mercury stayed together as they rolled around. This is because mercury has about seven times as much surface tension as water.

Oil and water don't mix, largely because water's molecules attract one another so much. Ducks make use of that fact. A duck floats high on the water not just because of its low density, but also because there is oil on its feathers. The oil keeps the water from seeping in and saturating them. Lightweight air stays trapped in the feathers, making the duck very buoyant. The duck preens regularly to distribute oil that is produced by special glands. If a baby duck tries to swim in a bathtub of soapy water, the oil dissolves and the baby duck sinks! The snakebird, or anhinga (Fig. 11–19), lacks oil glands to provide water repel-

**FIGURE 11–18**
Ducks are so buoyant that they have a difficult time ducking underwater. Their well-oiled feathers hold pockets of air well. Give a duck a bath in soapy water, which removes the oil, and it would sink like a rock, however.

**FIGURE 11–19**
An anhinga has no oil for its feathers, and so it easily feeds underwater. But on the surface, it manages to keep only its long neck out of the water, which earns it the popular name of snake bird. Out of water, the anhinga needs to dry its wet feathers in the sun.

lency.* Each time an anhinga dives to feed on fish, its feathers saturate. When it surfaces, it can manage to stick only its neck out of the water. Struggling out of the water and flying heavily away, it must take the time to spread its wings and dry them in the sun.

## Capillary Action

Sometimes water doesn't "seek the lowest level." In a teacup, tea climbs up the string of the teabag and over the edge, against the force of gravity. Dip the corner of a paper towel into a spot of water on a countertop and watch the water climb. The forces that make this happen are the forces between the molecules. Molecules of all kinds attract each other. The molecules within a liquid attract each other, and if the liquid comes in contact with a solid, there's an attraction between the liquid's molecules and the solid's molecules. The attraction between molecules of the same kind is called **cohesion.** The attraction between molecules of a different kind is called **adhesion.**

When a drop of water is put onto a horizontal plate of glass, the water curves outward at the edges of the drop where the water meets the glass. That's because the water's molecules are attracted more to the molecules of glass than they are to each other. We say the water *wets* the glass. Were water's surface tension less, the drop would creep over the plate's surface to cover it with a very thin film, as a drop of oil or gasoline would do. But if a drop of mercury is put on the same glass, the mercury turns inward where it touches the glass. In mercury the forces of cohesion are greater than the forces of adhesion, and the mercury is *nonwetting* (see Fig. 11–20).

**FIGURE 11–20**
Because water's adhesion to glass is greater than its cohesion, a water droplet wets glass. Mercury's molecules are more attractive to themselves (cohesion), however, than they are to the molecules of glass. Mercury is called nonwetting, and the figure shows why.

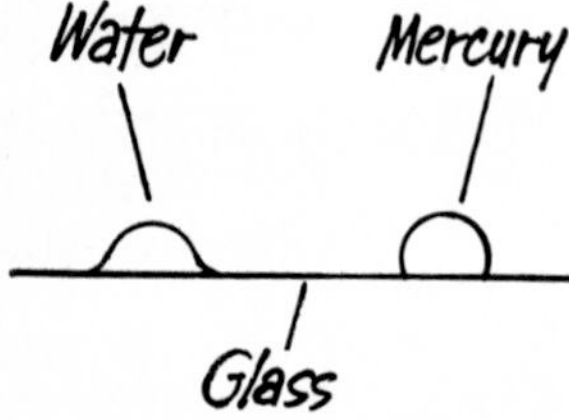


*This ancient species of bird has changed little over the eons. Before its young grow their feathers, they actually look more like lizards than birds.

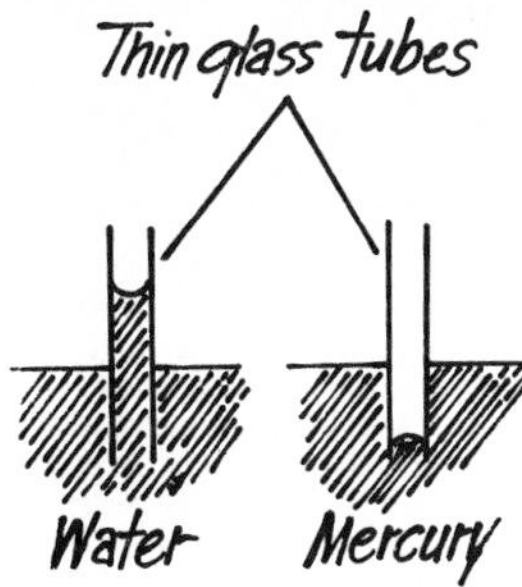


**FIGURE 11–21**
The behavior of water and of mercury on glass leads to capillary action of two kinds in a fine glass tube. Water's greater attraction to glass causes it to rise, while mercury's greater attraction to itself causes it to sink in the tube.

Strong adhesion gives wetting liquids the ability to rise against the pull of gravity into small vertical spaces. For example, if a small, hollow glass tube is held vertically so that its lower end dips into water, the water rises into the tube. The water is more attracted to the glass, and so where it meets the glass, it rises a bit. Then the water's surface inside the tube is curved, and surface tension at the curved edges pulls *up* on the rest of the surface layer. The surface moves upward, pulling (by cohesion) a column of water behind it. This is called **capillary action** (Fig. 11–21).

When you strike a match and light a candle, the heat of the candle's flame melts the wax at the top. Then capillary action pulls the melted wax up through the wick, where it burns in the flame. In a kerosene lantern the kerosene rises all the way up the braided cloth wick.

Capillary action, along with other lifting mechanisms in plants and trees, causes sap to rise against the pull of gravity. It also works with certain fabrics in clothes that wick perspiration from the body. There are other places where you might not expect to find capillary action, and yet it plays a major role. Working against gravity, capillary action holds water in the topmost layers of soil. When farmers cultivate fields of plants, it's not just to control weeds but to conserve water as well. Cultivation loosens the settled soil, adding large air spaces which interfere with the capillary action that brings water to the surface, where it would evaporate. Without those capillary pathways, the damp soil below the surface loses its moisture more slowly.

## CALCULATIONS

**TABLE 11–1**

**THE DENSITIES OF SOME LIQUIDS (GRAMS PER CUBIC CENTIMETER)**

| Liquid | Density |
|---|---|
| Gasoline | 0.66–0.69 |
| Ethyl alcohol | 0.791 |
| Kerosene | 0.82 |
| Turpentine | 0.87 |
| Olive oil | 0.918 |
| Castor oil | 0.969 |
| Water | 1.0 |
| Sea water | 1.025 |
| Milk | 1.028–1.035 |
| Mercury | 13.6 |

Just as for solids, a liquid's density (mass/volume) has the symbol $\rho$. This density determines the pressure in the liquid at a given depth and whether an immersed solid will float or sink. It also determines (see Example 4) how much of a floating solid remains above the waterline. The general formula for pressure is $P = F/A$; the pressure in a liquid (on earth) due to its weight at a depth $d$ is $P = \rho g d$.

EXAMPLE 1: In a hydraulic device on the orbiting space shuttle, a piston with an area of 1 square inch presses against a liquid reservoir, raising the pressure throughout the liquid. This pressure pushes on a piston of 3.3 square inches on the opposite side of the reservoir. Because the area is 3.3 times as large on this piston, the force is 3.3 times as great as the force exerted by the smaller piston. ($F = P \times A$, and the applied pressure $P$ is the same throughout the liquid.) If the small piston moves 4 inches, **how far must the larger piston move?**

The liquid is essentially incompressible, so its volume is constant. When the small piston moves along its cylinder, it displaces a certain volume of liquid. On the other side of the reservoir, an equal volume of liquid must move into the larger piston's cylinder, pushing that piston outward. The volume of liquid moved by that smaller piston is the piston's area times the distance the piston moves, or $V = 1\text{ in.}^2 \times 4\text{ in.} = 4\text{ in.}^3$. The larger piston has a larger surface area, so that volume of liquid *won't* push it 4 in. The large piston moves a distance $d$, and because its area is $3.3\text{ in.}^2$, the volume of liquid that moves into its cylinder is $A \times d = 3.3\text{ in.}^2 \times d$. Since that volume must equal $4\text{ in.}^3$, we see that $d = 4\text{ in.}^3/3.3\text{ in.}^2 =$ **1.2 in.**

(Conservation of energy tells us the work on the larger piston must be equal to the work done by the smaller piston if there is no friction. You can see that this is the case. Work $= F \times d = (P \times A) \times d$ for either piston. If the amounts of work are equal, then $(P \times A_1) \times d_1 = (P \times A_2) \times d_2$. The pressure $P$ cancels, and this equation becomes equivalent to the statement that the volume displaced by one piston is equal to that displaced by the other, so the numbers do agree. If the fluid were not incompressible, part of the smaller piston's work would go into compressing the fluid.)

---

EXAMPLE 2: **How deep must a scuba diver go to encounter water pressure equal to 1 atmosphere** ($14.7\text{ lb/in.}^2$ or $101{,}000\text{ N/m}^2$)?

$$P = \rho g d$$

The density of fresh water is $1\text{ g/cm}^3$, which is $0.001\text{ kg/}(1/100\text{ m})^3 = 1000\text{ kg/m}^3$. The value of $g$ is $9.8\text{ m/s}^2$, so

$$d = \frac{P}{\rho g} = \frac{101{,}000}{1000\text{ kg/m}^3 \times 9.8\text{ m/s}^2} = \mathbf{10.3\text{ m (or 33.8 ft)}}$$

---

EXAMPLE 3: A glass sphere used by commercial fishermen to buoy their nets in the ocean weighs 3 lb. If its diameter is 25 cm (about 10 in.), what upward force does it exert on the net when the sphere is completely submerged? $V_{sphere} = \frac{4}{3}\pi r^3$. The buoyant force on the glass sphere is equal to the weight of the seawater it displaces; $mg = \rho_{seawater} \times V_{sphere} \times g$. The density of seawater is $1.025\text{ g/cm}^3 = 1025\text{ kg/m}^3$, the volume of the sphere is $\frac{4}{3}\pi \times (0.125\text{ m})^3$, and $g = 9.8\text{ m/s}^2$, so

$$F_b = 82\text{ N (or about 18.5 lb)}$$

Subtracting its weight of 3 lb, we find that **the net upward force is about 15.5 lb.** If this seems like a lot, just try pushing a beachball completely under water. Water *is* heavy.

---

EXAMPLE 4: What percentage of an iceberg's volume is under water? The density of ice is $0.917\text{ g/cm}^3$, and the density of seawater is $1.025\text{ g/cm}^3$. The iceberg sinks until the volume of it beneath the water level displaces water equal to the entire weight of the iceberg. Then $F_b$ is equal to the iceberg's weight.

$$F_b = (\text{volume of iceberg under water}) \times (\text{density of seawater}) \times g$$
$$= V_{under} \times \rho_{seawater} \times g$$

The weight of the iceberg is $mg = \rho_{ice}V_{iceberg}g$. Setting these equal to each other and rearranging gives

$$\frac{V_{under}}{V_{iceberg}} = \frac{\rho_{ice}}{\rho_{seawater}} = \frac{0.917}{1.025} = 0.895$$

So **89.5** percent of an iceberg's volume is under water, and 10.5 percent is above water.

## REVIEW

**1.** Exactly how does a liquid exert pressure on a container?

**2.** How is the pressure applied to an enclosed liquid passed along through the liquid?

**3.** The pressure in water that's at rest changes with depth. What formula expresses this?

**4.** Define the buoyant force. When a solid object is immersed in water, what is the buoyant force on it?

**5.** An object that's held under water can rise to float only if the weight of the water it displaces is (a) greater than, (b) less than, (c) equal to its weight.

**6.** Because of friction between the molecules of the liquid and the solid, the speed of water at a stationary boundary is ______________.

**7.** Viscosity is a measure of the internal friction within a liquid. True or false?

**8.** Turbulence decreases the speed of flow for a liquid. True or false?

**9.** Surface tension gives small liquid drops their shape and resists any increase in the surface area. True or false?

**10.** What is the adhesive force between molecules? What is the cohesive force between molecules?

**11.** The ability of some liquids to rise against the pull of gravity into small vertical spaces is called ______________. In this action the molecular force of ______________ is greater than the molecular force of ______________.

## DEMONSTRATIONS

**1.** Iron's density is about 7.8 grams per cubic centimeter, and water's is 1 gram per cubic centimeter. So iron should sink in water, right? Not in this case. Fill a bowl with water and find a sewing needle. Tear off a strip of bathroom tissue that's large enough for the sewing needle to rest on. Then lay the tissue with the sewing needle on top on the surface of the water. In less than a minute the paper becomes saturated with water and falls to the bottom of the bowl. Not so the needle! It is supported by surface tension, just like the insects called *water striders* seen on ponds and slowly moving streams. Find the reflection of an overhead light on the surface of the water (you may need to bend over the bowl) and move your head until the light's reflection approaches the needle and watch carefully. The needle depresses the water, exactly as if the water had a stretchable skin, or membrane, over it.

Now remove the needle and stir a few drops of liquid soap into the water. Then place the needle on another piece of tissue and lay it on the surface again. This time the needle sinks with the paper. Soap reduces water's surface tension, and it can't support the needle. This is the property of soapy water that lets it penetrate the small spaces within fabrics to get clothes clean.

**2.** Prove that salt water is more dense than tap water. Put a fresh egg into a glass of water. It will sink (unless it is old enough to have formed some gas bubbles inside). Add a few tablespoons of table salt, stir, and watch the egg rise. The salt water exerts a larger buoyant force on the egg than the tap water does because the same volume of salt water weighs more.

**3.** Sprinkle some pepper over the surface of a bowl of water. Drop a small drop of cooking oil into the middle of the bowl. As the oil races outward over the surface of the water, the pepper particles are pushed to the side very rapidly by the spreading oil film.

**4.** Fluid pressure increases with depth, which is why when you are standing the blood's pressure at your feet is higher than in your head. Take off a shoe and look at the veins in the top of your foot. You can probably see those that stand out because of blood's pressure. Then sit down and slowly raise that foot (prop it on a desk). As your foot gets to the level of your heart or slightly higher, the veins deflate because the blood pressure in them becomes less. This also works with the veins on the back of the hand. Hang your arm by your side and clench your fist. The veins pop out. Release your fist and slowly raise your hand above shoulder level and watch the veins disappear.

**5.** The heart pumps blood into arteries that distribute the blood to your limbs. The blood returns to your heart through veins, some of which are exposed on the backs of your hands and feet (as mentioned in Demonstration 4). You can use the elasticity of a vein to see the direction of the blood flow, to or from the heart. Veins on the top of the foot are best for this purpose. Slight pressure from a fingertip placed on the vein stops the flow. Maintaining pressure, run the finger along the vein toward your heart. As the finger moves, blood fills the vein behind the finger. Now slide your fingertip away from your heart, keeping pressure on the vein. The collapsed vein stays flat because the blood doesn't flow through the vein in that direction.

**6.** Warning: This demonstration is not for the timid *or* the vainglorious! Rub the end of your nose with a tissue to get rid of any oil or moisture on the skin, and fetch a teaspoon from the kitchen. Now you're ready. Try to hang the spoon from your nose by pressing the end of your nose into the spoon's bowl. Most likely, it won't work. Then moisten the spoon with your tongue, and with a little wrist action, wave the spoon around in the air for a moment so it won't be *too* moist. You want only a *very* thin film of moisture on the spoon. It will now hang on your nose for quite some time, thanks to the forces of surface tension and adhesion.

# EXERCISES

**1.** Why do some people feel dizzy or faint if they ride in elevators that accelerate upward rapidly? Do they feel this way if the elevator accelerates downward?

**2.** People who snorkel know that after awhile the act of breathing becomes tiring. What causes this to happen?

**3.** If you go twice as deep in a liquid, do you double the pressure due to the liquid's weight? If $g$ were twice as large, what would happen to the pressure in a liquid? If one liquid is twice as dense as another, how do their pressures compare?

**$\vec{4}$.** *Suppose you double the depth of the water in the plastic beaker in Fig. 11–22. (a) Does the weight on the scale double? (Ignore the weight of the flask.) (b) Does the pressure on the scale double? (c) Does the pressure on the bottom of the flask double?

**FIGURE 11–22**

**5.** Identical holes are punched 1 centimeter from the bottoms of a large paint can and a soft drink can. If each is then filled with water to a depth of 10 centimeters, from which can should the water flow faster?

---

*Arrows indicate vector problems.

**6.** Will iron objects float on liquid mercury?

**$\vec{7}$.** Is the block shown in Fig. 11–23 propelled to the right by a larger pressure on its left face? Why or why not?

**FIGURE 11–23**

**8.** Ice floats in water. What does this tell you about the density of ice? Check Table 10–2 to prove it.

**9.** When an ice cube melts in a glass of water, does the water level rise?

**10.** Look up the average density of the planet Saturn in Table 10–3. What would happen if you could place Saturn in a large bathtub full of water in a gravitational field of 1$g$?

**11.** Why does your heart do less work when you lie down?

**12.** Have you ever stood up quickly after lying down for awhile and felt faint? What causes this? Should a giraffe experience the same effect?

**13.** Will an ice cube float in kerosene? See Tables 10–2 and 11–1.

**14.** Exactly what holds contact lenses on the eyes?

**15.** If ketchup pours too slowly, what can you do to decrease its viscosity?

**16.** When you pour ketchup (or salad dressing) from its bottle, at what part of the opening does it move fastest? Why?

**17.** Traveling downstream in a canoe, where should

you stay to take advantage of the current? Traveling upstream, where should you put the canoe if you want to avoid the current?

**18.** What's wrong in Fig. 11–24?

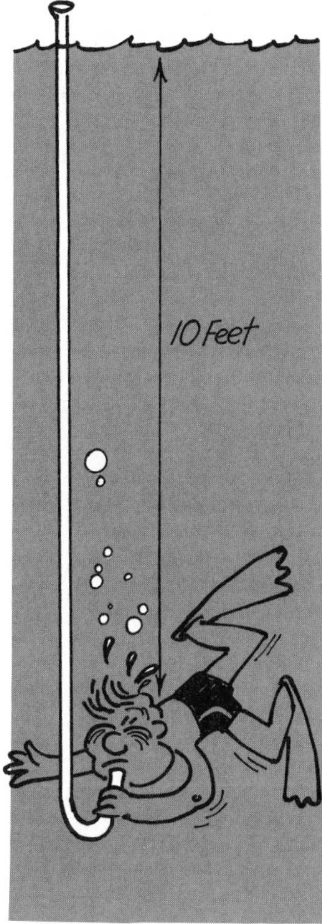


**FIGURE 11–24**

**19.** Why are towels dried on a clothesline stiff and scratchy while those dried in a clothesdryer are soft and fluffy?

**20.** After swimming in a pool or in the ocean, you will dry off by evaporation after awhile even if you don't use a towel. The next time you do that, notice the hairs on your arms. Why are they matted together in little clumps?

**21.** Liquid copper at 1131°C has a surface tension some 15 times greater than water at room temperature. How should this affect impurities that fall onto the surface of copper? (Anyone who has used a soldering iron has probably noticed dirt or dust or other impurities floating high on the melted solder.)

**22.** When a bathtub faucet is only partly open, a smooth stream of water may flow with no noise. But open it full blast and the gushing water is accompanied by much noise. Why?

**23.** In 1964 a freighter capsized in the harbor of Kuwait. Someone remembered an old 1949 Walt Disney comic book in which Donald Duck and his nephews Huey, Dewey, and Louie raised a yacht by filling it with ping-pong balls. It took 27 billion polystyrene balls, but the freighter was raised. Explain in terms of the freighter's density how this helped.

**24.** Why is hot soapy water better for cleaning than cold soapy water?

**25.** How would it help a swimmer if she or he could exert muscular tension to *expand* the lungs while underwater (without taking in water)?

**26.** A seamstress sometimes licks the end of a thread before pushing it through the eye of a needle. Why? What effect does this action use?

**27.** Explain why toothpaste comes out of the end of a tube even though you squeeze from the side.

**28.** Why should some patients exercise in water after operations or long illnesses?

**29.** Contrary to the rising of water in a small-diameter glass tube, liquid mercury is forced downward when such a tube penetrates its surface. Explain why.

**30.** Some plants wilt visibly on hot or windy days. What causes this?

**31.** Inspect Table 10-2 for the densities of fat, muscle, and bone and decide if people with larger percentages of fat than normal should float more easily than those with a smaller percentage of fat.

**32.** Why do you think soap bubbles blown by a child are so spherical?

**33.** Does capillary action help as you are drying off with a towel?

**34.** Should your blood flow faster through veins and arteries or through the capillaries in your body? Why?

**35.** If you drink a cup of coffee, do you think your average density will change?

**36.** A 1000-ton ship floats in the ocean. What weight of seawater has it displaced?

**37.** What property of oil could cause it to evaporate at a slower rate than water?

**38.** If you jump into a swimming pool, do you increase the pressure on the bottom of the pool?

**39.** Which should be built the strongest, a 10-foot dam to hold back a small pond or a 10-foot dike in Holland to hold back the ocean?

**40.** If fresh water weighs 62.4 pounds per cubic foot, how much does a cubic foot of seawater weigh? See Table 11–1.

***41.** Example 2 in "Calculations" found the depth in fresh water at which a scuba diver experiences 1 atmosphere of water pressure. The density of seawater is 1.025 g/cm$^3$. Show that a diver in salt water would experience water pressure equal to one atmosphere at only 33.0 feet rather than 33.8 feet as in fresh water.

***42.** A raft built from solid spruce ($\rho$ is about 0.6 g/cm$^3$) is 1 ft thick, 12 ft long, and 8 ft wide. What weight will it support without submerging completely in fresh water?

***43.** The Hoover Dam is 726 feet high. In June 1983 the runoff from heavy snows during the winter raised the level of Lake Mead, behind the dam, to the top, cresting the dam for the first time since an early test when the dam was new. What was the water pressure at the base of the dam?

***44.** Ships are charged according to their weights when they pass through the lock systems of some canals. The canal tenders read from markers on the side of a lock how many feet or meters of water the ship displaces in the lock, which corresponds to the volume of water the ship displaces. If a freighter displaces 1750 cubic feet of ocean water, how much does it weigh?

***45.** Estimate the total force water exerts on the Grand Coulee Dam, which is 550 ft high and roughly 4000 ft wide. (*Hint:* The average pressure is the pressure halfway down to the bottom. Presume the width is uniform all the way down.)

***46.** An average adult male has about 4.5 liters of blood and a heart rate of about 60 beats/minute. If his heart moves 80 cm$^3$ of blood each time it beats, about how long does it take for the blood to circulate through the body?

***47.** Explain why any variety of wood that would float in water on earth would also float in water on Mars.

****48.** The volume of the oceans is about $1.4 \times 10^{18}$ m$^3$. If the continents were smoothed out and the solid earth were a perfect sphere, about how deep would the water be over its surface?

***49.** When something's density is greater than water's, the object will sink when it is submerged. The net force pulling it down is $mg - F_b$. Show that this force is equal to $F_{net} = (\rho_{object} - \rho_{water})V_{object}g$. What does this formula suggest about how you might try to speed the settling of blood cells or bacteria (in a small sample of water) whose density was only very slightly greater than the water's? (Be sure to see the answer to this one!)

# CHAPTER 12
# GASES

For the holidays, the professor and his wife flew into Lima, Peru's largest city. The next day they boarded an old, British-made train dating back to the turn of the century and settled in next to a window for the trip. They would climb from Lima at sea level to an altitude of 15,600 feet on the world's only railroad to reach that altitude. They were glad they weren't traveling by bus—the narrow dirt roads along the way had washouts from the infrequent rains, and most railway cars, unlike buses, are off limits to livestock.

A hot, dry climate dominates the first 3000 feet of altitude. From there to 7000 feet, the average temperature during the year changes no more than 5°F. Between 7000 and 10,000 feet lies the region European settlers found comfortable, with temperature ranges similar to those of their homelands. Above this, the days are burning hot and the nights freezing cold. They'd go through all this in their 8 to 10 hour ride.

As the train approached 14,000 feet, the professor began to feel lightheaded. And he knew why. Each breath of air at high altitude brings in much less oxygen than a breath of air at sea level. He marveled at the Indians working on the terraced slopes far above the railway. Then he remembered the story about the sulfur miners in Chile who work at 19,000 feet or more during the day but come down several thousand feet each night to sleep. To live at 19,000 feet without interruption is past the limits of human physiology; the body, starved for oxygen, will lose muscle tissue, weaken, and die.

He looked around to see if any of the other passengers seemed as woozy as he; his wife was O.K. Just then the forward door of the car opened, and a porter stepped in. Under his arm was a long-necked goatskin bag with a stopper in the end; the porter was selling whiffs of oxygen. Feeling nauseous, he reached into his pocket for change. But while the porter made his way among the customers, the professor fainted. When he awoke, the porter and his bag had already moved on to the next car. This slight attack of altitude sickness left him with a headache and a feeling of weakness for a day. ■

## The Atmosphere

On a clear day in the western states, you might see a mountain range from a hundred miles or more away. Yet lift your eyes skyward from those mountains, and that same distance *up* is where the first space walks took place. (The deep blue sky really isn't very deep.) It was here that the suited astronauts, secured only by a safety tether, stepped outside their airtight capsules and ''flew'' over the earth's surface at almost 18,000 miles per hour—good proof that there is little air to cause friction. From there they could look down into the thin blue veil of the atmosphere below. Today, spacewalkers from the Space Shuttle move about without tethers, as in Fig. 12–1.

At the upper edge of this shallow sea, intense sunlight and fast-moving particles from the sun (the *solar wind*) and from interstellar space (the *cosmic rays*) hammer the sparse molecules of the air. Many lose electrons, becoming ionized. Aptly called the **ionosphere,** this region reaches downward toward the planet earth for about 50 miles. At the lower levels of the ionosphere, the air is dense enough to offer substantial friction to the incoming interplanetary debris we know as *meteors*. By the time they reach an altitude of 65 miles, these tiny bits of rock glow white-hot. The smaller ones burn up within a span of 10 miles; the somewhat larger (and brighter) ones might plunge to within 45 miles of earth. Those that reach the ground (the *meteorites*) have to be big to begin with because the atmosphere's resistance grows quickly as the piece of rock descends.

The region from 50 miles to 30 miles above the surface is called the **mesosphere,** literally ''in the middle'' (see Fig. 12–2). Immediately below the mesosphere, 30 miles over our heads, the region called the **stratosphere** begins. If you were there looking down, you could just

**FIGURE 12–1**
Astronaut Robert L. Stewart maneuvers outside the space shuttle in February 1984. High-speed bursts of nitrogen gas provide the small forces he needs to accelerate in this almost frictionless environment 150 miles over the earth's surface. Note the curved surface of the earth and its thinning atmosphere below him.

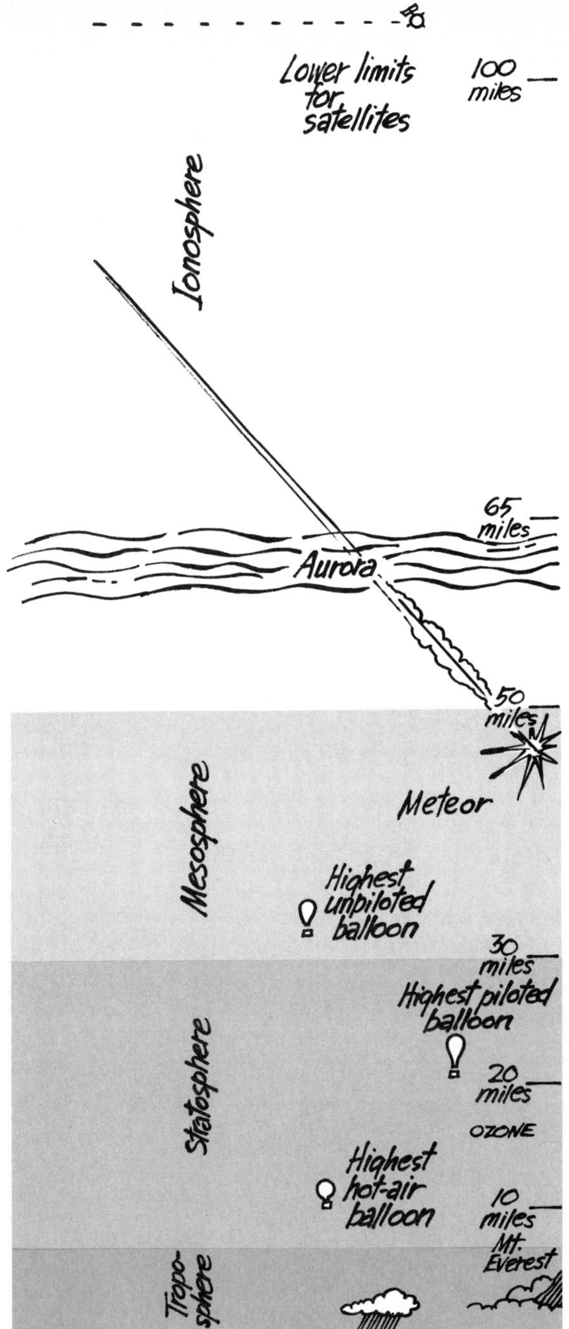


**FIGURE 12–2**
The regions of the atmosphere.

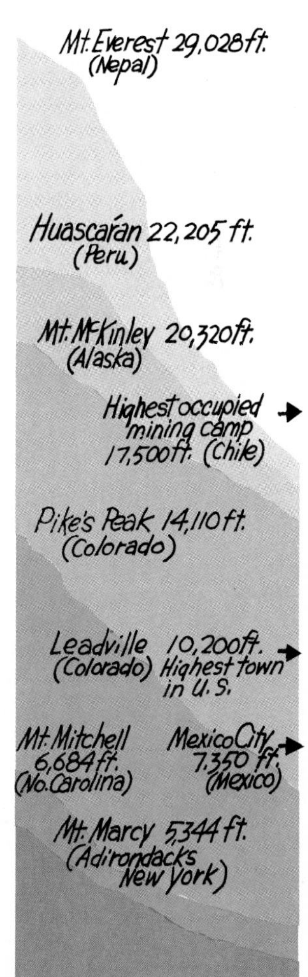


**FIGURE 12–3**
High elevations on the earth's surface.

make out things as large as football stadiums. The powerful sunlight throughout the stratosphere separates some of the $O_2$ molecules into two individual atoms of oxygen. Often these free oxygen atoms combine with undisturbed $O_2$ molecules to form $O_3$, or **ozone,** a molecule that absorbs the harmful ultraviolet radiation in the sun's light. Without this invisible guardian, ultraviolet rays could destroy the life on earth. The

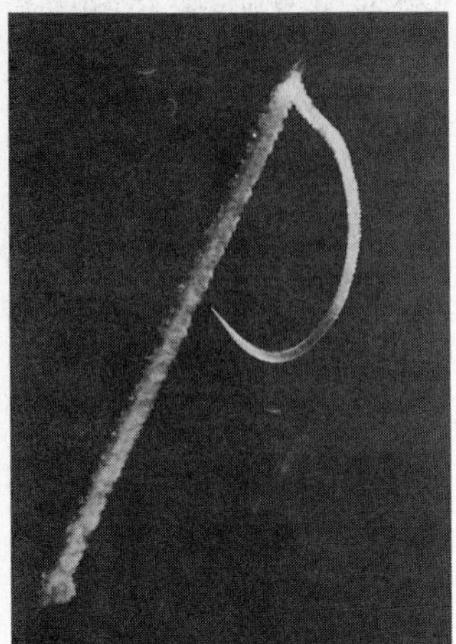

**FIGURE 12–4**
A message is left hovering in the clear air of the atmosphere. Such a cloud is only droplets of liquid large enough to reflect light well.

ozone is most dense at an altitude of 15 miles above the earth's surface. By comparison, the highest-flying commercial jets cruise at some 9 miles above sea level, at the stratosphere's lower edge.

Beneath the cold, thin stratosphere lies the **troposphere,** a shell 7 miles deep that harbors most of the molecules in our atmosphere. Our weather is produced here by the sunlight and the hot, cold, wet, and dry areas below. This region of rapid mixing is where water vapor rises, cools, and turns into clouds and ultimately into droplets of water that fall as rain. You are at home in this most dense layer, surrounded by the gaseous mixture of nitrogen and oxygen.

Hold your thumb at arm's length and take a look. A straight line between your nose and your thumb would intersect about a billion molecules at any instant. But you see through them—they are invisible—and we take little notice of the air, at least compared to the solids and liquids around us. Although we can't taste or smell clean air, we can feel it. Just take a breath and blow across the hairs on your arm. In each cubic centimeter of space at sea level some $10^{19}$ molecules move at bulletlike speeds. Each collides and rebounds with others at an average rate of a billion or more times a second. These fast-moving molecules hammer any solid in the air, as well as the surface of any liquid exposed

## PRESSURE IN A MIXTURE OF GASES

Air is a mixture of gases (see Table 12–1), so the pressure air exerts is due to molecules ($N_2$, $O_2$, $CO_2$, and some $H_2O$) and atoms (argon, neon, helium, and krypton) that collide with anything immersed in air. Each component gas in the air delivers a certain percentage of the total pressure air exerts, and the pressure exerted by a specific gas alone is called its **partial pressure.** The atmosphere's pressure is the *sum* of the pressures exerted by each species of gas in the mixture; *air pressure is the sum of the partial pressures of the different gases in the air.*

**TABLE 12–1**

**GASES IN THE ATMOSPHERE**

| | PERCENTAGE OF MOLECULES IN DRY AIR |
|---|---|
| Nitrogen ($N_2$) | 78.09 |
| Oxygen ($O_2$) | 20.95 |
| Water vapor ($H_2O$) | 0–4 |
| Argon (Ar) | 0.93 |
| Carbon dioxide ($CO_2$) | 0.03 |
| Neon (Ne) | 0.0018 |
| Helium (He) | 0.000524 |
| Krypton (Kr) | 0.0001 |
| Radon (Rn) | $6 \times 10^{-18}$ |

to them. However, from second to second the pressure of the air bathing our planet is very steady because the molecules are so numerous and their motions so fast.

(a)

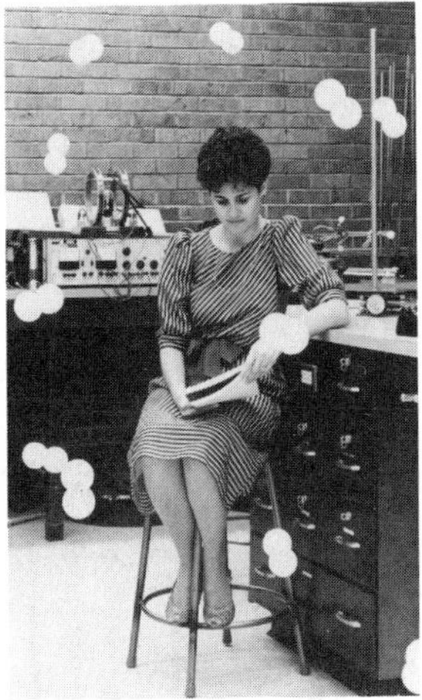

(b)

**FIGURE 12–5**
*(a)* A model of a nitrogen molecule, $N_2$, that is about 5 inches long. *(b)* Gail is surrounded by a stop-action model of air made up of "molecules" like the one shown in part *(a)* to show the relative density of nitrogen molecules in normal air. The air around us is mostly empty space.

## The Pressure and Density of the Atmosphere

The pressure our atmosphere exerts on us and the things around us begins with the solitary molecules 100 miles up. Widely separated, those atoms and molecules make long and graceful loops away from earth, turn, and fall back to collide with others below. In doing this they push downward, causing the particles they hit to push more on those beneath them. In turn, the descending molecules get bumped *upward.* This process continues all the way to earth's surface, so essentially the air below holds up the air above, supporting it against the pull of gravity. The collective weight of the atmosphere at sea level is enormous.

The air at sea level supports the overlying atmosphere with a pressure of almost 15 pounds per square inch. Since a square foot contains 144 square inches, this amounts to over a ton of force on every square foot of earth's surface at sea level. But this pressure doesn't push only down. The collisions of the air molecules, which transmit the pressure, take place in *all* directions, up, down, and sideways, just as in a liquid. *Gases flow* (like liquids) *when there is an unbalanced pressure.* The breeze you might feel on your face when you go outside comes from air that flows from regions of higher pressure toward those of lower pressure. Largely for these reasons, gases and liquids together are called **fluids.** But while a liquid is an incompressible fluid, a gas is compressible because of the space between its molecules. And the pressure caused by the weight of the air above compresses the air at lower elevations, pushing more molecules into each unit of volume. For this reason, the density of air closer to earth is greater than it is at higher elevations.

To discover how air's density changes, you could take a trip from sea level up to the summit of an 18,000-foot mountain. That train ride in Peru could take you almost that high! Take along some identical jars with rubber seals that make the lids airtight. At various altitudes open a jar, swish it around, and turn it upside down to pour out that thicker air from sea level. Then seal it again. Carefully weigh those closed jars on your return, and you'll find that the higher the sampling site, the lighter the jar. Then, by subtracting the weight of a completely evacuated jar (one pumped entirely free of air), you could deduce that 1 cubic foot of air at sea level weighs 0.08 pounds and the same volume of air at 18,000 feet weighs only *half* as much. The density of air at 18,000 feet is half the density of air at sea level.

## Measuring Air's Pressure

Awaiting you at any kitchen sink is a superb demonstration of air pressure. Run in a few inches of water, lower a drinking glass in sideways, and fill it completely. Now without letting the water run out, turn the glass upside down and ease it up out of the water until the rim is just below the surface. The water in the glass doesn't flow out even when the glass is mostly out of the water. Why doesn't it?

If you just dump a glassful of water into the sink, gravity pulls it down. It plunges in and raises the water level in the sink. In the glass you are holding, the water has weight too, but something counteracts that weight and keeps the water up in the glass. The water in the sink just below that column of water in the glass exerts pressure that won't let the water flow out.

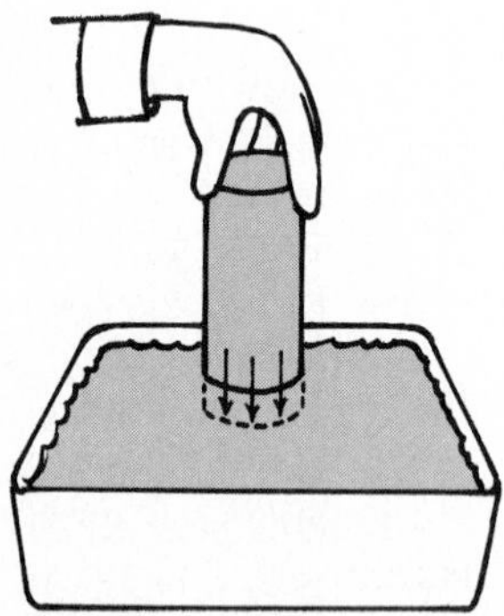

**FIGURE 12–6**
Why doesn't the water in the glass flow downward into the sink? What supports its weight? If the water in the glass descended, the water level in the sink would rise. But this doesn't happen because the air's pressure on the surface of that water is far greater than the pressure exerted by the water in the glass, and you see the result.

But where does that pressure come from? It can't come from the weight of the water in the sink, because its level is below the level of the water in the glass. The pressure comes from the air, which exerts almost 15 pounds of force on every square inch of the surface of water in the sink. That pressure is transmitted to every corner of the liquid, and it is more than enough to support the weight of the water in the glass. (See Fig. 12–6.) Actually the water would not empty from any glass with a height of up to 34 feet. This was known even in Galileo's day, when suction pumps raised water up to 34 feet but no higher. However, if you had a 35-foot glass for such an experiment and somehow raised it up to its rim, the water column in the glass would rise along with it—for 34 feet. But as you pulled it up that last foot, a space would appear above the water in the glass. (That space would be a very good **vacuum,** which would have some water molecules zipping through it. A *perfect* vacuum is a space with no atoms or molecules in it.) The air pressure of earth's atmosphere at sea level can hold up a column of water only 34 feet high.*

In 1643 Evangelista Torricelli, a student of Galileo, hit upon the idea of using mercury in a closed, upright tube (similar to the water in the inverted glass) to determine air pressure. Since mercury is 13 times denser than water, the column of mercury that balances the atmospheric pressure is not nearly 34 feet tall. In fact, it is only about 34 ft/13 high, or about 30 inches. Torricelli's **mercury barometer** was simple and convenient because of its size, and it immediately showed that air pressure varies somewhat from day to day at any location. Near sea level, the normal range of barometric pressure is 31 to 29 inches of mercury. Even today news programs report the local air pressure in *inches of mercury*. Warmer air is less dense than cold air—its molecules are farther apart. So when it gets hot outside, the local air pressure drops a bit; warm air weighs less. Likewise, the barometer climbs if cold air sweeps in; cold air weighs more.

The air pressure that supports a 1-millimeter column of mercury is called a **torr,** in honor of Torricelli. The pressure of 760 torr is equivalent to 29.9 inches on a mercury barometer. This is an *average* atmospheric pressure at sea level. The lowest air pressure ever measured in our atmosphere was 25.55 inches (649 torr) in the eye of a typhoon off Okinawa. The highest air pressure measured was 32 inches (813 torr)

*You can check this. Use $P = \rho gh$ to show that the pressure under 34 feet of water *due to its weight alone* is about 15 pounds per square inch, which is the pressure air adds to the water's surface to hold that enclosed column up. See "Calculations," Example 2, in Chapter 11.

## UNITS OF PRESSURE

The metric unit of pressure in a gas (or a liquid) is one newton per square meter ($N/m^2$), which is called a *pascal* (Pa). The English unit is the pound per square inch ($lb/in.^2$). Another common unit is the *standard atmosphere* (atm), which is 14.7 $lb/in.^2$ or 101.3 kPa (kilopascals). Air pressure measured with a mercury (Hg) barometer is indicated in *inches of mercury* or *millimeters of mercury*. Notice that inches of mercury isn't actually a unit of pressure in itself. In terms of a mercury column supported by air, 1 atm = 760 millimeters of Hg = 76 centimeters of Hg = 29.9 inches of Hg. A *torr* is the pressure that will support a millimeter column of mercury, so 760 torr = 1 atm. Another pressure unit used by meteorologists is called the **bar.** One bar is $10^5$ $N/m^2$, which is equivalent to 29.53 inches of mercury. A *millibar* is one-thousandth of a bar, so 1 atmosphere equals 1013 millibars.

on a bitter cold December day in Siberia. Fluctuations in air pressure are significant (as anyone with a sinus problem knows). For instance, a 1-inch drop in barometric pressure over the ocean allows the water level in that region to rise over a foot.

An **aneroid** (containing no liquid) barometer works another way. A partial vacuum is created in a small metal container with a thin metal side, and the container is sealed airtight. As the air pressure outside the container changes, its thin metal side bulges in or out very slightly. This movement is passed along to a dial that (if calibrated properly) shows the air pressure. If you carry an accurate aneroid barometer up one flight of stairs, you'll see a drop of about ⅓ torr in air pressure.

Rapid changes of just a few torr will pop your ears, as you may know from a road trip through the mountains or from a plane ride. This happens because the air space behind your eardrum is linked by tiny *eustachian tubes* to the air outside, and these are normally closed. You hear a pop when the tubes are forced open by a pressure difference between the outside air and the air trapped inside your middle ear. To prevent this popping, you can help to open the tubes gently by yawning or by holding your nose and mouth closed and blowing. A difference in pressure of about 60 torr causes pain because of stretching of the eardrums.

## Pressure in a Moving Fluid: The Bernoulli Effect

A moving fluid, gas or liquid, exerts pressure because of its motion. A solid surface directly in the path of the stream gets a backward push as it stops the forward motion of the fluid. That's called drag, as we've seen for airplanes and skydivers. But something else happens at the side of the moving fluid. The gas or liquid molecules at the side of a fluid

**FIGURE 12–7**
Any gas molecules in the straw that wander out into the fast-moving stream of air above the straw are swept away. This leaves a partial vacuum in the straw. The liquid in the straw then rises, since there is less pressure there than on the surface of the liquid in the container. Thus the atmosphere's pressure pushes the liquid up into the straw.

stream are caught up and swept along with the current. Because this leaves fewer molecules along the edges of the flow, the pressure there is less. *The pressure is reduced at the sides of a moving fluid*. This is called the **Bernoulli effect**.

The next time you sip a drink with a straw, blow sharply across the top of the straw. While the air rushes past, the pressure over the straw drops, and the liquid rises into the straw. Air molecules in the straw that bounce into the moving stream of air get caught up and swept along. This creates a partial vacuum in the straw. The liquid rises in the straw, moving into the area of lower pressure. If the stream of air over the straw is fast enough, the liquid rises all the way up and out of the top of the straw. The air stream becomes a stream of spray. Perfume atomizers, spray cans of paint or deodorant, insecticide bombs, airbrushes, and chimneys use this principle (Figure 12–7).

A strong gust of wind over the roof of a house makes the outside pressure less than the pressure inside. If the wind speed is very high, the roof can be lifted, not by the wind above but by the air pressure from below. In most houses metal strips connect the roof structure to the wall frames to keep this from happening. The Bernoulli effect occurs on large scales as well.

In a smoothly flowing stream, floating leaves or twigs or even people on innertubes tend to migrate toward the center, where the current is fastest. That's the Bernoulli effect at work. The water's pressure on the floating object is less on the side where the stream moves faster and greater on the slower moving side.

## Floating in Air

In the last chapter we saw that liquids buoy up immersed solids and that the buoyant force equals the weight of the displaced liquid. Air too is a fluid, so anything immersed in air is buoyed up with a force equal to the weight of the displaced air. Helium is less dense than air, so a balloon filled with helium floats. Similarly, hot air is less dense than cool air at the same pressure, so hot-air balloons float.

## More About Compressibility: Boyle's Law

Push on any trapped gas, and it gives; the molecules only crowd a little closer together, giving up some of the empty space around them. You can compress a gas until the molecules touch, at which point you've turned it into a liquid. (Liquid air, however, remains liquid only while you maintain the pressure. Relieve the pressure and it boils away.)

There's a very simple rule about compressing gases (such as air) called *Boyle's law:* **If you increase the pressure on a confined gas without changing its temperature, its volume decreases by the same factor,** and vice versa. This means that the product of the pressure $P$ and volume $V$ doesn't change.

$$\text{initial pressure} \times \text{initial volume} = \text{final pressure} \times \text{final volume}$$

**FIGURE 12–8**
The heated air in the balloon is only slightly less dense than the air around it, but a large volume of such air leads to a buoyant force capable of lifting both balloon and passengers.

## UP, UP, AND AWAY

In the summer of 1782 Spain had the British garrison at Gilbraltar under seige by land and sea. Joseph Montgolfier, a Frenchman who sympathized with the English, considered an alternate route of escape for the English—to wit, the air. He was aware that smoke and clouds of steam always rise upward, so he filled a large paper bag with smoke in his kitchen, and it promptly rose to the ceiling. The next year he and his brother launched a huge paper balloon almost 50 feet in diameter at Paris, its hot air provided by an iron brazier hung from chains beneath the balloon. During its short flight, two courageous passenger–adventurers kept busy dousing fires that sprung up in the paper balloon. (The Montgolfiers preferred to keep their own feet on the ground, and the dispute at Gilbraltar was settled without balloons.)

Two decades before, Henry Cavendish (the same physicist who first measured the gravitational constant, $G$) determined the density of hydrogen gas, and Joseph Black, a Scottish chemist (who discovered carbon dioxide), pointed out that an animal bladder filled with hydrogen should ascend because of buoyancy provided by the air. But no one knew how to make a lightweight material that hydrogen would not escape through. The French Academy of Science raised money by public subscription to solve that problem, and the physicist Jacques Charles (of Charles's law in Chapter 14) headed the project. Turpentine was used to dissolve natural rubber, and silk cloth was washed in the solution; when dried, it was rubberized and impervious to hydrogen. On December 1, 1783, Charles and another passenger triumphantly rose over Paris in a hydrogen balloon only 26 feet in diameter. After they descended, Charles went up alone so as to go even higher. The sun had just set, and as the balloon rose once more, he saw the sun again in the west—no doubt Charles was the first person ever to see two sunsets in one day. His balloon reached an elevation of 9000 feet and later landed by moonlight.

Ballooning soon became the rage. In 1785 a hydrogen balloon was used to cross the English channel; it sprung a leak, and the passengers had to throw out everything but the most essential clothing to stay aloft long enough to make the French coast. One cocky Frenchman ascended in a balloon while mounted on his horse!

Meanwhile, scientists were using balloons to test the properties of the atmosphere and their own biological responses at high altitudes (and those of birds they carried along). In 1804 the chemist and physicist Joseph Louis Gay-Lussac collected the first samples of air from elevations over *20,000 feet*. He found those samples, though rarefied, had the same percentages of gases as air at sea level. Flights to such high altitudes brought on bouts of altitude sickness, frostbite, and stories of thrilling adventures. A few years later, two Italians (one was an astronomer) took their balloon to the limit. As the air pressure dropped to 229 millimeters of mercury (indicating an altitude of 29,700 feet), the balloon expanded to capacity. The two were so incapacitated by altitude sickness, however, that they couldn't release hydrogen to descend. Up they went. When the air pressure decreased to 216 millimeters, the balloon exploded. As they fell, the ruptured balloon material acted like a parachute, and they landed hard but safe. Others weren't so lucky. In 1874 two French scientists were the first to carry oxygen along to breathe. The next year they made a try for the altitude record with a third scientist aboard, but their supply of oxygen ran out. Despite bouts of unconsciousness and the paralyzing effects of altitude sickness, they *kept throwing out ballast to go higher*. Soon all were overcome, and only one survived the flight. The modern record for high-altitude ballooning is 123,800 feet, but that balloonist too did not survive the descent.

In symbols,

$$P_iV_i = P_fV_f \qquad \text{(Boyle's law)}$$

Figure 12–9 shows an apparatus that exhibits Boyle's law. The molecular picture helps us understand this law. Gas pressure comes from collisions. All the molecules that hit a square centimeter of a solid container each second contribute to the pressure. If you halve the volume of a gas, you push all its molecules into half as much space, so that there are twice as many molecules per cubic centimeter as before. Since the space now has twice as many molecules, twice as many hit a given square centimeter each second, and twice the "hits" means the pressure doubles. If 60 cubic inches of air at 15 pounds per square inch absolute is squeezed into a space of 30 cubic inches by a bicycle pump, its new pressure will be 30 pounds per square inch.

There's one important point to understand about Boyle's law. The pressure $P$ is *absolute* pressure; that is, it is the amount of pressure a gas exerts as compared to a perfect vacuum that exerts no pressure at

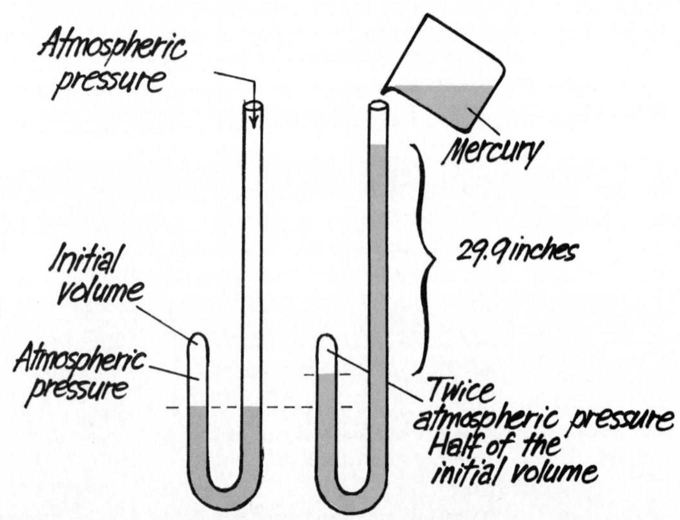


**FIGURE 12–9**
This U-tube experiment demonstrates Boyle's law. Notice what happened to the gas in the closed end of the tube when the equivalent of one atmosphere of pressure was added to it. (Remember, 29.9 inches of mercury is equal to 1 atmosphere of pressure.)

all. Absolute pressure is *not* the pressure that a pressure gauge, such as a tire gauge, measures. Almost without exception, a pressure gauge indicates only the gas pressure over and above the pressure exerted by the atmosphere at the gauge's location. When a tire gauge registers 22 lb/in.$^2$, the absolute pressure is 22 lb/in.$^2$ *plus the atmospheric pressure* wherever that tire is. Don't be misled. The psi (sometimes capitalized PSI) are pounds per square inch *over* atmospheric pressure. The few gauges that register absolute pressure indicate that by *psia*. And if you use anything but absolute pressure in Boyle's law, you've goofed.

## The Mixing of Gases and Liquids

"-THINK IT'S TIME TO BURP TH' BABY"

**FIGURE 12–10**

Whenever a liquid and a gas meet, some of the bombarding gas molecules penetrate the liquid's surface or **dissolve** into the liquid. At the same time, some of the liquid molecules evaporate into the gas. As you might expect, more gas dissolves if the gas is under a large pressure. Then if the pressure is reduced, the extra gas will come out of the liquid. When you uncap a carbonated drink or open a bottle of champagne, the dissolved carbon dioxide ($CO_2$) comes out in bubbles that you both hear and see.

Fish breathe oxygen that is dissolved in the water. Some of this oxygen comes from the air above the lakes, ponds, and oceans, and some is released by aquatic plant life. Tumbling mountain streams are especially rich in dissolved oxygen, since air is bubbled through the falling waters. In a home aquarium, an air pump keeps enough oxygen dissolved in a relatively small volume of water to support the fish. In some springs that come from deep underground, the water has very little dissolved oxygen. No fish can live in these oxygen-poor springs. But as the upwelling spring water flows away, air dissolves at its surface, and only a short distance down the stream, fish can thrive in the freshly oxygenated water.

## Breathing

When you take a breath, muscles expand your chest wall and pull your diaphragm down, allowing the elastic lungs (without muscles of their own) to expand. This brings the pressure inside your lungs to perhaps 10 torr below atmospheric pressure (a pressure that would draw water about 15 centimeters up a straw). When those same muscles relax, the stretched lung tissues relax too, and the air flows out, again without the use of muscles. During strenuous exercise, however, muscles come into play to force the air in and out more quickly.

With each breath of fresh air, oxygen comes into your lungs. It dissolves into the thin, moist walls of the *alveoli,* the hundreds of millions of bubblelike globules inside your lungs. To help transfer oxygen quickly, the alveoli have many protrusions that increase their surface area. In fact, your lungs present forty times as much surface area to the air as your entire skin does! At the same time, carbon dioxide is given up by the blood there to be expelled when you breathe out.

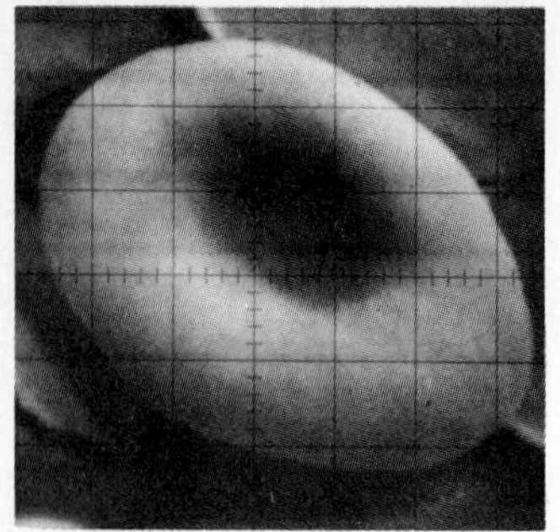

**FIGURE 12–11**
Red blood cells, which look like smooth disks with depressions at their centers on both sides, carry oxygen throughout our bodies to fuel the chemical processes of life.

Oxygen penetrates those walls in less than a second and enters the bloodstream via tiny capillary blood vessels. Surprisingly, the liquid portion of the blood cannot hold much dissolved oxygen (or nitrogen); 99 percent of the oxygen that reaches the capillaries combines chemically with *hemoglobin* molecules in the red cells of the blood (Fig. 12–11). They hitch a ride to the cells, where they are used in chemical processes to give energy. When you exercise rapidly and begin to tire your muscles, it's partly because your blood can't transfer oxygen to those muscles fast enough. Working muscles use up to ten times more oxygen per second than they do at rest, and that's why you have to breathe so much faster.

## Breathing under Pressure

Scuba divers breathe compressed air from tanks that are regulated to match the increasing water pressure as they descend. As the divers go deeper, the pressure in the air they breathe rises. More nitrogen and oxygen than normal dissolve in the bloodstream. Oxygen is a very reactive gas. Too much oxygen causes oxygen poisoning—the same effect would occur at sea level if you were to breathe pure oxygen at 15 pounds per square inch. Cells in the lungs swell from the adverse chemical reactions, but the divers are never down long enough for this to be dangerous. The danger comes from the dissolved nitrogen.

Nitrogen doesn't normally react in the body, but a condition known as *nitrogen narcosis* occurs at depths of 100 or more feet. The dissolved nitrogen at those pressure intoxicates divers, making them somewhat lackadaisical. It's precisely as if the diver is breathing small amounts of nitrous oxide, known as laughing gas, but the exact cause of the tipsiness isn't yet understood. Whatever the reason, it's not funny to be intoxicated when you're so deep beneath the surface.

A second danger from dissolved nitrogen comes if the diver stays a long time at such depths and comes up too quickly. The extra dissolved nitrogen in the tissues can come out of solution as bubbles of gas, not unlike the bubbles of carbon dioxide that fizz in a soft drink when you suddenly pop the top. Where the nitrogen bubbles form, they prevent the normal flow of blood and cause great pain, a condition known as the *bends*. To prevent the bends after a deep dive, the diver must hover at shallow depths for awhile, to give the nitrogen time to dissolve slowly back into the blood, where it will be exhaled through the lungs as the blood circulates. If the diver miscalculates and surfaces too soon, the bends are unavoidable. Then the only way to help is to take the diver to a *recompression* chamber, where the air pressure is raised to force the bubbles to dissolve back into the tissue. There the air pressure is very slowly relieved, giving the nitrogen time to diffuse into the blood without bubbling out. Navy divers who must dive deep get around these problems by breathing mixtures of helium and oxygen rather than air.

Surprisingly, astronauts share the possibility of getting the bends. U.S. spacecraft typically orbit with atmospheres of pure oxygen maintained at about ⅓ normal atmospheric pressure, or about 5 pounds per square inch. That's somewhat more than the partial pressure of oxygen in the air and in someone's lungs at sea level, about 3 pounds per square inch. The rocket ascents take place in only a few minutes, and the rapid drop in pressure could give the astronauts the bends *from the nitrogen normally dissolved in their bodies at sea level.* Before their trips to the moon, the Apollo astronauts suited up in advance and breathed pure oxygen for some time before boarding the Saturn rockets, denitrogenizing their bodies before liftoff.

## CALCULATIONS

EXAMPLE 1: A certain house at sea level has 1800 square feet of floor area. **Calculate the total force that the air inside that house exerts upward on the ceiling and the roof.**

One standard atmosphere is a pressure of 14.7 lb/in.$^2$, or 14.7 lb/in.$^2$ × 144 in.$^2$/ft$^2$ = 2100 lb/ft$^2$. Since $F = P \times A$, the total upward force is 2100 lb/ft$^2$ × 1800 ft$^2$ = **3,800,000 lb** of force.

EXAMPLE 2: If a tornado passed next to the front of a house, the pressure there could easily drop by 15 percent in less than a second. **Calculate the net force on the front door of a closed house if the outside pressure suddenly dropped 15 percent.**

A typical front door is about 6.5 ft by 3 ft, so its area is 19.5 ft$^2$. If the air inside is at a pressure of 14.7 lb/in.$^2$ and the air outside is 15 percent less, or about 12.5 lb/in.$^2$, there is a net pressure of about 2.2 lb/in.$^2$ outward on that door. In terms of square feet, 2.2 lb/in.$^2$ × 144 in.$^2$/ft$^2$ = 317 lb/ft$^2$, so the total force on the door is $F = P \times A$ = 317 lb/ft$^2$ × 19.5 ft$^2$ = about 6200 lb, or more than 3 tons.

EXAMPLE 3: A car tire is filled with 2.9 cubic feet (ft$^3$) of air at a tire gauge pressure of 30 lb/in.$^2$. **How many cubic feet of air** (at normal atmospheric pressure) **were forced into that tire?** The gauge pressure of a tire gauge normally begins at zero lb/in.$^2$. But that represents zero lb/in.$^2$ *over* the atmospheric pressure. So the absolute pressure in that tire is about 45 lb/in.$^2$ If air escapes, it will be at atmospheric pressure again, so

$$P_i V_i = P_f V_f$$

$$45 \text{ lb/in.}^2 \times 2.9 \text{ ft}^3 = 15 \text{ lb/in.}^2 \times V_f$$

Dividing by 15 lb/in.$^2$, we find

$$V_f = \mathbf{8.7\ ft^3}$$

Since 2.9 ft$^3$ of air was originally in the tire, the pump forced in 5.8 ft.$^3$

## REVIEW

**1.** Describe the atmosphere beginning at sea level and going upward.

**2.** Pressure in a gas is transmitted by molecular collisions. True or false?

**3.** Collisions between the molecules of a gas cause them to change speeds and directions constantly. True or false?

**4.** Is pressure in a gas transmitted in all directions? Why?

**5.** The collective weight of the atmosphere exerts a pressure of about ________________ lb/in.$^2$ on everything at sea level.

**6.** Unlike a liquid, a gas is compressible because of the space between its molecules. True or false?

**7.** Air, like liquids, flows when there is unbalanced pressure. True or false?

**8.** The column of mercury that balances the atmosphere's pressure at sea level is about ________________ inches high. A comparable column of water is ________________ feet tall. What is the most frequently used unit of barometric pressure?

**9.** Anything immersed in air is supported by a buoyant force equal to the volume of air it displaces. True or false?

**10.** Describe the Bernoulli effect in fluids. Give an example with air and an example with water.

**11.** State Boyle's law.

**12.** The amount of a gas that can dissolve in a liquid depends on the pressure the gas is under. True or false?

## DEMONSTRATIONS

**1.** With only a little practice you can defy gravity with this great wager-winner. Place a nickel (or dime or quarter) near the edge of a table. With a deep breath, blow sharply—not at the coin but *across the top* of it. Normal air pressure from *beneath* the coin pushes it upward into the region of smaller pressure, where it is swept along with the stream of moving air. (Remember, the stream of air needs a high speed to reduce the pressure over the surface.) With only a little practice, you can make the coin land in a tilted cup (or someone's hand) a few inches away.

**2.** Hold the back of a spoon under a moderate stream of water from a faucet. You'll feel the water press the spoon downward. With the very tip of the spoon's handle between your thumb and forefinger, hang the spoon parallel to that stream and ease the back over into the flow. Although it seems as if the spoon should be knocked away by the water's pressure, it isn't. Instead, it moves into the stream. That's Bernoulli's principle (with only a little help from surface tension). Now increase the flow from the faucet; the effect increases. Watch the stream of water below the spoon. Does it show action–reaction?

**3.** Turn your shower on full and watch the shower curtain move toward the stream of water. Daniel Bernoulli (1700–1782) probably never saw that happen. City water pressure today is typically 40 pounds per square inch or so.

**4.** Hold a ribbon or a piece of string near but not directly in the blast of a fan. Find the position *under* the stream where the Bernoulli effect will counteract the pull of gravity on the ribbon. Next put the fan on low speed and use a candle to see how the surrounding air near the stream is affected—the candle flame will bend.

**5.** It's easy to show that air is about 20 percent oxygen. Fix a large candle on the bottom of a pan with a few drops of melted wax. Then add water until only several inches of the candle remain above water. Light the candle, and place an inverted glass over it, letting enough water under the glass so the water level inside is at the level of the water in the pan. As the oxygen disappears from the air trapped in the glass, the water level will rise to fill about one fifth (20 percent) of the volume in the glass. (What happens to the hot $CO_2$? Carbon dioxide dissolves in water in much greater amounts than oxygen or nitrogen. Cool water will typically hold 20 to 30 times the weight of carbon dioxide compared to oxygen.)

**6.** Use a quart jar to measure the volume of a breath of air from your lungs. Run some water into a sink, *fill* the jar completely, and invert it, keeping the rim underwater. Then use a plastic tube or even a straw to exhale under the jar so that the air rises to displace water in the jar. Mark the point of the new water level on the jar. Take it out of the water, turn it right side up, and fill it with water to that mark. Then measure this volume with a graduated cylinder or a measuring cup.

**7.** Break off from a wooden match three small pieces of wood about 1 centimeter long. Fill a soft drink bottle with water and drop in the match sticks. Now put your thumb over the top and press down hard. (You can use your other thumb to help.) The extra pressure transmitted through the water compresses air spaces in the wooden sticks, and they become more dense and sink. Each stick will contain a slightly different proportion of air to wood, so if you release the pressure a little at a time, you may reach a point where the pressure causes one stick to rise, one to hover in the water, and one to be pressed to the bottom of the bottle. (This doesn't work as well if there is much air between your thumb and the water level in the drink bottle. Why?)

# EXERCISES

**1.** Match these characteristics of the atmosphere with appropriate elevations: (a) Where most meteors burn up, (b) the densest part of the ozone layer, (c) the maximum elevation of commercial flights, (d) where space walks can take place, (e) the level below which most of the air is found. Use elevations of about 7 miles, 9 miles, 15 miles, 50 miles, 100 miles or more.

2. We live at a boundary of our atmosphere where fluid meets solid. How does this help us with respect to winds?

**3.** At sea level there is almost 15 lb/in.$^2$ of air pressure pressing in on every square inch of your body. Why aren't you pushed around by it?

4. People not accustomed to elevation usually get sick from lack of oxygen at an altitude of 14,000 ft. So how can a commercial jet keep people at 38,000 ft or more for several hours on cross-continent flights?

**5.** Gases, it's said, fill all the space available to them. So why doesn't earth's atmosphere move out into space?

6. Explain how a suction cup is able to hold on.

**7.** A breath of air weighs about 1/100 lb. How can something that weighs almost nothing and consists mostly of empty space support such a huge pressure?

8. Why should a gasoline can have a small hole on the top of the can in addition to the spout? Why is it easier to pour chocolate syrup if you punch two holes on opposite sides of the top of the can?

**9.** Why should the blades of a windmill be placed high above the ground on a tower?

10. Where would you be safer during a hailstorm, in Leadville, Colorado (elevation 10,035 ft), or Boston harbor? (Assume each area has hailstones that are the same size.)

**11.** A skydiver who leaps from a plane at 14,000 ft reaches terminal velocity in 10 to 12 seconds. What happens to this terminal velocity as the skydiver approaches earth?

12. Why do soft drinks bubble when first opened?

**13.** How can dust stick on a car traveling at 55 mph?

14. Inspect the blades on a house fan that's in use. Invariably there are small dust particles that stay on even while the blades are moving. Why is this?

**15.** In a car engine the mixture of air and gasoline must burn very fast—you could say they almost explode—to power the engine. So the fuel first goes through an apparatus (carburetor or injectors) that turns it into very fine droplets and mixes it with lots of air. Why should that help?

16. A small bubble of air trapped in the brake lines of a car lessens the pressure the brakes apply on the brake shoes or drums. Why does that happen? (Car mechanics bleed the brake lines after working on brakes to eliminate any air bubbles.)

**17.** A bubble of air in the bloodstream can be fatal. (Physicians call that an *air embolism.*) Why should that be so?

18. Analyze what happens when you drink from a glass with a straw.

**19.** When a convertible with its top up travels at a cruising speed, why does the top balloon outward?

20. A box sealed airtight and sitting on a very sensitive scale weighs different amounts during the week. Why?

**21.** Explain what happens when you travel in the mountains and your ears pop.

22. If you cough to dislodge a foreign particle in your trachea, the air pushing through it can actually cause a partial collapse of the airway. Why? The smaller airway then helps increase the speed of the air that follows. How does this help eject the foreign matter? What principle is involved here?

**23.** A meteor that would weigh 100 tons on earth may or may not reach earth's surface. (If it does reach the surface, it is called a *meteorite.*) How would different paths (i.e., different angles) affect the meteor as it approached?

24. If you hold a very deep breath while standing on a scale, do you weigh more or less than when you exhale, or do you weigh the same?

**25.** For the Bicentennial celebration in 1976, the world's largest flag was unfurled on the Verrazano Narrows bridge in New York City. The stars were 11 feet in diameter, and the flag had a surface area of 2.5 acres. Unfortunately, because of the height of the bridge, the flag was in shreds within a few minutes. Why is the wind typically greater in high places that are exposed?

26. At a gas station, you put 25 lb/in.$^2$ of air pressure into the car's tires. Does it make any difference whether you are in Denver or Miami at the time? Discuss. (Presume the temperature is the same in both cities.)

**27.** Speculate on what eventually happens to a child's runaway balloon as it rises.

28. Would water in car tires work as well as air does? Would water work better than solid rubber tires? Which tires would probably produce the best gasoline mileage? Why?

**29.** A small percentage of the earth's land area lies over 19,000 ft above sea level. What would happen to us if all the area of the continents were above this elevation?

30. If overnight the acceleration of gravity near the earth's surface became ½$g$, give two ways this would affect the flights of birds.

**31.** Suppose the atmosphere of earth suddenly vanished. Would you (a) weigh more, (b) weigh less, or (c) weigh the same if you were standing on a scale at the time? What would happen to a hovering hot-air balloon?

**32.** If you go deep into a cave that takes a direction leading upward, should the air pressure increase or decrease as you climb?

**33.** The particles in an aerosol spray slow down quickly as they leave the nozzle of the can. Why? What would happen to the stream from a spray can on the moon?

**34.** The space shuttle's main engine generates 375,000 lb of thrust at sea level but 470,000 lb of thrust in space. Why should there be a difference?

**35.** If you want to throw a record fastball, should you try it in New York City or in Denver? On a hot day or a cold day?

**36.** Why do hot-air balloonists prefer early-morning or early-evening flights?

**37.** Insects are believed to have been the first fliers because of their large surface area to weight ratios. What advantage does that give?

***38.** The air pressure in the tires of a Boeing 747 is 200 lb/in.$^2$. How many square inches of tire must support a (fully loaded) 747 that weighs 600,000 lb?

***39.** Calculate the weight of sea-level air in a 10-ft by 12-ft bedroom with an 8-ft ceiling. Air weighs 0.08 lb/ft$^3$.

***40.** Some scuba divers take down double tanks that hold 142 ft$^3$ of sea-level air. What is the weight of the air in a filled double tank?

***41.** A standard scuba tank is filled with 71.2 ft$^3$ of air at sea level. The pressure gauge on the tank then registers 2250 lb. What is the volume inside that tank?

***42.** Show that a 1-in. drop in the barometric pressure (which can cause the local ocean level to rise 1 ft) is equal to a drop of about 70.6 lb/ft$^2$.

***43.** How much are you buoyed up by air? Air weighs 0.08 lb/ft$^3$. (First you need to know your volume. Your average density is about the same as water, 1 g/cm$^3$. One cubic foot of water weighs 62.4 lb, so if you weigh 135 lb, for example, your volume is 135 lb/(62.4 lb/ft$^3$) = 2.16 ft$^3$.) On Venus's surface, the air is 90 times as dense as on earth's surface. By how much would you be buoyed up on Venus?

***44.** Compare the buoyant force in air to the buoyant force in water.

***45.** The compression ratio of a car's engine is 8 to 1. (That means 8 volumes of air and gas vapor go into 1 volume when a piston is pushed all the way up into a cylinder.) When the piston is down, 16 cubic inches of air and vaporized gasoline at 1 atmosphere fill the cylinder. Then a valve closes, and the piston moves all the way up, as far as it ever goes in the cylinder. Find the new volume and the new pressure, assuming the temperature remains constant.

****46.** *Estimate* the air pressure in the mine at Carleton, South Africa, that goes about 12,000 ft below sea level.

***47.** The first hydrogen balloon that carried Jacques Charles aloft was 26 ft in diameter. (a) Assume it was a perfect sphere and calculate the buoyant force from the air on that balloon. (b) Calculate the lifting force of the hydrogen in that balloon, that is, the buoyant force exerted by the air minus the weight of the hydrogen itself. That is the force that's available to lift the weight of the balloon material, the people, and the gondola beneath the balloon. (Hydrogen weighs 0.07 as much as air.)

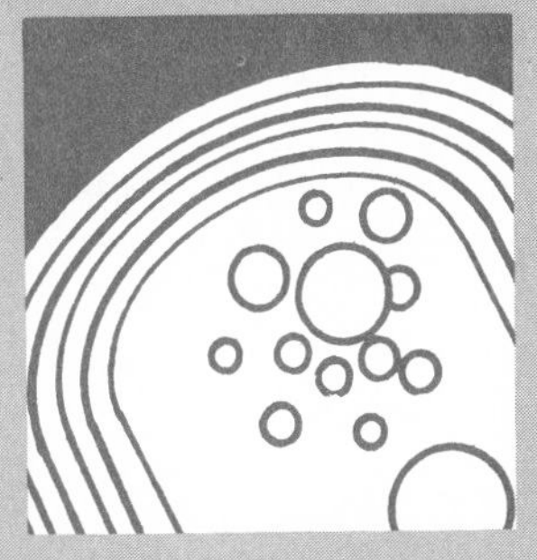

# CHAPTER 13
# TEMPERATURE AND MATTER

Each year in June the racing world comes to Le Mans, a 2 hour drive from Paris. There, over 8 miles of tight corners and straightaways command the attention of some of the world's foremost drivers. They lie in padded hammocks, only inches from the road. Catapulted by engines just behind their backs, they must endure noise, heat, and vibration while they guide their awesome machines around the course. In the straights they move at 220 miles per hour, and everything looks blurred except the roadway in the distance. Should a rock stick in a tire's tread, the driver doesn't hear the click-click-click that you can hear when a car moves from a gravel road onto pavement. Their tires rotate 45 times a second, and the clicks are a steady buzz.

The longest straightaway is the one before the sharp turn known as the Mulsanne corner. The drivers see it coming, and then they have to wait. To maintain the highest average speed, a driver waits until the last possible instant to apply the brakes. Hard. To stay on the road at the Mulsanne corner, a car can go no faster than 35 miles per hour. Spectators along the road get a brief glimpse of what happens then. Just before the corner, the metal discs of the brakes glow at 800°C (1500°F)—a bright cherry red. Unseen, the brake fluid boils in the lines next to that red-hot metal. The brakes of the race car deliver more power doing negative work to slow the car than the engine provides positive work to accelerate it down the straightaway. At that corner the car's brakes turn more than 95 percent of the car's kinetic energy into heat, on every lap, for 24 hours. ■

## Temperature and the Volume of Matter

The propane burners come on with a whoosh, sending heated air into the opening of the beautiful, giant nylon bag. It billows and grows and then lifts off the ground to hover over the basket as the burners continue at full blast. The balloonists climb aboard, and the ride begins. Heating the air makes its molecules move faster, and it expands. As the hot air becomes less dense, it rises, buoyed up by the cooler air around it. That's because the faster-moving molecules collide with the slower-

**FIGURE 13–1**
Filling a hot-air balloon.

moving molecules and push them outward. When the heated air expands, there are fewer molecules per unit of volume, so its density is less.

Liquids, too, generally expand when they are warmed because their loosely bound molecules move faster and stretch farther against the intermolecular attractions.* That's why you shouldn't fill the gas tank of a car to the brim. Gasoline from the underground tank at a service station is cool. If the air outside is warmer, that gasoline expands in the car's tank and will overflow if you've filled the tank too far. An automobile radiator that's been completely filled with coolant will overflow as the engine warms it for the same reason.

Solid matter also expands when the temperature goes up, but because the bonds are stronger in solids they don't expand as much as liquids do. For example, if the radiator and engine (including any spaces inside) expanded more than the liquid coolant did, the level of the coolant would *fall* as the engine warmed.

Various solids expand by different amounts per degree of heating because of differences in bond strengths and structures. When a heating system turns on in a cold house, the ducts and vents pop and snap because of expansion. Heated metal ducts expand unevenly, and they also expand more than the materials they are connected to; steel expands three times more than masonry, for example, for the same change in temperature. A house creaks and groans in the quiet of night as it contracts unevenly from the day's expansion. Some materials expand by different amounts in different directions. The wooden boards used in framing a house expand about 5 times more across the fiber (or grain) than parallel to it. If the heating or cooling is too uneven, a solid may

**FIGURE 13–2**
Underground pipes like these carry hot water between buildings that share a central heating facility, as is often the case on a large college campus. Periodic U-shaped bends in these metal pipes allow the long lengths between the bends to expand and contract without affecting the total system. Otherwise, the pipe could expand due to the heat of the flowing hot water and ram through the foundation of a distant building.

*Water, however, *contracts* as it is warmed from 0°C to 4°C. See the later section on freezing water.

(a)

(b)

**FIGURE 13–3**
In connected segments of large buildings, there are always expansion joints. Look for them in the floor; they are often covered by metal plates. Here in the hall of a university building, a large concrete arch over a lobby rests on steel rollers to allow for expansion and contraction due to changes in temperature. In part *(a)* the arrow shows the location of the roller; part *(b)* is a close-up of the roller.

crack, as a chilled glass fractures if you pour hot liquid into it or a warm glass from the dishwasher may crack if you fill it with an iced drink. The inside of the glass just expands or contracts too fast for the outside, causing a great (thermal) stress in the material.

Liquids also expand by different amounts upon heating, but gases don't. The molecules in a gas are too far apart on the average for their intermolecular forces to influence the expansion or contraction of the gas. For gases, then, the expansion occurs when faster-moving molecules cause higher pressure. If the pressure around any gas remains constant, it expands and shrinks by the same percentage as any other gas whenever its temperature increases or decreases.

## Measuring Temperature

A brand-new dad sprinkles a few drops of baby's formula on the underside of his wrist to test its temperature; mom feels a child's forehead to check for fever. A cook judges the temperature of the cooking oil by whether a drop of water sizzles in it, and ice in a bird bath in the morning means it was below freezing the night before. Rough estimates of temperatures are all we need for many purposes, but for more exact estimates we use thermometers.

The ordinary thermometer is one of the cleverest devices used in a household today. A bit of liquid enclosed in a glass bulb expands if the air warms it and contracts if the air cools it, but the percentage of change in its volume is very, very small. (The glass expands and contracts an even smaller amount.) Enter ingenuity. The rest of the thermometer is a thick glass tube with a very slender cavity. As the large pool of mercury or red-colored alcohol in the bulb expands, it has nowhere to go but into that skinny tube. An increase in volume of as little as one-thousandth of one percent sends the liquid scooting into it for a noticeable distance. A calibrated scale along the tube shows the temperature of the thermometer's bulb and hence the temperature of its surroundings.

An oral thermometer for taking a person's temperature has an even skinnier opening in its tube, so the mercury column moves even farther per degree of temperature change. The thick tube of an oral thermometer is specially curved to magnify the slim column at one particular angle of observation. When taken from the person's mouth, the mercury in the bulb cools quickly to air temperature. It contracts so quickly that it leaves the column of mercury in the tube behind. (The end of that column indicates the temperature of the mouth.) A bubble, a space of partial vacuum, forms between the mercury in the bulb and the column, and the mercury in the tube must be shaken down hard before the thermometer is used again.

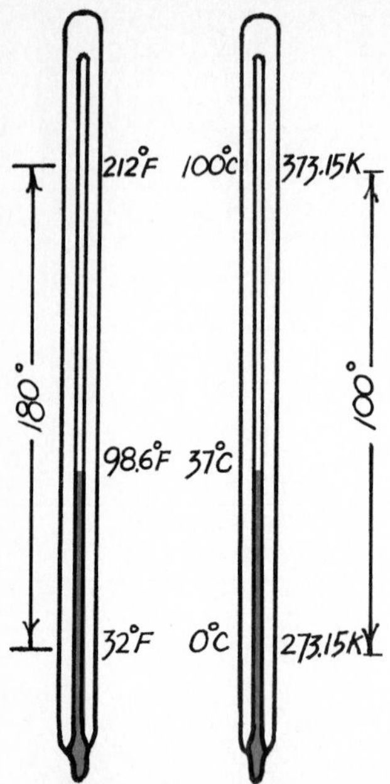


**FIGURE 13–4**
A comparison of the Fahrenheit, Celsius, and Kelvin scales. Note that the Celsius and Kelvin scales have equal-size degrees, while 180 Fahrenheit degrees equal 100 Celsius or Kelvin degrees.

A **Celsius** scale on a glass thermometer reads zero at the position of the liquid when the bulb is in freezing water and 100 when the bulb is in boiling water (at sea level). The liquid expands very evenly between these two points, so the distance between them along the glass tube is divided into 100 equal parts, or *degrees* (see Fig. 13–4). On the **Fahrenheit** scale water freezes at 32 and boils at 212, so 180 degrees separate the two points. The **Kelvin** scale, once called the *absolute* temperature scale, has degrees that match the Celsius degrees. The difference in scales is the choice of "starting" temperatures. The Kelvin scale has its zero point at *absolute zero,* a temperature we discuss in a later section, whereas the Celsius scale has its zero point at the temperature of freezing water. The result is simple enough. The Kelvin temperature is equal to the Celsius temperature *plus* 273.15 degrees. When

**FIGURE 13-5**
"Your temperature is 98.6 degrees Fahrenheit, 37 degrees Celsius, and 310.15 degrees Kelvin—all normal; you'll just have to take that exam!"

**FIGURE 13–6**
This disassembled wall thermostat controls an air cooling and heating unit. It has one coiled bimetallic strip for each function.

a Kelvin temperature is given, there is no degree symbol; 300 degrees Kelvin is written as 300 K.

The thermostat in most home heating systems operates another way. Thin strips of brass and steel (or two other dissimilar metals) are sandwiched together into a springlike coil. Because the two metals expand or contract by different amounts for the same temperature change, the coil winds or unwinds according to the temperature and, in doing so, triggers an off–on switch.

Temperature changes also affect the way metals and other materials conduct electricity. The modern thermometer used by hospitals and physicians uses that property to measure body temperature. Of course, the temperature they check is the core (internal) temperature, not the skin temperature, which accounts for why they must take it the way they do—with the thermometer underneath the tongue.

## What Temperature Actually Measures

When the very young begin to speak, two of their earliest words are "hot" and "cold" because they feel which temperatures are comfortable and which are not. They learn to avoid touching hot stoves and getting too close to fires, and later they learn to dress up against the bite of winter cold. We know about temperature from our sensory experiences—we feel it. But what exactly does temperature measure?

### CELSIUS, FAHRENHEIT, KELVIN

Anders Celsius (1701–1744) was a Swedish astronomer. On an expedition to the Arctic, he made astronomical measurements connected with the earth's curvature that proved Isaac Newton was right—the earth is flattened at the poles because of its rotation. However, the temperature scale that bears his name wasn't really his invention. Although he did publish a paper on the constant degrees of freezing and boiling, the 100 degrees between these two points had been in use earlier. Moreover, Celsius's own scale was upside down from the Celsius scale as we know it: He designated 0° the point of boiling water and 100° the point of freezing water. Three years after his death, his colleagues at the observatory in Uppsala, Sweden, inverted the scale, and for decades it was called the Swedish scale.

Daniel Gabriel Fahrenheit (1686–1736) was in the business of making scientific instruments in the Netherlands. To learn how to manufacture thermometers, he visited a scientist, Olaf Roemer, who had devised a temperature scale of his own. (Roemer was also the first to measure the speed of light.) In that era people thought the temperature of a mixture of ice and salt was the lowest temperature possible. Roemer, therefore, assigned that point 0°. Boiling water was assigned the point of 60°. Roemer had a healthy male put the thermometer under his armpit for awhile to determine human temperature. The temperature of 22.5° was called *blood heat*. (It was widely believed at that time that females had lower temperatures than males.) Freezing water without salt was at 7.5°. Fahrenheit refined the scale for convenience to get rid of the fractions and redid the measurements himself. Ice and salt was at 0°F, freezing water became 32°F, the underarm temperature was 96°F, and boiling water was 212°F. Modern measurements of core temperatures of humans average 98.6°F.

William Thomson, Lord Kelvin (1824–1907), an Irish-born physicist who settled in Scotland, was deeply involved with instrumentation in physics. In the 1850s, first one transatlantic cable and then another (to replace the first, which broke) were laid between Ireland and Newfoundland. As Kelvin had warned, the sending and receiving mechanisms were not adequate because of the long distance. He made a sensitive instrument that would enable rapid and sustained transmission along a third cable laid in 1865, saving an enormous British investment and in the process becoming a national hero (especially among the financiers). He made a fortune with it. The temperature scale that bears his name today honors him for being the first to establish an absolute scale based on physical laws of heat transfer rather than arbitrary points defined by physical processes. The modern Kelvin scale, however, evolved long after his initial suggestion.

You might rub your hands together to warm them on a cold day. When you pound a nail with a hammer, the nail gets warmer. Knocking and rubbing things together creates heat and raises the temperature. Even stirring a glass of iced tea increases its temperature by a tiny amount, an effect that James Joule studied on a larger scale in the 1840s. Joule was an English brewer whose hobby was physics, particularly heat and temperature, since these were important in the brewing process.* He placed a paddle wheel in an insulated tub of water and set about to measure the small temperature changes caused when the paddles stirred the water. (Insulation retards the flow of heat, as we'll see in the next chapter.) He let a falling weight attached to a cord turn the wheel. The slowly falling weight did work on the paddle wheel, and the paddles did work on the water. (How much? Force times distance, or $F \times d$, where $F$ was the weight and $d$ the distance of its descent.) *Joule found that the temperature of the water rose a precise fraction of a degree for every unit of work the paddle wheel delivered.* Work is energy, so turning the paddle wheel gave extra energy to the water, raising its temperature in proportion to the work done. We say the paddles added thermal energy to the water. When something's temperature changes (be it solid, liquid, or gas), the energy we call **thermal energy** comes in or leaves. *''Heat'' is just a shorter term for heat energy or thermal energy.*



*A thirsty patron can still buy ''a pint of Joule'' in the pubs of some areas in England.

Joule's discovery that equal amounts of work cause equal increases in temperature was the first step toward understanding what temperature measures, but the exact relation between molecular speed and temperature came from the molecular picture of matter. **The temperature of a substance is proportional to the average kinetic energy of its molecules, $\frac{1}{2}mv^2$.*** That is not to say that if Joule had stirred kerosene or oil or milk in his tub the same amount of work would have produced the same temperature change that it did in water. Not *all* of the work done by his paddles went into extra kinetic energy of the molecules. Molecules can store some added energy internally as potential energy and in various other ways, as we'll see in the next sections.

## HEAT ENERGY AND CALORIES

Heat energy is measured in **calories,** a hand-me-down term from the 1700s when philosophers thought a fluid (or "caloric") passed from one body to another whenever there was a temperature change. Even today we say heat flows from a warmer body in contact with a colder one. The hand you've held a cup of hot coffee with feels warmer, so you could say, "hot things warm up cooler things." If you are holding a glass of iced tea instead, you might say, "cold things cool warmer ones." But cold does not flow into your hand; heat leaves your hand to warm the glass of iced tea. The kinetic energy of the molecules in your warm hand spreads to the cold glass. Heat travels; cold doesn't.

**One calorie of heat energy can raise the temperature of 1 gram of water by 1 degree Celsius.** One thousand calories (1 **kilocalorie**) equals 1 **Calorie,** with a capital *C*. You find food energy measured in these Calories on diet and nutrition charts. The relation between calories and joules is

$$1 \text{ calorie} = 4.18 \text{ joules}$$

$$1 \text{ Calorie} = 4.18 \times 10^3 \text{ joules}$$

So if you consume 2400 Calories in a day, that's 10 million joules of energy that your body turns into heat (mostly) and muscular exertion.

### How Much Heat Does Matter Hold?

The total amount of heat energy or thermal energy in a quantity of water or anything else depends on its *mass* as well as its temperature. A second cup of coffee at breakfast means you've added twice the heat energy to your stomach as one cup brings. Suppose in his famous experi-

*For the exact relationship, see "Calculations."

**TABLE 13–1**

| APPROXIMATE SPECIFIC HEATS (CALORIES PER GRAM PER DEGREE K) | |
|---|---|
| Gold | 0.03 |
| Platinum | 0.03 |
| Lead | 0.03 |
| Mercury | 0.033 |
| Silver | 0.06 |
| Copper | 0.093 |
| Iron | 0.11–0.12 |
| Air | 0.2 |
| Clay | 0.2 |
| Granite | 0.2 |
| Aluminum | 0.21 |
| Wood | 0.4 |
| Ice | 0.55 |
| Steam | 0.5 |
| Alcohol | 0.58 |
| Paraffin | 0.7 |
| Water | 1.0 |

ment James Joule had used a tub with three times as much volume. A given amount of work by those paddle wheels would have raised the temperature of that larger mass of water only one-third as much.

Besides temperature and mass, there is a third property of matter on which its total heat energy depends: the *type* of matter itself. You can prove this in a kitchen. Add a cup of cold water from the faucet to a cup of water from the hot water faucet. The mixture's temperature almost exactly splits the difference between the temperature of the hot water and the cold water, as you can see if you have a kitchen thermometer. Next, put 80 pennies in a cup and run hot water over them for a minute to bring them to the same temperature as the hot water. Eighty pennies have the same *mass* as a cup of water. Dump the heated pennies into a cup of water from the cold water tap. This time the mixture of hot and cold matter won't be nearly so warm as the mixture of hot and cold water. *Kilogram for kilogram, copper contains less than 10 percent as much heat energy as water does at the same temperature*. The copper not only needs less energy than water does to raise its temperature but also loses less energy than water as it cools off. Consequently, the hot pennies don't warm the cup of cold water as much as the same mass of hot water did. Water stores more heat energy than copper does per degree of temperature increase.

The quantity that measures how much heat energy it takes to raise something's temperature a given amount is called the *specific heat*. **The specific heat of a substance is the quantity of heat that will raise the temperature of 1 gram by 1 degree Celsius.** One calorie of heat energy raises the temperature of 1 gram of water by 1°C (by definition), so the specific heat of water is 1 calorie/gram/°C. To raise the temperature of 1 gram of copper by 1°C requires only 0.093 calories of heat, however, so copper's specific heat is 0.093 calorie/gram/°C. The specific heat of a substance is a gauge of the heat energy a substance can store. Matter with a greater specific heat stores more energy for each degree of temperature change.

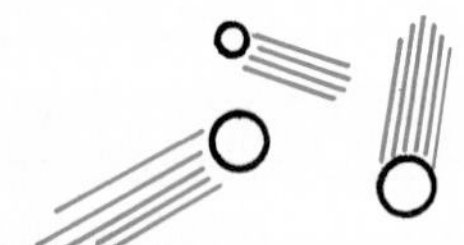

**FIGURE 13–7**
A gas composed of single atoms carries energy by translation.

## How Various Substances Store Different Amounts of Heat

The simplest kind of gas, such as a gas made up of single atoms of helium or argon or neon, also has the simplest manner of storing heat energy (Fig. 13–7). When a few calories are added to such a gas, *all* that heat energy goes to increase the average kinetic energy $\frac{1}{2}mv^2$ of the atoms (hence, its temperature, $T$). The next-simplest substance with respect to matter and heat is a gas of two-atom (diatomic) molecules, such as oxygen, $O_2$, and nitrogen, $N_2$ (Fig. 13–8). When a few calories are added to a gas of diatomic molecules like these, the temperature of the gas doesn't increase as much as for the single-atom gases. All the heat energy does not go into the kinetic energy $\frac{1}{2}mv^2$ of the molecules. These dumbbell-shaped molecules can *spin* when they are hit by other molecules. The energy they absorb by rotating doesn't translate the molecule and so does not give an increase in temperature. Thus the specific heat of a gas of diatomic molecules, that is, the quantity of heat needed to increase their temperature, is larger than the specific heat of single atoms.*

In solids and liquids, the molecular reaction to the addition of heat energy is more complicated because of the bonds. The molecules, atoms, or ions in many solids are tightly bound and can't rotate. But as the particles of a solid gain kinetic energy, they stretch against their bonds, storing energy as *potential* energy while the solid expands (Fig. 13–9). Therefore, it takes more heat energy to raise the temperature of a solid by a given amount, and how much more depends on the strength of its bonds. Liquids, too, store some of the heat energy they absorb in potential energy form as their molecules spread apart against their bonds. And because a liquid's molecules flow or move at the same time, a liquid molecule (consisting of two or more atoms) can rotate as well, storing some of the added heat energy in the form of rotational motion. An important example is water, $H_2O$. The two hydrogen atoms attached to the larger oxygen atom stick out like handles, and when struck at almost any point, a water molecule rotates. That energy added to the potential energy water stores during expansion is the principal reason why it takes so much heat energy to change water's temperature a given amount compared to most other substances. For each calorie of heat added to water, only a minor part goes to increase the temperature, because the temperature is proportional *only to the translational kinetic energy*.

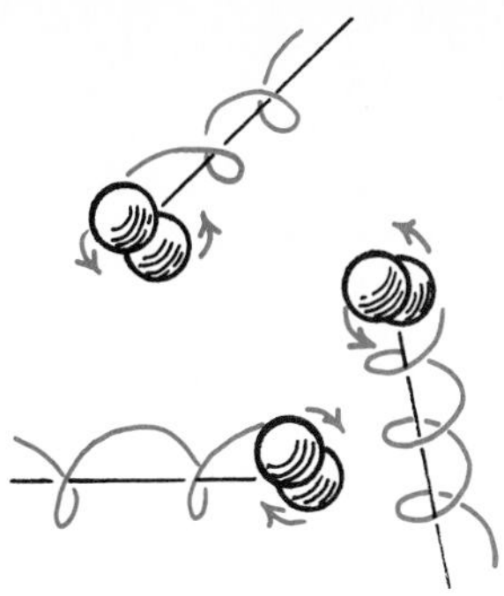

**FIGURE 13–8**
A diatomic gas carries energy by translation and also by rotation.

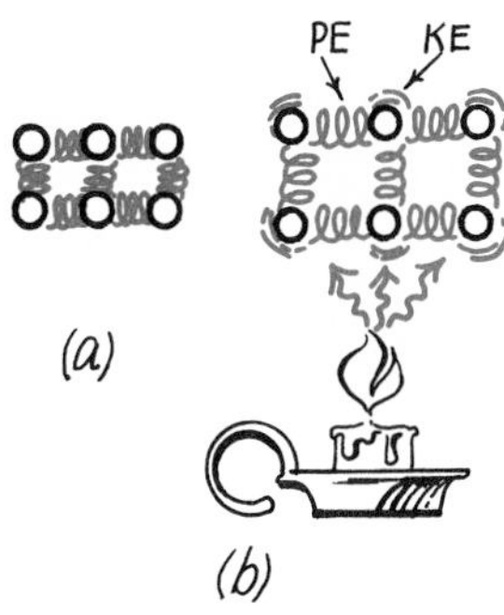


**FIGURE 13–9**
*(a)* Atoms or molecules have translational kinetic energy because they vibrate; they also have potential energy because of the bonds that hold the solid together. *(b)* As the atoms—for example, due to heating—vibrate faster, they pull out farther against those bonds.

## Melting and Freezing

In the 1850s Yankee sea captains with eyes for business carried ice cut from the winter ponds of New England around the tip of South America and back north to California. Packed in wood shavings for the trip, the ice brought a handsome profit from bustling saloons, where it was used to cool the summer drinks of goldrush miners.

Ice cools a liquid by absorbing heat energy from it. Like miniature cannonballs, the liquid's molecules slam against the solid ice, pounding its surface. Those molecules on the surface begin to vibrate faster as the frenzied beating continues. They take heat energy from the liquid, and the liquid becomes cooler.

The ice cubes in your refrigerator's freezer are probably at −25°C (−13°F) or even colder. Drop those cubes into a glass of water, and they begin to warm. Once their surfaces warm to 0°C, something special happens. The ice at the surface melts at that temperature, absorbing a lot of heat energy as it does. The interesting thing is that the ice turns into water at 0°C; *the average kinetic energy of the molecules doesn't change even though the ice absorbs a great deal of heat energy* (Fig.

*If struck hard enough, a diatomic molecule can vibrate as well, the atoms moving back and forth along the line between them. But at ordinary temperatures the collisions aren't strong enough to set them into vibrational motion. The details are explained by atomic physics, which we won't discuss here.

FIGURE 13–10
It requires 80 calories of heat energy per gram of ice to turn ice into water at 0°C.

13–10). Here's an example where heat energy seems to "disappear." But it doesn't, of course. The energy goes to break the bonds of the crystallized ice, which can't hold the jostling molecules at temperatures higher than 0°C. The extra heat energy changes the state of the water from solid to liquid.

The energy that goes to break a solid's bonds is called the **latent** *(hidden)* **heat of fusion** *(melting)*. For 1 gram of ice at 0°C to melt and become water at 0°C takes 80 calories of heat energy, so water's heat of fusion is 80 calories per gram (Fig. 13–11). That's a lot of energy. Suppose the water from a faucet is at 25°C. Ten grams of that water would have to drop 8°C to give up enough heat energy to melt only 1 gram of ice. That's why only a few ice cubes can cool a large drink.

The processes of freezing and melting are reversible. To freeze a gram of water at 0°C, 80 calories of heat energy must be taken from it. Whenever matter of any sort changes from the solid state to the liquid state, or vice versa, heat energy is transferred without a change in tem-

FIGURE 13–11
The water at 0°C gives up 80 calories of heat energy per gram of water as it changes into ice that is also at 0°C.

**FIGURE 13–12**
Semitropical plants can be protected from an overnight freeze with the latent heat of water if they are covered to prevent large heat losses by convection.

perature of the substance. *Any melting solid absorbs energy from its surroundings while its own temperature doesn't change.* Likewise, *any freezing liquid releases heat energy to its surroundings while its own temperature doesn't change.*

Some people in rural areas put tubs of water in their storage cellar or shed on a cold night. As the air temperature in the cellar or shed drops below freezing, the water doesn't. It remains at 0°C until it freezes throughout. All the time it is freezing, it gives up heat energy (its latent heat) to the cooler air and to the stored food, keeping the surroundings warmer than they would be otherwise. Fresh vegetables and fruit typically freeze at about −1°C, or 30°F, so this trick really works.

## FREEZING WATER

Heat something, and it expands; cool something, and it shrinks. That's generally how things behave. But there are exceptions, and one is water. Water at room temperature shrinks in volume if it begins to cool, just as you'd think it should. (See Demonstration 1.) Its molecules slow, and the intermolecular bonds (the forces of cohesion) bring them closer together. But at 4°C the shrinking stops; at 4°C water is as dense as it ever gets. If cooled further, it *expands*. So water at 3°C is not as dense as water at 4°C, and 2°C water is less dense than 3°C water. Water expands all the way from 4°C to 0°C. (See Fig. 13–13.)

**FIGURE 13–13**
Water expands when its temperature drops from 4°C to 3°C, or from 3°C to 2°C, and so on down to zero. Though ice can't exist at those temperatures, some of the molecules of water are moving slowly enough at those temperatures to bond together briefly. Buffeting by other water molecules quickly breaks them apart. Nevertheless, at any instant, enough of these small "crystals" exist in water that is at, say, 2°C to cause the water to expand. This expansion happens because the molecular structure of ice is more open than is the structure of water in the liquid state. Once at 0°C water will freeze solidly if more heat is removed to slow the faster-moving molecules to the point where they can join the solid structure, as shown in the figure.

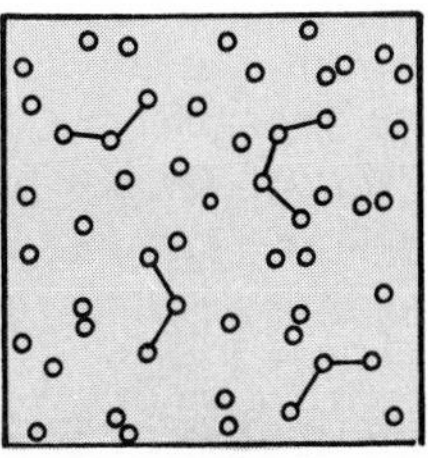


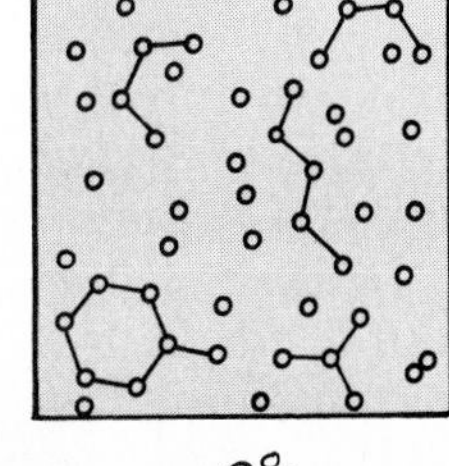


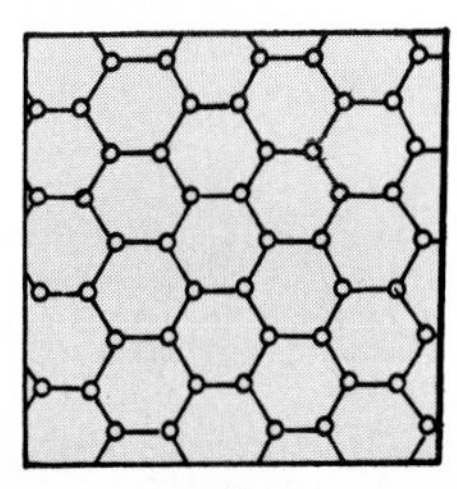

When water at 0°C freezes, the crystal called ice is made of water molecules that fit in a definite pattern because of the way in which the bonds form. That's not unusual, of course, but the structure of ice leaves a lot of empty space between the molecules; that is, they don't fit together as closely as they can in the liquid state. So as liquid water at 0°C freezes, it *expands* and takes up about 10 percent more volume. This is why ice is less dense than water, and why it floats.

The expansion of water between 4°C and 0°C affects how lakes and ponds cool off and freeze in fall and winter. Day after day, the water in a pond loses heat to the cool air at its surface. That cooled water is heavier and sinks to the bottom, and warmer water rises to the top to replace it. Then it too cools and sinks. This exchange goes on until the water that is sinking to the bottom is at 4°C. Water at that temperature is as dense as water ever gets, so each drop on the surface that cools to 4°C sinks, *until the entire pond or lake is at 4°C*. Then as water at the surface continues to cool (from 4°C to 0°C), it becomes lighter at each step and stays at the top. And that is where the ice forms, floating, because it is less dense than water at 0°C.

A layer of ice insulates the water below from the colder air above it, keeping deep ponds and lakes from freezing solid, which would kill any fish or other aquatic life. In the spring the ice melts, and the lake warms from the top down. The water at the top expands as it warms, and the cooler water at the bottom remains there because it is more dense than the water above it. That's why, if you wade into a lake in the summertime, the water at your toes is cooler than the water at your shoulders.

## Vaporization, Condensation, and Evaporation

Heat from a kitchen stove can quickly turn the water in a pot into water vapor, a gas of water molecules. Water at sea level and atmospheric pressure can get no hotter than 100°C, at which point any added heat energy goes into vaporizing the water. The energy that goes to break the liquid's molecular bonds and turn it into a gas is called the **latent heat of vaporization.** Like the heat that changes a solid to a liquid, *this heat energy doesn't change the temperature of the molecules that break away*. At sea level it takes a whopping 540 calories of heat energy to vaporize a gram of water that is at 100°C. In the reverse process, sea-level water vapor at 100°C can condense, turning into liquid water, at 100°C, but only if the vapor loses 540 calories per gram of water vapor. *Any vaporizing liquid (or solid) absorbs energy from its surroundings while its own temperature doesn't change*. Likewise, *any condensing gas releases heat energy to its surroundings while its own temperature doesn't change*.

Liquids don't always have to be heated for molecules to leave their

FIGURE 13–14
Steam from the surface of the water in the pot rises, and the large droplets quickly vaporize in the air. When the vapor-laden air (cooled to about 30°C) reaches the colder window, it gives up its latent heat of vaporization and condenses to form water droplets.

surfaces. It happens all the time in the process called **evaporation.** At the surface of a liquid, the molecules have a wide range of speeds and travel in every possible direction. Attractions from nearby molecules within the liquid keep most of the molecules at the surface from leaving. But when the most energetic molecules head almost straight out from the surface, they do escape. The rate of evaporation increases if the liquid is warmed because more of its molecules have the high kinetic energy needed to escape. No matter what its temperature, however, an evaporating liquid always absorbs the latent heat of vaporization from its surroundings. (The actual value of the latent heat of vaporization depends on the temperature. For water it is about 580 calories per gram at 30°C, as compared to about 540 calories per gram at 100°C. See Fig. 13–14.)

The surface area of the liquid also influences rate of evaporation. If you wipe a kitchen countertop with a wet dishcloth, the small droplets of water the cloth leaves dry up quickly by evaporation. Evaporation proceeds over all the surface that is exposed to air, and small drops have a lot of surface area for each unit of mass.

## Boiling

A pot of water boils at 100°C at sea level, and no matter how high you turn the heat after it's boiling, the temperature doesn't go any higher. The boiling happens as small regions of water close to the heat source absorb enough energy to vaporize even though they are under water. The liquid molecules break away from each other, bursting into bubbles of water vapor by pushing nearby water aside. The bubbles rise to the top, and the surface boils. If the heat beneath the pot is increased, the

**FIGURE 13–15**
The boiling point of water (and other liquids) depends on the air pressure at the surface of the liquid. Pressure cookers use this fact to cook with boiling water at temperatures higher than 100°C.

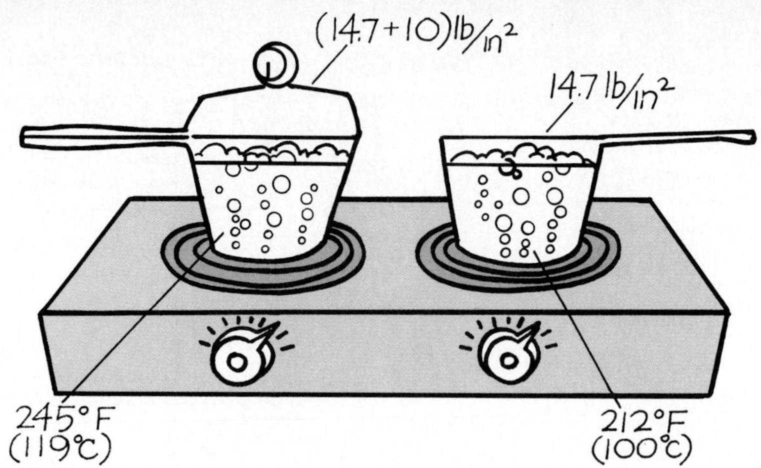


temperature of the water won't rise—but more bubbles will form per second, increasing the boiling action.

The temperature at which the boiling action takes place depends on the air pressure above the liquid. In a pot the water is only a few inches deep, so at sea level the pressure at the bottom due to the water's weight is much less than the atmospheric pressure that is transmitted to the bottom of the pot from the water's surface. At higher elevations, however, water boils at lower temperatures because the air pressure on its surface is less and it is easier for the prospective vapor bubbles to push back the water as they form near the pot's bottom. Many first-time backpackers in the mountains find this out when water that is boiling doesn't cook their eggs or rice in the usual amount of time. Because the water is boiling, the camper thinks the cooking process is going on, but it is the temperature of 100°C that cooks the food (at sea level) at a certain rate, not the boiling. In the mile-high city of Denver, Colorado, water begins to boil at about 94°C. Atop Mt. Mitchell, North Carolina, the highest mountain in the eastern United States (about 6600 feet), the atmospheric pressure is 0.78 atmosphere, and water boils at 92.8°C. On the summit of Mt. Everest, at almost 30,000 feet, the atmospheric pressure is about one-third of an atmosphere, and water boils at 73°C. That is about the partial pressure of oxygen maintained in Skylab in the 1970s, so even if there had been gravity to keep water in a pot, astronauts couldn't have made coffee by the normal methods. If you could go to the top of the atmosphere where there's almost no air pressure, water would boil at the temperature of your room. Perhaps your instructor can show you: A small cup of water placed under the bell of a vacuum pump boils when the air above it is removed.

## Sublimation

Not all solids become liquids as the temperature goes up. Some go directly from the solid state into the gaseous state, a process called **sublimation.** (In fact, all solids lose some surface molecules just as liquids do, and for the same reasons.) Solid carbon dioxide, the frozen form of the gas in carbonated drinks, sublimes rapidly when the surrounding air is at room temperatures. Napthalene crystals (moth balls) sublime more slowly. In dry air and direct sunlight, snow and ice sublime noticeably even at temperatures well below freezing. Most solids evaporate at much slower rates than these, of course. But even diamonds aren't really forever; occasionally a carbon atom at the surface gets enough speed to escape from the surface.

The reverse of sublimation works, too, as when snow crystals grow in cold winter air. The slow-traveling water molecules stick to each other as ice, not liquid water. On a cold winter's morning in the continental United States, you can often see frost (or hoarfrost) on lawns and cars. That's sublimation in reverse.

## The Coldest Temperature: Absolute Zero

Try this. Blow up a balloon and place it in the freezer compartment of a refrigerator. In just a few minutes, the balloon shrinks in size. As the air inside cools, the molecules have a lower average kinetic energy, so they move more slowly and push less on the inside of the balloon. Any water vapor inside the balloon soon becomes frost on its inner surface. If the refrigerator could cool to $-78.5°C$, even the carbon dioxide molecules of the air inside the balloon would become a frost of dry ice. At much lower temperatures the nitrogen and oxygen would become liquids and then solids as their heat energy decreased. The balloon, of course, would be completely flat by then.

Is it possible to cool the air in the balloon—or the balloon itself, or water, or a solid gold ring—to a point where there is no more heat in it? There is indeed a coldest temperature, at which the molecules of *any* substance have no more kinetic energy to give up.* At the temperature of $-273.15°C$ (or $-459.7°F$) matter is as cold as it can ever be. This coldest point is called **absolute zero.** Absolute zero is the zero on the Kelvin scale of temperature. Since temperature measures molecular kinetic energy, the Kelvin scale is the most natural temperature scale. That is, at 0 K, the zero point of the temperature scale, the molecules have no kinetic energy that can be extracted from them. Using sophisticated cooling apparatus, physicists can produce low temperatures that are very close to absolute zero. Temperatures as low as $5 \times 10^{-8}$ K, which is called 50 nanokelvins, have been produced experimentally. Thanks to the sun, matter on earth never gets so cold naturally.

## High Temperatures

The air temperatures most people think of as comfortable are between 20°C and 30°C. (For example, 25°C is 77°F and about 300 K.) Most solid things around you remain in solid form even if you dip them into

*In Chapter 25 we'll see that even then molecules have a tiny amount of vibrational energy it is *impossible* for them to give up. Molecules, it turns out, can never be made to be completely still and vibration free.

boiling water and raise their temperatures to 100°C (or 373 K). An iron frying pan and clay bricks can survive the 1000-K furnace of a full-blown fire that consumes a house. But at about 1400 K, sand melts in a glassblower's oven, and the temperature of molten steel flowing from a blast furnace might be over 1700 K. By the use of electric welding arcs, iron can be made to boil and vaporize at 2700 K. At the temperature of the sun's surface, 6000 K, matter can exist only as a gas of atoms. Within a vapor such as this, any molecule suffers collisions so violent that it splits apart, separating into component atoms. When a gas is at a temperature of tens of thousands of degrees Kelvin, there is a chance that any colliding atoms will suddenly splinter. At this point, gas becomes a plasma because the fractured bits of atoms have net electric charges.

When a stroke of lightning passes between the earth and a cloud, a plasma exists along that jagged, superheated path through the air. In modern street lamps, a tiny bit of plasma glows brightly and lights city streets, a kind of artificial lightning that has its temperature controlled electrically. Even the fluorescent lights in libraries and classrooms make use of thin, lower-temperature plasmas, although the light from these lamps comes from a fluorescent coating on the inside of the tubes rather than the plasmas themselves. When you strike a match, there are charged particles in the hot gases in the flame. Flames are one of the coolest plasmas of all.

Unlike the case at the low end of the temperature scale, there seems to be no upper limit in principle to the amount of kinetic energy an atom or molecule or ion might have. The energy we eventually see and feel as light originates near the center of the sun at a temperature of about 15,000,000 K. At the heart of exploding stars, the temperature briefly reaches billions of degrees Kelvin.

## CALCULATIONS

A simple formula converts temperatures from Fahrenheit to Celsius, or vice versa. If $T_C$ stands for the Celsius temperature and $T_F$ for the Fahrenheit temperature, then

$$T_F = \frac{9}{5}T_C + 32°F$$

or

$$T_C = \frac{5}{9}(T_F - 32°F)$$

**TABLE 13–2**

**COEFFICIENTS OF EXPANSION, α (IN UNITS OF $10^{-6}$ PER DEGREE KELVIN)**

| | |
|---|---|
| Oak wood | |
| Parallel to fiber | 4.9 |
| Across fiber | 54 |
| Pine wood | |
| Parallel to fiber | 5.4 |
| Across fiber | 34 |
| Platinum | 8.7 |
| Concrete | 9–14* |
| Glass | 9 |
| Brick | 9.5 |
| Steel | 11 |
| Gold | 14 |
| Copper | 16 |
| Silver | 19 |
| Aluminum | 22 |
| Ice | 50 |

*Depending on the mix of cement to aggregate and the type of aggregate

EXAMPLE 1: **What is the average internal temperature of humans on the Celsius scale?** That temperature is 98.6°F, so by the second formula,

$$T_C = \frac{5}{9}(98.6°\text{F} - 32°\text{F}) = \mathbf{37.0°C}$$

The volume of a solid, a liquid, or a gas changes as its temperature changes. If the temperature change is not too great, the change in volume, $V_f - V_i$, where $V_f$ is the final volume and $V_i$ is the initial volume, is proportional to the change in temperature, $T_f - T_i$, and is also proportional to the original volume itself, $V_i$. That is, if 1 liter of water expands to become 1 liter plus 1 $\text{cm}^3$, then 20 liters would expand to become 20 liters plus 20 $\text{cm}^3$ for the same temperature change.

If we wish to know how much a solid object expands in a certain direction, however, the amount of its expansion depends on its length in that direction, not its volume. The amount its length changes per unit of length for each degree (Celsius or Kelvin) of temperature change is called its **coefficient of linear expansion,** α (lowercase Greek letter *alpha*). This number has units of 1/(1 degree Kelvin), or $\text{K}^{-1}$. The formula that gives the change in length ($L$) is

$$L_f - L_i = \alpha\, L_i(T_f - T_i)$$

(In this formula *any* units of length may be used, since α doesn't depend on the units of length.)

EXAMPLE 2: Concrete sidewalks are poured in sections, and gaps are left between the sections. When the sidewalk expands on a hot day, the ends of each section move into the gaps for short distances. If the concrete has a coefficient of linear expansion of $11 \times 10^{-6}\ \text{K}^{-1}$, **how much will a 10-foot section of sidewalk expand** from a winter low of −7°C (20°F) to a summer high of 43°C (110°F)?

$$L_f - L_i = \alpha L_i(T_f - T_i) = (11 \times 10^{-6}\ \text{K}^{-1}) \times (10\ \text{ft}) \times (50°\text{C})$$

$$= 0.0055\ \text{ft}$$

or 0.066 in.

The gaps are more than adequate for such expansions. Notice a Celsius degree cancels a degree Kelvin, since they indicate the same temperature change.

Matter whose temperature changes either gains or loses heat energy. The amount of heat energy matter stores is proportional to its mass, temperature, and specific heat, symbol $C$. The heat energy that enters or leaves, called $Q$, when the temperature changes is

$$Q = mC(T_f - T_i)$$

The specific heat $C$ (sometimes called the *heat capacity*) takes into account all the ways in which the matter stores heat energy, not just the kinetic energy of its molecules.

EXAMPLE 3: A cup of hot coffee (mass about 237 g) fresh from the pot is at 90°C. **Estimate how much heat energy it loses when it cools to room temperature at 25°C.** Coffee is almost all water, and its specific heat is about the same as water.

$$Q = mC(T_f - T_i) = (237 \text{ g}) \times \left(1 \frac{\text{cal.}}{\text{g} \cdot {}^\circ\text{C}}\right) \times (25^\circ\text{C} - 90^\circ\text{C})$$

$$= -\mathbf{15{,}400\ cal} = -15 \text{ kcal (or } -15 \text{ Cal)}$$

The kinetic energy of the molecules in a gas, liquid, or solid is directly proportional to its temperature. The relation is $\frac{1}{2}mv^2 = \frac{3}{2}kT$, where $T$ is the Kelvin temperature and $k$ is **Boltzmann's constant** (not to be confused with the spring constant of Chapter 3). The value of $k$ is $1.38 \times 10^{-23}$ joule/degree Kelvin.

EXAMPLE 4: a) **What is the average kinetic energy of nitrogen molecules in air that is at 25°C?** A temperature of 25°C is about (273 + 25) K, or 298 K.

$$KE = \frac{3}{2}kT = \frac{3}{2}(1.38 \times 10^{-23} \text{ J/K}) \times 298 \text{ K} = \mathbf{6.17 \times 10^{-21}\ J}$$

b) **What speed on the average does a nitrogen molecule have in air that's at 25°C?** $\frac{1}{2}mv^2 = 6.2 \times 10^{-21}$ J. The mass of a nitrogen molecule is about (from Chapter 9) 28 amu = $28 \times 1.7 \times 10^{-24}$ g = $4.8 \times 10^{-26}$ kg. Then

$$\bar{v}^2 = \frac{6.2 \times 10^{-21} \text{ J}}{(4.8/2) \times 10^{-26} \text{ kg}} = 26 \times 10^4 \text{ m}^2/\text{s}^2$$

$$\bar{v} = \mathbf{5.1 \times 10^2\ m/s} \text{ (or about 1700 ft/s, which is about } \mathbf{1100\ mph}\text{)}$$

# REVIEW

**1.** Hot air rises in the atmosphere because heating increases the speed of its molecules, so it expands and becomes less dense. True or false?

**2.** Do liquids and solids generally expand when heated? Which generally expands the most?

**3.** Give the three temperature scales in general use and their freezing and boiling points for water.

**4.** What exactly does temperature measure?

**5.** To what form of energy does the term *heat* refer?

**6.** What did Joule discover about heat and temperature?

**7.** The temperature of a substance is proportional to the average kinetic energy of its molecules. True or false?

**8.** On which three properties of an object does its total heat energy depend?

**9.** Define specific heat.

**10.** Heat energy refers not only to translational kinetic energy of molecules but also to internal types of energy such as rotational energy, vibrational energy, and potential energy. True or false?

**11.** Define the latent heat of fusion of a substance.

**12.** When matter changes from the solid to the liquid state, heat energy is transferred without a change in the temperature of the substance. True or false?

**13.** Which is less dense, water at 3°C or water at 4°C?

**14.** Define the latent heat of vaporization of a substance.

**15.** Describe how a liquid boils.

**16.** Define absolute zero and contrast this point on the Kelvin scale to the Celsius and Fahrenheit scales.

# DEMONSTRATIONS

**1.** Fill a glass soft-drink bottle to the brim with water from the cold water faucet. (A 16-oz bottle works fine, but the larger, the better.) Make certain the water is level with the opening. Then run some hot water into the sink and place the bottle in it. In no time you can see the water overflow the top of the bottle as the warming water expands. Water (and other liquids) expands more than glass does, so the bottle of water is like a crude thermometer. The water in the bottle stops expanding when it heats to (nearly) the temperature of the water in the sink. (Then if you leave and come back in a few minutes, you'll find the level of the water has dropped *below* the rim. Warm water evaporates fast!)

Next, rinse the bottle with water from the cold water faucet to reduce its temperature and fill it to the brim with cold water again. This time place it in your refrigerator and check it in 15 minutes. The decrease in volume is very noticeable. Except for its behavior between 4°C and 0°C, water that is being cooled behaves just as you'd expect it to; it shrinks. (Some of the drop in level is due to evaporation; but not much, as you can show by placing the cold bottle of water into a sink of water at room temperature. The expansion brings it nearly back to its original level.)

**2.** Air, a gas, expands even more than water for the same increase in temperature. Stretch a balloon flat over the rim of the same glass bottle you used in Demonstration 1 to prove it. Tie it tightly to the lip of the bottle so no air will escape, but without stretching the balloon very tightly across the opening. Then immerse the bottle in hot water or just run hot water over it, and place the bottle in the refrigerator. The expansion and contraction of air is almost 20 times that of water, as you will see.

For a cute variation, put the empty bottle into the refrigerator and get it cold. Then take it out and immediately place a wet nickel over the opening. Wrap your hands around the bottle to provide heat energy, and the expanding air inside makes the nickel clank against the bottle.

**3.** For this demonstration you'll need a couple of meters of thin copper wire. (Perhaps your instructor can get this for you.) String it horizontally, (between two nails works fine), stretching it tightly. Hang a small object that weighs a pound or two from the center of the wire, and notice its height above the floor. You can prop a meter stick behind it, for example. Next, light a candle and move the flame along the length of the wire. The weight will drop as the wire grows longer.

**4.** Rubber when heated *contracts* rather than expands because of an unusual molecular structure. To show this, hang a rubber band on a nail and stretch it vertically a bit by hanging something from it with a piece of string. (Try some different kitchen utensils.) Bring the flame of a match close enough to warm the rubber band (but not too close), and you'll see the object rise as the rubber contracts.

**5.** Sneakers too tight? Don't throw them away without trying this. Shove a plastic bag deep into each shoe, fill them with water and close the tops tightly, making sure there are no air bubbles trapped inside. Put them in the freezer of your refrigerator for a day. The shoes will fit a foot that's 10% larger in volume!

# EXERCISES

**1.** Do all materials expand when heated?

**2.** A gas expands when heated if it can push back the matter that surrounds it. But in the kitchen device known as a pressure cooker, any gas inside is trapped in a space of constant volume. What happens as the gas is heated?

**3.** Why do ice cubes snap and crackle when put into a glass of warm tea?

**4.** Next time you're on an interstate highway, listen for the clicks the car's tires make as you go on or come off a bridge. What's involved?

**5.** How could the *height* of the Jefferson Memorial Arch in St. Louis (Fig. 13–16) be used as a seasonal indicator?

**6.** Iron rods are often embedded in concrete structures to strengthen them. What happens when these structures go through changes of temperature? (See Table 13–2.)

**FIGURE 13–16**
The Jefferson Memorial Arch, St. Louis.

**7.** If you ever want to freeze some apple cider in gallon jugs, you'd be well advised to drink a cup from each full jug first. Why?

**8.** Should the melting points of solids depend more on their molecular attractions or on the chemical bonds that hold the molecules themselves together?

9. Ancient miners (before the day of the iron pick) heated exposed copper ore inside the ground and then doused it with water. How could this help extract copper from the mine?

**10.** Why does it help to run hot water over the metal top of a jar of food when it's on too tight? (See Table 13–2.)

11. Don't leave your phonograph records in the sun. Discuss what will happen if you do and *why* it happens.

**12.** Rubber *shrinks* upon heating. If this was the largest effect of an increase of temperature on a car's tires, how would it affect the odometer's readings?

13. For what purpose did the Apollo 11 spacecraft rotate once every 20 minutes on its way to the moon?

**14.** James Joule found that a temperature increase of a substance was directly proportional to the energy added to it. True or false?

15. Bumblebees sometimes need to warm up the larvae and pupae in their nests. When they do, they first *decouple* the flight muscles from their wings. Tell how this can help them use body heat.

16. James Joule carefully measured with a thermometer the temperature at the top and bottom of one of the taller waterfalls in Europe. What was he up to?

**17.** A wise cook does not let a tightly fitted pot lid remain on a hot pot of stew that's left to cool. Why not?

18. The method for stir-frying food quickly over high heat was developed in China, apparently as a means to conserve fuel, which became scarce when China's large population decimated the forests several thousand years ago. In this technique, the food is chopped into small pieces and cooked in a single layer in a bowl-shaped vessel called a *wok*. Discuss how this helps cook vegetables in only a few minutes.

**19.** Once the water begins to boil in a pot of vegetables (or eggs to be hard-boiled), should you adjust the heat?

20. Exactly why does a drop of water thrown into a hot frying pan jump around?

**21.** During the construction of a large dam, pipes are laid a few feet apart throughout the concrete, and water is run through them while the concrete hardens. Why is this necessary?

22. Have you ever seen tennis players sweep the water puddles on a court after a rain? Exactly how does this help?

**23.** Compare what would happen if you boiled one egg at sea level, one in an airplane, and one on a high mountain.

24. If you lick a silver spoon that's been in a very hot cup of coffee, it probably won't burn your tongue. But a spoonful of that coffee dropped on your tongue could leave a blister. Why?

25. How does water's high specific heat make it a good coolant for car engines?

26. When a teapot containing water nears the boiling point, you can usually hear distinct popping noises. Speculate why, and then be sure to read the answer.

**27.** A car's coolant system is closed by a tightly fitting radiator cap that maintains pressure within the system. Why is this necessary?

28. The high-speed trains of Japan and other countries move on welded rails that don't have expansion gaps. This eliminates excessive wear and tear at the joints between rails. How, then, might such solid rails expand and contract without damage?

29. A wisecracker once quipped that if the glass bulb and tube of a thermometer expanded more than the liquid inside, it could be an original "Celsius" thermometer. What was meant by that?

**30.** Which could give you a more severe burn, 2 g of water at 100°C or 2 g of steam at 100°C?

**31.** Very cold snow, called dry snow, won't pack into snowballs. Why?

32. If you heat water to 100°C, does it necessarily boil?

33. You could know when a cooling liquid begins to freeze (or a warming liquid has reached its boiling temperature) by observing a thermometer in it. How would you know?

**34.** Can a thermometer placed in a ice tray in a freezer measure the latent heat of fusion? Can a thermometer placed in a pot of boiling water measure the latent heat of vaporization?

35. A spent lead bullet dug from a piece of wood it had penetrated is partially melted. Where did the heat to do that come from?

**36.** Why is cooking oil better for cooking at high temperatures than water is?

*37. Two cups of buttered popcorn give you 110 Calories. How much is that in joules?

*38. The highest recorded temperature for a location in the United States is 134°F. What is that on the Celsius scale?

*39. Isaac Newton proposed a standard thermometer scale for a bulb of alcohol connected to a glass tube. He marked the temperature of a mixture of ice and water as 0° and the temperature of the body as 12°. In this case, what would be the temperature in degrees "Isaac" of boiling water?

*40. Someone consumes 2400 kilocalories in a day; calculate how much water (at 100°C) that much energy could turn into vapor. (The heat of vaporization of water at 98.6°C is about 541 cal/g/K.)

*41. An oxygen molecule has 16 times the mass of a hydrogen molecule. If samples of each gas are at the same temperature, how much faster are the hydrogen molecules moving? Which would be more likely to escape from a planet's atmosphere? Why?

*42. Can two objects of the same mass but different composition have the same total heat energy if their temperatures are not the same?

*43. What is the change in height of a 1200-foot steel tower for broadcasting television signals if its temperature changes 20°C from day to night?

*44. What is the difference in length of a 500-meter steel bridge between a low winter temperature of −16°C and a summer temperature of 39°C?

*45. Five ice cubes of 9 grams each are used to cool 500 grams of water (about 2 cups) at 25°C. To what temperature does that water change just in changing that ice at 0°C to water at 0°C?

*46. Show that a 1100-kg racing car slowing from 220 mph to 35 mph dissipates more than 5 million joules of kinetic energy. (You consume twice that much energy in a 2400-Calorie day.)

*47. The molecules in a baseball move about because of their thermal energy. About how fast would the baseball have to be thrown to double its total energy (heat + kinetic)?

**48. A kilogram of copper is immersed in boiling water until its temperature is 100°C. Then the copper is dropped into a Styrofoam cup containing 1 kilogram of water at 0°C. What is the temperature of the mixture when the copper and water reach the same temperature?

# CHAPTER 14
# HEAT ENERGY IN MOTION

As Dave reached the lower deck and put away his six-pack of frozen soft drinks, his shirt was already soaked with sweat. Even the coolest spots in the engine room of the 20-year-old freighter were 100°F, and the heat raised stifling vapors of sea water, oil, and sludge. Two laboring steam turbines cranked around at over 3000 rpm. The sound level was at the threshold of pain, and the noise of rushing steam and water hammering in pipes added to the din that could slowly destroy unprotected eardrums. Behind this thunder was the hum of air blowers, placed everywhere in a futile attempt to ease the heat. Because the noise was steady, he could almost put it out of his mind—until he had to say something to a shipmate. To talk, he had to scream.

Two hours into his four-hour watch, a tiny leak sprung in a high-pressure superheater line from the boiler. Steam leaks were common, but a superheater leak sounds different. The thin, high scream of steam escaping at 800°F and 1200 pounds per square inch was unmistakable. All lights went out, and Dave froze in his tracks. Someone shut down the engines. This time it was bad. The invisible stream of water vapor had sliced a main electrical cable, and he would have to find the leak in total darkness. The ship slowed to a stop and rolled heavily with the waves. He groped for a flashlight and a broom and crept forward, waving the wooden broomstick up and down in front of him. There was no doubt when he came to the leak. He felt a firm jerk, and the end of the broom, surgically removed by the whistling vapor jet, tumbled to the floor. He knew it could have just as easily been his hand. The superheater cooled and was soon repaired. The ship's motion smoothed out as it gained speed, but the engine noise rose again. Dave wondered if it was worth it.

Steam turbines are the successor of the steam engine, which used the pressure of steam to push a piston back and forth. Perfected by James Watt (1736–1819) in the 1760s, the steam engine helped bring on the Industrial Revolution, replacing horse, oxen, and flowing water as the prime movers to supply power for machines. Steam locomotives, not railroads per se, ''built the West,'' and steam-driven paddle wheels meant large ships no longer needed to depend on catching favorable winds. The cast-iron blocks, cylinders, pistons, rings, valves, and crankshafts of today's automobile engines were first developed for steam engines. Today steam turbines propel ships, and more important, they provide the link between the heat energy from fossil fuels or nuclear reactors and the generators of the electricity we all use. Watt's name was a good choice for the unit of power.

This chapter is about the way heat moves around and some of the ways we move heat to do work. We begin by looking at the three ways heat moves: by *conduction, convection,* and *radiation.* ■

**FIGURE 14–1**
If you step barefooted onto an unheated tile floor or even a cool wooden one, heat energy rapidly begins to leave the bottoms of your feet.

## Conduction

When you throw back the covers on a cold winter morning and roll out of bed, you feel more than just the contact force as your feet hit the floor. The floor feels cool or even cold to your warm feet (Fig. 14–1). Heat flows from your feet into the floor as soon as the skin makes contact. The molecules of your skin are moving faster than the floor's molecules, vibrating with more energy. When they touch the surface molecules of the floor, the cooler molecules there are buffeted by the jittering and very quickly begin to vibrate faster themselves. Neighboring molecules beneath them are set into faster motion, and heat energy flows into the floor from your foot. When heat energy passes between molecules because of their contact, we say the heat moves by **conduction.**

You know by experience that different materials conduct heat at different rates. A tile floor feels colder than a wooden floor that's at the same temperature because it conducts heat faster. Granite and marble walls may look impenetrable to us, but not to heat! They conduct heat up to ten times faster than an equal thickness of concrete block, for example. But concrete block, plaster, and brick conduct heat too rapidly away from homes in winter, so buildings are *insulated* with extra materials. An *insulating material* is a poor conductor of heat; that is, it conducts heat very slowly. Heat flows through a material at a rate proportional to its *coefficient of thermal conductivity.* Values of this coefficient for various substances are given in Table 14–1, and their use is explained in the "Calculations" section at the end of this chapter.

**TABLE 14–1**

**APPROXIMATE VALUES FOR THE COEFFICIENT OF THERMAL CONDUCTIVITY, $k$ (W/m/°C)**

| Material | $k$ |
|---|---|
| Air (room temperature) | 0.026 |
| Styrofoam | 0.03 |
| Rock wool | 0.039 |
| Pine wood | 0.11 |
| Paper | 0.13 |
| Dry soil | 0.14 |
| Hard-packed snow | 0.22 |
| Plaster | 0.3–0.7 |
| Brick | 0.4–0.9 |
| Ice | 0.6 |
| Water (room temperature) | 0.61 |
| Glass | 0.7–0.9 |
| Concrete | 0.9–1.3 |
| Porcelain | 1.09 |
| Granite | 2.2 |
| Marble | 3 |
| Steel | 47 |
| Aluminum | 240 |
| Copper | 400 |
| Silver | 430 |

As you can see from Table 14–1, metals conduct heat far more rapidly than other solids do. When atoms of a metal bond together, many of the loosely bound outer electrons leave their parent atoms to wander throughout the solid. While the heavier atoms can only vibrate in place, passing heat from molecule to molecule, the free electrons can travel a long way between collisions with atoms or other electrons. They move very fast and conduct heat quickly through the metal. In silver, gold, and copper the electrons move more freely than in iron or nickel. But even though a steel spoon conducts heat poorly compared to a silver spoon, cooks know that steel spoons conduct heat rapidly compared to wooden spoons or plastic spoons.

The speed of heat transfer by conduction between two points also

**FIGURE 14–2**
Heat from the inside of this house has come through the roof to melt or sublime most of the snow. The snow on the eaves remained because the eaves had cold air beneath them, rather than a warm house. But why does that large square of snow on the roof remain? A phone call to the owner supplied the answer. A room had been added in the attic and insulated to modern standards, and the heat from within passed through only very slowly, as proved by the unmelted snow.

**FIGURE 14–3**
If the surface below your feet is warmer than your feet, heat flows into your feet. Whether the flow is in or out, the rate of heat flow depends on the difference in temperature between the two surfaces. If you "get used to" standing on a hot sidewalk (or a cold floor), it means your feet have become warmer (or colder), and the heat flow between surfaces decreases because the difference in temperature is less.

depends on how large the temperature change is from one point to the other. The larger the temperature difference, the more rapidly heat moves to the cooler area (Fig. 14–3). As an example, ocean water is almost always cooler than your skin temperature. So when you jump in, it feels cold because heat leaves your skin rapidly by conduction. Later, as your skin cools to a temperature close to the water's, the rate of heat flow is much less; you ''get used to'' the water.

## Convection

Solids, liquids, and gases all conduct heat, but conduction is usually most important in solid matter. In liquids and gases, the atoms or molecules flow, and they transport heat energy as they do. Currents of moving fluids such as air or water can move heat energy very effectively, as they do in home heating systems. When matter moves and carries heat energy along with it, we say the heat moves by **convection.** In fluids this is usually the fastest method of transporting heat. Air that is dry *conducts* heat very poorly, for example. It takes a relatively long time for the random collisions to transfer heat from one place to another. But a flow of that same dry air can move heat quickly.

Sunlight coming through a window warms whatever it touches—a curtain, a carpet, a waxed floor. Small currents of air rise above the warmed spots, carrying the smell of their origins. Because hot air is less dense than cold air, a parcel of relatively warm air is buoyant, and it rises. The rising current of air is a **convection current.** Outside, especially in summer, large convection currents mix the air in earth's atmosphere and cause winds. But air currents aren't caused just by differences in weight. Fans in heating systems force hot air from one place to another, and those currents too convect heat.

**FIGURE 14–4**
Usually the random motions of air molecules get them nowhere fast because they collide and change directions so often. It is convection currents that spread odors so effectively in our world. When the temperature differences in the air are large, as they are during cooking, convection currents are more apparent.

Home insulation slows the transfer of heat by convection. By isolating small pockets of air, insulation prevents large-scale convection currents (Fig. 14–5). Nevertheless, heat energy still moves through the insulating material by both conduction and smaller convection currents, but at a much slower rate.

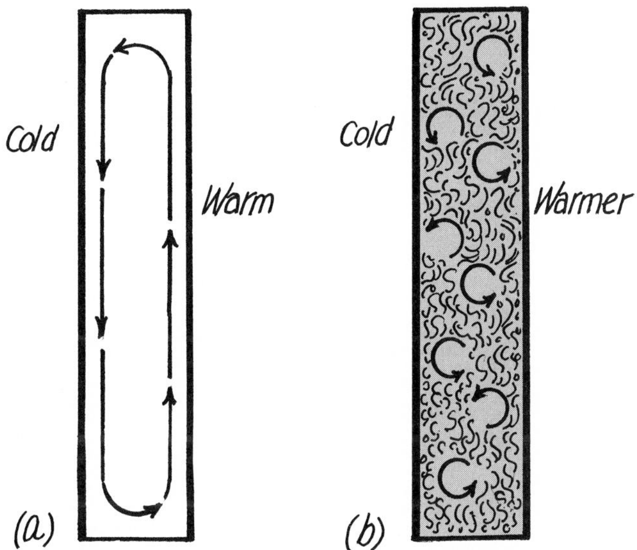


**FIGURE 14–5**
The air space between the walls in part *(a)* is not insulated, and large convection currents freely carry heat energy to the cooler surface. In *(b),* however, insulation prevents large convection currents, and the rate of transfer of heat energy between the warm and cold surfaces is less.

## Heat Radiation

When Joule did experiments to warm water with paddle wheels, he wanted to see precisely how much heat energy came from a given amount of work (Chapter 13). He insulated the tub carefully against convection and conduction, knowing that heat lost through the tub's walls would make his measurements inaccurate. But even if Joule's experiment had been done in a vacuum, with no way for atoms or molecules to conduct or convect heat away, *some heat energy would have escaped*. That's because there is another form of heat transfer, and this one doesn't use matter to carry the energy. Heat energy moves in the form of *radiation* called **heat rays** or **infrared radiation.** These waves are an invisible form of light, and like light, they can move through a vacuum. (Sunlight, including infrared rays, comes unimpeded through 93 million miles of vacuum to reach the earth.) We will see what light rays and infrared rays are in later chapters, but for now we can say that *the electrons in vibrating or colliding atoms and molecules produce these rays*. The faster the atoms vibrate or the harder they collide (that is, the higher the temperature of the material), the more heat energy they emit, or radiate.

**FIGURE 14–6**
While convection currents are carrying heat upward, the fire radiates heat in all directions. Infrared radiation, or heat waves, carry energy through solids, liquids, gases, and even through a vacuum. If the skewers for the hot dogs are metal, the picnicker will also feel the heat that conducts through them.

You can feel heat rays that come from matter that is warmer than your skin; if you sit beside a fireplace, the heat you feel is from those rays. (Convection currents from the fire don't warm you. They go straight up the chimney.) You can feel the radiation from a hot cup of tea or coffee on a cold day if you place your hands about a centimeter away from the sides of the cup. Amazingly enough, if that same cup were filled with cold water from your kitchen faucet it would still radiate about half as much heat energy as it does when it is filled with hot liquid. You don't feel those heat rays, however, because you are warmer than the cold water and you yourself emit more infrared radiation per second than you get from that cold water. The hotter an object is, the more heat energy per second it sheds, but every single atom of the matter around us both radiates and absorbs infrared radiation. *Only at absolute zero would atoms cease to radiate heat energy, and only if everything around them were at absolute zero would atoms cease to absorb radiant heat energy.*

Materials differ greatly in their ability to absorb or emit radiant energy. When you are in a parking lot on a sunny day, compare the temperatures of the paints on a light-colored car and a dark-colored car. Dark colors both absorb and emit more radiation per second than light colors do. Light colors reflect more radiation. Textured or rough sur-

**FIGURE 14–7**
The astronauts' face plates were coated with a thin layer of gold to reflect most of the intense sunlight they experienced on the moon, including both infrared and ultraviolet rays.

**FIGURE 14–8**
A concrete sidewalk gets very hot in the light of the summer sun. But it loses much of that heat after just a few minutes in the shade. Though convection and conduction help, most of the quick loss at the surface is by radiation.

faces, with their ups and downs, have more area to emit and absorb from, so they lose and gain radiant energy faster than smooth surfaces.

Since heat rays move just fine through a vacuum, there is no way to insulate something perfectly. Radiant heat energy always spreads throughout any matter or into empty space. The most that can be done to provide insulation against this kind of loss is to reflect some of the infrared rays back to the matter that emits them. A smooth coating of aluminum (or gold or silver) reflects infrared rays very well, and that is why the backing of house insulation often has shiny coats of aluminum. (The movable electrons in metals play a major part in this reflective property.) That's also why those lightweight but reflective space blankets used by outdoor sports fans and campers can keep them warm. Coffee stays hot (and lemonade stays cold) in a thermos because the walls of both the vacuum bottle and its inner liner have metallic coatings to slow the transfer of heat by infrared radiation. Convection and conduction is nearly eliminated by the partial vacuum between the walls.

When the infrared radiation that leaves an object's surface each second is greater than the amount of radiation it absorbs from its surroundings during that same second, it cools. Experiments show that the *rate of cooling or heating by infrared radiation is proportional to the difference in the temperature between the object and its surroundings,* just as for heat conduction. Isaac Newton discovered this fact, and it is called *Newton's law of cooling*.

## Thermal Equilibrium

Leave a glass of iced tea on a table in the summer, and in a while it warms to the temperature of the air in the room. The tea and glass absorb infrared radiation from their surroundings, and heat is conducted in from the warm air and the table. Air that's cooled at the surface of the glass is heavier and falls, to be replaced by warmer air, so convection currents play a part as well. Finally, water vapor in the air condenses on the outside of the glass, giving up its latent heat of vaporization and helping to warm the glass. Once the glass reaches the temperature of its surroundings, it is at *thermal equilibrium:* Then it radiates as much heat radiation as it receives each second, and because there is no temperature difference, no heat is transferred by conduction. However, thermal equilibrium, like Newton's first law, is really an idealization. Just as there are always forces everywhere to act on matter, there are always hotter and cooler regions, though the temperature differences might be slight.

Suppose the air in a house was in thermal equilibrium. There would be no convection currents at all. (This couldn't be the case if you were in the room; your breathing and uneven skin temperatures would cause convection currents.) The air molecules would still have average speeds well over 1000 feet per second, so it might seem that a lot of mixing should take place. But each molecule has some $10^{10}$ collisions every second. They all bounce off one another, constantly change directions, and as a result, get nowhere fast. Without convection currents an oxygen molecule in the room would need 8 or 9 hours to wander only 1 meter from its starting place. Of course the air, house, and surroundings are not in thermal equilibrium, and the convection currents that result are the reason you can smell food cooking in a kitchen even when you are in another room.

## Cooling by Expanding and Heating by Compressing

**FIGURE 14–9**
As a gas expands, it cools. You can feel the valve on a spray can get cooler as the gas escapes through it.

Open your mouth wide and breathe out on the palm of your hand. Then do it again with your lips nearly pressed together. This time the air expands as it escapes. Feel the difference? Expanding air cools.

Hot or warm air that expands into cooler air around it does work in pushing the cool air out of the way. That takes energy, and this energy comes from the kinetic energy of the molecules in the hot gas. The collisions that push the cold air back *slow the warmer molecules that do the pushing,* so the temperature drops as the hot gas expands.

The reverse takes place if a gas is compressed. To blow up a party balloon, you use your mouth like an air compressor, pushing each breath into the balloon at a pressure slightly higher than that of the air around it. The air molecules inside the balloon are crowded together a little closer than those outside. Work must be done to compress a gas, to push its molecules closer together. Pushing the molecules through a distance does work, it gives the molecules extra speed, and as their kinetic energy increases, so does their temperature. Compressing *heats* a gas.

The tires on a bike or car are tougher versions of the balloon and under higher pressure. A bicycle pump, a simple plunger (or piston) that presses on air that's trapped in a cylinder, can inflate them. The plunger pushes the air molecules head-on, and they pick up speed, like a tennis ball that's hit by a racket. A bicycle pump that's just been used is warm, and not only from friction between the plunger and the cylinder. The plunger does work on the air to compress it, which heats the air, and some of that heat escapes through the cylinder wall. If you let air out of a tire, you can feel the opposite effect: The tire valve cools off as the escaping air expands and cools.

**FIGURE 14–10**
In diesel and gasoline engines, fuel is burned rapidly to form a high-pressure, high-temperature gas. This gas pushes a piston downward, and some of the energy of the gas is converted to mechanical energy as the piston turns a crankshaft. When the gas expands, its pressure drops, and $P_iV_i/T_i = P_fV_f/T_f$.

## More about Temperature and the Volume of a Gas

Jacques Charles, the French physicist who flew the first hydrogen balloon (see "Up, Up, and Away," Chapter 12), discovered the relation between the temperature and the volume of a gas that expands or contracts while at a constant pressure. His experiments showed that if the Kelvin temperature increases by, say, 100 percent, the volume increases by the same percentage. In symbols, Charles's law for a gas under constant pressure is

$$\frac{V_i}{T_i} = \frac{V_f}{T_f}$$

where $T$ is the Kelvin temperature. Thus if the pressure remains constant and the Kelvin temperature doubles ($T_f = 2T_i$), the volume doubles as well ($V_f = 2V_i$).

Boyle's law for a gas at constant temperature (Chapter 12), $P_iV_i = P_fV_f$, can be combined with Charles's law to give a more general gas law. For any confined gas, the product of the absolute pressure and the volume, when divided by the Kelvin temperature, is a constant. In symbols,

$$\frac{P_iV_i}{T_i} = \frac{P_fV_f}{T_f}$$

This law tells what happens when air is compressed in a bicycle pump, when the fuel–air mixture is compressed in the cylinders of a diesel engine, or when a hot air balloon rises in the atmosphere.

## Evaporation as a Cooling Process

When water evaporates, it absorbs a lot of heat without a change in temperature. A single gram of water absorbs about 100 calories if it warms from 0°C to 100°C. Then it must further absorb *over five times that much heat* (540 calories) to turn into vapor at 100°C. Because the latent heat of vaporization ''disappears'' without causing a temperature increase, evaporation can be very effective at cooling.

By hanging wet mats in doorways and windows, the ancient Egyptians, Greeks, and Romans cooled the incoming air in the summertime. As water evaporated, it took heat from the air that drifted through the moist mats. Modern air conditioners and home refrigerators cool by the evaporation of a liquid too, though the liquid doesn't get away in the process.

## Moving Heat Energy to Cool Things Off: Refrigerators and Air Conditioners

In home refrigerators and air conditioners today, cooling takes place when a liquid called *Freon* vaporizes, taking its latent heat of vaporization from its surroundings.* The Freon is cycled through a closed cooling system that has two separate sides, a low-pressure region and a high-pressure region. The liquid enters the low-pressure side through a valve that separates the two sides of the system (see Fig. 14–11). The drop in pressure on the liquid lowers its boiling point, and it evaporates while traveling through the *evaporator* (or *refrigerator*) *coils*. In the

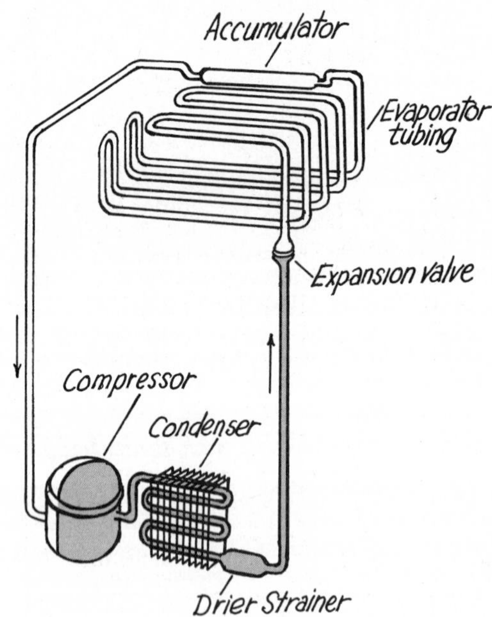


**FIGURE 14–11**
The cooling system of a refrigerator. The expansion valve is an adjustable plug that lets the pressurized liquid Freon pass at a slow rate into the low-pressure side of the system, where it evaporates in the evaporator tubing. The accumulator prevents any unevaporated Freon from entering the compressor, since liquid Freon is very incompressible and would damage the compressor.

*Freon* is the trademark of a family of gases made synthetically. Beginning with the natural gas methane ($CH_4$), the hydrogen atoms are partially or entirely replaced with fluorine and chlorine atoms. Gases of the Freon family are nonpoisonous, noncorrosive, nonflammable, and odorless and have favorable pressure–temperature characteristics to serve as refrigerants. They were widely used as propellants in aerosol cans (deodorants, paints, and so on) until research indicated they could destroy ozone, at which time most manufacturers replaced them in uses that released the gas into the atmosphere.

FIGURE 14–12
A central air-conditioning system for a building. Such a unit is only a large version of a refrigerator, with an arrangement to cool filtered air by passing it over the evaporator coil and a fan to push the cool air through a system of vents. The condenser, which sheds the heat gathered by the evaporator coil, is located outside.

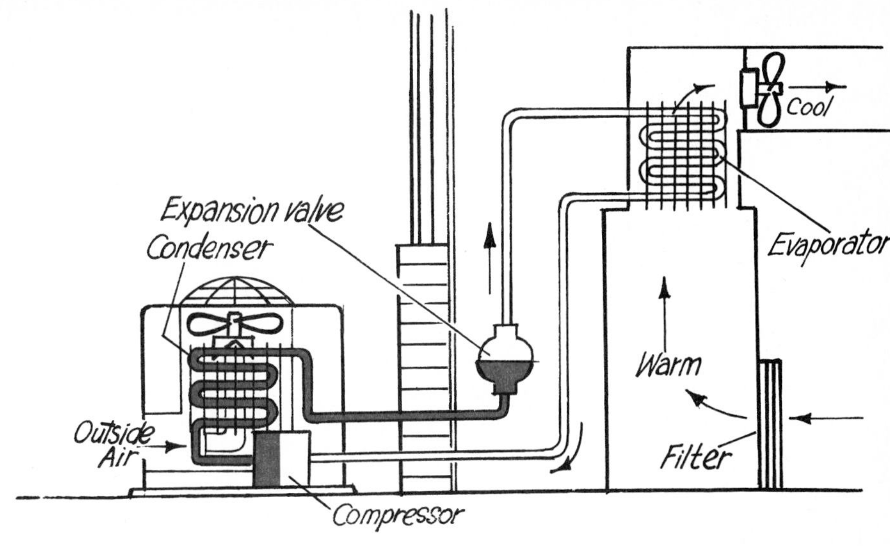


process, it absorbs heat from the air (in the case of an air conditioner) or heat from the food (in the case of a refrigerator). Then, still under low pressure, the gas enters a *compressor,* which is what you normally hear when a refrigerator or air conditioner is working. The compressor compresses the gas, which also heats it. The hot, dense gas emerges from the compressor into the high-pressure side of the system and passes into a *condenser,* which is something like a car's radiator. Copper tubing passing through metal fins conducts and radiates heat away from the hot Freon gas in the tubes. A fan moves air over the condenser fins to speed the cooling, and the gas condenses, giving up its latent heat of vaporization to the surrounding air. In the liquid form again, it is ready for another cycle through the system. The heat the gas absorbs on the low-pressure side is released on the high-pressure side. These cooling devices just *move* the heat from one place to another. (Many a pair of children's shoes have been dried out on the floor in front of the refrigerator where the heated air from the condenser exits.)

FIGURE 14–13
Blades of a steam turbine are made of special alloys to withstand the temperature, pressure, and vibrations associated with being struck by high-speed superheated steam. They must also be perfectly balanced.

## Moving Heat Energy to Do Work: Steam Turbines

When you turn on a light, it uses energy from a commercial power plant. Such a plant generates electrical energy by using fossil or nuclear fuel. (Only a few percent of commercial electricity in the United States and other industrialized countries comes from hydroelectric generators, geothermal plants, and windmills.) The small electric motors in every home, in vacuum cleaners, electric razors, fans, and even the motors that drive refrigerator compressors all use energy that begins as heat.

First, heat turns water into steam in a *boiler*. The steam's temperature is then boosted in a *superheater,* coils of tubes that pass directly through the heat source under the boiler. The additional heat energy the superheater puts into the steam is important for the next step. The high-

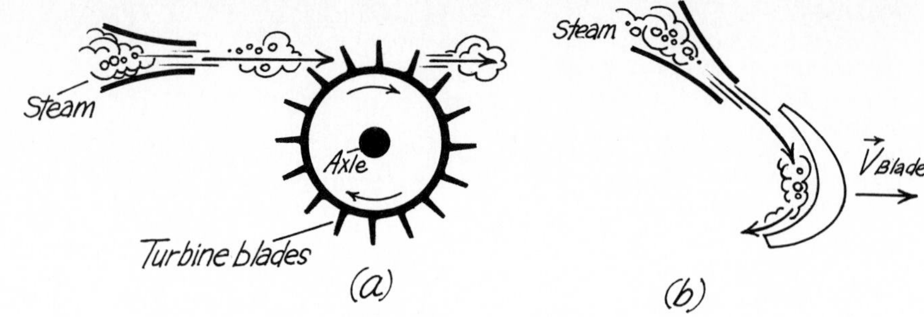


**FIGURE 14-14**
*(a)* A simple steam turbine. *(b)* Many steam turbines have blades shaped to get more work from the steam. As the steam presses against the surface, it is guided almost backward. At the same time, the blade gets the reaction force in its forward direction.

**FIGURE 14–15**
The cylinder of this motorcycle has fins to increase its surface area and promote cooling. The piston moves back and forth in the cylinder. The spark plug that ignites the air-gasoline mixture is visible, as are the exhaust pipes that take away the "spent" hot gas.

pressure steam has great heat energy in the form of the kinetic energy of its molecules, but its motion is random. By providing an opening into a lower-pressure region and letting the steam escape in channels or nozzles, its energy becomes more directed. The high-speed flow of steam then encounters sets of blades fixed around a movable axle, which is the working part of a *steam turbine* (see Fig. 14–14). The rushing steam spins the turbine blades around, like a house fan operating in reverse. The rotating shaft of the turbine then provides the mechanical energy to run other machines. The largest steam turbines produce commercial electricity by running electrical generators. The shafts of steam turbines aboard large ships go through gear boxes before turning the propellers.

## Thermodynamics Revisited: Efficiency

The first law of thermodynamics (from Chapter 6) has something to say about the operation of a steam turbine, an air conditioner, or the motor in a food blender. It tells us that **energy in = energy out + work done.** The total energy remains constant, and this is true for those devices or for any others. In the case of a steam turbine, this statement becomes *heat in* = *heat out* + *work done*. The **efficiency** of a turbine, or an engine or motor, is how much work it does compared to the energy that's put into it, or

$$\text{efficiency} = \frac{\text{work done}}{\text{energy in}}$$

For a steam turbine, for example, this is

$$\text{efficiency} = \frac{\text{heat energy in} - \text{heat energy out}}{\text{heat energy in}}$$

The second law of thermodynamics says even more. No machine can convert into work *all* of the energy that goes into the machine. Some energy always winds up as wasted heat, which moves into the surroundings of the engine. In a heat-driven engine like the steam turbine, the efficiency is limited by the *temperatures* of the heat that goes *in* and the heat that comes *out*. The maximum theoretical efficiency, expressed in terms of the final and initial temperatures of the heat-carrying medium, is called the Carnot* efficiency. In terms of Kelvin temperatures, the maximum possible efficiency of an engine that is driven by heat is

$$\text{maximum efficiency} = \frac{T_{\text{in}} - T_{\text{out}}}{T_{\text{in}}} \quad \text{or} \quad \frac{T_{\text{hot}} - T_{\text{cold}}}{T_{\text{hot}}}$$

An engine could have 100 percent efficiency only if the exit temperature ($T_{\text{cold}}$) were absolute zero. Then all of its heat energy would have been converted to work. But that's impossible. Most often the heat that leaves an engine (e.g., a car's exhaust gases or the spent steam from a turbine) is much higher than even room temperature (about 300 K). On the practical side, the equation for efficiency shows why it is important to superheat steam that's used to drive a turbine. To use heat most efficiently, the difference in temperature from the hot side to the cold side of the engine must be *as large as possible*. Diesel engines for cars, trucks, and tractors burn fuel at a higher temperature than gasoline engines, so their efficiency can be higher. But even with diesels, most of the energy from the burned fuel goes into waste heat. As we saw in Chapter 6, the quantity called *entropy* that physicists use to measure the *disorder* of a system always increases whenever energy changes forms. Heat energy always becomes more dilute.

## CALCULATIONS

EXAMPLE 1: A plastic balloon is *partially* filled with helium at sea level. Its initial pressure is $P_i = 760$ mm of mercury, and the helium is at a temperature of 70°F (21°C). Upon rising to just over 9 km in the atmosphere, where the temperature is −60°F (−51°C) and the air pressure is 229 mm of mercury, the balloon is fully expanded and encloses a volume ($V_f$) of 15,000 ft$^3$. **What volume of helium gas was put into the balloon at its launching?**

$$\frac{P_i V_i}{T_i} = \frac{P_f V_f}{T_f}$$

where $T$ must be the Kelvin temperature and $P$ must be absolute pressure, not gauge pressure. Rearranging, substituting, and cancelling units, we have

$$V_i = \frac{P_f V_f T_i}{P_i T_f} = \frac{229 \times 15{,}000 \text{ ft}^3 \times 294}{760 \times 222} = \mathbf{5986 \text{ ft}^3} \text{ of helium gas}$$

*Sadi Carnot (1796–1832), a French engineer, investigated operating temperatures while experimenting to improve the efficiency of steam engines. Although his analysis wasn't rigorous, it caught the attention of William Thomson, Lord Kelvin, who used Carnot's ideas to work toward a definition of an absolute temperature scale, now known as the Kelvin scale.

**EXAMPLE 2:** On large ships the superheaters aren't usually fired for operation until the cruising speed is reached in order to reduce wear and tear due to the high temperatures involved. Before its superheater is in use, steam comes out of a ship's boiler at 490°F (528 K), passes through the steam turbine, and condenses in a condenser that is cooled by circulating seawater at about 77°F (298 K). (a) **What is the turbine's maximum possible efficiency?**

$$\frac{T_{\text{hot}} - T_{\text{cold}}}{T_{\text{hot}}} = \frac{528\text{ K} - 298\text{ K}}{528\text{ K}} = \mathbf{0.44}$$

So at best, 44 percent of the energy put into the turbine could do work. (b) Later the superheater is used, and the steam enters the turbine at a temperature of 775°F (686 K). **What is the turbine's maximum possible efficiency then?**

$$\frac{686\text{ K} - 298\text{ K}}{686\text{ K}} = \mathbf{0.57}$$

(Friction, and heat lost by conduction, convection, and radiation, always prevent these maximum efficiencies from being achieved.)

You've had some experience with how fast heat flows between two regions of different temperatures. For example, to pick up a really hot pot, you need two potholders to protect your hand, not just one. The thicker the material that separates the two different temperatures, the slower the flow of heat. The area matters, too. Two windows in a room let out (or in) twice as much heat as one window. And, of course, *the greater the temperature difference, the faster the flow of heat*. The formula for heat flow through a substance is

$$\text{heat energy per second} = \frac{k \times \text{area} \times \text{temperature difference}}{\text{thickness}}$$

$$= kA\frac{(T_1 - T_2)}{d}$$

where $k$ is the coefficient of thermal conductivity of the material. If $k$ is large, the material conducts heat rapidly. For insulators, $k$ is a relatively small number. Table 14–1 lists values of $k$ for a variety of substances. The units of $k$ are watts/(meters × degrees Celsius), or W/(m · °C).

**EXAMPLE 3:** A Styrofoam ice chest has walls 2 cm thick. Suppose the chest is full of ice and canned drinks at 0°C and the outside air is 36°C. If the total surface area exposed to the air is 0.9 $m^2$ (if the bottom surface is on a wooden picnic table, that surface is better insulated), **about how long does it take to melt 2¼ kg (5 lb) of the ice inside?**

$$\text{heat per second} = \frac{0.03\text{ W/m}\cdot{}^\circ\text{C} \times 0.9\text{ m}^2 \times (36^\circ\text{C} - 0^\circ\text{C})}{0.02\text{ m}} = 49\text{ W}$$

or about 50 joules/second. (Notice that Styrofoam's coefficient of thermal conductivity is approximately the same as that of air, because the air bubbles in Styrofoam do the insulating.) Since 4.18 joules = 1 calorie, heat enters the chest at a rate of 12 calories/second. The heat energy needed to melt 2¼ kg of ice is

$$2250\text{ g} \times 80\text{ cal/g} = 1.8 \times 10^5\text{ cal}$$

**TABLE 14–2**

**EMISSIVITY, e (THE COEFFICIENT OF EMISSION OR ABSORPTION) AT ABOUT 20° C**

| | |
|---|---|
| Human skin (any color) | 0.97 |
| Lampblack | 0.95 |
| Green paint | 0.95 |
| Glossy black paint | 0.90 |
| Aluminum paint | 0.55 |
| Cast iron | 0.25 |
| Polished copper | 0.15 |
| Polished aluminum | 0.08 |
| Polished silver | 0.02 |

so

$$\frac{1.8 \times 10^5 \text{ cal}}{12 \text{ cal/s}} = 0.15 \times 10^5 \text{ s, or about } \mathbf{4.2\ h}$$

The rate at which a material gives off radiant energy is *not* something you've had much experience with. Experiments show how it happens, and the result is

$$\text{heat lost per second} = e \times (5.7 \times 10^{-8} \text{ W/m}^2/\text{K}^4) \times \text{surface area} \times T^4$$

This is known as the **Stefan–Boltzmann law.** $T$ is the Kelvin temperature of the substance; $e$ is a number that's always between zero and 1 (with no units) called the **emissivity** of the material, and it depends on the material's color, surface texture, and composition. For a perfect *absorber* $e$ is 1, and for a perfect *reflector* $e$ is 0. A perfect absorber is a material that absorbs all radiation that strikes it; then the energy it absorbs is reemitted in a way that depends on the temperature. *For temperatures near room temperature, all the heat lost is in the form of invisible infrared radiation.* For temperatures as high as those on the surface of the sun, most of the radiation is in the form of visible light.

EXAMPLE 4: The asphalt shingles on a 25-ft by 60-ft roof have a value of $e = 0.95$. On a summer day when the temperature of the shingles is about 130°F or 328 K, **how much heat energy per second does the roof lose by infrared radiation?**

$$\begin{aligned}\text{heat lost per second} &= 0.95 \times (5.7 \times 10^{-8} \text{ W/m}^2/\text{K}^4) \\ &\quad \times (7.6 \text{ m} \times 18.3 \text{ m}) \times (328 \text{ K})^4 \\ &= \mathbf{87{,}000\ W}\end{aligned}$$

As an object loses heat by infrared radiation, it also gains heat by absorbing infrared radiation given off by its surroundings: The Stefan–Boltzmann law works for both processes. If $T_{\text{environment}}$ is the temperature of the object's environment, the object *absorbs* heat at the rate of $e \times 5.7 \times 10^{-8}$ W/m²/K⁴ $\times$ surface area $\times T^4_{\text{environment}}$. In thermal equilibrium the object is at the same temperature as its environment, and it gains and loses heat at the same rate. Otherwise, when object and environment are at different temperatures, the net heat the object loses or gains per second $= e \times 5.7 \times 10^{-8}$ W/m²/K⁴ $\times$ surface area $\times (T^4_{\text{object}} - T^4_{\text{environment}})$. This expression can be shown (with a lot of algebra) to be approximately proportional to $(T_{\text{object}} - T_{\text{environment}})$, which is Newton's law of cooling.

## REVIEW

**1.** Describe conduction in a solid and convection in a fluid.

**2.** What makes metals good conductors?

**3.** How does house insulation work?

**4.** When an object is at the same temperature as its surroundings, is it at thermal equilibrium?

**5.** Will a gas that expands lose heat energy and become cooler?

**6.** To compress a gas, work is done on it, increasing its kinetic energy and its temperature. True or false?

**7.** State Charles's law. State the more general gas law that combines Charles's law and Boyle's law.

**8.** Are these statements true? The latent heat of vaporization turns a liquid into a vapor without a temperature change. The latent heat of vaporization is released by a vapor that condenses into a liquid without a temperature change in the substance.

**9.** When the gas Freon is cycled through a refrigerator or an air conditioner, it absorbs heat as it evaporates, and the vapor releases heat when it condenses. True or false?

**10.** A steam turbine is driven by the steam or the superheated steam that is directed onto its blades. True or false?

**11.** What general ratio defines efficiency for an engine or a motor or a steam turbine?

# DEMONSTRATIONS

**1.** Explore the convection currents in a house with a candle. The slightest motion of air bends a flame. If you hold a lighted candle near the top and the bottom of a partially opened door, you can see which of the rooms is warmer or cooler. Warmer air rises, spilling through the doorway at the top, and cooler air flows in the opposite direction near the floor to replace it. The flame points in the moving air's direction.

**2.** Peer across the top of a dark-colored car that's been sitting in a parking lot on a summer day. Objects in the background shimmer because convection currents rise unevenly from the hot roof and hood. That's because light bends as it travels through regions of varying density, and this bending causes the images to waver. (We'll see more about this in Chapter 23.)

The same effect is available in a kitchen any night. Turn on a burner of the stove and turn off the lights. Direct the beam of a flashlight through the air just *above* the burner. The light that hits the wall beyond the stove will shimmer. Next, set the flashlight down to one side, but aim it at some object in the room; position yourself so that the light you see reflecting from that object passes through the air above the burner on the way to your eyes.

**3.** Styrofoam is an excellent insulator. Hot coffee (or an iced drink) in a Styrofoam cup mainly loses (or gains) heat through the surface that's exposed to the air. Place a paperback book over the top to cover the cup, and the coffee will still be hot (or the iced drink cold) thirty minutes later!

**4.** Remove both ends from an empty soup can with a can opener. Use the yellow area of a candle's flame to blacken about half of the inside of the can, leaving the other part shiny. Then hold the can vertically with the flame centered between the blackened and shiny areas so that the convection currents rise through it without striking the inside surfaces. You'll feel the blackened area warm first as it absorbs infrared radiation from the candle's flame more rapidly than the light, shiny surface does.

**5.** About 90 percent of the heat from a fire in a fireplace goes up the chimney. Only about 10 percent radiates in the form of heat rays into the room. The glass or plastic lenses of an ordinary pair of eyeglasses let visible rays pass through them but block infrared rays. Sit with your face close to the fire and close your eyes gently. Your eyelids are very sensitive to the heat brought by the infrared rays. Now slip a pair of glasses on while your eyes are still closed and feel instant relief. Your eyelids cool immediately because the infrared radiation is stopped by the glass or plastic. Be sure to try this because the effect is dramatic.

# EXERCISES

**1.** Why do some cooks put long aluminum nails through potatoes to be baked?

**2.** Have you ever licked an aluminum ice cream scoop just after it's been used? What happened? Would the tip of your tongue stick to a wooden spoon that was at the same temperature?

**3.** Pick up a pair of scissors and a wooden pencil. Which feels colder even though they are both at room temperature? Why?

**4.** Astronomers who work with telescopes at high elevations learn to put mittens on their hands in subzero weather before touching any metal equipment. Why is this necessary?

**5.** Lighted candles wouldn't burn well in Skylab. Explain why and suggest a solution.

**6.** Exactly how does steam condensing in the pipes of old-fashioned radiators warm a room?

**7.** How does opening the cargo doors of the space shuttle help to cool it if the doors are on the shaded side of the shuttle?

**8.** Why does the air in a deep underground cave have a constant temperature all year?

**9.** The coefficients of thermal conductivity of ice and water are about the same. Then how does a layer of ice on a pond help to insulate the water below in winter?

**10.** A student has money for only one bag of ice. Would it last longer in a small ice chest or a larger one? Why?

**11.** Exactly how does turning down the thermostat in a house save fuel in winter?

**12.** On a hot day an elephant keeps cooler by throwing wet mud onto its back. Exactly how does this help?

**13.** Why were the faceplates of the Apollo astronauts coated with thin layers of gold? Why were their spacesuits made of white material?

**14.** A fire in a fireplace gives out more heat after it has been burning for some time. Why?

**15.** On a hot day a child once suggested leaving the refrigerator door open to cool the kitchen. Would this work? Discuss.

**16.** During the summer in Death Valley, some of the few residents turn off their hot water heaters. Then they use the *hot* water faucets for *cold* water and the *cold* water faucets for *hot* water. Explain how this works.

**17.** Explain why apartment buildings can be much more energy-efficient in the winter than single homes of the same size.

**18.** A naked person sheds infrared rays effectively from only about 85 percent of the skin's surface area. Why?

**19.** Explain why a glass of iced tea takes longer to warm up than a cold can of beer, which needs only a few minutes to become lukewarm.

**20.** Travelers of long ago discovered that a damp cloth cover on a canteen helps cool the liquid inside. Explain how.

**21.** Explain why you could stick your hand into an oven at 500°F for 15 seconds without harm, whereas you would never put your hand into boiling water at 212°F for even a second.

**22.** Astronauts who walked on the moon wore liquid-cooled underwear. Water was pumped through flexible plastic tubes to keep them cool. What factors made that necessary?

**23.** Faced with a bowl of thick, steaming hot vegetable soup, most people would stir it to make it cool faster. How would that help?

**24.** Exactly how does a lid on a soup pot help keep the soup hot?

**25.** Should it take a large turkey (a) more time, (b) less time, or (c) the same amount of time to cook *per pound* than a small turkey at the same temperature in a convection oven? (Cooks beware!) Explain your answer.

**26.** It takes a bit more runway for an airplane to take off at midday than early in the morning. Why?

**27.** The odor of onions and garlic won't spread through a kitchen if they are stored in a dark corner or a cupboard. Why not? What happens if you put them in a sunny window for an hour?

**28.** Why will a waiter or waitress spin a bottle of wine or champagne in its bed of ice before opening it for the diners? Will the bottle get colder faster in an ice-water mixture?

**29.** James Joule found no temperature difference between the top and the bottom of the waterfall in Exercise 16, Chapter 13. Can you guess why?

**30.** Heavy-bodied insects such as giant silk moths shiver for awhile before their first takeoffs of the day. Suggest why they do this.

**31.** Do you think a jogger on a hot day would prefer to run with a slight summer breeze blowing in the jogger's direction or to run directly into the same breeze? (If you jog, chances are excellent that you know the answer.)

**32.** When you cook rice or potatoes, the pot you fill too full of water will *always* overflow when boiling. Why? (The water level in a ship's boiler is critical also because of this effect.)

**33.** The rate of heat flow from an object depends on its surface area. What would be the best shape for a house in order to minimize heat loss in winter?

**34.** A road that goes north of the Arctic Circle in Alaska was opened in 1980. Much of it is built on high gravel beds to insulate the road from the permafrost below. Why should that be necessary?

**35.** Many of the metal pilings that support the Trans-Alaska pipeline have radiators on them so the heat from the warm oil traveling through the pipeline won't reach the ground. Why should that be a problem?

**36.** Just as compressing heats a gas, compressing a roadbed heats it too. From March to June 1, some Alaskan highways restrict trucks to 25 percent of their weight capacities. Why should they do this?

**37.** On a freezing night in Florida, would oranges on a tree freeze before or after grapefruit would? Why?

**38.** The leaves of rhododendron curl into cylinders when the temperature drops. How does that help protect the foliage against heat loss?

**39.** Why do bridges freeze over faster than roadways in a winter snowstorm?

***40.** (a) If the thickness of the insulation in the attic of a house is doubled, about how much should that reduce the rate of heat flow through that area? Why? (b) Suppose the temperature outside is −12°C (10°F) and the air in the house is kept at 24°C (75°F). To which temperature setting could the thermostat be lowered so the rate of flow of heat from the house would be about half as great? (Notice that this technique saves the same amount of energy per hour as doubling the insulation in the ceilings and walls of the house. However, the house will be colder if the thermostat setting is lowered and warmer if insulation is doubled.)

***41.** A full scuba tank stores air under pressure at 2100 lb/in.$^2$. In the trunk of a car on a hot summer day, the tank's temperature soars from 75°F (24°C) at the dive shop to 130°F (57°C). (a) What's the new pressure at the higher temperature? (b) Is the pressure more than 10 percent over the recommended 2100 lb/in.$^2$? (c) Suppose the tank was stored in a garage and a fire broke out, raising its temperature to 250°F (121°C). What would the pressure in the tank be at that point?

***42.** A tightly sealed glass jar of air is placed in the freezer of a refrigerator at sea level. As its temperature changes from 25°C to −30°C, the pressure from the air inside it drops. If $P_i$ = 14.7 lb/in.$^2$, what is $P_f$? (Assume the volume is constant.)

***43.** The windows and ventilators of a car parked on the beach are closed, sealing the air (at 80°F, or 27°C) inside it. Two hours later the sun has raised the car's temperature to 120°F (49°C). (a) What is the air's pressure inside the car? (b) What net force does the air's pressure exert on a rear window whose area is 5 ft$^2$?

***44.** The engine of an average-sized motorcycle uses about 1 gallon of gas in an hour's operation (i.e., it goes 50 miles on a gallon at 50 mph). A gallon of gasoline yields 32 × 10$^6$ calories of heat energy. If the motorcycle is 22 percent efficient, how much work does the engine do in 1 hour? How much heat energy is wasted?

***45.** At a commercial electrical power plant, superheated steam enters the turbine at 1000°F (811 K) and is condensed at about 110°F (316 K). What is the theoretical maximum efficiency of the turbine?

***46.** There are experimental projects underway to generate electricity from heat engines that use the temperature difference between the surface of the ocean (at about 25°C) and the ocean's depths (about 4°C). What is the possible maximum efficiency of such an engine? Why is this idea interesting despite the low efficiency?

***47.** A person's skin temperature varies over the body and depends somewhat on the temperature of the surroundings. At room temperature, an average skin temperature is about 33°C or 306 K. When a male model (surface area 1.8 m$^2$) and a female model (surface area 1.3 m$^2$) pose nude for an art class, each loses heat effectively from about 85 percent of the body's surface (see Exercise 18). (a) Calculate the rates at which each loses heat. (b) If the air and the walls of the room are at 24°C or 297 K, calculate the rates at which each absorbs heat from the room, and subtract those rates from the heat losses to find the *net* losses. (c) Convert the net losses to kilocalories/day to see how much food energy each model would expend as heat just by standing around naked.

# CHAPTER 15 HEAT, CLIMATE, AND LIFE

Before the time of steam locomotives, stagecoaches were the rapid transit systems of the world. On a good road, a lucky (and hardy) passenger might travel 50 miles in a single day. Have times changed! Between two human heartbeats, an orbiting space shuttle travels almost 5 miles.

Moving at over 17,000 mph, the shuttle has a lot of kinetic energy; its elevation of 150 miles means it has a lot of potential energy, too. Before it lands, the shuttle has to lose that energy. The first step is to slow the shuttle so it will angle down and catch the edge of the atmosphere. Turning the craft so the rocket engine points forward, the astronauts fire the engine briefly. When this *retrofire* has slowed the shuttle's speed by 200 mph, the engine is cut off, and the crew turns the shuttle back around. Then they tilt its nose upward and wait.

They feel the onset of deceleration about 90 miles up, as the shuttle's bottom becomes the world's largest brake shoe. At 60 miles up (where most meteors evaporate brightly), the nose of the fast-moving craft ionizes the air upon impact. If the shuttle is on the dark side of earth, the riders see an eerie reddish-pink glow from that plasma. Because the ionized air reflects radio waves, the crew is out of touch with the ground for 16 (long) minutes. When they are still 10 miles up, they begin to steer the shuttle into sweeping S-shaped turns and wait for air resistance to lower the speed enough for a safe landing. From retrofire to touchdown, the trip takes half an hour—not a bad commuting time from work to home, especially considering the mileage.

Each year since 1957 rockets have carried scientific instruments out of the atmosphere into orbits to study earth's air, oceans, lands, their thermal interactions, and more. Today the sharp images from weather satellites are an expected part of TV news programs. The vivid pictures from space showing the curving surface of earth, its beautiful blue atmosphere thinning outward into the black background, give us a magnificent perspective of our planet.

The atmosphere, the oceans, and the land surfaces share the heat from the sunlight they absorb, and together they are the home for life. Their thermal properties and the thermal interactions between them make the

(a)

(b)

**FIGURE 15–1**
Photos taken several hours apart on August 8, 1980, by two satellites in geosynchronous orbits over earth's equator. Over 26,000 miles from earth's center, the satellites revolve about the earth with the earth's own period of rotation, which keeps them stationary with respect to one point over earth's equator. *(a)* The western weather satellite monitors the eastern Pacific and the west coast weather. *(b)* The eastern satellite monitors the east coast weather. Note the two hurricanes. (In the summer of 1984 the eastern satellite went out of order, and the western satellite was shifted in its orbit to view the east coast as well.)

climates that determine which plants can grow where. That in turn determines where most species of animals can live. Earth's plants and creatures are uniquely adapted to the varieties of conditions its climates provide. These various topics are the subject of this chapter. ■

## The Origins of the Atmosphere and the Oceans

The earth's oceans and its initial atmosphere almost certainly came from its interior early in its history. Yet one sniff downwind from an active volcano confirms that the gases from earth's interior differ greatly from the air we normally breathe. (Such gases usually smell of hydrogen sulfide, the gas that gives rotten eggs their odor.) Volcanic gases contain almost 60 percent water vapor, more than 20 percent carbon dioxide, about 12 percent sulfur compounds, about 5 percent nitrogen, and *no* oxygen. Today's atmosphere of nitrogen and oxygen (99 percent) and only traces of carbon dioxide and sulfur compounds bears little resemblance to conditions some 4 billion years ago.

In a nutshell, here's what happened. The water vapor condensed and formed the oceans. The large volumes of carbon dioxide dissolved in the new oceans and reacted with calcium (the fifth most abundant element on earth by weight) to form a compound called calcium carbonate. This compound is taken up by marine animals and used in their shells. Over the eons shells accumulated on the ocean floor, compressing under their own weight to form the rock limestone. The sulfur compounds thrown out by volcanoes came down with rains—the first acid rains—and reacted with elements both in the oceans and on land to form more stable compounds. Nitrogen, however, is relatively inert, so most of the nitrogen that escaped from the earth remains in the atmosphere to this time. All the evidence indicates that oxygen didn't appear in the air until much later. Atmospheric oxygen comes from the chemical process called photosynthesis, which plants use to build carbohydrate molecules. In this process plants remove carbon dioxide from the air and release oxygen.

Things didn't work out this way for earth's sister planet, Venus. Venus is believed to be almost identical with earth in composition. Nearer to the sun than earth, however, Venus was (and is) too hot for water vapor to condense, so it has no oceans. When carbon dioxide outgassed from Venus's interior, it remained in its atmosphere. Both carbon dioxide and water vapor are more efficient at absorbing heat rays than nitrogen or oxygen, so Venus's atmosphere prevents heat from leaving the planet quickly. This entrapment of heat is called *the greenhouse effect,* and the planet Venus has a runaway case of it. Today its surface temperature is about 900°F. At such temperatures, water readily combines with iron compounds, so Venus's water is probably bound up in its rocks—if it didn't evaporate from Venus's atmosphere due to its relatively small mass. Instead of accumulating tremendous deposits of limestone from its carbon dioxide as earth did, Venus was left with an atmosphere of 96 percent carbon dioxide that exerts *90 earth atmospheres* of pressure at its surface. The high clouds that hide Venus's surface are droplets of sulfuric acid, very different from earth's clouds of water vapor and rain.

## Sunlight on Earth

When daytime skies are clear, sunlight strikes the oceans and the land. Some light is reflected there, but most is absorbed (except on snow covered areas.) The absorbed rays warm the matter, causing its molecules to move faster, collide harder, and give off more heat rays in all directions, including outward through the atmosphere into space. These warmed areas continue to radiate heat into space even after the earth's rotation has carried them into the shadow of earth. For that reason, night is a time of cooling.

As land and water are exposed to the sun's rays during the day, they don't warm by the same amounts. The specific heat of soil and sand is about ⅕ as large as water's. For the same amount of heat energy, then, the land's temperature can increase up to five times as much. Likewise, if the land were to lose the same amount of heat energy at night as the water (per unit of mass), its temperature would fall five times as much. What's more, sunlight penetrates deeper into water, so

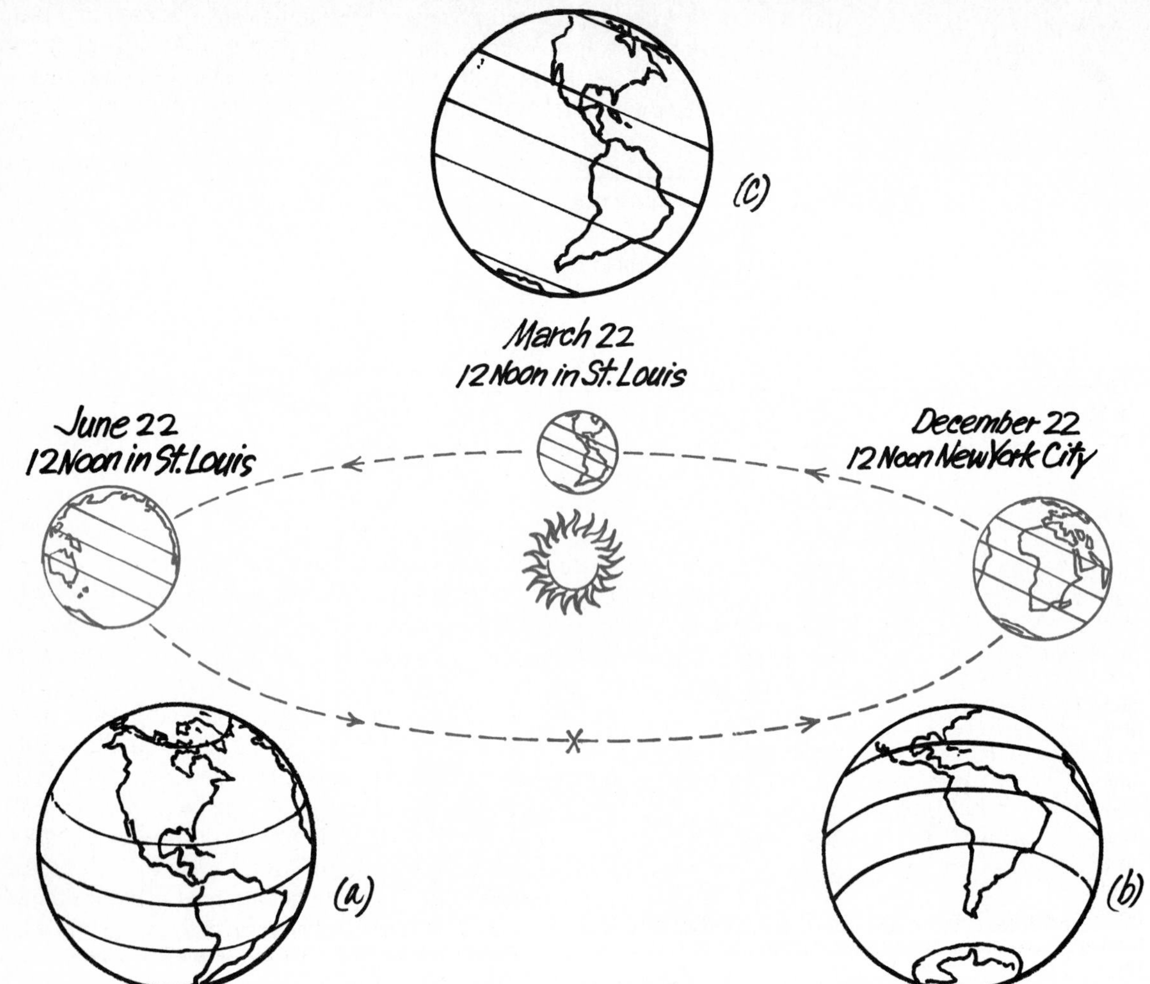


**FIGURE 15–2**
The tilt of earth's axis with respect to the sun causes the seasons. Here the earth is shown at three positions on its orbit. The expanded views *(a), (b), (c)* show the earth at those same dates and times *as seen from the sun.* In view *(a),* June 22, the Northern Hemisphere is tilted more toward the sun than at any other time during the year. As you can see from view *(a),* at that time the Northern Hemisphere faces the sun more directly than the Southern Hemisphere does. That's summer for North America, of course, and for Europe and Asia—all of the land masses in the Northern Hemisphere. Likewise, it is winter for the entire Southern Hemisphere. The situation is reversed in view *(b),* December 22. In view *(c),* about March 22, (and September 22, not shown), the hemispheres face the sun at equal angles. For more details on the cause of the seasons, see Figs. 15–3 and 15–4.

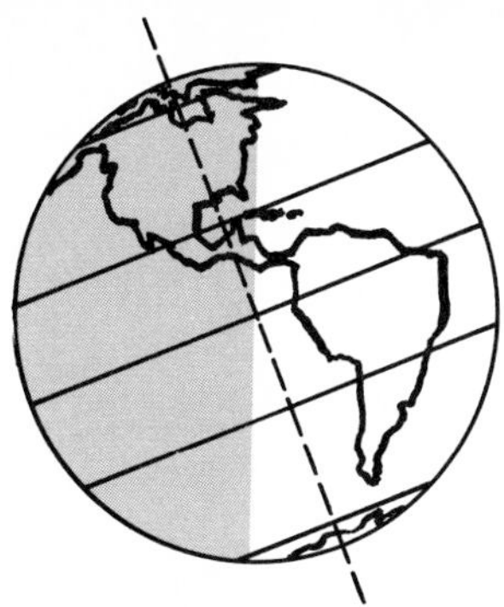

**FIGURE 15–3**
Another view of earth on December 22, arranged so that the sun is directly to the right. Points on the left horizon of earth (as you look at this figure) take about 6 hours to reach the dashed line and another 6 hours to reach the right horizon (the same for the trip on the back side). This shows why North America has fewer than 12 hours of daylight in winter whereas South America has more than 12 hours. Only points on the equator have 12 hours of day and 12 hours of night all year long. Notice that all of the region below the Antarctic Circle is exposed to the sun 24 hours on this day, while everything above the Arctic Circle is in darkness for 24 hours.

that heat energy is shared by a larger volume of matter than on land, where the light may be stopped within a centimeter or less. And currents in the oceans spread the energy even farther. With more mass to share the energy and a greater specific heat, the ocean's temperature is very slow to change compared to the land.

The orientation of the spinning earth as it orbits the sun is the major factor that determines how hot or cold a given place will be. The earth's axis of rotation tilts 23½° from the perpendicular to the plane of its orbit around the sun. Figures 15–2 and 15–3 show how the tilt causes varying amounts of day and night during a year at every location except the equator. That tilt of earth's axis causes the changes in solar heating that lead to our seasons.

In our summertime, earth's tilt points the Northern Hemisphere more toward the sun's direction, and each day locations north of the equator get more hours of daylight and fewer hours of dark. Meanwhile, the Southern Hemisphere is pointed more away from the sun, keeping the cities and countryside there in the dark longer than 12 hours and in the sunlight less than 12 hours each day. That's one reason summers are hotter and winters are colder. Another has to do with the angle of the incoming sunlight at a given location. In winter, for example, the sun's path in the sky each day as seen from the ground is low compared to its path in the summer. That means in winter sunlight strikes the earth more obliquely, and *a given amount of sunlight spreads over a much larger area* (see Fig. 15–4). The land and waters get less energy per second per unit of area in the wintertime, and the days are shorter as well. Both effects ensure lower temperatures in winter.

During any given day, summer or winter, the energy per second per unit of area from sunlight is greatest when the sun is highest in the sky—about 12 noon, local time. Yet the hottest time during a clear day is a few hours *after* 12 noon. Why? For several hours after the peak input of heat energy, the sun still adds more heat energy per second than the ground is losing by radiation, convection, and evaporation. So the temperature continues to rise into the early afternoon hours. The

**FIGURE 15–4**
Light that strikes a surface at an angle is spread over more area and so delivers less energy per unit of area. You can demonstrate this with a flashlight *(a)* just as it happens with sunlight on the earth *(b)*.

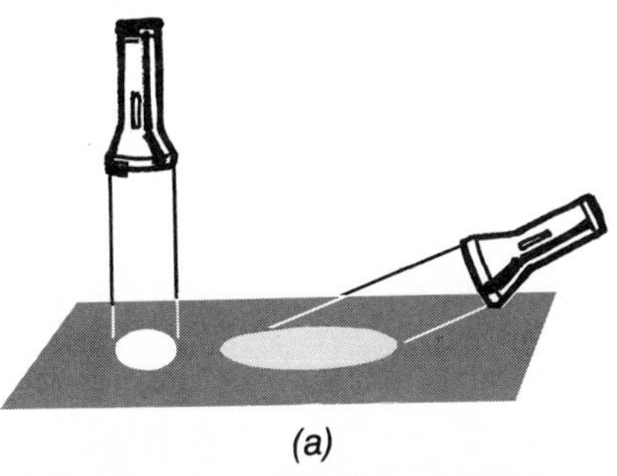

*(a)*

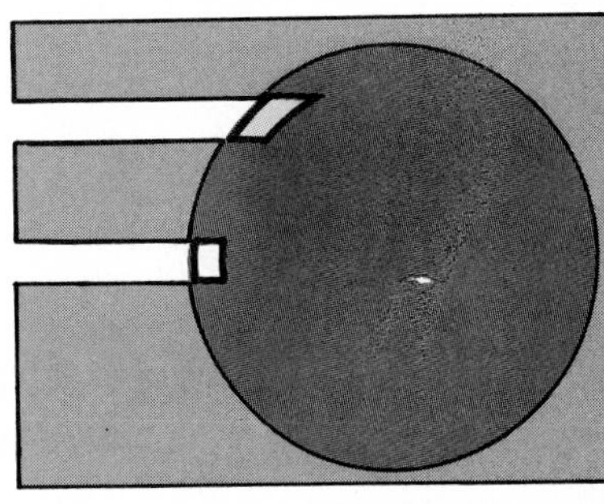

*(b)*

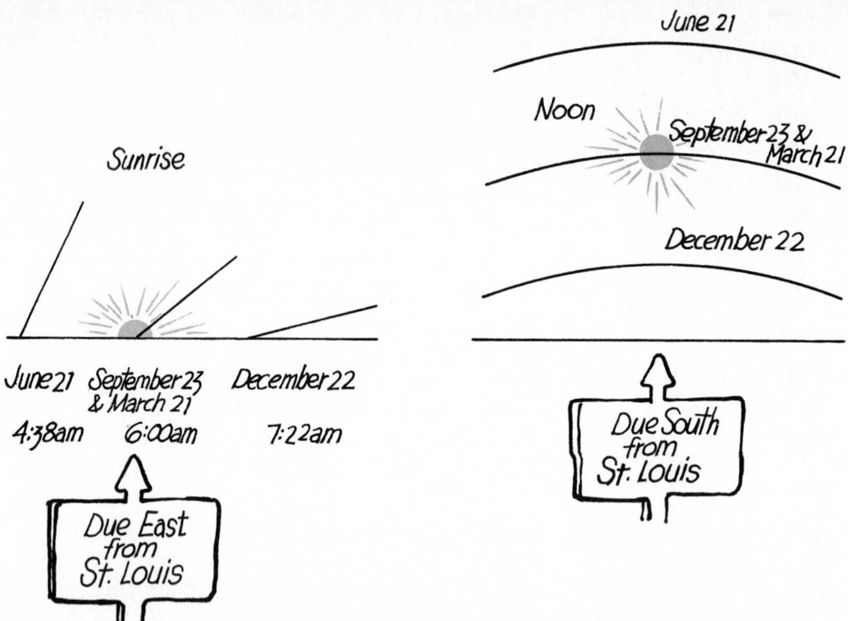


**FIGURE 15–5**
A view from a point on earth's surface that shows the daily path of the sun at different times of the year.

seasons show a similar lag in temperature change. From Fig. 15–5 you can see that the sun reaches its highest angle in the sky on June 22 for points in the Northern Hemisphere (above 23½° latitude). Summer's hottest months, however, are usually July and August. Though not at its highest angle in those months, the sun continues to add more heat energy per day than the hemisphere radiates away during 24 hours, so the Northern Hemisphere continues to warm. Similarly, December 22 is the lowest point for the sun in the skies of the Northern Hemisphere, marking the time of the least incoming solar energy per square meter. But the coldest months of winter are usually January and February. During those months the Northern Hemisphere still radiates away more heat than it receives each day, making those months colder than December.

## Convection Currents in the Atmosphere

Thanks to solar heating, earth's atmosphere abounds with convection currents. Warmed by sun-heated ground or parking lots or even sidewalks, parcels of air rise, buoyed upward by cooler air that comes in from the sides. They range in size from wisps that can rise between blades of grass to the giant updrafts that go on to make a thunderstorm. The larger ones cause gentle breezes or gusts you feel when you take a walk. Winds are caused when higher-pressure air moves into areas of lower pressure where the air has warmed and expanded.

Within a centimeter of the ground, however, convection currents can't move the air effectively because of friction with the solid surface. In this boundary layer, molecules must get around by diffusion, that is, they move randomly because of their collisions rather than being swept along in a current of molecules. As we saw in the last chapter, such diffusion is very slow—so slow, in fact, that bloodhounds in search of a lost child can sometimes follow a scent several days old by keeping their noses in or near the boundary layer close to the ground. (The molecular trail left by the child may slowly shift with the wind. The bloodhounds' paths, therefore, rarely follow the child's footsteps exactly. But contrary to popular belief, if both the air and water are still, bloodhounds *can* follow a trail of scent across water.) Fortunately for us, above that first centimeter and all the way to the upper atmosphere to roughly 100 kilometers (about 60 miles), the air is mixed and stirred by winds. It's difficult to appreciate this fact enough. Without that mixing, earth's atmosphere would be deadly. Quiet air is a poor conductor, so as the ground absorbed sunlight and got hot each day, the heat would accumulate in the air next to the ground and drive the temperature up. To make matters worse, the heavier gases (and pollutants) in the air would settle near the surface and the lighter ones would rise.

FIGURE 15–6

A cave (known as "the dog cave") in Italy gives us an example of what the environment might be like without convection currents. The air there is essentially in thermal equilibrium with the rock walls, ceiling, and floor, so no convection currents stir it. This cave is in an active volcanic region, and gases (mostly water vapor and carbon dioxide) slowly leak in from the surrounding rocks. The carbon dioxide is heavier than the other gases, and it settles low in the undisturbed air in the cave, hovering just above the cavern floor. Human visitors are well over that layer because of their heights, but if a dog wandered in, it could collapse and die from lack of oxygen.

Thanks to the sun's light, convection currents keep earth's air well mixed. Some of the carbon dioxide molecules you exhale today, or some of those emitted by a car you drive, will turn up at a beach in Hawaii in a year or less. And in another year some of them will be in the air a penguin breathes in Antarctica. Updrafts of air take some molecules from sea level in the morning to 5 miles up by noon. In a violent thunderstorm, molecules rise to the top of the troposphere in a matter of minutes.

## Prevailing Winds

Although local updrafts do a lot to mix the air vertically, the long-range horizontal mixing in our atmosphere is due to **prevailing** winds. These winds, like local breezes, arise from solar heating, only over a much larger area—the entire globe. We've seen that the equatorial regions get almost direct sunlight all year, and the polar regions get sunlight only at a low angle, so the energy per second per unit of area brought in by light is a great deal less at the poles. Warmed tropical air rises, cools, and spreads outward at a high elevation, while the cold polar air flows outward close to the ground. This evokes a picture of ground winds traveling toward the equator from the poles and high-altitude winds moving from the equator to replace the colder air. But earth's rotation brings the Coriolis force into play to turn the moving air, and this turning and other details of the heating process break the circulation pattern

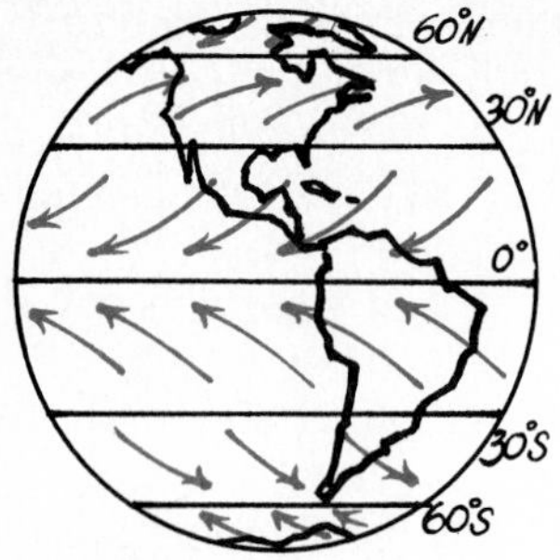


**FIGURE 15–7**
The zones or belts of the prevailing winds on earth's surface. The winds between the equator (0°) and 30° N latitude are called the northeast trade winds because they blow from the northeast. From 0° to 30° S, the winds are called the southeast trade winds. From 30° N to 60° N is the zone of the prevailing westerlies, which govern the climate to a large extent in the continental United States. The winds between 30° S and 60° S are also called westerlies.

in each hemisphere into three distinct cells, or zones, as shown in Fig. 15–7. On the average, the prevailing winds follow those patterns, though local winds may come from any direction.

Two hundred and more years ago, the sailing ships that were the only link between Europe and the New World would take the trade winds west, go up the east coast of the colonies, and take the prevailing westerlies back home. The times of cross-country bicyclists who race across the United States average days longer when they travel westward against the prevailing winds than when riding eastward with them. These winds also cause the principal features of the ocean's surface currents; compare Fig. 15–9 to Fig. 15–7 to see for yourself.

Prevailing winds are often a major factor in climate, too, especially where land meets ocean. We've seen that in winter the oceans are slower to cool off than the land. The pleasant winter climate in San Francisco is due to the westerly winds off the Pacific Ocean, which bring in relatively warm air from over the water. Because the prevailing winds over the continental United States come from west to east, they deal a poor hand to the East Coast. The current known as the Gulf Stream carries warm water from the Caribbean area along the East Coast all winter long. But the air warmed by the current doesn't usually get to Washington, D.C., or New York City because the prevailing winds are in the wrong direction. These cities receive chilled air from the interior of the continent instead. (Florida, whose westerly winds come from the Gulf of Mexico, is an exception.) Meanwhile, the Gulf Stream continues north, then east, then south to warm the British Isles, which are at the same latitude as Hudson Bay.

When summer arrives, the ocean is cool compared to the land, and once again that works to San Francisco's advantage. In the center of a large continent, however, the summers are relatively hot and the winters are relatively cold, winds or no winds. Siberia is infamous for those reasons.

**FIGURE 15–8**
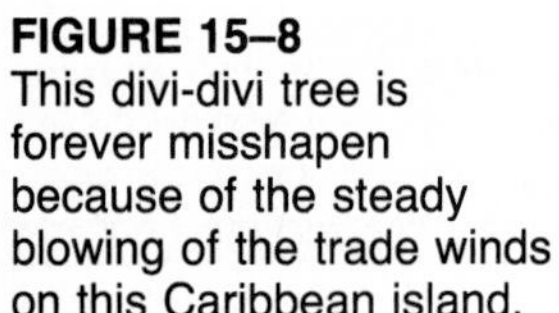
This divi-divi tree is forever misshapen because of the steady blowing of the trade winds on this Caribbean island.

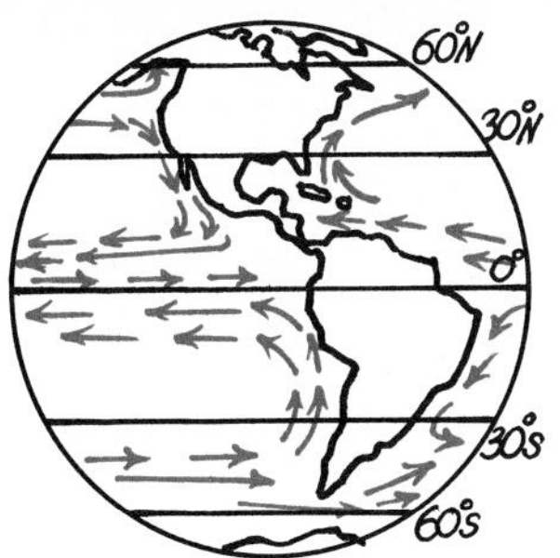


**FIGURE 15–9**
The major surface currents of the oceans. Except for a countercurrent directly on the equator and the interference of land masses, the surface currents follow the pattern of the prevailing winds.

## Water Vapor in the Atmosphere

Water vapor is never a major component of earth's air. The muggiest air is only 4 percent water vapor by weight, that is, 4 grams of water in 100 grams of air. But water vapor is crucial to life on land for the rain it brings. Water gets into the air by evaporation from the oceans (about 85 percent) and land masses (about 15 percent). *Over 100,000 cubic miles* of water rise into the air each year to fall back as rain, snow, and dew.

The portion of air pressure that's due to the water molecules in air is the *partial pressure* of the water vapor. But it's usually called the **vapor pressure** of water. Here's why. At the normal temperatures of our atmosphere, liquid water in the oceans and ice at the polar caps exist in *equilibrium* with the water vapor in the air above them. As molecules are evaporating from the liquid or subliming from the solid state, they are also condensing into or onto those surfaces at the same time. Because water and ice can exist at the temperature of the air, the water vapor the air can hold at normal temperatures is very limited, as the following example shows.

Suppose a dehumidifier was used to dry the air in a closed kitchen, that is, to remove all the water vapor. If the sink were full of water, evaporation would proceed. Suppose the temperature of the room was 26°C and the dry air registered a pressure of 760 millimeters of mercury on a barometer on the wall. As water evaporated, the barometer's reading would gradually rise because of the extra pressure from the water vapor. In a day or so the air pressure in that kitchen would stand at 785 millimeters of mercury, or 760 + 25. That's as high as it would go. No more water would evaporate from the sink, because *at that point just as many molecules would reenter the water's surface in that sink as would leave it each second.* The air in the kitchen would then be **saturated.** The **saturated vapor pressure** of water at 26°C is 25 millimeters of mercury. But if the temperature were raised, more water would evaporate: *The saturated vapor pressure depends on the temperature* (see Table 15–1).

**TABLE 15–1**

| SATURATED VAPOR PRESSURE OF WATER (MILLIMETERS OF MERCURY) | |
|---|---|
| −10°C | 2.1 |
| 0°C | 4.6 |
| 10°C | 9.2 |
| 20°C | 17.5 |
| 30°C | 31.8 |
| 40°C | 55.3 |

The saturated vapor pressure doesn't depend on the *pressure* of the rest of the air, however. If there had been a vacuum in that kitchen, the barometer on the wall would have risen only from 0 to 25 millimeters of mercury before the point when just as many molecules returned to the water in the sink as left it each second. (The barometer's rise would have been much faster, however, with no nitrogen and oxygen molecules to beat some of the would-be escapees back into the water. The *rate* of evaporation does depend on the air's pressure.)

## Relative Humidity

Most air contains some water vapor, but air is rarely saturated, especially near the ground. To describe the amount of moisture in the air, we can compare its *actual* vapor pressure to the *saturated* vapor pressure at that temperature. The ratio of these quantities is called the *relative humidity*.

$$\text{relative humidity} = \frac{\text{actual vapor pressure}}{\text{saturated vapor pressure}}$$

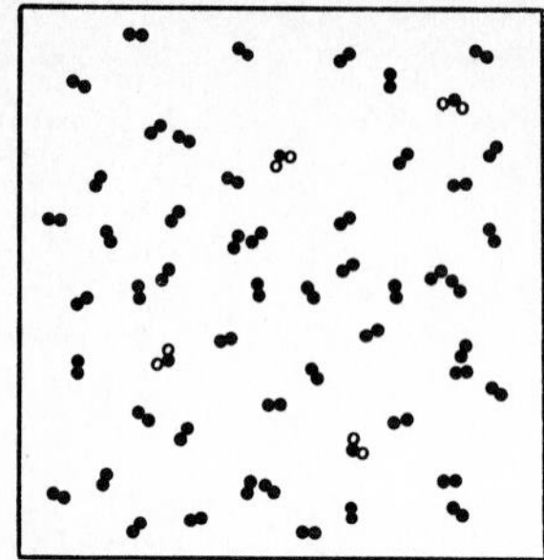

**FIGURE 15–10**
This representation of air saturated with water vapor at room temperature shows how little moisture the air can really hold. Saturated air at this temperature contains only 4 percent water by weight.

When multiplied by 100, that fraction is expressed as a percent. So if the actual water vapor pressure in a room at 26°C is 17 millimeters of mercury where the saturated vapor pressure is 25 millimeters of mercury, the relative humidity there is $17/25 = 0.68$, or 68 percent.

The relative humidity of a parcel of air depends on its temperature as well as its moisture content. Air whose relative humidity is 68 percent at 26°C can hold more moisture at 32°C, that is, its *saturated* vapor pressure is larger. So if that air is heated to 32°C while its moisture content remains the same, its relative humidity drops. Likewise, should that air's temperature drop below 26°C, its saturated vapor pressure would be less and its relative humidity would rise. If the temperature drops sufficiently, any parcel of air that contains water vapor will become saturated and its relative humidity will become 100 percent. The temperature of air when this occurs is called the air's **dew point.** This is the point where condensation of water vapor in cool air can begin.

## Clouds, Rain, Fog, and Dew

Picture a small water droplet floating in the air. Water molecules evaporate from its surface, and water molecules from the air enter its surface. One of three conditions will apply to this droplet:

**1.** If the relative humidity of the surrounding air is less than 100 percent, it loses more molecules to the air than it gains each second. The droplet shrinks.

**2.** If the air's relative humidity is 100 percent, just as many molecules go back into the droplet's surface as escape each second. The droplet's size stays the same; the air around it is saturated.

**3.** If the air's relative humidity is greater than 100 percent, the air provides *more* water molecules to the droplet's surface each second than escape it. The droplet grows. The air is said to be **supersaturated;** it contains more water than it can maintain indefinitely at that temperature.

It is supersaturated air that causes rain. When warm, moist air rises in the atmosphere, it cools. As it cools, its saturated vapor pressure drops and its relative humidity rises. If a parcel of moist air in an updraft reaches 100 percent relative humidity and continues to rise and cool, condensation is certain to occur.

Condensation mostly begins on small particles in the air that provide surfaces where the water molecules can stick and collect. Called **condensation nuclei,** these airborne particles have numerous sources. For example, many small water droplets are launched upward in the spray from the top of an ocean wave. The smallest of these evaporate and leave microscopic salt particles that the wind carries aloft. Over land, breezes pick up all sorts of microscopic particles, smaller even than dust. Forest fires, geysers, volcanoes, farming, and modern indus-

**FIGURE 15–11**
Forest fires contribute large numbers of particles that serve as condensation nuclei in the atmosphere.

trial activities all add particles that can serve as condensation nuclei for raindrops.

When growing water droplets in a parcel of air become large enough to reflect light well, they form a *cloud*. The clouds we see are rarely more than five miles above sea level, where the air is cold enough to cause the moisture to condense. Cloud droplets are tiny, nevertheless, so they have relatively large surface areas for their masses and have a very low terminal speed. Pushed along by the slightest breeze, they seem to have no tendency to fall. But if they grow, the larger drops have less surface area per unit of mass and the air affects their motion less. Then they can fall from the clouds as rain.

If a night is clear, the ground cools rapidly by radiating its heat into space. Air in contact with this ground also cools off. If that air is humid, a cloud can form just over the ground. We call this kind of cloud *fog*. Fog also forms when warm, moist air moves over cold water or cold land and cools.

Moisture often appears on blades of grass and other surfaces that cool quickly by radiation when the sun goes down. On a still night, the air temperature is lowest just before sunrise, so its relative humidity is highest then. Cooling by radiation, leaves of grass may well reach lower temperatures, lower even than the cool air's dew point. Water vapor then condenses on the cool blades of grass. This is called *dew*. A concrete sidewalk bordering the grass won't get wet, however, because as it radiates heat away, it also conducts heat up from the ground beneath it, keeping it warmer than the grass. A car parked outside might become wet with dew, though. A car's metal skin is thin, so there is lots of surface area to radiate from for each kilogram of mass. Metals also have small heat capacity, and a car's rubber tires insulate it from the warm earth, so the car's temperature drops quickly. As the car cools, the air next to the car cools, and the water vapor in that air condenses on the car.

If the formation of dew is heavy on some surfaces, such as the leaves of plants or even on a car, large drops of dew form and drip onto the ground like rain. In some nearly arid locations in the eastern Mediterranean region, dew contributes the equivalent of 8 inches of rainfall yearly, supporting certain crops that couldn't otherwise survive without irrigation.

If the night is cold, leaves of grass can radiate heat so fast they drop below freezing before the air does, and frost rather than dew forms. (See "Sublimation," Chapter 13.) In Alaska a winter fog is often made of ice crystals rather than water droplets. Called *ice fog,* it sparkles in the sunlight and sometimes causes a shadow, as will an ordinary cloud or fog if it is dense enough to block the light. These phenomena happen when the air temperature drops below freezing.

## Rainstorms, Large and Small

An ordinary summer raincloud is caused directly by solar heating. A buoyant plume of moist, warm air rises quickly, cools, and produces rain and pellets of ice called *hail*. Though hail is probably present in all

## TWENTY MINUTES IN A THUNDERCLOUD

Few have ever experienced a thunderstorm as Professor John Wise did from the basket of his balloon one summer afternoon in 1843. The professor had just lifted off from the town square in Carlisle, Pennsylvania, when he spotted a huge black cloud. His balloon drifted under it, and he "immediately felt an agitation in the machinery, and presently an upward tendency of the balloon, which also commenced to rotate rapidly on its vertical axis." Inside the enveloping cloud, snow and hail pummeled his balloon as he "watched everything around me of a fibrous nature become thickly covered with hoar-frost, my whiskers jutting out with it far beyond my face . . . . I supposed that the gas would rapidly condense and the balloon consequently descend and take me out of it. In this, however, I was doomed to disappointment, for I soon found myself whirling upward with a fearful rapidity, the balloon gyrating and the car describing a huge circle in the cloud." The unfortunate scholar was caught up in the circulation of the storm cell, hurled upward in the updraft only to fall breathlessly in the downdraft. "This happened eight or ten times . . . the discharge of ballast would not let me out at the top of the cloud, nor discharge of gas out of the bottom of it . . . the balloon had also become perforated with holes by the icicles that were formed where the melted snow ran on the cords at the point where they diverged from the balloon, and would by the surging and swinging motion pierce it through."

Finally the balloon fell clear of the updraft, much to the professor's delight, "after having been belched up and swallowed down repeatedly by this huge and terrific monster of the air for a space of twenty minutes, which seemed like an age, for I thought my watch had been stopped, till a comparison of it with another afterward proved the contrary. I landed, in the midst of a pouring rain, on the farm of Mr. Goodyear, five miles from Carlisle, in a fallow field, where the dashing rain bespattered me with mud from head to foot . . ."

thunderstorms, it melts (or sublimes) before it travels very far, so it rarely reaches the ground. Surprisingly, most raindrops don't make it to the ground either. Only about 20 percent of the rain in a typical summer thundershower falls to earth; most evaporates or disperses into less dense clouds.

The huge storms called *hurricanes* are another matter.* Born over the warm tropical oceans of late summer, these storms may be several hundred miles across with winds of at least 73 mph (by definition) and sometimes up to 200 mph. Their clouds often stretch much farther. The

*Hurricanes are called *typhoons* in the western Pacific. In the Bay of Bengal (Indian Ocean) they're called *cyclones*.

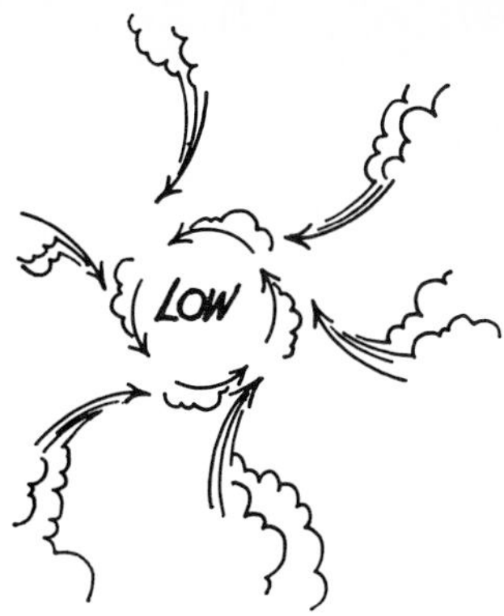


**FIGURE 15–12**
A cross-section of a hurricane showing its internal motion.

energy to drive the winds comes from the warm water beneath the storm. The center of the storm is an area of low pressure (called the *eye*), and winds come toward it from all directions. As they do, the Coriolis force due to earth's rotation turns them, giving the hurricane a counterclockwise rotation in the Northern Hemisphere and a clockwise rotation in the Southern Hemisphere. (See Fig. 15–12.) As a result the eye remains calm, but there is a wall of rapidly ascending air around the eye fed by warm, moist winds sweeping in over the choppy surface of the ocean. This ascending column of wet air soars straight up to heights of 40,000 feet. On its way the moisture in this air condenses into rain, releasing its latent heat of vaporization as it does. *That latent heat energy feeds the kinetic energy of its winds and drives the storm.* At the top of the rotating column, the air is thrown outward. If a hurricane travels onto land or over water that is cooler than 80°F (27°C), it loses its source of energy, slows, and disperses. Figure 15–13 shows two connected hurricanes, a rare occurrence; the clouds of the ascending columns around the eyes are easy to see, since they tower above the rest of the storm and cast shadows.

## Climate and Life

The sun warms our planet unevenly from the poles to the tropics; winds and ocean currents distribute heat energy and moisture or take them away. Ice caps and glaciers, steamy tropical jungles and swamps, grassy plains and barren deserts, hot seasons and cold seasons, and islands

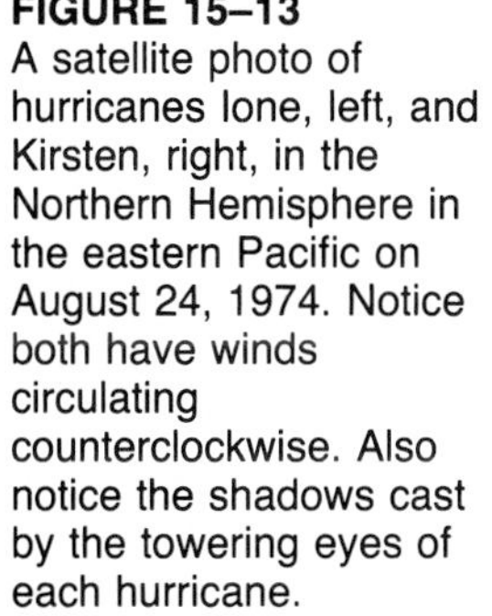

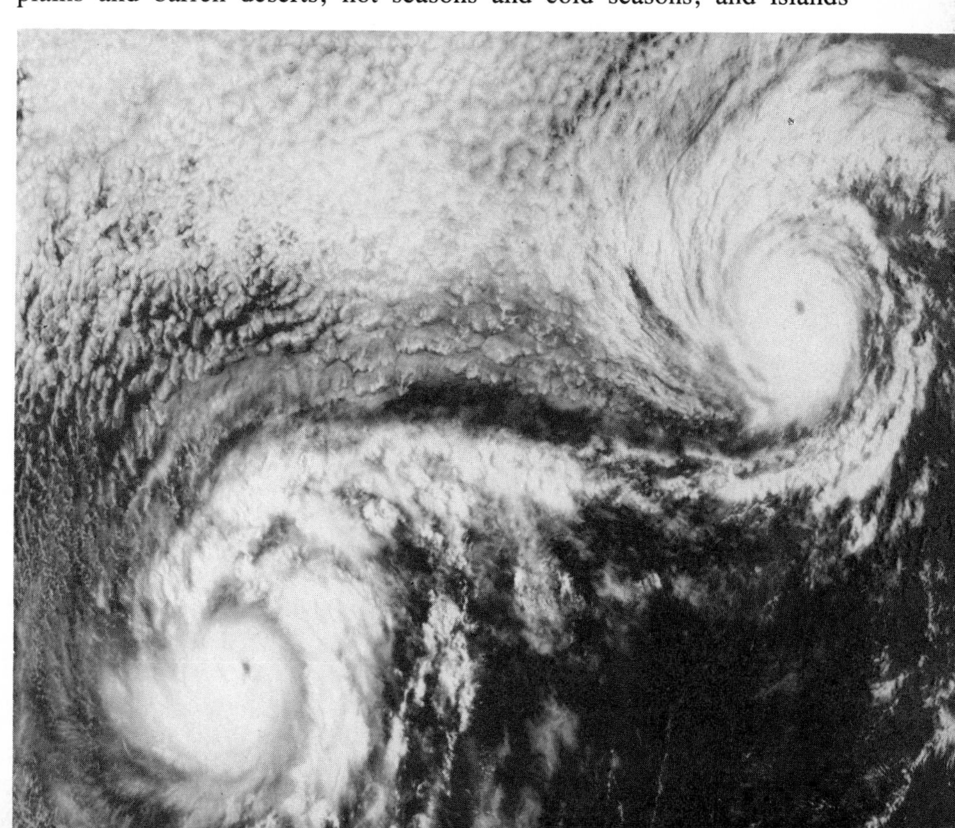

**FIGURE 15–13**
A satellite photo of hurricanes Ione, left, and Kirsten, right, in the Northern Hemisphere in the eastern Pacific on August 24, 1974. Notice both have winds circulating counterclockwise. Also notice the shadows cast by the towering eyes of each hurricane.

where there are no real seasons—such is the sun-driven climate of earth, home to life.

Because plant and animal life depends on chemical reactions that are very sensitive to temperature, living organisms exist only within a narrow temperature range. Plants take on the temperature of the air or water they live in; their chemical activities (whose rates depend on the speed of molecular collisions and the number of collisions per second) proceed when it's warm enough and cease when it is too cold. Small one-celled animals (with so much surface area per unit mass for heat to move through) do the same. Lizards and snakes, though larger, have internal temperatures that are usually only a few degrees above the temperature of their surroundings. These cold-blooded animals hibernate when it gets cold and their body temperature drops. Humans, other mammals, and birds maintain very steady internal temperatures, however. The body's nervous system, its most highly specialized network of cells, is a part of the machinery that regulates internal temperature. Because we are warm-blooded, we can function just as well at midnight as at noon, if we wish, and we can travel without harm to our systems from the equator to the poles, using only a little extra insulation. About 90 percent of the various kinds of reptiles on earth live in the tropics, locked into the climate, whereas humans and other warm-blooded creatures are found almost everywhere. But there's a price to pay. Warm-blooded creatures use more energy than cold-blooded ones, and this requires more food. And in the case of humans, we must take other steps, such as using more or less clothing, or even heating or cooling the air, if we want to live in certain locales. To go into space, we must take a bit of home—air, water, and climate—with us, as well as home cooking.

## Life and Heat: Keeping Cool

Your body both loses and gains heat energy continuously by radiation, conduction, convection, and evaporation, and the net rate of heat loss or gain depends on the temperature difference between you and your surroundings. Your internal temperature, however, must remain constant, so *your body is constantly reacting to adjust the heat flow to keep its core temperature at 98.6°F*. Even as you sleep, you shed heat that is the waste product of the energy-producing processes. Located in your brain is your body's thermostat, the hypothalamus, whose function includes turning on and off your body's mechanisms for control of heat flow.

When you do physical work like moving furniture or even jogging to get to class on time, your muscles produce extra heat energy that slightly raises the core temperature. This triggers the hypothalamus to act. The hypothalamus signals the blood vessels nearest the skin, and they relax and dilate (become larger), letting larger volumes of blood come nearer to the surface and radiate the extra heat away. The extra blood close to the skin gives you a flushed appearance. In the summer when the air temperature isn't very different from the skin's, it's not easy to lose heat quickly just by radiating, convecting, and conducting. To help you keep cool, the hypothalamus activates sweat glands in the skin. To evaporate, the perspiration draws much of its latent heat of evaporation from your skin, and you become cooler. Perhaps you've noticed when the relative humidity is high, the warm, moist air seems sticky or muggy against your skin. That's because the evaporation of perspiration is slower, and you don't get cooled as well as when the humidity is low.

The relaxation (called *vasodilation*) of the small blood vessels near the skin and evaporation of sweat let us tolerate hot air surprisingly well. In the late 1700s, Dr. Charles Blagden was an eminent London physician who served as secretary of the Royal Society for nearly 20 years. He took some friends, a pet dog, and a raw beefsteak into a room that was heated to 260°F. They all emerged unharmed three-quarters of an hour later; the beef steak, however, was cooked. Another group went into a room whose temperature stood at 320°F. They beat a hasty retreat in only 2 minutes, nursing scorched ears and noses.

Marathon runners generate heat more rapidly than their bodies can shed it with radiation and evaporation of sweat. At the end of a marathon some runners will have core temperatures in excess of 105°F. More rarely, however, a runner will run to complete exhaustion, depleting the body's energy resources almost entirely. The result is a dangerous drop in body temperature as normal chemical activities stall. (This apparently happened to the winner of the 1982 Boston marathon.) The marathon runners of the bird world, ostriches, have featherless legs to promote the cooling of their muscles as they lope along at speeds up to 45 mph.

Experiments conducted in Death Valley show a human can sweat over 12 quarts of water per day while drinking water and eating salt pills to replace the lost fluid and salts. Since an adult only has some 5 quarts of blood, you have to admire the efficiency of the sweat glands. When the sweat pours off in drops, however, it doesn't take away much heat from the skin. Only when sweat evaporates, *absorbing its latent heat,* does it carry away a lot of body heat.

## Keeping Warm

The same blood vessels that dilate if you get overheated also respond if your skin temperature falls. In that case, tiny thermoreceptors (organs that can sense hot and cold) in the skin relay the message to the hypothalamus, and the blood vessels constrict. Some people turn white or even blue as the flow of blood to the skin stops, making the outermost skin and flesh the insulating equivalent of cork! The outer layer gets cold and radiates much less heat per second to the surrounding air. If you curl up your arms and legs, you minimize the surface area that both radiates heat and gets cooled by convection of the air in contact with it. On a cool evening, puppies not only curl up but also huddle together to share their radiated heat with each other. The tiny pigmy shrew, a mouselike mammal, has the misfortune to be warm-blooded. Its small size means it has a tremendous surface area per unit of volume, and it

**FIGURE 15–14**
Judy's dog, Tut, has a thick layer of fur that serves as insulation. The fur prevents natural large-scale convection currents and slows the passage of heat radiation.

loses a lot of heat energy. It must eat some 2½ times its own weight each day to survive. Hummingbirds have the same sort of problem and must eat two times their weight per day. At night they hibernate to cut their energy consumption.

Your body radiates heat most rapidly when you are naked, of course. About 50 percent of this heat leaves as radiation, and most of the rest leaves by convection currents. Clothing can reduce the heat lost by radiation to only a few percent, however, and it reduces the rate of flow of heat by convection and conduction as well. Clothing breaks up the air space close to the body and eliminates large-scale convection currents. In air pockets approximately 1 centimeter or so across or even smaller, large-scale convection currents aren't possible, and still air is a very poor conductor of heat. For these reasons winter clothing should be loose-fitting rather than tight. A sleeping bag made of goose down works this way too. Goose or duck down compresses into a small volume for convenient carrying, but it lofts when you shake it, trapping a large volume of air inside with the small feathers. The many feathers prevent circulation of this air, which is a good insulator. The many feathers also have an enormous amount of surface area, which absorbs and reemits infrared radiation, slowing its passage. Hop into one of these bags, fluff it up around your body, and you get as warm as toast in only a few seconds. The source? Your own body heat.*

Have you ever gotten goose flesh when you were chilled? That's a reaction that raises the hairs on your body. If the hairs stand up, they provide more effective insulation for the skin. Of course, this action works much better for arctic foxes, kittens, and bears. One thing that does little to help warm us is thinking. In experiments where subjects volunteered to do intense thinking (performing difficult mental arithmetic), investigators found that an hour of hard thinking uses up only as much energy as you would get from eating one half of a salted peanut. (So much for all those trips to the fridge while you're studying—no excuses!)

When you venture out in cold weather, what gets cold first? Nose, fingers and toes, and ears all have a great deal of surface area per unit of mass, with no hair to provide insulation. Skinny people often get cold quicker than fatter people do for two reasons. First, thin people have more surface area per unit of mass, and second, fat is a good insulator. Whales are proof of this. Their internal temperature is the same as ours (37°C), and yet their environment, the arctic ocean water, may be at −2°C. (Their thick blubber actually provides much more insulation than they need, but it also serves as the energy source during migration through thousands of miles of food-poor ocean.)

---

*The insulating tiles on the bottom and sides of the space shuttle give spectacular proof of the heat-stopping abilities of small dead-air spaces. Made from 90 percent air and 10 percent sand, those tiles are almost as light as feathers. A popular demonstration at the Kennedy Space Center is to put one of those tiles in the flame of a propane torch until the surface of the tile is hot enough to glow red. Twenty seconds after the torch is removed, the demonstrator picks up the tile with bare hands on the *hot* side. The air has already cooled the tile's surface, because the heat could not penetrate far into the tile's material in such a short time. On the shuttle those tiles withstand 1200°C (2400°F) temperatures for several minutes during reentry.

## Cooling and Heating Effects of the Wind: Wind Chill

Cool air next to your skin becomes warmer by contact with it. The air molecules gain heat energy as they bump against your warmer skin. Once warm, this air slows any further heat loss from the body, since the speed of heat transfer is large only for great differences in temperature. But if there is a wind, it sweeps away the warmer layer of air next to the skin, and you chill more quickly. *The wind can't take the skin's temperature below the temperature of the air itself*—no heat flows when there's no temperature difference—but the wind does drain the heat away more rapidly. That effect is measured by the *wind chill factor*. For example, when it is 5°F in Fairbanks, Alaska, and the wind chill factor is −20°F, this means that *the moving air at 5°F cools as rapidly as quiet air would at −20°F*. It *doesn't* mean the air inside an unheated cabin, for instance, would drop below 5°F. It wouldn't. The wind chill factor is only a gauge of the *rate of cooling* of the moving air.

During a cold snap of −70°F on the Alcan highway in 1978, wind chill forced truckers to stop completely. Their trucks were equipped with wind shields *under* the engines and electric heaters for the batteries, fuel tanks, and even the differentials. Yet when they tried to move along through the cold air, the rapid heat loss caused freezing of the diesel fuel in the fuel lines between the engine and the fuel tank. The truckers parked and kept their engines running 24 hours a day. In *still* air the fuel did not freeze. Had the systems' heaters been shut off, the fuel would have frozen solid. The truckers got rolling again at the first warm spell, 7 days later.

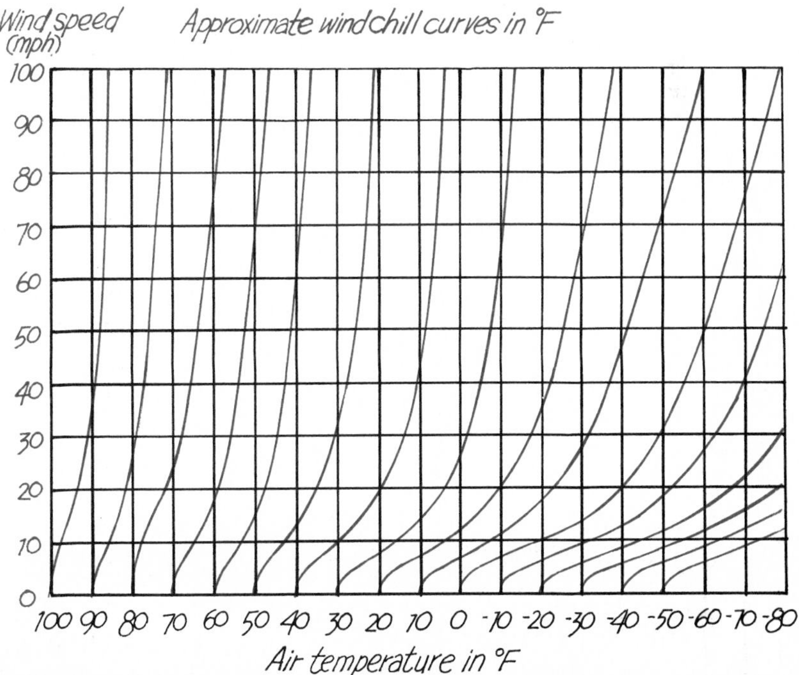


**FIGURE 15–15**
An approximate windchill factor chart.

The same effect that causes wind chill to lower temperature can work the other way as well, as it does in convection ovens for baking. A convection oven is designed so that a fan blows heated air directly onto the food. That eliminates insulation by a boundary layer of air at the food's surface and transfers heat faster, reducing the cooking time. Except for evaporation, air hotter than your body that's blown against you would heat you up too.

## CALCULATIONS

The black surface of a solar water heater absorbs sunlight and turns it into heat. To collect the rays of the midday sun, the flat panels of solar heaters in the Northern Hemisphere are tilted somewhat to the south. At noon the sunlight brings in more energy per second than at other times of the day when the sun is lower in the sky. Near sunrise or sunset, for example, the sunlight travels a greater distance through the atmosphere before it reaches the ground; when the light passes through more air, more of the light is scattered and absorbed.

The flat panel of a heater collects more solar energy when it directly faces the sun, that is, when the panel itself makes an angle of 90° with the incoming light. If the panel is tilted at an angle of less than 90°, it intercepts less light, as shown in Fig. 15–16. Table 15–2 lists the percentages (of its maximum amount) that a panel collects at various angles of tilt. At 45°, for example, the panel intercepts only 70.7 percent as much solar energy as it does if the panel makes a 90° angle with the direction of the sun's rays. An average midday value for the energy in sunlight at sea level is about 1000 joules per second per square meter, where the square meter of area is perpendicular to the incoming rays. Thus a 1-square-meter solar water heater panel tilted at 45° to the sun's rays collects 707 joules per second, or 707 watts.

The same situation, but from a different viewpoint, helps to govern the seasonal percentages of incoming sunlight enjoyed by all areas on earth. If in

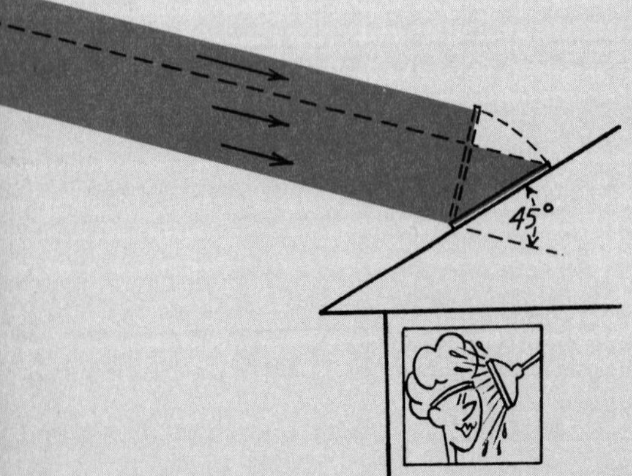


FIGURE 15–16

**TABLE 15–2**

**PERCENTAGE OF SOLAR ENERGY INTERCEPTED BY A PANEL (OR OTHER AREA) INCLINED TO THE SUN'S RAYS AT DIFFERENT ANGLES**

| ANGLE | | ANGLE | |
|---|---|---|---|
| 90° | 100% | 30° | 50% |
| 75° | 96.6% | 25° | 42.3% |
| 50° | 76.6% | 20° | 34.2% |
| 45° | 70.7% | 10° | 17.4% |
| 40° | 64.3% | 0° | 0% |
| 35° | 57.4% | | |

Refer to Fig. 15–16.

the winter the sun climbs to only 45° above some location's horizon on a given day, a square meter of ground there is warmed by only about 70.7 percent of the radiant energy that it would receive if the sun were directly overhead.

---

EXAMPLE 1: A solar water heater with 4 square meters of area faces the sun at noon on a clear day. **About how many calories of heat energy does it collect during an hour if its efficiency (energy trapped/energy in) is 80 percent? How much would it collect during that hour if it made an angle of 25° with the incoming sunlight rather than the 90° angle?**

About 1000 joules per second per square meter comes in, so 4000 joules per second falls on the panel, and (4000 joules/second × 0.8) = 3200 joules/second becomes heat. Converting joules to calories gives 765 calories/second, so in an hour (3600 seconds), the water heater collects **$2.75 \times 10^6$ calories.** If tilted at 25°, the heater would intercept only 42.3 percent of that energy, according to Table 15–2. That would be **$1.16 \times 10^6$ calories.**

---

EXAMPLE 2: In Washington, D.C., the sun is about 75° above the horizon at noon on June 22. On December 22 at noon it is only 27.5° above Washington's horizon. **About what percentage of the incoming solar energy that Washington receives on June 22 is received on December 22?**

Use Table 15–2 again. At 75°, the nation's capital city collects about 96.6 percent of the energy it would if the sun were directly overhead; at 27.5°, it collects about 46 percent, halfway between 42.3 percent (25°) and 50 percent (30°). In other words, on June 22 Washington intercepts 97 parts per hundred of energy it would receive if the sun were directly overhead, and on December 22 it intercepts only 46 parts per hundred. Thus 46/97 = 0.47, or **47 percent.** So at noon on the day when the sun is lowest in Washington's sky, the city gets less than half of the solar energy it receives on the day when the sun is highest, provided both days are clear. (The actual value is less, however, because less light comes in at the lower angle. This effect, along with fewer hours of sunlight, makes Washington's winters invigorating. For the same reasons San Francisco would be colder than Washington, D.C., except for those prevailing winds.)

---

EXAMPLE 3: Table 15–3 lists the hours of daylight at various latitudes in the Northern Hemisphere for the summer and winter *solstices,* the days with the highest and lowest points for the sun in the sky. **Estimate the lengths of**

**TABLE 15–3** APPROXIMATE MAXIMUM AND MINIMUM HOURS OF DAYLIGHT AT VARIOUS LATITUDES

| LATITUDE | JUNE 22 | DECEMBER 22 |
|---|---|---|
| 20° north | 13 hours 12 minutes | 10 hours 48 minutes |
| 30° north | 13 hours 56 minutes | 10 hours 4 minutes |
| 40° north | 14 hours 52 minutes | 9 hours 8 minutes |
| 50° north | 16 hours 18 minutes | 7 hours 42 minutes |
| 60° north | 18 hours 27 minutes | 5 hours 33 minutes |

**the longest day and the shortest day at Boston, Massachusetts, latitude 42.3° north.**

42.3° lies between 40° and 50°, so its length of day versus night is between the values listed for those latitudes. You can *approximate* the value by assuming that since 42.3° is 23 percent of the way from 40° to 50°, the value for the number of daylight hours should be the same as it is for 40° *plus* 23 percent of the difference between the 40° and 50° values. Then the number of daylight hours on June 22 at 42.3° north is about

$$14 \text{ h } 52 \text{ min} + 0.23 \times (1 \text{ h } 24 \text{ min}) = \mathbf{15 \text{ h } 11 \text{ min}}$$

Likewise, the daylight hours in Boston on December 22 are about **8 hours, 48 minutes.**

## REVIEW

**1.** What are the two most abundant gases in volcanic emissions? What are the two most abundant gases in earth's atmosphere?

**2.** Sunlight that falls on water is absorbed by a larger mass than sunlight that falls on land. True or false? A given amount of heat energy from sunlight will raise the temperature of sand more than water, gram for gram. True or false?

**3.** Is the portion of the air's pressure due to water molecules what we call the vapor pressure of water?

**4.** A parcel of air is saturated with water vapor when just as many molecules return to the surface of liquid water that's exposed to that air as leave the surface of that liquid each second. True or false? (Assume the liquid and air have the same temperatures.)

**5.** Define relative humidity.

**6.** Both the moisture content of the air and its temperature determine the relative humidity. True or false?

**7.** Does the rate of evaporation of a liquid depend on the pressure of any gases over its surface? On the pressure of its own gas over its surface?

**8.** Define the dew point for air.

**9.** Describe how fog forms over oceans, lakes, and wet ground.

**10.** What are condensation nuclei? Give some examples.

**11.** Dew is water vapor that condenses on cool surfaces. True or false?

**12.** Describe the cycle where air becomes moist and produces rain.

**13.** From the chart of the wind chill factor (Fig. 15–15) estimate the wind chill of 45°F air moving at 15 mph.

## DEMONSTRATIONS

**1.** Here's a way to feel cooling by evaporation and wind chill. The cool feeling you get if you use cologne or perfume or after-shave lotion comes from the evaporation of alcohol, which absorbs latent heat of vaporization as it leaves. Alcohol has weak intermolecular attractions and evaporates even faster than water does. If you splash some cologne on your forearm and *blow over it,* the cooling effect is even greater. The wind speeds the rate of evaporation, and heat leaves the skin much faster.

**2.** Like many other organic materials, human hairs swell and expand when there is moisture in the air. From 0 percent to 100 percent relative humidity, the length of a hair increases about 2½ percent. Using a single hair, you can make a *hygrometer,* a device for measuring relative humidity in the air.

Take a long hair, rub it with alcohol to strip off the oil, and tape one end to a wall. Tie the other end near one end of a soda straw. With a pin, loosely tack the other end of the straw to the wall. The end near the hair will be free to act as a pointer, or marker, and the weight of the straw keeps the hair stretched. Check the radio or TV weather report for the relative humidity in your area. Mark this percentage beside the straw pointer's position. Repeat this several times during that day, or the next, and mark subsequent readings above or below this mark according to the reports. As the hair lengthens or shortens, the hair-and-straw hygrometer can be calibrated in this way.

**3.** You can feel the equivalent of wind chill in water. The next time you take a dip in a cold spring, pond, or even the ocean (out past the breakers), stand still so you don't stir the water around you. Your body soon warms the layer of water next to your skin. Then move around a bit. Heat begins to leave your skin in a hurry. The layer of water that had been warmed by your body is swept away and replaced by a colder layer, and you feel the chill immediately.

**4.** To simulate the formation of clouds and rain, boil water in a teapot with a spout. The steam leaving the spout is invisible, but as it expands, it cools to its dew point, where the relative humidity is 100 percent. It then condenses and becomes visible as a cloud. (Most people call that cloud *steam* also.) To produce rain from this cloud of water vapor, hold a large metal soupspoon or metal spatula in the cloud. Its water vapor will condense on the metal surface (which conducts heat away fast). As the droplets condense and collect there, they grow large and fall as "rain."

**5.** Air conditioning units always drip, whether in a home or a car. They drip because the air that passes over and around the evaporator coils cools to its dew point and condensation occurs on the cold metal surfaces. The relative humidity of the cold air that comes from the air conditioner is close to 100 percent. If you stand in this blast on a hot day, there will soon be an invisible layer of moisture on your skin. If you then step outside into warmer air, more than likely you will shiver for a split second as that layer of moisture evaporates almost instantly and takes a lot of heat from your skin. It's especially noticeable if the relative humidity of the warmer air is low. This effect is common in warm-weather areas for people leaving well-chilled department stores or supermarkets.

## EXERCISES

**1.** The carbon dioxide pumped into the atmosphere by car engines, electrical generating plants that use fossil fuel, forest fires, volcanoes, and other sources, is increasingly raising the carbon dioxide percentage in our atmosphere. (The oceans seem to be dissolving only about half of each year's total output.) Speculate what effects this might eventually have on earth's climate.

**2.** Explain why the air temperature drops more on clear nights than on overcast, cloudy nights.

**3.** Explain why the southern exposure of a house in the Northern Hemisphere can get *warmer* during the winter months than during the summer. (*Hint:* A 2-foot overhang of the roof doesn't interfere with the sun in winter.)

**4.** A builder is aware that a house heats more when the sun is to the east and west and given a choice will place the narrow ends of a house in those directions. Is this idea correct? Discuss.

**5.** About 50 percent of earth's surface is covered with clouds in the daytime. What would happen to earth if that percentage were greater? If it were less?

**6.** On a typical night the average temperature in New York City, Washington, D.C., or Los Angeles drops about 20°F. Speculate what would happen to the atmosphere if the sun didn't shine tomorrow morning, and the next, and the next, etc. (Breathe deeply! You're enjoying solar energy.)

**7.** Why does it help to fan yourself on a hot day? (By the way, about 1500 A.D. Leonardo da Vinci made a water-powered fan, perhaps the first practical powered fan.)

**8.** Most animals stay in the shade on a hot, sunny day. Camels don't have that option. In fact, when they rest, they lie down facing the sun. Why? Those camels also have layers of fat on the tops of their bodies but none on the bottoms, which rest on the sand. How does that help?

**9.** The bees in a beehive can keep the inside air at 38°C by the use of water and a bit of wing power even though the hive is exposed to air at 50°C. How can they accomplish that?

**10.** In the Arctic the sun is never far above the horizon, so the sunlight comes in obliquely and brings little energy per square meter per second as compared to equatorial regions. Yet the Arctic is where many species of whales go to feed in the summertime and store up thick layers of blubber. Why should there be an abundance of plankton (plant life) in arctic waters in the summer?

**11.** In the summer you might be very comfortable in 70°F air at a too-cool movie theater, but if you jump into a spring whose water temperature is 70°F, it feels like ice water. What exactly is the difference?

**12.** Experienced campers know goose down bags can't keep them warm if the bags get wet. Why not?

**13.** In a thunderstorm the large drops of falling rain are accompanied by a strong downdraft of air, sometimes called the *raindraft*. What causes this?

**14.** A philosopher tells you the heat in your blood really comes from the sun. Discuss.

**15.** When you breathe out, the air coming from your warm, moist lungs is at 100 percent relative humidity. Explain why it turns into a white cloud on a cold day.

**16.** In winter ground squirrels, woodchucks, and big brown bats hibernate, their metabolic rates dropping to less than 1/50 of normal and their core temperatures falling to only 1 or 2 degrees above their surroundings. What does this do?

**17.** Seeding a cloud with tiny particles of a chemical compound can actually induce rain. What takes place?

**18.** In some places temperature inversions (areas where the air above is warmer than the air below) are common. Local air pollution can then become a serious problem because only minimal convection currents occur. Why is that?

**19.** To tell which way the wind is blowing, wet a finger all around and hold it up in the air. How does this old trick work?

**20.** In the western deserts of the United States, summer thunderstorms form and rain falls, but very often it never reaches the ground. How can that be?

**21.** Spheres have the smallest ratio of surface area to volume for any shape, and the larger the sphere, the less surface area per unit of volume. Does this fact benefit round-shaped penguins in cold climates or elephants in hot ones?

**22.** The earth is closest to the sun during our winter season (January 4 or thereabouts) and farthest during our summer (July 4 or close to it). Wouldn't this seem to make the Southern Hemisphere's summers hotter and winters colder? A check of average temperatures versus latitude and altitude shows little difference between hemispheres, however. To find out why, pick up a globe at your library. Look down on the North Pole, then turn the globe the other way and look down on the South Pole. Tell why climates in the Southern Hemisphere can be and are much the same as in the Northern Hemisphere.

**23.** Key West, Florida, is at the same latitude as Karachi, Pakistan, and Saudi Arabia, yet its climate is more moderate. Why?

**24.** The leather shirts worn by Indians and cowboys often had leather tassels across the chest and down the sleeves. Actually these were practical not just decorative. Guess their purpose.

**25.** Speculate on what might happen to the climate in the United States if the prevailing westerly winds we now have become prevailing *easterly* winds. (Florida's climate wouldn't change too much.)

**26.** How might small particles released into the atmosphere by industrial activities influence the amount of sunlight the earth gets?

**27.** Breezes most often come up the slope of a mountain at noon and go down the slope after midnight. Why?

**28.** You've probably seen someone use moisture from a puff of breath to clean eyeglasses. How does that moisture condense on the lens? Does it make a difference if the person huffs or purses the lips to blow? Why or why not?

**29.** At the beach, where does the breeze most often come from in the daytime? Why? At night? Why?

**30.** *Cold-blooded* is a misnomer. On a sunny day a lizard or a snake can have a blood temperature greater than a human's. Explain how, and suggest ways reptiles can regulate their temperatures on sunny days.

**31.** You may have seen jumping beans from Mexico. They come from seed pods of arrow plants that have been visited by small moths that lay eggs in them. After the eggs hatch, the worms they produced eat the seeds and line the inside with silk to await metamorphosis. These beans jump around if you hold them in your hand. Guess what's happening.

**32.** Look at Fig. 15–7 and explain how every year some Monarch butterflies from the United States wind up in Europe. (It's about a 3-day trip for them.)

**33.** Keeping eggs warm in the Antarctic isn't easy. An Emperor penguin couple does it by keeping their single egg between the male's legs for two months. When the chick hatches, it stays on papa's feet until it is big enough to sustain the heat loss caused by standing on ice. Interestingly, the veins in a penguin's webbed feet route the cold blood back past the arteries coming in with warm blood. Explain how that feature helps the penguin's feet lose less heat per second.

**34.** A rabbit feels warm to you because its body temperature is about 103°F. If you look at its big ears, you'll notice many small blood vessels near the surface. What purpose do they serve?

**35.** The lunar lander Eagle landed on the moon when the sun was only 10 degrees above the horizon. What was the purpose of choosing that condition?

**36.** Above 60 miles elevation, the air is very thin and convection currents cease to play the role they do below. Gases above this *turbopause* mix mostly by diffusion. What gases, then, would you expect to find at the very highest elevations, that is, at the edge of the atmosphere?

**37.** As an insulator, fur works in two ways; for instance, polar bears have fur, and so do desert camels. Explain how it helps camels to have fur.

**38.** Explain why dew forms on a cold can of soda pop from a vending machine.

**39.** In a convection oven, the actual cooking temperature often can be up to 50°F *less* than a conventional oven for the same purpose. Why?

**40.** A traveler driving farther and farther north notices that the various species of wild rabbits and foxes seen in the car lights at night have smaller and smaller ears. Why should they?

**41.** A canopy of leaves keeps the ground under a tree shady and cool. Grass does the same for soil. So after a hot, sunny day, should breezes flow into or away from a concrete–asphalt town surrounded by countryside?

**42.** From Fig. 15–16 speculate how the curves will turn when the airspeed grows to higher and higher values. Explain your guess.

**43.** Suppose you find you are 25 pounds overweight, and you diet and jog and lose weight. Give two reasons why it is easier for you to become chilled after a weight loss.

***44.** Using the information in Example 1 in "Calculations," find how many gallons of water could be heated by the solar heater from 15°C to 70°C at each angle.

# CHAPTER 16
# WAVES IN MATTER

When a bird soars in to perch on a telephone wire or a clothesline, it starts a chain reaction. The wire under its feet gives a little, which affects the parts of the wire to each side of the bird's feet, and so on—you can see the disturbance move along the line. One molecule's action affects the next, which affects the next, and though those molecules don't move far at all, some of the energy from the original action travels a long way through the matter. Such a disturbance is called a **wave.** A wave in matter carries energy through the matter without taking the matter along with it. (The matter, be it solid or liquid or gas, is called the *medium* of the wave.)

Sometimes you can see the waves, sometimes you can feel them, and sometimes you can hear them. Other times they can be detected only with sensitive equipment. But waves generally occur wherever macroscopic matter is pushed. Waves abound in the water of a baby's first bath, and most people have been close enough to a moving train or a heavy truck to feel waves that travel outward through the ground. This chapter is about the waves in matter, especially those we can see or feel. Those we hear, sound waves, are the subject of the next chapter. ■

## Some Facts about Waves

Close the windows in a room, but leave one door slightly open. Then quickly open a closed door on the opposite side of the room. In a split second the far door left ajar will move. (It opens or closes depending on how the door is hinged.) A wave travels through the air in the room to move that door. Air molecules pushed by the door you moved push into other air molecules that bump into others, and some of the energy you put into the air when you moved the door travels across the room in a wave. The wave moves the individual molecules of air in the same direction as the wave moves. Such a wave is called a **longitudinal** wave.

You can see such a longitudinal wave on a toy spring called a Slinky. Have a friend hold one end on a smooth floor, or just tie one

**FIGURE 16–1**
Compressions and rarefactions of a longitudinal wave on a long Slinky. Though the molecules in the spring only move back and forth slightly from their positions at rest, the compressions and rarefactions move away from the hand that caused them.

end to the leg of a piece of furniture. Then stretch the spring along the floor. Next, gather some of the coils together at your end and release them. They'll snap back, creating a disturbance that moves along the spring, passing from coil to coil. If you jerk your end of the spring back and forth in a regular (periodic) manner, you'll send a series of uniform disturbances along. As seen in Fig. 16–1, such a wave consists of regions called **compressions,** where the coils are pushed closer together, alternating with regions called **rarefactions,** where the coils are stretched farther apart. For that reason a longitudinal wave is also called a **compressional** wave.

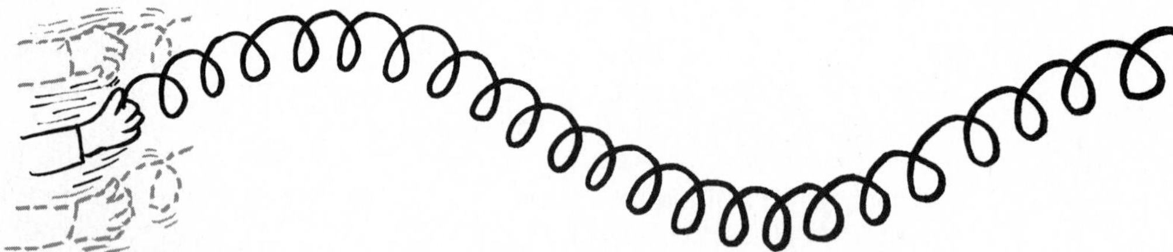

**FIGURE 16–2**
A transverse wave on a long Slinky. In this kind of wave, the back and forth motion of the molecules in the spring is perpendicular, or transverse, to the direction of the wave's motion. As with the longitudinal wave, the wave pattern travels away from the hand that caused it.

You can make another type of wave on the stretched Slinky. Shake your end of the spring from side to side with a regular motion, perpendicular to the spring's length. A different wave pattern, where the movement of the molecules in the spring is *perpendicular* to the direction of the wave's motion, will travel along the spring (Fig. 16–2). This kind of wave is a **transverse** wave. Transverse waves move through matter only if there are forces that tend to restore the molecules to their original positions when they are pushed to the side by a wave. Should molecules in a liquid or gas move to one side as a wave passed, they

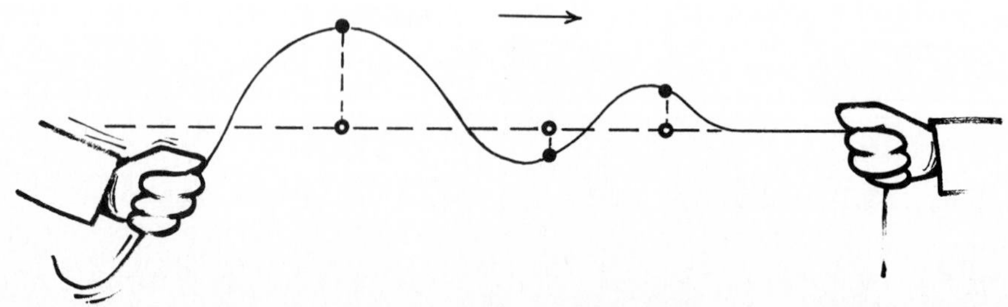

**FIGURE 16–3**
In a transverse wave, the particles move perpendicularly to the direction of the wave itself.

won't usually return to their former places because of the rapid rearrangement of the surrounding molecules. Consequently, such transverse waves cannot travel through liquids and gases. Longitudinal waves, however, can travel through matter in any of its states.

## Frequency, Amplitude, and Wavelength

Shake one end of a stretched clothesline with your hand, and a pattern moves along the line. When your hand moves the line up and then back to its undisturbed position, a hump (or "hill") travels along the line. When your hand moves the line down and then back to its undisturbed position, a dip (or "valley") moves along the line. Flipping the line up and down with regular motion sends a succession, or train, of hills and valleys along it. One complete up-and-back, then down-and-back motion is called a *cycle,* or an *oscillation*. Each complete oscillation makes one "hill" and one "valley" on the clothesline. The top of the hill is the wave **crest;** the bottom of the valley is the **trough.** The distance from the undisturbed position of the clothesline to the crest or to the trough is called the **amplitude** of the wave. (See Fig. 16–4.)

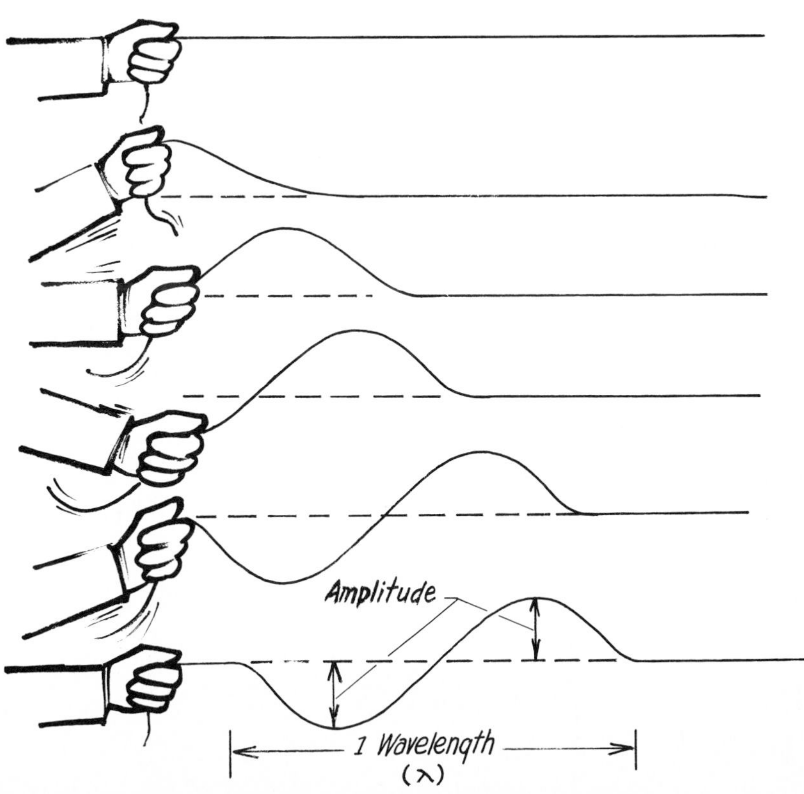


**FIGURE 16–4**
Making a wave on a stretched string. A tug up and back puts a crest on the line, and a tug down and back makes a trough. One complete up-and-back and down-and-back motion is called a cycle. The distance from undisturbed line to a crest or trough is the amplitude of the wave.

**FIGURE 16–5**
To measure the frequency of a passing wave, you only need to count the number of oscillations the medium makes in a certain amount of time and divide the number by the time.

If you furiously jerk the end of a clothesline back and forth, you can probably perform 3 oscillations in a second, or perhaps even 4 if you don't move the line very far (small amplitude). The number of oscillations per second is called the **frequency** of the wave. The frequency is the number of up-and-down trips a particle on the line makes in 1 second as the wave passes. A friend at some position along the clothesline could count the oscillations per second and tell how fast your hand was moving back and forth at your end of the line. A **wavelength** is the length of the wave that is made by one oscillation of the hand (see Fig. 16–5). A wavelength is also the distance from one crest to the next or from one trough to the next.

The wavelength of a wave and its amplitude are measured in meters, centimeters, feet, or any other convenient unit of length. The frequency is usually given as a number (of oscillations) per second, so its units are 1/s. A frequency of 1 (cycle) per second is known as a **hertz** (Hz) in the metric system. So a frequency of 3 oscillations per second is 3 hertz.

## The Speed of a Wave

Suppose you make a wave on a stretched clothesline by performing 3 oscillations per second with your hand. In 1 second, 3 wavelengths move down the line, one after the other. That means the front of this wave train (the leading point) travels a distance of 3 wavelengths along the clothesline in 1 second. Its speed, then, is 3 wavelengths/1 second, and all the successive hills and valleys of the wave pattern follow the front at this same speed. In other words, the frequency of oscillation multiplied by the wavelength gives the speed of the wave.

$$\textbf{wave speed} = \textbf{frequency} \times \textbf{wavelength}$$

For example, if you oscillate the end of a clothesline at 3 hertz and the speed of the wave measures 9.6 meters/second, the wave's wavelength is

$$\text{wavelength} = \frac{\text{wave speed}}{\text{frequency}} = \frac{9.6 \text{ meters/second}}{3/\text{second}} = 3.2 \text{ meters}$$

In symbols, the equation for the speed of a wave is

$$v = \lambda \times f$$

where $v$ stands for the wavespeed, $\lambda$ (Greek letter lambda) for the wavelength, and $f$ for the frequency. This equation holds for all types of wave motion.

The speed of a wave depends on the strength of the restoring force or action and the inertia of the medium. For transverse waves on a clothesline, the restoring force is the tension in the line. (See Fig. 16–6.)

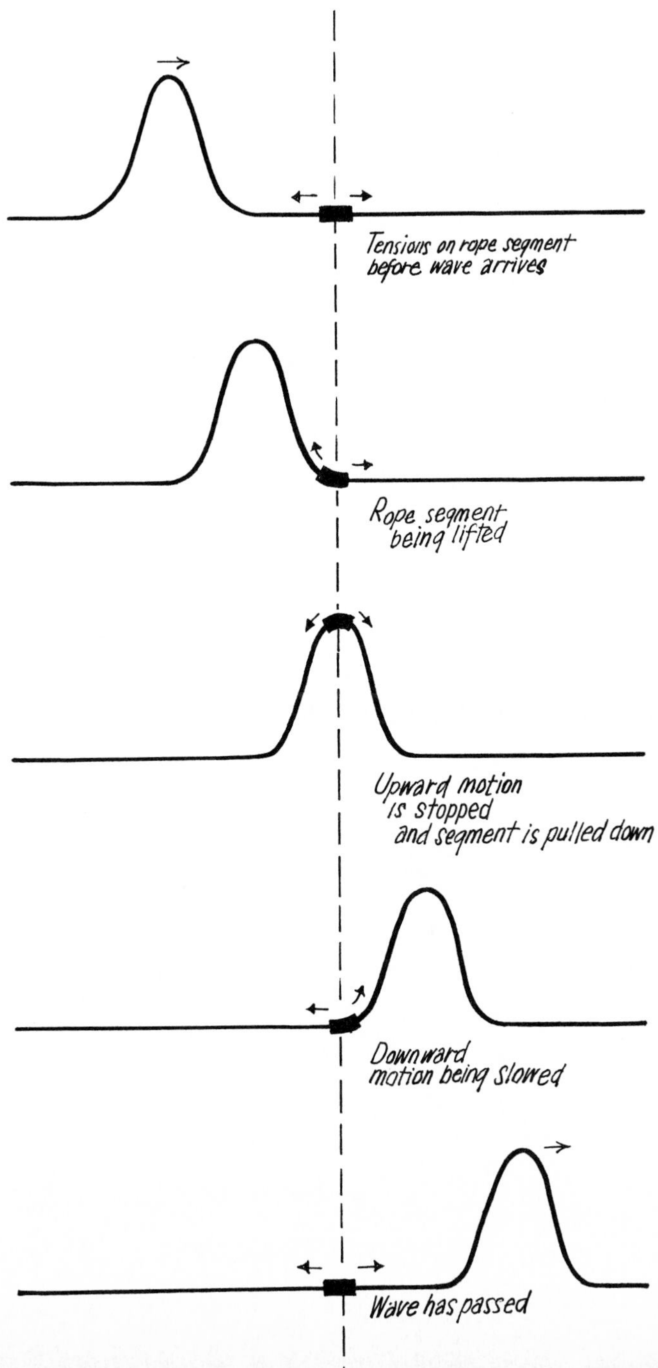


**FIGURE 16–6**
Whenever a wave travels along a string, the tension on any segment of string that's raised or lowered by the pulse tends to return that segment to its original position. The tension, then, acts as a restoring force. If there were no tension in a string, a wave couldn't move along it.

Just stretch that same clothesline more tightly, and you can actually see the waves travel faster. Or to see the effect of inertia use a heavier line and see the wave move more slowly.

There's another important fact about the speeds of the waves on a clothesline. If the tension remains unchanged, experiments show the waves that move on the line have the same speed *no matter what their frequencies*. This fact is very important for waves that we will study later, those of sound and light.

Compressional waves move through solids faster than through liquids or gases. That is because the strong bonds holding the atoms or molecules in place return them quickly to their normal positions when they are displaced. A compressional wave of 1000 hertz, say, could travel through a steel rail at 11,000 miles per hour, through water at 3300 miles per hour, and through air (at sea level and 20°C) at only 770 miles per hour. Transverse waves travel through solids at less than half the speed of compressional waves. This is so because a sideways displacement doesn't bring as much force to an atom; it doesn't stretch the bonds along a chain of atoms as much as the same displacement in the longitudinal direction does. Less force means less acceleration, and the particles (and hence the wave) move more slowly. See Demonstration 5 for an easy proof of this effect.

**FIGURE 16–7**
To generate a high-frequency wave requires more energy per second than to generate a low-frequency wave. Thus, a high-frequency wave carries more energy than a low-frequency wave of the same amplitude.

## Energy in Waves

The work you do if you shake a clothesline provides the energy that the wave takes with it. Because the work you do is force times distance, the amount of energy the wave carries depends (somehow) on how far you pull the line from its resting position as you make the wave; that is, the energy depends on the wave's amplitude. Also, if you shake the clothesline faster, you put more energy per second into making the wave, and the wave delivers more energy per second to the particles farther down the line when it moves past. If you try making waves on a clothesline, you'll see for yourself. Making short-wavelength, high-frequency waves is tiring, but making long, lazy, looping waves takes little effort. See "Calculations" for more details.

Not all the energy used to shake a clothesline makes it to the far end, however. Some is left behind as heat due to friction within the medium as the wave moves through. Any wave through matter eventually dies out because of the loss of its energy to heat.

## Reflections, Standing Waves, and Resonance

Tie one end of a rope to something substantial, a tree or a heavy piece of furniture, stretch the rope out, and give your end of the rope one hard flip. Then close your eyes and keep that rope taut. In a second or so, you'll feel a jerk at your hand. The wave has *reflected,* or turned around, at the far end of the rope. The returning wave didn't carry all the energy it left with. Besides the thermal energy it loses to the medium, some of its energy travels on into whatever matter is anchoring the far end of that rope. However, if the rope is anchored to something rigid, most of the energy is reflected, and the returning wave's form is almost identical to the initial shape of the wave. But there is one change: The returning wave pattern is upside down! Try it and see.

Now if you flip the end of that rope back and forth with a regular period, a train of waves travels down the rope. When the initial part of the wave reflects from the end, the rope plays host to two waves at once, one traveling from your end and the other from the reflecting end. Wherever the two waves merge, the particles on the rope get a push from each wave and move according to the *net* force they receive. The result can be fascinating.

If two crests of the same amplitude, one coming from each direction, meet at some point, the particles at that point rise twice as high as if only one wave were there; *the amplitudes of the crests add.* This is known as **constructive interference** of the waves (see Fig. 16–8). On the other hand, if a crest from one wave meets a trough of the same amplitude from another wave, particles at that point *do not move at all.* The upward pull from one wave is canceled by the downward pull of the other. Though two energy-carrying waves are simultaneously passing that location, there is *no* visible response by the medium at that instant. That is called **destructive interference** of the waves. Despite interference, the rope is a perfect host: The waves pass through each other and afterwards their forms are the same as before they met! Waves interfere only while they share the same part of the medium they are moving through, of course. And even when waves of different amplitude and frequency cross, they add or subtract at every point in the action we call interference.

Most often two interfering waves on a rope create a tumultuous, rapidly changing pattern. However, if you move your end of the rope

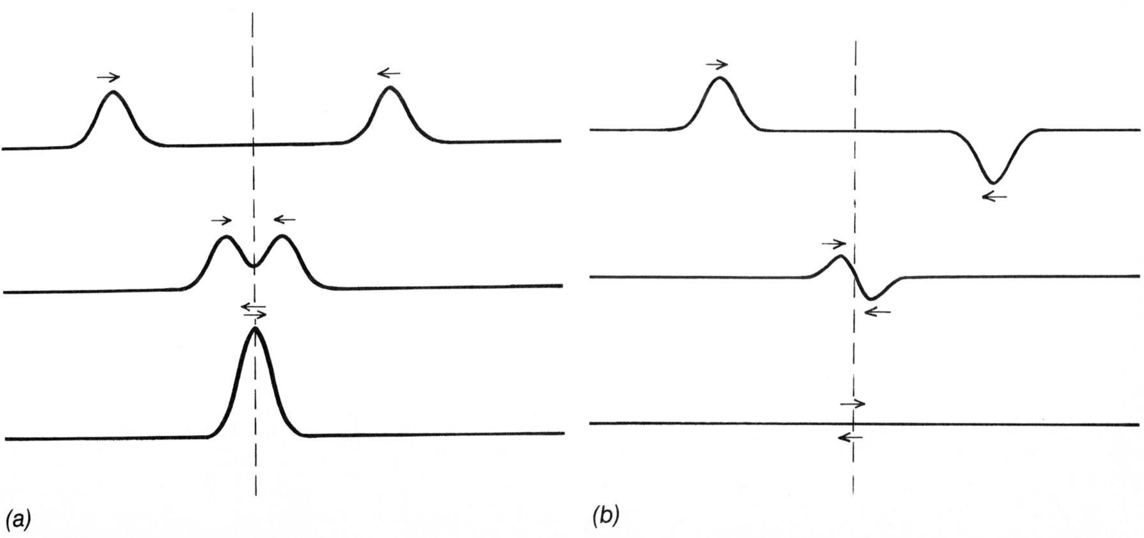


**FIGURE 16–8**
*(a)* Two crests add together in constructive interference when they meet and then pass on unaltered. *(b)* A crest and a trough of equal amplitude merge and then disappear for an instant in an act of perfectly destructive interference. For that instant, an observer could not tell there were waves present on that stretched string. Immediately afterward, of course, the interfering waves pass and look as if nothing has happened.

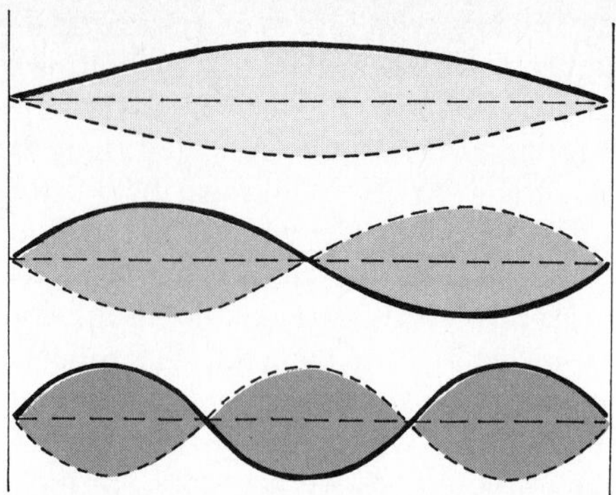

**FIGURE 16–9**
The three lowest-frequency standing waves on a stretched string. Each has a node, a point of zero motion, at each end. The greater the frequency of the standing wave, the more nodes it has. (Notice that for the very lowest-energy standing wave only one-half of a wavelength occupies the entire string. The next standing wave has two half-wavelengths, or one whole wavelength, and so on.)

with a few different frequencies, you'll see something else. For certain rates of oscillation, the reflected wave at the far end interferes with the wave you are sending to cause a steady (or *stationary*) pattern on the rope. That is, the rope moves up and down *in a pattern that doesn't travel along the rope*—you can't see a shape moving in either direction. Figure 16–9 shows three such patterns on a stretched cord whose ends

**FIGURE 16–10**
There is no motion of the medium at a node of a standing wave.

are fixed. These are called **standing waves.** Perhaps you've fiddled with a telephone's cord during a conversation that was dragging. If so, you may have excited one or more of these standing wave patterns on the cord. Notice the points in Fig. 16–9 where the cord *never* moves. At those points the two waves moving in opposite directions cancel; they interfere destructively *at all times*. Those points are the **nodes** of the standing wave. The points where the cord flips up and down with maximum amplitude are the **antinodes.**

The frequencies of oscillation that give rise to standing wave patterns are the **natural frequencies** of the medium. Any object that supports waves has such natural frequencies of oscillation. A disturbance at the frequency of one of the natural frequencies will excite that standing wave. If the disturbance continues, energy is added to the standing wave, and its amplitude grows. That process is called **resonance,** and the natural frequencies are called *resonant* frequencies.

## Standing Waves in More than One Dimension

A bow drawn across the strings of a violin excites standing waves on the strings, and a guitar string that's plucked vibrates in a standing wave pattern. But standing waves aren't found just on strings, ropes, or cords, traveling only in one dimension. When struck, a drum, cymbal, or gong vibrates in two-dimensional standing waves (see Fig. 16–11). Such patterns can be seen if a plate of glass is clamped down, sand sprinkled on it, and a violin bow drawn over the edge of the plate of glass to cause vibrations. The glass will vibrate in a standing wave, and the sand is set into motion, moving away from the antinodal areas, where the vibrational amplitude is largest, toward the nodal areas, where the sand isn't disturbed. When clamped at varying points, the plate vibrates in different patterns as the various reflections cause different standing waves. Another two-dimensional standing wave pattern forms on the surface of tea or coffee in a Styrofoam cup sitting on a countertop. The fuller the cup, the better. Tap the countertop with the heel of your hand, and circular patterns will appear on the liquid's surface. The circular

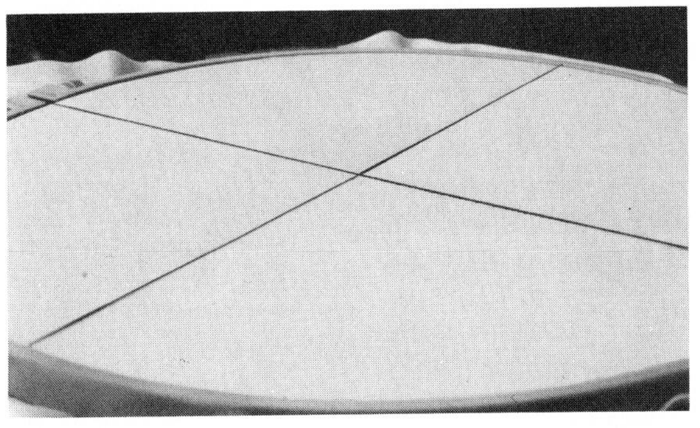

(a)

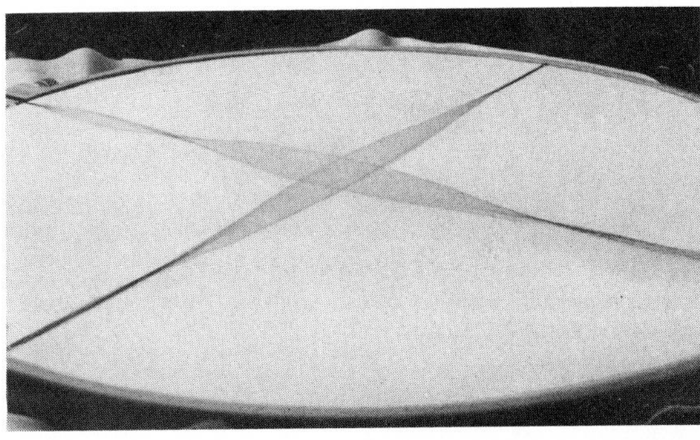

(b)

**FIGURE 16–11**
A stereo speaker vibrates back and forth to produce the longitudinal waves in the air that we call sound. Such waves can excite the natural frequencies of some objects. *(a)* A rubber membrane is placed over a speaker that's emitting steady low-frequency sound waves. *(b)* Standing waves are excited by the sound.

standing waves arise because of the circular boundary, from which all the waves are reflected. This brings us to our next topic, waves on the surface of water.

## Surface Waves on Water

Before the sun comes up in the morning, the surface of a lake can be as smooth as glass. But the sunlight heats the land and water, which causes varying temperatures in the air above them and brings breezes to life. A gentle breeze causes wrinkles on a pond, but a persistent breeze in one direction over a large lake pushes up larger crests of water and sends them along. The ultimate size of the waves depends on the speed and duration of the wind as well as the distance it blows over the water, called the *fetch* of the wind. Calculations show that a wind of 90 mph blowing in the same direction for 24 hours over the open ocean could cause waves 220 feet high. (Waves over 100 feet high have actually been observed at sea.)

Once waves are in motion, gravity acts as a restoring force, much like the tension in a stretched cord, tending to flatten any wave or wrinkle on the surface. Indeed, a cross section of ocean waves offshore looks just like the patterns that travel down a stretched rope (so long as no local winds disturb their shape). But a water wave isn't a transverse wave, nor is it longitudinal; it is a combination of the two. As a wave train passes through deep water, the water at the surface moves in a *circular* pattern. A cork floating on the surface shows the action of the water as a wave passes (see Fig. 16–12). Swimming in the ocean, you can feel this back and forth motion if you float just beyond the point where the waves begin to break up near the shore.

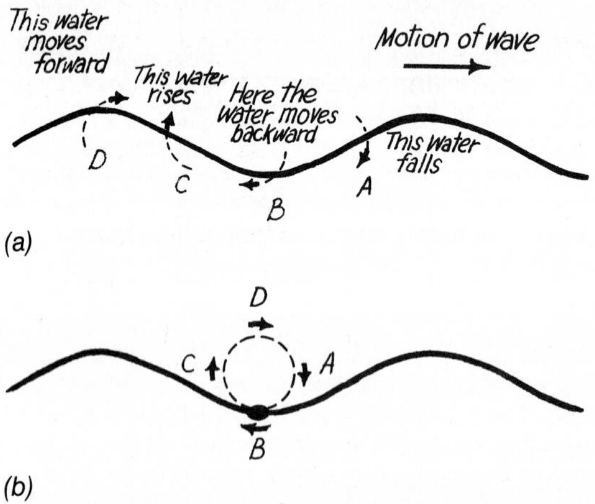


**FIGURE 16–12**
*(a)* The molecules at the surface of deep water perform a circular motion when a surface wave passes their positions.
*(b)* A small floating cork moves with the water and shows the same pattern of motion.

Although the cross section looks like a periodic transverse wave on a string, a water wave's pattern has an extra feature because it moves on a two-dimensional surface. When a surfer catches an ocean wave and scoots along the front side of the crest, he or she is riding a wavefront. A **wavefront** is a surface where the passing wave is affecting the

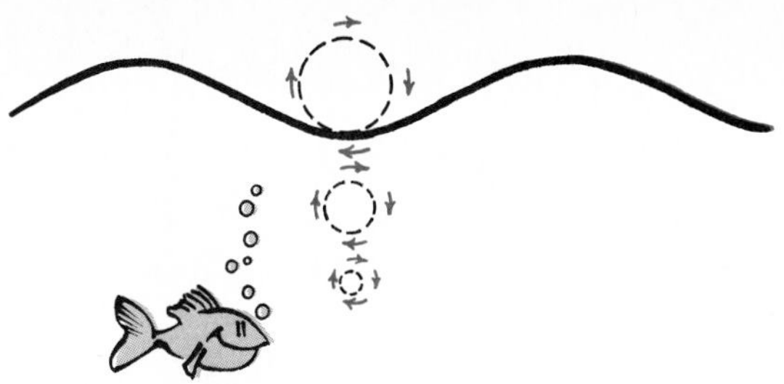

**FIGURE 16–13**
Though molecules of water at the surface sweep through large circular paths as a wave passes, those beneath them move on smaller circles. At a depth beneath the surface that is as great as 1 wavelength of the surface wave, the disturbance of the water is quite small.

particles of the medium in the same way at the same time. Examples of wavefronts can be seen in any pond or lake. Toss a pebble into a quiet pond, and surface waves move out in all directions from the point of disturbance. The circular patterns made by the crests and troughs are the wavefronts. As the waves move away, the circular wavefronts grow in circumference, but the amplitudes of the waves diminish. The energy the waves carry (that is, their ability to do work on the water they come to) spreads out as the wavefronts grow. It is like that on the ocean too. Giant waves made in a typhoon in the western Pacific Ocean spread out and lose amplitude as they move across the water. Those that hit the beaches of Hawaii may still be substantial in size, but by the time they lift surfers along the coast of California, they may have spread their punch over too many miles of wavefront to be really spectacular.

Surface waves don't affect the water that is far below the surface. *More than one wavelength below the surface, wave motion is negligible.* (See Fig. 16–13.) When a wave comes from deep water up to a beach, friction with the shore's bottom soon interferes with the wave's motion. The crest may then *break,* or fall over in front of the portion of the wave below it, which has been slowed by friction at the bottom.

## The Speed of Water Waves

Unlike the case for waves on a stretched rope or spring, the speed of waves on the surface of water depends very much on the wavelength (or on the frequency). For large water waves such as those made by large boats, waves with *longer* wavelengths travel faster. The next time you see a speedboat on a lake, notice the waves as they come to the shore. The first to arrive have the longest wavelengths. But at the other end of the range of wavelengths, like the tiny water waves that you can make by wiggling a finger in a glass of water, the reverse is true. For those waves surface tension acts as an additional restoring force. In Chapter 11 we saw that the force of surface tension is strong (as compared to gravity) only over small areas, such as for drops of water a centimeter or less across. Likewise, the smaller the wavelength, the

**FIGURE 16–14**
The small wavelength capillary wave outruns the current. Those with the smallest wavelengths have the greatest speed. Surface tension contributes much more to the restoring force for such small waves than gravity does.

more effective the surface tension is as a restoring force. For tiny water waves, the *smaller* wavelengths travel faster. Waves that move with the help of surface tension are often called **capillary** waves.

Figure 16–14 is a photograph of a small twig breaking the surface of a slowly moving stream. The fast capillary waves run upstream from this stick; the slower waves, such as those trailing behind the twig, are of the gravity type. If you look for them, you'll find them in any smoothly flowing stream or brook.

## Interference and Diffraction of Water Waves

Maybe you've seen a ski boat pull a skier around in a tight circle. Waves spread from both sides of the boat's path. Not long after the boat leaves that area, the show begins. Near the center of the boat's curved path, waves from the boat meet. As they do, the water's surface jumps up and down furiously. The water waves interfere to make a wave whose amplitude is sometimes equal to the sum of the amplitudes of the two merging waves, and likewise for the troughs (Fig. 16–15). Because the relative speed of the two waves is twice the speed of an individual wave, these larger oscillations occur twice as fast as for one wave alone. Once they've separated, however, the waves go their ways unchanged.

Whenever water waves pass a boundary on the surface, such as a boat dock on a lake, waves that pass its edge spread out somewhat into the region that's blocked to the direct entrance of the waves. This phenomenon is called **diffraction.** Figure 16–16 shows what happens when straight wavefronts come to an opening in a barrier, for instance, a jetty that protects a harbor in bad weather. The incoming wave disturbs the water in the opening, and this disturbed water emits a circular wavefront into the region that the straight wavefront could not reach. Diffraction is most pronounced when the barrier's opening is the same size or

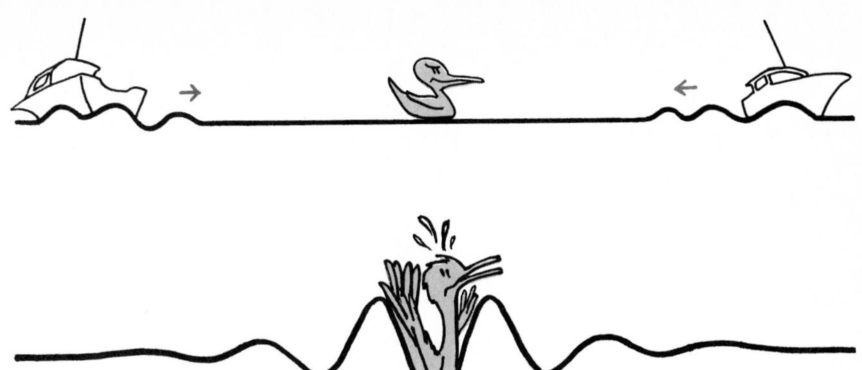

**FIGURE 16–15**
Wave interference on lakes or oceans can be hazardous, as for this duck. Watch for such merging waves from circling ski boats. If waves of equal amplitude merge from opposite directions in the region of interference, the waves are twice as high, twice as deep, and oscillate twice as fast as a single wave does. For this reason, canoers should be careful not to get caught between a passing large boat and a reflecting shoreline, such as a rock face in a lake. In 1979 a huge low pressure area coming eastward took an abrupt left turn near the British Isles and went north. The winds changed, too, and the waves made by the storm before it turned met the ones formed after it turned, catching many of the yachts in the Fasnet Race, the finale of the five-race Admiralty Cup Series. Twenty-three yachts were sunk or abandoned.

smaller than the wavelength of the incoming wave. Nevertheless, when any kind of wave passes a sharp boundary, some diffraction occurs.*

## Refraction

When a wave that moves in two or three dimensions comes to a region where its speed changes, its direction of travel may change. If it enters that region straight on, no turning takes place. But if the wave enters at any other angle, the wave's direction of travel changes in an action called **refraction.** You can get an idea of how refraction works from experience you may have had while driving a car. Picture this.

---

*In the 1670s Christian Huygens (a Dutch scientist who first defined centrifugal force and who was the first to observe the rings of Saturn) offered the first explanation of diffraction. He thought each point in the medium vibrated with the frequency of the passing wave and, in doing so, emitted a circular wave (in two dimensions) or a spherical wave (in three dimensions). But as you can see in Fig. 16–16, only *half* of a circular wavefront appears, so Huygen's idea wasn't completely correct. The full explanation of diffraction is best understood mathematically, and we won't discuss the details.

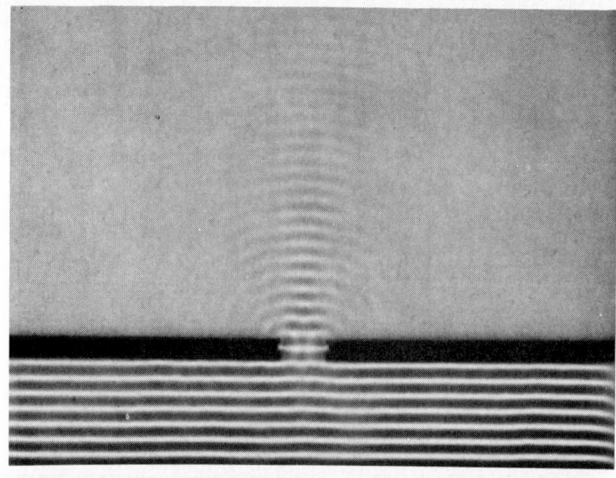

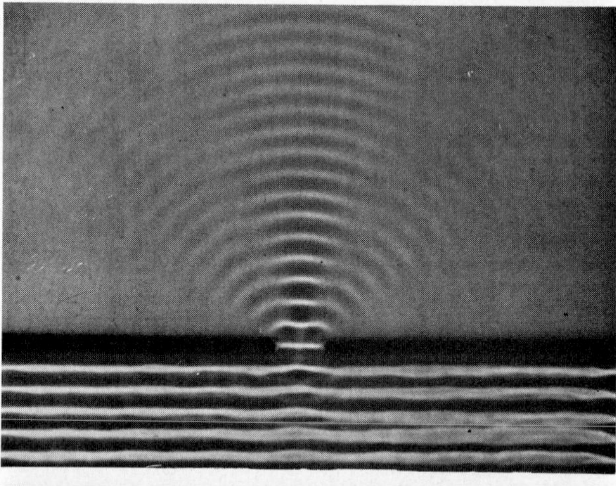

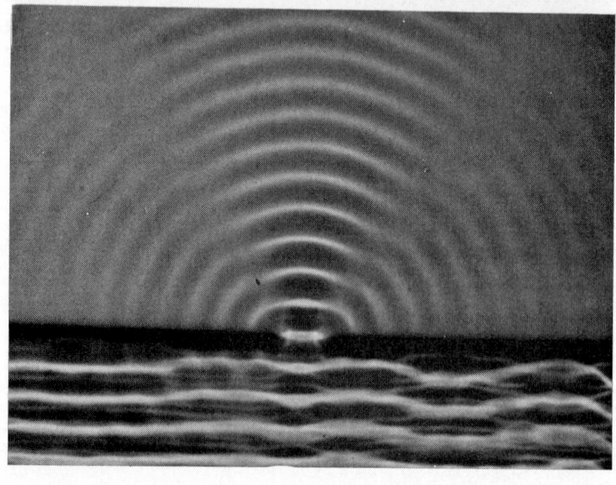

**FIGURE 16–16**
A very general property of the diffraction of waves that pass through openings can be seen with these water waves. As a wave passes through the opening in a barrier, more diffraction occurs if the wavelength of the wave is about the same size or larger than the size of the opening itself. If the wavelength is smaller than the hole in the barrier, a portion of the wave goes through in a straight line and little diffraction occurs.

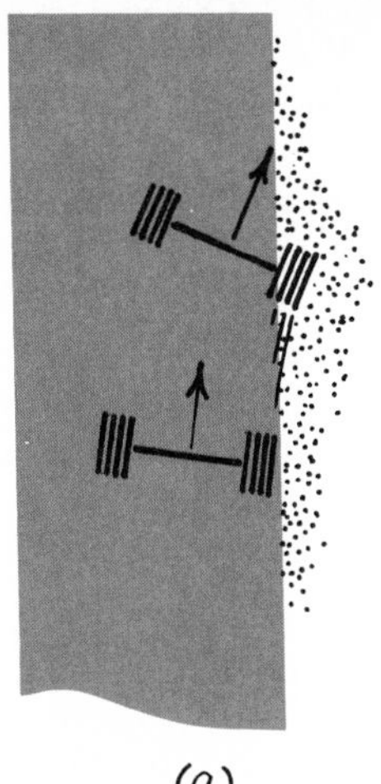

(a)

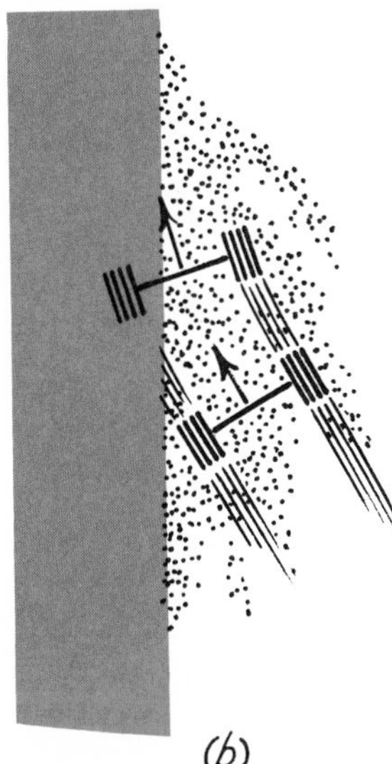

(b)

**FIGURE 16–17**
*(a)* A car angles into a softer portion of a road that offers more resistance. Likewise, a wavefront angles into a medium where its speed is slowed; this change in direction of a wave is called refraction. *(b)* A car angles away from a firmer portion of road that offers less resistance. In much the same way, a wavefront angles away from a medium where it will add speed.

A car with good steering alignment travels straight along a level road with very little steering. But watch out if the right wheels hit a deep water puddle on the side of the road during a hard rain, or if they slip off the pavement into some soft sand or snow. The right wheels will get extra resistance to their motion while the left wheels do not. That causes the car to swerve to the right. To describe what happens, you could say the car turns *into* the medium (water or sand) where it travels more slowly. Something similar happens if the car is in soft sand and angles left to get onto the pavement; the left wheels clear before the right ones. As that happens, the slower right side angles the car back toward the edge of the road. (See Fig. 16–17.) It's as if the car turns somewhat *away from* the faster medium.

Water waves do the same type of thing as they approach a beach. Ocean waves often come toward a beach obliquely. As they approach, their interaction with the rising ocean bottom slows them, and they turn into a shallow area where they move more slowly. This explains the long parallel rows of waves coming onto the beaches in Florida. Waves angling in turn as they meet the gradual slope of the ocean bottom and then travel almost at a right angle to the shoreline. For a contrast, watch movies of surfers in Hawaii. Large waves there often come onto the beach at a considerable angle. The ocean bottom near those beaches falls off so fast that there is no chance for refraction to turn the waves straight toward the shore.

## Interacting with Waves

Floating on a raft out past the breakers in the ocean, you bob up and down and back and forth as waves pass, just as a cork does. It's relaxing because the waves give you a smooth ride. That's not true, however, for a small boat or skiff whose length is about the size of one wavelength of the passing waves. Such a boat falls from a crest into a trough and is pushed upward again before it stops moving downward. The boat takes a pounding from the same waves that give you a smooth ride. A larger boat, like an ocean liner, is many wavelengths long, however, so the up-and-down pushes of the waves more or less cancel; the ship's large mass means it won't accelerate much anyway. Incoming waves reflect from its side, and there is a quiet area next to the other side. Interestingly, farther behind the ship, the waves that diffract from each end of the large craft fill in, and the wave's pattern is soon the same as before. The physical size of an object, both its dimensions and its mass, determine how much energy it can absorb from a wave. We'll encounter this fact again in later chapters.

## Earthquake Waves and Tidal Waves

The solid earth, like any other solid medium, supports wave motion. The movement of a railroad train or a semitrailer truck causes small-scale waves in the earth; the waves that arise during earthquakes are more significant. Every day tiny slippages along cracks, called *faults,* in the earth cause waves, or tremors. This happens at depths of only several miles to as deep as 400 miles, and most tremors are too small when they reach the surface to be felt by humans. But more than half a dozen times a year, there is a major shift of crustal rock somewhere in the zones where the earth's crust is beset with faults. As a result, tremendous pressure is released almost instantly at that point, which is the *focus* of the earthquake. The moving rock crashes into a new position, and the motion stops, but waves carry much of the energy of this action in all directions throughout the earth.

As earthquake waves spread outward from the focus, their energy does likewise. For that reason the waves of greatest amplitude at the earth's surface are at the *epicenter,* the point on the surface directly over the focus and therefore closest to it. The waves made by earthquakes are of two types, longitudinal and transverse. Transverse waves move the earth from side to side and are called **S waves.** Longitudinal waves push and pull the earth and are called **P waves.** Instruments called *seismographs* detect these waves as they reach the surface. A seismograph is basically a mass that is suspended so that it can swing in a single direction. The mass tends to remain at rest while the earth vibrates beneath it. When the seismograph shakes because of a passing wave, a pen connected to the mass moves across a moving strip of paper (Fig. 16–18). In this way, it records the relative size of the waves in its vicinity.* A cluster of three seismographs whose masses move in mutually perpendicular directions records all the motion from passing compressional or transverse waves.

The energy released in the form of waves during a very large earthquake may be as high as the energy of a hundred 100-megaton hydrogen bombs. That energy appears in only a few seconds, so the power of an earthquake is awesome. As the waves reach the surface, the motions they cause level buildings, make landslides, and more. If the epicenter is located on the ocean floor, the motion there can lift and then drop an enormous weight and volume of water. The surface level over a large area may rise or fall as much as a meter, creating a surface wave with a small amplitude but a very long wavelength. This kind of

---

*The standard for measuring earthquakes is the *Richter scale.* A slight earthquake rated as 1 isn't noticeable to us, but earthquakes rated from 5 to 8 can be very destructive. Normally only six to twelve earthquakes a year rate 7.5 or larger. The destructive power of an earthquake wave is related both to its amplitude and to its frequency. If the frequency matches a resonant frequency (or natural frequency) of a building, the structure is more likely to be damaged. The Richter scale is as follows:

| MAGNITUDE | |
|---|---|
| 8 | Great earthquake |
| 7–7.9 | Major earthquake |
| 6–6.9 | Destructive earthquake |
| 5–5.9 | Damaging earthquake |
| 4–4.9 | Minor earthquake |
| 3–3.9 | Smallest generally felt |
| 2–2.9 | Sometimes felt |

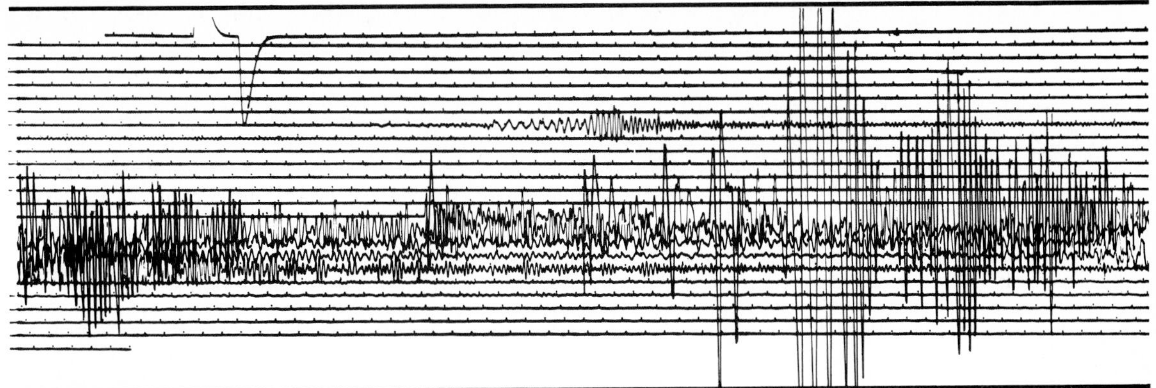

**FIGURE 16–18**
A printout from a seismograph. This one was recorded on October 11, 1975, at Mundaring, Australia, and the large fluctuations show the motion of the earth there due to a large earthquake at Tonga Island.

wave, called a *tsunami* (pronounced "soo-nah-mee"), can travel across the open ocean at speeds over 600 miles per hour.* When the leading edge of a tsunami hits shallow water somewhere, such as at the shore of an island or a continent, the rising ocean floor slows its passage. In effect, this allows the back of the wave to catch up with the front. The wave's amplitude, although initially small, grows as the wave slows. As a result, when it hits a beach, the wave may be traveling at only 30 miles per hour, but it can be as high as a 10-story building (Fig. 16–19). Such a tsunami hit Hawaii in 1946. The devastation was watched in horrified silence by the crew of a freighter a mile offshore. Because they were anchored in deep water, the sailors felt nothing at all as the tsunami passed beneath them on its path to the island.

**FIGURE 16–19**
An earthquake in the Aleutian Islands on April 1, 1946, caused a tsunami to hit the island of Hawaii. This photo shows the wave in the process of destroying an oceanfront building.

*Although these waves are called tidal waves by most people, the tides have nothing to do with them. The Japanese word *tsunami* means harbor wave, which seems only a little more appropriate because one of these waves can funnel into a harbor and destroy it.

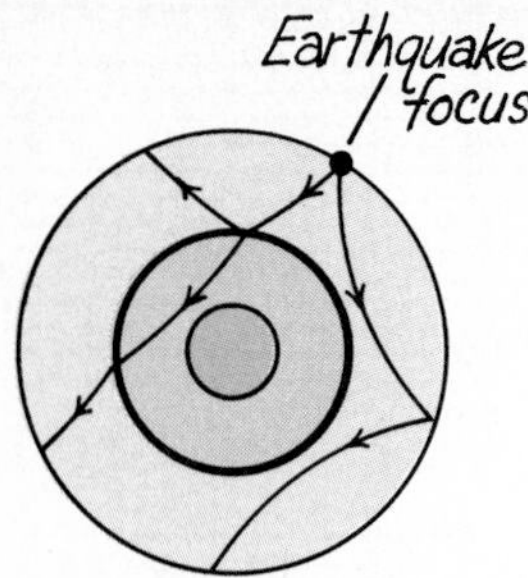


**FIGURE 16–20**
A cross-section of the earth showing the paths taken downward by two segments of a compressional wave generated by an earthquake near the surface. The wave moving to the left (from the focus) strikes the boundary between the mantle and the outer core of earth. Part is reflected there, and part is transmitted. The wave moving down the page and to the right rises to strike the surface, where it is reflected. The waves curve upward toward the surface because the deeper they travel the more compressed the matter is and the faster they go. All waves refract away from the faster medium.

Most of what geophysicists know about earth's interior comes from the study of earthquake waves. A major earthquake sends S and P waves to most points on earth's surface. Figure 16–20 traces some of the paths taken by earthquake-generated compressional waves. The waves reflect when there is an abrupt change in their speed due to a change in the properties of the rock they are passing through. The waves also curve upward, because the rock at greater depth has greater density and the waves will travel faster there. That is refraction at work. The paths followed by S waves are similar, except that there is always a shadow zone on the far side of the earth where the S waves never arrive. Tracing backward through the earth with the equations that govern wave motion, geophysicists have shown that S waves stop abruptly in the earth's outer core, about 2900 kilometers (1800 miles) below our feet. Since S waves cannot travel through fluid matter, it is thought that the outer core is in the liquid state.

The majority of earthquakes come from sites under the great fault zones, called *rifts,* in the crust. Not coincidentally, that's also where the world's active volcanoes are. Volcanic eruptions usually come on the heels of an earthquake, though not always. These catastrophic events are as yet largely unpredictable. In 1556 in Shensi, China, approximately 800,000 people died in a severe earthquake. In 1920 an earthquake in China took 150,000 lives, and 3 years later another in Japan caused 150,000 deaths. The 1970 earthquake in Peru killed 70,000 people, and the 1976 earthquake in Tangshan, China, killed over 250,000 and injured another 750,000 people.

**FIGURE 16–21**
Damage to a department store in Anchorage, Alaska, from the March 27, 1964, earthquake.

## CALCULATIONS

EXAMPLE 1: On a stretched spring waves of any wavelength (or any frequency) travel at the same speed so long as the tension isn't changed. **Show that when such waves have smaller wavelengths, they also have higher frequencies, and when such waves have longer wavelengths, they have lower frequencies.**

Let $v = \lambda_1 f_1$ be the equation for the speed of one wave and $v = \lambda_2 f_2$ be the speed of another. Since the speeds $v$ are the same, $\lambda_1 f_1 = \lambda_2 f_2$. Thus, if $\lambda_1$ is longer than $\lambda_2$, $f_1$ must be smaller than $f_2$ for the products of each to be equal. Likewise, the wave with the higher frequency must have the shorter wavelength.

---

The energy a wave carries away is equal to the energy used to make it. Making waves on a stretched clothesline can give an idea of how the energy in a wave depends on the wave's frequency $f$ and amplitude $A$. Suppose that with your hand you make two oscillations per second on a clothesline, using a stopwatch (or the "locomotive" counting technique of Demonstration 1, Chapter 1). Next, suppose you *double* the frequency, making four oscillations per second and being careful to keep the amplitude the same. As you double the frequency, you can tell that you put more than twice as much energy per second into the wave. In fact, you must use *four times* as much energy to create a wave with the *same* amplitude but *twice* the frequency of another. In general, the energy in such a wave a proportional to the square of the frequency $f$. Or

$$E \sim f^2$$

where $\sim$ means "is proportional to."

The energy in a wave also depends on the amplitude $A$ of the wave. If you create a wave with *twice* the amplitude of another but with the same frequency, you once again do *four times* as much work. (Try it with clothesline, and you'll believe it.) In general again, the energy in a wave is proportional to the square of the amplitude $A$.

$$E \sim A^2$$

Together, the dependence of energy on amplitude and frequency is

$$E \sim A^2 f^2$$

---

EXAMPLE 2: Two compressional waves in air created by musician's tuning forks have exactly the same amplitudes. One has a frequency of 256 hertz, and the other has a frequency of 440 hertz. **Find the ratio of their energies.**

All factors other than the frequencies being equal, the ratio of the energies is

$$\frac{E_{440}}{E_{256}} = \frac{(440)^2}{(256)^2} = \mathbf{2.95}$$

The wave of frequency 440 hertz has almost three times the energy of the wave of 256 hertz, although they have the same amplitude.

---

When a wave spreads out in three dimensions, its wavefronts grow in size, and the energy carried in each wavefront is then spread over a larger area.

An object close to the wave's source intercepts more energy per second than an object farther away. Hold your hand close to a 100-watt light bulb, then slowly pull it backward to feel this effect. Or just back away from a blasting stereo speaker; your ear then intercepts less energy per second. A wave source (someone's voice box, a firecracker's explosion, a light bulb) emits a certain amount of energy per unit of time (or power). At a distance $r$ from that source, the waves have spread over a surface area equal to $4\pi r^2$, the surface area of a sphere, just like the surface of an expanding balloon. The energy released at the source is spread over that area.

How much energy an object receives from such a spherical wave at distance $r$ from its source depends on the energy per second that crosses a unit of area *perpendicular* to the wave's direction of travel: That quantity is called the **intensity,** $I$, of the wave. The equation for the intensity is

$$I = \frac{\text{energy/second (released by the source)}}{4\pi r^2}$$

As $r$ increases, the wave's intensity becomes smaller by the factor $1/r^2$. This equation is often called the *inverse-square* law.

---

EXAMPLE 3: **Compare the intensity of a wave at a point 3 meters from the source to its intensity at a point 1 meter from the source.** Use the equation for intensity to make the ratio $I_{3\text{ m away}}/I_{1\text{ m away}}$. You'll see that everything cancels except the factors

$$\frac{r^2(3\text{ m away})}{r^2(1\text{ m away})} = \frac{(1\text{ m})^2}{(3\text{ m})^2} = \frac{\mathbf{1}}{\mathbf{9}}$$

The intensity of the wave three times farther from the source is 1/9 of the intensity closer to the source.

## REVIEW

**1.** A wave in a medium is a disturbance that transports energy over a distance without displacing the matter very far by comparison. True or false?

**2.** If the particles in a medium move in the direction of the wave's motion, it is a ________ wave.

**3.** If the particles of the medium move perpendicularly to the wave's direction, the wave is called a ________ wave.

**4.** Define compression and rarefaction.

**5.** The distance between two successive peaks or two successive troughs is the ________ .

**6.** Is the frequency of a traveling wave the same as the number of oscillations per second of the particles as the wave passes?

**7.** What is the amplitude of a wave?

**8.** Describe a standing wave.

**9.** A point that doesn't move in a standing wave is called a ________ .

**10.** The peaks on a standing wave are called ________ .

**11.** The spreading of a wave that passes a sharp boundary is called diffraction. True or false?

## DEMONSTRATIONS

**1.** Borrow a child's marbles to show a wave. Make a line of marbles along a straight crack between two sections of concrete sidewalk, making sure each marble touches the one next to it. Then they can react to a displacement, much as a row of molecules in some solid will. Thump the end marble of the line with your finger and almost *instantly* the other end marble will roll. A longitudinal wave with a very high speed moves through the marbles.

**2.** Fill a soft plastic rectangular ice tray with water until it overflows. Then tap the counter it's resting on and watch the standing waves that appear because of the straight boundaries. In both the long direction and across the width of the tray, these standing waves intersect and make grid-like patterns.

**3.** Stand a row of dominoes on end and gently tap the end one. The tiny bit of energy you use to topple the first one is transferred to the last one in the line. None of the dominoes moves very far, yet the wave travels from one end to the other, stopping only when the medium comes to an end.

**4.** Make circular standing waves in a Styrofoam cup full of water, as discussed earlier. Those waves of small wavelengths are surface tension waves. Then mix a few drops of dishwashing liquid into the water. The amplitude of waves in the mixture will be much smaller because the soap decreases the surface tension of the water.

**5.** Hook identical rubber bands onto both ends of a paper clip and have someone pull the rubber bands in opposite directions, exerting tension. The paper clip is like an atom in a chain of atoms in a solid, and the stretched rubber bands represent the "bonds" holding it to its nearest neighbors. Move the paper clip with your fingers, first along the direction of the rubber bands (the longitudinal direction) and next in the direction perpendicular to the rubber bands (the transverse direction). The difference in effort needed points out why longitudinal waves travel faster through solids than transverse waves do: The restoring force in the longitudinal direction is larger.

**6.** Lay one end of a pencil or a ruler against a tabletop or desk and press down. Flick the end that is off the surface and watch the vibrations. That is a standing wave with a node at one end (where you're holding it down) and an *antinode* at the other end. The length of the pencil or ruler, then, is one-fourth the wavelength of the standing wave. (Draw a sketch to check that.)

**7.** Hold the base of an empty thin-rimmed wine glass against a tabletop with one hand. Moisten a finger of your other hand and press it to the rim, and slowly move it around the circle of the rim. The loud sounds you hear are from standing waves the friction from your finger excites in the glass, much like the waves the friction from a violin bow makes on the strings of a violin.

## EXERCISES

**1.** Define a wave in your own words.

**2.** If you make a compressional wave on a stretched Slinky by moving one end back and forth with your hand, how does the frequency of your hand's motion correspond to the frequency of the wave on the Slinky?

**3.** When rain falls into a lake, each raindrop creates small waves. Why don't the numerous waves add to make other waves with huge amplitudes?

**4.** Define what is meant by the wavefront of a water wave.

**5.** When there are tall trees surrounding a small pond, the choppy water caused by the wind is located mostly in the middle of the pond. Explain why.

**6.** Explain why an earthquake is felt most severely directly over the focus, the point where the wave begins.

**7.** When someone begins to pull on one end of a stretched rope, does the other end begin to move at the same instant? Is this also true for a steel cable or for a steel rod? Discuss any differences in these situations.

**8.** Why can't transverse waves travel through a gas or a liquid?

**9.** Visitors to Florida never see large alligators (12 to 16 feet long) on the lakes when the wind is causing large-amplitude waves on the water. Why not?

**10.** A bird takes off from a utility wire. What happens to the energy of the wave created in the wire by the bird's leaving?

**11.** Give some reasons why two tin cans with a string stretched between them could be a better way to communicate than merely shouting through the air.

**12.** The Waldorf-Astoria Hotel in New York City is built over a subway tunnel. The hotel's foundation rests on several inches of cork and lead. Comment on the function of that material.

**13.** Sketch the three longest-wavelength standing waves on a stretched cord that's clamped firmly at both ends. Identify the nodes and antinodes of the waves and give the wavelengths of each in terms of the length of the cord, $L$.

**14.** Two identical clotheslines are stretched side by side in a yard. You pluck each of them and notice that the disturbance travels faster on one than on the other. What does this tell you about the tensions in the clotheslines?

**15.** A seismograph that oscillates to the maximum with a P wave coming from some direction cannot be used to detect an S wave coming from that same direction. Why not?

**16.** Explorers who have crossed the ice packs off Antarctica have noticed that ice shifts occur up to 2 days before a storm arrives. What could be happening?

**17.** Every waitress and waiter knows from experience how easy it is to excite the lowest-energy normal mode of water waves in a cup of coffee or a glass of beer or a wineglass of wine. (It's often called the *sloshing* mode, and it is easy to excite in a bathtub or on a waterbed, too.) Discuss the shape of that standing wave.

**18.** The destructive earthquake waves that raise and lower the land or the ocean are transmitted to some extent to the air when the surface of the earth or the water moves. Why are those air waves harmless?

**19.** When the earth slips along a fault, the motion may be horizontal, vertical, or at an angle. Which type of dislocation do you think would be most likely to cause a tsunami?

**20.** The snout of a mole is covered with hairs that connect to sensitive nerves. What purpose could this feature serve? Blind cave fish can evade a biologist's net because of nerve endings on their heads and sides. How could this feature help?

**21.** Discuss this statement: Cooks intuitively avoid resonances while beating eggs.

**22.** The sort of vibrations made in the earth by a rumbling train is familiar to most of us. But even in rural locations away from highways and railroads, earth's surface moves enough with another kind of wave action to disturb the most sensitive seismograph. These devices must be lowered into holes drilled several hundred meters deep to escape the disturbances at the surface. What could be causing those waves at the surface?

**23.** Explain why a compressional wave that travels through air at 1090 feet/second should travel 19 times faster through aluminum.

**24.** Explain why you can't stir sugar at the bottom of a full cup of coffee or a full glass of iced tea just by jostling the glass to make waves on the surface. (Try this; if you pour out some of the liquid a little at a time, the sugar will eventually move as the container is jostled.) What shape of glass or cup would better lend itself to that action?

**25.** Should you be able to hear for a greater distance in the desert during a bright summer day or during a summer night when the stars are out?

**26.** A solitary swimmer jumps into a swimming pool and climbs out. What becomes of the energy in the waves created by the swimmer?

**27.** Why do scuba divers 30 feet beneath the ocean's surface feel little motion from the waves passing overhead?

**28.** In 1831 troops marching in step across a suspension bridge in England unwittingly set the bridge into an oscillation (they excited a normal mode) that grew until the bridge exceeded its elasticity and collapsed. Tell what must have happened. How could this situation have been prevented?

**29.** Is it possible to ride a pogo stick by jumping at any frequency you want? Would a child's frequency of jumping vary from an adult's?

**30.** Comment on the rings around the craters on Callisto, one of Jupiter's moons (Fig. 16–22). (Such craters are also found on the moon, Mars, and Mercury.)

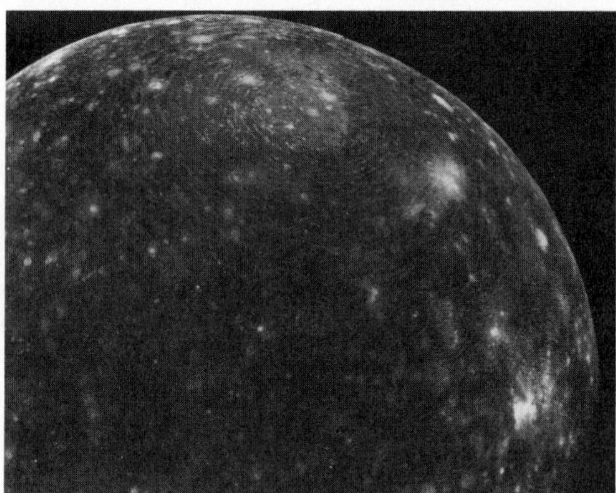

**FIGURE 16–22**

**31.** When a breeze blows into shore on a fresh-water lake, you can see an interesting phenomenon if there are weeds along the shore. The short-wavelength waves on the lake don't penetrate a weedbed, yet the long-wavelength waves do. Discuss.

**32.** Hold a meter stick at both ends and shake it to excite the normal mode that looks just like the lowest-energy, longest-wavelength standing wave on a stretched string. Draw a sketch of this standing wave. Then take the meter stick with one hand in the middle, leaving both ends free, and shake it again. This time the lowest-energy normal mode has antinodes at both ends of the stick. Sketch this mode; decide which of these two modes has more energy or if they both have the same energy. You can check your answer by estimating the frequency of the wave. If they have the same frequency, for example, their energies will be the same. (*Hint:* Your hand won't be at a node when you hold the meter stick at its center.)

***33.** Sound in sea-level air at 25°C travels at about 1090 feet/second. How many seconds does it take to go 1 mile?

*34. One day when the fish weren't biting, a fisherman noticed that the waves lifted his boat about once every 2 seconds. Then he glanced at the waves and estimated their wavelength to be 6 meters. What was the speed of the water waves? Draw a sketch of this and show the waves that will hit the boat in the next 4 seconds.

*35. Explain how you could stand at one end of a stretched rope and measure the speed of transverse waves on it if you knew the length of the rope.

*36. During a nighttime thunderstorm, the thunder arrives 8 seconds after you see the clouds light up. About how far away was the lightning strike?

*37. Jupiter is five times farther from the sun than earth is. What is the intensity of the sun's light at Jupiter's position compared to the sun's light at earth's position?

*38. In its elliptical orbit around the sun, the earth moves as close as 0.983 AU and as far away as 1.017 AU. (1 AU = 1 astronomical unit = 149,600,000 km.) Compare the minimum intensity of sunlight on earth to its maximum intensity.

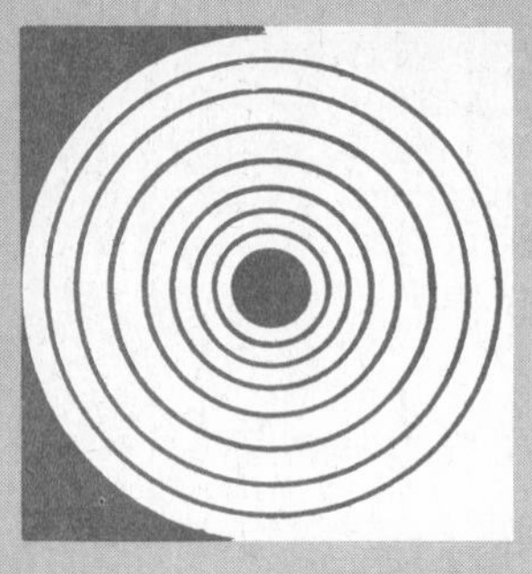

# CHAPTER 17
# SOUND

In 1915 bad luck struck a British expedition that had hoped to cross Antarctica. They became icebound in the Weddell Sea, just off the continent, and the moving ice floes slowly crushed their wooden ship. Trapped in this frozen solitude, the crew listened to the wind and the noises made by the shifting ice. The ice spoke a language all its own, as recorded in a passage from the book *Endurance, Shackleton's Incredible Voyage:**

> There were the sounds of the pack in movement—the basic noises, the grunting and whining of the floes, along with an occasional thud as a heavy block collapsed. But in addition, the pack under compression seemed to have an almost limitless repertoire of other sounds, many of which seemed strangely unrelated to the noise of ice undergoing pressure. Sometimes there was a sound like a gigantic train with squeaky axles being shunted roughly about with a great deal of bumping and clattering. At the same time a huge ship's whistle blew, mingling with the crowing of roosters, the roar of a distant surf, the soft throb of an engine far away. . . . In the rare periods of calm, when the movement of the pack subsided for a moment, the muffled rolling of drums drifted across the air.

No doubt the physicist aboard, Reginald James, had opportunity to recall the questions the expedition leader had asked about his qualifications for the venture. Was he good-tempered? Were his teeth good? Did he have varicose veins? And that was that. ■

## The Nature of Sound: Sound Waves

When you walk, you disturb the air around you. You push the air in front of you, and behind you air moves in to fill the space your body just left. At walking or even running speeds this takes place rather

---

*Alfred Lansing, *Endurance, Shackleton's Incredible Voyage* (New York: McGraw-Hill Book Company, Inc., 1959), p. 5. Trapped without hope of rescue at the southernmost point in a million square miles of chaotic ice, the expedition had an escape that makes a terrifying and heroic story. Antarctica was not crossed for another 43 years.

**FIGURE 17–1**
Compressions and rarefactions move out in *all* directions from the source of a sound. The vibrating string of this guitar sets up oscillations not only in the air but also in the wooden body, providing a louder sound than the string alone would make. Listeners at all angles around the guitar hear the sound.

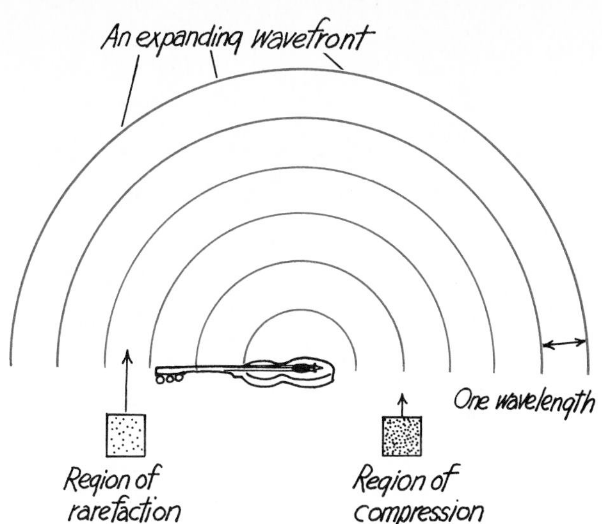


smoothly. But faster motions affect the air differently. A string, stretched tight with its end fastened rigidly, can be plucked at its center. As the string whips through the air, it shoves the molecules in its way, pushing them closer together. At the same time, a partial vacuum is left in the wake of the string. In other words, the vibrating string creates compressions and rarefactions in the air (Fig. 17–1). Because it is under pressure from its own weight, the air around the region of partial vacuum rushes in from all sides. The air springs out in every direction from the region of compression. When these molecules move, they influence the neighboring molecules; the air that snaps in or out of a region changes the pressure on the neighboring air, causing it to move too. Just like a disturbance that moves from molecule to molecule over a stretched spring, the disturbance caused by the vibrating string spreads through the air. Though an individual air molecule moves only a small fraction of a centimeter, the regions of compression and rarefaction continue to travel outward. Such a disturbance, as we saw in the last chapter, is called a compressional wave.

When the waves coming from a vibrating string enter the ears of someone nearby, the person hears a sound. That is, the mechanisms in the ear (discussed in the last section of this chapter) sense the vibrations in the air caused by the string and trigger nerve impulses that signal the brain. A sound you hear is only your brain's interpretation of how the invisible molecules in the air jiggle as such a wave passes. Like any instrument of detection, our ears have limits: We can hear waves only if they oscillate the air's molecules from as few as perhaps 20 times a second to as much as 20,000 times a second. Thus the frequencies of the compressional waves we call sound waves lie between 20 hertz and 20,000 hertz. Compressional waves with frequencies higher than 20,000 hertz are called **ultrasonic waves.**

Any sound wave of a single frequency above about 100 hertz is called a *tone*. The highness or lowness of a tone is called its **pitch.** Pitch relates to the frequency of the sound wave, the number of vibrations per second it imparts to the air's molecules as it passes. A high pitch means a high frequency, etc. The loudness of a sound depends on how much energy per second the wave brings to the air next to the ear's detection system. That is dependent on the amplitude of the wave, which is how far the wave moves the air molecules it passes, as well as the wave's frequency. (Loudness is discussed again in "Calculations.")

Compressional waves with frequencies of 20 hertz to 20,000 hertz can also move through liquids and solids, which is why we can hear sound underwater in a swimming pool or in a bathtub, or by pressing an ear against a door. Not only could the train robbers in early cowboy movies hear the approaching train through the rails, they could get the news faster that way than through the air. As we saw in Chapter 16, such waves travel more quickly through solids, where the molecules are bound by strong, springlike bonds, than through gases, where the wave travels only by molecular collisions.

## Details of Sound Waves

As with other waves, there is a relation between a sound wave's speed, its frequency, and its wavelength. One complete oscillation of an air molecule takes place when each complete wavelength passes, so if a wave vibrates a molecule back and forth 300 times in a second (the wave's frequency is 300 hertz), a total of 300 wavelengths move past in one second. In other words, the wave's speed is

$$\frac{\text{distance}}{\text{time}} = \frac{300 \times 1 \text{ wavelength}}{1 \text{ second}}$$

or

$$\text{wave speed} = \text{frequency} \times \text{wavelength}$$

In symbols,

$$v = \lambda \times f$$

where $\lambda$ is the symbol for the wavelength.

The measured speed of sound in sea-level air at 20°C is 343 meters per second (or 767 miles per hour). The lowest frequency we can hear is 20 hertz, so

$$\text{wavelength} = \frac{\text{speed}}{\text{frequency}} = \frac{343 \text{ m/s}}{20 \text{ Hz}} = 17 \text{ m (or 56 ft)}$$

Likewise, the wavelength of the highest frequency we can hear is about 0.017 m (about ⅔ in.).

A remarkable and important property of sound waves is that they all have the same speed as they move through a parcel of air. That is because the energy in a sound wave is transferred by the same molecular motions that are responsible for the pressure in air. The speed of sound depends on the *average speed of the air molecules,* not on the frequency of the passing compressions and rarefactions. *Long-wavelength* (low-frequency) *sound waves take the same amount of time to cross a sym-*

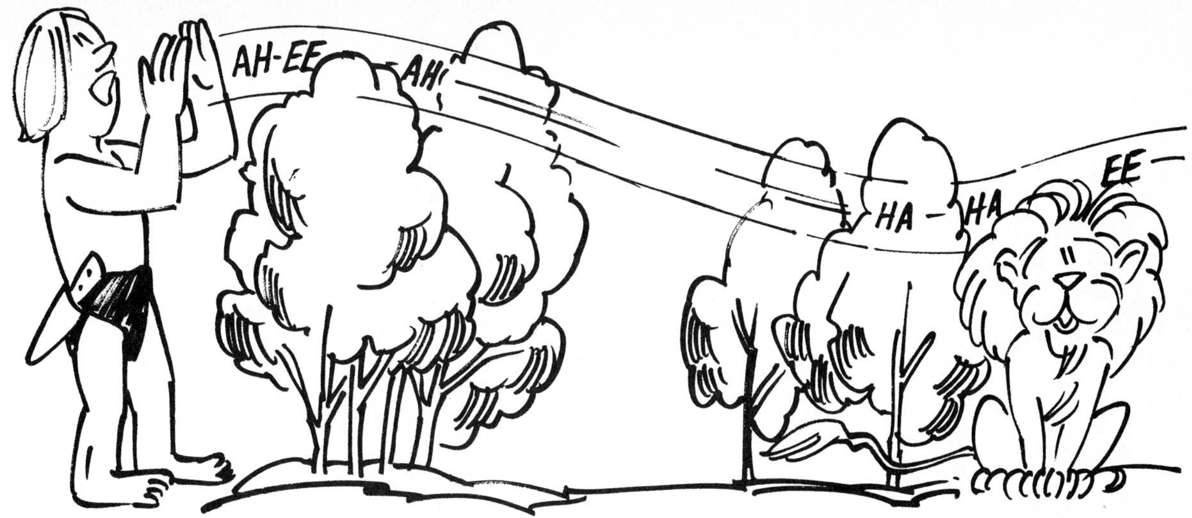


FIGURE 17–2
If the different frequencies that make up sounds traveled at different speeds through the air, the original sounds would be garbled as the wave patterns changed during travel. Fortunately, that isn't the case; all sound waves moving in air have the same speed.

*phony hall as short-wavelength* (high-frequency) *sound waves do*. Imagine the confusion if this weren't so! Each word you utter is made from a mixture of different frequencies, or wavelengths. If sound waves of different frequencies moved across a room at various speeds, the words (the patterns of the waves that your brain will interpret as words) would become garbled as the waves traveled.

Snap your fingers, and compressions and rarefactions ripple out through the air in all directions. As the waves spread, the energy they carry (which moves the molecules of the air the waves come to) is spread over larger and larger areas (Fig. 17–3). Just like the circular waves that leave the point where a raindrop falls into a water puddle, sound waves decrease in amplitude as they spread and the molecules of the air they pass move over smaller distances. That finger snap could startle someone sitting next to you, but a person a hundred feet away might not hear it.

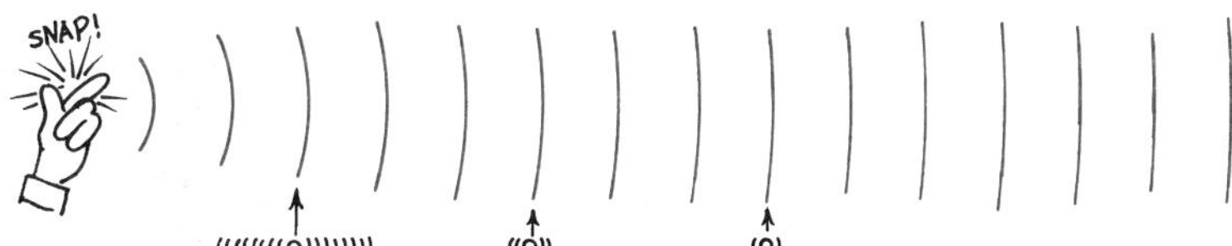


FIGURE 17–3
As a sound wave spreads, so does the energy it is carrying. The molecules of air farther from the source vibrate back and forth much less as the sound wave passes than do the molecules closer to the source.

## Properties of the Speed of Sound in a Gas

Sound waves travel through air, or any other gas, when displaced molecules run into others and transfer energy. It follows that the speed of sound depends on the temperature of the gas, since the average molecular kinetic energy, $\frac{1}{2}mv^2$, in a substance is directly proportional to its absolute temperature (Chapter 14). In warmer air, then, the molecules move faster and sound travels more quickly. At 10°C sound travels at 337 meters per second. At 20°C it moves at 343 meters per second, and at 30°C it travels at 349 meters per second.

Experiments show that sound waves travel a little faster through pure nitrogen gas than through pure oxygen gas at the same temperature. Here's the reason. The molecules of nitrogen have the same average kinetic energy as do molecules of oxygen at the same temperature.* A nitrogen molecule has a mass of 28 atomic mass units, while an oxygen molecule's mass is 32 atomic mass units. The nitrogen molecule, then, has a larger $v^2$ (hence a larger speed $v$) to make up for its smaller $m$, since $\frac{1}{2}mv^2$ is the same for both the nitrogen and the oxygen. Air is a mixture of 78 percent nitrogen and 21 percent oxygen, and sound waves move through it at a speed *between* the speed of sound in pure nitrogen and its speed in pure oxygen at the same temperature, in accordance with the average molecular speed. Thus at 0°C sound travels through nitrogen at 334 meters per second, through oxygen at 317 meters per second, and through air at 331 meters per second.

The relative humidity of air also affects the speed of sound. The presence of even a small percentage of lightweight water molecules (18 atomic mass units) allows sound waves to move a bit faster. Sound moves through pure water vapor (at temperatures over 100°C) at speeds close to 500 meters per second.

## Reflection and Absorption of Sound Waves

It can be an eerie feeling to step inside an unfurnished house or apartment. Your footsteps, your voice, and the snick of a closing door sound hollow and harsh. Flat walls, floors, and ceilings make good reflectors of sound. As sound waves come in, they bounce off these surfaces with little distortion. That is, the waves that reflect keep the same shape, and the reflected sound, or *echoes,* resemble the original sound. On the other hand, furniture, lamps, and pictures normally in a room *scatter* the sound waves in many different directions. We say they reflect sound **diffusely.** Less rigid, porous materials such as draperies and rugs do more than just diffuse sound. They *absorb* a large percentage of the energy of the sound waves as well, quieting echoes and softening noises.

The ceiling and walls of large concert halls can be specially designed to reflect sound to the audience in the back. Otherwise, it is disturbing if echoes of the same sound come from different directions at different times. To prevent such unwanted reflections, acoustic technicians place panels of absorbing materials at key areas on the ceilings and walls. In the tuning of a concert hall, they sometimes place sand-

*See ''Calculations,'' Chapter 13, Example 4.

FIGURE 17–4
*(a)* In the open air, oratory is more difficult because so much of the energy put into a sound wave never reaches the listeners. *(b)* In a properly designed lecture hall, much of the energy is focused by reflection to the listeners farther away.

FIGURE 17–5
When rock climbers are out of sight of one another on high cliffs, they can hear one another only by means of diffraction.

bags in the chairs to simulate the absorption and diffusion effects of the people that will be sitting there during a performance.

Sound reflections can be used to detect things. A sightless person can sense the nature and size of close objects by listening to the echoes from someone talking, or from footsteps or the taps of a cane. During a routine physical examination of a patient, a doctor gets a crude picture of the body's internal spaces by tapping on the patient's back and chest, a method known as percussion. The reflected sounds reveal the size of internal organs and the presence of fluids in cavities. For pictures of greater detail, a doctor may beam ultrasonic waves into the patient, which reflect from organs, bones, air, etc. The reflected sound waves appear as electronically generated pictures on a TV screen. This painless, noiseless method involves far less risk for patients than do diagnostic X rays. Sonar gear on surface ships or submarines can detect other submarines through the water by emitting pulses of sound and listening for echoes.

## Diffraction of Sound

As you know if you've ever had a somewhat vocal roommate, you hear sound from another room much better than you would think you should even if a door is only slightly ajar. Like other kinds of waves, sound waves spread, or *diffract,* somewhat as they come through an opening or past a corner. (See Chapter 16.) *But diffracted sound waves are effective only along paths that angle just slightly away from a straight line to the original source.* If someone talks to you from around a corner, *most* of the sound you hear is reflected from nearby surfaces. Out-

**FIGURE 17–6**
Someone standing near the center of this geodesic dome hears echoes from his or her voice simultaneously returning from all directions. The result is strong constructive and/or destructive interference of the sound waves at that position.

side, even trees and bushes reflect sound, helping us to hear things better. Few people have been in places so isolated that reflected sounds are absent and diffraction alone lets them hear around corners. Rock climbers are one exception. High on a rock cliff away from any other nearby surfaces that could reflect sound, climbers not in direct sight of each other must shout mightily to be heard. Even then, if the angle of diffraction is too large, the climber listening for signals hears nothing. (Often climbers must resort to another kind of wave motion for signaling, such as tugs on the rope used between them for protection.)

## Interference of Sound

Sound waves that arrive at the same point at the same time interfere with each other. For instance, if a compression of one sound wave occurs at some point when the rarefaction of another sound wave is there, some cancellation takes place; the molecules respond only to the *net* force they get when the two waves push them. The crossing of sound waves goes on around you all the time, but you rarely notice. That's because one sound is louder than the others, or the adding and canceling of frequencies is not complete because of the different frequencies involved. On the other hand, if a pair of stereo speakers emit the same sound (such as when you turn the amplifier to monaural), you can sit in various locations in the room and find dead spots where the sound gets softer because of destructive interference and live spots where the sound is louder because of constructive interference. If there is an observatory dome or a geodesic dome on your campus, try this. Stand in the center of the structure and say something. All sound waves reflect uniformly from the angled walls and come together where you stand. The interference is unmistakable—in fact, you may become momentarily disoriented trying to talk while your own words come back to you.

**FIGURE 17–7**
A variety of household items resonate distinctly if placed in front of a speaker when music is playing loudly.

## Soundboards, Sympathetic Vibrations, and Resonance

A stretched string that's plucked or struck sharply emits sound waves, but that sound is neither loud nor especially musical. The tones we hear from musical instruments such as guitars, violins, cellos, or pianos don't come primarily from the stretched strings or wires, but from *soundboards* that are set into motion by the strings' vibrations. Guitar strings, for example, vibrate against a bridge that transmits the vibrations to the guitar's wooden body, which is its soundboard. This large surface area then vibrates in sympathy with the strings and emits certain tones, or musical notes.* In Chapter 16 we saw that any object has natural patterns of vibration, or normal modes. A vibrating guitar string excites one or more of the natural frequencies (see Chapter 16) of the thin, solid soundboard. The shape, thickness, and elastic qualities of the wood determine the natural frequencies and hence the musical quality of the sound it produces.

Alternating pushes and pulls cause any object to vibrate in sympathy to some extent. However, if a certain frequency of disturbance produces a large vibration in an object so that it absorbs a lot of energy, the object is said to **resonate** with the disturbance. An empty cardboard shoe box with a lid or an empty metal coffee can resonate when certain low-frequency (bass) sound waves strike it. The air pressure on the object fluctuates as compressions and rarefactions arrive. Put the container in front of stereo speakers, and you can feel the resonant vibrations with your fingertips. If you ever feel a loud bass sound in your chest, it's because the air cavity there resonates with the sound waves of those particular low frequencies. That is, the transmitted sound of those frequencies reflects in the cavity in just the right way to form standing waves that absorb a great deal of energy from the sound waves striking you. A soprano's high note can excite a standing wave in the thin wall of a champagne glass. If that oscillation grows by resonance until the elasticity of the solid is exceeded, the glass shatters. Hold a finger against that glass, however, and the finger absorbs some energy from the vibrating surface; then there's no chance the glass will shatter. Likewise, a cloth packed into a bell deadens the oscillations of its surface if it is rung.

**FIGURE 17–8**
No matter what its size, shape, or composition, every object has natural resonance frequencies.

## The Turning of Sound Waves: Refraction

When you talk, sound waves radiate from your mouth, the compressions and rarefactions spreading through the air in front of you. The expanding region of a single compression or rarefaction is similar to the crest or trough of an ocean wave in that the particles throughout that region undergo the same type of motion at the same time. In other terms, the region of a single spreading compression (or rarefaction) is like a wavefront of an ocean wave. (See Fig. 17–1.) As a result, a person a little to your right hears the same message as someone to your left.

The wavefronts of sound waves bend and turn if they angle into a region where the speed of sound changes for some reason. Sound waves, like water waves and all other waves, *refract* (Fig. 17–9). A sound wave always turns *into* a cooler region of air, where it will travel more slowly, and it turns *away* from a warmer area, where it would

---

*Now is a good time to review the sections on standing waves in Chapter 16.

FIGURE 17–9
*(a)* During the day the sound bends upward, away from the warmer air near the ground. This limits the distance that sound travels. *(b)* At night, sounds can often be heard over longer distances. Sound waves traveling over the ground bend (refract) downward into the air cooled by contact with the ground, where they travel more slowly.

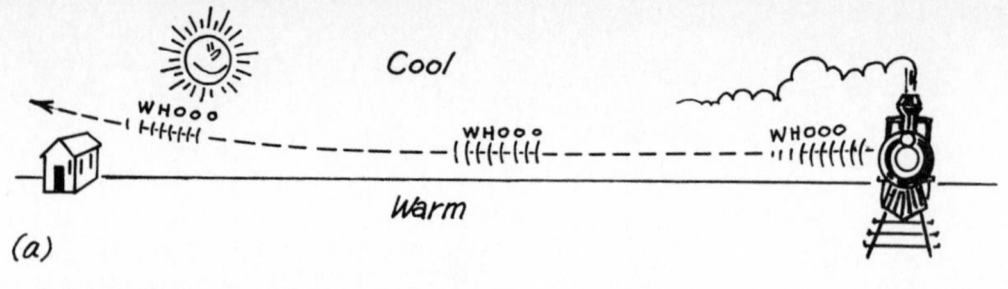


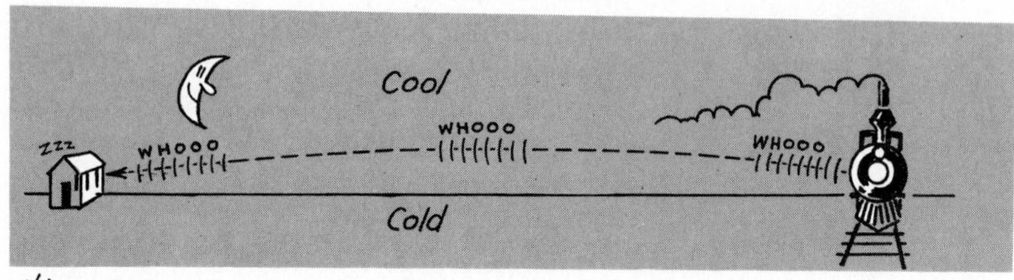


travel faster. Someone within earshot of a refracted sound wave hears the sound coming not from its direction of origin but from a different direction. During daylight hours the air just over the ground is normally warmer than the air higher up. Sound waves traveling parallel to the ground slowly bend upward into the cooler air. At night the ground sometimes radiates its heat quickly and by contact cools the air just above it. Sound waves from voices or from traffic noises that would go up in daylight hours instead bend downward into the cooler air, hugging the ground for a longer distance than in daytime. People who fish often notice how far sound carries over a quiet lake. If the water is cooler than the air temperature, the air next to the lake is also cooler. Sound that starts to travel upward at a slight angle bends back down into that cool layer of air before traveling far.

## The Doppler Shift

When a car or motorcycle roars past you, the pitch of the sound you hear changes abruptly. The sound you hear as the vehicle comes toward you has a higher frequency than the sound you hear as it recedes in the distance after passing you. These frequency changes are easy to hear on sports telecasts of motorcycle or car races as the machines pass the sportscaster's microphone. The loudness of the sound has to do with how close the car or bike is, but the change in frequency (pitch) is due to its motion. (The change in loudness is often more impressive than the change in pitch.) If a parked car's horn were stuck and you passed it while moving in another car, you'd hear the same sort of change in frequency. When you move toward a source of sound, the pitch is higher, and as you recede, the pitch is lower. Any such change in frequency because of motion is called the **Doppler shift.** Here's why it happens.

Suppose a parked car's horn vibrated 400 times a second; that is, 400 compressions alternating with 400 rarefactions left the horn each second. If you were standing some distance away, a sound wave of frequency 400 hertz would reach your ears. But if you moved toward

**FIGURE 17–10**
An observer who is moving toward the source of a sound intercepts more wavelengths per second. As a result, the frequency of the sound heard is higher than if the observer were standing still.

that car, you'd run into *more* than 400 wavefronts per second (just as a rowboat that's rowed into the waves gets more slaps per second against its hull than if it were anchored). That means you'd hear a higher-pitched sound with a frequency greater than 400 hertz. Likewise, should you move away from the horn, fewer wavefronts would catch up to you each second. The frequency of the sound you'd hear would be less than 400 cycles per second. That's how the Doppler shift in frequency takes place when the *listener* is in motion (Fig. 17–10). In contrast, here's what would happen if you were sitting by the side of the road as the car with its stuck horn moved away from you. The horn, of course, oscillates at 400 hertz even though it is moving. At the end of each oscillation, the horn would be farther from your position than it was when that oscillation began. The wavelength of the sound wave is "stretched out" compared to its length if the horn were at rest. Fewer crests (or troughs) reach you per second; the sound wave at your position has a frequency lower than 400 hertz. If the car were moving toward you, the wavelengths would be compressed instead, raising the wave's frequency to over 400 hertz. Figure 17–11 gives a visualization of how the Doppler shift comes about in these cases where the *source* of the sound is in

**FIGURE 17–11**
When the source of a sound is moving toward an observer, the wavefronts are closer together than they would be if the source were still. That observer then hears a higher-frequency sound. Likewise, an observer left behind by a moving source receives sound waves of longer wavelength (lower frequency).

**FIGURE 17–12**
When a bee or other large insect passes by, the pitch during its approach is noticeably higher than the pitch after it passes.

motion. (As an example, if a 400-hertz horn comes toward you at 55 miles per hour, you hear a sound whose frequency is 431 hertz.)

If you've been around flowering plants outside, you've no doubt heard honeybees and bumblebees buzzing around. When one passes near to your head during a straight-line dash, the Doppler shift is hard to ignore. Perhaps you've been stuck at a railroad crossing as a train rumbled by, blasting its air horn all the way. The Doppler shift is unmistakable, as it is with the siren of a moving police car or an ambulance that passes you at high speed.

## Shock Waves in Air: The Sonic Boom

Earlier we saw that someone moving through the air parts the air and pushes it to the sides and it fills in behind the person in a smooth motion. We also saw that faster motion such as that of a vibrating guitar string causes a compression in front of the string and a rarefaction behind it. If the speed of an object through the air is very great, such as for a high-speed rifle bullet, the change in pressure fore and aft is very abrupt, and the resultant sound is very sharp. If a rifle bullet travels faster than the speed of sound, sound waves made by the bullet cannot travel in front of the bullet. The sharp disturbance still travels outward at the speed of sound, but its wavefront forms a *cone* behind the bullet. We call this a **shock** wave. There are two shock waves for a bullet, a compression from the front end and a rarefaction from the back, but they are very close together. Therefore, when the compression and rarefaction arrive at someone's ear, the person hears only one distinct crack. When a jet aircraft or the space shuttle flies faster than sound, however, people below its path hear two explosive sounds, the first from the compression by its nose and the other from the rarefaction at its tail (Fig. 17–13). These sounds are called **sonic booms.**

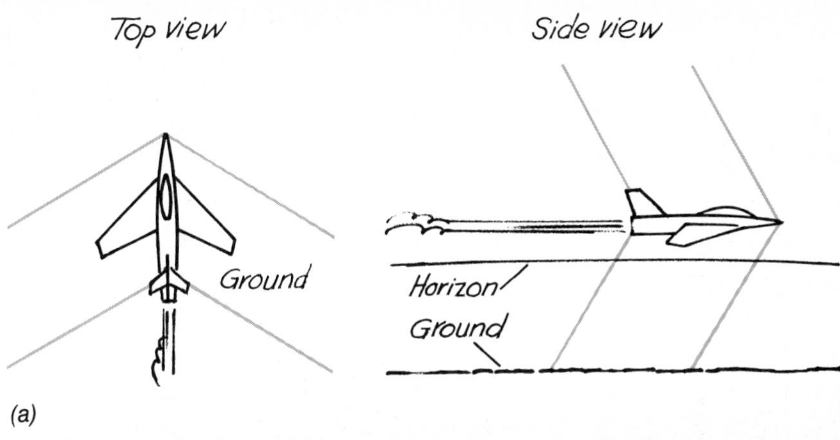


(a)

(b)

**FIGURE 17–13**
*(a)* The shock waves from a plane traveling faster than the speed of sound. The compression at the nose and the rarefaction at the tail spread out in cone shapes and cause explosive sounds, or sonic booms. *(b)* Just as an airplane creates a shock wave by traveling faster than the speed of waves in air, this duck and ducklings create one by traveling faster than the speed of waves in water.

A shock wave from a plane contains a lot of energy as it travels outward at the speed of sound in air. But like any wave, it diminishes in amplitude as it spreads. The sonic booms of the shock wave cause no problem to anything on earth's surface if they begin at elevations of 70,000 feet. A sonic boom from a plane at 40,000 feet can be an annoyance to people on the ground. If shock waves originate at 25,000 feet, however, they break windows in homes and buildings below the plane's path; at 8000 feet the energy in the waves can cause structural damage in buildings.

## Thunder

Most of us have counted the seconds between seeing lightning and hearing the thunder. Sound travels about a mile in 5 seconds, so you can find the approximate distance to the unlucky spot. The noise you hear as thunder is generated all along the lightning's path. The stroke superheats the air around it, raising the air temperature by thousands of degrees Kelvin, and that heated air expands explosively, compressing the air around it. This creates a shock wave that travels outward and loses energy as it does. The result of all this is what you hear: a piercing thunderclap if you are very close or ordinary thunder if you are farther away. Sound waves are unleashed at all points along a stroke of lightning at essentially the same time. The rolling thunder you hear is the succession of sound that reaches you from higher and higher points along a stroke of lightning. If there are hills nearby, reflections also play a part in the sound. And that's not all. Refraction is at work too. The waves bend as they pass through hotter or cooler, or even wetter, parcels of air, and this plays a part in the drum rolls of thunder. Whenever you hear thunder, you can bet the lightning stroke was less than 15 miles away. At that distance or greater, the sound waves moving almost parallel to the ground curve upward into the cooler air of the atmosphere and miss tickling your ears.

## The Sounds from Trees and Brooks

Listen to the sound of the wind in a pine tree. The needles cause small ripples in the stream of air, much like a twig that breaks the smooth surface of a fast-moving brook. The air's internal friction, or viscosity, prevents it from parting and flowing smoothly around to close in behind the needles. Like the twig in the brook, swirls (or eddies) appear behind the needles, causing pressure fluctuations in the air that travel outward in every direction (Fig. 17–14). For pine needles this action (and its soft sound) begins at wind speeds of 4 to 5 miles per hour. If the breeze picks up speed, it causes more ripples per second, and the pitch of the sounds from the tree becomes higher.

In trees with broader leaves, the breeze flutters the leaves about, causing a rustling noise similar to the one you can make by flipping the pages of a book, or flapping a single page as you turn it. These sounds are often pitched lower than the sound from pine trees, because the smallest objects cause the highest-pitched sounds in the wind. When the boughs of some trees lose their leaves in the autumn, the wind still makes a sound. The empty branches are larger in diameter and stiffer than the leaves and create fewer of the small eddies per second. The sound from the bare trees becomes low in pitch and somber compared to the sounds of the trees in full leaf.

In a slow brook where the water flows smoothly around stones or logs, practically no sound stirs the air. But if you hear a quiet murmur, look closely at its source. The sound doesn't come from smooth waves or even from eddies in the water. If the water flows over or around an irregular object fast enough to break its surface, it can trap air. Then bubbles burst at the surface or even underwater, sending vibrations in all directions. In waterfalls, the water collides violently with the water or rocks below, trapping air and compressing it. These violent, large-scale compressions create the roar. But when raindrops fall on bushes and trees, the leaves give with the impact and compressions are less violent, making a much softer sound.

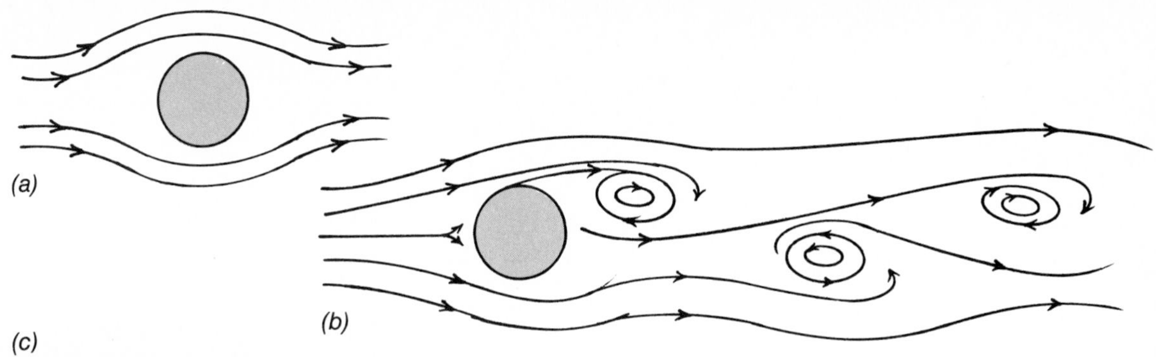

**FIGURE 17–14**
*(a)* At very low speeds wind can flow smoothly around a pine needle without making noise, just as water flows around a stone in a quiet brook. That's called laminar flow. *(b)* At higher speeds the parted air can't flow smoothly; eddies or swirls in the fluid air form behind the pine needle (or any other object). This rippling causes compressions and rarefactions, and that's the sound of the wind we hear. At even higher speeds, the needle or object will have a very ragged, turbulent wake behind it. When the compressions and rarefactions are closer together, the sound rises in pitch. *(c)* When a canoe paddle pushes against the water, it causes eddies like those in part *(b)*.

## Sounds from Insects and Animals

Have you ever *heard* a butterfly fly? Most flap their wings at only 8 to 12 cycles per second or even less, causing oscillations that are below the frequencies we can hear. Nevertheless, if you are close, you can hear the air rippling from the edges of a butterfly's wings as they beat, much like the sound of wind going around a leaf. Big bumblebees in flight move along with the help of 130 wingbeats per second, making a sonorous drone that's easily heard. The honeybee performs 225 beats per second to make a higher-pitched hum. The mosquito owes its annoying high-pitched whine to 600 wingbeats per second. When you hear these insects, the sound comes from the compressions and rarefactions caused as their wings beat against the air.

Many rodents, birds, and insects emit ultrasonic sound waves. The ancestors of domestic dogs once hunted those creatures for a living, and today's canine pets still hear high-frequency squeaks up to 50,000 hertz. Those silent (to us) dog whistles put that ability to practical use. Cats hear even higher frequencies—up to 70,000 hertz. Bats sense ultrasound so well that they use their ears like we use our eyes. A bat sends out pulses at ultrasonic frequencies and locates obstacles and insects according to the direction of the echoes and the time it takes the reflections of the sounds to return. These ultrahigh-frequency sound waves (50,000 to 120,000 hertz) let them fly and hunt in total darkness. Some species can actually locate a flying insect to within several millimeters. But even this natural sonar isn't a perfect hunting instrument; apparently some moths can hear the high-pitched sounds of the bats. Studies have shown that whenever bat sounds are produced, the moths swoop down to land, seeking safety on the ground.

Porpoises and whales use underwater sound. Porpoises send out ultrasonic frequencies and listen for reflections. They can easily sense schools of fish within a half mile of their position. Killer whales have been recorded talking to each other over distances of 20 miles. (Water, remember, conducts sound more effectively and faster than air.) Other species of whales are thought to hear one another at a distance of 500 miles.

The human voice comes from the vocal cords in the larynx, with the mouth and nasal cavity working together as a resonant chamber. To speak, you must force air from the lungs through the narrow passage between the tightened vocal cords. This causes the air in the acoustic cavity to resonate. Varying the shape and opening of the mouth and the placement of the tongue gives the amazing variations of sound in the human voice. Usually a person's voice in a tape recording sounds flat and "not like myself," even when the recording is of high quality. That's because the voice on the tape does not include the low-frequency resonances from the voice box that the ears get through the surrounding bone structure and tissue of the speaker.

## How We Hear Sound: The Ear

Audible frequencies are those heard by people with normal hearing. Usually only the young hear sound waves as high as 20,000 cycles per second; as people get older, their eardrums lose elasticity and don't vibrate as well in sympathy with high-frequency sound waves. Studies indicate that during a person's forties the hearing capability on the high-frequency side loses about 150 cycles per second each year. That's not too serious, however; most frequencies in normal speech are between 100 and 4000 cycles per second.

Although ears look something like funnels, their shapes contribute little to the process of hearing. The sound waves that enter the ear canal are guided to the eardrum with little change in amplitude. It is at this thin, conical, skin-covered membrane that the hearing process actually begins (see Fig. 17–15). Between the brain and the eardrum is a menagerie of miniature equipment capable of detecting motions of the eardrum with amplitudes as small as *one atom*. In any case, sound waves don't move the eardrum far. A movement of even 0.25 millimeters brings pain and/or damage to the mechanism of the ear. Sounds from close, violent explosions, for instance, cause deafness, as does repeated close exposure to the amplified sound from speakers at rock concerts. Behind the eardrum is a space containing air, a region known as the middle ear. The air within this space isn't used to transmit the sound, however; solid matter takes the sound on the next step of its trip. Three tiny bones (the tiniest in your body) called the hammer, anvil, and stir-

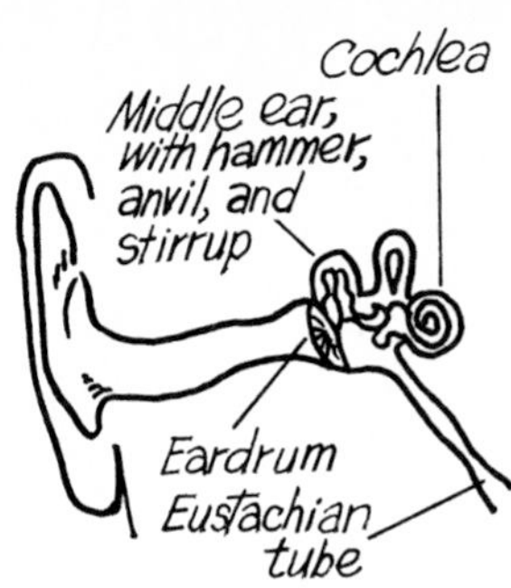


**FIGURE 17–15**
The mechanisms of the human ear.

rup occupy this air chamber. The hammer has a handle that rests on the interior side of the eardrum. When that membrane vibrates, so does the hammer, which in turn moves the other two bones. The three bones are in a lever arrangement. The stirrup rests its footplace on another membrane, oval in shape. Behind this oval membrane is the cochlea, or inner ear, which is filled with fluid. The oval membrane has about 1/20 the area of the eardrum membrane. Consequently, the minute amount of pressure on the eardrum is increased about 20 times as the wave travels through the smaller oval membrane into the fluid of the cochlea. The cochlea is a snail-shaped chamber built into the bone of the skull for protection. The waves that travel around its waterway move membranes that rest on cells that can trigger nerve impulses. Not an easy path for a sound wave, from gas to solid to liquid, and finally to sets of electrical impulses that the brain must interpret as sound. But for most of us, it works just fine.

## Calculations

The speed of sound through air is the same for all frequencies. As for Example 1 in "Calculations" in Chapter 16, that fact means the higher the frequency of a sound wave, the smaller its wavelength, and vice versa.

EXAMPLE 1: The tone called *middle C* on a musical scale is a sound wave of 262 hertz. **What is its wavelength in air that is at 10°C?** ($v$ = 337 m/s.)

$$\lambda = \frac{v}{f} = \frac{337 \text{ m/s}}{262/\text{s}} = \mathbf{1.29\ m}$$

When a sound wave goes from one medium to another and its speed changes, what should happen to its frequency and its wavelength? The molecules of the first medium move back and forth at a rate equal to the frequency of the wave, directly causing the molecules of the second medium to oscillate. The molecules of the second medium, then, will oscillate at the *same rate f* as the molecules in the first medium; the frequencies in both regions are the same. Since the speed changes, however, the wavelength must change so that $v = \lambda \times f$ holds in each region. If the speed increases, $\lambda$ must increase; if the speed decreases, $\lambda$ must decrease.

EXAMPLE 2: A 500-hertz sound wave is emitted by the burglar alarm in Skip's house. He hears it even though he is under water in his backyard pool. **What is the wavelength of the sound in the air and also in the water?** (Use $v_{\text{air}}$ = 343 m/s; $v_{\text{water}}$ = 1480 m/s.)

When the wave is traveling through the air, $\lambda_{air} = v_{air}/f =$ **0.69 m.** When it strikes the water, the transmitted wave has the same frequency, 500 hertz, since its surface molecules move "with" the molecules of the air. The wavelength of this sound in the water, however, is $\lambda_{water} = v_{water}/f = (1480 \text{ m/s})/(500/\text{s}) =$ **2.96 m.**

---

As they spread, sound waves, like the other waves of Chapter 16, also lose intensity, $I$, or energy per second per unit of area perpendicular to the wavefronts. The equation for this is (as in Example 16–3)

$$I = \frac{\text{energy/second (released at source)}}{4\pi r^2}$$

---

EXAMPLE 3: On takeoff a DC-10 airplane produces about 1500 watts of audible sound. **What is the intensity of this sound at a distance of 10 meters? At 100 meters? At 1000 meters?**

$$I_{10 \text{ m}} = \frac{1500 \text{ W}}{4\pi(10 \text{ m})^2} = 1.2 \text{ W/m}^2\text{, or about } \mathbf{1 \text{ W/m}^2}$$

$$I_{100 \text{ m}} = 0.012 \text{ W/m}^2\text{, or about } \mathbf{10^{-2} \text{ W/m}^2}$$

$$I_{1000 \text{ m}} = 0.00012 \text{ W/m}^2\text{, or about } \mathbf{10^{-4} \text{ W/m}^2}$$

Look at Table 17–1 to see if the sound levels at 100 meters and 1000 meters fit your own experiences at airports.

**TABLE 17–1**

**THE DECIBEL SYSTEM**

| | WATTS/METER$^2$ | dB |
|---|---|---|
| Threshold of unaided hearing | $10^{-12}$ | 0 |
| Your own breathing | $10^{-11}$ | 10 |
| A whisper, or rustling leaves, or standing on the rim of the Grand Canyon | $10^{-10}$ | 20 |
| Room in a house when no one is talking (perhaps an air conditioner is on) | $10^{-9}$ | 30 |
| Library's sound level | $10^{-8}$ | 40 |
| A quiet office | $10^{-7}$ | 50 |
| Someone talking normally 1 meter away | $10^{-6}$ | 60 |
| Traffic, New York City (as heard on second floor of a tenement) | $10^{-5}$ | 70 |
| Third-floor apartment next to Los Angeles freeway | $10^{-4}$ | 80 |
| Very noisy street | $10^{-3}$ | 90 |
| Inside a subway | $10^{-2}$ | 100 |
| Intolerable sound for long | $10^{-1}$ | 110 |
| Car horn at 1 meter | ⅓ | 115 |
| Threshold of pain (stick-fingers-into-ears-noise) | 1 | 120 |
| | $10^{1}$ | 130 |
| | $10^{2}$ | 140 |
| Standing *next* to a runway when jet takes off | $10^{3}$ | 150 |
| Standing near the space shuttle at takeoff | $10^{6}$ | 180-up |

Intensities of sound are most often compared by using powers of 10. One reason for this is the wide range of intensity of sound the human ear can detect—from about $10^{-12}$ watt/meter$^2$ to more than 1 watt/meter$^2$, or 12 powers (or factors) of 10. For such comparisons the barely detectible $10^{-12}$ W/m$^2$ is called the *reference* intensity. A sound that has 10 times that intensity (or $10^{-11}$ W/m$^2$) is 1 **bel** or 10 **decibels** (dB) more intense. Table 17–1 gives typical sounds and their intensities.

Loudness is *how our ears detect and our brains perceive the intensity of sound waves*. A sound wave of 3500 hertz at 80 dB of intensity sounds about twice as loud to our ears as a 125-hertz sound wave at 80 dB, for example. The ear is particularly sensitive to sound between 250 and 500 hertz, and even more so in the range 3000 to 4000 hertz, dropping off significantly at higher frequencies. The intensity of a sound wave does not measure loudness by itself.

Television commercials often seem loud compared to regular programs. A commercial's monitored intensity is equal to or below the sound intensity level of a sports program—believe it or not! But if the commercial's sound is concentrated at frequencies where the human ear is more sensitive, it will seem to be much louder.

## REVIEW

**1.** Are sound waves in the air compressional or transverse waves?

**2.** Audible waves in air have frequencies between _____ hertz and _______ hertz.

**3.** The speed of sound in a gas depends not on the frequency of the wave but on the average molecular speed of the gas it moves through. True or false?

**4.** Which properties of a gas affect the speed of sound waves through it?

**5.** Sound waves travel through air at 0°C at a speed of about _______ meters per second.

**6.** If sound waves reflect from smooth, hard surfaces, do they keep the same shape when they are reflected? Does the reflected sound resemble the original sound?

**7.** When a sound wave spreads out after passing an opening, we say it is diffracted. True or false?

**8.** Does constructive or destructive interference of crossing sound waves affect the sound you ultimately hear from those waves?

**9.** When an object absorbs significant energy from the sound waves striking it, we say it resonates. True or false?

**10.** A refracting sound wave turns away from cooler air, where it travels more slowly, but into warmer air, where it travels faster. True or false?

**11.** Explain how a Doppler shift in frequency can occur.

**12.** Describe a shock wave and a sonic boom.

**13.** Does the sound of thunder originate from a shock wave produced by lightning?

**14.** Do the hearing capabilities of many animals and insects exceed ultrasonic frequencies?

## DEMONSTRATIONS

**1.** Excite some standing waves by tapping an empty drinking glass with a spoon. The sound it makes comes from oscillations of the glass itself, not from standing waves in the air column inside it. Prove this by putting a finger to the glass before you tap it again—there's not much sound then. Put some water in the glass, and the pitch changes; only the portion of glass above the water level oscillates enough to give much sound. The shorter the standing waves, the shorter their wavelengths and the higher their frequencies.

Using a soft drink bottle, you can prove it is the glass surface that makes the sound when you tap it and not the air column inside. To excite the air column, blow over the bottle's opening. As your breath oscillates in and out of the bottle, it excites the lowest-mode standing wave, and you will hear a low hum. Then tap the glass bottle with a spoon at the same time to hear the two very different tones.

**2.** The next time you are outside on a windy day, turn your head until you hear the wind whistling past your ears. That sound comes from the air rippling around the edges of your ears, just like the noise from the wind in the trees.

Inside a building, listen to the sounds the air makes as it comes from a heating or cooling duct; that's the same effect.

**3.** Depress the pedals of a piano. This lifts the dampers from the strings. Then sing out loudly. Certain of the piano's strings will resonate with the sounds of your voice and sing back to you. In a similar way, if you hang some narrow strips of aluminum foil and whistle very loudly, they'll sometimes resonate in response. If they are hanging down the side of a solid surface, you may hear a buzz from their back-and-forth motion against that surface.

**4.** Press one end of a ruler against a kitchen countertop, letting three-fourths of the ruler extend over the edge. Then twang the ruler by plucking the end that's off the counter. The noise you hear is mostly from the countertop, not the ruler, as you can quickly see if you place something absorbent between the ruler and the countertop. Test other surfaces. Glass and steel are especially efficient at being sounding boards, and so is wood, but the sounds they all make are very different.

**5.** With a firecracker and a friend, you can estimate the speed of sound. Go to a large field where the two of you can get far apart, 1000 feet or more. When one of you sets the firecracker off, the other must watch the flash and begin to count the seconds, or use a stopwatch.

# EXERCISES

**1.** Can sound originate in still air without a disturbance?

**2.** The maximum displacement of a molecule from its undisturbed position as a sound wave passes is called the ________ of the wave.

**3.** Does sound travel faster or more slowly if the air (or gas) pressure is less? Discuss.

**4.** Does sound in a gas travel faster or more slowly if the temperature is less? Does sound in a solid or liquid travel faster or more slowly if the temperature is less?

**5.** Both the temperature and air pressure are much greater at Venus's surface than at earth's. How would these factors affect the speed of sound there? Venus's atmosphere is mostly carbon dioxide (molecular mass 44 atomic mass units). How would this affect the speed of sound there compared to earth's nitrogen–oxygen atmosphere?

**6.** Would you expect sound to travel a greater distance in the desert or over a cool lake? Discuss each situation.

**7.** If you submerge one ear into the water in a bathtub, you can often hear sound coming from other parts of the house or apartment building that you couldn't otherwise hear. Discuss why.

**8.** Why must the volume of a stereo in a room with a wall-to-wall carpet be turned higher than in a room with tile or wooden floor?

**9.** The building of the wooden bodies of violins and guitars is an art. How are these bodies important to the sounds of the instruments?

**10.** A motorcycle rider in a box canyon approaches a distant wall with the horn of the cycle blowing. Discuss the frequency of the sounds a bystander near that wall would hear. Discuss the frequency of the sounds the rider would hear.

**11.** Why is the sound of a harp so soft compared to the sound of a guitar or a piano?

**12.** Why does tightening a guitar string raise the pitch of the tone it makes?

**13.** Why should the sound of raindrops on grass be softer than raindrops on a sidewalk? Why should the sound of rain in a puddle be softer than the sound it makes on a car's roof?

**14.** The air is quite thin at an altitude of 70,000 feet. If a supersonic airplane breaks the sound barrier there, the shock wave it creates carries away less energy than if the plane flew through air at 10,000 feet at the same speed. Use the law of conservation of energy to tell why that is true.

**15.** An *infrasonic* sound wave is one whose frequency is too low for us to hear. Tell how you make and detect such a wave in a room with two doors just by moving one door back and forth in a regular motion.

**16.** The feathers under an owl's wing are very soft compared to those of a vulture or an eagle. What advantage does this give to the owl?

**17.** If builders could evacuate the air from the open spaces in the walls of apartment buildings, the rooms would be effectively soundproofed, since no sound could move where there are no molecules to carry it. Why wouldn't this be practical? Discuss.

**18.** Why did the astronauts in Skylab have to shout to each other to be heard? The gas pressure in the cabin was one-third of an atmosphere. (They also discovered they couldn't whistle in Skylab.)

**19.** When a sound wave strikes a closed door, the incoming wave pushes and pulls on it. However, little sound is heard on the other side. Why?

**20.** Using stereo speakers to find "dead spots" in a room works better with long-wavelength sounds. Can you suggest why?

**21.** Lord Rayleigh (1842-1919) noted that sharp sounds made at the edge of pine woods returned an echo with a higher pitch than the original sound, while woods containing trees with broad leaves returned an echo more like the original sound. Explain why.

**22.** The foam on the top of beer in a mug kills the sound if the glass is tapped with a spoon. Why is that so?

**23.** An arrow from a bow might travel as fast as 60 meters per second. Bow hunters know there is little chance to hit a deer at 100 meters even if the aim is accurate. Why is that so?

**24.** A rock outcropping (known as Stoney Point) on the Hudson River was removed in a construction project. As a result, riverboat captains found the river much more difficult to navigate in foggy conditions. Guess why.

**25.** Explain how the Doppler shift might be used with an ultrasound generator and detector to measure the speed of blood flowing through an artery of the body.

**26.** Why do you think cold winds howl but warmer winds do not?

**27.** A passing car that blows its horn gives you a chance to hear the Doppler shift. Is the change in frequency greater, hence more obvious, if the car is traveling in the same direction as your car or if it is traveling in the opposite direction from your car? Why?

**28.** Sometimes when you hear a plane that is far overhead, its sound isn't steady. Instead, the muffled sound rolls, fading out, then rising in loudness, over and over again. What's happening?

**29.** Would you expect the space shuttle to make a double or a single sonic boom?

**30.** Listen to the sounds of a fan, an electric mixer, and an electric razor as each is turned on and speeds up and then as it is turned off and slows down. Why does the sound of each correspond to the revolutions per unit of time of its motor?

**31.** The lowest resonant mode of the human ear canal is excited with the least energy from sound waves. Depending on the individual, the frequency of that mode is generally from 2500 hertz to 4000 hertz. What does this tell you about the sensitivity of human hearing for sound waves of these frequencies?

**32.** The propellers of merchant seagoing vessels push so hard against the water that it separates as the propellers cut through it. This separation creates regions of partial vacuum called cavities. This *cavitation process* makes noise that is easily picked up by submarine sonar devices. What causes the noise in the cavitation process?

**33.** A bat called the horseshoe bat uses the Doppler shift to locate moving insects. It emits pulses at different frequencies so it can receive an echo of 83,000 hertz. How could reception of that single frequency be beneficial to the bat?

**34.** Why does an eavesdropper place a glass between ear and wall in order to hear what is said in the next room?

**35.** Although the speed of sound is over 700 mph near the ground, it is about 650 mph at 36,000 feet. Explain how that change in the speed of sound affects the cone of a sonic boom as it approaches the ground.

***36.** Suppose the speed of sound in a certain room is 330 meters per second. If you wanted to adjust a technician's sound generator to give waves that are 20 centimeters long, to which frequency should you set the dial?

***37.** A small motorcycle engine revs up to 10,500 revolutions per minute. What insect does it sound like? (Refer to the section on "Sounds from Insects and Animals.")

***38.** Show that an ultrasonic sound wave of 120,000 hertz emitted by a bat might well locate a flying insect that is about one-fourth of a centimeter in diameter. (Refer to "Interacting with Waves" in Chapter 16, p. 331.)

***39.** Show that the wavelength of a 20,000-Hz sound wave is about 1.6 cm. Show that the wavelength of a 20-Hz sound wave is about 16 m.

***40.** The sound 10 meters from the stage at a rock concert is at the 90-dB level. What is the sound level 100 meters away? (*Hint:* See Example 3, "Calculations.")

***41.** A person talking normally emits at most $10^{-4}$ watts. If such energy could be used to power a light bulb, how many persons would have to talk to turn on a single 100-watt bulb?

***42.** Studies have shown that if a truck passing along a road at night makes a noise of intensity 75 dB in someone's bedroom near the road, 80 percent of the people tested would wake up. Advertisements for popular commercial smoke alarms for homes, which should wake up almost everyone in case of fire, state their noise level is at 85 dB. About how much more intensity do those waves carry than those from the noisy truck? Do you think that should be sufficient? Discuss.

# CHAPTER 18
# ELECTRIC CHARGE

Summer, 1980, in Colorado

It was cold in the shadows of Eldorado Canyon. Paul shivered. Staring at the huge rock above his head, he visualized his route, and when Jim gave the signal, Paul began to climb. Eight consecutive lengths, or pitches, of their 150-foot rope and they would be at the top. After three pitches, he felt the first warm rays of sunshine and noticed some clouds rolling in. At the end of the fifth pitch, he waited on a narrow ledge; it was Jim's turn to climb again. There wasn't much room, and Paul wondered if they could both fit even their feet there. The climber's guidebook had accurately described this place as a ''spectacularly exposed airy perch'': One step over and it was 700 feet straight down. As Jim neared the ledge, the rain began. The climbers pressed their backs against the cold rock wall and waited for it to stop.

The shower passed, and Paul was soon leading the last pitch, following an age-old crack in the rock's surface. His mouth was dry. This part was difficult, and the wet rock was slippery. With a final, determined, arm-straining move, he was up on top. In a burst of exhilaration, Paul gave a hoot and raised his arm over his head. As he did, *snap!* A row of bright arcs some 2 inches long stung his upstretched fingertips. In one unbroken motion, he pulled his arm back and climbed down from the top of the pinnacle, fast. Jim's turn to reach the summit would have to wait for another day. Later, Paul's friends joked about the ''world's smallest bolts of lightning'' and called him an ''enlightened'' climber.

The long electrical sparks at Paul's fingertips were streamers, a severe form of the colorful phenomenon called *St. Elmo's fire*. This electrical discharge was known hundreds of years ago to sailors who watched green, blue, or even violet ''fire'' crackle along the tall masts of their wooden ships at night. One of Columbus's sailors wrote of the ''ghostly flame which danced among our sails.'' Today airplane pilots sometimes see St. Elmo's fire dancing on the wings and even over the windshields of their aircraft at night.

If the hair on your arms ever tingles and stands up, or if you see St. Elmo's fire or streamers, beware! They may be signals that lightning is close at hand. ■

## Charges

Shred some tiny pieces of tissue or newspaper and lay them on your desk. Bring a plastic ballpoint pen close to them. Notice anything happening? Now rub one end of the pen vigorously against your clothing and wave it just over the bits of paper. They will momentarily fly up to the pen, defying gravity. Ordinarily, if you press an inflated party balloon against a wall, it won't stick there. But stroke one side of the balloon against your clothing, hold that side against a wall, and gently let it go. If the relative humidity of the air isn't too high, the balloon will most likely cling there for quite some time.

Do these things, and you join the company of the ancient Greeks who rubbed pieces of amber* (the yellow-brown fossilized resin used for jewelry) and wrote about its powers of attraction for small bits of matter. Of course, those scraps of paper and the balloon don't really defy gravity. Another force, stronger than gravity, is at work. This attraction comes from a force that we call the *electric force*. The electric force doesn't depend on the mass of the paper, pen, or balloon but on a different property of matter called *charge*. Usually, the things around you are not charged, and we say they are neutral (or uncharged). The pen and the balloon became attractive only because the friction from rubbing "charged" their surfaces.

**TABLE 18–1**

| **(+) Positive Charge** |
|---|
| Rabbit's fur |
| Glass |
| Wool |
| Quartz |
| Cat's fur |
| Silk |
| Your skin |
| Cotton |
| Wood |
| Amber |
| Rubber |
| Hard plastic (combs or ballpoint pens) |
| India rubber |
| **(−) Negative Charge** |

Long before the days of plastic pens and rubber balloons, investigators found that charges, once summoned by friction, could be transferred from one material to another. For example, a glass rod rubbed with a silk cloth becomes charged, and if it is pressed to a tiny scrap of paper, the paper takes charge from the rod. A hard rubber rod, too, becomes charged if it is stroked with cat's fur, and small bits of paper will take some of its charge in the same way. Those experimenters discovered that two scraps of paper, each charged by a hard rubber rod, *repelled* one another, as did two bits of paper each charged by a glass rod; but if a bit of paper charged by rubber was brought close to one charged by glass, they *attracted* each other. From these kinds of observations, they deduced that there are *two* types of charge and that *like types of charge repel while unlikes attract*.

Today we know that any material can become charged to some degree by bringing it into firm contact with a different material. (Rubbing helps by bringing more of the surfaces together.) Look at the list in Table 18–1, and try some of these simple experiments. What is charge? No one knows. Charge, like mass, simply names a property of

*The root word *electr* comes from the Latin word *electrum*, which means *amber*. Today *static electricity* means charge that doesn't move; *electricity* most often refers to moving charge.

matter. But we *do* know how charge behaves. The two types of charge are fundamental properties of electrons and protons, and these particles are the carriers of charge in normal matter.*

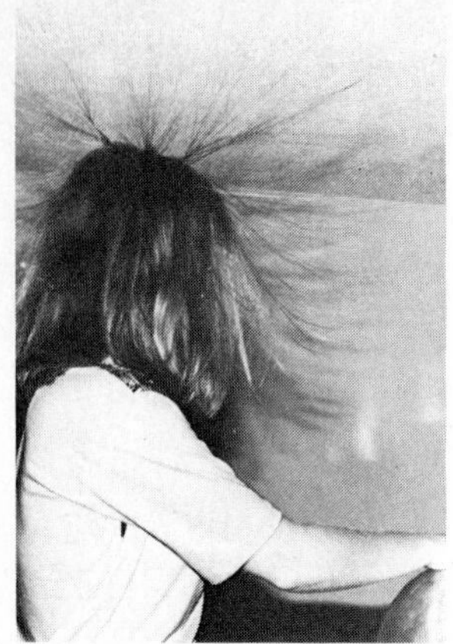

**FIGURE 18–1**
This girl's hair is experiencing the effect of like charges repelling one another. The apparatus she is touching is a Van de Graaff generator. It separates electrical charges by means of friction and deposits them on a metal dome. When the girl touches that dome, some of the charges move onto her body, and since like charges repel, they spread out. Much of the charge runs onto the strands of her hair, which lift and spread apart because of the repulsive force between the like charges.

In Benjamin Franklin's time, public demonstrations with charge became popular. Enthusiasts built huge spinning wheels that rubbed continuously against a surface, producing so much charge it would arc through the air to escape (because like charges repel one another). You can do this on a smaller scale by shuffling across a nylon carpet on a dry day and touching a metal doorknob. The sparks you see and hear and feel are the same kind of thing. Such electrification displays were seen by most as mere parlor tricks, but people like Ben Franklin gave electrical phenomena more attention. He knew that rubbing two different substances together produces both types of charge—the silk cloth gets the opposite charge from the glass rod—and, moreover, when like amounts of oppositely charged matter come together and touch, the charges disappear; the matter is once again neutral. Because of observations like these, Franklin decided that the charges were simply an unseen, movable fluid that could be drawn from normal matter by friction. Accordingly, any body with an excess of this fluid was charged one way and any matter with a deficiency was charged the other way. With this in mind, he decided to give the symbol (+) to the fluid and (−) to the deficiency; equal amounts of (+) and (−) could cancel if all the fluid flowed back to the deficient material, leaving the matter once again in a normal, uncharged state.

There was only one thing left to decide, and that was quite a riddle. Which type of charge was the plus, the unseen fluid? The physical appearance of the material didn't change when it became charged, so it didn't seem to matter. Franklin made an arbitrary choice. He assigned the type of charge on rubbed glass to be positive and that on a rubbed rubber rod to be negative. His idea was right, but his choices were wrong. It is the rubber that has the excess fluid (today we know it has the transferred electrons) and the glass that has the deficiency. In spite of this, the convention persists to this day, which is why we call electrons negative and protons positive.

The negative charge on a piece of rubbed amber or rubber is due to an excess of electrons. The cloth or fur that rubbed those substances is left with fewer than normal electrons; it is positive. Likewise, the positive charge on glass rubbed with silk means that electrons are missing from the glass, having been taken up by the silk cloth.

---

*The numbers of electrons and protons in ordinary matter are equal; for a review of the facts about atoms and their charges, see Chapter 9.

## ELECTRON AFFINITY

Amber rubbed with wool always gets a negative charge, and glass rubbed with silk always comes away positive. The reason has to do with the structure of molecules and atoms. Each species of atom holds onto its outermost electrons with a different strength. We say they have different **electron affinities.** Likewise, molecules hold onto their outermost electrons with different strengths. Solids, collections of atoms or molecules bonded strongly together, also have different affinities for their least tightly held electrons. If an atom or molecule holds onto its outermost (hence most loosely held) electrons well, it has a large electron affinity. If it loses its outermost electrons relatively easily, it has a small electron affinity. When two materials come into contact, the one with the greater electron affinity comes away with extra electrons; thus it has a negative charge. The material that is left with fewer electrons because of its smaller electron affinity is positively charged.

**FIGURE 18–2**
Benjamin Franklin could only guess which of two materials electrified by rubbing together had the excess of electrical fluid.

## The Electric Force: Coulomb's Law

In 1785, when Ben Franklin was 79 years old, Charles Coulomb, a Frenchman, managed to show that the electric force between two charges is consistent with a law that looks in form a lot like the law of gravity. This is **Coulomb's law*:**

$$F = K\frac{q_1 q_2}{d^2}$$

where $d$ is the distance in meters between the quantities of charge $q_1$ and $q_2$. The unit of charge is a **coulomb** (C), and by definition it is equal to the charge of about $6 \times 10^{18}$ electrons. $K$ is a constant that is found by experiment, just like $G$ in the law of gravity. The force points along a straight line drawn through the locations of the charges $q_1$ and $q_2$. If both charges have the same sign, the force repels them, but if the charges have opposite signs, the force between them is attractive. The value of $K$ is about $9 \times 10^9$ newton·meter$^2$/coulomb$^2$. For numerical comparisons of the electric force and the force of gravity, see ''Calculations.'' The electric force is amazingly strong compared to gravity; for example, the electric force between an electron and a proton is about $10^{40}$ times as great as the gravitational force between them (see Example 2). Coulombs are a large unit of charge when compared to the charge on an electron or a proton. But coulombs relate easily to electrical devices and home electrical use, where newtons are appropriate for force and watts for power, and where we are never concerned with single electron motions, as we'll see in the next few chapters.

**There are . . . Agents in Nature able to make the Particles of Bodies stick together by very strong Attractions. And it is the Business of experimental Philosophy to find them out.**

Isaac Newton, about half a century before Coulomb's experiments.

## What Being Neutral Really Means

The matter around us normally is electrically neutral, exerting no net electric force. But matter is made of atoms, and all atoms contain charged protons and electrons, so we should recall what being neutral

*Historians of science have claimed that Coulomb's experiment wasn't accurate enough to prove the law now called Coulomb's law, and that the fame should belong to others who verified it at later dates. Nevertheless, it is still called Coulomb's law.

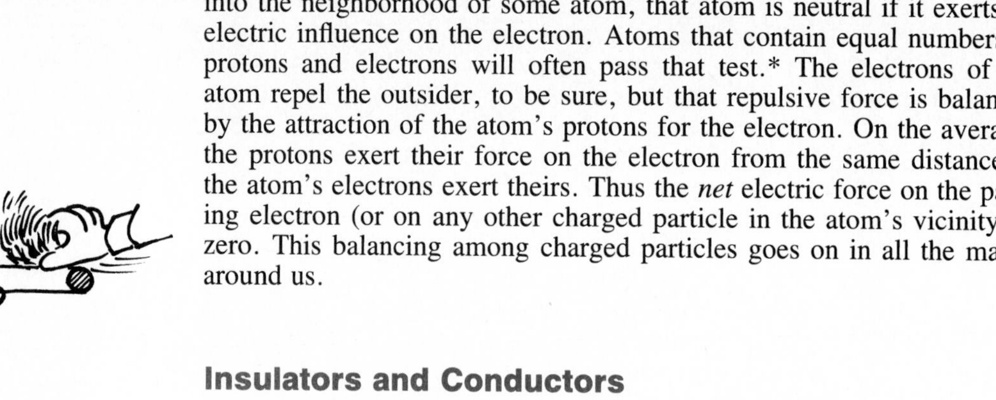


means. From the point of view of an unattached electron that wanders into the neighborhood of some atom, that atom is neutral if it exerts no electric influence on the electron. Atoms that contain equal numbers of protons and electrons will often pass that test.* The electrons of the atom repel the outsider, to be sure, but that repulsive force is balanced by the attraction of the atom's protons for the electron. On the average, the protons exert their force on the electron from the same distance as the atom's electrons exert theirs. Thus the *net* electric force on the passing electron (or on any other charged particle in the atom's vicinity) is zero. This balancing among charged particles goes on in all the matter around us.

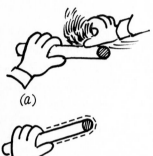

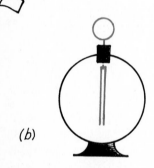

## Insulators and Conductors

If you rub a plastic pen on one end, the other end won't pick up any scraps of paper. The charge more or less stays at the point you stroked. It's the same for a hard rubber rod or a glass tube. The charge stays put unless something else touches the charged area. Because any charge on them remains pretty much immobile, glass, rubber, and plastic are called **electrical insulators.** In fact, most solid materials are insulators. Since a charge on such a material doesn't readily move about, we say it is a **static** charge.

(b)

Materials such as aluminum and other metals are **conductors.** Excess charges readily move on (or in) conductors. To demonstrate this point, carefully peel the thin aluminum foil from the wrapper of a stick of gum, cut the foil lengthwise into two identical strips, and hang the strips side by side from a paper clip or a clothespin so they touch at their top ends. Then charge a glass rod with vigorous strokes of a silk cloth and touch it to one strip of the foil. As you do, the dangling leaves push slightly apart. The charge one metal leaf got from the glass spreads out, and some of the charge moves onto the opposite leaf. Because both now have excess positive charge, they repel each other and separate. A two-leaf aluminum foil arrangement like this is a crude **electroscope** (see Fig. 18–3). An electroscope of gold foil is a simple and reliable charge detector that once was commonly used to estimate quantities of charge. The more charge the gold foil received, the greater the repulsion between the leaves and the farther they separated. (Very thin gold foil is even lighter than aluminum foil and separates more easily.)

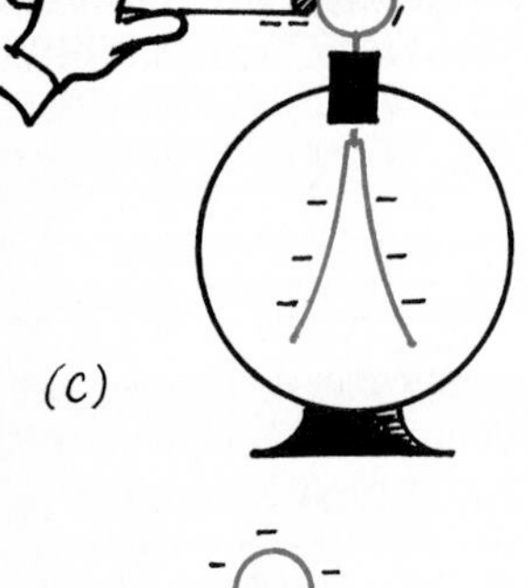

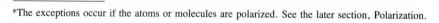

*The exceptions occur if the atoms or molecules are polarized. See the later section, Polarization.

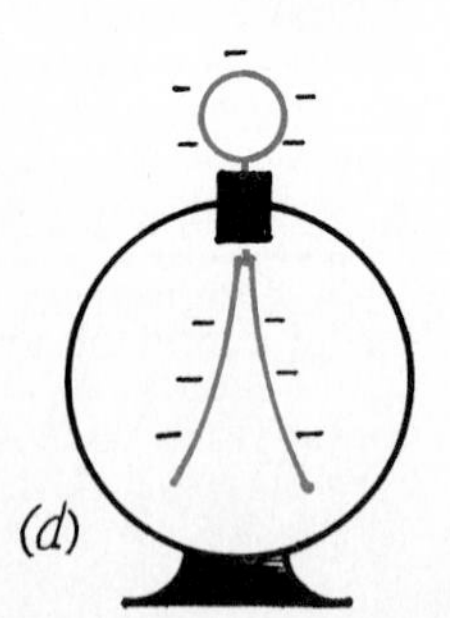


**FIGURE 18–3**
The steps in the charging of an electroscope with negative charge. Charge is separated *(a),* then a negatively charged rod is brought to the electroscope *(b),* where contact is made with the conducting knob. Electrons transferred to the metal knob spread to all regions of the conducting metal, and the like charges on the thin conducting leaves push them apart *(c).* The charged rod is withdrawn, and the electroscope is left with a net negative charge *(d).*

**TABLE 18–2**

| FAMILIAR CONDUCTORS* | FAMILIAR INSULATORS* |
|---|---|
| Silver | Amber |
| Copper | Hard rubber |
| Gold | Nylon |
| Aluminum | Porcelain |
| Brass | Beeswax |
| Iron | Glass and wood |
| Lead | Shellac |
| Mercury | Very pure water |
| Graphite | Air |
| In addition, water is a good-to-excellent conductor when it contains dissolved salts (or acids or bases). | |

*Each substance in the list is a slightly better conductor or insulator than the one below it.

If a negative charge is given to an electroscope by a charged rubber rod, the extra electrons have no trouble moving over the metal, which has free electrons of its own, not bound to any of the metal atoms. The extra electrons from the rubber rod repel each other and move as far apart as possible. Even if they were placed inside the metal, they would flee to its surface (but not leave it) because of their mutual repulsion. Thus, *any net charge on a metal resides on its surface.*

When an electroscope is given a positive charge by a stroked glass rod, its leaves separate just as if free positive charges had repelled one another and spread over the leaves. But that doesn't happen. Even when the metal has a net positive charge, it is the electrons that do the moving. The protons never leave the nuclei at the center of the atoms, and the bonded atoms don't move either. Here is why it appears that the positive charges move. When the positively charged glass rod touches the metal, it takes some electrons from that point on the metal's surface, leaving that area positive, or with a deficiency of electrons. Other free electrons in the metal are attracted to the positive area, and some move in there. When they make this shift, they leave their own areas slightly positive and the two leaves spread apart. The result is that the *positive area seems to spread out.* The effect is the same as if *free* positive charges were present and repelled each other to every point on the metal's surface.

Liquids and gases, too, can be either conductors or insulators. To be a conductor, a fluid must have ions, charged atoms or molecules or even free electrons, to carry charge along. Most of the water in our environment is a good conductor of electric charge because it contains dissolved minerals and salts that have broken into ions. For example, if you stir a cup of water with a finger, salt particles from your skin dissolve into the water. Salt (NaCl) breaks into sodium ions ($Na^+$) and chlorine ions ($Cl^-$) as it dissolves. Salt water is a good conductor. On the contrary, chemically pure water has relatively few ions, making it a good insulator. Likewise, air is an insulator, since it also has very few ions compared to the number of neutral molecules.

When there is a lot of moisture in the air, any solid object exposed to the air usually has a thin coating of water molecules on it. That water

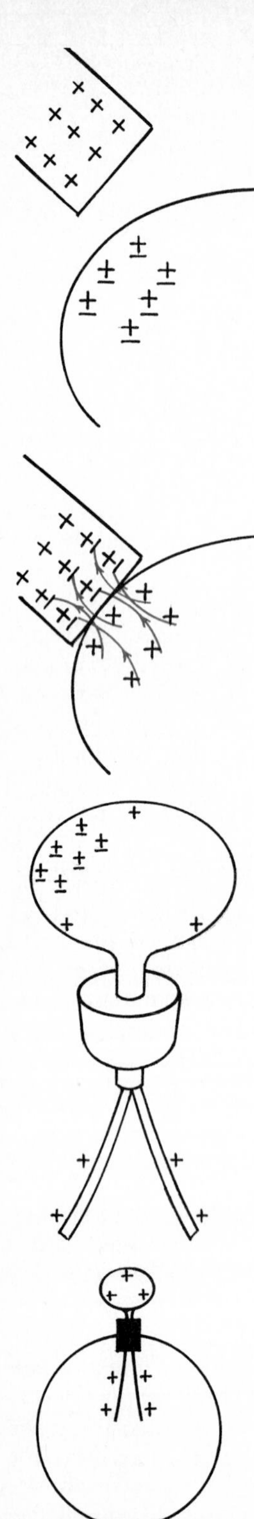

**FIGURE 18–4**
This is how an electroscope can get a net positive charge: A positively charged rod is brought into contact with the electroscope. Some electrons jump to the positive rod, leaving the electroscope with a net positive charge. Other mobile electrons in the electroscope are then attracted to the positive areas and move to them; the areas they leave then become positively charged. At this point, it appears, however, as if the net positive charge had repelled itself to spread over the metal. But in fact, the positive charge, belonging to the protons in the atomic nuclei, never moves. The electrons do all the moving around. The electroscope leaves are left with a net positive charge and spread apart.

dissolves ions from the solid surface and turns the liquid layer into a good conductor. A net charge on that surface can spread out and leak off to the ground or even into the air. That is why a rubbed balloon won't stick to the wall for long on a warm and humid day. On the other hand, that's also why after a stroll across a rug you summon sparks at a metal doorknob more easily in winter than in summer. Colder air can't hold nearly as much water vapor as warm air, so the surfaces in a cold room generally have less moisture on them. To get the rubbing experiments to work on a humid day, first use a hairdryer or a heat lamp to evaporate the unseen layers of wetness from the solid surfaces.

## Induction

An electroscope has a metal knob on top that is connected to the fragile gold leaves hanging in a glass case below. If someone brings a charged rod of glass or hard rubber near the knob of an uncharged electroscope, the leaves spread apart as the rod approaches before the rod can touch the knob to transfer charge. Suppose it is a rubber rod with a negative charge. The movable electrons in the metal knob of the electroscope are repelled by the electric force from that negative charge as the rod draws near. The electrons are pushed away by the negative charge on the rod and move down onto the leaves, making them both negative (which causes them to spread apart) while leaving the knob itself positive. If the charged rod is then taken away without touching the metal knob, the gold leaves fall together again as the electrons return to the positive knob.

The rod's charge made the charges within the electroscope move, so we say the external charge **induces** the positive and negative areas on the electroscope. *Induced charges in matter are brought about by the presence of an electric force*. The net charge on the electroscope is still zero, however, since no charge has been added or subtracted. The electroscope is still neutral overall, despite the separation of the charge.

A simple trick lets an experimenter give an electroscope a net charge by induction. The first step is to make those movable electrons go not to the gold foil but to somewhere outside the electroscope. To do this, touch a conducting wire to the electroscope, one that leads away

**I understand everything in this book except what is meant by an electrically charged body.**

A remark made by a philosopher to mathematician Henri Poincaré, 1885.

to some large conductor such as a metal cabinet or desk. Then bring up the negative rod, but don't touch it to the electroscope. Once again the electrons flee from the electroscope knob, but this time most of them run all the way onto the large conductor. If you remove the negative rubber rod at this instant, the electrons come back into the positive electroscope. But if you disconnect the wire *before* you take away the negative rod, the displaced electrons cannot return to the electroscope as the rod retreats. That leaves the electroscope with a net positive charge. This process is called **charging by induction** (Fig. 18–5).

**FIGURE 18–5**
The steps in giving an electroscope a net charge by induction. *(a)* A charge is brought close to the electroscope's neutral knob, whereupon it repels like charges in the metal. *(b)* Some of those mobile charges move onto the electroscope's leaves, causing them to spread apart. *(c)* When wire that is connected to a large conductor touches the metal, those charges repel even farther by going to the larger conductor, such as a metal desk or even the ground. (See the next section, Grounding.) *(d)* When contact with the conducting wire is broken, the electroscope is left with a net charge of the opposite sign, even though no contact with the charged rod was made. Then, when the charged rod is removed, the net charge spreads over the electroscope, and *(e)* the leaves repel.

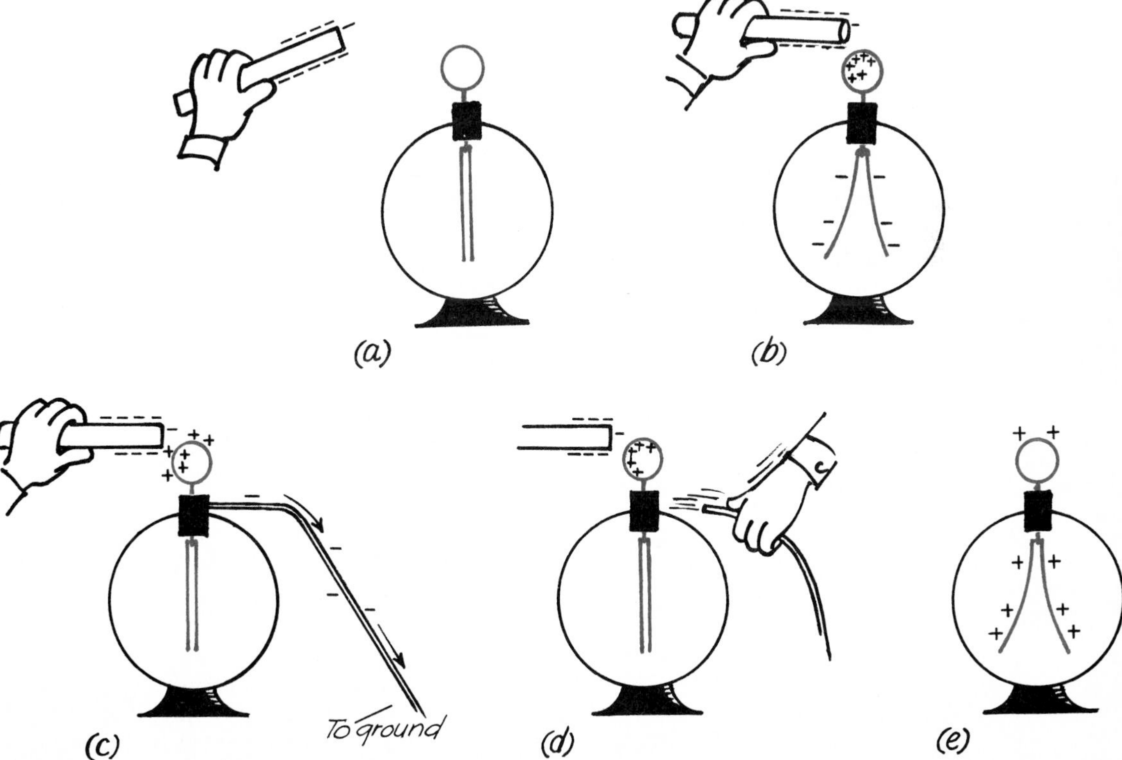

**The Parts of all homogeneal hard Bodies which fully touch one another, stick together very strongly. And for explaining how this may be, some have invented hooked Atoms, which is begging the Question; . . . I had rather infer from their Cohesion, that their Particles attract one another by some Force, which in immediate Contact is exceedingly strong, at small distances performs the chymical Operations . . . and reaches not far from the Particles with any sensible Effect.**

Isaac Newton, on bonds in neutral matter.

When a comb you've run through your hair (or a plastic pen you've rubbed on your clothing) picks up tiny scraps of paper, it is because of induction. The negative charge on the comb exerts an electric force on the charges in the scraps of paper. Paper normally picks up moisture from the air, making each scrap a conductor. Negative charges run to the far side of a scrap (away from the comb), leaving the side nearest to the comb positive. Because the positive side is closer to the comb, its attraction to the comb is stronger than the far side's repulsion (according to Coulomb's law). The bit of paper flies to the comb. Once the scrap touches the comb, however, some of the comb's negative charge can transfer to the paper. If this happens, the paper suddenly has a net negative charge and is just as suddenly pushed away from the comb.

## Grounding

To charge an electroscope by induction, a large conductor was used to hold the induced charge. Though a metal cabinet works fine to take on the charge induced in an electroscope, often a much larger quantity of charge needs to be disposed of, such as when a lightning strike brings charge to earth. That calls for a larger conductor, and all around us is the largest available conductor—the earth. Below our feet lie soil, minerals, and organic material, a vast collection of damp matter capable of holding an immense quantity of charge. A metal wire or cable that leads underground or even a metal water pipe that runs into the ground can serve to drain off charge from a faulty household electric connection, from a lightning rod, or even the induced charge on an electroscope. Furnishing such a conducting path to the ground is called **grounding.** In the next chapter we will see why ordinary appliances should be grounded to the huge conductor beneath your home.

## Polarization

When either a negative or a positive charge is brought close to any metal object, electrons on the surface move in response. But in an insulator the atoms hold onto their electrons. Nevertheless, the electrons do respond: They shift a little bit toward an external positive charge or away from an external negative charge. This shifting causes the atoms to become ever so slightly elongated in shape. As seen in Fig. 18–6, *the surfaces of the insulator become charged,* but the induced charge doesn't move around as it would on the surface of a conductor. When an atom or a molecule has a negatively charged side and a positively charged side, it is said to be **polarized.**

Some molecules, such as water molecules, are *naturally* polarized. When the two hydrogen atoms join with the oxygen atom to make a water molecule, the hydrogen atoms become somewhat positive because the oxygen atom has a large affinity for electrons, and it pulls their electrons toward itself. The oxygen end of the molecule becomes rela-

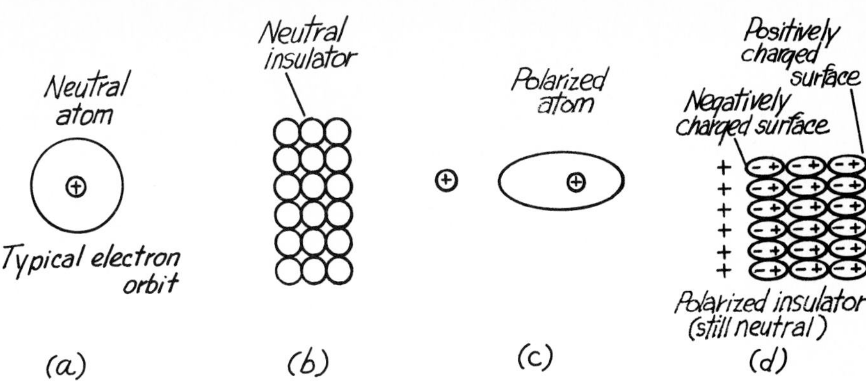


**FIGURE 18–6**
*(a)* Neutral atoms tend to have symmetrical charge distributions. That is, the average position of all the electrons of the atom is at the center, which is also the center of the positive charge in the atom's nucleus. *(b)* An insulating material with no free electrons ordinarily has no displacement of charge in any direction. *(c)* However, if a charge is brought close to a neutral atom, its electrons shift in response to the external attraction or repulsion, and the average position of the electrons is displaced. This leaves one side of an atom relatively negative and the other relatively positive. The atom is then polarized. *(d)* Likewise, the atoms of an insulating material are collectively polarized if charge is brought close by, leaving one surface of the material relatively positive and the opposite surface relatively negative.

**FIGURE 18–7**
The strong chemical bonds between the oxygen atom and the two hydrogen atoms that combine to make a water molecule cause that molecule to be permanently polarized. The oxygen atom has strong affinity for electrons, causing its region of the molecule to be relatively negative and leaving the hydrogen regions of the molecule relatively positive.

tively negative at the same time. Overall, the water molecule is still neutral, but its natural polarization plays a large role in determining which substances dissolve in water and which do not. The water molecule is called a **polar molecule,** whereas a molecule with no strongly charged areas is *nonpolar*.

## The Electric Force Field

The earth's huge mass gravitationally attracts the moon (and vice versa) through the emptiness between them. The Apollo capsules, coasting through the gap between earth and moon, felt pulls from them both at every point. To describe this influence, we say that any mass changes the properties of the space around it so that any other mass is attracted to it. This condition of the space around a mass is called a *gravitational force field*.

A proton repels another proton and attracts any electron, though nothing tangible bridges the distance between them. Charges, like mass, change the properties of space, and this stress too is a force field, the *electric force field*. As an illustration, suppose a quantity of positive charge is spread evenly over a small glass sphere. Coulomb's law tells

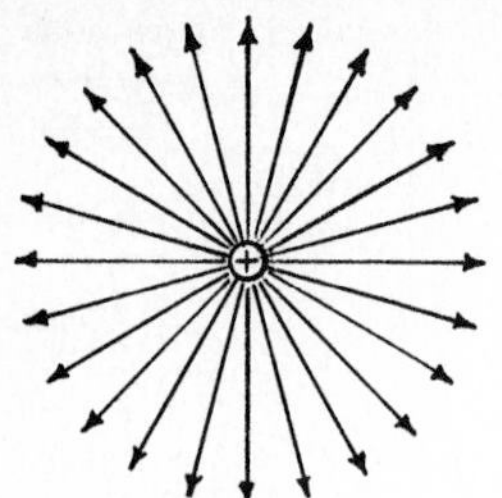

FIGURE 18–8
A positive charge has an electric force field that is represented by lines radiating in every direction. The field lines around a negative charge of equal value are the same except the arrows point inward rather than outward.

us the electric force field of that positive charge on the sphere acts this way: If an external positive charge $q$ approaches, it is pushed straight away from the sphere. A charge twice as large at the same place gets a push that is twice as big. To make things simpler, it makes sense to define an electric field, $\mathscr{E}$, where $\mathscr{E}$ gives the force per unit of charge.

$$\text{electric field} = \frac{\text{electric force on charge } q}{\text{magnitude of charge } q}$$

$$\mathscr{E} = \frac{F}{q}$$

where $\mathscr{E}$ is measured in newtons/coulomb and $F$ is the net force (as found from Coulomb's law). Just as the electric force $F$ is a vector, so is $\mathscr{E}$. Here's how the electric field makes things simpler: If you know the electric field in some region of space and a particle of charge $q$ is introduced, the electric force on that particle is the value of the electric field times the amount of charge on the particle. $F = \mathscr{E}q$

To visualize the electric field around an object such as the charged glass sphere, we can draw lines that show the directions of the force on another positive charge brought into the space near that sphere (Fig. 18–8). These imaginary lines are called **field lines.** The field lines from a uniformly charged sphere radiate straight away from the surface of the sphere. This means that a positive charge near the sphere would get a push outward along the field line passing through the positive charge's position. The arrows of the field lines pointing outward show the direction of the electric field and thus the direction of the force on a positively charged particle. (Electrons near the sphere would experience a force in the other direction, or toward the charged sphere.) Where the field lines are close together (near the sphere), the force is strong. Where the lines are far apart, the force is smaller. Note that these field lines aren't real. They are just imaginary aids to let us visualize the stress in space around the charges, to let us see where the field is strong or weak, and also the direction of the push or pull on another charged particle.

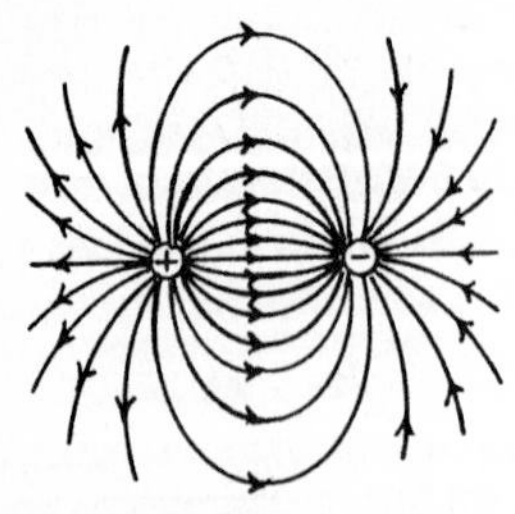

FIGURE 18–9
If a negative charge is brought close to a positive charge, their fields add like vectors since the forces they exert are vectors. The *net* electric field around equal but opposite charges is portrayed here. This arrangement of charges is called a dipole. A dipole's field occurs often in nature.

Figure 18–9 shows the field lines around a pair of positive and negative charges of the same magnitude that lie some fixed distance apart. A (+) and (−) charge arrangement like this is called a **dipole.** The electric field of a polar molecule has much the same shape as the electric field around a dipole.

## Electrical Shielding

We'll see in later chapters that it is possible to make an electric field so strong it will tear the electrons away from the nuclei of atoms; any matter immersed in that field would be ionized. But most electric fields don't pack such wallops. In fact, even the strong electric field between a thundercloud and earth that gives rise to a bolt of lightning isn't strong enough to cause a net electric field *inside* a metal object as large as a car, which is why a car is a good place to be during an electrical storm. As we've seen, any extra charge on a metal object stays on its surface, as does an induced charge. When a metal object becomes immersed in an electric field from some external source, an amazing reaction takes place. The marvelously mobile electrons in the metal respond at once. Electrons scoot around on the surface of the metal, jamming up in some places and vacating from others in response to the external field. As they move, their own individual fields go with them. Those charges stop moving when their own electric fields add to cancel the part of the electric field from the external source that could make them move. That is, they stop accelerating across the surface when there is no net electric field *tangent* to the surface. Under the surface of the metal, the electric field from the external source is canceled *completely* by the induced charges on the surface. If that weren't so, more of the metal's movable electrons would move to the appropriate place on the surface; if there is a net field inside a conductor, the free charges move.

In effect, then, the safest place to be when there are large electric fields around is inside a conductor. Even with its glass windows, a metal car frame shields the passenger area from penetration by an electric field. Maybe you've noticed metal boxes or cannisters covering some of the parts in a stereo set or a radio. Those metal covers serve as electrical shields to keep out unwanted electric fields that could interfere with performance. To guard against damage from lightning, some computers are housed in rooms that are entirely lined with metal, including the floor and ceiling, with the metal shield grounded to the earth.

## Potential Energy from the Electric Force

A bowling ball on a closet shelf has potential energy. If it rolls off, it gains kinetic energy equal to its loss in potential energy and is able to do work on something in its way. The ball has potential energy due to the gravitational field of earth. We say the ball has *gravitational potential energy*. A charged particle that is in an electric field also has potential energy. The electric field exerts a force on the particle, and if it is free to move, it gains kinetic energy. We say the charged particle has **electrical potential energy** when it is in an electric field.

The amount of potential energy a charge has because of an electric field $\mathscr{E}$ depends only on its position in the field and the size of $\mathscr{E}$. If the charge moves from one place to another and the electric field does work on it, the *difference* in the potential energy of the charge because of its change in position is equal to the work done on it by the electric field, just as for gravity. To describe such changes in electrical potential energy, we use a quantity called the **electric potential,** which is the potential energy per unit of charge due to the electric field. The units of electric potential are joules per coulomb, which are called **volts** (V).

When a charged particle in an electric field moves from a point that has a given electric potential to a second point that has a different electric potential, work is done on the particle by the electric field.

**work done on charge = change in electric potential energy = difference in electric potential between two points (in volts) × charge (in coulombs)**

That is, the energy given to the charge is

$$\text{energy} = V \times q$$

**There is an immensity of facts which justify us in believing that the atoms of matter are in some way endowed or associated with electrical power to which they owe their most striking qualities . . . .**

Michael Faraday, about half a century before the discoveries of the electron and the nucleus.

where $V$ is the difference in electric potential between the two points and $q$ is the charge in coulombs. $V$ is often called the *potential difference* or the *voltage difference* between the two points in the electric field. For numerical examples using this equation, see "Calculations."

When you comb your hair, the comb takes electrons from your hair. The negative charge on the comb can induce charges on a small scrap of paper, causing it to fly to the comb. The difference between the electric potential of the comb and the electric potential of the bit of paper is about 5000 volts. And just as the force of gravity does work on a bowling ball that rolls off a shelf, the electric force does work on the paper as it moves to the comb.

Shuffle your feet across a nylon rug on a cold day, and you become positively charged. The difference between your electric potential and that of a metal doorknob that gives you a shock is at least 5000 volts; if a spark jumps between the doorknob and your hand, the voltage difference is perhaps 10,000 to 15,000 volts. Normally, dry air is a good insulator. However, if a potential difference of 15,000 volts occurs at points only a centimeter or two apart in the air, electrons can leave one point and smash into air molecules hard enough to ionize them. That ionized air offers a good conducting path that huge numbers of electrons can then follow. The spark you feel and hear is a miniature lightning bolt.

## Very Large Potential Differences: Lightning

Thunderstorms are rare in some areas, but not in central Florida. The lightning belt there takes first place in the United States for days per year when thunder is heard: 90. Worldwide, however, there are an average of about 100 lightning bolts every second. Lightning comes from a storm cloud where separated charges abound because of friction within rising and falling parcels of air, rain, and hail. If a lot of negative charge accumulates in one area and the corresponding positive charge collects in another area, a large electric field results, along with a great difference in electric potential energy. A bolt of lightning occurs when electrons jump from a negative area to a positive area.* After the charge moves, the positive and negative areas are less charged than before. The electric potential between the areas is less, and the electric field is smaller. Such a bolt is called an **electric discharge** for that reason.

*St. Elmo's fire sometimes occurs just before the potential difference builds up to the point where lightning will occur. Pilots can hear a static growing on their radio headsets and watch the dancing lights on their aircraft just before seeing lightning.

**FIGURE 18–10**
A strong negative charge near the base of a cloud induces a positive charge on the ground beneath it. That is, mobile electrons on the ground are repelled by the negative charges in the cloud and move out, leaving the area beneath the cloud positively charged.

When lightning is mentioned, most of us think of bolts that strike the ground. But for every one that comes to earth, four others jump between clouds or between parts of the same cloud; only 20 percent of all lightning bolts come to the ground. A cloud usually becomes negatively charged near its base, and the negative charge *induces* a positive area on the ground and all conducting objects beneath the cloud. When the potential difference between the cloud and the ground is large enough, the strong electric field ionizes molecules in the air, and the bolt comes down. Just before the strike, the potential energy difference is typically 10 million to 100 million volts.

Mountaintops bring the conducting earth closer to the region of the charges in thunderstorms, and a storm cloud induces static (nonmoving) charge on the peaks. If you visit such a place and your hair stands up, make sure you're not on the top, the closest place to the cloud and the most attractive spot (because of induction) for its ultimate discharge. In fact, any tall objects bring the ground closer to the cloud's charge. Solitary trees are especially good targets for lightning, as are the tallest trees in a grove. If the weather threatens and you are caught out of doors, it is better to crouch in a field than under a tall tree. Crouch, but don't lie down, because when lightning strikes, the charge can run along the ground. If such a charge travels from your head to your feet, it delivers more energy than if it travels between two feet close together. Unless wet, your shoes are insulators and can help provide some protection from charges that move along the ground. Stay away from railroad tracks, too. The rails are often insulated from the earth by the wooden crossties. Not only can they carry a charge well, but induction makes them likely candidates for a strike.

In general, it is better to be in a house than out of doors during a thunderstorm. Most people who are struck are outside when it happens. In the house, stay away from any window with a tall tree nearby. Because the charge that accompanies a bolt moves most easily along con-

**FIGURE 18–11**
The lightning rod to Doreen's right is only a foot or so higher than the roof of the building it protects. The thick copper cable connects all the other lightning rods on the roof and terminates on a thick copper rod that is driven into the ground to a depth of 10 feet or more.

ductors, stay away from metal appliances and out of the bathtub; the metal plumbing of a house has a vent up to the roof, and often it is a metal pipe. Cars are safe, as mentioned before, and so are metal airplanes, and all metal buildings. Buildings with lightning rods are safe as well. Lightning rods, invented by Ben Franklin, should be a foot or more taller than any chimney or vent on the house, and they should be connected by a conducting cable to a copper rod that goes at least 10 feet into the (moist) ground, which is a good conductor. The charged cloud overhead induces charge in the tip of the brass or copper rod. If lightning strikes the area of the roof, the rod provides the easiest path to the ground, and the charge runs along that path, averting damage to the home.

## CALCULATIONS

EXAMPLE 1: A single electron has an amount of charge equal to $1.60 \times 10^{-19}$ coulombs. But it is a negative charge, so we write it as $q_{\text{electron}} = -1.60 \times 10^{-19}$ coulombs. The proton has the same amount of charge but with a positive sign. In a hydrogen atom the solitary electron is found at an average distance of about $0.50 \times 10^{-10}$ meters from the single proton in the nucleus. **Calculate the electric force between them.**

$$F = K\frac{q_{\text{electron}}\, q_{\text{proton}}}{d^2}$$

$$= \left(9.0 \times 10^9 \frac{\text{N-m}^2}{\text{C}^2}\right) \times \frac{-1.6 \times 10^{-19}\,\text{C} \times 1.6 \times 10^{-19}\,\text{C}}{(0.5 \times 10^{-10}\text{m})^2}$$

$$= \mathbf{-9.2 \times 10^{-8}\ N}$$

The negative sign here means the particles are drawn together, whereas a positive sign would mean the force is repulsive. (Notice that two negative charges would have a positive value for the electric force, since two negative quantities multiplied together give a positive number. Like charges repel.)

---

EXAMPLE 2: (a) **Calculate the gravitational attraction between an electron and a proton in a hydrogen atom as in Example 1.** (b) Compare the force of gravity between them to the electric force between them; that is, **form the ratio $F_e/F_g$.** The mass of an electron is $9.11 \times 10^{-31}$ kilograms, the mass of the proton is $1.67 \times 10^{-27}$ kilograms, and $G$ is $6.67 \times 10^{-11}$ newton-meter$^2$/kilogram$^2$.

(a)

$$F_g = G\frac{m_1 m_2}{d^2}$$

$$= \left(6.67 \times 10^{-11}\frac{\text{N-m}^2}{\text{kg}^2}\right) \times \frac{9.11 \times 10^{-31}\text{ kg} \times 1.67 \times 10^{-27}\text{ kg}}{(0.5 \times 10^{-10}\text{ m})^2}$$

$$= \mathbf{4.05 \times 10^{-47}\ N}$$

(b) Using $F_e$ from Example 1, we have

$$\frac{F_e}{F_g} = \frac{9.2 \times 10^{-8}\text{ N}}{4.05 \times 10^{-47}\text{ N}} = \mathbf{2.27 \times 10^{39}}$$

To the nearest power of 10, the electric force between an electron and a proton in a hydrogen atom is some $10^{39}$ times as great as the gravitational force between them. The strength of the electric force is why atoms are so tough. Electrons and protons ''hang together'' tenaciously.

---

EXAMPLE 3: **Calculate the electric field, $\mathcal{E}$, at a distance of 1 centimeter from a positive charge of $10^{-6}$ coulombs that's on the very tip of a comb's tooth.**

From Coulomb's law, $F = K(q_{\text{comb}}q)/d^2$, gives the force on a charge of amount $q$ at a distance $d$ from the point of the comb's tooth. The definition of the electric field is $F/q$, the electric force per unit of charge. Just rearranging Coulomb's law to find $F/q$, we see that

$$\frac{F}{q} = K\frac{q_{\text{comb}}}{d^2} = \mathcal{E}\text{ (due to charge on comb)}$$

Using the values given, we find

$$\mathcal{E} = \left(9 \times 10^9\frac{\text{N}\cdot\text{m}^2}{\text{C}^2}\right) \times \frac{10^{-6}\text{ C}}{(0.01\text{ m})^2} = \mathbf{9 \times 10^7\ \frac{N}{C}}$$

---

EXAMPLE 4: As we'll see in Chapter 19, a battery uses chemical reactions to maintain a potential difference $V$ between two points, its terminals. Most car batteries are 12-volt batteries. If a total of 4 coulombs of charge leaves one terminal of a 12-volt battery and flows through the conductors of a car's electric system to return to the other terminal, **how much energy did the battery give to that charge?**

$$\begin{aligned}\text{energy} &= \text{voltage difference (in volts)} \times \text{charge (in coulombs)}\\ &= 12\text{ V} \times 4\text{ C} = 48\text{ V}\cdot\text{C}\\ &= 48\,\frac{\text{J}}{\text{C}} \times \text{coulomb} = \mathbf{48\ J}\end{aligned}$$

---

EXAMPLE 5: Professor Phate noticed Coulomb's law in a physics book. "Aha!" thought Phate, who was in search of a levitation device, "If I eliminate some electrons in my shoes and some more in an insulator beneath my feet, the positive areas will repel and I'll be uplifted." In no time he had invented a machine to dispose of one electron per atom, a feat that leaves a kilogram of ordinary matter with a positive charge on the order of $10^7$ coulombs. Pointing his invention downward, in an instant he charged his boots (about 1 kilogram) and a like amount of wood on the floor beneath them. (a) **Estimate the instantaneous force Phate felt.** (b) **Presume those like charges somehow managed to stay together on the floor and in his boots and estimate the acceleration of Phate's takeoff, assuming his mass to be 70 kilograms.**

$$F = K\,\frac{q_{\text{boots}}\,q_{\text{floor}}}{d^2}$$

where $q_{\text{boots}} = q_{\text{floor}} = 10^7$ coulombs and $d$ should be the average distance apart of the charges in his boots and floor. Use $d = 1$ centimeter $= 0.01$ meter.

(a)

$$F_{\text{Phate}} = \left(9 \times 10^9\,\frac{\text{N}\cdot\text{m}^2}{\text{C}^2}\right) \times \frac{10^7\text{ C} \times 10^7\text{ C}}{(0.01\text{ m})^2} = \mathbf{9 \times 10^{27}\ N}$$

(b) Ignoring the force of gravity, which is only about 700 newtons, we have

$$F_{\text{Phate}} = m_{\text{Phate}}\,a_{\text{Phate}}$$

so

$$a_{\text{Phate}} = \frac{9 \times 10^{27}\text{ N}}{70\text{ kg}} = \mathbf{1.3 \times 10^{26}\ m/s^2}$$

Since $g$ is about 10 m/s$^2$,

$$a_{\text{Phate}} = \mathbf{1.3 \times 10^{25}}\boldsymbol{g}$$

Though impossible, *this example points out the tremendous strength of the electric force that holds ordinary matter together.* That same force tends to reunite charges that are separated, to keep matter electrically neutral. To be sure, $10^7$ coulombs is an enormous amount of charge. A typical lightning bolt brings only 25 coulombs of charge to earth. (Also, though today we know of reactions that can destroy electrons, those reactions also destroy an equal amount of positive charge. In every reaction known today, the *net* charge remains the same: We say that the total amount of charge is constant, or that *charge is conserved*–just as energy is.)

## REVIEW

**1.** There are only two types of charge in all matter, and they exert forces on each other. The like types of charge ________________; the unlike types ________________.

**2.** The symbol (−) represents a ________________ charge, which is an excess of ________________ in the matter.

**3.** Most matter on earth is in a neutral state; that is, it has no net charge. True or false?

**4.** Any material can become charged to some degree by bringing it into firm contact with some other material. True or false?

**5.** Give the equation for Coulomb's law and identify each variable and constant.

**6.** What is the metric unit of charge?

**7.** A thermal insulator is a material that conducts heat poorly. Define an electrical insulator as discussed in this chapter.

**8.** Define induction.

**9.** Define grounding and give an example.

**10.** Why does any net charge on a metal reside on its surface?

**11.** A charge in an electric field has potential energy. Its potential energy divided by the quantity of the charge is called the electric potential. It has units of joules per coulomb, which are called ________________.

**12.** What is the difference in electric potential between two points in an electric field called?

**13.** When a molecule has a negative side and a positive side, it is called a ________________ molecule.

## DEMONSTRATIONS

**1.** On a cold winter night, go into a dark room and quickly pull your wool or nylon sweater over your head. The snap, crackle, and pop and the flashes of light you see mean that potential differences of some 15,000 volts were generated by the friction.

**2.** If your refrigerator has a metal ice tray with a handle to release the cubes, try this. Take the tray of ice cubes into a closet or a dark room. (Better still, sit in the dark for 5 or 10 minutes to get your night vision and have someone bring you the frozen tray.) Then jerk the handle quickly; you will see sparks as the ice breaks away from the metal separators. The bonds holding the solids together are electrical in nature, and when those bonds are broken, electrons are displaced in the solids. When the negative and positive charges get back together, it's like a miniature bolt of lightning—light is given off. (Photographers who work in darkrooms see sparks if an ordinary piece of tape is ripped from a metal film canister. The explanation is the same.)

**3.** Charge a plastic pen or comb by rubbing it briskly against a dry garment. Then hold it close to a slow stream of water from a faucet. The stream is attracted by the charged object. Yet if you tried this with some rubbing alcohol, or a can of automotive oil, or even gasoline, nothing would happen. (You shouldn't do this with gasoline because of the danger of sparks.) It is the polar molecules of water that are responsible for this phenomenon. The plastic's negative charge repels the negative side of the water molecules and attracts the positive side. The tumbling water molecules respond and orient themselves with the negative side away from the plastic pen or comb. The closer, positive side of the molecules is tugged more strongly than the far side is repelled, and the stream of water moves toward the negatively charged object.

**4.** Here's a great demonstration for Halloween in a cold, dry climate. Rub a leg of a pair of nylon pantyhose briskly over a wool sweater that's held down on a table. Turn the leg over and rub the other side. If the relative humidity of the air is low, the hose expands just as if someone's leg were in it! Like charges (on both surfaces) repel.

**5.** Using a bottle, a rubber or cork stopper, a paper clip, and aluminum foil, follow the diagrams in Fig. 18–12 to make an electroscope. Then experiment with the substances in Table 18–1, as discussed in the chapter.

**FIGURE 18–12**

# EXERCISES

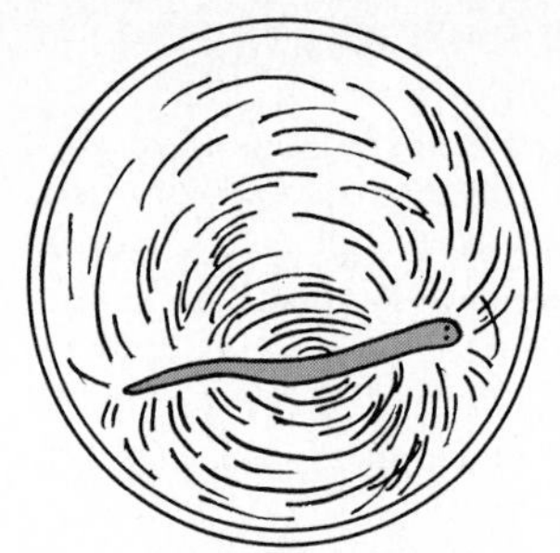

**FIGURE 18–13**
The electric field of an electric eel as sketched by Michael Faraday. He used his bare hands to estimate the intensities.

**1.** Early Germanic tribes imported glass from the Mediterranean countries and referred to it as *glesum,* their word for amber. Tell how an educated tribesman could have shown that the substance was not amber. (*Hint:* Refer to Table 18–1.)

**2.** Sparks are extremely dangerous around grain storage bins where there is a great deal of grain "dust" in the air. Explain why this is true. It is the same reason medical personnel guard against sparks in operating rooms where alcohol and ether vapors are in the air.

**3.** Explain why you are grounded when you are in a bathtub of water.

**4.** Why do you think vinyl phonograph records attract dust so avidly? Why is it impossible to clean dust from them by rubbing with a dry cloth?

**5.** When someone with long hair washes, dries, and then combs it with a plastic comb on a dry day, some of the hairs stand up and apart. What is the sign of the charge on the hair? On the comb? Consult Table 18–1.

**6.** Rubber tires get charged from friction with the road. What sign is their charge? Perhaps you have seen a gasoline truck trailing a metal chain beneath it. What purpose does the chain serve? (Modern tires for such trucks are made to be conductors of electricity.)

**7.** Describe how you could put a net charge on a metal sphere without actually touching it to something that is charged.

**8.** Over 80 percent of all cloud-to-earth lightning brings negative charge (electrons) to earth. What about the rest? The positive area of the cloud is involved in these bolts. Do you think they bring positive charge to earth? Discuss.

**9.** Lightning bolts took a great toll on tall-masted wooden sailing vessels. As soon as Ben Franklin invented the lightning rod, sailors began to hoist copper chains from the water to the mast top when lightning threatened. Sailors were killed on at least one occasion when lightning struck while they were lifting the chain. The metal ships that came onto the scene in the 1860s had no such problems. Why not?

**10.** Induced charges in metal containers shield their interiors against strong electric fields on the outside. Why can't we have a shield against the gravitational field in the same manner?

**11.** Explain why it is difficult to keep your hair really clean if you brush it often.

**12.** Figure 18–13 is a facsimile of a sketch made by Michael Faraday in a lab notebook in 1838. (You'll learn of his contributions to the physics of electricity and magnetism in the next few chapters.) This sketch showed Faraday's estimate of the lines of force around an electric eel in a dish of water at the time of discharge. Faraday estimated the intensity of the electric field by dipping his bare hands into the dish at various points! Of what in this chapter does this field remind you?

**13.** Michael Faraday (see Exercise 12) built a metal cage and charged it until sparks flew to objects held nearby. Then he entered the cage and performed experiments to detect electric fields. He found none. Explain.

**14.** Explain how a neutral molecule or atom can be attracted to a charged particle.

**15.** In Chapter 14 we saw that metals were good conductors of heat. In this chapter we see they are good conductors of charge. Is there a connection between these facts?

**16.** The gravitational force law for mass is similar to the electric force law for charge except for the great difference in the magnitude of the proportionality constants $G$ and $K$. There is another great difference between mass and charge: There is no *mass* dipole that is analogous to an electric dipole. Discuss.

**17.** Explain how induction can make a charged balloon stick to a wall.

**18.** Explain why humid air takes away static electricity better than dry air. (*Hint:* Remember that water molecules are polar molecules.)

**19.** If you have access to a black and white television, turn up the brightness and gently lay a piece of notebook paper on the vertical screen. Why does the paper stick to the screen?

**20.** Plastic wrap is a thin organic film that's designed to electrify easily so that it adheres to solid objects and to itself as well. Explain how, using items made of substances listed in Table 18–1, you could go about determining the sign of the charge on the plastic wrap.

**21.** Have you ever bitten down onto an aluminum foil wrapper of some candy or gum so that it made contact with a metal filling in a tooth? The small but sharp shock you felt was from a transfer of charge between two different metals. Discuss.

**22.** Explain why those movable electrons inside a copper penny that all repel one another don't rush out to the surface of the penny.

**23.** You should never attempt to free a kite that's snared in local power lines by using a long pole. Why not?

***24.** If two like charges are taken *twice* as far apart, by how much does their repulsive electric force decrease?

***25.** Following Example 5, show that if Phate had eliminated only one electron per 10 billion atoms in his shoes and in the kilogram of wood beneath his feet, his acceleration would have been only 130,000 $g$.

***26.** Find the (approximate) number of electrons in one coulomb of negative charge.

***27.** A flashlight battery has a voltage difference of 1.5 volts between its ends. If 0.001 coulombs of charge passes from one of the ends to the other, how much work does the battery do on those electrons?

***28.** A certain television has a potential difference of 30,000 volts in its picture tube to accelerate electrons toward the screen. Assuming an electron starts at rest, calculate its energy after passing through that potential difference and also its speed. ($q_{electron} = -1.6 \times 10^{-19}$ coulomb; $m_{electron} = 9.1 \times 10^{-31}$ kilogram.) Compare this speed to the speed of light.

***29.** A lightning bolt 1000 meters long discharges across a potential difference of 100 million volts. What was the average value of the electric field along that stroke?

***30.** A 1-centimeter spark jumps from your finger to a metal doorknob on a cold winter's day. Assuming a 15,000-volt potential difference between your finger and the doorknob, find the electric field and compare it to that of a lightning bolt (Exercise 29).

***31.** A lightning bolt brings 25 coulombs of negative charge to earth through a potential difference of 100 million volts in only 0.005 seconds. What was the *power* expended?

# CHAPTER 19
# MOVING CHARGES

A child who becomes tall enough to reach the light switch on a wall gains a certain measure of independence. Security too. Chances are that as a kid you knew exactly where to stand in the doorway to reach around the facing and find the switch to turn on the light before you stepped into your darkened room. Even before then, you were a user of the instant energy available from the electrical outlets in your home. Today you may wake to the buzz of an electric alarm clock and shower in water heated by electricity. The milk in the refrigerator is electrically cooled, since an electric motor drives the compressor. Breakfast rolls in the toaster oven get piping hot because of electrical energy. The list goes on. This chapter and the next two give details of how all this is done.

On the average, each man, woman, and child in the United States makes daily use of electrical energy in a quantity that equals the energy an adult derives from the normal daily intake of food. It is as if there were a silent genie, channeled invisibly through thin copper wires in the walls where you live, at your command 24 hours a day. If the total electrical energy used annually in the United States (including industrial and commercial use as well as residential) is divided by the population, that energy amounts to eight times the food energy an adult needs during a year. So each of us benefits in goods and services from the energy equivalent of eight adults laboring full time. We are accustomed to the comfort and convenience of this energy, but we should remember that in less developed countries it is not this way. An estimated two-thirds of the world's family dwellings don't have electricity. ■

## Electric Current

Anywhere there is a net electric force, a charged particle that is free to move *will* move. However, most electric charge is bound in neutral atoms. In these tiny structures the attractions between the electrons and the nuclei are normally hard to overcome. But in metals the atoms bond together in a way that frees some electrons to wander. If there is a net electric field, the free electrons in a metal will move along in the direction in which the field pushes them. In the same fashion, many liquids (such as salt water or even blood) have ions that move in an electric field. Any flow of charged particles, whether electrons or ions, is called an **electric current,** or, more simply, a **current.** The measure of a current is *the quantity of charge that passes by per second.* In other words, current is the *rate* at which charge moves past a location. Current has the symbol $i$, so

$$\text{current} = i = \frac{q}{t} = \text{rate of flow of charge in coulombs per second}$$

If 1 coulomb of charge passes by in 1 second, the current is 1 **ampere,** or 1 **amp** (A). A current of 1 ampere might mean (1) there are lots of charged particles per unit volume moving slowly or (2) there are a few charged particles per unit volume moving quickly. The current is positive if the charges are positive and negative if the charges are negative, but for our purposes the sign of the current will rarely matter.

## Making Charges Move

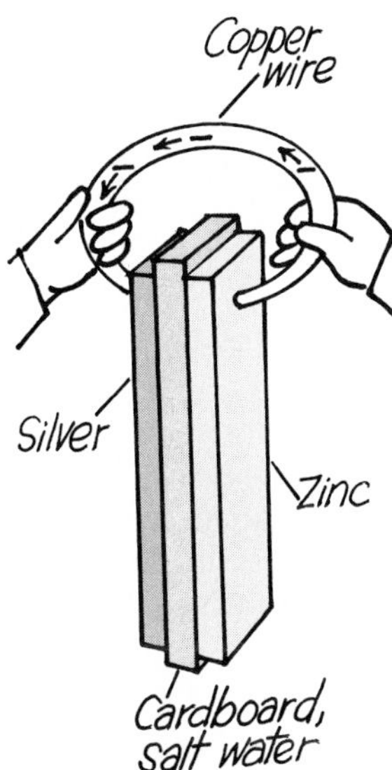


**FIGURE 19–1**
A simple electric cell, like the one made by Volta in 1800. When a conducting wire is connected to the unlike metals, a chemical reaction proceeds in the cell, and a current flows through the wire.

When one material makes contact with another, electrons can transfer because of the difference in affinities those materials have for their electrons. The large "rubbing" machines mentioned in Chapter 18 work in this way. But if the difference in electron affinity between the two substances is great enough, the transfer (or rearrangement) of electrons is automatic; that is, a *chemical reaction* takes place. Some chemical reactions release electrons and others take up electrons. A pair of such chemical reactions can serve as a source of electric current, or a source of electric potential difference.

For example, stick a straightened paper clip and a copper wire into an ordinary lemon. Knead the lemon to make the inside more liquid, and stick the other ends of the wires close together on your tongue. A tiny current will move through the moisture on your tongue, activating your taste buds. This happens because of chemical reactions between the lemon juice and the metals. At one wire, a reaction with the liquid takes electrons from the metal. At the other, electrons are given up to the metal from the liquid. The moisture on your tongue provides a conducting path for the electrons to move *from* the metal that is adding electrons in the lemon juice reaction *to* the metal that is losing them there. (See the section "Are Batteries On All The Time?" for more details.)

In 1800, shortly after doing a similar two-metals-to-the-tongue experiment, Giuseppe Antonio Anastasio Volta made the first electric cell. In place of the lemon, he used a strip of cardboard soaked with salt water. He sandwiched the wet cardboard between a plate of silver and a plate of zinc. Chemical reactions with the salt water made one plate positive and the other negative, and when a copper wire connected the zinc to the silver, a current flowed in the wire. That was the first chemical electric **cell** (see Fig. 19–1).

Volta then made the first **battery** of cells by stacking identical cells in a cylindrical arrangement (see Fig. 19–2). An electron moving

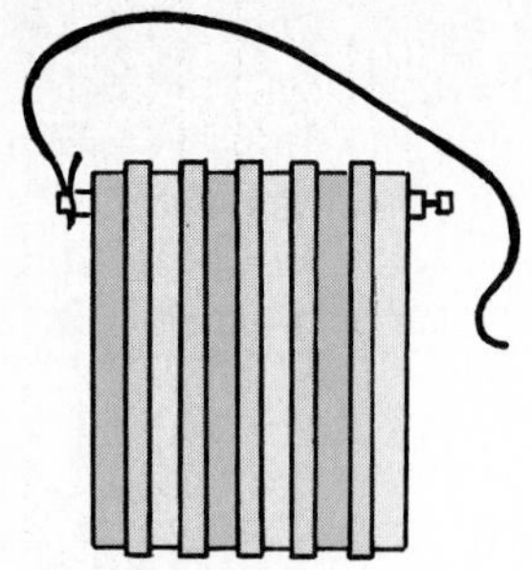

**FIGURE 19–2**
A simple battery (a battery of cells). If each cell contributes 1.5 volts of electric potential, this battery will have a total electric potential of 4.5 volts across its ends, or terminals.

from the negative plate to the positive plate in a single cell has work done on it. That is, it moves from a point of higher electric potential (for a negative charge) to a point of lower electric potential. If it then travels through another cell, it has more work done on it, and the total change in its electrical potential is even greater. Cells connected like those in Volta's battery give a larger potential difference between their outermost plates, or *terminals,* because they collectively do more work on a passing charged particle than a single cell does. An ordinary 12-volt car battery which has 6 cells connected in line uses plates of lead and lead oxide immersed in sulfuric acid solutions to generate electric potential. Each cell contributes about 2 volts to the total potential difference across the terminals of the battery. In the chemical reaction in each cell, some of the water of the solution is broken down into hydrogen gas and oxygen gas. Each cell has its own cap so the owner can replenish the water lost in the reaction. (The holes in the cell caps let the potentially explosive gases escape harmlessly.)

What most people call a flashlight battery is actually a single cell—the standard-size D–cell. When new it has a potential of about 1.5 volts between its terminals. The quantity of chemicals in such a cell is small compared to a car battery, so it produces a smaller total charge during its lifetime. The various types of batteries are rated in *amp-hours* to give an indication of how much charge they can push through a circuit before they are completely discharged, or dead. (One amp-hour is the charge drawn from a battery if a current of 1 amp is used for 1 hour, or if ⅓ amp is used for 3 hours. Charge = current × time.)

Car batteries and certain flashlight batteries can be recharged by connecting them to an external voltage difference that forces the chemical reactions inside the cells to go backward. That is, the materials are for the most part returned to their original states and can proceed to generate more charge as before. That is the function of the alternator on a car. As the engine runs, the alternator supplies a reverse electric potential to the terminals of the battery to restore its state of charge.

## Electric Charge in Motion

From Ben Franklin's time, people have thought of electric charge as a fluid. In many ways charged particles do respond the way a liquid does. Liquids are almost incompressible; moderate pressure on the water in a pipe doesn't change the number of molecules in a unit of volume. Similarly, in a moderate electric field the electrons in a conducting wire are incompressible. That is, there are just as many in every cubic millimeter of the wire as there are when no field is present. Otherwise, any region of the wire that was left positive would exert a *tremendous* electrical force to return the lost electrons. So electrons move through a conducting wire only when just as many electrons come in at one end as leave at the other—just as water flows through a pipe.

Acting under the force of gravity, the water in a creek flows downstream; it naturally moves from a place of high gravitational potential energy to points of lower potential energy. In conductors, electrons too naturally move from places where they have a high electric potential to

## ARE BATTERIES "ON" ALL THE TIME?

In a fresh battery, be it for a flashlight or a car, two chemical reactions stand ready to proceed—whenever they can. At the positive end of each cell, there's a reaction that will take place if electrons are available to add to the reactants. At the negative end of each cell, there's a chemical reaction that will occur if the reactants can shed electrons. But neither reaction is strong enough to proceed without a mechanism to shed or absorb those electrons. When the ends of a copper wire touch those terminals, the electrons given up in the chemical reaction at the negative terminal can move onto the wire, *but only as electrons in the copper wire are taken up by the chemical reaction at the positive end.*

The instant a copper wire is connected between the terminals, a storm of chemical activity inside the battery releases electrons at the negative terminal and takes up electrons at the positive terminal. Electrons are drained from the end of the wire at the positive terminal and pop into the wire at the negative end. During all this, the wire remains absolutely neutral (as does the battery). Because of the strong attraction of the metal ions in the wire, electrons never pile up at any point; each cubic millimeter of the copper wire must remain neutral even though a current of electrons is moving through it.

A large current can flow through a copper wire, because there are about $10^{22}$ free electrons per cubic centimeter. But if the wire is disconnected, those chemical reactions stop—*almost*. Left unused, all batteries eventually go dead. One reason is that some current flows even without a conducting wire between the terminals. Though air is a poor electrical conductor, each cubic centimeter of air has from a few to a few hundred ions (principally due to cosmic rays that ionize air molecules), and if positive ones strike the negative terminal and negative ions strike the positive terminal, the reactions creep along. Under humid conditions, when a thin layer of moisture bathes the battery's surface, even more current flows over the battery's surface. Pure water doesn't conduct well, but add a trace of impurities, even salt from fingers or fingerprints, and it's a different story. Ions are very mobile in water, and the battery can discharge through water that has ions in it. Then too, the reactants in batteries are corrosive. In time, acids can eat through most containers, ruining the cells. But until this happens or until the reactions have completely discharged, the battery gives a potential difference *all* the time, and a current of charge takes place whenever there's a conducting path between its terminals.

places where their electric potential is less. We say a current of electrons flows. We also say a conducting wire draws a certain current when it is connected to points of different potential. (Similarly, people used to speak of drawing water from a well or a faucet. In this sense, to draw is not to attract but to *obtain*.) In many circumstances, then, we can think of a current of electrons that move in a conducting wire as a current of water moving through a pipe. And rather than saying "charged particles move," we can just say "charge flows."

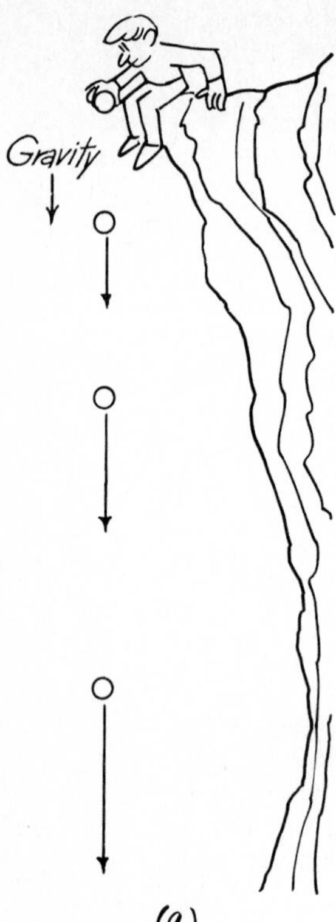


## Resistance to an Electric Current

Electrons, like water, can move through a potential difference in various ways. The simplest is an unrestrained free-fall, such as when water passes over a precipice and becomes a waterfall. At first there is little air resistance to its fall; the water accelerates rapidly. By analogy, this is what happens when electrons move in the vacuum of a television tube between points of different potential with not even air to retard their motions. They accelerate freely.

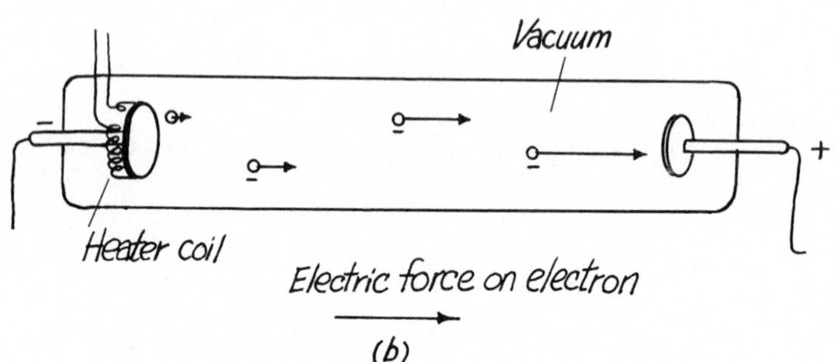


**FIGURE 19–3**
*(a)* An unrestrained acceleration due to a force. The heavy ball accelerates continuously. *(b)* An unrestrained acceleration due to a force. The electrons accelerate continuously.

But sometimes a mountain stream stays within its banks and descends gently along an incline, its contact with the ground preventing much acceleration. Because of friction, it might arrive at the base of the mountain with no greater speed than it has at the top. That is the way an electric current flows in a conducting wire. Confined to move within the wire, the movable electrons meet with friction: They often crash into atoms in the wire, causing them to change directions. Though on the average they move in the direction of the current, they cannot gain much speed. A collision may stop an electron's forward progress entirely or even send it backward against the flow at any time. A typical current in a copper wire creeps along at perhaps a millimeter a second. We say the conducting wire *resists* the flow of the electron stream.

Of course, the movable electrons in a conducting substance have motion even when there is no electric field supplied to push them along. They have thermal energy, just as the atoms do. In fact, in a conductor that is at room temperature, the electrons move with a high average instantaneous speed, much faster than a rifle bullet. But because of their numbers and the random nature of their motions, the average electron current past a point is zero. No more go in any one direction than in another. Yet if you connect that conductor across the terminals of a battery, in a manner of speaking, you create a small incline, a hill. As the mobile electrons drift downhill from the negative terminal to the positive terminal, each bounces from one atom to another, much like

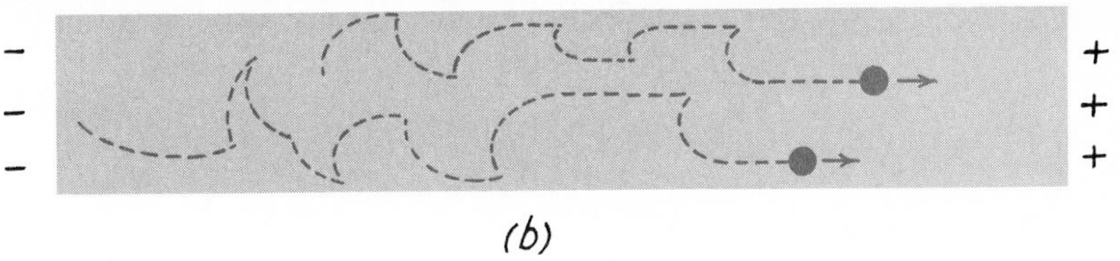


(b)

the ball in a pinball machine. No matter how many collisions change its direction, it eventually winds up at the bottom of the slope.

## A Measure of Electrical Resistance: Ohm's Law

A copper wire connected to a battery conducts a certain current, $i$. Another wire of the same length and diameter but made of silver would conduct a larger current from the same battery. An aluminum wire of the same size and length allows less current to flow than the copper wire does. And yet the difference in electric potential, $V$, between the battery terminals is the same for each wire. Each metal has a different resistance to the flow of charge, for two principal reasons: One, in different conductors a different percentage of bonding electrons are free to move. Two, the distance such electrons can travel before having collisions and changing directions depends on the electronic properties of the conducting material.

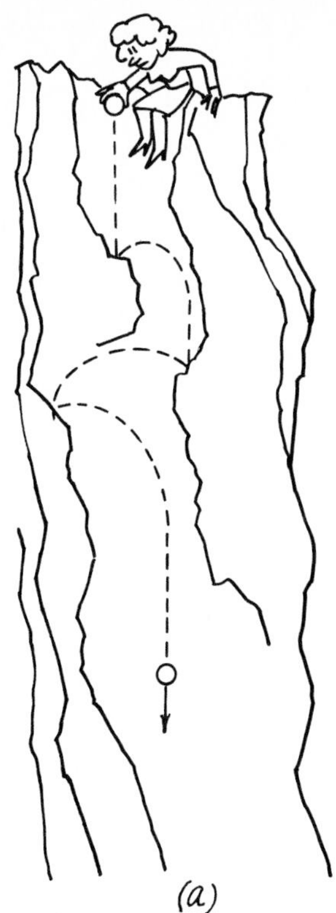

(a)

**FIGURE 19–4**
*(a)* Although the force acting on this ball is the same as that acting on the ball in Fig. 19–3*a*, this ball experiences resistance that keeps it from gaining speed continuously. Collisions with the ledges slow the ball. *(b)* Although a force is acting to push these electrons through the wire, collisions with atoms slow them and keep them from gaining speed continuously.

The current in a wire of any conducting material increases if the potential difference between the ends of the wire increases; but the larger the resistance, the smaller the value of the current. Georg Ohm discovered about 1820 that the voltage across the ends of a conducting wire is directly proportional to the current that results in the wire.* The constant of proportionality for the wire is given the symbol $R$. So

$$\text{potential difference} = \text{current} \times \text{resistance}$$

or

$$V = iR \qquad \text{(Ohm's law)}$$

where $R$ is defined as the **resistance** of the conducting path to the current $i$. The unit of resistance is the **ohm.** Rearranging Ohm's law as $R = V/i$ leads us to the definition of an ohm. A conducting path has a

*Georg Simon Ohm (1789–1854) was educated at home by his self-taught father. He showed great promise in math, but in college apparently Ohm spent too many hours at the billiard tables and at partying. Forced to leave school by his angry father before he could get his Ph.D., Ohm took a job as a high school math teacher. By working long and hard, Ohm won his Ph.D., but he couldn't find a university job. He eventually took a high school job that forced him to teach physics in addition to math. Eight years later he started the research that would lead him to Ohm's law. While the importance of his findings was accepted immediately in England and France, his German colleagues ignored it. In 1852, Ohm was given a university professorship, but he died less than 2 years later.

resistance of 1 ohm if a potential difference of 1 volt across the ends of the path draws a current of 1 ampere.

## The Resistance of a Conducting Wire

The type of material is only one factor that governs the resistance of a conducting wire. Another is its length. A long copper wire connected across the terminals of a flashlight cell draws a certain current. Cut that wire in half, and one of the shorter lengths connected across the cell will draw approximately *twice* as much current for the same potential difference. This is why. An electron that moves from one terminal of a 1.5-volt flashlight cell to the other gets the same amount of work done on it, no matter what distance it has to go. Work = force × distance, so if the distance increases because a longer wire is used, the electric force must decrease. For a longer wire, then, the electric force is less, and the electrons don't accelerate as fast. Consequently, the rate at which charge passes a given point is smaller, meaning the current $q/t$ is less.

One more factor determines the resistance of a conducting wire. A thicker copper wire, one with a larger diameter, offers less resistance than a thinner one of the same length (Fig. 19–6). A copper wire has a

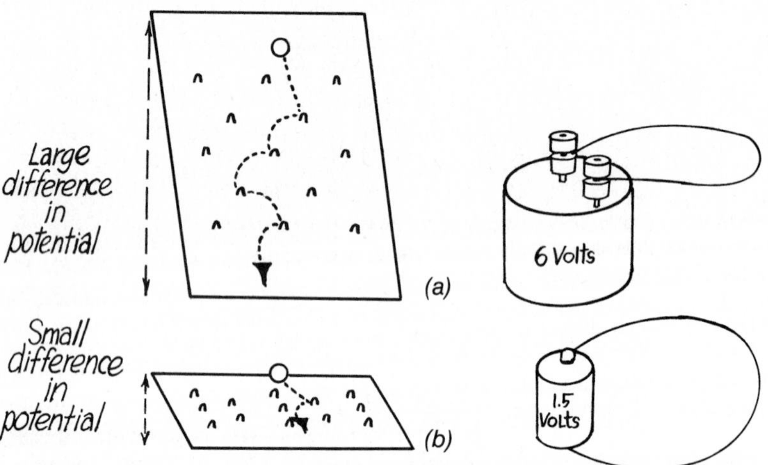


**FIGURE 19–5**
Identical wires are connected across two batteries. In *(a)*, the battery has four times as much voltage difference between its terminals as in *(b)*. The electrons move similarly to marbles sliding down inclined planes that have pegs to slow their motions. In *(a)*, the marbles lose four times as much potential energy as those in *(b)*. In the wires connected to the batteries, the electrons move through differences in electric potential rather than through differences in gravitational potential energy, but the effect is the same. That is why the 6-volt battery draws a current four times larger than the 1.5-volt battery: The electrons move faster in a wire when the potential difference is greater, meaning more charge per second passes a point.

**FIGURE 19–6**
More charge per second can pass through a large diameter wire than a small one, even when the electric force is the same. A thicker wire, therefore, offers less resistance (or draws more current) than a thinner one of the same length.

certain number of electrons in each *unit of volume* that can move. The thinner wire has a smaller volume than the thicker one because of its smaller cross-sectional area. If the same potential difference is applied to each wire, the current $i$ will be less in the thin wire because there are fewer electrons to move past a point in a unit of time. As seen from Ohm's law, if $V$ is the same yet $i$ is smaller, the thinner wire has a larger resistance $R$ than the thicker wire does. We'll return to the topic of resistance in conducting wires in "Calculations."

## Electrical Circuits: DC and AC

The potential difference between the battery terminals in a car can start the engine, light the way, clear the rain from the windshield, and bring you music. The potential difference between the wires in the walls of a house can do an even wider range of tasks. What does the work? Energy that comes from the *source* of the potential difference. In the case of a battery it is energy from the chemical reactions. For a home circuit it is energy from a power plant, which comes from burning coal or oil, or from water falling at a dam, or from nuclear fission.

To operate any electrical device, all you need to do is provide a connected conduction path for electrons between the potential difference that is supplied by the battery or the home circuit. Any such conducting path is called an **electrical circuit.** An electric switch lets you turn the appliance on or off. If the switch is on, the circuit is complete or *closed,* and a current flows in the appliance. If the switch is off, the circuit is *open,* and there is a break in the circuit; with no connected conducting path between the potential difference for electrons to move along, no current flows.

When electric current flows in one direction through a circuit, as it does when a battery supplies the potential difference, it is called **direct current,** or *DC*. The current from the electrical outlet in a home, however, is **alternating current,** or *AC*. Its voltage is not steady; instead, it oscillates, changing the direction of the electric force first one way, then the other. (The reason has to do with efficiency. We'll see why in Chapter 21.)

The alternating current used by power companies throughout North America is a 60-cycle-per-second current. Responding to the alternating electric field in a wire, electrons sway back and forth a very short distance some 60 times every second. The electric field is caused by an alternating voltage applied to the wires at the power station. So when an appliance is connected at an outlet, the electrons in the appliance don't really go anywhere; they just surge back and forth under an *average* potential difference of 110 to 120 volts. This points out an important feature of AC electricity. It is the energy of the moving electrons that turns on the lights and runs the refrigerator, the hair dryer, or the electric razor. The electrons in the power station never leave, nor do the electrons in a home circuit. They merely swing back and forth a very short distance, quicker than your eye could follow. An alternating current is just a vibration of the sea of electrons superimposed on their faster, random thermal motions.

## Electrical Power

Because the voltage difference is the potential energy change per unit of charge, when a charge of $q$ coulombs travels through a voltage difference of $V$ volts, the work done on the charge is $q$ (coulombs) times $V$ (volts), or $qV$ joules. In other words, the total energy a charge $q$ receives when it goes through a potential difference $V$ is $qV$.

$$\text{electrical energy} = E = qV$$

From this we can also see that twice as much charge moving through half as much voltage requires the same amount of energy.

To operate electrical appliances requires not just energy but a certain amount of *power,* which is the rate at which energy is delivered. Power is energy divided by time:

$$\text{power} = P = \frac{E}{t}$$

Let's put electrical power into electrical terms. If $E = qV,$ then

$$P = \frac{E}{t} = \frac{qV}{t} = \left(\frac{q}{t}\right)V = iV$$

Power is measured in joules per second, or **watts,** while the current is in coulombs per second, or amps, and the voltage difference is in joules per coulomb, or volts. So **electrical power** (in watts) = **amps × volts.** When lots of energy is transferred per second, the process is powerful, meaning the value of $P$ is large. A light bulb that gets 60 watts of power from the potential difference of 110 volts in a home's electrical wiring draws a current of $i = P/V =$ 60 watts/110 volts = 0.54 amps. A vacuum cleaner that uses 650 watts of power draws a current of 650 watts/110 volts = 5.9 amps. With patience you could get the same amount of energy from single flashlight cells as you can from a wall socket, but you need that energy *fast* to operate an iron or a dishwasher or a refrigerator compressor (see Fig. 19–7).

**FIGURE 19–7**
When you are using an appliance, it's not just the voltage that matters; it's also the rate at which electrical energy can be delivered.

## Electrical Heating

A current of electrons moves along a copper wire that's connected between the terminals of a battery. In human terms, these electrons seem to have a frustrating journey. Each time they move off toward the positive terminal, they run into a copper ion in the wire. On the average, that slows them down. Of course, for every action there is an equal but opposite reaction, and the ion gets a quick push that, on the average, speeds it up; that is, the ion vibrates faster in its place in the atomic structure of the wire. As a consequence, *the temperature of a wire goes up when a current passes through it.*

Retarded by its collisions in the metal, an electron that reaches the positive terminal arrives with the same speed (hence kinetic energy) on the average that it had when it left the negative terminal. This means *all the work done by the battery as the electrons travel between the terminals turns into heat as the electrons move through the wire.* It follows that the faster a circuit uses energy, the faster it heats up. That is, *the power used by a circuit of simple resistance is equal to the rate at which the circuit generates heat.* In Chapter 21 we'll study electric motors, which turn a part of the energy used by a circuit into mechanical energy.

At this point, it is tempting to make a guess about resistance and heating in a circuit: "The more resistance, the more heating." But that's not so. A short copper wire connected to the terminals of a battery gives off more heat than a long copper wire does. The shorter wire has less resistance, so it draws more current, and power = current × voltage. The voltage would be about the same for the longer wire, but the current would be smaller.

## High and Low Resistance

A toaster is connected to the 110-volt potential difference in a wall socket by an electrical cord. Inside the cord are two copper wires, each wrapped in an insulating material to keep them apart and connected to metal prongs in the plug. These copper wires lead to opposite ends of a strip of high-resistance conducting wire in the toaster. That high resistance means the current flowing through the toaster is small. Exactly the same current flows in the copper wires in the electrical cord too, but because those wires have less resistance, much less heating occurs. *Essentially all of the energy the toaster circuit draws goes into heating the toaster wire (or element).*

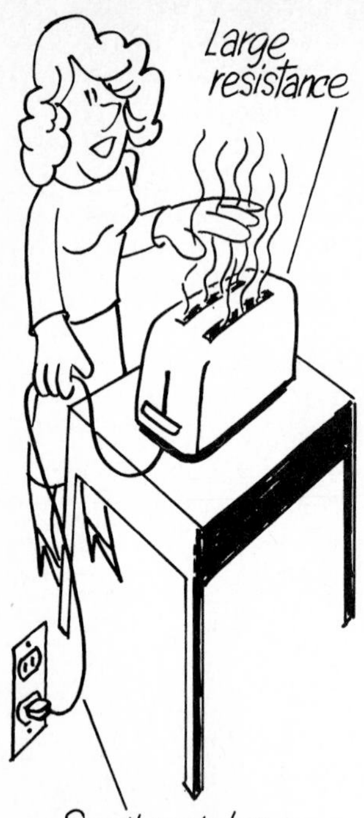


**FIGURE 19–8**
The high resistance wire in a toaster gets hot enough to cook toast, but the copper wire with less resistance only gets warm even though the same amount of current flows through both of them.

Suppose the insulation between the wires in the toaster cord becomes damaged, and the two wires at different electric potentials touch while the toaster is on. Suddenly the electrons have an easier path to follow. The new path, through that point of contact, would offer almost *no* resistance compared to the toaster element. Ohm's law, $V = iR$, tells us that if $R$ gets very small while $V$ remains constant (110 volts), the current in the wires and across that point grows. As the current goes up, the rate of heating increases, and that heat can melt the copper wires, usually at the point of contact. Such a place of low resistance in a circuit is called a **short circuit,** or just a **short.** Shorts are dangerous because where there is enough heat to melt copper, there's ample heat to ignite nearby combustible material.

A safety feature is used in home circuits to protect them from overheating in case of a short. These devices also work in case too many appliances are connected to some part of the household circuit, raising the current to high levels and overheating the wires in the wall. In older homes a meltable *fuse,* encased in glass, is part of each circuit (Fig. 19–9). If an appliance shorts out, that circuit draws a large current. In a matter of seconds, the fuse melts (at a predetermined current noted on the fuse) and breaks the circuit before the wires can overheat. The dangerous short should be repaired before the melted fuse is replaced.

Another type of device is used in the circuits of all newer homes, a bimetallic strip that curls up when it gets hot from carrying too much

**FIGURE 19–9**
A fuse limits the amount of current that a circuit can carry.

current. (It works the same way as the bimetallic thermostat described in Chapter 13.) As it heats and curls, it snaps away from a contact to break the circuit. This kind of device, called a *circuit breaker,* works a little faster than a fuse, so it's safer (Fig. 19–10). And once an overload is corrected and the open bimetallic strip cools, it only has to be reset to become functional again, usually a matter of flipping a switch.

You can demonstrate the action of a fuse or a short circuit with a flashlight cell. From a pad of steel wool, select the thinnest strand you can find. Then attach two copper wires to the cell, one at each terminal. Use the thread of steel wool to complete the circuit. If you touch the copper wires to the steel wire at points about a centimeter apart, the steel gets warm. If you move those copper wires very close together along that thin steel thread, the length of steel between the copper wires gets really hot. A smaller length of the steel wire offers less resistance, so the current increases. If the points of contact of the copper wires come to within a millimeter or so of each other, the steel will glow white hot like the filament of a light bulb and melt, breaking the circuit.

**FIGURE 19–10**
A circuit breaker acts more quickly than a fuse to break a circuit and is easier to restore to operation.

## Electrical Circuits: Experiments You Can Do

### Series Circuits

With a few small lengths of copper wire, a couple of flashlight cells (or batteries), and some flashlight bulbs you can show a great deal about how electrical circuits perform. For example, make the two arrangements shown in Fig. 19–11.

Touch the loose wire to the end of the bulb.

*Question 1:* In which circuit does the bulb grow more brightly? Why?

*Question 2:* What happens if you turn one of the cells around in circuit (b)? Why? What happens if you turn the cell around in circuit (a)?

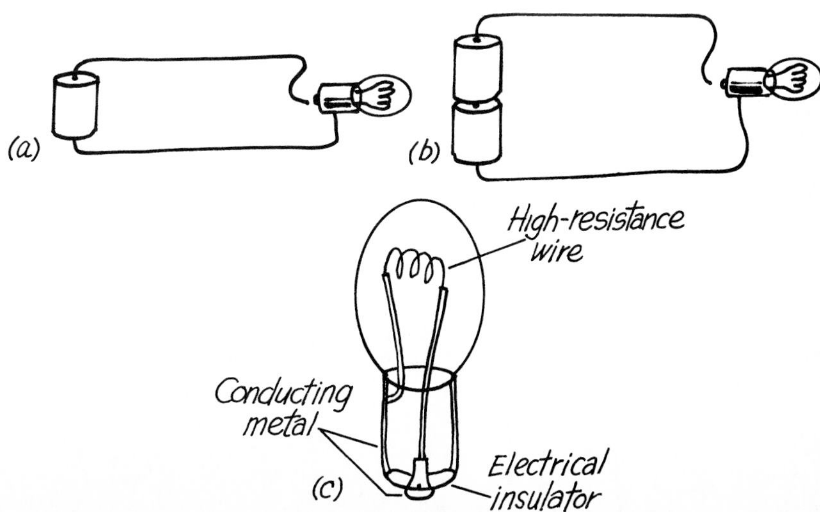


**FIGURE 19–11**
*(a)* A bulb lighted by one cell. *(b)* A bulb lighted by two cells. *(c)* The details of an ordinary flashlight bulb.

## A PUZZLE

Telephones are one of those things taken for granted in developed nations today. From the time we are young, we use the phone system to talk with friends much as if they were in the next room. Ask a question, and the answer comes right back with no hesitation or pause. But someone who makes an overseas phone call usually notices a difference. There is a fraction of a second between the time a question is asked and the time the other party hears it. Then there's an additional lapse between when the party answers the question and when the questioner receives the answer. Though short, the pause can be disturbing. (It's caused in part by the electronic signal being beamed almost 22,000 miles out to a satellite and back. Even though it travels at the speed of light, the signal takes a good fraction of a second for the round trip.)

Most of us never think about what happens when we use the phone to talk to someone who is, say, 2 miles away. If we could shout and be heard at that distance, our words, moving at the speed of sound, would need almost 10 full seconds to reach the person. Then allow another 10 seconds for the reply to reach us—even a very short conversation would become a very long one. Since electrons typically move only a millimeter or so a second along copper wire, how does a telephone signal travel that distance over the phone lines so fast? The answer to this puzzle is found if we ask what causes an electrical current. Electrons in a wire move because of the electric force that is caused by applying a difference in electric potential to the ends of the wire. Any change in the current, likewise, comes from a change in the applied potential. So the real question is, *how fast does the change in the force field travel when the potential across the wire changes?* The change in the electric force field through a copper wire travels only somewhat slower than the speed of light in a vacuum, *186,000 miles per second.* It's true the electrons move slowly as they deliver power to operate the phones. But they get their energy from the push and pull of an electric force field (or *signal*) whose changes move through those wires extremely fast.

1: Circuit (b) makes the bulb glow more brightly because an electron completing the circuit gets the benefit of the electric potential of two cells—in effect, 3 volts rather than 1.5 volts. When two cells or batteries are connected as in circuit (b), they are said to be in **series.** In a standard flashlight the cells are in series, as are the individual cells in a car's battery. When parts of a circuit are in series, there is only one path for the current of electrons to take.

In both of these circuits, essentially all the energy of each electron goes into thermal energy in the bulb's filament, the only place of high resistance in the circuit. The hotter the filament, the more it glows. For an even brighter light, use three or four cells in series. But use a heavy-duty bulb, or you'll melt the filament in a second or two.

2: No current flows. A cell's potential difference pushes electrons from the negative terminal through the external circuit to the positive terminal and *only* in that direction. When two similar negative terminals touch, there's no potential difference, *hence no electron motion*. The same is true for two equally positive terminals.

Next, try this. Turn the single cell in circuit (a) around. There's no difference; the bulb lights as before. No matter in which direction the current flows, the resistance of the light bulb filament (and the copper wire) is the same.

Now make the three arrangements shown in Fig. 19–12.

**FIGURE 19–12**
A single cell used to light one, two, and three bulbs.

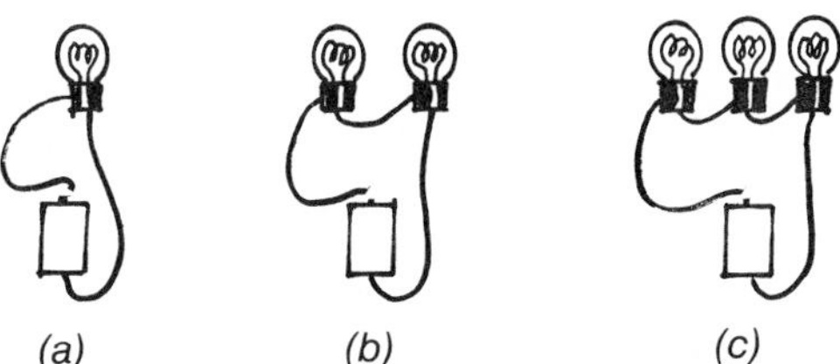


*Question 1:* When you touch the copper wire to the cell's terminal, in which circuit do the bulbs glow most brightly? Most dimly? Why?

*Question 2:* In circuit (b) do the bulbs seem to emit just about equal amounts of light and warmth? Why? In circuit (c)? Does it matter which way the cell is turned?*

1: First let's recount how the bulb glows. The filaments of the bulbs offer much more resistance (typically about 10 ohms) to the current than the wires (typically about 0.1 ohm) do. In fact, if you connect the bulbs with longer wires, you hardly see a difference in brightness. The heating that occurs when the current passes through the filaments makes the bulbs glow. The single bulb in (a) glows most brightly because there is less total resistance in the circuit, hence more current flows. The bulbs in circuit (c) glow most dimly. There the three bulbs offer about three times the total resistance of (a). This means the electrons each lose about one-third of the cell's potential difference as they move through each bulb. Once again, the direction of the current makes no difference.

2: If the two bulbs in circuit (b) or the three bulbs in circuit (c) give the same brightness when they are connected to a single cell, they'll have the same brightness in these circuits. The same number of electrons per second must leave the negative terminal of the cell as enter the positive terminal during every second, so the current through each of those bulbs has to be the same. Though the electrons lose energy at each filament they travel through, *they lose the same amount of energy in each bulb.*

*For this experiment you need to find flashlight bulbs that have nearly the same resistance. In a box of new bulbs, there will be some with noticeable differences in brightness. If in circuit (b) or circuit (c) any bulbs have a different brightness, interchange their positions to convince yourself that the difference arise from the bulbs' properties and *not* from the positions in the circuit (closer to one terminal or the other).

These circuits are also series circuits, with only one path for the electrons to take. Given time, each charge moves through every bulb and cell in the circuit. When the filament of only one bulb burns out, the circuit is broken and all the bulbs go dark.

Note that the voltages of the separate cells in series are added to give the total electrical potential across the circuit, and the resistances in series are added to give the total resistance to the circuit.

## Parallel Circuits

There is a more convenient way of connecting electrical appliances to a voltage source than in series. Figure 19–13 shows how. Wires lead from a voltage source to the region of the bulbs. Each bulb then is connected directly to the (+) and (−) wires. This is called a **parallel** circuit. In this circuit there is more than one path that can lead electrons back to the cell. Parallel circuits have two distinct advantages over a series circuit. (1) You can disconnect one bulb without interrupting the current flow to the other bulbs. (2) Every bulb that's connected taps the full voltage of the source: No matter which route a particular electron takes, it uses the full electric potential of the cell in its trip. This avoids the loss of current, and hence of power, that occurs when additional bulbs are placed in a series circuit. Typical household circuits are parallel circuits; many appliances use the same 110-volt potential difference.

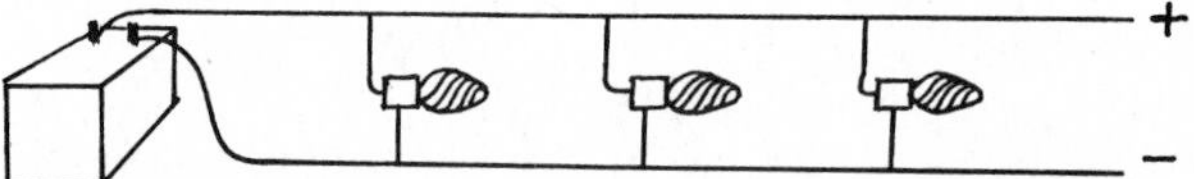


**FIGURE 19–13**
In a parallel circuit, every bulb (or appliance) gets the same voltage.

If you connect two or three bulbs in parallel with a flashlight cell, you may notice some uniform dimming as current is drawn in larger amounts from the cell. That's because there is a limit to how fast the chemical reactions in the cell can proceed. The voltage source from the local power company that you use in home circuits is much more powerful than these cells, and you don't usually notice any dimming as extra appliances are plugged in—except maybe when the refrigerator compressor turns on or you use an iron.

Next make the circuits shown in Fig. 19–14.

*Question 1:* What is the potential difference between the two parallel wires in circuit (a) and in circuit (b)?

*Question 2:* In which circuit do the bulbs burn more brightly when you connect them to complete the circuit? Why?

1: For circuit (a), the potential difference between the conducting wires is the same as the potential difference between the terminals of the cell, about 1.5 volts. For circuit (b), the voltage difference between the wires is the voltage difference across *one* flashlight cell, not *two*. Electrons traveling through the circuit can only pass "through" one cell or the other during one trip around, but not both. That means the current of electrons is moved along by a 1.5-volt potential difference only. The two cells are connected in *parallel* rather than in series as before.

FIGURE 19–14

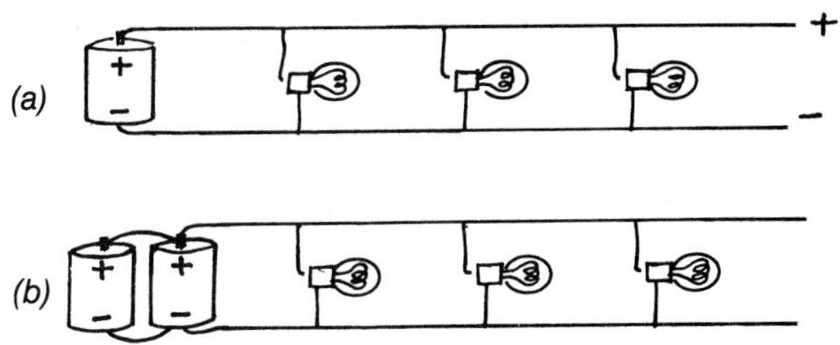


If the two positive ends of any two similar cells are connected by a conducting wire, no current flows because there is no potential difference across the ends of the wire. So there is no boost in the voltage when batteries are in parallel. But there *is* a boost in capacity for delivering power to the circuit, as we'll see next.

2: Very likely the bulbs in circuit (b) burn more brightly than those in circuit (a). That's because the chemical processes that maintain the potential difference in a cell as current is drawn don't have to proceed so fast when two cells contribute to the process. The three bulbs put twice the demands on the single cell in circuit (a) as they do on each of the two cells in circuit (b).

You should know how to connect batteries in parallel if you ever need to ''jump-start'' a car whose battery is too weak to start its engine. This is done by connecting another similar battery (fully charged) to the weak one with jumper cables, thick copper wires with strong grips to make the contacts. The fresh battery than lets the starter draw the current it needs. It is very important that you *don't* connect the batteries in series, because that would increase the voltage to a level the car's electrical circuits aren't designed to handle. Connecting the extra battery in parallel with the car's battery gives the proper voltage to the starter and the extra capacity for delivering current to the starter. So here's the rule: The positive terminals on the two batteries should be connected with a cable first. Then with the other cable, connect the two negative terminals; or better yet, connect the negative terminal of the fresh battery to the engine block or the chassis of the car.* That avoids a spark at the battery which might result when you connect the two batteries' terminals and close the circuit. This would be dangerous if there is hydrogen gas emerging from the discharged battery (see the section ''Making Charges Move'' earlier in this chapter).

Figure 19–15 is a diagram of a home's parallel circuit wiring. The third prong on the electrical plugs of many appliances leads to the grounding device of the home for the users' protection. Should the electrical insulation around a ''live'' wire in an appliance fail, the wire might touch a conducting part of the appliance. Someone who then touched the appliance could provide a conducting path to the ground, and the current would flow through that person's body. A ground wire

*You should check to make certain the car has a negative ground. Some older cars have positive grounds.

**FIGURE 19–15**
A diagram of a house circuit. If connected, this lamp would draw a current equal to $i = P/V = 100$ watts/110 volts = about 1 amp. This refrigerator would draw $P/V = 600$ watts/110 volts = about 6 amps of current.

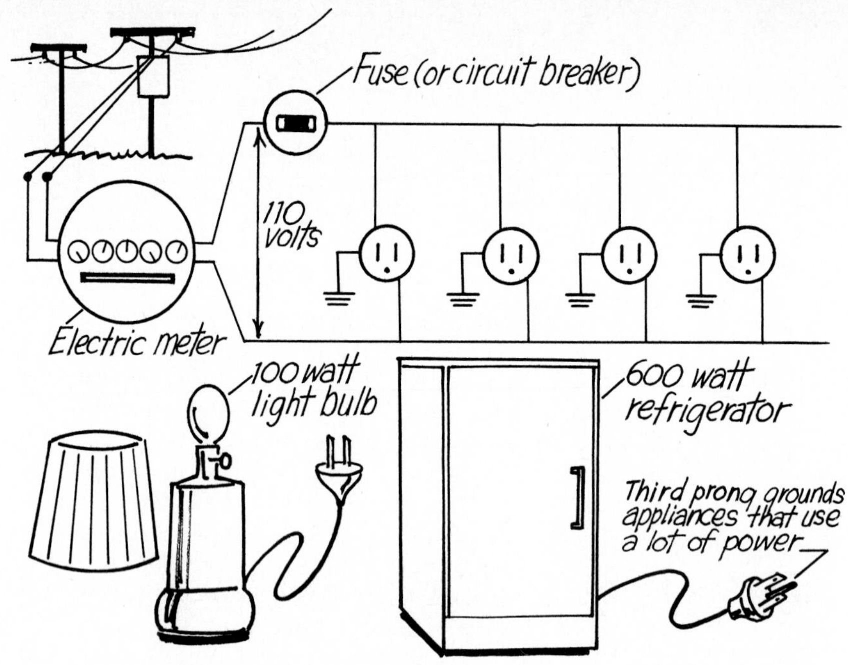


provides an easier conducting path for the current, thereby protecting the user from a severe shock. Likewise, any static charge that might build up from friction as the appliance operates goes to the ground, preventing a surprise for the operator who must touch the device.

## CALCULATIONS

EXAMPLE 1: **Calculate the number of joules in a kilowatt-hour. Show that the energy provided by 1 kWh could lift a 70-kg person to a height of over 5000 m.**

A power of 1 watt used for 1 second gives 1 joule of energy, since $P = E/t$ and $P \times t = E$. One kilowatt used for 1 hour gives

$$1000 \text{ W} \times 3600 \text{ s} = \mathbf{3.6 \times 10^6 \text{ J}}$$

The work required to lift a person (mass = 70 kg) a distance $h$ is given by $W = F \times d = mg \times h$. The amount of energy $3.6 \times 10^6$ J will lift that person a distance equal to

$$h = \frac{W}{mg} = \frac{3.6 \times 10^6 \text{ J}}{70 \times 10 \text{ m/s}^2}, \text{ or about } \mathbf{5100 \text{ m}} \left(\text{since } 1 \text{ J} = 1 \frac{\text{kg}\cdot\text{m}^2}{\text{s}^2}\right)$$

A kilowatt-hour from a power company might cost 9¢ or less. Electrical energy (circa 1984) is a real bargain.

---

EXAMPLE 2: A student studies by the light of a 100-watt bulb for 4 hours one evening. **If the power company charges 9¢ per kWh, how much did this cost?** $P = E/t$, so

$$\begin{aligned} E &= P \times t \\ &= 100 \text{ watts} \times 4 \text{ hours} \\ &= 400 \text{ watt-hours} = 0.4 \text{ kilowatt-hours} \end{aligned}$$

Cost equals 0.4 kWh × 9¢/kWh = **3.6¢.**

---

EXAMPLE 3: Some appliances don't list the power they use but instead give the current they will draw. If a toaster is listed at 5 amps when connected to a wall circuit whose potential difference is 115 volts, **how much power does it use?**

Using $P = iV$, we find

$$P = 5 \text{ amps} \times 115 \text{ volts} = \mathbf{575 \text{ watts}}$$

---

EXAMPLE 4: A 1-horsepower motor (at 120 volts) drives an attic fan for 8 hours per night. **About how much does this cost per month if the utility rate is 11¢ per kWh?**

One horsepower is 746 watts. $E = P \times t$, so

$$E = 746 \text{ W} \times \left(8 \frac{\text{h}}{\text{night}} \times 30 \frac{\text{nights}}{\text{month}}\right) = 1.8 \times 10^5 \text{ W·h/month}$$

$$= 1.8 \times 10^2 \text{ kWh/month}$$

The cost per month is 180 kWh × 11¢ = **$19.80.**

## Convenient Formulas for the Power Used in a Circuit

Using the definition of power together with Ohm's law gives two convenient equations for finding the power used by an electrical circuit. Since $V = iR$, the power $P$ is equal to $iV = i(iR)$, or

$$P = i^2R$$

This gives the rate of heating in an electrical circuit (or part of a circuit) if the resistance $R$ and the current $i$ are known. Sometimes, however, the voltage and the resistance might be the known quantities in a circuit. Ohm's law gives $i = V/R$, and the power becomes $P = iV = (V/R) \times V$, or

$$P = \frac{V^2}{R}$$

Once again, from this formula we can see that a short wire across a battery's terminals will draw more power (and generate more heat) than a long wire.

Likewise, if the voltage in a circuit is held constant and the resistance increases, the power consumption goes down.

---

EXAMPLE 5: Some appliances list their resistances but not the currents or power they draw. An electric stove element connected to 240 volts in a household circuit has a resistance of 48 ohms. **How much power does it use? How much does it cost to use for 2 hours each day for a month if the cost per kilowatt-hour (kWh) is 10¢?** Use $P = V^2/R$ to find the power. The answer is **1200 W.**

If this element is on for 2 hours, the energy used is

$$E = P \times t = 1200 \text{ watts} \times 2 \frac{\text{hours}}{\text{day}} = 2400 \frac{\text{watt-hours}}{\text{day}} = 2.4 \frac{\text{kilowatt-hour}}{\text{day}}$$

For a month, then, 2.4 kilowatt-hour/day × 30 days/month = 72 kilowatt-hour/month. The cost would be

$$72 \frac{\text{kilowatt-hour}}{\text{month}} \times \frac{10¢}{\text{kilowatt-hour}} = \mathbf{\$7.20 \text{ a month}}$$

---

In this chapter we saw how the resistance of a copper wire depends on its length $L$, its cross-sectional area $A$, and the specific material it is made from. Experiments show the following formula gives the resistance for such a wire:

$$R = \rho \frac{L}{A}$$

where $\rho$ stands for the **electrical resistivity** of the material, not its density. $R$ is measured in ohms, whose symbol is $\Omega$ (Greek capital omega). $L$ is measured in meters, $A$ in square meters, and $\rho$ in ohm-meters, or $\Omega \cdot$m.

---

EXAMPLE 6: **Calculate the resistance of 1 meter of copper wire** that has a cross-sectional area of about $2 \times 10^{-2}$ cm$^2$, and **compare it to the resistance of a flashlight bulb.** (See Table 19–1.)

$$R_{\text{wire}} = \rho_{\text{copper}} \frac{L_{\text{wire}}}{A_{\text{wire}}} = (1.7 \times 10^{-8}\ \Omega\cdot\text{m}) \times \frac{1\ \text{m}}{2 \times 10^{-6}\ \text{m}^2}$$

where we've used 1 cm$^2$ = $10^{-4}$ m$^2$. Doing the calculation you'll find

$$R_{\text{wire}} = \mathbf{0.85 \times 10^{-2}\ \Omega}$$

Let's compare this to the resistance of a flashlight bulb that has a power rating of 1 W and operates at 3 V (two flashlight cells in series). $P = iV$, so 1 W = $i \times 3$ $V$, so $i$ = ⅓ A. Using Ohm's law, $V = iR$, we see that the resistance is

$$R_{\text{bulb}} = \frac{V}{i} = \frac{3\ \text{V}}{^1/_3\ \text{A}} = \mathbf{9\ \Omega}$$

Here's the point: The resistance of copper wire is very small compared to the resistance of other devices in a circuit that uses electrical energy. The copper wire's resistance can for most purposes be *neglected* when calculating the power or current an electrical appliance draws from a circuit.

**TABLE 19–1** ELECTRICAL RESISTIVITY AT ROOM TEMPERATURE IN OHM-METERS (Ω·m)

| Material | Resistivity |
|---|---|
| Silver | $1.6 \times 10^{-8}$ |
| Copper | $1.7 \times 10^{-8}$ |
| Gold | $2.4 \times 10^{-8}$ |
| Aluminum | $2.7 \times 10^{-8}$ |
| Tungsten | $5.6 \times 10^{-8}$ |
| Iron | $9.7 \times 10^{-8}$ |
| Graphite | $8 \times 10^{-8}$ |
| Nichrome (toaster wire, an alloy of iron, nickel, and chromium) | $1 \times 10^{-6}$ |
| Glass | $10^{10}$ to $10^{14}$ |
| Hard rubber | $10^{13}$ to $10^{16}$ |
| Amber | $5 \times 10^{14}$ |

## REVIEW

**1.** The quantity of charge that passes a point in a unit of time is called a current. Give the formula that expresses this fact, and also give the units for current and charge.

**2.** When electrons in a cell or battery move, it is the result of chemical reactions inside it. True or false?

**3.** Give three factors that affect how much resistance a conducting wire offers to a flow of electrons.

**4.** State Ohm's law and define the unit of resistance, the ohm.

**5.** The conducting path electrons can take between a supplied potential difference is called an electrical circuit. True or false?

**6.** Define direct current and alternating current in a circuit.

**7.** State the equation for electrical energy in terms of the electric potential and the charge that goes through that potential.

**8.** Power is energy divided by time. In electrical terms, the equation to find the amount of power used is $P =$ ___. Fill in the blank, and give the units for each quantity in this equation.

**9.** Discuss the reason why the temperature of a wire increases when a current begins to pass through it.

**10.** Is a short in a circuit a point of high or low resistance?

**11.** Describe a fuse and a circuit breaker, and state their purposes in circuits.

**12.** A __________ circuit has only one conducting path for the electrons; otherwise, it is a __________ circuit.

## DEMONSTRATIONS

**1.** Figure 19–16 is a photo of a meter a local power company provides to measure the electrical energy used at a home or apartment. This meter registers the energy consumed in kilowatt-hours (kWh), a unit that is convenient for typical consumer use. (See Example 2 in "Calculations.")

The counting mechanism of an electric meter is similar to the odometer of a car in that it continually counts up to a certain number and then returns to zero before starting over. An odometer typically returns to all zeros after 99,999 miles; an electric meter registers all zeros after counting 99,999 kilowatt-hours. There are five dials on an electric meter, one for each digit. The reading on each dial is taken separately, according to the position of the hand on the dial. If the hand lies exactly on a number,

**FIGURE 19–16**
An electric meter at a residence.

that's the reading for that dial; otherwise, it is the last number that the hand has *passed*. The five digits, recorded from left to right, indicate the number of kilowatt-hours used since the meter was last reset to zero. To find the number of kilowatt-hours used during an intermediate period of time, two readings must be taken, and the numbers subtracted. The difference between them is the number of kilowatt-watt hours used between the dates of the two readings. The interval between readings (often 30 days) and the power company's charge per kilowatt-hour (typically 8¢ to 12¢ per kilowatt-hour in 1984) is indicated on the monthly bill to the consumer. The cost per kilowatt-hour multiplied by the number of kilowatt-hours used is an estimate of the amount due. (Tax and other charges may be extra.) Figure 19–17 gives an example of a meter reading in kilowatt-hours.

# EXERCISES

**1.** Explain how turning on a flashlight is like turning on the faucet at a sink. (Electricity on tap?)

**2.** Exactly what does an electric current always transport from one place to another? (Careful.)

**3.** Can you compare drinking a milk shake with two straws rather than one to using a thicker wire rather than a thinner one to draw current from a battery? Discuss.

**4.** What do you think would happen if someone dropped a large screwdriver across the terminals of a car's battery?

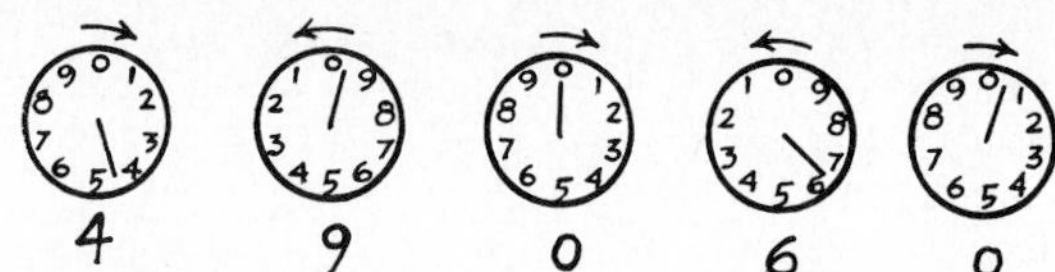


**FIGURE 19–17**
The reading on this meter is 49,060 kWh. To find the energy consumed between two dates, you must record the meter reading on each date and subtract the first reading from the second.

**2.** The current in a circuit can be adjusted with a device called a *rheostat*. In fact, a rheostat is a mechanical device that lets someone vary the resistance in a circuit with only the turn of a knob, and as the resistance changes, so does the current. Rheostats are commonly used in dimmer switches for home lighting, for the house lights in movie houses and theaters, and in radios and televisions for controlling the volume of the sound.

You can make a rheostat with a flashlight cell, a number 2 pencil, and some copper wire. Split the pencil into halves to expose the graphite lead. (As seen from Tables 18–2 and 19–1, graphite is a conductor, but not a very good one.) Connect a circuit as shown in Fig. 19–18. By varying the contact to use more of the graphite rod in the circuit, you increase the resistance and decrease the current, as you can see by the dimming of the bulb.

**FIGURE 19–18**

**5.** When you turn on an electrical appliance connected to the wall sockets in a room, do any new electrons come in or leave from the appliance? Discuss.

**6.** Two points on an object are at different electric potentials. Does a current flow between them? Discuss.

**7.** Exactly why will a small copper wire melt if connected between the wires in a home electrical socket while it won't melt if connected across the terminals of a flashlight battery?

**8.** Why will a 3-volt flashlight bulb quickly burn out if used with a 6-volt flashlight battery?

**9.** What does a fuse in an electric circuit limit, the potential difference or the current?

**10.** In the United States, accidental electrocutions occur at the rate of about 1000 per year. Many of these involve electrical appliances placed on the edge of a bathtub, on a shelf over a bathtub, or in a shower stall. Discuss the dangers there.

**11.** Discuss this statement: ''Power companies get all their electrons back, so they should just charge you a rental fee.''

**12.** In 60-hertz alternating current, how many times per second does an electron change its direction?

**13.** What purpose is served by connecting batteries in series? In parallel?

**14.** A certain flashlight can use a 10-ohm bulb or a 5-ohm bulb. Which bulb should be used to get the brighter light? Which bulb will discharge the battery first?

**15.** What is an open electric circuit? What is a closed electric circuit?

**16.** If a conductor is heated, its resistance usually increases, since the vibrating atoms interfere more with the electrons' motions. In the wire of an extension cord, this heating doesn't matter so much. Think of several cases where a change in the resistance due to an increase in temperature could make a difference.

***17.** Make an analogy with inclined planes to show why a wire 100 times as long as another will have an electric field 1/100 as large if the wires are connected to a potential difference of the same amount.

**18.** If an electron's mass were less but its charge were the same, would it accelerate to even higher speeds in a television tube?

**19.** What is the difference between a cell and a battery?

**20.** The 60-cycle-per-second alternating current stops only instantaneously as it reverses direction. Why don't you notice a flickering of an incandescent lamp as this takes place? (Recall from Demonstration 4, Chapter 1, that you can see the flicker that occurs in a mercury vapor or sodium vapor streetlamp.)

***21.** If you made a monstrous flashlight by connecting 73 flashlight cells in series, each of 1.5 volts, what sort of bulb should you use with it?

***22.** An electric clock that requires 4 watts to operate uses ________________ joules per second.

***23.** Calculate the resistance of a 60-watt light bulb that is connected to a 120-volt source.

***24.** How does it help to use 220 volts rather than 110 volts to operate an electric clothes dryer?

***25.** Compare the current used by a 1-watt flashlight bulb that runs on 3 volts and a 40-watt light bulb that runs on 115 volts. Discuss your answer.

***26.** If 2 coulombs of charge pass a point in 10 seconds, what is the current in amps?

***27.** If a television set uses 300 watts, how much current does it draw?

***28.** A bulb that has a resistance of 30 ohms is connected by copper wires across the terminals of a 6-volt battery. (a) How much current passes through the bulb? (b) How much electrical power does it use? (c) How much electrical energy does it use in one minute?

***29.** How many amps would flow through a household circuit if the following appliances were being used at the same time: two irons (1100 watts each), a hair dryer (1300 watts), and a vaccuum cleaner (600 watts)? Why would the wiring rather than the appliances heat up dangerously? Would a 30-amp circuit breaker break the circuit under this load?

***30.** Each headlight on a car uses about 3 amps of current. How long would it take to discharge a fully charged 12-volt battery rated at 80 amp-hours if the lights were accidentally left on?

***31.** An adult's food energy intake is about 3000 Calories, which is equal to about 3.5 kilowatt-hours of energy. Show that if you leave a 100-watt bulb and a 60-watt bulb on for 24 hours, they use more (electrical) energy than an adult uses in the same period.

***32.** Look up the utility charge per kWh in your area (use anyone's electric bill). Then see if electrical energy is cheaper than food energy. In other words, could you feed yourself for a day on the amount 3.5 kWh (3000 Cal) costs?

***33.** The United States generated about $2.37 \times 10^{12}$ kWh of electrical energy in 1982, and the U.S. population was about 232 million people. (a) Show that the share of electrical energy per person was about 10,200 kWh. (b) Using 3000 Cal per day as the food energy required by an average adult, show that a year's food energy amounts to almost 1300 kWh. (1 kWh = $3.6 \times 10^6$ J; $4.18 \times 10^3$ J = 1 Cal.) (c) Show that the average U.S. resident's share of the annual (electrical) energy amounts almost to the (food) energy needed by eight adults.

# CHAPTER 20
# MAGNETISM

The hopeful visitors converging on the fashionable Paris address were from the upper ranks of society, circa 1778. Each new arrival suffered from some type of medical complaint, from rheumatism to skin blemishes. Inside, the sober group gathered around an oaken tub some 5 feet across. Beneath its heavy wooden lid, dark, motionless water concealed layers of iron filings, powdered glass, and other artifacts. Those sitting closest to the tub grabbed iron rods jutting through holes in the cover; those in the second row clasped hands.

Suddenly their host appeared. Franz Anton Mesmer, dressed in a lilac-colored robe, and his young assistants made their way among the patients, pointing iron rods in the direction of each guest's affliction. Mesmer himself pressed his fingers on the offending area for as long as 15 minutes or more. Occasionally a patient underwent a "crisis" from this "treatment," becoming as animated as a small magnet in the presence of a larger one. Later some claimed they remembered nothing that had taken place. Mesmer thought that an invisible fluid essential to health acted on the human body and could be reinforced by ordinary magnets. His tub of iron filings and the iron bars were meant to focus extra magnetism on his patients and cure their ills. Of course, what was really happening is called hypnotism today, though it was first known as mesmerism.

By 1784 Mesmer was popular and rich, but he was also controversial. King Louis XVI appointed a distinguished committee of scientists, including Ben Franklin and Antoine Lavoisier, to investigate Mesmer's claims. They found no magnetism and no indisputable benefits in his treatments and attributed the crises to the patients' imaginations. But Mesmer wasn't deterred, and soon he turned up in London. England, too, had its skeptics. Hannah More wrote a letter to Horace Walpole in hopes that he would "be as severe as you please on the demoniacal mummery . . . . Mesmer has got a hundred thousand pounds by animal magnetism in Paris." But the disciples seemed to outnumber the skeptics. Mesmeric salons sprang up, selling both services and magnetic pharmaceuticals, including magnetic snuff. The idea was still popular in the 1840s and 1850s, and Alfred Lord Tennyson and Charles Dickens were among the practitioners. Dickens once wrote: "I have the perfect conviction that I could magnetize [hypnotize] a frying pan,'' but he seems to have concentrated his practice on his wife and a certain Italian duchess. And in hospitals of that time, before the days of anesthesia, doctors began to induce Mesmer's trances to help patients through surgery.

Discoveries in science come in many ways. Sometimes an investigator suspects what a particular result might be, looks for it, and finds it. But sometimes a discovery is purely accidental. *Serendipity*—a chance discovery of something you're *not* looking for—is a word Horace Walpole coined in the 1700s before Mesmer came on the scene. Mesmer's wild analogies of animal magnetism didn't work, but they did lead to the discovery of hypnotism. We'll discuss animal magnetism of a more realistic kind in the last two sections of this chapter. ■

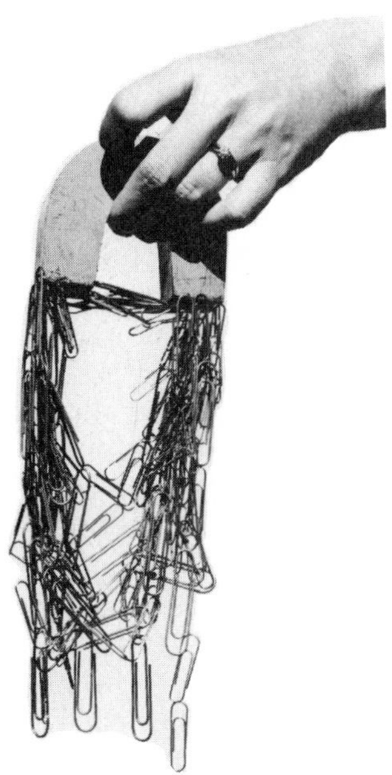

**FIGURE 20–1**
A magnet attracts bits of iron—here, paper clips—and causes those bits of iron to attract others. This magnetic attraction is stronger than the attraction of gravity for these objects, so they cling to each other.

## Magnetic Force

Legend has it that, once upon a time, a shepherd by the name of Magnus was pasturing his flocks in a province near Macedonia. The iron tip of his wooden staff, so the story goes, and his shoes, cobbled with iron nails, stuck to the ground (no doubt to his amazement), and the province became known as Magnesia. In this region rocks of the mineral magnetite ($Fe_3O_4$) abound and had been mined from early times. These rocks, called *lodestones,* exert a force that attracts bits of nearby iron. Moreover, attraction from a lodestone travels through a bit of iron to attract a second piece of iron–an act recognized in the following passage from a poem by the Roman poet Lucretius (96?–55 B.C.):

> Now sing my muse, for 'tis a weighty cause,
> Explain the Magnet, why it strongly draws,
> and brings rough Iron to its fond embrace.
> This men admire; for they have often seen
> Small Rings of Iron, six, or eight, or ten,
> Compose a subtile chain, no Tye between;
> But, held by this, they seem to hang in air,
> One to another sticks and wantons there;
> So great the lodestone's force, so strong to bear!*

Today manufactured lodestones—we call them **magnets**—are used in all kinds of ways. In refrigerator doors they provide the last little tug that snaps the door closed; they are used in radio and stereo speakers, and in children's toys. If you can find a couple of small magnets, you can observe how they act. (Try the local hardware or discount store, or ask your instructor.) Dip a small magnet into a bag of iron filings, and you'll find there are two spots on the magnet where many filings gather.

*Lucretius, *De Rerum Natura (On the Nature of Things),* trans. T. Creech, London, 1714.

**FIGURE 20–2**
The places of greatest attraction on a magnet are called its poles. One end of this bar magnet was dipped into a bag of loose iron filings, which cling to the area around its pole on that end.

These spots are called the magnetic **poles.** The magnet's force is strongest at these poles (see Fig. 20–2).

A common type of magnet is a *bar* magnet, a rectangular piece of iron that has been magnetized. Its two poles are near the ends of the bar. If you bring one bar magnet close to another, you can see that the approaching poles either *repel* one another or *attract* one another. Magnetic poles come in two kinds and obey the rule "opposites attract, likes repel." Since each magnet has a pole of each type, you could label them positive and negative, just as Ben Franklin labeled charges to keep them straight. But historically, they're called north and south poles for the following reason: Float a small bar magnet on a small piece of wood in water, and the magnet (and wood) will slowly turn so that one end of the magnet points somewhat toward the geographic north. That end, the magnet's north-seeking end, is the magnet's **north pole.** The other end, the magnet's **south pole,** points roughly in a southerly direction. (They align that way because the earth itself is a weak magnet, as we'll see later, with its south and north magnetic poles at very high and low latitudes.) A compass needle is just such a magnet suspended in a case of nonmagnetic material.

The magnetic poles of magnets, then, exert forces on each other. The force is called the *magnetic* force, and it follows a law very much like the force of gravity and the electric force. That is, the mutual force between two magnetic poles is much stronger when the poles are close together; and as the poles are pulled apart, the force gets weak very quickly. Some magnets are very strong compared to others. A bar magnet is normally much stronger than a compass needle. If $p_1$ represents the strength of one pole of a magnet and $p_2$ the strength of another magnet's pole, the **magnetic force** $F$ between them is proportional to their product, divided by the square of the distance between them. Or,

$$F \propto \frac{p_1 p_2}{d^2}$$

where $d$ is the distance between the poles. The poles push straight apart or pull straight together. (We won't bother with defining the pole strength $p$, since we won't use it in calculations. Likewise, we won't discuss the constant of proportionality corresponding to $G$ or $K$ in the formulas for the gravitational force and the electric force.) One interesting thing about this magnetic force law is that it lets you *feel* a force that, like gravity and the electric force, varies in strength inversely as the square of the distance between the interacting objects. That is, when the strengths of the poles $p_1$ and $p_2$ don't change, the value of the force depends only on the factor $d^2$ in the denominator of the equation. If you align two identical bar magnets end to end (north pole to north pole) and let them push apart for a short distance, say, less than half the length of the magnets, you feel an inverse-square *repulsion* with your hands. Then turn one of them around and feel an inverse-square *attraction*. (For more details about this, see Example 1 in "Calculations.")

So each magnet has a north and a south pole. What if you cut a bar magnet through the middle? You don't wind up with the north pole isolated on one piece and the south pole on the other, but you have two smaller magnets, each with a north and a south pole. You can prove this. Magnetize a nail by stroking it in one direction with one end of a bar magnet. Then test it with a compass to show that it has a north pole and a south pole. Now cut the nail in half with a pair of pliers and test the two new pieces. Each will have a north and a south pole. In a later section we'll see why magnets act this way.

## Magnetic Fields

Magnets, like mass and charge, somehow put a selective stress on the space around them. Any nearby magnets have forces exerted on them just because the others are there. We say a magnet has an invisible **magnetic force field.** Just like an electric field or a gravitational field, a magnetic field can be visualized by drawing *field lines* with arrows that show the direction in which a north pole is pulled by the force field. A south pole is pulled in the opposite direction.

The pattern of a bar magnet's field comes into view if iron filings are sprinkled on a sheet of paper and the bar magnet is brought up beneath the page (Fig. 20–3). Tap the paper, and the individual filings rotate to make a clear pattern, one that looks like the force field of a (+) and a (−) charge, a dipole. So the magnetic field around a north and a south magnetic pole resembles the electric field around a positive and negative charge. Turn the bar magnet beneath the paper, and tap

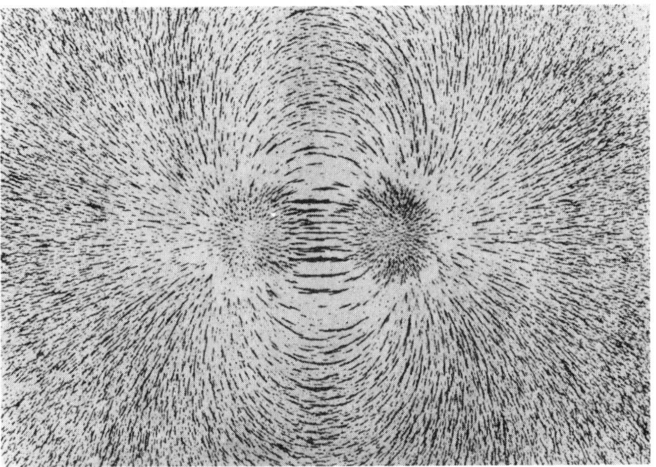

**FIGURE 20–3**
The loose iron filings on a sheet of paper responded to the magnetic field of a bar magnet that was placed beneath the paper. The filings simply turned, aligning their longest dimensions with the direction of the magnet's magnetic field. The direction of the field lines? The convention is they point away from north poles and toward south poles.

the paper again to see the iron filings shift and outline the field's new orientation. The needle of a small compass lines up with the direction of the filings when the compass is placed on the paper. Notice that the iron filings behave like compass needles; in fact, they become small compass needles in the presence of a magnetic field. (We'll see why in the section on magnetic induction.) Another way to trace magnetic field lines is to place a single compass in the magnet's field and move it all around, recording the needle's orientation at each new position Figure 20–4 shows how a group of compasses can be used to trace a magnetic field.

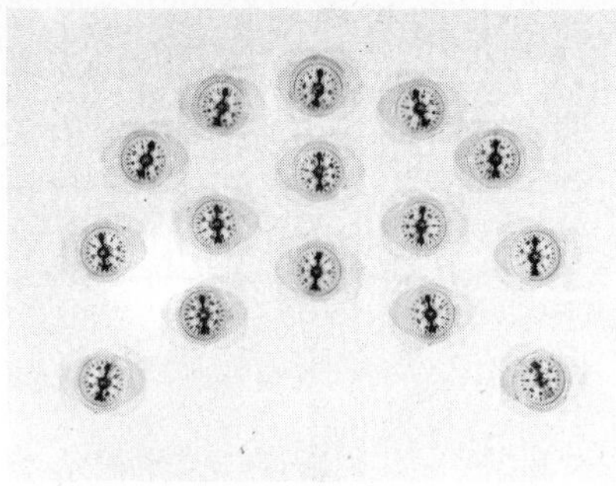

(a)

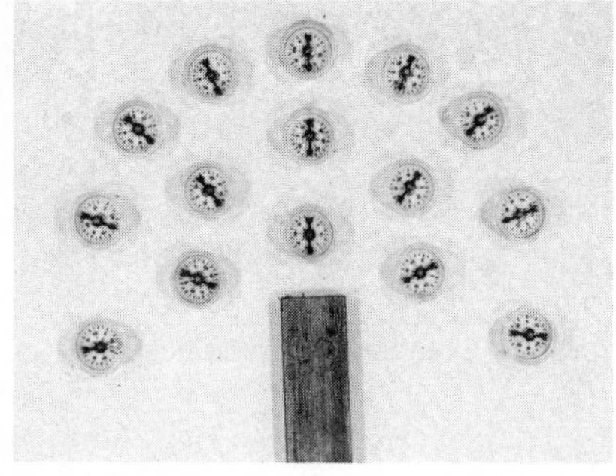

(b)

**FIGURE 20–4**
*(a)* The magnetized needles in these inexpensive compasses align (more or less) with the earth's magnetic field. *(b)* A magnet placed nearby produces a much stronger magnetic field than the earth's; as a result, the compass needles are seen to align with the magnet's field lines. If a compass is moved within the field, its needle responds to the change of direction in the field lines. Only a few feet from the magnet, however, its field decreases, and the earth's field dominates again. A compass that is taken only 5 feet or so from a magnet like this one will point toward the magnetic north pole of earth.

Magnetic fields pass undiminished through most materials, including most metals. As the demonstration with iron filings on paper shows, a magnetic field penetrates paper easily. You could do the same demonstration on a sheet of plastic or a thin piece of wood, or even a piece of aluminum or brass. These materials don't noticeably modify a magnetic field, nor does the field modify these materials. But iron behaves differently, as those filings show, and so do the metallic elements nickel and cobalt.

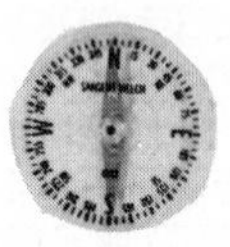

(a)

(b)

**FIGURE 20–5**
A pair of compasses pointing to the magnetic north and south of earth, abruptly realign when a natural lodestone is brought near them. Such lodestones floating in water served as the first compasses.

If a small piece of iron is placed in a magnetic field, examination of the field lines around the iron show they bend somewhat toward the iron. The lines of force are to some extent gathered in by the iron. If there is a cavity in the iron, the lines of force miss passing through it because they draw together or contract within the metal. Such a concentration of magnetic lines of force by iron is the way an antimagnetic watch works. A soft iron cover surrounds the moving parts, gathering in the lines of force of any local magnetic fields. The moving steel parts of the watch, then, are protected against magnetization and the resulting forces that would interfere with their actions.

## The Magnetic Field of an Electric Current

In 1681 an English ship sailing to Boston was struck by lightning. After the storm had passed, the sailors noticed that the ship's compass no longer pointed north. The north pole of the magnet pointed south and the south pole pointed north! Somehow the lightning bolt had reversed the magnet's poles. Nevertheless, using the wrong end of the compass for orientation, they came safely into Boston Harbor. The incident seemed to show that there was a connection between electricity and magnetism.

Many of the features of magnetic behavior resemble electrical behavior: Opposite magnetic poles attract, likes repel; the force laws are similar; a pair of opposite magnetic poles have a magnetic force field that looks like the electric field of a pair of opposite electric charges. As soon as these basic facts were known, natural philosophers began to search for a link between electricity and magnetism. Their experiments using electric charge and magnetic poles didn't show any forces or other changes between them when they were brought close together. Nonetheless, they wondered about *currents* of charge, such as in a bolt of lightning. Volta's invention of the battery in 1800 provided investigators with a source of steady currents, and the search continued. Some of the ideas were less than inspired. One group of amateurs set a battery afloat in hopes it would align itself north–south like a compass. (Oddly enough, none of them connected a wire between the battery terminals so that a current could flow.) Other attempts were more credible. Some investigators used batteries to send a current through a wire lying on a table. Then they placed a compass on the table *beside* the wire and watched for a deflection from the north–south direction that would tell them an additional magnetic force was involved. However, no matter

### A QUESTION

If magnetic poles are so like electric charges, could they be one and the same? A quick experiment gives the answer. Hang a small piece of glass or rubber from a string, charge it by rubbing, and bring up either pole of a bar magnet. There'll be no twisting or turning; there are no attractive or repulsive forces between the magnet and the charged substance. Likewise, a charge placed in the neighborhood of a compass needle doesn't deflect the needle. In this sense at least, magnetism is not electricity, and magnetic poles are not charges.

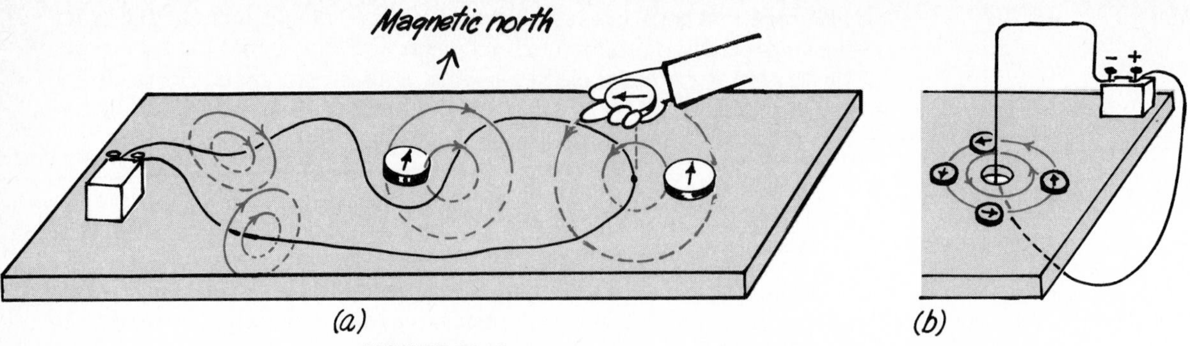


**FIGURE 20–6**
*(a)* Magnetic field lines form a circular pattern around a current-carrying wire (here the solid segments of each circle are above the surface of the table and the dashed segments are below the surface). The needles of the compasses lying on the table are not deflected by the field because to point in the direction of the field the needles would have to point into or out of the surface of the table, and they can't move in that direction. However, the needle of a compass held above the surface of the table and parallel to the field will be deflected. *(b)* If the wire goes through the surface of the table, the field lines are in the plane of the table surface, and the needles of compasses on the surface align themselves parallel to the field.

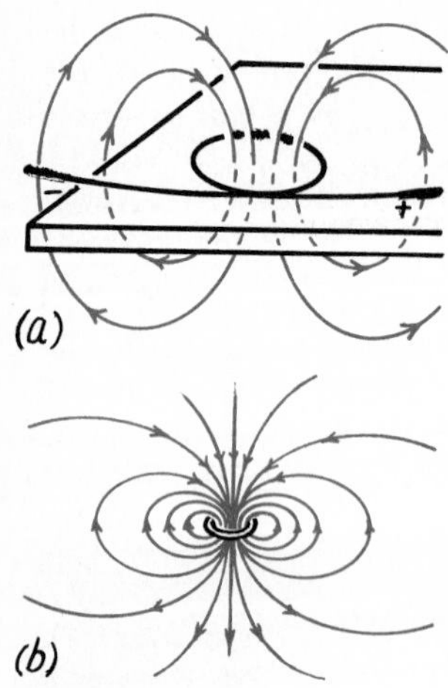


**FIGURE 20–7**
*(a)* When Ampère made a loop in a current-carrying wire, he saw that the loop's magnetic field was a dipole field like that of a bar magnet. *(b)* A more detailed view of the magnetic field of a current loop.

where they placed the compass or how the wire was arranged on the table, there was no deflection.

Then in 1820 there was an accidental discovery during a physics lecture in Copenhagen. Knowing the result of that experiment we just mentioned, Professor Hans Christian Oersted was showing his students that moving charges did not make a magnetic field. With the wire and compass both on the table, he showed there was no effect on the compass. But then he picked up the compass and inadvertently held it directly *over* a north–south segment of wire attached to a battery (Fig. 20–6*a*). He was amazed to see the needle twist toward the east–west direction. Serendipity! There *was* a magnetic field associated with the current of charge after all, and a strong one.

Before the discovery by Oersted, investigators had (quite logically) looked for magnetic lines of force pointing straight out from the current-carrying wire the way that electric field lines point out from charges. But the magnetic field made by an electric current is a field *without poles;* the magnetic field *circles* the current (see Fig. 20–6). A compass on the table wasn't deflected left or right because the magnetic field pointed straight up or down. Shortly after hearing of Oersted's findings, a French mathematician–scientist named Ampère made a loop in a current-carrying wire and discovered that the magnetic field inside and around that loop looks much like the field of an ordinary bar magnet.*

*We'll meet André Marie Ampère's (1775–1836) important contributions to electricity and magnetism in the next chapter. The unit of current, the ampere, is named for him.

He had discovered that *circulating electric currents have magnetic fields like those of dipoles* (see Fig. 20–7). Then Ampère speculated that currents in magnetic atoms (like iron) caused their magnetic fields. (This was 80 years before the discovery of the electron and about 100 years before models of the atom had electrons orbiting the nucleus.)

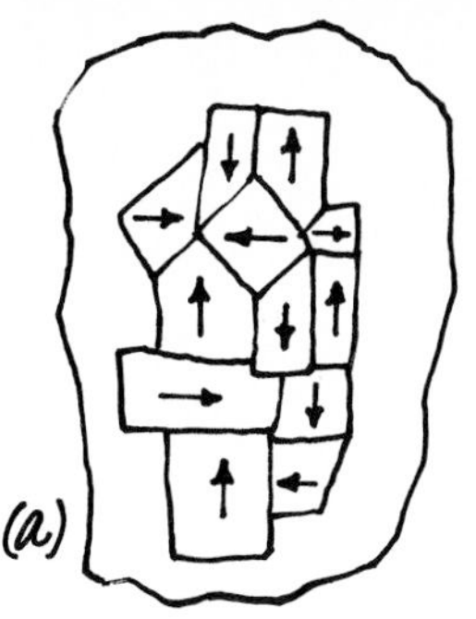


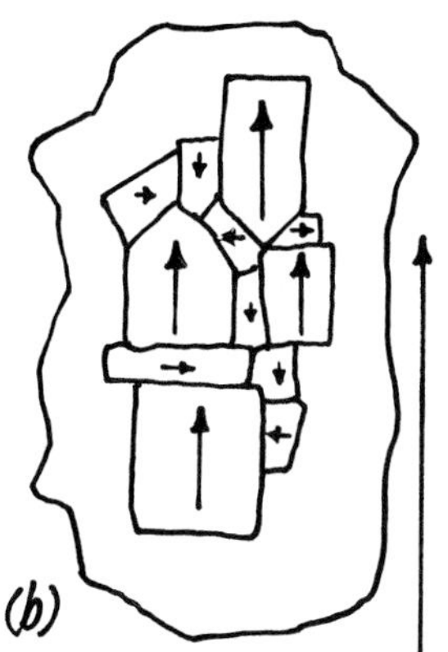


**FIGURE 20–8**
In an unmagnetized piece of soft iron *(a)*, the magnetic domains have random sizes and directions. The fields of the domains more or less cancel. But when the iron is in a magnetic field *(b)*, the domains whose fields are in the direction of the applied field actually grow, increasing in strength, while those in other directions shrink and decrease in strength.

## Magnets: An Atomic View

Today we know that electrons are the subatomic basis for the magnetism in permanent (or induced) magnets. The magnetic fields they cause come about in two ways. Electrons in orbital patterns around the nucleus can constitute a loop of current, and, as Ampère found, any current loop gives rise to a magnetic field like a bar magnet's field. Also, electrons (and protons and neutrons too) have magnetic dipole fields of their own. These **intrinsic** magnetic fields are unalterable fundamental properties of these particles. Whether in an atom's electron cloud or drifting freely, the electron has a magnetic field around it. It is this intrinsic magnetic field that is mostly responsible for permanent and induced magnetism in iron and other elements and compounds. Iron, for example, has a majority of its electrons' intrinsic dipole fields aligned in the same direction, giving each iron atom a net magnetic field. That's why you can cut a magnet in half once, then again, and again, and there will still be smaller magnets, all the way down to the atomic level. Each iron atom is itself a tiny magnet. However, not every piece of iron is magnetized.

In ordinary iron, small groups of perhaps a million iron atoms align naturally, their magnetic fields adding together. These magnetized groups of atoms are called **magnetic domains** (Fig. 20–8). But iron has a net magnetic field only if many of these domains are aligned, pointing in the same direction. Stroking a piece of soft iron, such as a nail, over and over in the same direction with a magnet aligns many of the domains and magnetizes the iron. Just tapping a nail in the presence of a strong magnet magnetizes the nail. However, the nail isn't a very permanent magnet. You can demagnetize it (or disalign the domains' directions) by a sharp blow or two with a hammer or even by just throwing the nail on the floor. But some iron alloys hold the domains' direction much better. If these metals are first taken to very high temperatures and cooled while in the grip of a strong magnetic field, the magnetic domains settle into place, aligned with the external magnetic field as a compass needle aligns with the earth's magnetic field. The permanent magnets used in household cabinet latches and refrigerator doors are made from magnetic grains of iron alloys that are frozen in place in epoxy that hardened while in a strong magnetic field. The resulting solid is a magnet that can be machined into any shape.

On the other hand, the magnetic domains in magnetized soft iron don't stay aligned so well. If the soft iron is heated or jolted by an impact, the direction of many of the domains will change, decreasing the net magnetic field. If you shake a box of small bar magnets, you can see how this might happen. They attract one another and tend to come together in pairs, side by side, north pole to south pole and south pole to north pole. This doesn't eliminate the magnetic fields of the pair entirely, but it does reduce their strength dramatically (Fig. 20–9). If you keep shaking the box (gently), pairs of such pairs will join, further reducing the net magnetic field of the box of magnets.

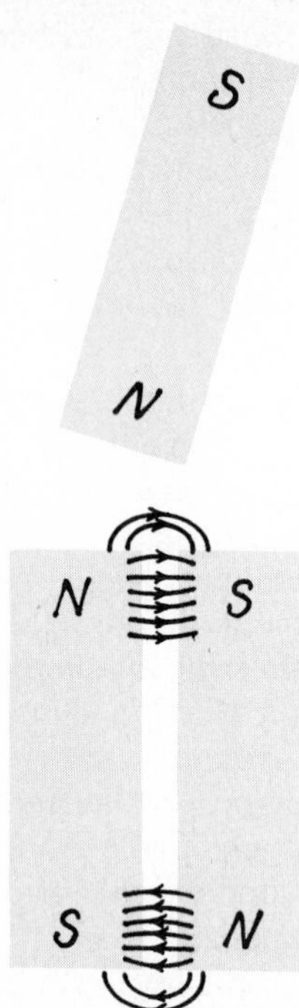


**FIGURE 20–9**
When two similar bar magnets align side by side but in opposite directions, they attract each other strongly, and the net field of the two almost cancels at points around the magnets. If a third magnet is placed above the two aligned magnets, it gets nearly equal and opposite forces from them.

## Magnetic Induction

Iron or steel in a magnetic field becomes induced; that is, the magnetic domains temporarily turn so that their magnetic fields add. This is how long strings of iron rings or nails or paper clips are able to hang from the pole of a magnet. Each becomes a temporary magnet. Remove the magnet, and their domains disalign immediately. This reduces the magnetic field of each, and the iron objects separate.

Devices called electromagnets use magnetic induction to create very strong magnetic fields with variable intensities that can be turned on or off in an instant. An **electromagnet** is a length of insulated copper wire wrapped around a soft iron core (Fig. 20–10). When current flows through the wire, each loop of the wire has a magnetic field like that of a small bar magnet, and the fields of all the loops add together. By itself this field is rather weak. However, the tiny but powerful magnetic domains in the iron core are induced by that field to align with the fields of the loops. In this way the *net* field becomes many times as strong as the field of the current loops alone. The more loops there are, the stronger the field is. Likewise, the greater the current through the wire, the greater the strength of the magnetic field.

The earth's magnetic field is weaker than the field of a toy magnet brought close to a compass (see Demonstration 2). Nevertheless, it does induce a magnetic field in iron, especially if the iron remains undisturbed for a long period of time. Jostled by normal thermal vibrations, a small percentage of the iron's magnetic domains respond to the earth's weak field and align themselves north-south. Stroll through your kitchen

**FIGURE 20–10**
An electromagnet uses the combined magnetic field of many loops of a current-carrying wire to induce a core of soft iron to become magnetic, thereby intensifying the net magnetic field (of the iron plus the coils).

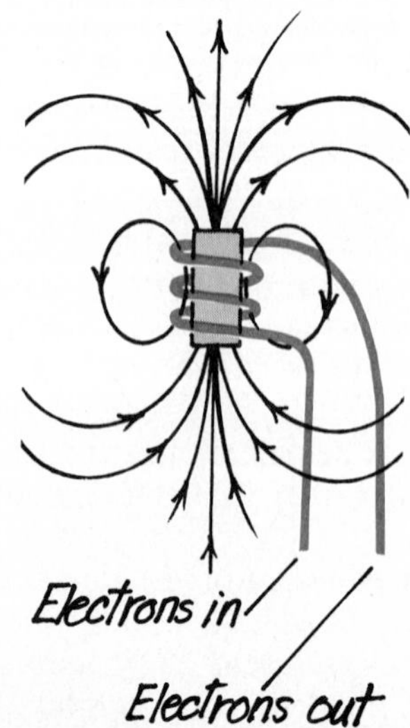

with a compass. Check the refrigerator, kitchen stove, and other iron or steel appliances whose positions don't often change. Chances are you'll find evidence of magnetic fields in these objects.

If an ocean-going vessel with a steel hull travels north, its bow temporarily becomes a weak north pole and its stern a south pole just because of induction by the earth's magnetic field. The mere size of the iron hull of a tanker makes its induced magnetic field strong enough to affect its own or any nearby compass. If the tanker heads south, the induced poles immediately reverse.

## MEASURING CURRENTS AND VOLTAGES WITH ELECTROMAGNETS

Electromagnets are the working parts of the standard instruments used to measure currents and voltages. The idea is simple. An electromagnet is mounted on an axle directly between the two poles of a permanent magnet. If a current passes through the electromagnet's coil (meaning that a voltage difference exists across the ends of the coil), it becomes a magnet, and its north pole is attracted to the permanent magnet's south pole and repelled by its north pole. That causes the electromagnet to turn on the axle, where a spring resists its motion. The stronger the electromagnet's field becomes, the farther it twists against the spring. A needle that rotates with the axle indicates the current (or voltage) on a dial.

The electromagnet that rotates in the field of the permanent magnet is called a *galvanometer* (Fig. 20–11). A single galvanometer

**FIGURE 20–11**
The tiny wire coil and iron core of a galvanometer temporarily become a magnet when current passes through the wire. Because this coil and its attached needle are in the field of a large permanent magnet, they twist according to the strength of the magnetic field induced by the current. An attached spring exerts a force to restore the needle to its original alignment, and the indicator stops moving when the opposing forces balance. If the current is increased, the temporary magnet is stronger and twists farther against the spring before the spring force counteracts it. The dial is calibrated to correspond to the current through (or the voltage across) the needle's coil of wire.

can measure currents or voltage over different ranges. The user merely flips a knob that adds or subtracts various resistors in parallel or series to the electromagnet's circuit. When a galvanometer measures currents, it is called an *ammeter,* and when it measures voltage, a *voltmeter*. By the way, the name galvanometer is a tribute to Luigi Galvani, who, while dissecting frogs about 1791, accidentally discovered that if two dissimilar metals touched a frog's leg muscle, the muscle twitched. Galvani thought he was activating animal electricity and he proceeded to investigate, like Mesmer with magnetism, the use of animal electricity to cure disease. But, as we've seen, Volta created an electric cell without animal parts in 1800, and this proved the downfall of Galvani's ideas.

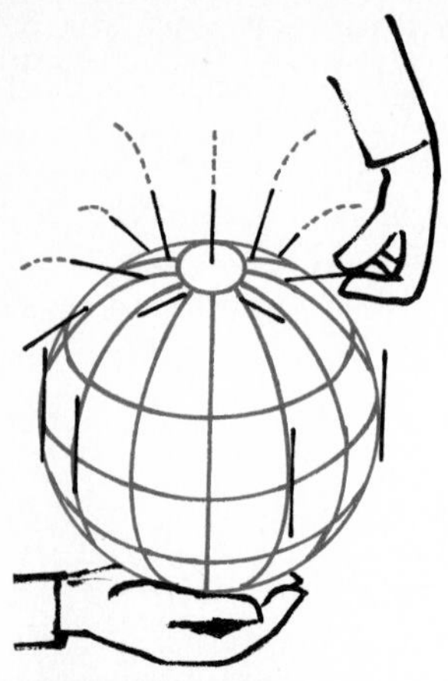

**FIGURE 20–12**
A terrela, or spherical magnet, illustrates the magnetic field of the earth. The iron needles show the directions that freely hanging compass needles at similar locations in the earth's magnetic field would have.

**FIGURE 20–13**
The magnetic poles of the earth are not quite aligned with the poles of the earth's rotational axis, and the magnetic "equator" doesn't quite coincide with earth's geographical equator.

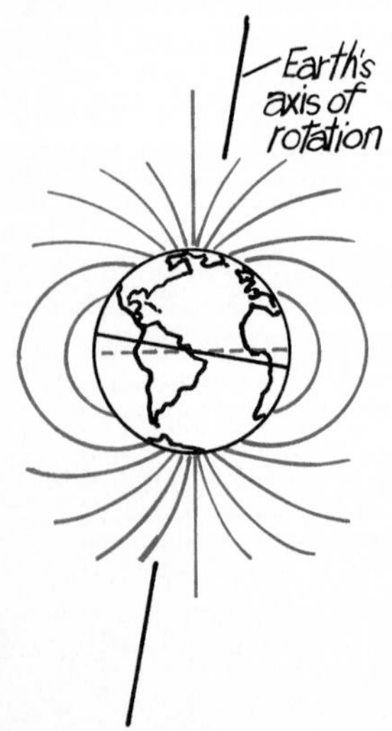


## The Magnetic Field of the Earth

Figure 20–12 shows how steel needles respond to the magnetic field at the surface of a *terrella,* which is a spherical lodestone. By induction they become small magnets and lie along the terrella's magnetic lines of force. Near the magnetic equator of the sphere, the needles lie flush against the surface, but nearer the poles the needles stand out against the lodestone at slight angles to the surface. At the magnetic poles of the terrella, a needle stands straight up on one end.

A compass needle that's held only by a string from its center of mass behaves in the same way in the earth's magnetic field; in fact, it's just as if the earth had a great bar magnet at its center. Figure 20–13 shows the orientation of the earth's magnetic field. The magnetic poles of the earth are more than 1000 miles from the earth's geographic poles, which are on the earth's axis of rotation. Just as with a terrella, the earth's magnetic field rarely runs parallel to the horizontal. The field lines usually dip toward or come out from the surface at some angle, which is called the **magnetic inclination,** or the **magnetic dip,** at that location. You can see the magnetic dip in your own geographical area by suspending a bar magnet with a string at its center of mass. Figure 20–14 shows an instrument for determining the magnetic dip.The magnetic dip anywhere in the continental United States and Canada is very large, as you can see from the field lines in Fig. 20–13.

The tendency of a compass needle to dip is a nuisance for compass users. To eliminate this motion in a compass made for use in North America, the needle is suspended off center, or even counterweighted on the southern end, so that it moves only in the horizontal plane of the compass. Even then, a compass needle points roughly toward the north *magnetic* pole, whereas a backpacker or a sailor needs to know the direction of the north *geographical* pole in order to coordinate a position on a map. The deviation of a compass reading from the direction of true geographical north is called the **magnetic declination** (Fig. 20–15). To plot an accurate course with a map, a person must know the magnetic declination in that area. At Daytona Beach, Florida, and Chicago, Illinois, the magnetic declination is close to 0°, so a compass needle at either city points nearly to true north. In Maine the needle points about

**FIGURE 20–14**
A magnetized pointer, free to rotate about its center, indicates the direction of earth's magnetic field lines at the location of the magnetized pointer. The magnetic dip is about 62° in Orlando, Florida, where this photo was taken.

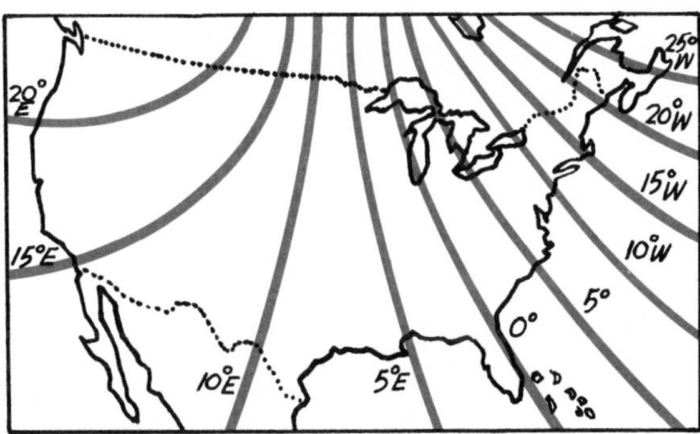


**FIGURE 20–15**
These lines of constant magnetic declination indicate the approximate deviation from true north that a compass reading would have at any point along the lines.

20° *west* of geographic north, while in Oregon it points about 20° *east* of the direction of the north geographic pole.

## Changes in the Earth's Magnetic Field

At any location on the earth's surface, the strength of the magnetic field, its declination, and its dip slowly change as time goes by. Variations in magnetic declination were noticed as early as the 1600s. In Paris the magnetic declination was 9° *east* of true north in 1580, 0° in 1665, and 22° *west* of true north in 1820. At present the magnetic declination at Paris is approaching 0° again. Such fluctuations are connected with the origins of the earth's magnetic field itself.

It is thought that the earth's magnetic field arises from electric currents in the molten iron region next to its solid core. The spinning earth carries any separated charge in its interior in circular paths, making current loops. Currents of charge far beneath our feet could be the source of the dipole magnetic field in and around the earth.

By way of comparison, space probes have discovered that the moon, Mercury, and Venus have no significant magnetic fields. Though Venus has about the same size and mass as earth, it rotates only once in 243 days, much too slowly to make a significant current of any separated charge deep inside. The moon, smaller than earth, is thought to be cooled enough to be solid throughout. Mars rotates much as earth does, but it, too, is small compared to earth and likely to have cooled to the solid state all the way to its center. Jupiter, on the other hand, is massive and rotates much faster (about 9.8 hours) than earth. As passing space probes proved, it has an intense magnetic field, much stronger than earth's. Jupiter's interior is thought to be largely liquid hydrogen, which becomes a good conductor when it is under great pressure.

Magnetic minerals are abundant in the rocks of earth's crust today. Many have retained the effects of the magnetic field at the geological time when they solidified. Such rocks show that the earth's magnetic poles have wandered extensively. The north magnetic pole strayed south of the equator in the Cambrian period, 500 to 600 million years ago, and another time even before that. In addition, they have actually reversed, the north magnetic pole becoming the south magnetic pole and vice versa. Over the past 5 million years a compass left at one location more often would have pointed south than north! In recent centuries, however, the average position of the wandering north magnetic pole has been very close to the present north geographical pole. A reversing of the magnetic field isn't exclusive to earth. The sun's magnetic field reverses regularly, with a period of 22 years.

## Magnetism and You

Anywhere there is an electric current, there is a magnetic field: in a bolt of lightning, in the circuits of a hand calculator, and even in the extremely weak ion currents that travel along the nerve cells in your body. The magnetic fields of the currents along these nerves are very weak, about one-billionth as strong as the earth's magnetic field, which turns compass needles. Whenever you reach out to touch something, nerves carry an impulse to the muscles you need to use and cause the contraction. This electrical impulse creates a temporary magnetic field. Two major sources of magnetic fields in the body are the heart (whose powerful contractions are triggered by large synchronous electrical impulses) and the brain. Strangely enough, if you drink a glass of cold water, the magnetic field around your middle increases slightly. Apparently there is an ion flow stimulated by cold liquid in your abdomen.

Because the mineral asbestos has magnetite in it, asbestos miners often have magnetic fields some 10,000 times larger than normal around their chests. The asbestos dust is harmful to the miners, but not because of its magnetism. The magnetite in the dust is insoluble in the body—it is inert. Therefore, asbestos in human lungs can be detected by its magnetic field before the concentrations are large enough to show up on X rays. The lungs of foundry workers, too, sometimes have large mag-

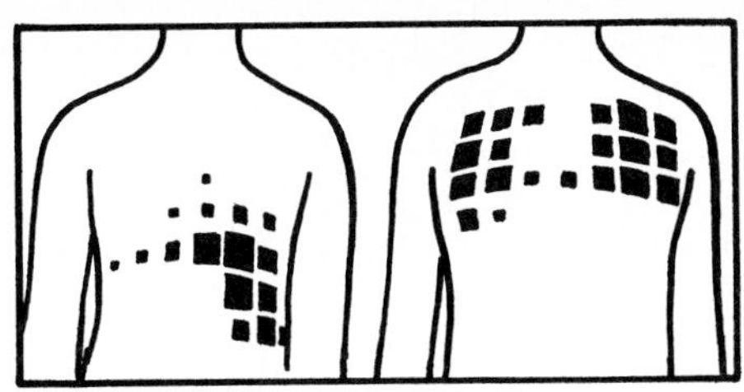

**FIGURE 20–16**
The magnetic profiles of two people just after an external magnetic field has aligned the iron particles in their bodies. Special magnetic detectors were used to pick up the particles' small magnetic fields which were about 7 factors of 10 smaller than the earth's magnetic field. The dark squares represent the intensity of the magnetic fields, which relates to the amount of iron present. The subject on the left had just eaten canned beans, which contained about 100 micrograms of iron oxide absorbed from the interior of the can. The lungs of the subject on the right, a welder, had about 500 micrograms of iron in them.

netic fields due to inhaled iron particles, as do those of arc welders. To detect such particles, the patient is painlessly magnetized in a medium-strength magnetic field (shades of Mesmer!), and any magnetic particles in this field align. The external magnet is turned off, and special detectors quickly measure the magnetic field of the aligned particles. The same technique is used to detect iron levels in the liver. With these devices, a medical physicist could tell if you'd just eaten canned beans by the trace amounts of iron in your stomach from the can (see Fig. 20–16).

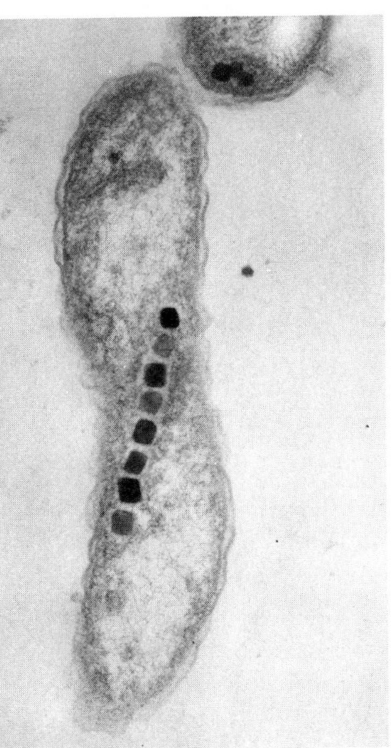

**FIGURE 20–17**
A closeup of the chain of magnetic domains in a magnetic bacterium.

## Other Animal Magnetism

During 1974 several species of aquatic bacteria were discovered that swim almost entirely along the earth's magnetic lines of force. These bacteria have their own built-in lodestones, perfect little crystals of magnetite put together from iron (taken from the water) and oxygen. A minute chain of these crystals, apparently of one magnetic domain each, points along the length of the cell. This strong internal compass responds to the earth's field and orients the bacteria along the field lines *through no action of their own*. In the laboratory, researchers have reversed the local net magnetic field, and the bacteria's lodestones adjusted the bacteria's direction of travel within a single second. Dead specimens also responded to the direction of the field, proof that it is the magnetic force that governs their direction of travel.

In the species of magnetic bacteria native to the Northern Hemisphere, the magnets' north poles point forward, away from the bacteria's tails. The best guess for the reason concerns survival. If a bacterium is stirred up from the bottom (its natural habitat), traveling north along the magnetic field lines in the Northern Hemisphere takes it back down to its home in the sediments of the pond or the saltwater marsh.

More precisely, the magnetic dip of the earth's magnetic field steers it down as it paddles. Because of its microscopic size and neutral buoyancy, the bacterium would otherwise wander aimlessly; gravity would not help it return to the bottom. Similar magnets have been found in algae and in shrimp, and magnetite crystals have been discovered in the abdomens of honey bees and in the brains of homing pigeons. We may be only the newest species on earth to use a compass.

## CALCULATIONS

Magnetic poles attract or repel with a force proportional to $p_1p_2/d^2$. But magnets, which consist of *pairs* of magnetic poles, exert net forces on one another that depend very much on their distance and angles of inclination to each other. The net force between magnets is seldom proportional to $1/d^2$. However, if two identical bar magnets are placed lengthwise along the same line and brought close together, they attract or repel each other with a force proportional to the inverse square of the distance between their closest poles.

---

EXAMPLE 1: Use a drawing of two bar magnets close together on a line (Fig. 20–18) to estimate the appropriate force vectors between the poles. **Sketch those vectors to convince yourself the *net* force on each magnet is (approximately) inversely proportional to $d^2$.**

FIGURE 20–18

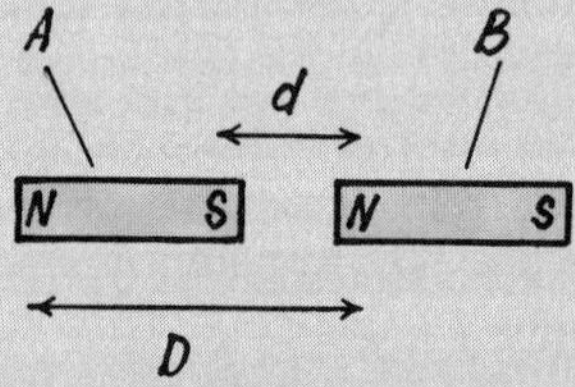


Each pole on magnet $B$ feels two forces, one from each pole of magnet $A$. There is, then, a net force on each pole of magnet $B$ to be found by adding those two forces. (The sum of the two net forces on the poles of magnet $B$ give the net force on that magnet. By Newton's third law, magnet $A$ gets an equal but opposite net force.) We'll first estimate the net force on the north pole of magnet $B$, and then we'll do the same for the south pole.

Here's how to estimate the net force on the north pole of magnet $B$. It gets an attraction from $A$'s south pole that's proportional to $1/d^2$. It gets a repulsion from $A$'s north pole that is proportional to $1/D^2$. If you estimate $D$ on the diagram with a ruler, you'll find $D$ is about $2.25d$, so

$$\frac{1}{D^2} = \frac{1}{(2.25d)^2} \text{ which is approximately } \frac{1}{5}\left(\frac{1}{d^2}\right)$$

The repulsive force is about ⅕ as strong as the attractive force; therefore, the vector force to the left is about 5 times as long as the vector force to the right:

and the net force on the pole is

The south pole of magnet $B$ is repelled by a south pole a distance $D$ (or $2.25d$) away and is attracted by a north pole that's about $3.5d$ away, as you can see with a ruler. There is a force to the right that is proportional to $1/(2.25d)^2 = \frac{1}{5}(1/d^2)$, and a force to the left that is proportional to $1/(3.5d)^2 = \frac{1}{12}(1/d^2)$. Drawing these on the same scale as before, we see the components

give a net force equal to

The net force on the magnet is the vector sum of the net forces on the north and south poles.

From these forces we see the dominating influence on magnet $B$ is the attraction between its north pole and magnet $A$'s south pole. That attraction is inversely proportional to $d^2$, the distance between them, so for small distances the two magnets attract one another with a force very nearly inversely proportional to the distance between the closest poles.

Here's an inverse square force *you can feel with your hands*. The force you feel when the magnets' closest poles are a distance $d$ apart is proportional to $1/d^2$. If that distance is doubled, the force is one-fourth as large. $1/(2d)^2 = \frac{1}{4}(1/d^2)$.On the other hand, if the distance $d$ is halved, the force is quadrupled. That is,

$$\frac{1}{\left(\frac{d}{2}\right)^2} = \frac{1}{\frac{d^2}{4}} = \frac{4}{d^2} = 4\left(\frac{1}{d^2}\right)$$

If you've never felt this force, it's worth finding two magnets to do it. It's very different from a spring's force, where if you double the displacement you only double the force.

## REVIEW

**1.** Like magnetic poles ________________, and ______ magnetic poles repel.

**2.** Give the formula that expresses the mutual magnetic force between the poles of magnets, and define the terms.

**3.** What other type of force field does the magnetic force field of a bar magnet resemble?

**4.** The magnetic field of a current in a straight wire has no poles; instead, it circles the current. True or false?

**5.** The magnetic fields of some of the electrons in iron atoms add to give iron magnetic properties. True or false?

6. What is a magnetic domain in iron?

7. Describe a method for aligning magnetic domains in a bar of iron, magnetizing the iron.

8. Electromagnets use magnetic induction to create magnetic fields that can be varied in intensity. True or false?

9. Sketch the magnetic field around the earth.

10. Define magnetic inclination, or dip, and magnetic declination.

## DEMONSTRATIONS

1. Put a paper clip into a jar of water and bring a large magnet up to the rim to see if the water or glass interferes with the magnetic field. Put another paper clip on a table and hold your hand an inch or less above it. Then hold the strong magnet over the top of your hand to see if your hand shields the magnetic field.

2. With a compass and a small magnet, you can compare the magnet's strength to the horizontal component of the earth's magnetic field. With the compass lying flat on a table, note the needle's direction. Then bring the south pole of a bar magnet along the direction perpendicular to the needle's direction. (How can you tell which is the south pole of the bar magnet? The south pole attracts the north-seeking pole of the compass; the north pole of the bar magnet would repel the compass's north pole.) When the magnet is close enough to deflect the needle to midway (45°) between the magnet's direction and the direction of the local magnetic field of the earth, the strengths of the two fields are about equal at that point. By the way, this demonstration shows that the magnetic pole in the earth's Northern Hemisphere is actually the *south* pole of a magnet—or else the north-seeking pole of the compass needle should point south.

3. Experiment with the strength of magnetic induction. (a) Attach a paper clip to a rubber band. Holding the end of the rubber band as shown in Fig. 20–19, you'll feel the force of attraction increase as the distance to the magnet decreases. (b) You can also hang weights from a paper clip that's suspended from a magnet (or combination of magnets) to see how much force is needed to break the magnet's grip on the paper clip.

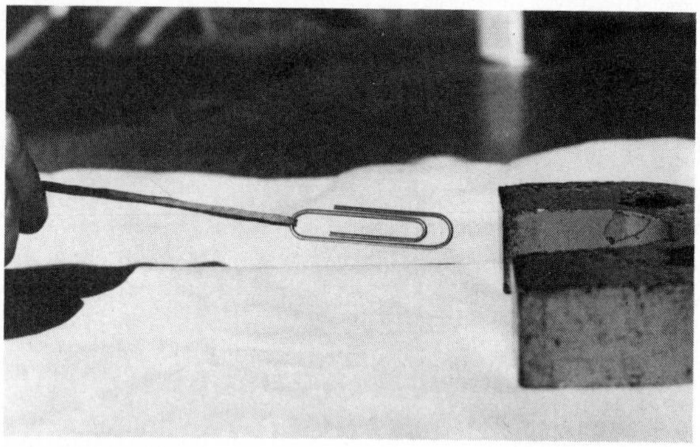

**FIGURE 20–19**

4. With a 1.5-volt cell, a short length of very thin insulated copper wire, and a large nail, you can make an electromagnet. First, connect the ends of the wire to the cell terminals to see if the current-carrying wire will pick up a paper clip. (It probably won't.) Don't leave the wire connected for long; the cell will run down quickly. Next, shave about a centimeter of the insulation from the wire somewhere near the middle and try to pick up a paper clip by touching it to the exposed wire. (It usually will. The magnetic field is stronger closer to the wire.) Now make a coil of the wire by wrapping it around your finger a few times, making sure the bare centimeter of wire lies at the very end of the coil. This time, when you connect the wire to the cell, the bare spot should be able to attract several paper clips at once. Then coil the wire tightly around a large nail. When this electromagnet is connected to the cell, the nail will lift more paper clips than the coil without the nail can.

## EXERCISES

1. Where are the lines of force most dense around a bar magnet? Why?

2. Where in the field of a bar magnet do the lines of force point straight from the north pole to the south pole?

3. Suppose someone handed you three similar iron bars and told you one was not magnetized but the other two were. How would you find the one that wasn't (without using the earth's magnetic field)?

4. Why will *either* pole of a strong magnet attract an iron nail?

5. Huge magnets at junk yards are useful for moving heavy steel junk around. These are always electromagnets and never permanent magnets. Why?

6. A steel garbage can more than likely becomes slightly magnetized between pick-up days. How could you reverse its magnetic poles?

7. If you performed Demonstration 1 in Chapter 11, you may have noticed something interesting about the

floating needle. If oriented in the east–west direction when the tissue paper sank, the needle rotated to the north–south direction. Why are sewing machine needles magnetized?

**8.** If a large steel bolt is held pointing north and downward at a large angle, it becomes temporarily magnetized by the earth's magnetic field when struck a few times with a hammer. (Try it! It will pick up several paper clips.) Why should it be angled downward as it is struck?

**9.** Using Fig. 20–12 again, decide if the north magnetic pole induced in a kitchen stove in Orlando, Florida, will be near its *top* or at its *base*.

**10.** Shortly after the discovery of magnetic bacteria in the Northern Hemisphere, a similar species was located in the Southern Hemisphere. However, the north poles of those bacteria were near their tails, opposite the orientation for those in the Northern Hemisphere. Explain this fact.

**11.** Tell how you can use an inexpensive compass to find the magnetic dip in your area.

**12.** Why can't you feel the $1/d^2$ attraction of gravity as you can with a magnetic force?

**13.** The path of the wandering north magnetic pole of earth for the last few centuries has been traced in studies of annual deposits of clay from river banks. Explain how such layers might reveal the north magnetic pole's direction.

**14.** The clappers of many doorbells are driven by electromagnets. What takes place between an electromagnet and a clapper?

**15.** Sensitive *magnetometers* (devices that detect and measure magnetic fields) can be carried in airplanes to search for ore deposits beneath the flight path of the plane. Which of these ores are likely to be detected in this way: Titanium? Copper? Iron? Magnesium? Nickel? Gold? Silver?

**16.** At which locations on earth would you *not* expect to find magnetic bacteria?

**17.** Saturn rotates in 10 hours. It has a diameter of 120,000 kilometers and a mass 95 times as great as earth's, and its magnetic field is at least twice as intense as earth's. Uranus rotates in about 17 hours. Its composition is thought to be much like Saturn's, and its mass is about 14 times as great as earth's. Would you expect it to have a magnetic field?

**18.** The magnetic dip in Florida is about 60° down from horizontal. From the looks of things in Fig. 20–12, would you expect the dip to be greater or less in Chicago, Illinois?

**19.** Magnetic stirrers are used in chemistry labs and hospital labs. First, a plastic-covered magnet is dropped into a beaker of liquid that needs to be mixed. The beaker is put on a small platform, a switch is turned on, and the stirrer begins to rotate rapidly. Explain how the device works.

**20.** Even the strongest permanent solid magnets are demagnetized if their temperatures exceed a certain value (different for each magnetic material). At that particular temperature, known as the *Curie temperature,* the effect of the forces holding the magnetic atoms in alignment is overcome by the thermal motion of the atoms. This normally happens well below the melting point of the magnetic material. For example, the Curie point of iron is 770°C, while its melting point is 1535°C. Considering this, you can see one reason why scientists think the earth's magnetic field is not due to permanent solid magnets inside the earth. Discuss.

**21.** Explain why a compass needle tells you the magnetic pole of the earth in the Northern Hemisphere is actually a south magnetic pole.

**22.** When steel ships are built in a shipyard, the metal plates of the hulls vibrate as rivets are inserted to hold them on. The jarring aligns some of the domains in the iron, which a ship can keep permanently. Likewise, the heated rivets cool in the earth's magnetic field at the shipyard, and they become magnetized as they lose their heat energy; the extra molecular jostling from the heat lets domains align more with the magnetic field. During World War II this permanent magnetization was removed at the shipyard before launch. The procedure began when workers wound coils of copper wire around the ship. Speculate on what the rest of the procedure was.

**23.** During World War II the Germans were able to take advantage of the magnetic field induced in any large steel ship by the earth's magnetic field. They mined English harbors with magnetic mines that would explode whenever the magnetic fields around them fluctuated, which occurred whenever a ship passed over them. Think of a way to neutralize a ship's magnetic field to avoid setting off a magnetic mine.

**24.** Figure 20–20 is a schematic of a speaker such as those in radios or TVs. Tell how a changing voltage across the wire of the electromagnet can make a sound wave.

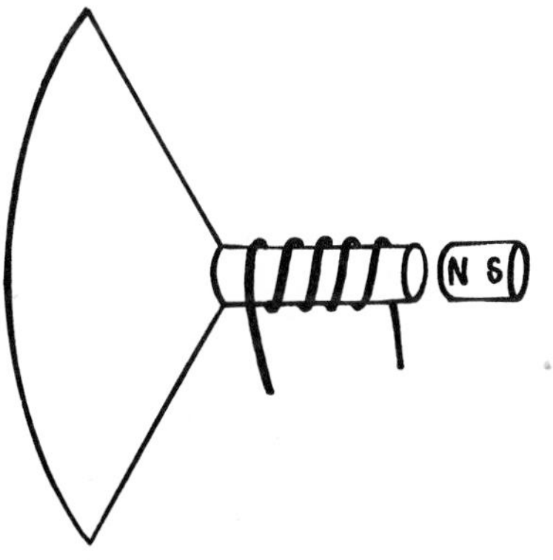


**FIGURE 20–20**

**25.** After Ampère's discovery that loops of conducting wire made magnetic dipole fields, he suggested that internal currents in iron probably caused the magnetic properties. Another scientist, Augustin Jean Fresnel, pointed out in several letters to Ampère that the ''Ampèrian'' currents must be on the atomic or molecular level because of the lack of heating. Discuss.

**26.** Will the north-seeking end of a compass needle still point north if the compass is taken to the Southern Hemisphere? Describe the magnetic dip of a compass needle in the Southern Hemisphere by observing Fig. 20–12.

**27.** If a bar of iron is bent into the shape of a horseshoe and then magnetized, its north and south poles are close together. What advantage does that offer?

**28.** What is a galvanometer? Describe how it works.

***29.** Repeat Example 1 in ''Calculations,'' using $d$, the distance between the two closest poles, equal to the length of one of the magnets. Convince yourself that the force at that distance is much less than a force proportional to $1/d^2$.

***30.** Describe how, with a small spring scale, a ruler, and two identical bar magnets, you could make a graph of the force between the magnets compared to the distance $d$ between the two closest poles. Could you use this graph to see if the force varied according to $1/d^2$ for small distances $d$?

# CHAPTER 21 ELECTRICITY AND MAGNETISM

News traveled slowly in horse-and-buggy days, so the message that came to Paris in September of 1820 was already 2 months old. But the timing, it seems, was perfect. Hans Christian Oersted's announcement that electric currents indeed produced magnetic fields was read aloud to the faculty at the Institute of France, and in the audience was the person who would make the next important steps connecting electricity and magnetism. Oersted discovered the key, and André Marie Ampère would use it to open many doors. Ampère left the meeting and began to experiment at once. One week later, he presented his own discoveries in a paper to the French Academy of Sciences. Knowing only that a current in a straight wire affects a compass, Ampère began with a simple experiment. He formed a loop in a wire, connected it to a battery, and investigated the magnetic field around it. He learned that the field looks and performs like the magnetic field of an ordinary bar magnet with a north and a south pole. Carrying this investigation further, he made a tight *coil* of such loops, sent a current through the looping wire, and found a similar but much stronger magnetic field because the fields of the loops added together. Then, with two separate coils, he managed to demonstrate that while current is flowing through them they act in every way like two bar magnets; a north pole repels a north pole and attracts a south pole.

At this point, Ampère made an even more important discovery, one that would become the basis for electric motors. He knew that forces behave according to Newton's laws. Applying the third law to Oersted's discovery that electric currents exert forces on compass needles or other magnets, Ampère reasoned that magnets (or magnetic fields) exert equal and opposite forces on currents. More than this, he concluded that the magnetic field from a current-carrying wire could exert a force on a current in another nearby wire. To test his idea, he strung two straight wires side by side and passed currents through them at the same time. It worked! When the currents were in the same direction, the wires drew closer together; if the current in one wire was reversed, the wires repelled each other. (This discovery gave people the means to produce physical pushes or pulls with electrical power, yet electric motors weren't invented until more than 50 years later.) Today we can explain Ampère's discovery in terms of the motions of the electrons in the wires. ■

## The Effect of a Steady Magnetic Field on a Charged Particle

As we saw in Chapter 20, a steady magnetic field exerts no force on a charged particle at rest in that field. If that same charged particle moves in a direction that takes it parallel to the magnetic field lines, there is still no force, no interaction between it and the magnetic field. But if that charged particle moves perpendicularly to the field lines and goes straight across those lines, the magnetic field deflects the charged particle, it causes that particle to turn. The charged particle gets a push that is perpendicular to the field lines *and* perpendicular to its velocity. Figure 21–1 shows the details. Because the force from the magnetic field is perpendicular to the particle's velocity, the force does no work on the particle and cannot change the particle's kinetic energy. (Remember that work = force × distance, where the force is parallel to the displacement of the object.) The magnetic field exerts only a *turning* force on the charged particle.

Figure 21–2 shows what happens when a charged particle has a velocity that carries it obliquely across the magnetic field lines. The particle's component of velocity parallel to the magnetic field is not affected, while its component of velocity perpendicular to the field lines turns continually. The particle follows a helical path.

This effect of a magnetic field on electrons explains why parallel wires push or pull on each other when currents flow through them. Each current generates a magnetic field that penetrates the other wire. The moving electrons deflect, turning to one side. In their collisions with metal ions, they give the wire a sideways push. Mere featherweights, whose mass accounts for less than one-tenth of 1 percent of the mass of the copper wire, the electrons collectively push the wire around nevertheless.

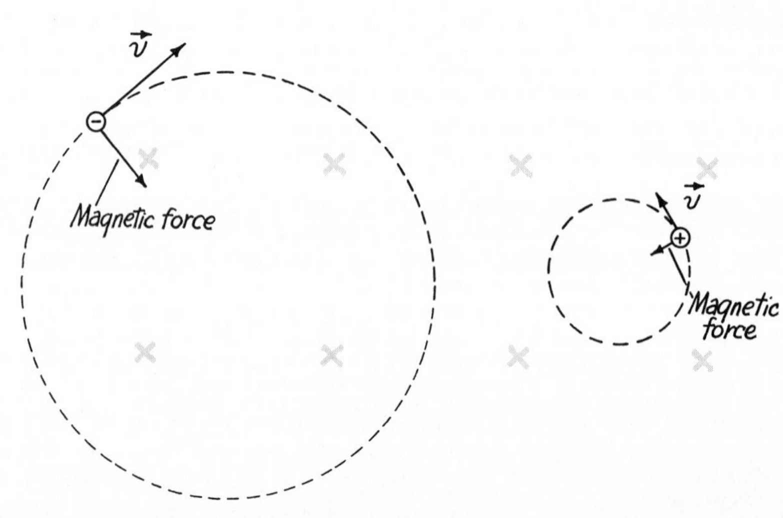


**FIGURE 21–1**
In a uniform magnetic field (here it is perpendicular to the page and going *into* the page, as indicated by the crosses), a charged particle moving perpendicular to the field lines receives a centripetal push from the magnetic force and moves in a circle. The greater the particle's speed, the larger the circular path. Opposite charges revolve in opposite directions. Here a positive particle has ⅓ the speed of a negative particle of the same mass, and its circular path has ⅓ the diameter of the negative particle.

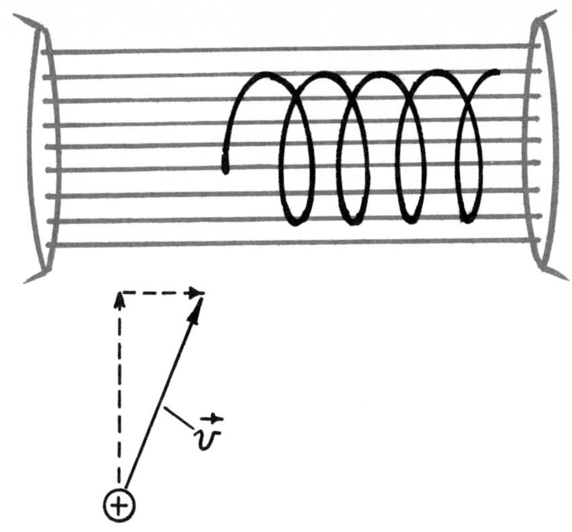


**FIGURE 21–2**
This particle has a velocity with a component parallel to the magnetic field lines in addition to one that is perpendicular to the field lines. The component parallel to the field lines is not affected by the magnetic field; the perpendicular component of the velocity turns in a circular fashion, as before. The result is a helical path shown in black.

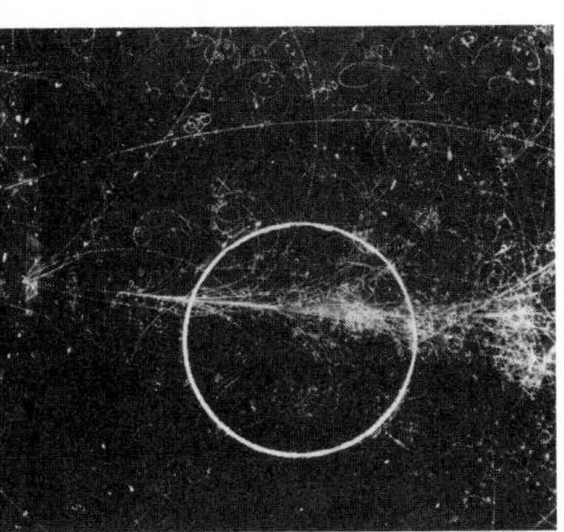

**FIGURE 21–3**
Subatomic particles leave tracks of bubbles in a liquid "bubble chamber." The charged particles travel in curved paths because of a strong magnetic field in the chamber. As they lose energy (and speed) they spiral inward.

## THE VAN ALLEN RADIATION BELTS

The strength of earth's magnetic field weakens with distance. Even so, it has a strong effect on charged particles from space that happen to approach it. Most of these particles are electrons and protons from the sun's atmosphere. Whizzing outward from the sun in every direction, they make up the solar wind. However, about 10 percent of the charged particles that come toward earth come not from the sun but from interstellar space; these are called cosmic rays.

As speeding charged particles approach, they encounter our planet's magnetic field. Any that cross magnetic field lines get a push perpendicular both to their own velocities and to the direction of the magnetic field. Their corkscrew motions are complex but result in many being trapped in two doughnut-shaped regions centered on earth's magnetic equator. They are called the **Van Allen radiation belts.** (James Van Allen, an American physicist, deduced the existence of these belts in 1958 from data sent to earth by Explorer 1.) The inner one is centered about 2000 miles above earth's surface; the outer one is some 10,000 miles overhead. The concentration of electrons and protons in these belts is thousands of times greater than that of the solar wind in earth's vicinity. Some of the particles in the inner belt have enough energy to penetrate several inches of lead. No space station meant for human occupation will ever be sent to orbit in that region.

As you can see in Fig. 21–4, the portions of the belts near the magnetic equator have higher altitude than the portions that dip close to the surface near the magnetic poles. At the magnetic poles, incoming charged particles aren't confined to the radiation belts; they have a clear path into the atmosphere itself. As a consequence, the number of charged particles from space that reach the earth's surface is

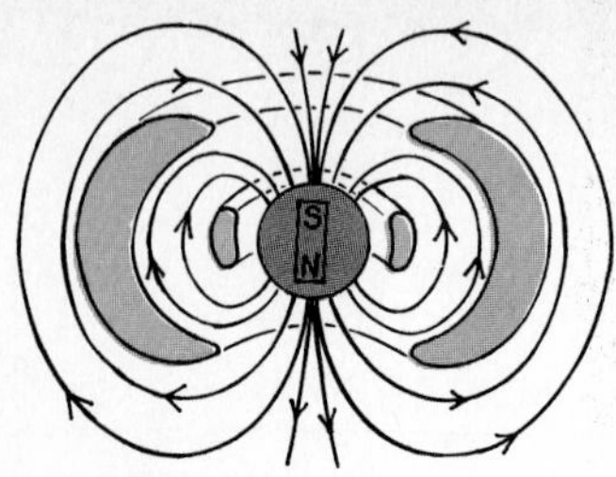


**FIGURE 21–4**
The Van Allen radiation belts around the earth. Incoming charged particles from space are trapped in these regions by earth's magnetic field. The motions of these particles aren't simple circles or helixes, however, since earth's magnetic field isn't uniform, as it is in Figs. 21–1 and 2.

greater at latitudes near the north and south magnetic poles and less in equatorial regions.

The moving charged particles of the solar wind, like any electric current, carry their own magnetic field. This field, in effect, compresses the earth's magnetic field lines on the side that faces the sun, while stretching out the field lines on the side away from the sun (see Figs. 21–4 and 21–5). The intensity of the solar wind picks up whenever a magnetic "storm" breaks out on the sun's surface. Such storms are signaled by the appearance of many sunspots or a nearly invisible solar flare (see Fig. 21–6). When this occurs, earth's magnetic field compresses even more, and the belts dip even lower at the magnetic poles. If the compression is severe, the belts actually spill charged particles into the upper atmosphere. At 60 miles up, those particles (mostly electrons) stimulate atmospheric atoms and molecules to emit light in an action called *luminescence* (we'll study this in Chapter 24). That light is called an *aurora*. If it happens near the north magnetic pole, it is an aurora borealis (or northern lights); near the south magnetic pole, it is an aurora australis.

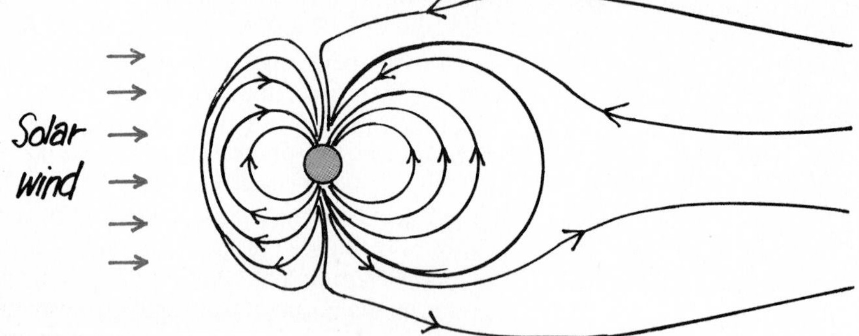


**FIGURE 21–5**
Farther out into space, past the Van Allen belts, the earth's magnetic field is influenced by the solar wind, in effect resulting in a magnetic "tail" that sweeps outward, away from the direction of the sun.

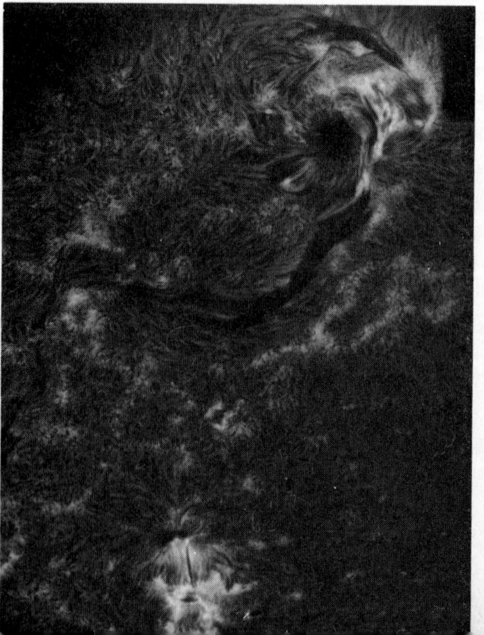

**FIGURE 21–6**
A sunspot (dark area at upper right) on the surface of the sun. Sunspots appear where the sun's magnetic field is especially intense. When sunspot activity is at a peak, the solar wind is also intense.

## Electricity from Magnetism: Michael Faraday

We now return to the story of the quest to relate electricity and magnetism. While Ampère continued to work in Paris, the next great step was taken in London by Michael Faraday. Through his teens, Michael Faraday had been a bookbinder's apprentice. But after listening to some public lectures given by the famous chemist Humphrey Davy, he decided at age 21 to change trades. He made careful notes of Davy's lectures, bound them in leather, and sent them as a gift to Davy at the Royal Institution. Soon he was working in Davy's lab as a bottle-washer.

Faraday had no formal education and never acquired a working knowledge of mathematics; he has been called a mathematical illiterate. Instead of using mathematics to describe physics, he composed physical pictures to help him understand things. It was Faraday who first used imaginary field lines to describe the magnetic field and the electric field. And he is regarded as one of the finest experimenters the world has ever known. His notebooks reveal that as early as 1822 he had sought to "convert magnetism to electricity." He even carried a small magnet and a coil of wire in his pocket to tinker with at odd moments. But it wasn't until August 1831, that Faraday made a breakthrough: *He discovered how to use magnetism to produce, or generate, electric currents.* This is how he found what he had sought for so long.

He coiled an insulated wire around one side of an iron ring, and on the other side he coiled another. Faraday connected the first coil to a battery. Current flowed through this coil, bringing forth a magnetic field in the iron ring. The iron amplified the magnetic field (billions of tiny magnetic domains in the iron snapped into alignment with the coil's field), and the ring guided it through the region surrounded by the second coil. Faraday noticed that when he touched the ends of the wire of the first coil to the battery, the needle of the galvanometer attached to the second coil jiggled. It jumped quickly to one side and just as quickly returned to zero, indicating a brief flow of charge in the second coil. When he disconnected the first coil from the battery, he noticed that the needle skipped again, but to the other side, indicating a flow of charge in the opposite direction in the second coil. In between, when the current through the first coil was steady, the galvanometer needle showed no current flow in the second coil.

Faraday reasoned that the first jiggle took place while the current was increasing in the first coil; that is, before it reached its steady rate.

**FIGURE 21–7**
Faraday's experiment. Only when the current *changes* in the first coil does the magnetic field change across the second coil and induce a current there. If the current in the first coil is steady, its magnetic field is steady also, and no current is induced in the second coil. In such an arrangement with a battery, the current changes only briefly whenever the circuit is closed or opened.

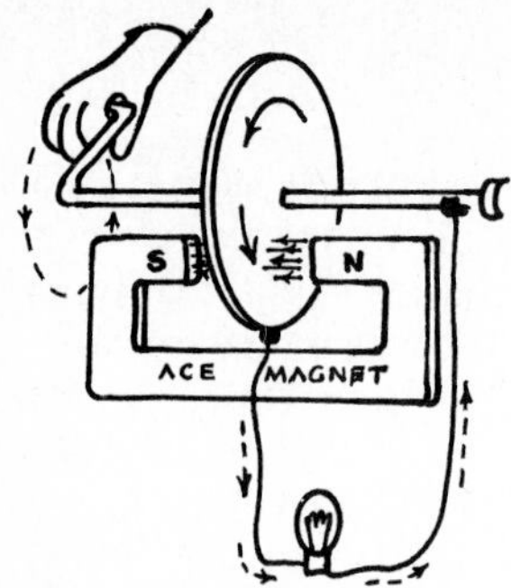


**FIGURE 21–8**
If a metallic disk is turned in a magnetic field, the mobile electrons in that disk get a sideways push as they are carried perpendicular to the magnetic field lines. If a complete circuit is formed, a current will flow. Such a device is called an electric generator.

As the current grew, so did the magnetic field, and the growing magnetic field spread through the second coil. This was when the needle jumped, showing a flicker of current in that coil. When the current in the first coil reached a steady value, the magnetic field in the second coil was no longer increasing, and the needle returned to zero. Then as he disconnected the battery, the magnetic field of the first coil collapsed to zero and retreated across the second coil. It was this *change* taking place that caused the second jump of the needle, which showed there was a flash of current in the other direction. Faraday had discovered that *a changing magnetic field induces a current in a conductor*. His discovery is called **electromagnetic induction,** and he later "explained" this action with his picture of magnetic field lines.

As we saw earlier, Ampère's findings revealed that when a charged particle crosses magnetic field lines it gets pushed to one side. Faraday's findings revealed that if magnetic field lines (as in a changing field) cross a charged particle (as the electrons in the second coil of wire), the particle gets a similar push. In summary, then, any *relative motion* between magnetic field lines and a charged particle causes a force on the particle. It doesn't matter whether the particle moves across the magnetic field or the magnetic field moves across the particle. Either way, the particle gets a push.

Change the magnetic field in any way across a closed conducting circuit, and a current will flow. In fact, if you merely wave a coil of copper wire in the air, as Faraday once did, the weak magnetic field of the earth induces enough current in that coil to deflect the needle of a sensitive galvanometer. Push a magnet into a coil of wire, and a current flows through that coil. Or just take a class ring or a gold wedding band and use one finger to stand it vertically on a tabletop. Flick it on one side with another finger to set it spinning like a top. You guessed it—a current goes around the circle of the ring because of Mother Earth's magnetic field.

## ABOUT ELECTROMAGNETIC INDUCTION

If magnetic field lines move across a charged particle, it gets pushed. *An electric field pushes it; an electric field appears whenever a magnetic field changes in time*. Recall that the moving charges of a current create a magnetic field. That magnetic field arises because of the motion of the electric field carried by those charges. So a change in a magnetic field produces an electric field, and a change in an electric field produces a magnetic field. That is what is meant by *electromagnetic induction*. In the 1860s, some 30 years after Faraday's discovery, another physicist, James Clerk Maxwell, put together everything that was known about electric and magnetic fields and came up with an amazing insight. As we'll see in the next chapter, the stuff we call light can be explained as coming from electromagnetic induction.

## Generators of Electricity

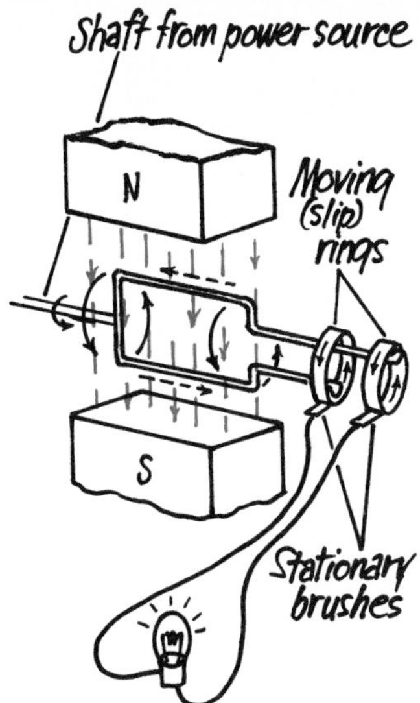


**FIGURE 21–9**
A schematic of a modern generator (showing only a single loop of wire) that generates an alternating current and voltage. Halfway through each revolution of the loop, the current reverses direction as the sides of the revolving loop change their directions across the magnetic field. Stationary bits of graphite, or brushes, press against the rotating rings to provide a complete circuit so current will flow. A steam turbine usually provides the power to turn the coils through the magnetic field, although falling water or wind can be used.

Faraday saw right away how he could use his new discovery to produce an electric current. He mounted a solid copper disk on a conducting axle and brought a magnet up close beside the disk. When he turned the disk, the free electrons in the metal passed through the magnetic field lines of force and received a sideways push. To provide a closed circuit for the moving charges, he touched one end of a copper wire to the axle and the other to the edge of the spinning disk. The moving electrons in the disk then had a complete conducting path and a push along that path; hence, a current flowed through the wire.

Faraday's first simple, hand-cranked generator (Fig. 21–8) soon evolved into larger and more complex machines to convert the energy of falling water or high-pressure steam into electricity. But they all work in the same basic way. Rather than a spinning disk, one or more coils of wire circle an axle that rotates in the magnetic field(s) of one or more magnets. The falling water (or rushing steam) pushes the blades of a turbine to turn the axle and its attached coils. Figure 21–9 shows how a single loop of a hand-cranked generator works. Turning on its axle, one side of the conducting loop has a velocity that is in exactly the opposite direction from the other side of the loop. Electrons on one side of the loop get a push one way, while those on the other side get an opposite push. If the ends of the loop are connected to a circuit outside the generator, a current will flow. (Electric potential is still generated in the loop even when the circuit is open, but electrons won't move around the loop. The circuit must be closed for current to flow.) Notice what happens after half a turn of the loop. Each side then travels in the opposite direction from before, and the direction of the current through the bulb reverses. This means the generator gives an alternating voltage and current (AC). In commercial power plants in North America, electrical generators deliver an alternating voltage that oscillates 60 times a second. If you ever visit Europe, you will most likely need an adapter for your hair dryer or electric razor; commercial electric power in England, for example, is delivered at 220 volts rather than 110. The generators there and in Europe oscillate at 50 times a second, rather than the 60 hertz that's the standard in North America. Commercial clocks that use 60-hertz voltage to keep time in the United States register only 50 minutes for every hour when connected to European power lines.

Years after the breakthroughs of Faraday and Ampère, other scientists and inventors learned to make practical use of the power produced by electrical generators. The first commercially produced light bulb appeared in 1880, and in no time, New York City had the world's first electric power utility plant to sell electrical energy to the public. As conducting wires from that plant were taken to farther and farther points, the longer lines meant more resistance. The wires turned electrical energy into heat as a current passed. Soon much of the electrical energy produced by the generators was wasted in transmitting the power. An efficient method of transmission was needed, especially to take advantage of more remote natural energy resources such as Niagara Falls. Its falling water could drive electric generators at minimal cost, but the great distance from population centers was a drawback. In the

### THE REWARDS OF SCIENTIFIC RESEARCH

Faraday could never have imagined the ultimate effect his generator of electric current would have. Indeed, he knew he couldn't foresee its consequences, as we can tell by his replies to inquisitive people who asked the question, "Of what use is it?" He startled one matron with the answer, "Madam, of what use is a newborn baby?" At another time, the prime minister of England saw Faraday's demonstration and asked what use such an invention might possibly have. Faraday replied, "Why, Sir, someday you will be able to tax it." That was a prophetic answer, as anyone who pays electricity bills knows.

The same question is often asked of people who do research in science today. "Why should the government (or industry, or private foundations) support the explorations of scientists that may well never pay off (especially by fiscal year 1987 or 1988)?" "What could society miss by not sponsoring scientific research?" is a better question. Historically the answer is clear: A *lot*.

next two sections we'll see how alternating currents make possible an efficient method of transmitting electric power. You can play an album, fry an egg, or wash clothes with the help of electricity only because electrical power can be transmitted in conducting wires for long distances.

## Transmitting Electricity: Transformers

When a current encounters resistance, it turns electrical energy into heat, which removes that energy from other use. There is nothing a power company can do at present to reduce the resistance of the long copper wires (called powerlines or transmission lines), but it can and does reduce the current the transporting lines carry to minimize the energy loss in the wires.* How does a generating plant deliver power with low currents? The answer lies in the formula for electrical power, power = current × voltage. To deliver power at *low* current requires a *high* voltage.

Commercial AC electric generators produce electrical power at a modest voltage, perhaps 2200 volts. Then the voltage is increased to 120,000 volts or more before it is applied to the long-distance transmission lines. (Remember, electrons aren't transmitted from the power plant to a home; only energy is transmitted.) In the vicinity of the con-

*In "Calculations" of Chapter 19, we saw that a resistance turns energy into heat at the rate $P = i^2R$. The $R$ for long transmission lines is large, but if $i$ is made very small, the power loss is still manageable. For example, see Example 3 in this chapter's "Calculations" section.

**FIGURE 21–10**
*(a)* In a modern utilities plant, fossil or nuclear fuel is used to make steam to turn the turbines, and the turbines turn the electrical generators. The voltage output from these generators is immediately stepped up (increased) for economical transmission via transformers. *(b)* High-voltage power lines connect these transformers to transformers at substations near the localities where the electrical energy will be used. *(c)* The electric lines on neighborhood power poles convert that high-voltage electricity to lower voltages suitable for home use with another step-down transformer. *(d)* From that transformer, electric wires lead directly to homes and businesses.

sumers' homes, the voltage from the high voltage (or *high-tension*) wires is lowered to 2200 volts or so. Then even closer to home or dorm, usually at a power pole a block or less away, it is converted to lower voltage that is safer to use, normally 220 volts or 110 volts for use in the wall receptacles. Both the upward and the downward voltage conversions are done with *transformers,* devices much like the one Faraday used to discover electromagnetic induction (Fig. 21–10).

A **transformer** is two separate coils of insulated conducting wire wrapped around an iron core that intensifies and guides any magnetic field that pierces it. If one coil (the **primary coil**) carries a current, the magnetic field it produces is shared almost totally with the other (**secondary**) coil. The primary coil carries an alternating current, which surges back and forth creating an alternating magnetic field. The moving magnetic field lines cut across the second coil and induce a force on the movable electrons there. The net magnetic field that goes through each loop on the primary also passes through each loop on the secondary. Faraday and others discovered that *this equal magnetic coupling of all the loops causes each loop to have the same voltage across it.* No matter what the voltage is across the primary coil or how many individual loops it contains, the voltage per loop in the primary is the same as the voltage per loop induced in the secondary coil. (Notice a constant direct current wouldn't generate a changing magnetic field, so the voltage from a DC generating plant couldn't be stepped up or down with transformers unless the direct current changed frequently.)

Here is what happens in a transformer that has 5 primary and 20 secondary loops. Suppose there are 10 volts across the primary coil, giving it 2 volts per loop. The secondary coil will have 2 volts per loop induced for each of its 20 loops, for a total of 40 volts across it. The voltage coming from the output side of this transformer is greater than the voltage entering the primary coil. In electrical jargon, the voltage has been stepped up. In a step-down transformer, the secondary coil has fewer loops than the primary coil.

Even though transformers can multiply voltage in this way, they cannot multiply energy. Because energy is conserved, the power out can't be any more than the power in. Transformers waste very little electrical energy. In commercial transformers the electrical energy coming from the secondary coil each second is more than 90 percent (sometimes up to 99 percent) of the energy going through the primary each second. Since electrical power is equal to current times voltage, this means

$$\text{current}_{\text{primary}} \times \text{voltage}_{\text{primary}} \approx \text{current}_{\text{secondary}} \times \text{voltage}_{\text{secondary}}$$

where $\approx$ means "approximately equal to."

This equation shows how voltage can be stepped up by a transformer without "multiplying energy." If the voltage is stepped up in the secondary, the current is stepped down by the same factor: A voltage in the secondary that's four times as great as the voltage in the primary means the current in the secondary is about one-fourth the current in the primary.

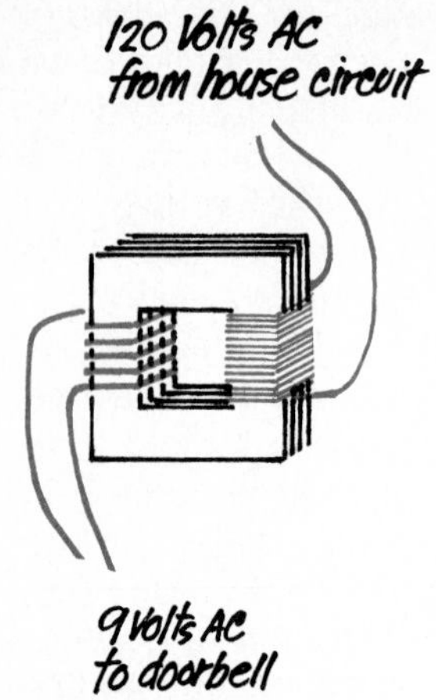

120 Volts AC
from house circuit

9 Volts AC
to doorbell

**FIGURE 21–11**
In a transformer such as this one, each loop of wire experiences the same changing magnetic field and each loop gets the same voltage difference through it. This transformer has 4¾ loops in its secondary coil, so if the voltage output from that coil is 9 volts, each loop contributes about 1.9 volts. The primary coil, which has 120 volts applied across it, must therefore have about 120 V/1.9 V, or 63 loops.

The next time you press the button on a doorbell, think of a transformer (Fig. 21–11). A small transformer reduces the voltage between doorbell and button so that a wet finger on it is less likely to receive a serious shock. Likewise, a toy electric train runs at a reduced voltage to protect the children's fingers (and the dog's nose) should they complete a circuit by touching both rails of the track while the switch is on.

## Motors

Washing machines, refrigerators, vacuum cleaners, blenders, dishwashers, fans, and record turntables all use electric motors that convert *electrical* energy into *mechanical* energy (Fig. 21–12). This is how they do it: A coil of wire that's wrapped around an axle sits in a magnetic field. You flip a switch, and the ends of this coil complete a conducting path with the electric circuit in your home. A current then flows in the coil. As electrons in the wire move through the magnetic field, they get a sideways push that also gives the wire a sideways push. That push causes the coil to rotate and turn the motor's axle. In a sense, a motor is an electrical generator that works in reverse. Whereas an electrical generator turns mechanical energy into electrical energy, an electric motor turns electrical energy into mechanical energy. There's a surprising bit of history behind this. Faraday's first generator was made in 1831, but the electric motor did not come into being until 1873, at an exhibition in Vienna. A worker there made a mistake in connecting some conducting wire from one electrical generator to another. When the first generator was switched on, the second came into action, its coils whirling around at a great speed, transforming the electrical power of the first generator back into mechanical power in the second. Needless to say, it was a significant accident!

## ELECTROMAGNETIC LAUNCHERS

Physicists and technicians continue to find new ways to use electrical energy. Examples in communications and computers are everywhere, but there is one use of electrical energy that for sheer power sounds like something out of science fiction. Electromagnetic *launchers* use the interaction of electricity and magnetism to propel objects up to extreme speeds, about 25,000 miles per hour (or 40,000 kilometers per hour).

The idea is simple. Two parallel conducting rails are bridged by a freely sliding metal bar. A direct current along one rail crosses the sliding bar and travels back along the second rail in the other direction. The magnetic field from *both rails* pushes on the current of electrons as it travels across the moving bar. As Fig. 21–13 shows,

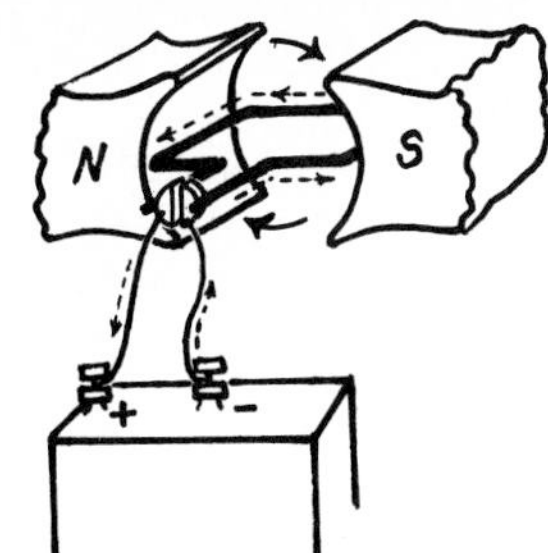


**FIGURE 21–12**
A schematic of one loop of a coil in an electric motor of a household appliance. The current moving through the loop of wire gets sideways deflections from the magnetic field, which cause the loop to turn. The coils of such a motor are wound around an axle (not shown) that is used to drive the blender, vacuum cleaner, refrigerator compressor, and so on.

this pushes the conducting bar along the rails. Accelerations approaching 1 million *g* are achieved by pouring millions of joules of electrical energy into the rails in just a few thousandths of a second.

The rails repel each other as the current flows through them, so they are mounted in a gun barrel to keep them from flying apart. The barrel also serves to keep the projectile from disintegrating under its great acceleration. Because a sliding bar can melt with the electrical heating that takes place, a newer version of electromagnetic launcher replaces the conducting bar with a plasma arc between the rails. An electrical discharge ionizes the air, which conducts electrons quickly across the gap; the magnetic field from the rails pushes those moving electrons and, hence, the entire plasma arc along the rails. A nonconducting projectile surfing in front of this traveling lightning bolt accelerates at up to 5 million *g* before emerging from the gun barrel. Present versions are limited by the projectile's elastic strength. After only a 2-meter ride, it has the speed it would need to escape from earth's gravitational pull if the atmosphere weren't around to interfere (25,000 miles per hour), and the air friction is great enough to cause the small projectile to disintegrate just after leaving the barrel.

Gerard K. O'Neill, of Princeton University, has proposed lifting raw materials from the moon to build large space stations nearby, using electromagnetic launchers (of a somewhat different design) as transporters. With the proper launchers, material could even be sent from earth! Calculations have shown that a ton of material shaped much like a telephone pole could escape from earth's surface when propelled by an electromagnetic launcher. The missile would be in the atmosphere such a short time (only a little more than 1 second) that only 5 percent of its mass would be stripped off because of friction with the air it ripped through. The great accelerations would limit the cargo to ore and the like. Higher forms of life could not withstand the acceleration of such an extremely-fast mass-transportation system.

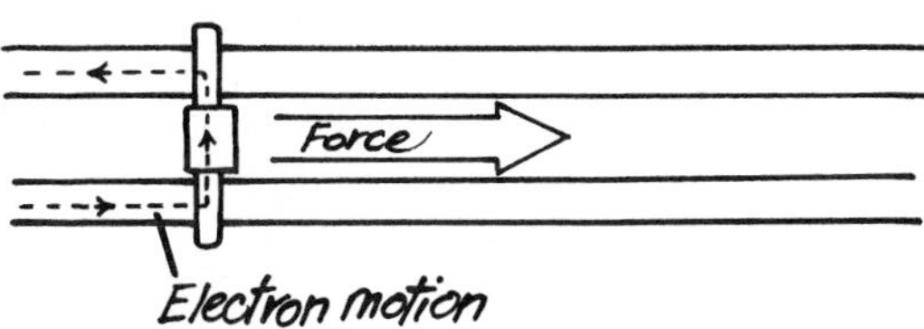


**FIGURE 21–13**
An electromagnetic mass accelerator, or launcher. A sliding conducting bar completes a circuit across two parallel rails at greatly different electric potentials. The current that passes through the bar gets a sideways deflection from the magnetic fields of both rails, and the bar is propelled along the rails.

## CALCULATIONS

In a transformer the induced voltage per loop is the same for the primary coil and the secondary coil. In other words,

$$\frac{\text{voltage across the primary coil}}{\text{number of loops in the primary coil}} = \frac{\text{voltage across the secondary coil}}{\text{number of loops in the secondary coil}}$$

is the expression that shows the relation between input voltage and output voltage for a transformer. In symbols,

$$\frac{V_p}{N_p} = \frac{V_s}{N_s}$$

---

EXAMPLE 1: A single transformer raises a voltage from 2200 V to 120,000 V. **How many loops does the secondary coil have for each loop in the primary coil?**

From the formula for transformers,

$$\frac{N_s}{N_p} = \frac{V_s}{V_p} = \frac{120{,}000}{2200}, \text{ or about } \mathbf{55}$$

There are about 55 loops in the secondary coil for every 1 loop in the primary coil.

---

EXAMPLE 2: In the transformer of Example 1, **how many amperes of current are produced at the higher voltage for each ampere in the primary coil?**

The power in must be about equal to the power out, so $P_{in} = P_{out}$. But $P = i \times V$, so

$$i_{in}V_{in} = i_{out}V_{out}$$

Rearranging, we find

$$i_{out} = \frac{V_{in}}{V_{out}} i_{in}$$

Substituting $i_{in} = 1$ amp, $V_{in} = 2200$ V, and $V_{out} = 120{,}000$ V, we find

$$i_{out} = \mathbf{0.018\ amp}$$

which is quite a drop in current.

---

EXAMPLE 3: A voltage is to be applied to a circuit of resistance $R$. Using the formula $P = i^2R$, **compare the energy that disappears as heat when the current is 0.018 amp rather than 1 amp, as in Example 2.**

$$\frac{P_{0.018\ \text{amp}}}{P_{1\ \text{amp}}} = \frac{(0.018)^2R}{1^2R} = \mathbf{0.00032}$$

This answer shows why high voltage is so important when power is put onto a commercial network (of resistance $R$). Stepping up the voltage from 2200 to 120,000 volts means the company wastes only 0.032 percent of the energy the circuit would waste at the lower voltage.

## REVIEW

**1.** A charged particle moving perpendicularly to magnetic field lines receives a push that is perpendicular to the field lines and to its own velocity. True or false?

**2.** Faraday discovered that a changing magnetic field induces a current in a conducting circuit. Briefly describe his experiment that led to electromagnetic induction.

**3.** As Ampère found, a charged particle crossing a magnetic field gets a sideways push; Faraday saw that magnetic field lines in motion across a charged particle produce a similar push. Both of these are examples of ____.

**4.** An electric generator is a device that converts mechanical energy into electrical energy. True or false?

**5.** In commercial power plants in North America, electrical generators deliver an alternating voltage that oscillates 60 times a second. True or false?

**6.** What does a transformer transform?

**7.** Why do power companies use high-voltage lines to transmit electrical power into residential neighborhoods? How is this high voltage reduced to safe levels for the home?

**8.** An electric motor is a device that converts electrical energy into mechanical energy. True or false?

## DEMONSTRATION

This demonstration is worth your while even if you have to ask your instructor for the materials. You'll need two coils of thin copper wire with numerous loops, two magnets, and two short lengths of insulated copper wire.

Connect the two coils with the wire. Place one coil on a table and suspend the other over the edge of the table. Use a book to pin the wires to the table so the hanging coil doesn't pull on the other one. Bring a magnet up close to the hanging coil (which is at rest), and hold it steady. Have a friend move the second magnet back and forth past the coil on the table, as close as possible without touching. The changing magnetic field this causes induces a current in the coil on the table. Electrons move in the lengths of copper wire and in the hanging coil as well. The current in this coil gets a push from the magnetic field of the magnet you are holding next to it, and the suspended coil swings back and forth as a result.

The coil on the table and the moving magnet represent an electrical generator. The relative motion between the magnetic field and the movable electrons in the wires cause an electric current to flow. (A changing magnetic field induces an electric field.) The lengths of copper wire are the transmission lines. The hanging coil and stationary magnet represent an electric motor that uses the push from a magnetic field on a current to convert electrical energy into mechanical energy.

## EXERCISES

**1.** What is so unusual about the magnetic field around a long, straight, current-carrying wire?

2. Did the copper disk that Faraday used to make the first generator make AC or DC?

**3.** How can a magnetic field be used to move a charged particle that's initially at rest?

4. If Ampère had mounted two straight current-carrying wires perpendicular to each other rather than parallel, would the wires have attracted, repelled, or had no effect on each other?

**5.** Is the current that flows through the coil of the generator in Fig. 21–7 AC or DC?

**6.** Could you hold the coil in a generator stationary and rotate the magnet to obtain a current from a generator?

7. Explain (as if to someone who is uninformed) how energy can be transmitted in a power line that sits still day and night with nothing visible going on. Why don't the wires move? Why don't electrons flow into one end and out of the other?

8. Would a transformer work if a direct current were used rather than AC? (Careful!)

9. In a television tube, electromagnets deflect a fast-moving beam of electrons that sweep across the fluorescent screen, where their energy produces visible light. Do these electromagnets change the speed of the electrons, or only their direction? Exactly how do these electromagnets guide the beam?

**10.** Any coil of wire in the orbiting space shuttle has a small current in it, even if no potential difference is applied to its ends. Explain why.

11. Why is it more accurate to say electrical power is *transmitted* in conducting wires rather than transported?

**12.** A small device mounted next to the rim of some bicycle wheels powers a headlight for use at night. What's that device called? Where does it get its energy?

**13.** How does coiling a wire (making loops that lie on loops) intensify the magnetic field produced by the wire?

**14.** Higher voltage lowers the current for long-distance transmission of electrical power. So why don't power companies step up the voltage from 120,000 volts to, say,

100 million volts? (*Hint:* Glance over the last section of Chapter 18.)

**15.** In the late 1800s, there was a great debate about whether AC or DC should be generated and distributed by power companies. DC lost. What arguments can you think of that would support this historical decision?

16. Look at Figures 21–1 and 20–12 and tell in which direction an eastward-moving beam of electrons at the earth's magnetic equator would be deflected.

**17.** Would a class ring that is spun like a top in the earth's magnetic field have a current in it if a small slice is cut from the ring so that is doesn't make a complete circle?

18. The cassettes used for recording music and messages contain a plastic ribbon (the tape) coated with iron oxide or chromium oxide, both of which have magnetic domains. To record, sound waves create a varying electrical current in a microphone. An amplifier magnifies this current, which is used to operate an electromagnet in the recording head. As the tape passes through that head, the fluctuating magnetic field aligns the domains in the iron or chromium oxide. The magnetic pattern thus created can be used to make a sound very much like the original sound. Tell how the tape recorder can replay the tape, using the same magnetic head.

**19.** A ski lift carries a girl to the top of a ski slope. The energy to do this comes from the swarm of ultralight electrons that sway back and forth a fraction of a millimeter even as they jitter around with their normal thermal motion within the copper wires. Explain how such tiny motions can deliver so much energy.

**20.** Not far from San Francisco there is a geyser field where ground water passes through heated rock and turns and uses its energy to turn electrical generators. Meanwhile, a 750-horsepower electric motor in San Francisco uses some of that energy to move underground cables along the streets. Cable cars clamp onto that cable to be hoisted to the tops of the city's hills. Trace all the steps in the harnessing, transmission, and use of the heat energy from the earth's interior.

21. Which of these planets don't have radiation belts like the Van Allen belts: Mercury? Venus? Mars? Jupiter? Saturn? (*Hint:* Recall from Chapter 20 which have magnetic fields and which do not.)

22. In car radios there are power converters that use higher voltages than the 12 volts supplied by the car's battery. A small transformer steps up the voltage—but wait! Battery current is *DC,* not *AC,* so how can a transformer possibly work? Think about this for a moment, and then read the answer at the back of the book.

**23.** The small lamps that provide light for the dials on stereos, radios, and TVs may require only 4 or 6 volts for operation. Using the 110-volt AC current from a home circuit, you could light these lamps in two ways. A step-down transformer could provide the voltage they need, or a large resistor could be placed in series with the lamps. Most of the 110 volts across the circuit would be used to push the current through the large resistor, leaving only 4 or 6 volts across each bulb. Which method is more energy efficient?

**24.** Maxwell was quoted in a popular magazine in 1883 as saying the reversibility of the electric generator into a motor was "the greatest scientific discovery of the last quarter of a century." Take the view that this was probably a misquote and explain why.

25. Do the copper wires in your home lead to other wires that form an unbroken conducting link all the way to the power company that supplies your home with electricity? Explain.

*26. The input voltage to the transformer in Fig. 21–14 is 110 volts. What is the output voltage? If 1 amp is the current in the primary coil, about what is the current in the secondary coil?

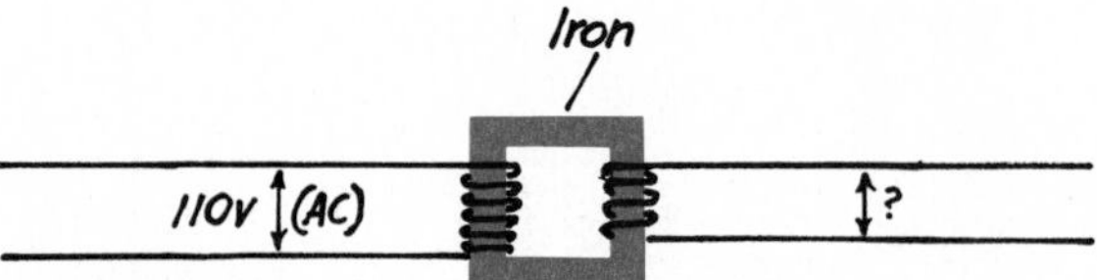


**FIGURE 21–14**

*27. The voltage used to cause a spark across the spark plug of a car is about 15,000 volts. However, the source of electrical power is a 12-volt battery. The car's "coil" converts 12 volts to 15,000 volts. How many loops does the secondary coil in that transformer have for every loop in the primary coil?

*28. If the voltage in the secondary coil of a transformer is 4 times the voltage in the primary coil, show that the current in the secondary coil is ¼ the value of the current in the primary.

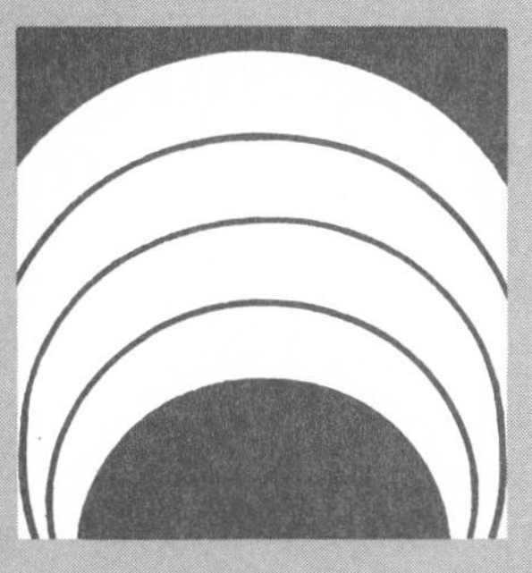

# CHAPTER 22 ELECTROMAGNETIC RADIATION

Next to almost every dentist chair there's an evacuated glass tube partially shielded with lead at the end of a flexible arm. Inside the tube are two bits of metal (called electrodes) a centimeter or two apart. Next to one lies a heating coil. With the press of a button, a high voltage power supply is turned on, causing the heated element to become negatively charged, and the other positive. Just like that, there is a potential difference of 60,000 volts between electrodes. Thermally agitated electrons evaporate from the surface of the heated element into the grip of a strong electric field between the charged metal pieces, and they accelerate through the vacuum toward the positive metal surface. The very strong electric force gives them enormous accelerations. After only a centimeter or so of travel, they move at over one-third of the speed of light!

As they arrive and collide with the positive electrode, 99 percent of their energy turns into heat. But a few penetrate close to the positive nuclei of the metal atoms, where they get a very strong pull indeed. They turn sharply around the nuclei and, as they do, emit radiation of a very high energy called X rays. We'll see in this chapter that an accelerated charged particle always emits radiation of some type. (The electrons in the appliances and wiring of your house or dorm don't accelerate nearly that fast. Nor would you want them to. The last thing anyone needs is X rays from a toaster.) ■

## Light

The morning sun brings us images of our world that are hidden in the darkness. When the white sunlight bathes the trees and buildings and roads, we are treated to rich greens, reds, oranges, yellows, the bright blue of the sky, and hundreds of other hues. All of these colors come from sunlight that reflects from the matter around us. We can also see colors when sunlight comes through a transparent substance, such as the cut-glass pendant of a chandelier. In such a piece of glass we can glimpse a splash of the colors of the rainbow. It was this phenomenon that led Isaac Newton to begin investigating the nature of light.

**FIGURE 22–1**
White light enters the face of a prism at an angle and separates into the rainbow spread of colors we call the spectrum of white light.

Newton closed the curtains to his study and arranged a small opening so that a narrow shaft of sunlight cut into the darkened room. In its path he placed a prism, a triangular cut of solid glass, turning it around in the sunbeam until a rainbow spread of colors (the **spectrum** of white light) fell on the wall. Red, orange, yellow, green, blue, and violet, one color gradually fading into the next, appeared on the wall. He then took a second prism and placed it in the beam of each color, one at a time. All the individual colors passed through the second prism unchanged; there was no further separation. This indicated that the colors that come from the spectrum of white light are more basic than white light itself. But just how one pure color was different from another remained a mystery. He later used a lens to recombine the colors separated by a prism, and the united colors appeared as white sunlight on the wall. White light, he showed, was the combination of colors of light all traveling together.

Light's speed of travel was another early mystery. If you watch your own shadow in the sunlight, it seems to follow you instantly. As soon as you move, sunlight strikes the place your shadow had been, or so it seems. In Galileo's time, some people thought light traveled instantaneously. But Galileo didn't agree. He proposed (and apparently performed) an experiment to see if light required time to move. Holding covered lanterns, he and an assistant stood some distance apart (less than a mile) in the dark of night. Galileo uncovered his lantern, and his assistant, upon seeing Galileo's lantern, uncovered his own. The time light took to travel to his assistant and then back to him, divided by twice the distance to his assistant, would give the speed of light. This experiment failed because the experimenters' own reaction times far exceeded the time light needed for the trips. Then Galileo proposed repeating the experiment at greater and greater distances, until finally telescopes would be needed to see the lanterns. Surely at some great distance the travel time for light would be large compared to the human reaction times. However, there is no evidence that Galileo followed his own advice, and the results would not have been any different anyway. Over such distances, light's time of flight is too small for human timekeepers to observe in such an experiment. In modern experiments to determine the speed of light, human reaction time has been eliminated. These show that the speed of light in a vacuum, called $c$, is about 186,282.03 miles per second (or about 299,792.46 kilometers per second). This great speed is why light appears to us to move instantaneously. Such experiments also show that in a vacuum all the colors of light travel at that same speed, $c$. The beautiful ingredients that blend to make white light keep pace with each other as they move through space.

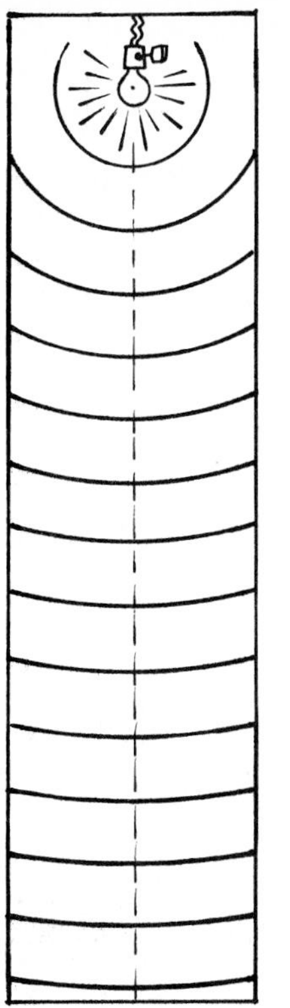

**FIGURE 22–2**
Wavefronts of light from the sun or a light bulb spread out in all directions. At great distances from the source, they move almost parallel, like lines of waves approaching a beach.

## The Wave Nature of Light

As early as Newton's time, there were two ideas about what light actually is. Newton argued that light was composed of tiny, fast-moving corpuscles, or particles; others thought that light was vibrations, waves of some sort. In 1802 an Englishman, Thomas Young, did an experiment to try to settle this question.

What Young proved is that two similar beams of light can meet and cancel one another at certain points, a difficult thing to imagine with particles. (The word *particle* brings to mind permanence; how could particles meet and disappear?) Recalling from Chapter 16 what takes place when waves on a string or on the surface of a lake interfere will help us see why Young set up this experiment as he did: First, imagine the meeting of two similar waves that are traveling in opposite directions on a string. When a crest passes some point just as a trough arrives from the opposing wave, they cancel perfectly, and the string lies flat for that instant (see Fig. 16–8). The waves are interfering *destructively*. The same thing happens with sound waves and water waves. Next, imagine ocean waves coming parallel into a jetty with two small channels through it. The waves that enter these openings spread out (diffract) into the quiet waters behind the jetty, where they eventually reach a straight line of buoys floating parallel to the barrier. We can predict the motions of the buoys due to the interference of the waves. The buoy that is an equal distance from both openings will rise and fall dramatically. Crests from each channel meet there and add; then the following troughs meet and combine to give a trough twice as deep. This is *constructive* interference. But in either direction to the side of this buoy, there will be a buoy that's becalmed. It never rises or falls, because each time a crest arrives from one wave, a trough arrives from the other and they cancel. This is *destructive* interference. Along the line of buoys, then, you could see points of constructive interference alternating with points of destructive interference. This is called the *interference pattern* of the waves.

Let's apply the behavior of these waves to Young's experiment with light. Young made certain that the light waves coming into the screen had parallel wavefronts like the ocean waves approaching the jetty (see Fig. 22–2). To do this, he sent a bright beam of light through a small opening in an opaque screen. Behind this he placed another screen with two parallel slits to let the light through. He reasoned that the waves from the opening in the first screen would spread out and become more parallel as they approached the second screen. When he looked *behind* the double slits on the unlit side of the screen, he found an interference pattern, that is, a bright band of light in the center with alternating dark and bright bands to either side. Even though there was just as much light falling onto the areas of the dark bands as in the

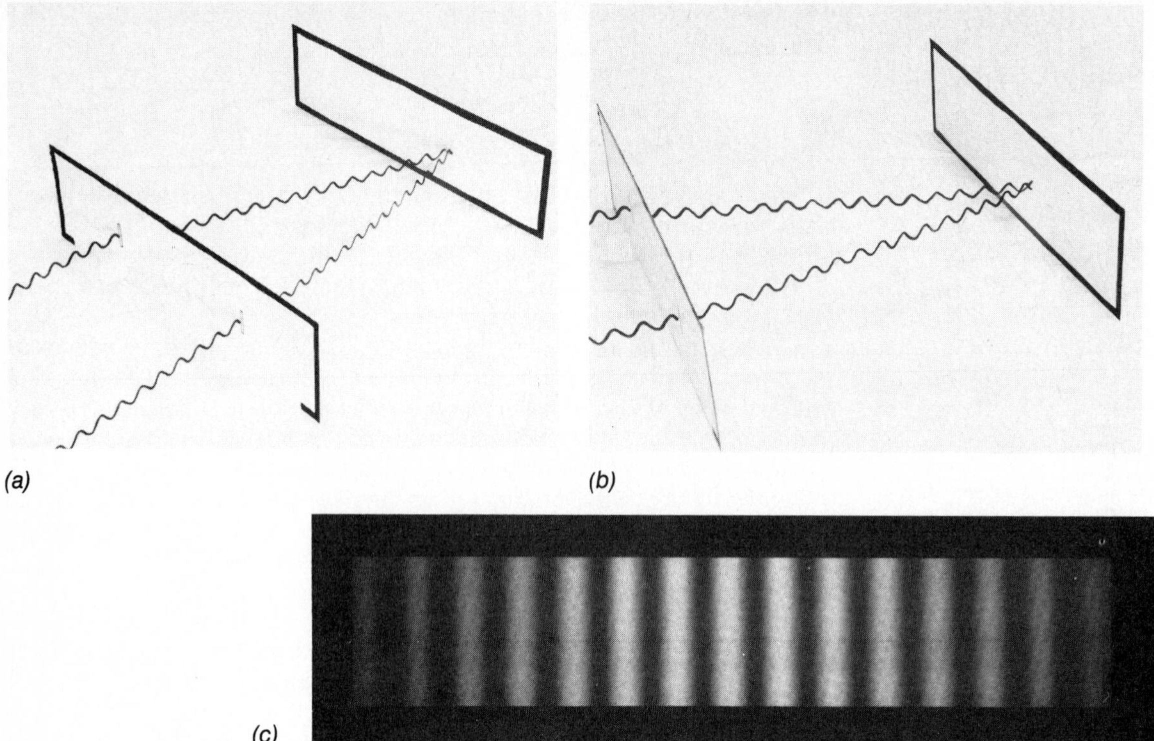

**FIGURE 22–3**
A partial schematic of Young's experiment. Young first sent a bright beam of light through a small opening in an opaque screen (not shown here). Parallel wavefronts of this light strike a second screen that has two thin parallel slits to admit the light. The light waves then strike a third screen. To reach a given spot on the third screen, the light waves must travel different distances from each slit, and a different number of wavelengths fit on each path. In *(a)*, the crests (and troughs) of the two waves arrive at the screen at the same time. The waves add constructively, and a bright band will appear on the screen. In *(b)*, the crest of one wave arrives at the screen at the same time as the trough of the other wave. The waves add destructively and cancel; this area of the screen will be dark. *(c)* The interference pattern revealed by Young's experiment.

bright bands, the light waves in the dark bands canceled almost perfectly. With this experiment, Young had established that light has the properties of waves.

## Maxwell's Waves

In the 1860s a gifted young Scot, James Clerk Maxwell (1831–1879), put together some of the discoveries of Ampère, Faraday, and others, along with contributions of his own. In four equations (see Fig. 22–4)

**FIGURE 22–4**
Christina, a physics major, is wearing a shirt that displays Maxwell's equations. Some physics instructors ban such shirts on examination days.

he expressed in mathematical terms the connections between electricity and magnetism. Together these equations explained the actions and behavior of the electric and magnetic forces. And then Maxwell made a great discovery. He found that his equations predicted that the electric fields of charged particles could carry ripples on them, oscillations that looked (in the mathematical sense) exactly like the transverse waves that travel down a stretched string. As these ripples move along, the changing electric field produces an accompanying magnetic field. So the waves Maxwell predicted came to be known as **electromagnetic waves,** or **electromagnetic radiation.** He calculated how fast these waves would travel and noted that their speed was very close to the estimate (in those days) of the speed of light. In this way, Maxwell discovered what Thomas Young's waves of light are. *They are traveling oscillations in the electric field of a charged particle. Such waves arise when* (and always when) *charged particles accelerate*. The acceleration of the charged particle puts kinks in the field lines that are at right angles to the acceleration vector, and these kinks move out along the field line at the speed of light. Light from the sun, from fire, from electric bulbs, and from fireflies comes only from charged particles that have accelerated for some reason.*

Maxwell's equations showed even more. Like ocean waves, waves on a stretched string, and sound waves, electromagnetic waves of all frequencies reflect, refract, interfere, diffract, are absorbed, and are scattered when the conditions are right. About 8 years after his death, Maxwell's theory of electromagnetic waves was confirmed by Heinrich Hertz, who first produced and detected Maxwell's waves by purely electrical means. For his tremendous accomplishment in explaining electromagnetic behavior, Maxwell is remembered as one of the giants in the history of physics.

**I have a paper afloat, with an electromagnetic theory of light, which til I am convinced to the contrary, I hold to be great guns.**
J.C. Maxwell, 1865.

When we use the word *light,* we usually mean the particular wavelengths of electromagnetic radiation that we can see. The colors in white light are only the different wavelengths or frequencies of this electromagnetic radiation; our eyes and brains interpret the various wavelengths as individual colors. Yet Maxwell's equations placed no mathematical bounds on the frequency or wavelengths of such waves. Every frequency and wavelength was possible, even frequencies of electromagnetic radiation that we cannot see.

## Light We Cannot See

In 1800 William Hershel, the famous English astronomer, was experimenting with the spectrum of white light. Since the light from the sun warms anything it touches, Hershel placed a thermometer in the colors of the spectrum made by a prism to see how each affected the thermom-

*One problem Maxwell had with his electromagnetic waves was explaining what medium they moved through. The only waves known in his day moved through a solid, liquid, or gas. He postulated an "ether" in space that waves of the electric and magnetic fields could move in. It turned out that no ether was necessary, that the waves moved through a vacuum just fine. In Chapter 26 we'll see that Albert Einstein did away with the idea of the ether in 1905.

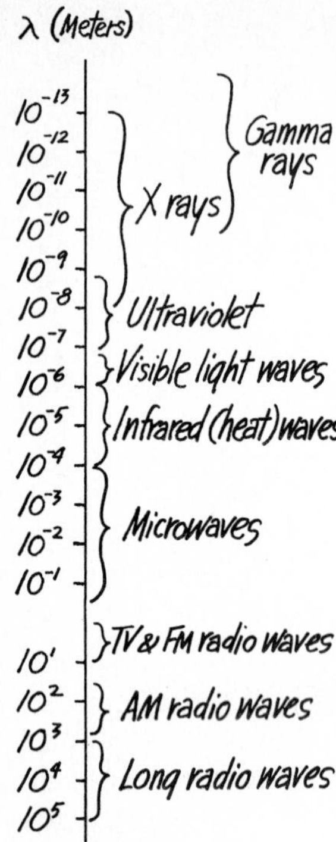


**FIGURE 22–5**
The electromagnetic spectrum, of which the visible spectrum of Fig. 22–1 is only a small part. To obtain the frequency of any of these waves, use $f = c/\lambda$, where $c$ is about $3 \times 10^8$ meters/second.

## LIGHT'S OTHER NATURE

In Chapter 24 we will see that between 1899 and 1905, physicists discovered another property of light, namely that individual atoms and molecules absorb or emit light only in lumps of energy. These individual quantities of light energy are called *quanta* (singular, *quantum*), and the bit of light that carries one such lump of energy is called a *photon*. Although the wave properties of light explain many of the effects we will study, others need the quantum viewpoint for their explanation.

eter. He was startled to see the temperature rise the most when the thermometer was off to the *side* of the red band of the spectrum, where there was no visible light at all. The prism directed invisible rays to this area, rays that were highly effective in heating ordinary matter. These rays are called **infrared** rays, and their wavelengths are longer (and their frequencies lower) than the waves of the visible spectrum. Most solid matter absorbs infrared rays extremely well, as opposed to reflecting them, and their energy warms the matter. Since they are so effective at heating matter, they are also called **heat** rays. The next year, J.W. Ritter, of Germany, detected other invisible rays to the side of the blue–violet region of the spectrum. (He exposed paper that was soaked with a silver compound to the spectrum of sunlight that passed through a prism. Wherever light fell, the silver compound turned black. Ritter found the greatest blackening occurred just beyond the violet end of the visible spectrum.) These **ultraviolet** rays have shorter wavelengths (and higher frequencies) than visible radiation.

## The Electromagnetic Spectrum

Today we know there can be any frequency of electromagnetic wave, as Maxwell suspected. The field lines of a charge, though imaginary, give us a visualization of these waves. Jiggle an electron or proton by any means, and kinks appear on its field lines. However fast the charge wiggles, the field's adjustment to the charged particle's motion is even faster; those kinks whip out along the field lines at the speed of light. The spectrum of visible light as seen with a prism is only a very small part of the spectrum of all possible electromagnetic waves (see Fig. 22–5). The electromagnetic waves with the longest wavelength are called *radio* waves. Next down the ladder in wavelength size are *microwaves,* which carry TV and telephone signals and cook food. Then, with even shorter wavelengths, come *infrared, visible,* and *ultraviolet rays,* followed by *X rays* and *gamma rays,* which have the shortest wavelengths of all.

Accelerating electrons create most of these waves, although protons too could bring them about. That's because electrons have such a

## ELECTROMAGNETIC WAVES YOU CAN MAKE

Stroke a rubber rod with cat's fur, and it becomes charged. If you then shake the rod back and forth a few times per second, invisible electromagnetic waves travel out from it at almost $3 \times 10^8$ meters per second. Their frequency of oscillation is the frequency of the rod's oscillation, probably 3 or 4 hertz. As with waves on a string, a small frequency ordinarily means there is little energy in the waves. (The exception would be if the waves had tremendous amplitudes.) If you flipped the rod back and forth 4 times a second, the wavelength of the electromagnetic waves from the charged rod would be some 60 thousand miles across. But if you put the charged rod on the edge of a table and twanged it, the rod might oscillate at as much as 20 times a second, making waves with wavelengths comparable to the diameter of the earth. If you could oscillate such charges at the rate of $10^{14}$ to $10^{15}$ times a second, the electromagnetic waves you'd make would have wavelengths of perhaps 5000 atomic diameters, and they would be visible. Those electromagnetic waves are light waves.

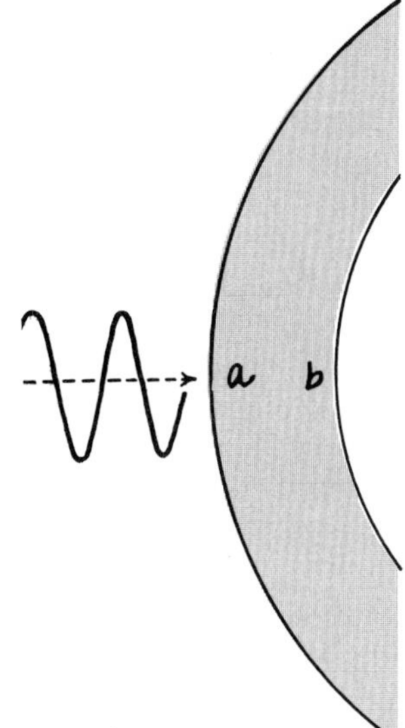


**FIGURE 22–6**
When light strikes the film of a soap bubble (or an oil slick), some of the light reflects from the front surface at *a*. Some light also enters the bubble and reflects from the back surface *b* to emerge at the front surface *a*. As these two reflected waves come to an observer's eyes, some wavelengths add constructively (the observer sees these colors) and some add destructively (the observer doesn't see these). The thickness of the film determines whether a given wavelength inside emerges in step with the similar wave reflected at the film's surface.

small mass compared to protons and the nuclei most protons belong to. If the same force is applied to an electron and a proton, the electron's acceleration is almost 2000 times as great as the proton's, and it is the acceleration that causes the wrinkles on the field lines of a charged particle. For example, the molecules of any solid, liquid, or gas at ordinary temperatures emit infrared rays just because of the thermal motion that makes every molecule jostle with its neighbors. Every bump means electron clouds collide, and electrons accelerate and radiate. If the bumps are really violent, as in the heated filament of a light bulb, the accelerations are harder and the electrons send out shorter wavelengths, those of visible light. In an arc or a plasma the collisions are even more violent, and the light is blue–white. In an X-ray tube the electrons that penetrate atoms almost to the nucleus have extreme accelerations and emit X rays, extremely high frequencies of light. Gamma rays, with even greater frequencies than X rays, come from the nucleus. We'll study the modern understanding of emission and absorption of electromagnetic waves in Chapters 24 and 26; in the next four sections, however, we'll look at some of the effects around us that can be explained with the wave nature of light.

## Interference of Light Waves

The shimmering colors you can see in soap bubbles come from the interference of light waves. When light strikes a bubble, some reflects from the outside surface of the thin soap film and some penetrates into the film (Fig. 22–6). Once inside the film, the transmitted portions of the waves partially reflect against the inside surface of the film, and return to the outside surface and escape from the bubble. (Remember that when a wave changes mediums abruptly, there is always a reflected portion.) Among these escaping waves, some wavelengths will meet waves of the same wavelength reflecting from the front surface in just the right fashion for them to cancel. That is, a crest leaving the thin film cancels a trough of the same wavelength reflecting from the front surface. Other wavelengths from inside the film exit in such a way as to add constructively. Those added waves are the bright colors you see on the glimmering bubble.

Dip a circular ring into a soap solution, hold it up vertically, and watch the colors in the film. As the liquid in the film drains downward, leaving the top of the film thinner and thinner, the colors change. That's because as the film becomes thinner, a different number of wavelengths of each color of the light fit along the path the light travels inside the film. Thus one color of light after another will meet the criterion for cancellation (or addition) as the light emerges. Similarly, a tiny bit of oil or gasoline on water spreads over the surface to make a very thin transparent film. The colors you can see from such a thin film of oil show that interference is at work.

Makers of optical equipment must work around the interference behavior of light in a lens. For example, when light enters a camera lens, some is reflected from the front surface of the lens, and some passes in. When the transmitted light reaches the back interior surface of the lens, most passes through to make the desired image on the film. But some of the light is reflected there. The reflected portion poses a problem. It returns to the front interior surface, where some of the light reflects again. At the back surface, some passes through and interferes with the light that's making the image on the film. To avoid this unwanted interference, a thin, transparent coating is applied to the front surface of the lens. The internally reflected light reflects from *both* surfaces of the coating, just as with the film of a soap bubble. The lens maker can make the coating just the right thickness to cancel a given wavelength of light, corresponding to the color green, for example. Another thickness would cancel red, and another blue. Separate coatings, one on top of the other, eliminate much of the unwanted reflections by getting them to cancel. That's why you should never rub the front surface of a camera lens to clean it: The coatings are thin and they aren't as hard as glass.

## Polarization

If you shake one end of a clothesline straight up and down, the wave that scoots away wiggles only in the vertical direction, which is in one plane. Shake the same line horizontally, from side to side, and the wave you make wiggles only in the horizontal plane. Waves whose oscillations are in a single plane are called **plane polarized,** or often just **polarized** (Fig. 22–7). If you jerk a clothesline in random directions, the wave that moves away from your hand and along the line isn't polarized.

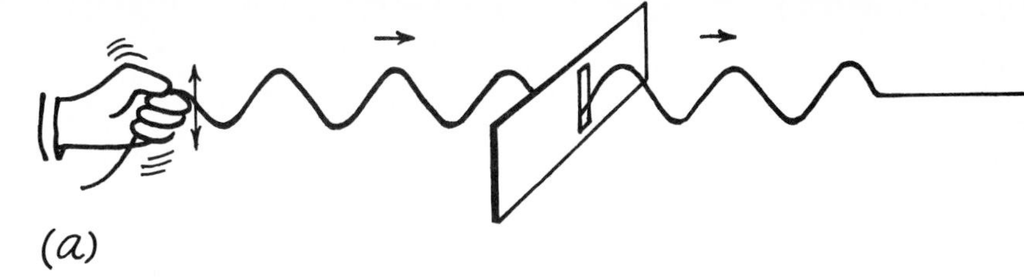


(b)

**FIGURE 22–7**
Two polarized (vibrating in a single plane) transverse waves on a string. A slit whose length is in the plane of the polarized wave lets it pass, as in *(a),* while a slit perpendicular to a wave's plane stops it, as in *(b).*

Most light isn't plane polarized. That is because the electrons that emit the light, whether from the sun's surface or from the hot surface of a light bulb's filament, accelerate randomly in every direction. Some substances are able to absorb those light waves that vibrate in a certain direction while letting others vibrating in a perpendicular direction pass. An example is the material *Polaroid,* invented by Edwin Land in 1938. Embedded in Polaroid are long molecules in alignment, with electrons that can move along the length of the molecules. When electromagnetic waves travel through the material, the electrons move and absorb energy from the part of the wave that pushes them in the direction along the molecules. The oscillations of the electric field perpendicular to the

**FIGURE 22–8**
A polarizing material for light works like the slits in Fig. 22–7. Sunlight is composed of light waves vibrating in every possible plane. When it strikes a polarizing material, only those waves whose oscillations are in one certain plane get through; the rest are blocked. These two oversized Polaroid sunglass lenses have their planes of transmission marked by the arrows. In *(a)* the light that passes through one lens is transmitted by the second lens when its plane of transmission is in the same direction. But in *(b)* the second len's plane of transmission is perpendicular to the the first, and the light is blocked.

*(a)*

*(b)*

length of these molecules cannot move the electrons far, however, so these electromagnetic waves pass through the material with less absorption.

Sunglasses made of polarizing material use this property to cut out some of the strong reflections of sunlight. Direct sunlight is not polarized. But if sunlight reflects from a road or a lake (or any non-metallic surface), it becomes partially polarized. The bright reflected sunlight we call glare is mostly light whose waves oscillate in the horizontal plane. Polarizing lenses block those horizontal glare waves, while vertically oscillating light waves pass through them.

## Making and Detecting Radio Waves

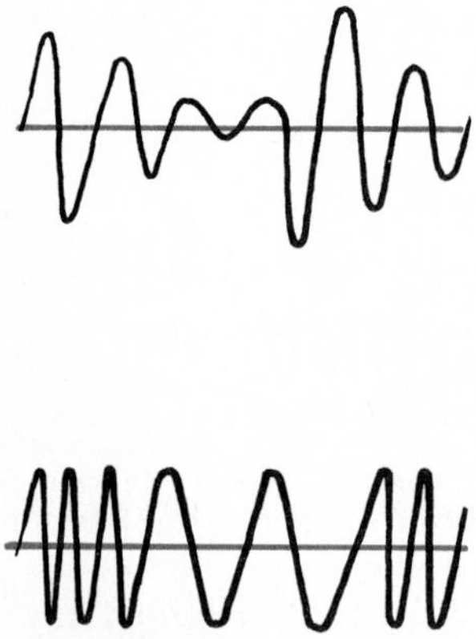

When James Clerk Maxwell died in 1879, he couldn't have known how the waves he predicted would revolutionize the technology of communications. Speed-of-light telephone communications are taken for granted in many countries today. Available to anyone with a radio or a television set are news, entertainment, and educational programs, courtesy of electromagnetic waves generated by commercial and public radio or television stations.

Here's a brief explanation of how a radio wave is broadcast at one location and received at another. A sound wave from a voice or muscial instrument enters a microphone at the radio station. The variations in air pressure the wave causes within the microphone mechanically generate an electrical current called the *signal*. That signal is used to modify, or **modulate,** an intense carrier radio wave that is made by the station's transmitting antenna. The carrier wave comes from fast oscillations of charge within this antenna. In essence, one end of the antenna (a conducting rod) is positive and the other is negative, setting up a dipole electric field around the antenna. When the charges in the station's antenna are reversed hundreds of thousands of times a second, the dipole field follows suit, and oscillating field lines move outward over the local landscape at the speed of light. If the signal at the station modulates the amplitude of the carrier wave while its frequency remains constant, the station is broadcasting an AM (amplitude-modulated) radio wave. If the carrier wave's amplitude is steady but its frequency is modified by the signal from the microphone, the broadcast wave is FM (frequency modulated). See Fig. 22–9.

**FIGURE 22–9**
Compare both the frequencies of these waves and their amplitudes. An AM radio wave (such as the top wave) has a constant frequency, but the amplitude is modulated (modified). The pattern of modulation carries information to a receiver, amplifier, and speaker, which reconstruct the original sound. An FM radio wave (such as the bottom wave) has a constant amplitude but a modulated frequency.

As it leaves the transmitting antenna, a radio wave spreads out in all horizonal directions. Its energy too spreads out along the expanding wavefronts; we say the signal weakens. However, any radio that's within range of the station can pick up the distance-weakened signal, make a large copy of it (the amplifier in the radio does this), and use that copy to reconstruct the original sound from the radio station (the speaker takes care of this part). Here's how the radio receives the signal. The movable electrons in the radio's metal antenna are pushed about by the electric field of the incoming radio wave. These electrons oscillate back and forth, duplicating the radio wave's pattern precisely, both in frequency and in relative amplitude. In other words, the elec-

## HEATING WITH MICROWAVES

Rather than "a chicken in every pot," some future politician may well promise "a chicken in every microwave"; the current new wave in kitchen technology is the microwave oven. A conventional oven cooks by conducting heat from the outside of the food to the inside, but a microwave oven cooks by simultaneously heating the food throughout its volume.

A water molecule, remember, is a permanent electric dipole, more negative on one end and more positive on the other. In the presence of a microwave's oscillating electric field, the positive end is pulled along the field line, and the negative end is pulled in the opposite direction. These actions cause the molecule to rotate—it has work done on it. As the electric field oscillates, the molecule turns back and forth, gaining rotational energy. Through collisions with neighboring molecules, its rotational energy is changed into translational kinetic energy, heat. That heat is conducted through the food. Because microwaves penetrate food very well, this heat-producing process goes on throughout the food wherever there are water molecules. A portion of microwaves that enter the food travel all the way through it, but the oven's metal liner reflects them back for another pass and another. Each time more of the microwave's energy is absorbed. Often in just a matter of minutes, heat from the flipping water molecules cooks the food. (The dry paper or Styrofoam plates or containers recommended for use in microwave ovens contain no water, and the microwaves pass on through with no effect on the material.)

There are other situations where heating of matter by microwaves is useful. For instance, in *diathermy,* the heating of tissue for medical therapy, low-intensity microwaves penetrate outer layers of fat, which contain little water, and warm the underlying muscles, which have a higher percentage of water content.

trons in the radio antenna perform a jig that's a miniature copy of the electron motion in the station's transmitting antenna. (Incidentally, each metal pot and pan in the kitchen is a receiving antenna for these waves, too, but these utensils have no amplifiers to enlarge the waves and no circuits and loudspeakers to use the electrical signals to reproduce the sound.) Because radio waves come in from every nearby station at the same time, a radio's receiver must be adjusted, or *tuned,* to follow the particular carrier signal you want it to copy. When the set is tuned to a given station, it amplifies only that one part of the various signals picked up by the antenna. Sometimes, however, when an electrical storm is in the area, accelerated charges in the lightning bolts send out bursts of radio waves of all frequencies. Some will be in the frequency range of the station the set is tuned for, and the radio amplifies that signal (called *static*) along with the commercial signal it is receiving.

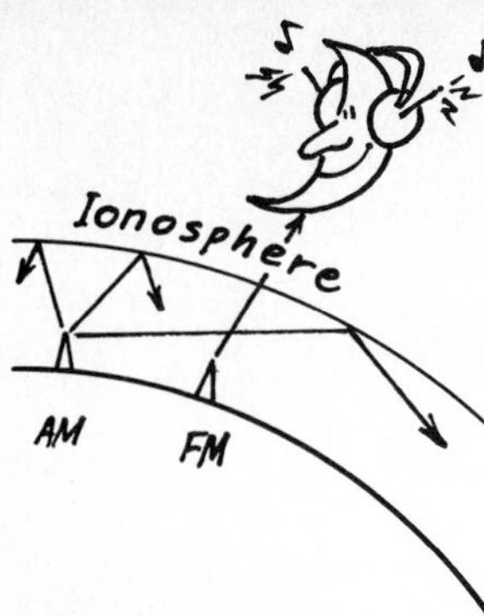


**FIGURE 22–10**
FM radio waves aren't reflected effectively by the ionosphere although AM radio waves are.

**FIGURE 22–11**
The greater the frequency of a light wave, the more energetically it oscillates the electrons of the atmosphere's molecules. Because blue light has the highest frequency of the visible frequencies, the molecules scatter the blue light from sunlight as it passes much more effectively than they scatter the lower-frequency red light, for example. That is why the sky overhead is blue. (Ultraviolet is scattered even better than blue light since it is of a higher frequency. However, there is less ultraviolet light in sunlight than there is blue light, and our eyes don't sense ultraviolet anyway.)

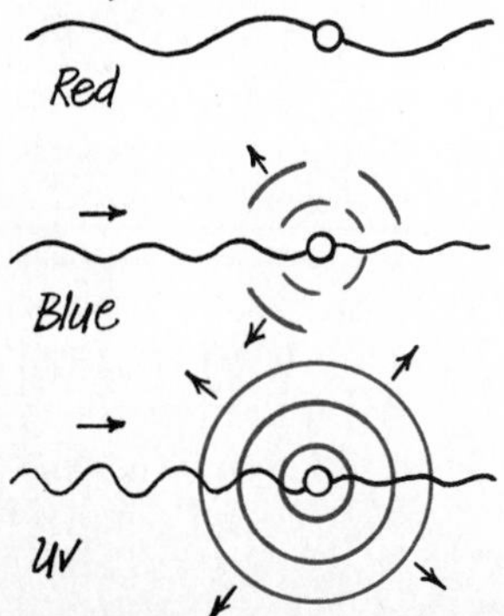


Radio waves move electrons only at the surface of metal; electric fields don't penetrate metal, as we saw in Chapter 18. Instead, the oscillating electrons at the surface absorb a wave and then, because of their own accelerations, reemit it in the action called reflection. Metals reflect radio waves just as a mirror reflects light. A radio won't play inside a closed metal box. And yet waves pass right through our bodies all day, every day. (Don't believe it? Put a small portable radio on the ground and lie over it—does it still play?) Radio waves don't greatly disturb the electrons in our molecules because they carry very little energy. Even though their oscillations push the positive nuclei in the molecules one way while pushing the electrons in the opposite direction, the effect is too small to count. By the same token, nonconducting matter doesn't disturb the incoming waves much, and this is why they can pass through a roof or a wall or a person without much absorption.

In the thin gas of earth's ionosphere above the ozone layer, some of the atoms and molecules are ionized by strong ultraviolet radiation. These ions respond to the oscillating fields of AM radio waves rising from earth's surface. As the ions accelerate, they send out a perfect copy of the radio waves as if the ions were miniature antennae. Part of those signals return to earth. This is why shortwave radio transmissions and some AM radio stations may be received over great distances without a straight line of sight to the transmitters. FM radio signals and TV signals, however, are not reflected by the ionosphere. Their frequencies are so high that the electrons in the ionosphere can't respond fast enough to the changes in the electric fields of those waves to reflect them back to earth.

## Blue Skies and Red Sunsets

Radio waves have almost no effect on the molecules of air they pass through. The electrons, bound tightly to their atoms, barely move when long-wavelength electric and magnetic oscillations pass (tens of millions of atoms span one radio wavelength). It's as if those electrons were corks that rise and fall gently on a lake as a smooth wave passes. But shorter wavelengths of similar amplitude carry more energy, as we saw in Chapter 16, and their more rapid oscillations can cause greater disturbances of the electrons. Visible light has higher frequencies and shorter wavelengths than radio waves. (Only a few thousand atoms will span one visible wavelength.) Because of this, the electrons in the molecules of the air absorb some of the energy from the radiation and then reemit it. The neutral molecules themselves become miniature antennae and send the light waves out in all directions. We say the molecules **scatter** the light. The greater the frequency of the light (or the shorter the wavelength), the more the electrons respond.*

*Unless the oscillations are so fast and powerful that they tear the outer electrons from the molecule, wrecking the antennae. Radiation of such high frequency, like ultraviolet light, X rays, or gamma rays, is called ionizing radiation.

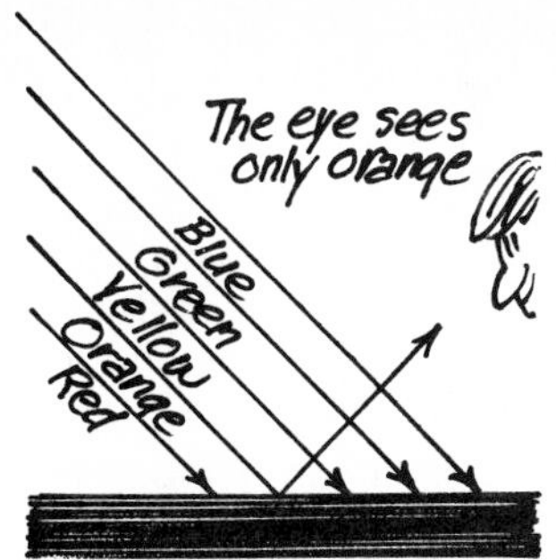


**FIGURE 22–12**
When light strikes the surface of a solid, certain wavelengths are absorbed by the material, and others are reflected. In some instances, most wavelengths are absorbed, leaving only one narrow range of wavelengths to reflect. That range of wavelengths gives the object its color when viewed with our eyes. In other cases, several small ranges of wavelengths reflect, and the eyes see the combination of these fundamental colors as yet another color, one not in the spectrum of sunlight (such as brown, grey, aqua, pink, and so on).

**. . . the colors of all natural bodies have no other origin than this, that they reflect one sort of light in greater plenty than another.**

Isaac Newton

This scattering of light is why the sky is blue. It is not the color of the gases in our atmosphere that we see; those gases are colorless. Sunlight is a mixture of all frequencies, but blue light, with a higher frequency (and shorter wavelength) than green or yellow or red light, is scattered more effectively than those other wavelengths. It is the scattered blue light that we see in the sky. (See Fig. 22–11.)

This scattering effect also explains why the sun is red at sunset or sunrise. When the sun is low on the horizon, its light must pass for many miles through the densest part of the atmosphere in order to reach your eyes. In this case, the blue frequencies (and the yellows and greens) are effectively removed by scattering all along the light's line of travel. When the sun's rays reach your eyes or reflect from the nearby clouds, almost all that is left of the visible frequencies of light is the least scattered (or longest wavelength) which is red.

Ultraviolet light at the very end of the visible spectrum is scattered even more than blue light. Of the ultraviolet waves that get through the ozone layer, half are in the sun's direct rays, and the other half are from the blue sky. This is why on a sunny day fair-skinned people can get a mild sunburn even though they stay in the shade, out of the direct sunlight.

## The Color of Things

When sunlight strikes someone's favorite yellow shirt, all the colors of the rainbow shine on the material; yet the yellow wavelengths are the ones we see when we look at it. The other wavelengths (or frequencies) of light are absorbed by the material—to a large degree, if not entirely. Just which frequencies are absorbed by any material depends on how the electrons are bound by the molecules and atoms of the substance. The electrons in the molecules of the yellow dye in a yellow cotton shirt absorb most of the blue, green, and red frequencies of light. The energy of that absorbed light raises the temperature of the material, and some of that energy is reemitted from the shirt as infrared radiation. The yellow frequencies are another matter. Some of the electrons in the yellow dye, behaving like miniature antennae, first absorb and then reemit the yellow light in all directions. Those particular electrons *resonate* at the frequencies of the incoming yellow light, absorbing its energy and then giving it up at the same frequencies. This reradiation of only certain frequencies is what we call *selective reflection* (Fig. 22–12). We see a yellow shirt only by reflected light, and it has that color because the dye reflects mostly the yellow frequencies of sunlight.

For another example of selective reflection, a carrot is orange in sunlight (or under electric light) only because blue, green, and dark red light is largely absorbed while orange light (and smaller amounts of yellow and red) is reflected by it. If that carrot were held under a source of pure blue light with no other light reaching it, the carrot would look black. It would absorb the blue light, and there would be no orange light to be reflected. What we call black is actually the absence of visible wavelengths of electromagnetic radiation.

Any color not in the spectrum of white sunlight is a mixture of different frequencies from the rainbow spectrum. When an artist mixes two or more paint pigments to get a desired hue, the color the artist sees is a mixture of whatever frequencies of visible light those various pigments reflect. The grand variety of colors we see around us when the sun is shining is due to the selective absorption by matter of colors from the visible spectrum of white light. Our brains perceive the remaining frequencies, those reflected from trees and flowers and the paints on cars, as colors.

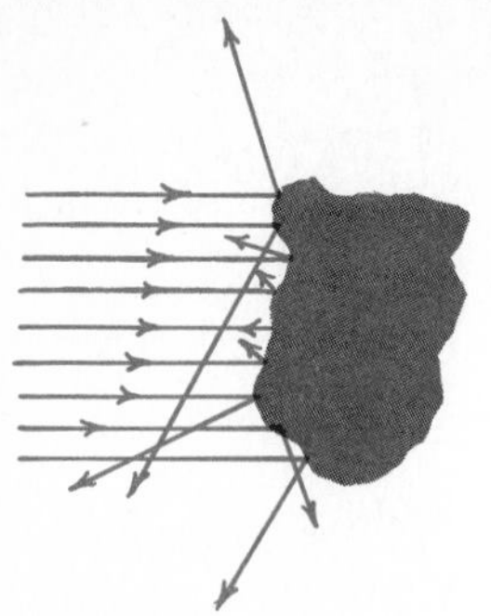

**FIGURE 22–13**
Large particles of dust or smoke in the sky scatter light selectively from their surfaces in all directions. If enough of those particles are in the air, the sky takes on the color (or mixture of colors) they reflect.

Reflection of selective colors can also play a part in the light we see from the atmosphere. Particles from volcanic emissions, forest fires, and air pollution in cities often reflect light that colors local skies. One fire in a Canadian forest of resinous trees caused purple sunsets for the people who were close by. Selective reflection is also what colors Mars's skies. The atmosphere of Mars is very thin compared to earth's. Nevertheless, before Viking 1 landed there on July 20, 1976—only 7 years to the day after the first lunar landing, and during the U.S. bicentennial year—many scientists suspected that Mars would have a pale blue sky from molecular scattering, just as earth has. But Viking found it to be pinkish to orange because of microscopic dust particles that reflect light just as the Martian surface does.

## CALCULATIONS

Electromagnetic waves, like all other types of waves, have a basic connection between their speeds, wavelengths, and frequencies, namely:

$$\text{speed} = \text{wavelength} \times \text{frequency}, \quad \text{or} \quad c = \lambda \times f$$

(See the section "The Speed of a Wave" in Chapter 16 for a discussion of this formula.) For electromagnetic waves in a vacuum, all frequencies from radio to gamma rays move at the speed of light, $c$. For reference, $c$ is equal to about 300,000 kilometers per second, or $3 \times 10^8$ meters per second. In the English system, $c$ is about 186,000 miles per second.

---

EXAMPLE 1: **What is the frequency of a 10-centimeter microwave?**

$$c = \lambda f$$

so

$$f = \frac{c}{\lambda}$$

$$f = \frac{3 \times 10^8 \text{ m/s}}{0.1 \text{ m}}$$

$$= 3 \times 10^9/\text{s} = \mathbf{3 \times 10^9 \text{ Hz}}$$

EXAMPLE 2: An AM radio station broadcasts at "1250 on your dial." Look on a radio's dial and you may see a notation *kHz*. That means 1250 on the dial stands for a frequency of 1250 kilohertz. (Often you'll see 125 instead, with " × 10 kHz" by the dial.) **What is the wavelength of that station's radio wave?**

$$\lambda = \frac{c}{f}$$

$$= \frac{3 \times 10^8 \text{ m/s}}{1250 \times 10^3 \text{ s}} = \mathbf{2.4 \times 10^2 \text{ m}} \text{ (or 240 m)}$$

# REVIEW

**1.** What is the speed of light in a vacuum? Do all colors of light travel at the same speed in a vacuum?

**2.** Do visible light waves, heat rays, X rays, and radio waves all travel at the same speed in a vacuum?

**3.** Thomas Young's experiment showing interference of light proved that light has the properties of waves. True or false?

**4.** Waves of light, according to Maxwell's equations, are traveling oscillations in the electric field of a charged particle. They are known as electromagnetic waves or electromagnetic radiation. True or false?

**5.** Define the word *light* in terms of the spectrum of electromagnetic radiation.

**6.** Which type of wave has the shortest wavelength? Which has the highest frequency? (a) radio waves (b) infrared rays (c) ultraviolet rays (d) green light

**7.** Define an AM radio wave; define an FM radio wave.

**8.** What is a polarized wave?

**9.** Which has the higher frequency, TV signals or AM radio signals? (See Fig. 22–5)

**10.** What natural phenomenon sends out radio waves that a radio can pick up?

**11.** The sky appears to be blue because the other colors have been scattered from our view. True or false?

**12.** Our eyes and brains perceive the mixture of frequencies of light reflecting from an object as its color. True or false?

# DEMONSTRATIONS

**1.** Here's a way to make your own blue skies and sunset. When a flashlight beam penetrates a pitcher of clear water, there's little change in the color of the beam. But add a few drops of milk and you'll see the beam of light turn orange (Fig. 22–14). The milk's molecules scatter the blue light (and some green and yellow, too) in all directions before it can reach your eyes, just as the air's molecules do for the rays of sunlight at sunset. Now look through the side of the pitcher, perpendicular to the beam. Presto! There's the blue light, scattered to the sides (and in all directions), just as the air scatters blue light from sunlight to give us blue skies.

**2.** The light that comes down from the blue sky is partially polarized. Hold Polaroid sunglasses a few feet in front of you and look through them at an angle of 90° away from the sun. Slowly rotate the glasses until you find a position where the polarizing material stops much of the light.

**FIGURE 22–14**

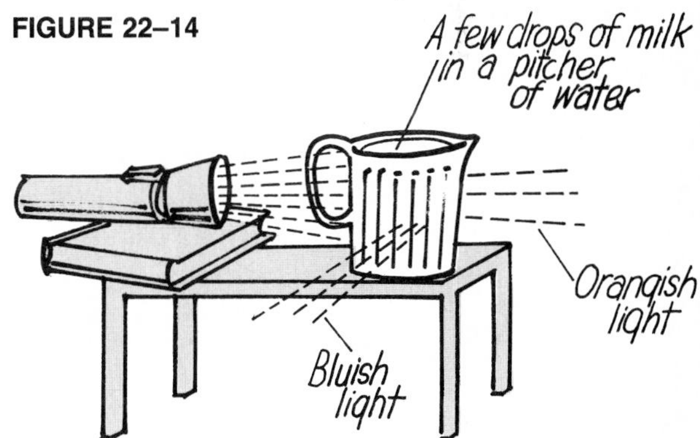


**3.** Motion pictures, we've seen, take advantage of the persistence of human vision so that 24 pictures flashed before our eyes each second can simulate motion. The same thing happens with colors. Your brain can't instantaneously turn off an image *or* its color. When you look at a color and

turn away quickly, there is a very brief afterimage. As a matter of fact, you can trick yourself into seeing white even though you are gazing at colors. Cut a circle from posterboard about 10 centimeters in diameter and color it so there are three pie-shaped segments, one red, one green, one blue. Punch two holes on either side of the center, pass a long cord through them, and tie the cord so about 30 centimeters of cord is on each side of the disk. Holding opposite ends of the cord, twirl the disk in one direction so that the cord winds around itself; then pull sharply against the ends of the wound-up cord, and the colored disk will twirl as the cord unwinds. As it does, your brain is stimulated by all three colors simultaneously, and the twirling disk appears to be white. (Because red light, green light, and blue light merge to give white light, they are called primary colors.)

# EXERCISES

**1.** We call black and white colors, but neither is found in the spectrum of sunlight. Are they really colors? Discuss.

**2.** Arrange the following electromagnetic waves in order of *increasing* wavelength: ultraviolet rays, X rays, infrared rays, AM radio waves, microwaves, green light.

**3.** Why isn't direct sunlight polarized?

**4.** Explain why the light from a candle's flame comes from electrons rather than protons.

**5.** Often it is possible to tune in AM radio stations from a distant city, yet you can't pick up TV broadcasts from that far away. Why?

**6.** If a radio station's signal is weak in your area, you can sometimes improve the reception by placing the radio by an open window. Why should that have an effect?

**7.** Why is the sunlit surface of a black car so much hotter than the sunlit surface of a white car?

**8.** Why does the color of cloth sometimes depend on whether it is viewed in fluorescent light, incandescent light, or in sunlight?

**9.** Communications with astronauts in space have always been carried on using microwaves rather than standard radio waves. What makes this necessary?

**10.** The space between the stars in our galaxy, the Milky Way, contains dust in varying amounts. As a result, the light from stars farther away from our solar system appears more red than the light from similar stars that are close. Discuss.

**11.** If you charge a plastic comb by rubbing it and wave it back and forth in the air, do you create an electromagnetic wave?

**12.** The ozone layer in the earth's atmosphere blocks not only most of the ultraviolet radiation and X rays from the sun but also the small amounts of X ray and gamma radiation that come to earth from interstellar space. Why is this important for life?

**13.** What color would a red car appear to be if it were under a lamp that emitted only blue light in a dealer's showroom at night?

**14.** At fairs, circuses, and promotionals for shopping centers, searchlights are sometimes used, throwing their beams into the night sky so they will be seen far and wide. Would this work on the moon? Discuss.

**15.** In terms of the energy an electromagnetic wave carries, explain how the light from a large, luminous star in the night sky pales against the light from your desk lamp at night.

**16.** The U.S. National Park Service has used infrared detectors mounted on helicopters to spot the hidden entrances to caves and caverns in the area around Carlsbad, New Mexico. Discuss how that is possible.

**17.** Discuss this statement: "Venus is a world with no shadows."

**18.** Because we are warm from the heat released by chemical processes in our bodies, we emit infrared radiation. That is why infrared detectors can reveal a person's presence at night. How do you think clothing should affect a person's visibility where an infrared detector is concerned?

**19.** Snow, sand, and ice are excellent reflectors of ultraviolet light, the electromagnetic waves that cause tanning and sunburn. Discuss the implications for beachcombers and alpine skiers. (Small doses of ultraviolet light are beneficial to humans, however, producing vitamin D precursors in the skin that are essential for survival.)

**20.** After an extensive forest fire in British Columbia, the full moon looked blue to people in the area; and after the volcanic explosion of Krakatoa, many people reported a green moon. Discuss.

**21.** A long time ago, an atomic bomb test on an island in the Pacific Ocean was telecast to the United States. TV viewers in the United States heard the explosion before observers stationed nearby heard it. Explain how that happened.

**22.** Why is someone high in the mountains in more danger of sunburn than at sea level?

**23.** The haze in the atmosphere over industrial centers and large cities comes from particles much larger than molecules and closer to the size of the wavelengths of visible light. How would you expect those particles to scatter the various colors of light? Does this correspond to what you've observed or what you've seen in photographs?

**24.** Why does the sky sometimes appear so much bluer after a rainstorm?

***25.** Suppose a beam of light could travel in a circle around the equator of the earth. Show that it would pass someone standing there 7 times in a single second. (The circumference of the earth is about 24,900 miles, or 40,070 kilometers.)

***26.** A hydrogen atom is roughly $1 \times 10^{-10}$ meters across; how many hydrogen atoms would fit across one wavelength of green light of frequency $5.5 \times 10^{14}$ hertz?

***27.** The longest wavelength electromagnetic wave broadcast by TV stations has a frequency of 54 megahertz ($1\text{MHz} = 10^6$ Hz). Find its wavelength.

***28.** The current in an appliance that's in use in your room accelerates electrons back and forth with a frequency of 60 cycles per second, or 60 Hz. These accelerated charges give off electromagnetic waves of frequency 60Hz. (a) Calculate their wavelength. (b) Explain why they are harmless to us.

***29.** FM radio stations broadcast on frequencies between 88 and 108 megahertz. What are the longest and shortest FM radio waves? AM radio stations use carrier waves with frequencies between 550 and 1500 kilohertz. What are the maximum and minimum sizes of the wavelengths of those waves?

# CHAPTER 23 PROPERTIES OF LIGHT

Ten stories above the floor of the civic center, George had a good view. He watched the 7-foot-long barrel of the spotlight as the floor crew hoisted it toward him. When it was within reach, George leaned out to bring the spotlight barrel onto his narrow platform. There was no rail on the front side of his perch, and, without any leverage, pulling the light in was dangerous. Once on its base, however, the huge spotlight was perfectly balanced. George adjusted the tension on its pivot until he could move the barrel with his little finger. But to follow the performer later that evening, he would tense his arms and use his whole body to guide the spot. This extra inertia helped to make the motion much smoother; otherwise, a tiny jerk at the lamp would jiggle the bright circle on the stage.

After dinner, as people began to fill the auditorium, George climbed the ladder to the catwalk and crossed over to his platform. An earphone in one ear would bring him split-second cues from the lighting director below, while his other ear took cues from the music. For the next 3 hours he'd wish for two pairs of hands and eyes to watch the spot on stage, focus and follow with it, change the colored filters to color the beam, and tend the carbon arc that made the light.

He pulled on the trigger to bring the two carbon rods together and then flipped the power switch. The flow of current between the rods in the lamp made them sizzle where they touched, and he slowly released his grip. As the smoking rods eased apart, a current of electrons shot the gap, ionizing the air between the rods and producing blinding blue-white light in its arc. A 12-inch mirror at the rear of the barrel sent the light to a clear lens at the front, where it emerged in a narrow beam. A small motor pushed the rods closer as they wore down in the intense heat. He adjusted the motor drive by the crackling sound of the arc. Too much or too little distance between those rods, and the arc would go out. It wasn't much fun changing those hot rods during the concert, but he always had to. The rods lasted only 30 minutes, while the concert sets were 45 minutes each.

The lead performer on the stage sweated in the intense light of the lamp, and George sweated from the lamp's heat. The heat was so great that any dampness on the front lens or mirror would cause them to shatter. "At least this is an indoor show," he thought to himself. He would never forget the nighttime water shows in Florida, where his job was to find the fallen skiers with his spot so the ski boats could avoid them. The problem had been the rain. With 10 amps flowing just below his fingertips, he'd had cause to worry. The biggest worry here was his one uncovered ear, which would still ring the next day from the loud music reflecting off the ceiling 10 feet above him. This was a crazy job. ■

## Light Rays and the Pinhole Camera

You can see the path of the sharp beam of a spotlight in a theater or at a concert because smoke, dust, and the air itself scatter a small portion of the light from the beam as it travels. Notice the edges of such a beam appear sharp, indicating that the light in that beam goes in straight lines. You may have watched as a lighting technician in a theater narrowed the width of a spotlight beam until it disappeared. Early philosophers decided that even the thinnest bundle of light travels in straight lines, so they spoke of light in terms of **rays,** infinitely thin beams of light. Although there is no such thing as an infinitely thin beam of light, the idea of such a ray is useful in tracing the path light takes through space. Try the following experiment, which is easily explained with rays of light.

Take a large cardboard box outside on a sunny day and punch a small pinhole in one side. (Use a sharp pencil or an icepick. A hole that's 0.5 millimeters to about 0.75 millimeters across works well.) In the bottom, cut a hole that's large enough to admit your head. Don't

**FIGURE 23–1**
The clear images we see with light we owe to the fact that light normally travels in very straight lines through the ocean of air around us.

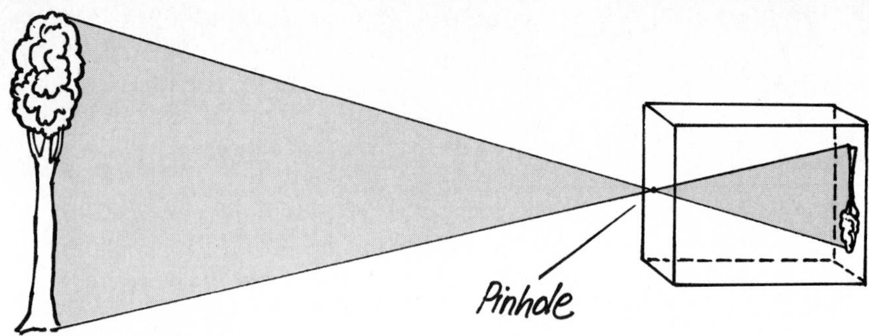


**FIGURE 23–2**
A diagram of the image produced by a pinhole in a box. Rays of light from the object travel in straight lines and enter the pinhole. A ray leaving the top of the tree, for example, goes straight through the pinhole and strikes the lowest point at the image's position on the back of the box. A ray from the base of the tree travels straight through the pinhole to strike the highest point at the image's position. The image, then, is upside down relative to the object.

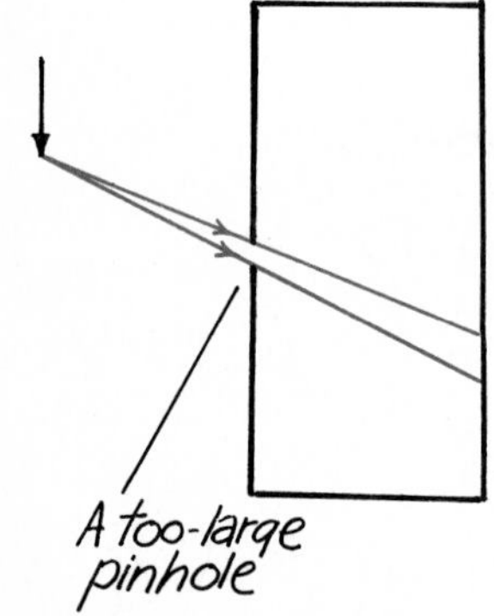


**FIGURE 23–3**
Here's how the diameter of a too-large pinhole affects the sharpness of the image. Each point on the object (here, an arrow) emits rays of light in all directions. Because of the width of the pinhole, rays traveling in slightly different directions enter the box. Two such rays are shown here. These rays go on to strike the back wall of the box in different locations, and that causes the image of the tip of the arrow to be smeared. The entire image of the arrow will appear blurred in this same way.

make the hole in the center. Place it off to one side so one eye will be about 25 centimeters from the wall directly across from the pinhole. Stick your head into the opening, and use a cloth around your neck to keep light from entering there. If light comes in other than through the pinhole, use tape to seal the leaks. Look at the wall opposite the pinhole. On that wall you'll see a faint *upside-down* **image** (a likeness) of the scene outside the box, as Figure 23–2 shows. The rays of light traced from the scenery through the pinhole show why this image appears as it does. That image could be recorded on photographic film; a *pinhole camera* is just such a light-tight box with a pinhole opening.

Images recorded with a pinhole camera aren't quite as sharp as those taken with standard cameras, however. Rays from each point on the outside scenery pass through every point in the pinhole opening, as shown in Figure 23–3, and smear the details of the image slightly. Although a smaller pinhole would produce a sharper image, it would let in less light and the image would be too faint to see. And if the pinhole is *very* small, the wave nature of light reveals itself, and a greater portion of the light diffracts, or bends, as it passes through the opening, as we saw in Chapter 22.

## Reflection of Light from a Smooth Surface

From the time we open our eyes in the morning until we go to sleep, light enters our eyes to form images of our world. When you look at a scene in front of you, what do you see? Usually not light "straight from the source," but light reflected by the objects around you. You see this page because light from the sun or a lamp or a fluorescent light strikes the page and reflects in all directions, and some of that light comes to

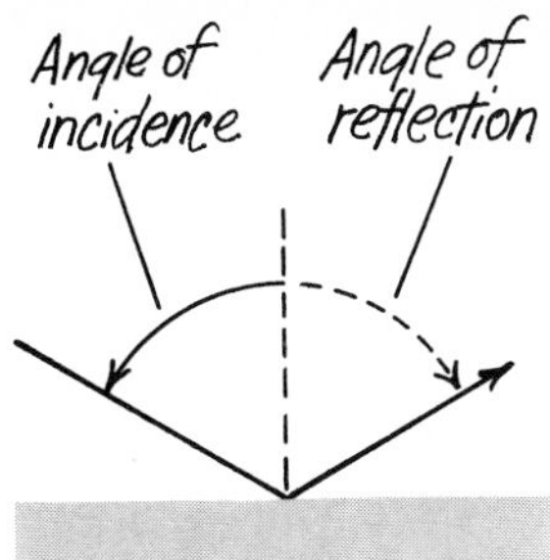


**FIGURE 23–4**
The angles of incidence and reflection for a light ray are measured from a line that is perpendicular to the reflecting surface.

your eyes. The printing on the page reflects much less light (it absorbs more) than the white paper does, and so the letters appear darker.

This page reflects more than half of the visible light falling onto it. By comparison, ordinary sand reflects about 30 percent and snow about 60 percent. But the surface of this white page and sand and snow are microscopically uneven and so don't reflect light uniformly. Such surfaces **diffuse** the light; that is, the light reflected from them is scattered in many directions.

In contrast to these materials, the smoother surface of a mirror reflects light more uniformly. Those rays that come into the mirror perpendicularly reflect straight back along their incoming paths. Any rays that come in at an angle to a perpendicular line drawn from the mirror's surface reflect at an equal angle, as shown in Figure 23–4. The **angle of incidence** made by the path of the incoming ray with the perpendicular line is equal to the **angle of reflection** made by the outgoing ray with the same line. Rays of light, then, bounce off a mirror as a pool ball might bounce from the cushion of a billiard table. Because of this behavior, the reflected rays can make an image that to some extent resembles the object those rays came from.

When you stand in front of a mirror, you see an image formed by light reflected from the mirror's surface. However, the image seems to peer at you from *behind* the mirror itself. Such an image, formed by rays that don't actually come from the location of the image itself, is called a **virtual image.** The image produced in a pinhole camera, on the other hand, is formed by rays of light that actually reach the image's location; that sort of image is called a **real image.**

When you look into a bathroom mirror, your image looks almost like a perfect picture of you, but it isn't. Blink your right eye, and the image that faces you blinks its left eye. The mirror's image exchanges the left side for the right side, and the rays traced in Fig. 23–6 show why.

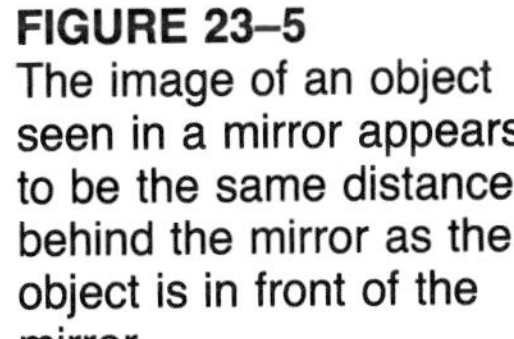
**FIGURE 23–5**
The image of an object seen in a mirror appears to be the same distance behind the mirror as the object is in front of the mirror.

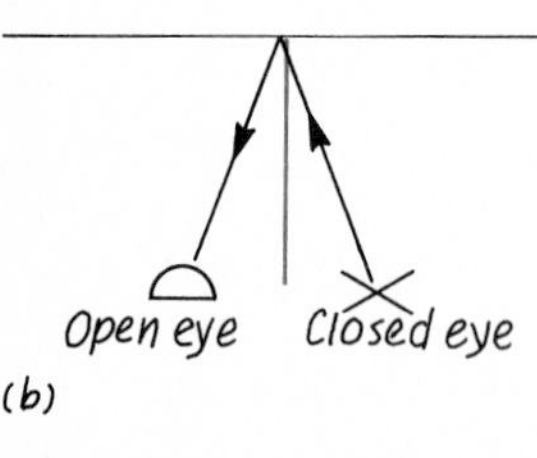


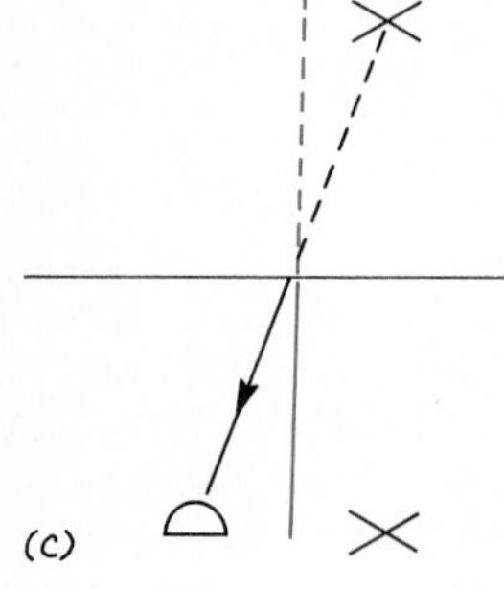


**FIGURE 23–6**
*(a)* Light from his closed right eye reaches the man's left eye after reflecting from the mirror at an angle. Because the angle of incidence equals the angle of reflection, the left eye sees the light from the closed right eye coming from the *left* eye of the image. *(b)* A schematic of the person's eyes, as seen from above. A ray from the closed eye travels a distance to the mirror, reflects at an angle equal to its angle of incidence, and travels an identical distance to the open eye. The open eye forms an image of the closed eye from many such rays, and the image it receives has a size that depends on the total distance the rays travel to meet the eye. The greater the distance, the smaller the image. *(c)* The image of the closed eye appears to be behind the mirror at a distance equal to the distance traveled by the incoming ray in *(b)*.

Reflections from a curved mirror give distorted images, as you can see by looking into the curved surface of the bowl of a metal spoon. Curving a mirror in a predetermined way can make such distorted images useful. Curved glass or metal mirrors in a spotlight or searchlight or flashlight direct the rays of light from a bulb or an arc into a parallel shaft of light. When light from a star falls onto the curved mirror of a reflecting telescope, the mirror directs all the light to a single point. Concentrating the light this way makes stars that are too dim to be seen with the unaided eye become visible through the telescope.

## The Slowing of Light by Matter: Refraction

On a clear day, your sharp shadow on the ground agrees with the old saying, "light travels in straight lines." But this saying should have a second part—"unless it doesn't." Dip one end of a straw or a pencil into water at an angle to the surface. The image you see of the submerged part angles out abruptly, as if the straw or the pencil were bent (Fig. 23–7). It's not bent, of course; its image is only displaced. This happens because the light coming from the underwater portion of the straw changes direction as it leaves the water in a process that is called **refraction.** The refraction of that light can be traced to the fact that light takes more time to travel through water than through air.

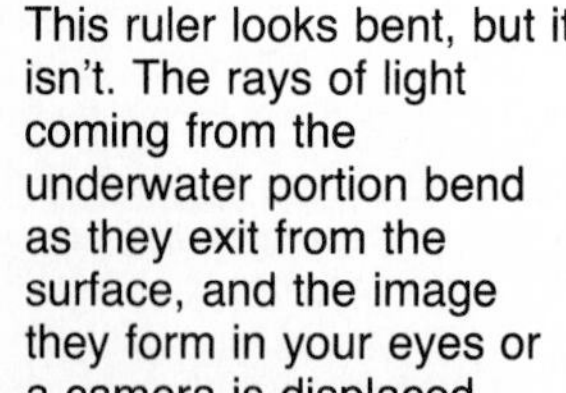

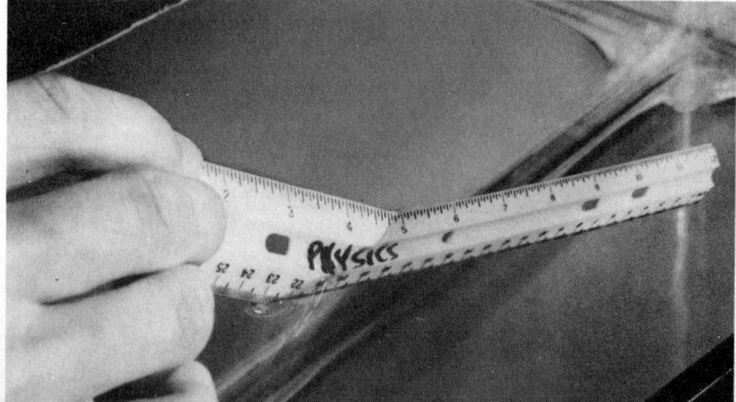


**FIGURE 23–7**
This ruler looks bent, but it isn't. The rays of light coming from the underwater portion bend as they exit from the surface, and the image they form in your eyes or a camera is displaced compared to the actual position of the ruler.

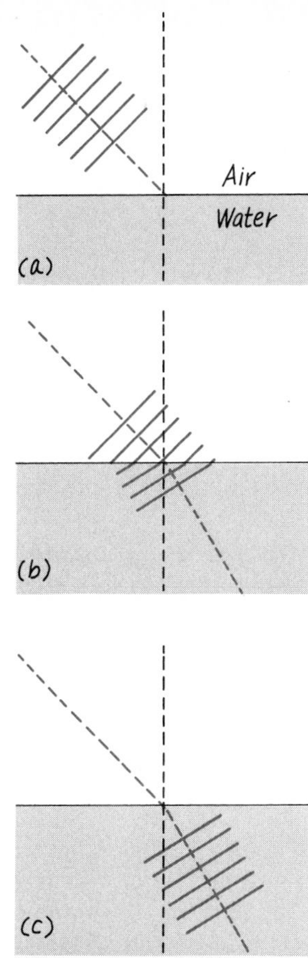

**FIGURE 23–8**
This sequence of illustrations shows how the wavefronts of light behave when light from the air enters water at an angle. For more details, see "Experiencing Refraction."

The wave nature of light provides a picture of how light is slowed as it passes through various kinds of transparent matter. The oscillating electric field of a light wave affects the electrons of any molecules or atoms the wave comes to. It moves the electrons, especially the outermost ones that are less tightly bound. Those electrons oscillate and absorb the light's energy, becoming radiators in the process. The light they emit in turn has the same frequency as their frequency of oscillation, which is the same as the frequency of incident light that caused the electrons to oscillate. So light that travels through matter is continually absorbed and reemitted—and these processes take time. The light itself always travels with speed *c,* but light's passage through the matter is delayed as its energy is held briefly between absorption and reemission, giving light a smaller *apparent* speed. In air at sea level, light's speed of transmission is about 99.97 percent of its speed in a vacuum; in water it is about 75 percent as fast, and in glass or quartz, 65 percent. The molecules in a diamond slow light even more, to 42 percent of its speed in a vacuum. But if light rays leave a transparent solid or liquid to move again through air where there are fewer molecules to slow them, they once more travel at 99.97 percent of *c,* light's speed in a vacuum.

Light refracts, or changes directions, when it angles out of one material into another where it travels with a different speed. Figure 23–8 shows what happens when wavefronts of light traveling through air enter a flat surface of water at an angle. Remember, a wavefront is perpendicular to a ray's direction. The first edge of the wavefront that enters the water slows down as the absorption and reemission processes take place. That end drags its feet while the part of the wavefront still in the air continues to scoot along; this causes the wavefront to bend at the surface. After the entire wavefront has entered the water, the wavefront is seen to travel in a different direction. The rate of light's travel in the new medium determines the amount of refraction. The slower the apparent speed, the greater the refraction.

The reverse happens if light leaves glass or water to travel in air. If the wave comes out at an angle, the emerging edge of the wavefront picks up speed while that part of the wavefront still in the glass pokes along. The result? The emerging light bends *away* from the perpendicular line to the surface.

## EXPERIENCING REFRACTION

You can feel and see this sort of change in direction in your car (as we noted in Chapter 16). If the right tires hit a deep water puddle near the curb, the car pulls to the right. As the tires hit the water, there is drag, the friction that comes when the tires push through the water. This drag slows the right side, and the car turns *toward* the slower medium. Although it isn't friction that slows light, the effect of refraction is the same. (See Fig. 16–17.)

If the car is on a sandy shoulder and pulls onto a paved road at an angle, the last wheel (or wheels) leaving the soft sand lags while the wheels on the road break free of the drag. The car turns sharply *away from* the faster medium. Light, too, bends away from the normal to a medium where it picks up speed.

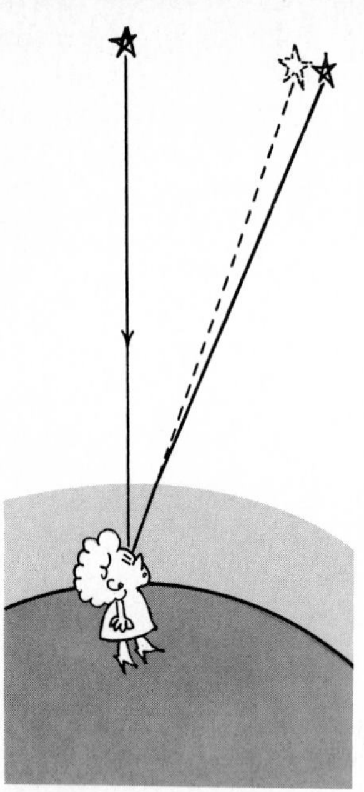

FIGURE 23–9
The star directly over this stargazer is seen in its true position. However, the atmosphere refracts (bends) starlight that comes in at an angle, and we see those stars slightly out of position, that is, higher above the horizon than they really are. (The dashed line shows the direction of the displaced image.)

## Refraction in the Atmosphere

The atmosphere of our planet is very thin at its outer edge and relatively dense at the ground. Incoming sunlight or starlight travels at the speed of light, *c*, until it strikes the outermost molecules; as it travels downward into thicker gas, its speed drops more and more. The loss in speed is slight, less than one-tenth of 1 percent at sea level. But even this small difference in speed causes refraction when light angles into the earth's atmosphere.

If the light from a star comes down from straight over your head (the point astronomers call your *zenith*) at some time during the night, you see its image in its true position. But a star at some lower angle than the point directly over your head is seen slightly out of its true position because its light refracts downward as it comes through the air. You see the image of that star a little closer to the zenith than it really is (see Fig. 23–9). This effect is slight, but astronomers must correct for it when they measure the position of stars. The greater the angle of incidence of the entering starlight, the more the bending. Rays coming in nearly horizontally are bent most. This happens, for instance, each time the sun rises or sets. At sunset you can sometimes watch the red globe of the sun flatten a bit and finally touch the horizon. It flattens because the rays from the bottom of the sun pass through a bit more of the thickest part of the atmosphere; thus they bend more toward your zenith than do the rays from the top of the sun. In fact, when the sun is

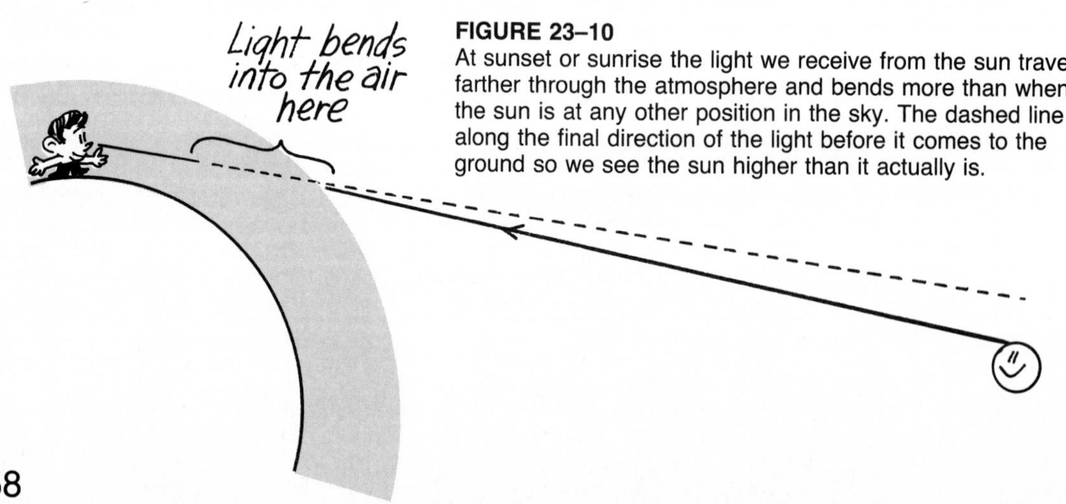


FIGURE 23–10
At sunset or sunrise the light we receive from the sun travels farther through the atmosphere and bends more than when the sun is at any other position in the sky. The dashed line is along the final direction of the light before it comes to the ground so we see the sun higher than it actually is.

**FIGURE 23–11**
When the sun's lower edge is touching the horizon at sunset or sunrise, the true position of the sun is one diameter below the image you are seeing.

at such a low angle, *all* its rays are bent quite a lot during their trip through the atmosphere. As a result, the sun itself is just *below* the horizon when you are seeing its image just *touch* the horizon. On Venus, if there were ever a clear day, the refraction would be much more pronounced. (Venus's atmosphere is 90 times as dense as earth's.) Figure 23–12 compares sunsets on earth and Venus.

**FIGURE 23–12**
Some actual positions of the sun (dashed color spheres) and its images (in black) as they would be seen in the sky of Venus (if it was clear) and as they are actually seen in the sky of earth. Notice the flattening of the setting sun on both planets. (Since Venus is closer to the sun, the sun looks larger there, in the horizontal direction at least. The flattening of the sun is more severe when viewed from Venus.) Refraction of light on Venus is much greater because of its denser atmosphere. The view of Venus here shows only the nearby landscape; the image of Venus's distant horizon would be displaced upward because light rays angling upward from points on the horizon would bend down and over to an observer's eye. In fact, the observer would seem to be standing at the center of a dish whose rim would be near the most flattened image of the sun shown in this illustration.

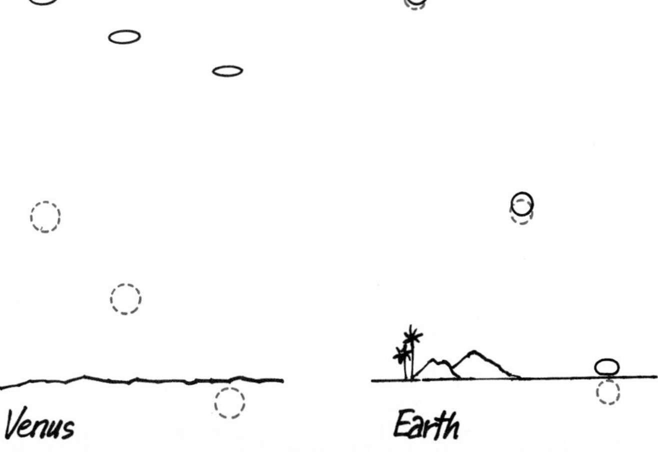

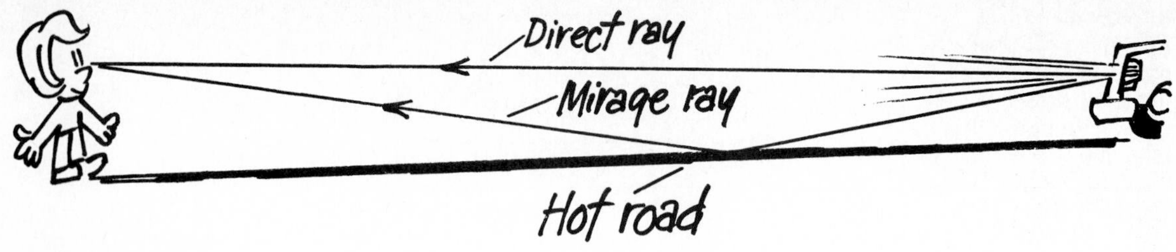


**FIGURE 23–13**
Light that comes nearly tangent to the boundary layer of heated air next to a hot road can be refracted upward, keeping it from striking the road. (The angles are normally much smaller than those shown here.) At bumps and dips, rays of light can pass more nearly tangent to the road for a short distance, increasing their refraction and causing a mirage.

## Mirages

**Mirages** are images of light that are misplaced by refraction. They can be seen on any sunlit highway, where they appear at a distance and look like water puddles, but they grow thin and disappear as you draw close to them. In a mirage you see the light from trees, clouds, or sky, or even car lights, anything that you might see over the mirages and on beyond them, as if a mirror were lying on the road. These mirages are caused not by reflection of light on the road surface but by refraction. A road heated by sunlight warms the air that touches it, forming a thin boundary layer of hot air next to the pavement. Light that enters this hotter air travels faster than it travels through the cooler air above it, because there are fewer molecules to slow the light. If light angles into this hot layer, it picks up speed and bends *away from* the road. The bending is slight, but light rays that almost graze the road can refract and turn upward enough to travel to your eye (see Fig. 23–13). The mirage comes from the light from the sky or from any objects just above the road from your line of sight. A mirage forms over any smooth dip or bump in the hot road where it finds a nearly parallel area to graze. If the surface of a road is wavy, you might see a series of puddles, as in Fig. 23–14.

**FIGURE 23–14**
Most often a mirage will be the image of the sky near the horizon. Those mirages usually look like puddles of water on the highway. But this photo shows a series of mirages from the running lights of a sports car. Rays from the car's lights travel nearly tangent to the road at a series of dips and each time refract upward in the hot boundary layer of air there.

Look for a mirage over a road and walk toward it. The puddle will get narrow and finally disappear. Since the light from a mirage is only slightly bent, the rays drop just below your line of sight as you approach the image. If you stoop down while looking at such a mirage, you see more of the refracted rays, and the mirage gets wider.

Mirages over roads appear even in winter when there is only a few degrees of temperature difference between the road and the air. You can see mirages above any flat surface that has been warmed enough by the sun, even a vertical surface. A dark-colored wall of a building absorbs sunlight well. To view a mirage on such a wall, stand at one corner and look directly along the surface. Any object that's just to the side of your line of sight along the wall will appear as a mirage on its surface. Move your head slowly side to side, and you'll see a clear image of the object appear and disappear.

**FIGURE 23–15**
This photo reveals that the bending of light that makes a mirage over a hot road takes place very close to the road's surface. The safety reflectors to the left of the dashed center line remain visible and are not a part of this huge mirage occurring at the intersection of two roads. Even though the reflectors are only a centimeter or so above the road's surface, the light traveling at their level is not refracted and doesn't form a mirage. The dashed center lines disappear into the "puddle," however.

## Total Internal Reflection

To see refraction work, take a flashlight to a swimming pool at night. Holding the light at arm's length just above the surface of the water, tilt it sideways so you can view the beam from the side. When the angle of incidence is large (when the light is grazing the surface), you cannot miss seeing the light beam change direction. It angles downward more steeply as it enters the water.

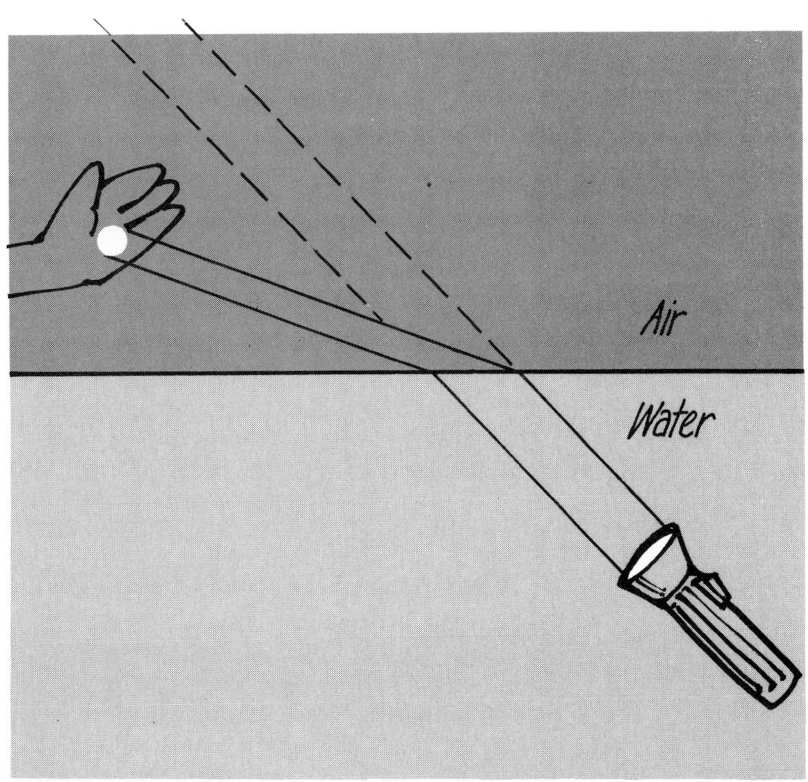


**FIGURE 23–16**
The path of a refracted beam of light from a flashlight in a swimming pool. The exiting beam picks up speed and angles away from the air, back toward the water's surface. (Not shown is the portion of the beam that's reflected at the water/air surface and goes toward the bottom of the pool.)

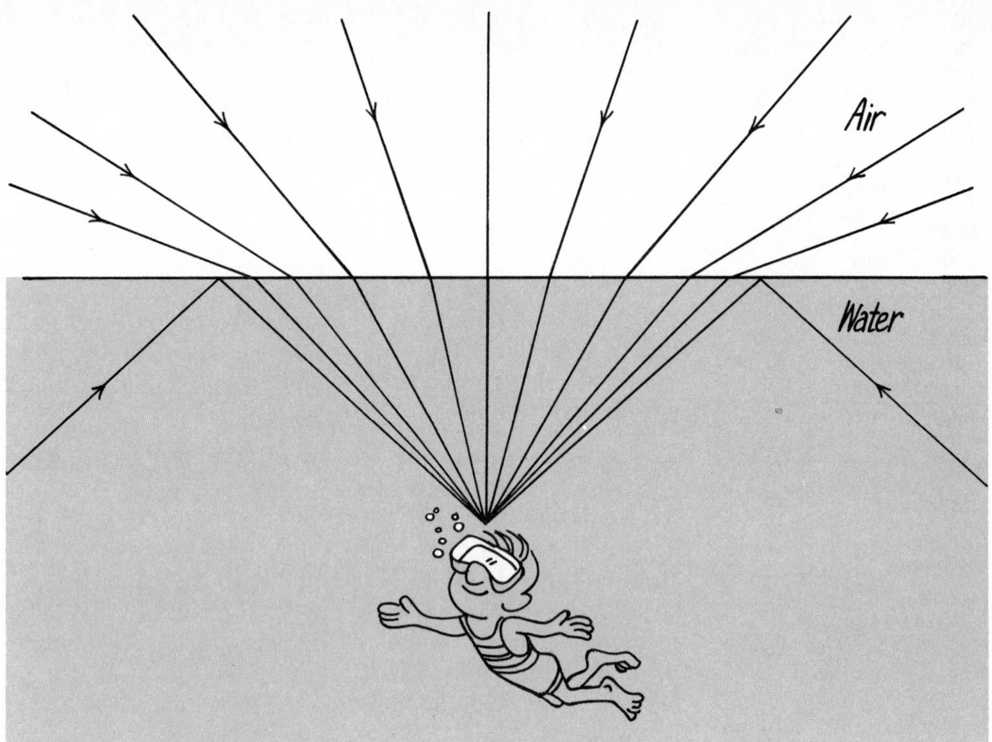


**FIGURE 23–17**
The critical angle of light at an air/water surface is 48.6 degrees. Light traveling through air almost tangent to the water is transmitted at a 48.6 degree angle when it enters the water. If light coming from the water hits the surface at an angle of incidence greater than 48.6 degrees, it is totally reflected.

If you wrap your flashlight in a clear, waterproof plastic bag, you can watch refraction work in the opposite direction. Holding the light just under the water, tilt the beam upward at an angle. Place your other hand above the water's surface, directly in front of the flashlight. There's no light there! Then slowly lower that hand until the reflected light hits it. You'll see that the beam is traveling close to the water's surface. At the bottom of the pool, you'll see part of the light that *reflects* from the water's surface to travel downward.

If you tilt this underwater flashlight even more, so the beam strikes the surface at a larger angle of incidence, you'll eventually have to put your hand down to the water to find the refracted beam as it travels nearly parallel to the surface. Tilt the light a tiny bit more. Your hand goes dark. The refracted light vanishes. At that *critical* angle, or at a greater angle, no light escapes from under the surface (Fig. 23–17). The light is totally reflected within the water. This is called **total internal reflection.** The water–air surface acts like a perfect mirror.

Divers have no trouble finding the critical angle for total internal reflection when they look up at the surface. Light coming from straight

**FIGURE 23–18**
Part of a diver's view of a bright circle that ends at the critical angle. The dark area that is farther from the diver's zenith (overhead point) transmits no light from above (unless waves are present to vary the angles of incidence for the light from above). The light from that dark area is only the light from the bottom that is reflected back down at the water's surface.

above their heads comes straight down, while light that comes in at an angle is bent. Even the light from the horizon angles in and bends downward. But if the surface is smooth, at angles greater than 48.6° from the vertical, divers can see only reflected light from beneath the surface. Because this happens in any direction around a diver's position, north, east, south, west, and all directions between, each sees a bright circle centered above his or her head. This circle's edge is at the critical angle. See Figure 23–18.

## Light Pipes

Electromagnetic waves are excellent message carriers. Radio waves and microwaves (used for TV transmission and long-distance telephone calls) pass signals along at the speed of light. Their oscillations are fast, and transmitting equipment can crowd a lot of information onto their waves per second by changing either their amplitude or their frequency. Satellites in geosynchronous orbit match the earth's rotation rate, and microwave signals beamed up to them are returned in broader beams that can cover about half the earth's surface. Communications have never been so good.

Of course, there's always room for improvement. Waves of visible light have a much higher rate of oscillation, so more information can be sent per second with light beams than with radio waves or microwaves. And although light travels in straight lines through a uniform medium, light can be *guided,* made to turn sharp corners or even go in loops. Mirrors could also make it do this, but each time light reflects from a mirror, a percent or so of its energy is absorbed. Light travels with less absorption through the thin glass or plastic fibers known as *light pipes,* or *optical fibers*.

Total internal reflection makes light pipes possible. Light comes in at one end of the pipe and moves straight along until it hits the inner wall. Because it moves more slowly inside the transparent solid than it would in the air outside, it is totally reflected back into the pipe if its angle of incidence on the wall is greater than the critical angle. Then the reflected beam travels in a straight line until it hits the inner wall again.

Figure 23–19*a* shows typical reflections in a small-diameter light pipe. A thin optical fiber can be twisted into loops or even knots and still carry the light as shown in Fig. 23–19*b*.

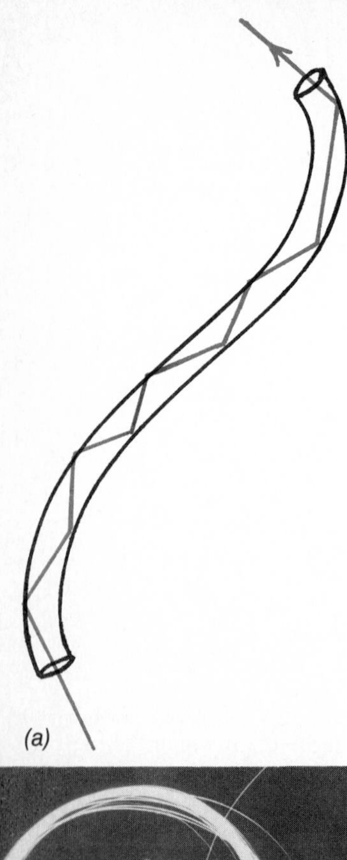

(a)

(b) **FIGURE 23–19**
*(a)* Reflections in a thin light pipe are of the total internal reflection variety. Nevertheless, some light is scattered in the glass, and some light is lost at tiny imperfections on the surface of the glass fiber, where some rays strike it at less than the critical angle of reflection and escape. *(b)* Light emerges at the end of a coiled optical fiber many meters long.

## Prisms and Rainbows

When white light enters a facet of a cut diamond, brilliant splashes of color may emerge from some other point. This happens because of a phenomenon called **dispersion**; the different frequencies of light that make up white light travel at somewhat different speeds in the diamond. For that reason they bend (or refract) by different amounts when entering or leaving at some angle through a face of the diamond. All transparent matter—diamond, glass, quartz, clear plastics, water—is dispersive to some extent. When Newton passed white light through a prism, dispersion caused the spray of colors on his wall.

As the visible frequencies of light pass into diamond or glass, red light is slowed less and blue light is slowed more. As a result, the blue light bends (or refracts) more than the red light if they both enter at the same angle. Figure 23–20 traces some of the rays of white light through a prism. Notice that refraction takes place where the rays leave as well as where they enter; the waves are spread even more after the second refraction.

A rainbow is a spectrum you see when sunlight falls on a distant rain cloud or on the mist of a lawn sprinkler or a waterfall. When sunlight strikes a spherical drop of water, some passes into the drop. The rays that enter at a certain angle to the drop's surface (see Fig. 23–21) undergo total internal reflection at the back of the drop and return to pass through the front surface, bending once more as they do. Because of dispersion, the colors separate inside the drop and reappear as a spectrum. Red light exits at an angle of about 42° from the direction line of

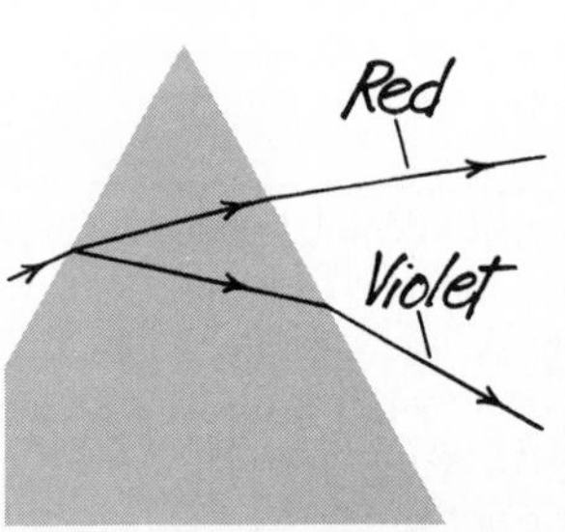


**FIGURE 23–20**
When visible frequencies of light enter glass, the lower frequencies lose less speed and are refracted less. Violet, the highest visible frequency, slows the most and is bent at a larger angle both when it enters and when it leaves the glass.

FIGURE 23–21
The dispersion of a beam of white light in a spherical raindrop.

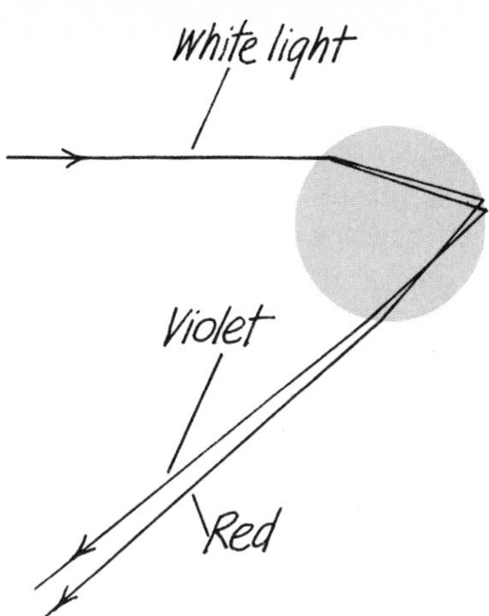


the sun's rays, while violet leaves each drop about 40° from that line. The other colors fall in between.

You see the internally reflected light from a drop only if a straight line from your eye to the drop makes an angle of 40°–42° with the sun's incoming rays. The rainbow appears as an arc for that reason; see Fig. 23–22. Someone standing next to you will see a rainbow too, but from different drops of water. If you are on the ground, the rainbow's arc ends at the ground, where there are no more droplets to reflect the light. From an airplane flying past this same cloud, however, you'd see a *full circle of rainbow*. Sometimes you can see a companion rainbow that circles farther out than the bright *primary* rainbow. This one, called the *secondary* rainbow, comes from rays that undergo two total internal reflections at the back of each raindrop and exit at a wider angle, about 50° from the incoming sunlight. The second internal reflection changes the order of the emerging colors, so the secondary rainbow's colors have their order reversed in relation to the colors of the primary rainbow.

## Lenses

Flat panes of glass let light come through your windows almost undisturbed (Fig. 23–23). But when light passes through a clear glass bottle or jar, the curved surface sends incoming rays of light in different directions. The purposely curved surfaces of eyeglasses put such refraction to good use, improving the wearer's vision. Crude eyeglasses came

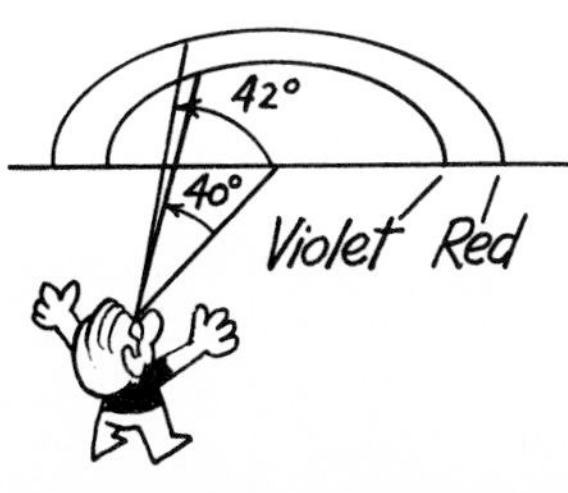


FIGURE 23–22
A rainbow produced by a rising or setting sun at the back of the observer, so the rays that strike the raindrops are traveling parallel to the ground. The red part of the arc is everywhere 42 degrees from the arc's center, and the violet arc is everywhere 40 degrees from the arc's center. This rainbow corresponds to horizontal incoming sunlight, as in Fig. 23–21. If the sun is at a higher angle, the arc of the rainbow will be nearer to the ground for this observer, as you can see by tilting Fig. 23–21 somewhat clockwise. In fact, if you rotate Fig. 23–21 until the incoming sunlight is almost vertical, as it nearly is at noon in the summertime, you'll see that the only rainbow you are likely to see at that time of day is in the mists of a lawn sprinkler near your feet.

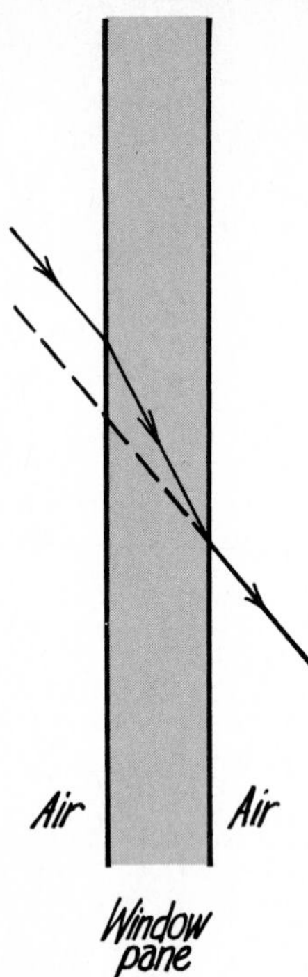


**FIGURE 23–23**
An image of an outside object is not distorted, but it is slightly displaced when viewed at an angle through a window. The ray in this diagram appears to come from the direction indicated by the dashed line when viewed by an observer on the right.

into use in the thirteenth and fourteenth centuries, and today more than half the adults in the United States depend on eyeglasses or contact lenses for good vision.

A simple lens is a thin, smooth piece of clear glass or plastic with its surfaces shaped either to spread the passing rays (a **diverging** lens) or to bring them together (a **converging** lens). The surfaces are normally polished into a spherical or cylindrical shape; that is, the lens surface is a ''piece'' of a perfect sphere or cylinder. All eyeglasses have such surfaces, because more complex surfaces, such as parabolic surfaces, are difficult to shape accurately.

A converging lens, also called a **convex** lens, (Fig. 23–24*a*), is thicker at its center than at its sides, while a diverging lens, called **concave,** is thinner at its center. Parallel light rays passing through a converging lens meet on the other side of the lens at a point called its **focal point.** The distance from the center of the lens to the focal point is the **focal length** of the lens. A diverging lens has a focal point, too, as shown in Fig. 23–24*b*. It is the point from which parallel rays passing through the lens seem to originate.

Because of the sun's distance, the rays of sunlight that fall on the earth are quite parallel. So if you've ever used a magnifying glass to focus sunlight to a point and burn a hole through a piece of paper, you've found the focal point of that particular converging lens. To use a converging lens to magnify, the object of interest must be closer to the lens than the focal point of the lens itself. To see how the magnified image appears, look at Fig. 23–25. In the figure, three rays (the solid lines) from the top of the tack head and one ray from its sharp point are followed. *These special rays are chosen because they are easy to follow, or trace, through the lens.* Ray A, which runs from the point of the tack through the geometrical center of the lens, goes through the lens unaffected. Ray B, from the top of the tack through the center of the lens, is bent only a little going in and is bent backward by the same amount going out because of the symmetry of the lens. Ray C, from the top of the tack, goes to the lens at the same angle as would a ray coming from the left focal point. (See the dashed line behind that ray.) Therefore, when ray C emerges from the lens, it travels parallel to the axis of the lens. Ray D, from the top of the tack, travels parallel to the axis before entering the lens. It, therefore, refracts to pass through the focal point on the right side of the lens. The dashed lines at the left of the figure show the directions from which the rays appear to come after they are refracted by the lens. The top of the image is located where the three rays from the top of the tack seem to come from, that is, where the dashed lines converge. The image you see isn't ''real,'' in the sense that light rays do not come from (or pass through) the image. Just as with mirrors, such an image is called a *virtual* image.

Converging lenses can also form *real* images, like the images slide projectors or movie projectors focus on a screen. Real images appear only when the object is at least one focal length from the lens. Figure 23–26 shows how this works. A candle that's exactly two focal lengths from a convex lens has a life-sized real image two focal lengths behind

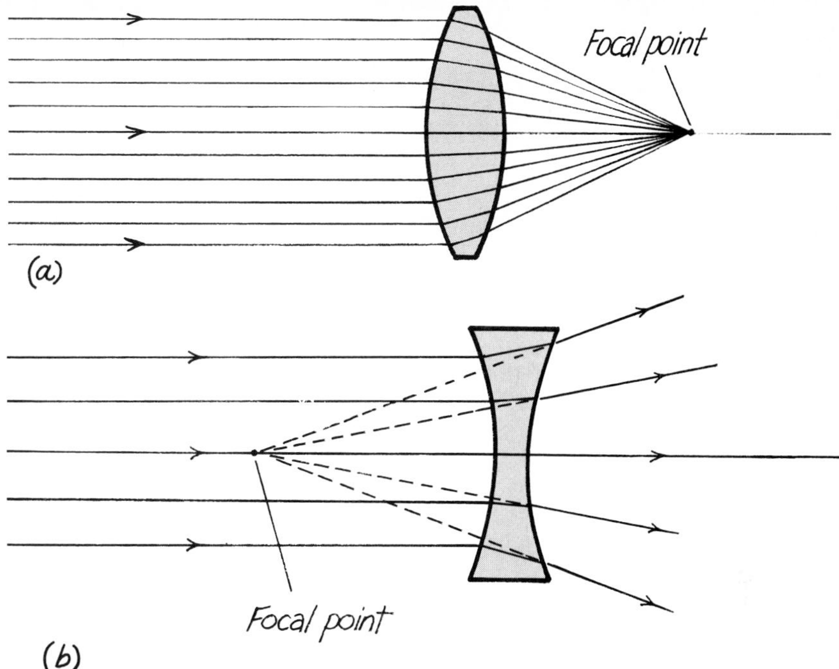


**FIGURE 23–24**
*(a)* The light rays on the left are traveling parallel to the axis of a convex lens, the center line perpendicular to the face of the lens. The lines refract at both surfaces of the lens and converge at a point on the opposite side of the lens called the focal point of the lens. *(b)* Parallel light rays coming directly into a concave lens refract and diverge so that they appear to come from a point that is located in front of the lens. That point is the focal point of the concave lens.

**FIGURE 23–25**
Using a convex lens as a magnifier for a small thumbtack. The large thumbtack at left is the image of the small tack as seen by an observer on the right. It is called a virtual image since the light rays don't really pass through it or come from it. Notice the tack is nearer to the lens than the focal point of the lens.

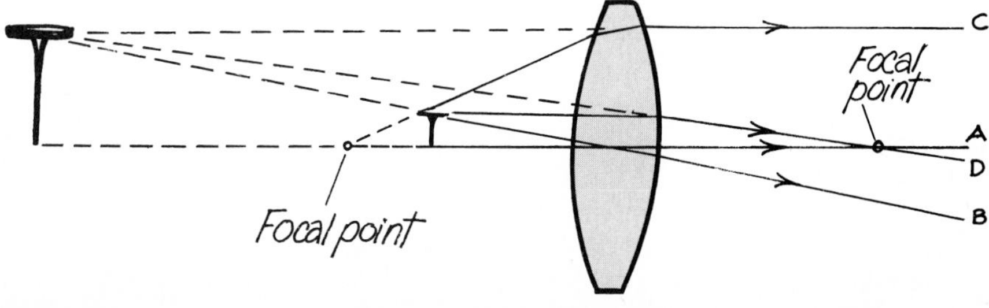

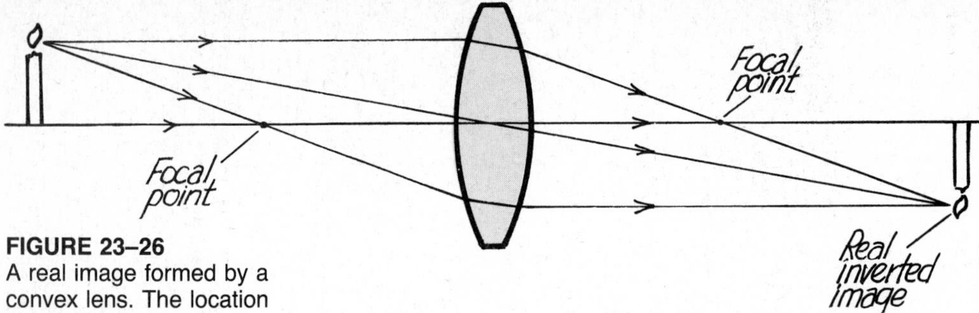


FIGURE 23–26
A real image formed by a convex lens. The location of the candle's flame is found by using rays similar to those used in Fig. 23–25.

the lens, but it is upside down, or inverted. (This is why the slides must go in upside down in a slide projector.) Move the object farther away from the lens, and the position of the inverted image moves in toward the lens. Move the object closer and the image is larger. As the position of the image moves in, the image shrinks in size. As shown in Fig. 23–27, a scene on the far horizon forms an image at the focal plane of the converging lens. In a typical 35-mm camera the upside-down image made by the lens falls on the film to expose it. To take a picture of a distant scene, the lens must be one focal length from the film. To focus on something closer, the lens must slide forward away from the film to keep the image in focus on the film.

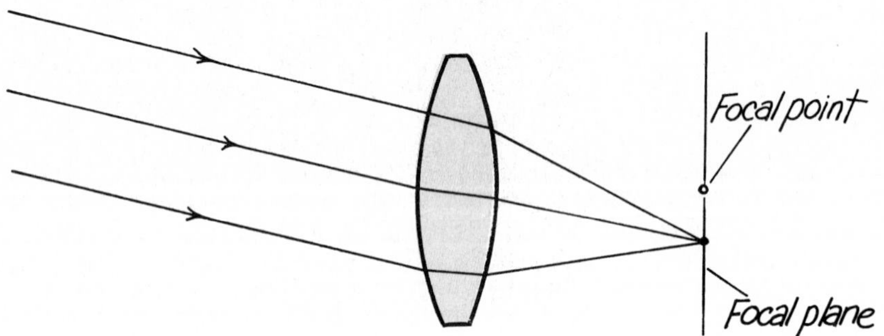


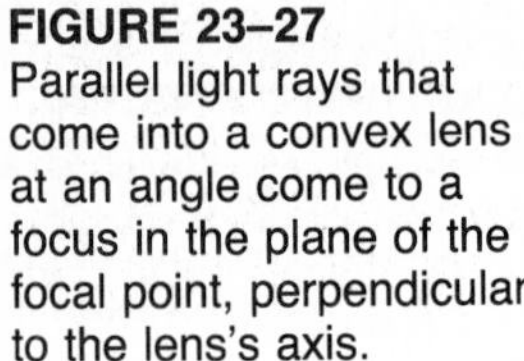
FIGURE 23–27
Parallel light rays that come into a convex lens at an angle come to a focus in the plane of the focal point, perpendicular to the lens's axis.

If we compare a slide projector to a camera, the role of image and object are reversed; the small slide is the object, and the larger image is projected on the screen. The light moves in the opposite direction from the way it moves through a camera lens to fall on the film.

Light coming through the edge of a spherical lens focuses a little closer to the lens than the light passing through its center, distorting the image. Modern cameras have specially designed series of spherical lenses, some convex and some concave, that when glued together compensate for this *spherical aberration* through multiple refractions. These lenses, usually made with different kinds of glass, are designed with the help of computers. Guided mathematically by the known refractive properties of the glass in the lenses, a computer sends imaginary rays to bend and turn through a set of hypothetical lenses. Adjusting the curves of the lens surfaces with the computer lets the designer minimize aberrations without having to experiment with the actual lens.

## Eyes

Your eyes resemble a camera in several ways. Your eyeballs are curved, and the front surfaces, the corneas, serve to converge the rays of light. Further, just beyond the surface of each eyeball, there is a real convex lens that converges the light a little more. When light passes into the eye, a tiny inverted real image is formed on the curved retina (see Fig. 23–28). The hair-thin tissue of the retina serves as more than a projection screen for the brain. Its light-sensitive cells continually sample the light of the image and send electrical signals to the brain; thus your eyes act more like a color TV system than a simple camera.

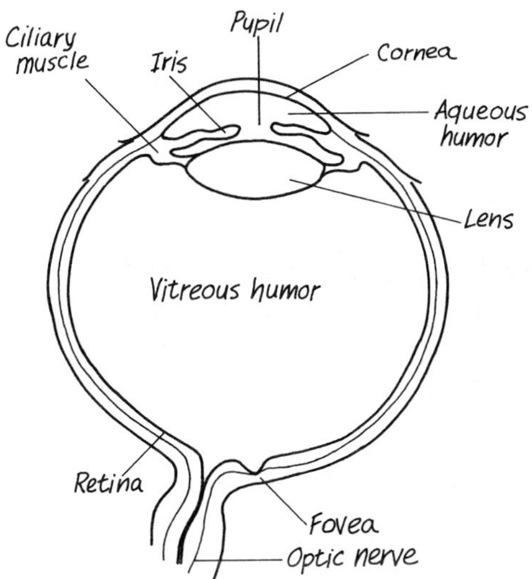


**FIGURE 23–28**
The human eye.

The lens of the eye does not move in and out to focus at different distances as a camera does. Instead, a muscle around the elastic lens squeezes down to change the shape of the lens. When the muscle is relaxed, the lens is thinnest and your eyes are focused on things far away. For you to read this page, the muscles contract, giving the lenses more curvature to keep the images on the retinas of your eyes.

Nearsighted or farsighted people often have eyeballs that are too long or too short, respectively, from front to rear. The light through the lens then forms an image that falls in front of (nearsightedness) or behind (farsightedness) the retina, causing the eye to see a blurred image rather than a well-focused one. Either condition is correctable with a glass or plastic lens, a diverging one for nearsightedness and a converging one for farsightedness). However, the shape of the cornea is more often responsible for bad vision. Eyeglasses or contact lenses usually solve either problem or both problems together.

The retina itself is several layers of tissue that are the bed for the cells that sense light, the **rods** and the **cones.** Seventy percent of the body's sense receptors are there, packed in enormous numbers (more than 125 million nerve cells per eye). Cones are concentrated at the eye's axis, a spot called the *fovea centralis,* to increase the eye's resolving ability. For example, if you see some friends a block away, their images on your retina are a mere two-hundredths of a centimeter across. But the concentration of receptor cells at the retina's center lets you see enough detail to tell them apart clearly, or to **resolve** their images. At the fovea itself there are about 250,000 cells per square millimeter.

The nerve cells of the retina don't merely pass signals to the brain. These complex photoreceptor cells and nerves "weigh" the flashes of light and decide (somehow) what to send to the brain and what to ignore. The accepted flashes of light are absorbed; then the rods or cones signal the nerve complex behind them, which sends electrical impulses that travel along the fibers (there are over a million) of the optic nerve to the brain. While the image on the retina may change in a continuous fashion, there is a lag in the brain's reception and interpretation, which is why 24-frames-per-second movies look so real to us. A part of this lag is because the rods and cones flash an electrical response to the light and then must "recharge" briefly before they flash again.

The cones are concentrated at the central portion of the retina, and they give us color vision. Although color vision is not yet fully understood, some cones seem to respond best to blue light, some to red light, and some to green.

The rods outnumber the cones about 20 to 1, and they account for night vision. Their density and sensitivity allow them to pick up the faintest gleam of light. A rod can fire with as little as $1/100{,}000$ of the light energy required to fire a cone. The pupil opens wide in the dark, and almost all of the areas of the lens and retina collect the light. The rods are mostly concentrated away from the retina's central axis. Try looking to the *side* of a faint star rather than directly at it, and you'll actually see it better.

At times, the light from a full moon casts sharp shadows and is bright enough to read by. Moonlight is reflected sunlight, and so contains red, green, and blue frequencies. Nevertheless, anything we see by moonlight alone appears in black, shades of gray, and white. By itself, moonlight on the earth isn't intense enough to fire the cones in our eyes, leaving our night vision to the rods. That's why you see strictly black and white from reflected moonlight.

## REVIEW

**1.** When light focuses to form a likeness of an object, that likeness is called an image. True or false?

**2.** What fact about light makes a pinhole camera workable? Can the pinhole be made too small?

**3.** If light reflected by a surface is scattered in many directions, it is diffused. True or false?

**4.** Draw the angle of incidence and the angle of reflection for a ray of light reflecting from a flat surface.

**5.** What is the difference between a virtual image in a mirror and a real image in a pinhole camera?

6. What property of light is responsible for refraction?

7. How does the speed of light through air, water, and glass compare to its speed through a vacuum?

8. Describe the effect of refraction on a star's position as we see it in the sky.

9. What causes a mirage?

10. Define dispersion. Name several dispersive media. Is a vacuum a dispersive medium?

11. How does dispersion produce a spectrum in a spherical, sunlit raindrop?

12. Define converging and diverging lenses, focal point, and focal length.

13. In what ways do our eyes resemble an ordinary camera? A television camera?

## DEMONSTRATIONS

1. Where the optic nerve in your eye is connected to the retina, there is a blind spot in your vision. You can find it with the help of Fig. 23–29. Close your left eye and look straight at the black square. As you stare at the square, move the book toward you. You'll see the cat disappear as its image falls on your blind spot. Move the book even closer as you continue to stare at the square, and the cat reappears—but the mouse disappears (but not its tail).

**FIGURE 23–29**

2. Starlight can show that earth's atmosphere not only refracts light but disperses it. With a pair of binoculars on a clear night, look at a bright star near the horizon. The spectrum of starlight you see shows that air slows the various frequencies of light by different amounts, causing them to bend at slightly different angles as they come through the atmosphere. Look at the stars overhead with the binoculars and you won't see a spectrum.

3. Your brain retains an impression of the image that falls on the eye for about 0.1 seconds after that image is removed. That is called persistence of vision, and it is why the moving images in movies (24 frames per second) and on television (30 frames per second) look so real to us. To test your persistence of vision, you need only to blink your eyes. Ordinarily you don't notice a gap in the image when you blink because your eyelid is shut for only a tenth of a second or less. A lazy blink of a half second could be risky in a game of catch or in very fast, close traffic.

4. You can make your own rainbow—and even a moonbow, when the moon is full. It's easy! Experiment with a garden hose that makes a fine spray. (If you want to read about it, look up Helen Perry's article in *The Physics Teacher,* **13,** p. 175, 1975, in a library.)

5. On a still, sunny day, find a shade tree and stand in its shadow. Occasionally a small beam of sunlight may emerge from between the leaves to strike the ground. If the opening it comes through is small, a small circular image appears on the ground, even though the opening isn't circular. The tree, the hole, and the tree's shadow are like a natural pinhole camera, and the image you see is an image of the sun. Because the image is round or slightly elliptical, depending on the time of day, it isn't especially striking. You might try looking for an image of the moon the same way—a crescent moon should give a crescent image. Investigate!

## EXERCISES

1. Why does a high-flying plane cast no shadow although a low-flying plane does cast one? Draw a ray diagram as part of your explanation.

2. Flat mirrors reverse the image of an object, exchanging right for left and vice versa, as you can see from Fig. 23–6. Why, then, don't mirrors exchange the top of your face for the bottom? Draw a ray diagram to show why a flat mirror doesn't make an image that is upside down.

3. The archer fish (sometimes called the rifle fish) squirts water with its tongue and pallet. The volley explodes through the water's surface to knock unsuspecting bugs into the water. The range is about 4 feet. It almost always fires from straight beneath its prey. Why does this technique make sense?

4. Discuss what happens when a pinhole camera has a too-small pinhole. When it is too large?

5. Kingfishers most often dive into the water parallel to the rays of the sun; in the morning hours they angle in from the east, near noon they dive straight down, and in the afternoon they tend to angle into the surface from a westerly direction. How do these methods help these birds fish for food?

6. An entire classroom of students can see an instruc-

tor's face at the same time. Would you say that fact derives mostly from (a) refraction, (b) transmission, (c) reflection, (d) diffusion, or (e) absorption of light?

7. Backpackers in high mountains can see a very different sunset or sunrise than people at sea level do. How is the sunset different? Why does the difference come about?

**8.** The fact that we normally see our environment with very little distortion depends on (a) the wave nature of light, (b) the speed of light, (c) diffusion, (d) the fact that light travels in straight lines through the air, or (e) all of the above.

**9.** The image made by a camera lens on the film is inverted, but the image in a typical camera's viewfinder is not. A prism of glass in the viewfinder is responsible. Draw a ray diagram to show what must happen. (*Hint:* The light enters one edge of the prism perpendicularly and exits at an angle of 90° both to the original beam and to another side of the prism.)

**10.** Explain why a shaft of sunlight streaming through a break in storm clouds seems to have such sharp edges, whereas light passing through a small slit diffracts.

**11.** Light travels more slowly through transparent matter than through a vacuum because it is absorbed and reemitted by the matter's electrons and so does not spend all its time traveling. True or false?

12. If you go from a well-lit room into a dark room at night, it takes a moment or so for your eyes to become dark adjusted. In fact, even though you can see pretty well in the dark room after 5 minutes, it takes about 15 minutes for your night vision to reach a really good level. It takes that long for the eye to furnish photosensitive chemicals to increase the sensitivity of the rods and cones. Turn on the light, however, and a moment of bright light causes you to lose most of that night vision. Think of a way to make your way safely back to bed after a late-night trip to the fridge without turning on lights in each room as you pass.

13. The park rangers for New Cave, near Carlsbad, New Mexico, take visitors to a location far into the dark cave and have everyone extinguish their flashlights. Ten minutes later, the visitors are still blind. Why don't their eyes adjust to the dark—or do they?

**14.** Have you ever seen a part of a rainbow, that is, only part of the normal arc? Explain why it's possible.

15. Look at Fig. 23–22 and decide where the sun must be in the sky to create a rainbow of the greatest possible height.

**16.** A landscape seen only by the light of the moon has no color, only black and white and shades of gray. Review the properties of cones and rods and tell why that is true.

**17.** Explain that the water-puddle mirages you see on a hot road are light that never touches the ground.

18. The first electric light bulbs were made of clear glass. Today most of them have frosted glass. Why? (The first frosted light bulbs, by the way, were frosted on the outside of the bulb. Why is interior frosting more desirable?)

19. The best mirrors are made by coating very smooth surfaces with a thin layer of aluminum or silver. Why is metal a better reflector of light than most materials?

20. The smoother the surface is before it is coated, the better the quality of the mirror. The best flats of glass are smooth to within 20 atomic diameters over appreciable areas. Explain why these flat pieces of glass reflect visible light so uniformly even though they contain variations of as much as 20 atomic widths.

**21.** Explain why water that is being heated (as in a glass teapot) produces wavering images if you look through the water.

22. The cleaved surface of a single crystal of graphite may be perfectly flat for distances of several thousand atomic diameters. Why can't this material be used to make the most perfect mirrors in the world?

**23.** Home movies of the 8-mm variety flash 18 different still pictures, or frames, onto the screen each second, while the 16-mm films shown in many classrooms project 24 frames per second. Which makes the better use of persistence of vision and produces less flicker in the motion?

24. Laser light from high-powered lasers can be focused with a lens and used to weld delicate metal parts. Such lasers can also drill (or melt) smooth holes in ultrahard metals to a precise one-millionth of an inch; explain why this powerful laser beam doesn't crack the lens as it travels through the glass.

*25. The speed of light is about $3 \times 10^5$ kilometers per second (or 186,000 miles per second), and the moon's average distance from earth is about 384,000 kilometers (240,000 miles). When you look at the moon, how old is the image you are seeing?

***26.** Make a drawing of a person standing in front of a full-length mirror that is just as tall as the person. Then use a ruler and a protractor to draw rays that reflect from the shoes to the eyes. (Remember the angle of incidence equals the angle of reflection.) Use that drawing to convince yourself a mirror needs to be only half as long as the person is tall to be a full-length mirror and show the person his or her entire image.

**27. Light travels through air at almost $c$. If it then enters a diamond, it travels at only $0.42c$. The electrons in the diamond vibrate because of the incident light wave, which has a frequency $f$; so as the light wave moves into the diamond, its frequency of vibration remains the same. This tells us something about the wavelength $\lambda$ of the light while it is in the diamond. The speed of a wave is always $v = \lambda f$, so outside: $\lambda_{air} f = c$

inside: $\lambda_{diamond} f = 0.42c$

Use these equations to find $\lambda_{diamond}/\lambda_{air}$, and check Fig. 23–30 to see if it matches your answer. (*Hint:* Divide each side of the inside equation by the corresponding side of the outside equation to form the ratio $\lambda_{diamond}/\lambda air$.)

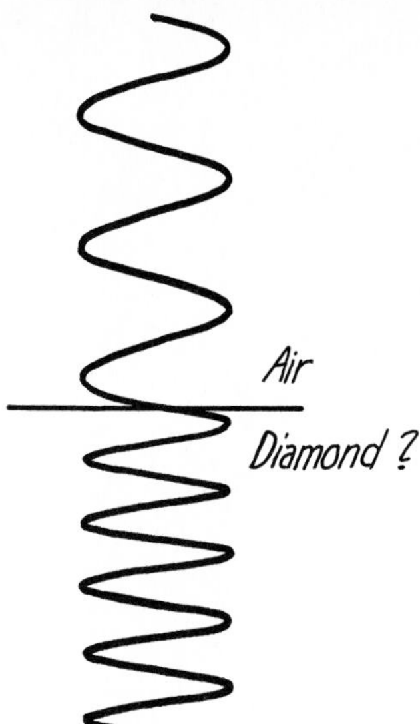


FIGURE 23–30

## New Beginnings: Planck, Einstein, and Bohr

Early investigators of electricity learned how to make a spark jump between two pieces of metal, or *electrodes.* Today anyone who has jumped-started a car has seen such sparks move through the air, caused by the passage of an electric current. But in the early days the nature of the current was unknown. When scientists put two electrodes in a sealed glass tube and evacuated the tube with a vacuum pump, their ammeters showed that a current traveled through the wires leading to the electrodes as the air pressure dropped. Studies with these *discharge* tubes led to the discovery of the nature of the electrical current, which flowed better through a vacuum than through air.

In 1839 Michael Faraday wrote of a phosphorescent glow in a discharge tube, and, quick to follow, other investigators soon discovered more. When the vacuum was very good, the end of the glass tube near the positive electrode glowed with a soft green light. Apparently something was striking the tube to produce that light. Moreover, solid obstacles placed between the two electrodes cast shadows upon the glowing end of the tube. Such sharp shadows meant that the invisible cause of the green glow traveled much as rays of light—that is, in straight lines. Named *cathode rays,* because they travel from the cathode (negative terminal) to the anode (positive terminal) and never backward, these radiations were soon found to have a property not shared by light: When a magnet was brought up to the discharge tube, the shadows moved, proving that the cathode rays were some sort of charged particles in motion.

By 1895 physicists had shown that the cathode-ray particles could penetrate thin metal foils, meaning they were probably smaller than atoms. In the years 1897 to 1899, the English physicist J. J. Thomson managed to determine an important property of the charged particles. He found that the ratio of the amount of charge the particle carried to its mass to be about 1000 times greater than the same ratio for ionized hydrogen (hydrogen is the lightest of the atoms). Presuming that the amount of charge on the cathode-ray particle is not much different from that of ionized hydrogen, this indicated that the cathode-ray particles had very little mass. For this evidence that cathode-ray particles (now known as *electrons)* were the first of the particles to be discovered that were smaller than atoms, Thomson was later given the Nobel Prize. (Despite the progress Thomson made with his investigations, it has been said that he always left the second decimal place to someone else to discover, as he did in this case: The more accurate ratio of charge to mass for the electron, $e/m$, is about 1837 times the same ratio for ionized hydrogen, or protons as they are now called. But the assumption about the amount of charge was correct. Protons and electrons have equal but opposite charges.)

In the wake of this development, discoveries by Max Planck (see "Max Planck's Model for Blackbody Radiation" in Chapter 24), Albert

Einstein ("Another Puzzle: The Photoelectric Effect," Chapter 24) and Neils Bohr ("The Bohr Model of the Hydrogen Atom," Chapter 24) were the beginnings of the understanding of the atom's structure. Their findings are in Chapter 24, but here we'll take a brief look at the scientists themselves.

Max Planck was quiet, very religious, and musical. His views were typical of many physicists at the end of the 1800s, especially in Germany; he thought all of natural behavior could be traced in principle to Newtonian mechanics. Many of his colleagues were of the opinion that all the great discoveries of physics had already been made. When his revolutionary discovery came along, he could not quite believe it himself. He was over 40 years old in 1900 when he discovered the radiation law that predicts how matter radiates because of its temperature. It earned him the Nobel Prize in physics in 1918. Unfortunately, Planck's personal life was rather tragic. His wife died early, one son died in World War I, and another, a member of a group that almost succeeded in assassinating Adolf Hitler, was executed by the Nazis. Most of a lifelong collection of books, manuscripts, and correspondence was destroyed by fire during an Allied air raid on Berlin.

A scientific contemporary of Planck's, Albert Einstein was only 26 in 1905 when he explained the details of the photoelectric effect, the knocking of electrons from the surface of a metal by light. He won the 1921 Nobel Prize in physics for that work. According to his sister, at his birth his mother was frightened because of his large head. His grandmother's first words when she saw him were, "much too thick, much too thick." Between 2 and 3 years of age, the not-as-yet-talking Albert decided to speak—in entire sentences, quietly rehearsing each before saying it out loud. When he was 4 or 5 his father showed him a drafting compass, and he was so struck by it he trembled. He later said of the event, "I experienced a miracle . . ." When he was 12 someone gave him a geometry book, and he soon called it his "holy geometry book." For the next 4 years he studied math on his own, and in time he became an excellent student.

Einstein graduated from college with a diploma that allowed him to teach in high school. After a short time as a substitute teacher he settled in Bern, Switzerland, and tutored students until a friend's father got him a job at the Swiss patent office. A student during that period wrote of him: "about five feet ten, broad-shouldered, slightly stooped, a pale brown skin, a sensuous mouth, black moustache, nose slightly aquiline, radiant brown eyes, a pleasant voice." He seemed to be able to be deep in thought without appearing to be aloof from those around him. It was while he was at the patent office, where he could easily do his regular duties and still have time to think about physics, that he published three papers (in 1905) in different areas of physics, any one of which was worthy of a Nobel prize.*In the same year, he completed a Ph.D. thesis. His first, submitted in 1901, had been rejected.

Einstein remained at the patent office until 1908, when he left to become a junior, then an associate, then a full professor at various Swiss, German, and Austrian universities. During this period he worked on the general theory of relativity, which deals with motion in accelerated frames of reference (see Chapter 26). Early in 1919 Einstein divorced his first wife, and he agreed that she would have the proceeds of a Nobel prize should he win one. He promptly married a cousin, Elsa, herself divorced. He was awarded the 1921 Nobel Prize in physics and his former wife received all the money, as promised.

Einstein continued to work on fundamental problems of physics for the rest of his life. He attempted to unify the forces of nature into one grand theory, the unified field theory. In 1932, several months before the Nazis came to power in Germany, he was appointed professor at the Institute for Advanced Study at Princeton, New Jersey. There his only duty other than research was to attend faculty meetings (he disliked teaching courses). Despite his position as a pacifist, first stated during World War I in Germany, he sent a letter to President Roosevelt in August 1939 to draw attention to possible military uses of nuclear energy. This helped influence Roosevelt to set into motion the effort that resulted in the development of nuclear weapons. Upon Einstein's death in 1955, his ashes were scattered by friends in an undisclosed location.

Niels Bohr was born in Copenhagen in 1885, and his parents were progressive and took part in every phase of their children's education. Though overshadowed as a student by his brother Harald, who became an eminent mathematician, Bohr earned early fame as a thoughtful, thorough scientific investigator. Upon finishing his Ph.D. thesis in 1911, Bohr went to England to study with J. J. Thomson. Because his thesis had been on the actions of electrons in metals, Bohr wanted to work with the discoverer of the electron. But Thomson had other interests by then, so a disappointed Bohr went to Manchester where Ernest Rutherford had established an active laboratory and only the year before had discovered the nucleus. There Bohr attacked the problem of how the electron behaved about the nucleus of the atom. After 6 months he returned to Copenhagen, where he worked out a model for hydrogen that explained how the atom absorbed and emitted light with the motions of its electron. This model of the hydrogen atom brought Bohr the Nobel Prize in 1922. Several years later the Danish government built a laboratory for Bohr on land donated by some of his friends. The Institute for Theoretical Physics

*Chapter 26 is devoted to his special theory of relativity, which explains motion as viewed from frames of reference with different speeds, the subject of one of those 1905 papers. Yet another, concerning Brownian motion, is claimed by some to have settled once and for all the question of whether there really are atoms and molecules.

became a focal point for a revolution in physics over the next few decades.

The revolution resulted in the *quantum theory* of atoms and molecules. It explained the structure and the behavior of those tiniest units of matter. First introduced by Planck, Einstein, and Bohr, the Quantum Theory would come into full bloom in 1925 and the years afterward. This theory overturned the classical notions of the behavior of matter (see Chapter 25), and in doing so it divided the opinions of the three who made the earliest contributions to its development. Bohr became the champion of the quantum theory, and Einstein and Planck, among others, became its antagonists. Again and again, Bohr and Einstein argued the theoretical points, with Bohr and the quantum theory the ultimate winners. Bohr came to use these arguments with Einstein as private daily mental exercises, providing a constant identity with the new theory. Even after Einstein died, Bohr was heard to argue against Einstein as though he were still alive.

When the Nazis occupied Denmark during World War II, Bohr refused to leave Copenhagen. He stayed to exert whatever influence he could for exiled scientists in Denmark. He was also concerned about retaliation against his family, friends, and colleagues at the Institute should he leave. At some point, however, Bohr and his brother Harald were warned that they were on a list to be deported to German concentration camps (their mother was Jewish). Bohr arranged for his family and himself to escape to neighboring Sweden. Once there, he was invited by Winston Churchill to come to England to help the war effort, and he accepted. On a moonless night, he was flown out of Sweden over occupied Norway in a one-person military aircraft. As the extra passenger, he had to ride in the empty bomb bay; the flight itself was not without additional risk. Because the plane was to fly at high altitude, Bohr was given, just before takeoff, an oxygen mask and a helmet with an intercom. The flight was underway before Bohr discovered that the helmet was too small for his head! He quickly passed out because of the lack of oxygen. Unable to get a response from Bohr on the intercom, the pilot figured out what had happened, and once past Norway, promptly descended to a lower altitude over the North Sea. When the plane touched down in Scotland, Bohr had revived without harm. Later he came to the United States to act as a consultant for the international team of scientists involved in the development of the nuclear bomb.

# CHAPTER 24
# LIGHT AND ATOMS

In 1871 Dr. David Livingstone, a Scottish missionary, disappeared in the African Congo. An Englishman, Henry Morton Stanley, led an expedition to look for him. The rainforest where Livingstone had vanished was larger than France, covered with trees almost 200 feet tall, and so dense in most places that sunlight never reached the ground. The hot, humid air trapped beneath the canopy of trees was stale with the odor of decaying vegetation. Many in Stanley's party suffered from dysentery and fevers, and as the expedition pushed farther into the jungle, the native bearers grew sullen and quietly deserted in twos and threes. The natives believed that far from their villages the jungle became darker and darker, until somewhere—at the end of the earth—a serpent lay in wait.

To ease these fears, Stanley told the natives he was from a far-away land that was completely different. He described for them a cool, spring morning in a sunny English meadow. The natives exchanged puzzled looks, then emphatically shook their heads. *All* the world was like *this,* a spokesman said, sweeping his arms wide at the dark jungle canopy. Everywhere there was only trees and more trees. Nothing in their experience could let them relate to a description of springtime and an English countryside.

During the same period in the late 1800s, some physicists were taking part in a very similar performance. Flushed with the success of Newtonian mechanics and Maxwell's equations and all their predictions, one well-known physicist proclaimed there was little left to do but determine the physical constants (such as the speed of light, $c$) to more and more decimal places. Yet, unknowingly, the physicists of that day faced a world as remote to them as Stanley's world was to those natives. Eventually some of those physicists pushed on to discover a realm so different from everyday experience that happenings there go against all common sense. Anyone can relate to the pushes and pulls and motions that Newton analyzed, but no one can ever personally experience what goes on in the interior of the atom, or the strange space and time of relativity. Developed largely in this century, this physics is usually called **modern** physics, while pre-1900 physics is often called **classical** physics. The next five chapters describe the world of modern physics. This part of nature provides a great test for human investigation and imagination and has been the scene of great triumphs of understanding. This understanding has led to progress in other sciences such as chemistry and medicine, and in technology, manufacturing, communications, and energy production. The physics developed in the twentieth century is having a major impact on our world. ■

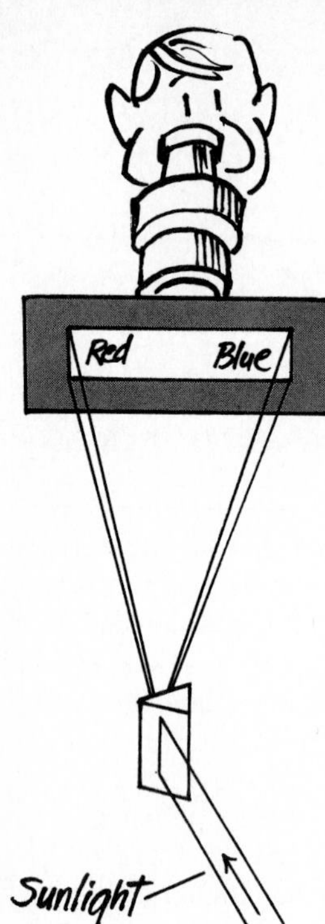


**FIGURE 24–1**
A spectroscope is an arrangement to enlarge small portions of the spectrum of light (here, sunlight).

## Spectra: The Fingerprints of Atoms

In 1800 William Wollaston, a wealthy London physician, became partially blind and decided to retire from medical practice. His hobby was science, and he set up a laboratory at the Royal Society with his own funds so he could study light, electricity, mineralogy, and chemistry. Two years later Wollaston devised an instrument to get a closer look at the spectrum of sunlight, the rainbow span of the visible frequencies of light.

Like Newton a hundred years earlier, Wollaston passed sunlight through a prism, but where Newton only projected the spectrum onto a wall, Wollaston sent the light through a small telescope to enlarge the spectrum. (An instrument that spreads the colors or wavelengths of light for visual observation through a telescope is called a **spectroscope.** If the instrument has a camera arrangement to photograph the spectrum, it is called a **spectrograph.**) When Wollaston put his eye to the telescope, he saw the same type of rainbow spectrum (the continuous spectrum) that Newton had seen. But the magnification revealed dark spaces in this rainbow where some wavelengths of light seemed to be absent or very dim. However, these ''missing'' parts in the sun's spectrum didn't seem to arouse his curiosity much, perhaps because of his poor eyesight and the crudeness of his spectroscope. Wollaston decided these dim, fuzzy regions might be some sort of boundaries between the colors, and he then turned to other experiments.

In 1814 an apprentice optician, Joseph von Fraunhofer, made a spectroscope far superior to Wollaston's. He found that Wollaston's fuzzy regions of the spectrum were actually very sharp, distinct *dark lines.** Furthermore, he managed to measure the wavelengths of the light that was missing in almost 600 of the darkest lines of the sun's spectrum. As to why these wavelengths were missing, he couldn't say.

In the 1840s and 1850s, the German physicist Gustav Kirchhoff managed to produce and investigate dark spectral lines in a laboratory. The dark lines appeared when he passed white light (composed of all wavelengths) through windows in a container of gas and sent the light that emerged into a spectroscope. For each different element he used in the container—hydrogen gas, oxygen gas, and others—different dark lines appeared; each element left its own fingerprint on the spectrum of

*Most spectroscopes pass the light through a slit, and the spectral ''lines'' are just images of the slit, which appear as sharp lines.

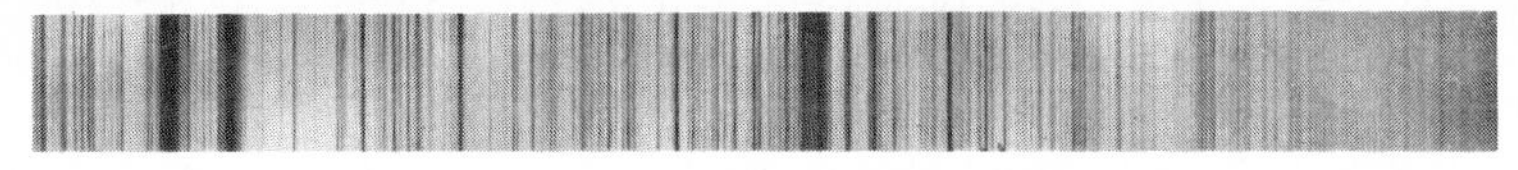

**FIGURE 24–2**
A portion of the spectrum of a beam of sunlight that has passed through a slit. The dark lines show that certain precise wavelengths (or frequencies) of light are missing in the sunlight.

white light. Somehow the atoms of the gas in the container absorbed certain precise wavelengths of light, letting all the others pass. Such rainbow spectra with dark lines, or missing wavelengths, came to be called **absorption spectra.**

Kirchhoff next placed a spectroscope to the side of the container of gas (see Fig. 24–3) so that none of the white light from the beam would enter the instrument. When he peered through the instrument, he saw lines of color in an otherwise dark field of view. These *bright lines* of light had the same wavelengths that the gas was absorbing from the white light to cause the absorption spectrum. The gaseous atoms absorbed light energy of certain wavelengths, which added energy to the gas, and then they lost that energy almost instantly by emitting light of identical wavelengths randomly in all directions. The bright-line spectra are called **emission spectra.** But why the atoms in a gas emit and absorb light this way remained unknown for awhile.

**FIGURE 24–3**
Photographing the absorption and emission spectra of a gas. Mercury vapor absorbs some frequencies from the white light, causing dark lines in the spectrum of the beam when it exits from the container. A spectrograph at the side of that gas gets none of the rainbow spectrum of the direct beam, however. Instead, the spectrum has bright lines on a dark background. These are the frequencies absorbed from the direct beam.

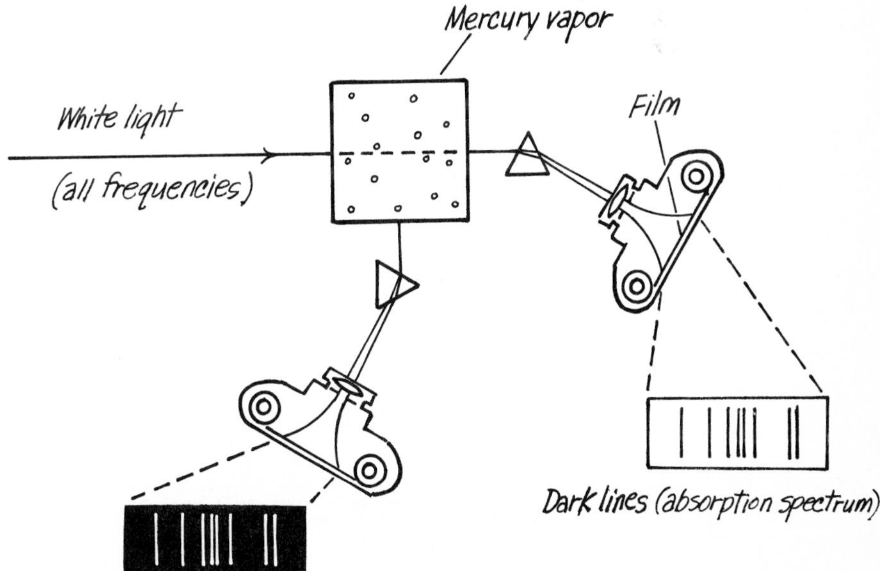

## AN IDEAL BLACKBODY

At least a small fraction of any radiation that strikes a solid is reflected. In reality, then, not even the darkest solid can be a perfect blackbody. But if a closed box has a tiny hole punched in it, any radiation that strikes the hole will pass into the interior. The hole, which reflects no light, figuratively ''absorbs'' the radiation. Once inside, the radiation is absorbed by the material inside the box (sometimes after several bounces). The material on the interior surfaces emits radiation itself, however, as does any matter that's above the temperature of absolute zero. Detectors on the outside can analyze the spectrum of the radiation as some of it escapes from the hole in the box. Using a similar arrangement, physicists in the 1800s studied the emission spectra of various materials this way. They determined that *no matter what material is lining the box, the spectrum of light emerging from the hole is the same*. And that spectrum changes only if the temperature of the box's interior changes. A hole in a box, then, is an ideal blackbody, absorbing all radiation that strikes it and emitting radiation in a way that depends only on temperature. The spectra of blackbody radiation at several different temperatures are shown in Fig. 24–4.

## Radiation and Solids

Unlike a thin gas that absorbs and emits only special wavelengths, solids both absorb and emit radiation over a wide range of wavelengths. Certain solids, like glass, diamond, or quartz, transmit some wavelengths of light. Those frequencies that aren't absorbed or transmitted are reflected, and it is this light that gives solids their colors. In general, darker-colored bodies absorb more of the white light that falls on them than lighter ones do. A black-colored body absorbs light best of all. An ideal **blackbody** reflects no light, so that it has the simplest absorption spectrum of all solids: It absorbs all the light that strikes it.

Blackbodies also emit radiation. If a blackbody is in thermal equilibrium, it emits as much radiant energy as it absorbs (unless conduction and/or convection is taking place at the same time). The emission spectrum of a blackbody was known from experiments in the late 1800s. At that time its origin was as much a mystery as the spectral emission and absorption lines of radiation by gases. The blackbody emission spectrum for several temperatures is shown in Fig. 24–4. Physicists tried to predict the blackbody emission curves, presuming the vibrating atoms were emitting the light. The atom's thermal vibrations, according to the classical ideas, should be random. They presumed the charges in the atom that emitted electromagnetic radiation would vibrate randomly as well. But they couldn't get the answers to come out right; the laws of classical physics weren't predicting the facts found by the experiments. Like the

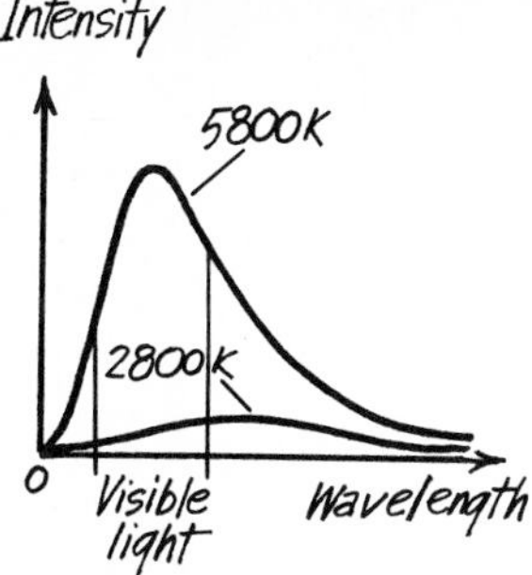


**FIGURE 24–4**
Emission spectra for blackbodies at 5800 K and at 2800 K. A blackbody emits some light at almost every wavelength, and the change in intensity from one wavelength to the next follows smooth curves like the ones on this graph. A blackbody at 5800 K (the temperature of the sun's surface) emits most of its radiant energy in the wavelengths of visible light. At 2800 K, the temperature of a lightbulb's filament, most of the energy is emitted at longer wavelengths than visible light—that is, as infrared, or heat, rays. (A typical incandescent lamp emits less than 10% of its radiant energy as visible light.)

spectral lines of gases, the blackbody emission spectrum was a piece of a puzzle awaiting a better understanding of nature at the atomic level.

## Max Planck's Model for Blackbodies

Just before the turn of the century, Max Planck was pondering the riddle of how a hot blackbody glows. He decided to work backward to a solution. The curves in Fig. 24–4 were his answers, so he looked for a mathematical formula that when plotted on the same graph would match those curves at every point. After much effort, he succeeded.

Planck found a family or series of curves that would fit the blackbody emission spectra for any temperature the blackbody could have, provided that a constant number appearing in the formula was chosen properly. In his formula (which we won't discuss) he called that constant number *h,* and it is now known as **Planck's constant.** The formula is called **Planck's radiation law.**

Planck still had to explain how the molecules or atoms gave up this light, and this turned out to be the hardest part of the riddle. Electrons were discovered only in that same year (1899), and as yet no one knew about the atomic nucleus. Radiation, he knew, occurred when charged particles accelerate or decelerate. Planck, like other physicists, first assumed that any charged particle in the solid would bounce around randomly because of the thermal motion of the atoms. But the light emitted by randomly accelerated charges did not match the light emitted by the blackbody. Nevertheless, Planck pressed on, and several weeks later he discovered a model that showed how charged particles in solids could give off light according to his formula.

Planck assumed the charged particles oscillated in place much as atoms in a solid vibrate in their positions. The oscillators would swing back and forth, their accelerations causing them to radiate. Their frequency of oscillation, called *f,* would be the same as the frequency of radiation they emitted. What Planck found was this: Whichever frequency *f* an oscillator had, *it gave up or absorbed radiation only in definite amounts of energy that were equal to a whole number times Planck's constant times the frequency of the oscillator*. It was as if the energy were emitted in lumps, or bundles. The size of the energy bundles from an oscillator with a frequency *f* could only be *hf,* or 2*hf,* or 3*hf,* and so on, any whole number times *hf,* where *h* is the constant from Planck's radiation law and *f* is the frequency. Planck called the quantity of energy *hf* a **quantum.**

energy of a quantum of light
= a small constant (*h*) times the light's frequency *f*

$$E = hf$$

The value for *h* in the metric system is $6.626 \times 10^{-34}$ joule·seconds.

Planck's model implied that the blackbody's oscillating charges couldn't swing back and forth like a mass oscillating on a spring. That is, their motion couldn't have just any amplitude. Apparently there was a definite *set* of oscillations the charged particles could have, and they could not move with other amplitudes. If light of some frequency were shone on an oscillator, an oscillator could absorb energy, but only if its amplitude increased to equal a larger amplitude in the set. Then the oscillating charge would have an extra quantum or so of energy. If an oscillator lost a quantum of energy, it would abruptly move with a smaller amplitude. But that amplitude, too, had to be in the set. The difference in energy for the oscillator as it moved from one amplitude to the next possible amplitude (up or down) was exactly equal to $hf$, where $f$ is the frequency of the emitted (or absorbed) light.

Imagine how it would be if a child's swing followed such a rule when you gave it a push. Rather than swinging as it normally does from the vertical up to an angle of 15 degrees or so, it would jump abruptly from, say 0° to 3°, then from 3° to 6°, and so on, never to be found at any angles in between. Nothing you've ever seen or experienced moves that way! Yet Planck's oscillating charged particles seemed to. (The very tiny value for $h$ means the jumps in energy are very small indeed, and perhaps real motion in our world would take place in such jumps if we could detect such tiny energy differences for such a large object as a swing. We'll return to this point in the next chapter.)

Planck's model made no physical sense to him or anyone else at that time. But because it explained the blackbody radiation curves so perfectly, he published it in 1900. For the next five years no more was heard of the quantum, though Planck spent a lot of time trying to find another way to explain his radiation law. He couldn't bring himself to believe that his strange model really described what was going on in a radiating solid.

## Another Puzzle: The Photoelectric Effect

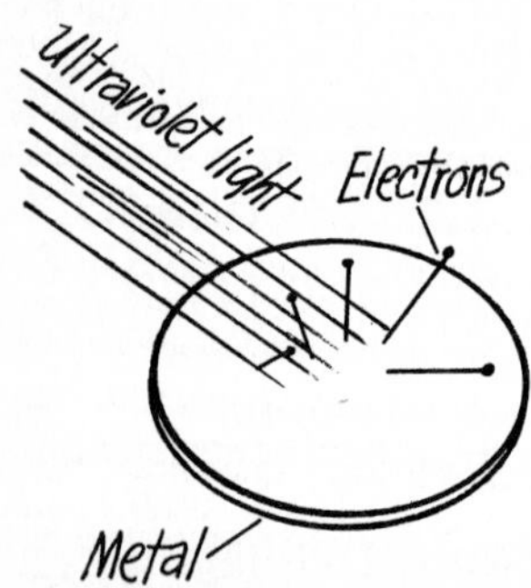


**FIGURE 24–5**
Electrons in metals such as zinc are able to absorb energy from ultraviolet light and can escape from the metal's surfaces.

About the same time that Planck was wrestling with his model for blackbody radiation, other physicists were experimenting with another effect between light and solids. Ultraviolet light shone on a plate of zinc caused electrons to pop from the surface with various speeds, much like water molecules evaporating from a sunlit lake. This action was later named the **photoelectric effect.**

The experimenters sent in a single frequency of ultraviolet light and varied the wave's amplitude. In other words, they used different intensities of ultraviolet light, corresponding to different brightnesses in visible light. Much to their surprise, the speeds of the electrons that came out of the metal were unchanged. More electrons came out of the metal per second if the intensity was greater, but none moved faster than before.

If you push something through a distance, you do work, you give it energy. The harder you push, the more energy it gets. But these electrons didn't act that way. A light wave with extra amplitude did not give them any more energy. It just pushed more of the electrons, giving each the same energy as before. This was quite a puzzle, and it took an exceptional physicist to discover what was happening.

## Einstein Answers This Puzzle

In 1905 Albert Einstein studied the photoelectric effect and decided that the known facts of physics couldn't explain the actions of the electrons. When an ultraviolet wave passed, the electrons weren't being pushed around as a cork is by an ocean wave. Instead, Einstein saw that the electrons reacted to these light waves exactly as if the wave were a hail of *pellets of radiant energy*. That is, the electrons absorbed energy from the electromagnetic wave only in precise lumps of energy, or quanta of energy.

Einstein named these pellets or ''particles'' of light **photons.** Each photon carried a quantum of energy in the amount $E = hf$, where $f$ is the frequency of the light. The bundles of energy that Einstein found in light waves corresponded exactly to the difference in energy that Planck's blackbody oscillators could take on. Together these findings of Planck and Einstein revolutionized the understanding of light.

For the hundred years before 1905, light was thought of as a smooth and continuous wave. The interference experiments of Young (Chapter 22) proved that light acts like a wave on a scale that we can see; the interference patterns in Young's experiment are visible to the eye. But Einstein found that when light interacts with electrons, it has a different character. On the scale of atoms, light is a hail of photons carrying energy. The oscillators that Planck had described with their curious jumps in energy levels are only electrons absorbing and emitting photons. The electrons in matter cannot absorb or give up energy in any amounts other than whole quanta of energy. To an electron, a light wave of greater intensity only brings more photons per second, and not a greater push or pull. We'll soon see that these findings were only the first revelations about the very strange world of electrons and atomic structure.

## The Bohr Model of the Hydrogen Atom

The spectacular discoveries of Planck and Einstein offered no help toward explaining the absorption and emission spectra of gases. The origin of those neat patterns of precise wavelengths was still a mystery. Then in 1911 Ernest Rutherford announced the discovery of the nucleus, and all the volume in the atom around the tiny nucleus was seen to be the domain of the electrons.* The next year a young Danish physicist, Niels Bohr, took on the problem of how the electrons behaved. Their motions, he thought, would explain how an atom emitted and absorbed light.

Bohr made no progress until a remarkable formula came to his attention. Hydrogen, the lightest and simplest element, has a spectrum of visible wavelengths that get progressively closer together as the

*Recall the descriptions in Chapter 9, or see the beginning of Chapter 27.

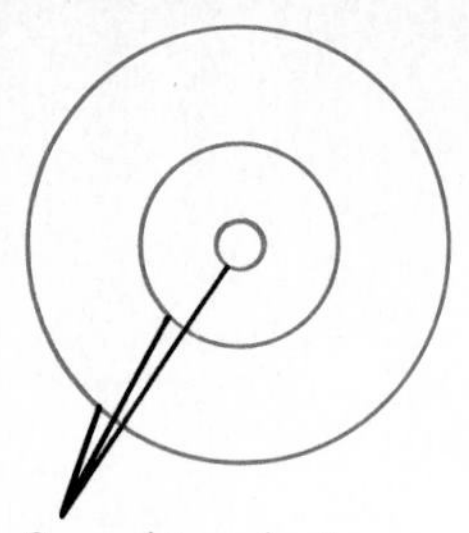


**FIGURE 24–6**
Bohr's model of the hydrogen atom; the electron circles the nucleus (invisible in the scale of this drawing) on any of a certain series of concentric orbits. The three orbits closest to the nucleus are shown here.

wavelengths get shorter. Some years earlier a Swiss schoolteacher, Johann Jakob Balmer, had, by pure guesswork, found a numerical formula that represented the wavelengths for the spectrum of the hydrogen atom. So just as Planck had worked backward to find a model that explained the blackbody radiation formula, Bohr set out to find a model for hydrogen that could explain Balmer's formula.

Since the charges of an electron and a proton are of opposite sign, they attract each other, and Bohr assumed correctly that the electric force is strong enough to keep them together as a unit, the hydrogen atom. Thanks to Rutherford's discovery of the nucleus, Bohr knew the electron has most of the atom's space to itself. The electron, Bohr decided, must be in orbit about the proton, falling around the proton rather than falling straight toward it, something like a planet in orbit about the sun. The proton, after all, is almost 2000 times more massive than the electron, so the proton moves relatively little as the electron sweeps around it. This helps to account for the relatively large volume of the atom as compared to its nucleus.

Next, Bohr tackled the problem of how the hydrogen atom absorbs or emits light. When a photon of light strikes a hydrogen atom and is absorbed, the atom takes on the photon's energy. The electron, being the lighter particle, is affected the most. When the electron's energy increases, the electron pulls away from the proton to orbit at a greater distance (just as an artificial satellite moves to a higher orbit around the earth if its energy is increased by booster rockets). Likewise, if that hydrogen atom emits light, the electron must give up the energy taken away by the photon and fall toward the proton to orbit at a closer distance. But a hydrogen atom, Bohr knew, absorbs or emits only the frequencies of light found in its absorption and emission spectra. To account for those facts, Bohr decided the electron must move only on a certain *set* of orbits. Moving from one of these orbits to another, the electron absorbs or emits energy in the amount $E = hf$ for precisely those frequencies found in the hydrogen spectra. Bohr deduced where the set of orbits would be so the electron's spectrum would fit the Balmer formula perfectly.

To visualize the electron energies in Bohr's model, think of walking up or down a flight of stairs. It takes energy to go up a step, and you lose energy if you step down. A step down (toward the proton) means a loss of energy, and a step up (away from the proton) means the electron gains energy. The steps of hydrogen's electron aren't of uniform size, which means a different amount of energy is lost or gained between any two steps. But just as you can't stand between two steps, neither can the electron land between its energy steps, or **levels.** The electron, like you, can take steps two or three at a time (or even more), skipping one or more energy levels; but while it is part of the atom, it must always stop on an allowed orbit with its definite energy level. There are no ramps for the electron in the hydrogen atom, only a stairway.

Despite its grand success at predicting the spectral wavelengths of the hydrogen atom, there were serious problems with Bohr's model. Why couldn't the electron orbit anywhere between Bohr's allowed or-

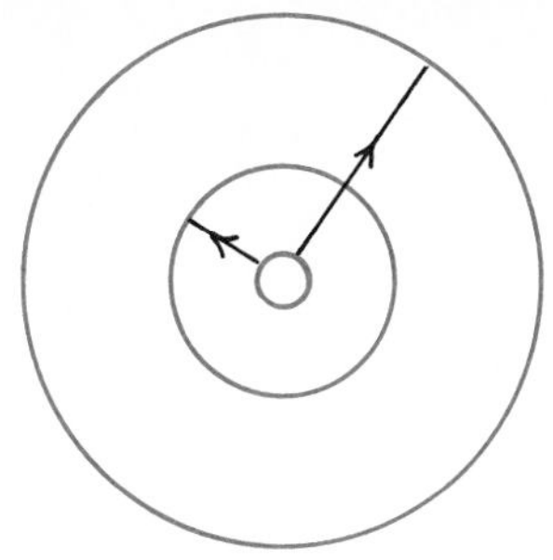

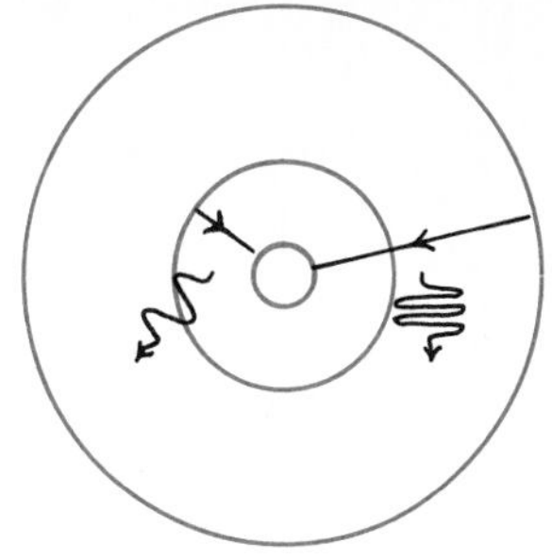

*(a) Absorption* *(b) Emission*

**FIGURE 24–7**
*(a)* In Bohr's model, the hydrogen atom absorbs energy when the electron jumps from a closer orbit to a farther orbit. Two possible outward transitions for the electron are indicated here. The farther the orbit from the nucleus, the more energy is needed to make the transition. *(b)* Emission takes place when an electron moves to an orbit closer to the nucleus. Here two inward transitions that end on the closest orbit to the nucleus are shown. The longest "fall" loses the greatest energy, so the photon that accompanies that transition has the highest frequency.

bits (as a planet could)? Also, Maxwell had shown 50 years before that any accelerated charge radiates (see Chapter 22), and an electron that moves along a circular path accelerates toward the center of the circle (see the discussion of circular motion in Chapter 4). Yet Bohr's electron did not radiate while in its orbit; it radiated only when it made a transition, that is, by jumping from one orbit to another. Clearly some part of the explanation was missing. Another problem surfaced when Bohr tried to use his model to find the spectrum of helium, an atom that has two electrons, or the spectra of other atoms with even more electrons. His model couldn't predict the observed spectral lines for any atoms but hydrogen. A dozen years later the reasons for these failures would be known. We'll fill in the missing parts of the explanation in the next chapter.

In the meantime, Bohr's model is easy to visualize, and by thinking of planetlike electrons hopping back and forth on imaginary tracks around the nucleus, a good many physical phenomena may be understood without the misrepresentation being too great. So for the remainder of this chapter, we'll look at some of these atomic effects using Bohr's model.

## Why We Get White Light from the Sun

A hot, thin (rarefied) gas in the chamber of a spectroscope sheds energy by giving off photons. The frequencies of those photons are the emission frequencies of the gas, and in the spectroscope that light appears as a few sharp, bright lines. That is, the gas emits the light of only a few narrowly defined wavelengths, or colors. If more gaseous atoms are stuffed into the chamber, however, the density and pressure of the gas increases. Then its spectrum begins to change. Once-sharp spectral lines that signify precise wavelengths begin to broaden, spreading out to become small *bands* of color. The pressure can be increased until those bands of color spread to overlap completely. The light from any gas under these conditions gives a **continuous** (rainbow) spectrum when seen through a spectroscope. In simpler terms, a hot, dense gas gives off white light.

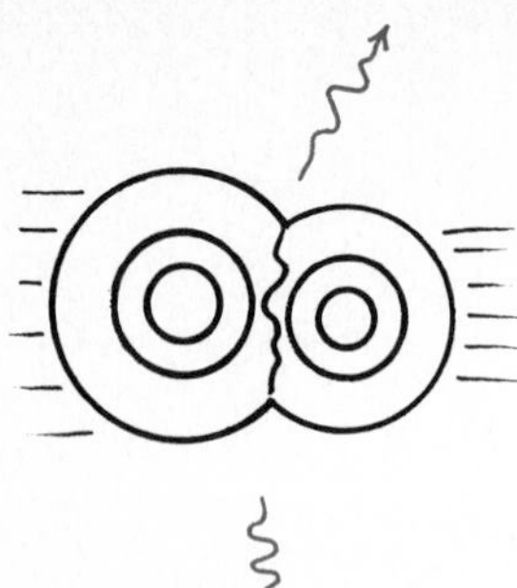

**FIGURE 24–8**
Atoms under pressure often have collisions while emitting or absorbing radiation. The pressure during a collision affects the energy of the outer orbits of the atoms, changing the frequencies that will be emitted or absorbed, making them different from the normal emission or absorption frequencies.

Collisions between atoms are one reason for the broadening of their emission lines. In a thin, hot gas the atoms rarely collide while in the process of emitting photons. The energy levels in such atoms are all the same, and the spectral lines are very sharp. But in a dense, hot gas the chances are good that any excited atom will collide with another atom while it is emitting a photon. During the collision, the excited electron comes under pressure from the other atom, which influences its orbit a bit and alters its energy. When such an electron jumps, the emerging photon has a different energy (and therefore frequency) from photons emitted by atoms that are not under pressure from collisions. The slightly different frequencies and wavelengths emitted by colliding atoms are seen through the spectroscope: A once-sharp (almost a *single* wavelength) spectral line has become broadened. The hotter the gas, the more violent the collisions can be, and the denser the gas, the more collisions per second. Both *temperature* and *density* influence how wide the spectral lines will be.

High temperature broadens spectral lines in another way, too. Because atoms have great speeds if the gas is at high temperature, the Doppler effect (Chapter 17) comes into play. If an atom moving in a direction away from your position emits a photon, its frequency relative to you is lower—much like the lower pitch you hear from a receding train whistle. If the atom is approaching as its photon emerges, the light you see has a higher frequency. Because in a hot gas some atoms have great speeds of approach and recession, the photons that appear have a spread of frequencies, and the spectral lines you can see in a spectroscope will be broadened. This effect is called **Doppler broadening** (Fig. 24–9).

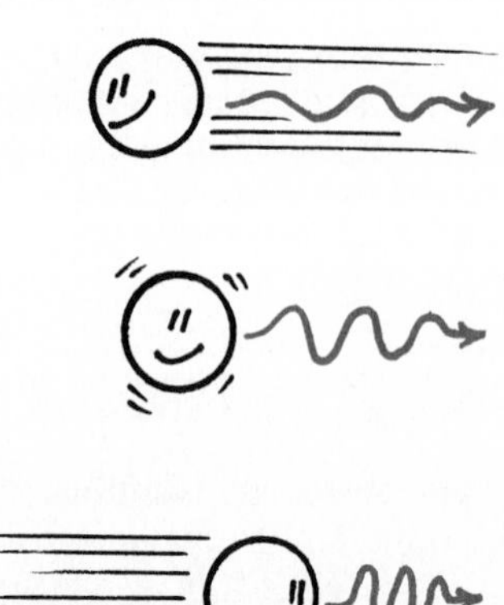

**FIGURE 24–9**
The effects on the frequency of a photon due to the emitting atom's motion. The frequency of a photon is reduced if an atom is speeding away from the observer (here to the right) during an emission. An atom that is relatively still during an emission emits a photon whose frequency belongs to the atom's normal spectrum. If an atom approaches the observer while the emission takes place, the photon's frequency is higher. The shift in frequencies due to the relative motion of the atoms is called the Doppler effect.

Sunlight is white light for all these reasons. The gas (mostly hydrogen and helium) at the sun's surface is hot, excited by light that comes from the sun's hot interior. The incessant collisions at high speeds and the Doppler effect spread the spectral frequencies of the atoms to the extreme until they overlap completely and become a continuous spectrum.

Sunlight has all the frequencies of the rainbow as it leaves the sun's surface. But it then travels through the sun's rarefied atmosphere. There, atoms of many elements absorb their respective spectral wavelengths from the passing light, while letting the others through. This is why, when he turned his spectroscope toward the sun, young Fraunhofer saw thousands of dark absorption lines within the rainbow spectrum of white light.

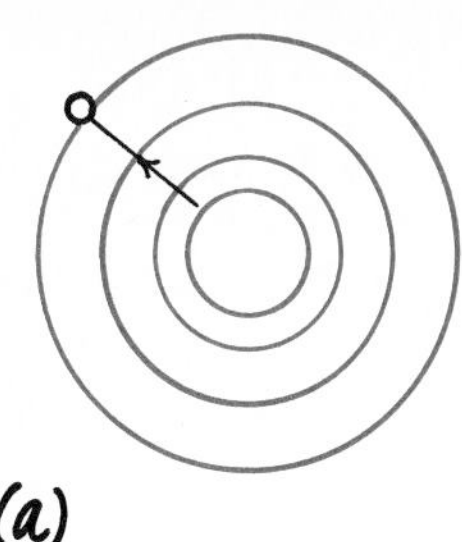

(a)

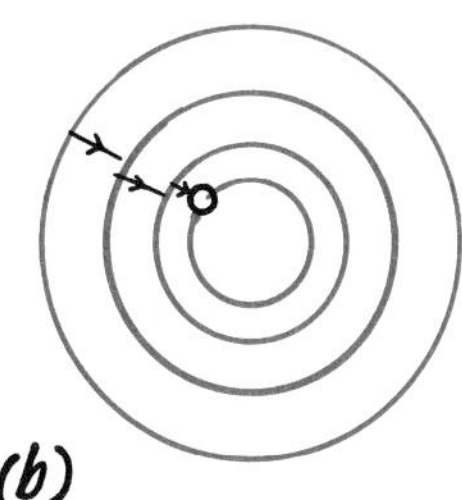

(b)

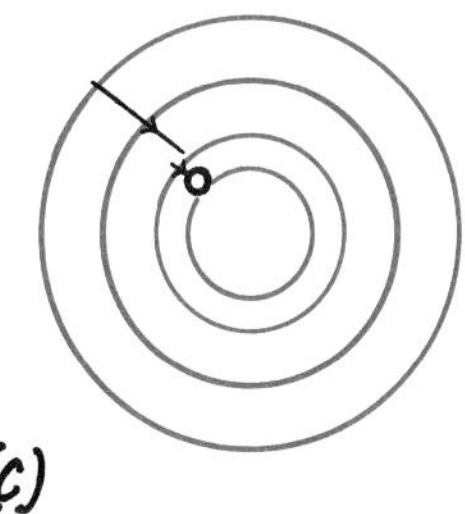

(c)

**FIGURE 24–10**
*(a)* The electron leaves the ground state to orbit in an excited state. *(b)* This electron returns to the ground state by making three successive transitions, each of which is accompanied by a photon of a different frequency. *(c)* This electron returns to the ground state after two transitions. The energy emitted during the first transition equals the energy lost during the first *two* transitions in *(b)*.

## Fluorescence and Phosphorescence

Suppose an atom has been excited. Perhaps another atom has struck it, causing the outer electron to move out to an orbit higher in energy than the lowest-energy orbit it normally occupies, which is called its **ground state.** Or perhaps the atom absorbed a photon from an incident lightbeam to cause the transition. However it happened, the excited atom will eventually lose the extra energy. It does this spontaneously.

The electron may return to its lowest energy state with one transition, emitting a photon that carries away all its extra energy. Or by emitting a photon of lesser energy, it may jump to an intermediate orbit and later, by emitting a second photon, make another jump and return to its ground state. Some of the possibilities are shown in Fig. 24–10.

Once an atom has been excited, it usually emits one or more photons within a time interval of about $10^{-8}$ second. This particular way of giving off light energy is called **fluorescence.** But sometimes the electron finds itself in a state (often an intermediate state) where the transition to a lower level takes more time. It may need anywhere from $10^{-3}$ second to days or even weeks to make the light-producing jump to the lower-energy state. This slower process is called **phosphorescence,** and the state where the electron is hung up temporarily is called a **metastable state.**

The soft red glow from a neon light in the window of a store comes from a gas of neon atoms. These atoms are struck by oscillating electrons in the (colorless) tube and become excited. Excited neon atoms have no transitions that would give photons of blue or green light, but they emit red light intensely by fluorescence.

In a fluorescent light, such as a desk lamp, oscillating electrons excite mercury atoms that give off intense ultraviolet light, photons that we cannot see. But the inside surface of the lamp is covered with a **phosphor** (a phosphorescent material) that first absorbs the ultraviolet photons and then emits photons of visible light. The excited electrons in this phosphor take several jumps to come back down to their original energy level, splitting the energy of those ultraviolet photons into several parts; these less energetic photons have frequencies in the range of visible light.

The glow-in-the-dark numerals on some watches come from phosphorscence. The phosphor is excited by the photons in sunlight or artificial light. Because its electrons take some time to drop back down to their ground states, the phosphor glows in the dark long after its exposure to light. In the screen of your television set, there are phosphors that lose extra energy more rapidly. But even then, if the TV camera focuses on a very bright light source such as a stadium light, a brief afterimage from the overactivated phosphors remains on the screen for a second or two after the camera moves away.

## Lasers

In the light from the sun or a light bulb, the photons are of all frequencies and go in all directions. They are emitted independently from one another and arrive at any surface as do the raindrops in a shower. Though most photons are emitted spontaneously by atoms, in 1917 Albert Einstein determined that not all photons were spontaneous. Excited atoms can also be *stimulated* to emit photons.

An excited atom is unstable and will lose its extra energy in due time. But if an excited atom is struck by a photon of exactly the same frequency as one it can eventually emit, the chances are very good that it will emit that photon immediately. The presence of the stimulating photon makes it easy for the emission to occur. Stimulated radiation is different from ordinary, spontaneous emission of radiation in several ways. The stimulated photon goes off with the stimulating photon *in the same direction*. They also travel in step so the peaks and valleys of their electric fields coincide exactly. That means their *amplitudes add perfectly;* the two photons essentially have perfect constructive interference. This sort of radiation is called **coherent.** (The photons from spontaneous emissions are **incoherent;** the photons occur independently and at random.) Once two photons are moving together, should they happen upon another excited atom of the same type, the chances are even better that this atom will be stimulated to add its photon to these two. The more stimulations that occur, the better that group of photons are at stimulating other photons. With each stimulation, the light gains energy—the light is amplified.

An incoherent beam of light resembles a rain shower falling on a surface, but a coherent beam is something like all of the drops falling *at once*. The energy per second delivered by a brief burst of stimulated radiation can be remarkably large. Lasers are devices that amplify light by stimulated emission. The name is an acronym, formed by the first letters of the words that describe it: **l**ight **a**mplification by **s**timulated **e**mission of **r**adiation. The first laser was made by using a ruby, a crystal that is mostly colorless aluminum oxide. The pink-to-red appearance of the gem comes from a small percentage of chromium atoms that give up photons of red light when excited. When bright white light strikes a ruby crystal, the photons excite the electrons in the chromium atoms. The excited electrons quickly drop down into a metastable state where they are briefly trapped. Chromium atoms are phosphorescent, so their

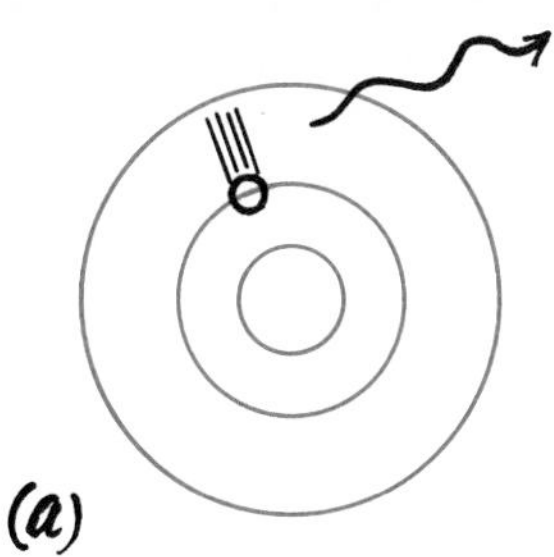


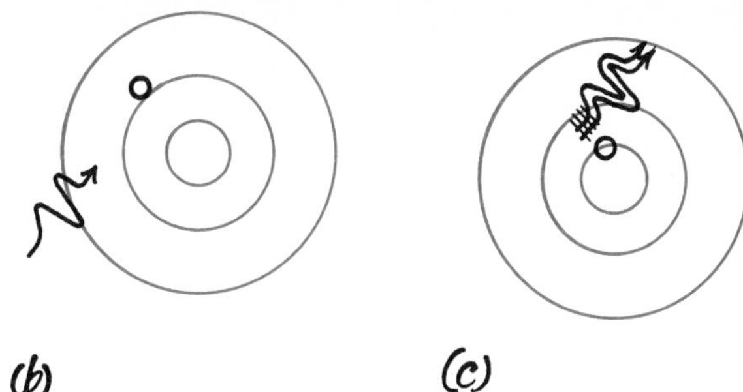


**FIGURE 24–11**
The steps in stimulated emission. *(a)* An electron in an excited atom first makes a transition to an intermediate state where it is metastable and where it will remain for a while before losing more energy. *(b)* While the atom is still excited, a photon passes by that was emitted when a similar atom made the transition from this metastable state to its ground state. *(c)* The electron is stimulated by that photon to release its own energy and also make the transition. The photon it releases moves off in step with the stimulating photon. If enough atoms are in their metastable states at the same time, this process can go on to give a light beam of tremendous power, since all the photons' electric fields add constructively.

electrons have a longer stay on the intermediate level. This property of phosphorescence is used to produce coherent radiation.

A powerful lamp (called a *flash lamp*) that gives off a brilliant, brief blaze of light is wrapped around a cylindrical rod of ruby. When the flash lamp is turned on, the intense flood of photons excites most of the chromium atoms. As the first electrons drop from the metastable states to their ground states, photons of red light appear. If one of these photons happens upon another excited chromium atom, a stimulated emission usually occurs. (Some of the photons, however, will find unexcited chromium atoms that *absorb* them. If the flash lamp has done its job, there are fewer unexcited chromium atoms than excited ones.) Many of the stimulated photons just pass through the side of the crystal, of course. But a small percentage travel straight to the ends of the ruby rod. Each end of the cylinder is polished and plated to function as a mirror. Those photons that reach either end strike these mirrors. One surface reflects almost all of the light that reaches it. The other end reflects about 99 percent, letting 1 percent of the light pass through. The mirrors cause the photons to reflect repeatedly through the cylinder, stimulating more emissions with each pass and increasing the intensity of the light. The small percentage of light that emerges through the partially reflecting end is the laser beam, highly amplified coherent light.

Much more energy is sent into the ruby crystal by the flash lamp than appears in the laser beam. Many of the absorbed photons are transformed into heat, and most of the chromium atoms' red light escapes from the sides of the crystal. But because the laser's beam is coherent,

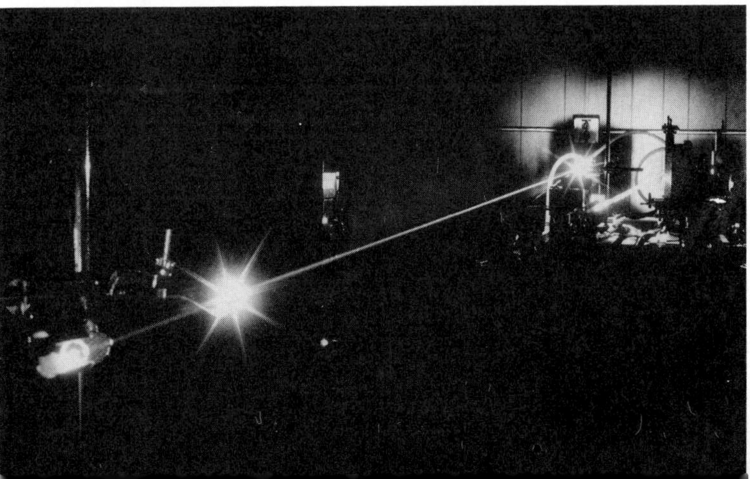

**FIGURE 24–12**
The beam of this laser carries about 50 milliwatts of energy emitted in 30 pulses each second.

its electric force field is astonishingly strong. And because the stimulated photons travel together in direction, a laser beam spreads very little as it travels. These unique properties make laser light invaluable for many applications in science, medicine, and communications.

**FIGURE 24–13**
McDonald Observatory in Fort Davis, Texas. When the powerful laser beam sends a burst of light to the moon, some of the photons are returned along their original path by the reflector left by astronauts.

## Lasers in Action

Today many varieties of lasers are made using different atoms and molecular compounds in the solid, liquid, or gaseous states. These various lasers produce their special wavelengths of light throughout the range of visible, infrared, and ultraviolet frequencies. There are lasers that fit into a shirt pocket and operate on a single penlight battery; others are large and powerful enough to be used as weapons against airplanes, tanks, or guided missiles.

Unlike light from a bulb that spreads out quickly as it leaves, the cross section of a typical laser beam might expand less than half an inch for every mile it travels. As a consequence, the compact laser beam serves as an instant straight line for use in the construction of skyscrapers and pipelines and also for the precise alignment and calibration of machinery. For example, engineers used lasers for precisely aligning the cutting tools that drilled an 18-foot-wide tunnel through 4 miles of solid rock, and at the exit the center line was only 2 inches from perfect alignment.

In an early, dramatic exhibition of a laser's power, brief bursts of laser light containing $10^{20}$ photons were directed at the moon. In less than 2½ seconds some of these photons had reflected from the moon's surface and returned to a telescope at an observatory. The actual time of flight was measured with an atomic clock to one part in a hundred million. Multipying this time ($t$) by the speed of light ($c$) revealed the round-trip distance to the moon ($d = ct$) with unprecedented accuracy. It was an impressive experiment. About $10^{18}$ photons from the original burst reflected from the moon's surface to travel out in all directions, but the sensitive photon detectors at the focal point of the telescope mirror received only about *10* photons. From this, however, the distance to the moon was found with an uncertainty of only a few tens of meters. Astronauts later left a special reflector on the moon. The distance from the observatory to that reflector can now be measured with an uncertainty of *under 10 centimeters* (about 4 inches).

Closer to home, lasers are used by computerized cash registers at the checkout counters in many grocery stores. The beam illuminates (and the computerized detector reads) the universal pricing code placed by the manufacturer on almost all grocery products. Some nationally circulated newspapers are printed simultaneously in a dozen or more cities across the country with laser assistance. Guided by a satellite transmission, the laser exposes the photographic film that is used to make the printing plates each day.

Lasers have been used in medicine since the 1960s. A surgeon working on the human eye can send a perfectly controlled laser beam through the cornea and lens and bring the beam into focus at the edge of a rip in the thin retina. At the point of focus, the concentrated laser

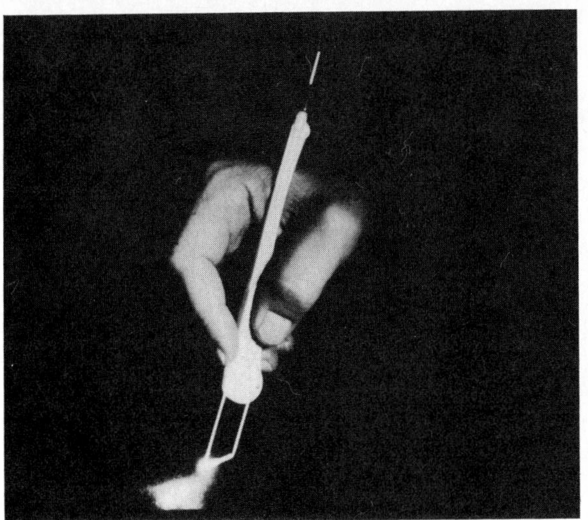

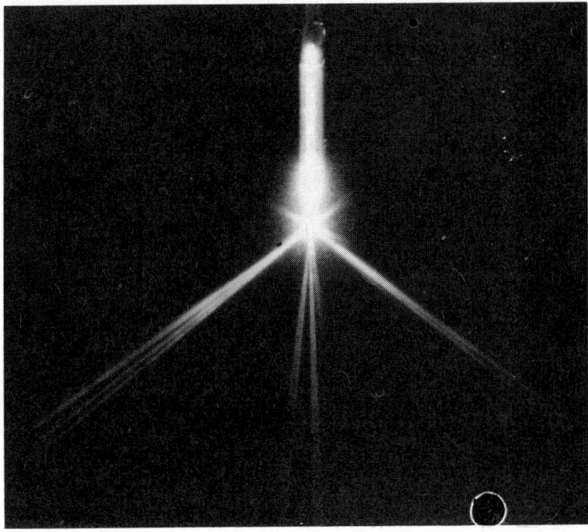

**FIGURE 24–14**
*(a)* A laser scalpel. Light from a laser is brought to this transparent scalpel via a light pipe and exits at the cutting edge. This scalpel which sterilizes and cauterizes as it is used, is especially valuable for skin grafts on burn patients and other skin surgery where infection is a primary problem. *(b)* An edge-on view of the scalpel as laser light emerges at the cutting edge.

light increases the temperature, coagulating the proteins there and welding the retina into place. Other lasers designed as surgical knives have an advantage over traditional scalpels. They sterilize and seal off blood vessels as they cut, minimizing infection and bleeding.

## CALCULATIONS

EXAMPLE 1: A ruby laser produces a 20-million-watt light pulse for about one twenty-millionth of a second. If the frequency of the red light is $4.32 \times 10^{14}$/second, **about how many photons are in that pulse?**

Each photon has an energy equal to $hf$, where $h$ is $6.6 \times 10^{-34}$ joule·second. If there are $n$ photons in that pulse, then $nhf$ is the total energy. Power equals energy divided by time, so the energy of the pulse is equal to the power output multiplied by the time that power is delivered.

$$E = P \times t = 20 \times 10^{6} \text{ watts} \times \frac{1}{20} \times 10^{-6} \text{ seconds} = 1 \text{ joule}$$

Setting $nhf = E$, we find

$$n = \frac{1 \text{ J}}{hf} = \frac{1 \text{ J}}{(6.6 \times 10^{-34} \text{ J·s}) \times (4.32 \times 10^{14}/\text{s})}$$
$$= 0.035 \times 10^{20} = \mathbf{3.5 \times 10^{18} \text{ photons}}$$

## REVIEW

**1.** A spectroscope produces a magnified spectrum of a light beam for visual observation. True or false?

**2.** Dark spectral lines appear in a continuous spectrum when certain wavelengths of light are absorbed as white light passes through a gas. True or false?

**3.** Gaseous atoms absorb light energy of certain wavelengths and emit light of identical wavelengths. True or false?

**4.** The light emitted by a thin gas of excited atoms appears in a spectroscope as bright lines of color on a dark field. This is called an ______________ spectrum.

**5.** Unlike a gas, a solid generally absorbs and emits light over a wide range of wavelengths. True or false?

**6.** Define a blackbody in terms of the light it absorbs.

**7.** Does an atom give off light when one of its electrons makes a transition to a lower energy level? To a higher level?

**8.** In Bohr's theory, a transition from one orbit to another means the electron has lost or gained a quantum of energy. True or false?

**9.** What is the photoelectric effect?

**10.** What is a photon? What formula gives the amount of energy carried by a single photon of light that has a frequency *f*?

**11.** Broadening of the characteristic spectral lines of an element's atoms takes place because of (a) collisions, (b) high temperatures, (c) the Doppler effect, (d) all of these.

**12.** Name two processes that describe the ways an atom can emit photons.

**13.** A device that amplifies light by stimulated emission is a ______________.

## DEMONSTRATION

**1.** Some night, turn off the lights in a kitchen that has an electric stove. Then turn one of the stove element controls to the setting for simmer or warm. In daylight you wouldn't be able to see the element glow at this setting, but in the dark you can because of your night vision. Hold a hand close to the side of the element; what you feel is light that you cannot see, infrared radiation. Slowly turn the control to the low, medium, and high heat settings, feeling the infrared radiation at each step and noting the colors of the element. As the temperature increases, the color of the most intensely emitted wavelength goes from dark red to lighter reds to red-orange; in other words, to shorter wavelengths, just as the peaks of the blackbody radiation curves do in Fig. 24–4. The different infrared intensities you feel also correspond to the curves of blackbody radiation. As the stove element becomes hotter, the infrared intensity rises, just as the blackbody curves indicate. The stove element isn't a perfect blackbody, but its emission spectrum follows the same general pattern as that of a blackbody.

## EXERCISES

**1.** A star that appears red emits most of its visible light in the red region of the spectrum, and a blue star emits more visible light in the blue region. Which is hotter?

**2.** Ultraviolet radiation, X rays, or gamma rays that pass through a gas can ionize some of the atoms or molecules and make the gas conduct electricity. Would you expect the greatest ionization from (a) a small intensity of high-frequency light or (b) a large intensity of low-frequency light?

**3.** If you turn off a TV in a dark room, why does the screen continue to glow for a few seconds?

**4.** How can a technician in a crime lab tell the atomic and molecular composition of a flake of automobile paint found at the scene of a hit-and-run accident?

**5.** In a photography darkroom you can find red lights of very low power that let the photographer see to work yet don't expose the light-sensitive chemicals in the photographic paper. Comment on whether any other color would do just as well.

**6.** What might the spectrum of hydrogen look like if the electron could move between orbits of *any* radius in Bohr's model?

**7.** Astronomers today use the absorption lines in the spectrum of the light from a star or planet to tell about its atmospheric composition. Explain how that works.

**8.** If an astronomer gave you photos of the spectra of a dozen different types of stars, some of the spectral lines you would see would be precisely the same in each spectrum, provided the exposures were identical. Guess the origin of those lines.

**9.** How could the photoelectric effect be used to open doors when someone approaches?

**10.** An astronomer can use a spectroscope to observe the sun's spectrum, an absorption spectrum. What would happen if he or she were watching this spectrum when a total solar eclipse occurred, where the moon moves between the sun and the observer, leaving only the sun's atmosphere (called the corona) visible? The continuous part of the solar spectrum would suddenly disappear, but that's not all. The dark lines would become bright lines. Explain.

11. In museums and geology exhibits, you can sometimes find specimens of rock on display that seem to glow in the dark with brilliant reds, greens, and even violets. Look closely, and you'll see a lamp is present. What sort of radiation must the lamp be emitting—could it be infrared? What property of the minerals in the rock causes them to glow?

12. Lasers can be used to detect flaws inside solid objects. A larger laser beam is focused on the solid's surface and literally blows off a small amount of the material, causing only minor cosmetic damage. As the vaporized material blows off the surface, it exerts a strong, sudden pressure on the surface. It creates a sound wave that begins at that point and travels through the solid. Tell how that large-amplitude, short-width sound wave might be used to detect flaws.

**13.** Suppose you are looking through a spectroscope at the lines of the emission spectrum of hydrogen gas. Describe what will happen to the emission lines if you turn on a heater that greatly increases the temperature of the gas.

14. How would sunlight be different if the collisions between atoms and their speeds of approach or recession from the observer had no influence on the photons the atoms emit?

**15.** Think of a solid as a condensed gas of atoms or molecules, and explain why a solid's emission spectrum is continuous, containing almost all frequencies of light (although different frequencies are emitted with different intensities).

**16.** Broadcasts via microwave transmissions let people watch the astronauts walking and riding on the moon and sent pictures to earth from Mars, Jupiter, and Saturn. Microwaves spread out too much to be detectable from Pluto's orbit, the edge of the solar system. How might we be able to get clear signals back from that distance?

17. Think of a reason fluorescent lamps are more efficient than incandescent lamps in turning energy into light. That is, fluorescent lamps give equal amounts of light with much less energy lost as heat.

**18.** Ultraviolet and infrared laser beams cannot be seen with the eye. Why does that make them more dangerous to use in a laboratory than visible beams are?

19. How do some materials and paints visibly glow under a black-light, an electric bulb that emits invisible ultraviolet light?

**20.** Could materials without metastable states produce coherent beams of light by laser action?

21. A neodymium laser can emit powerful pulses of green light. Could a flash tube that emitted only red light be used to power that laser?

22. High-pressure sodium vapor lights emit an amber-pink glow. Low-pressure sodium vapor lamps emit a deep yellow light and are brighter and less expensive to operate. Both are used as street lamps that scatter light into the atmosphere in every direction. Astronomers want the cities close to the observatories to use the type of light that interferes less with their spectral analysis of starlight. One of these lights emits a broad continuum of light, while the other emits its light mostly at two discrete frequencies, allowing the astronomers to work in the remainder of the spectrum with their spectographs. Which light is the one the astronomers prefer?

*23. About how many photons per second are emitted by a 100-watt bulb if the *average* frequency of the photons is $5 \times 10^{14}$/second? (Typically 7 percent of the energy each second goes away as visible light. Heat takes away most of the energy.)

*24. The wavelengths of the light found in the spectral lines of atoms can often be measured to an accuracy of 1 part in $10^{13}$. They provide the means for some of the most accurate measurements in physics and astronomy. Compare this to the accuracy which laser-ranging provides for the moon's distance to earth, 10 centimeters (0.1 meter) to about 380,000 kilometers.

*25. A laser and a photon detector, mounted in an airplane, have been used to chart the depths of shallow coastal waters. A brief burst of light from the laser strikes the water's surface, where some of the beam is reflected and returns to the detector. However, a portion of the beam enters the water and travels to the bottom, where once again a small amount of light is reflected. Even in 10 meters (about 30 feet) of turbid water, a few photons from the light that made it to the bottom of the water will emerge from the surface and enter the photon detector on the plane. The time lag between the light reflected from the surface and the light reflected from the bottom can be used to find the depth. What would the time lag be for a water depth of 10 meters? (The speed of light in water is $0.75c$.)

# CHAPTER 25
# THE WAVE NATURE OF PARTICLES

"Is there any point to which you would wish to draw my attention?"
"To the curious incident of the dog in the night-time."
"The dog did nothing in the night-time."
"That was the curious incident," remarked Sherlock Holmes.
From "The Silver Blaze," *Great Stories of Sherlock Holmes,* by Arthur Conan Doyle

Like the dog that did nothing in the nighttime, the inaction of the electron in Bohr's model for hydrogen was a clue that something was wrong: The classical laws of physics said it should radiate light, lose energy, and spiral inward to crash into the proton—but it didn't. And although his model explained the spectrum of hydrogen, neither Bohr nor anyone else could say why the electron should move only on a certain set of orbits.

It took an entirely new theory to solve this mystery, one that shattered established notions of what particles were like and how they behaved. The radical ideas needed came from a number of physicists, and, like Planck with his theory of blackbody radiation, some of them could not believe in the end what they had collectively discovered. The first step was taken by none other than a prince of the French nobility, Louis Victor de Broglie (born in 1892). ■

## De Broglie's Idea

As a graduate student in physics at the University of Paris, de Broglie became fascinated with Albert Einstein's revelations about light. The fact that light acts like a wave in some experiments and like particles in others led de Broglie to pose this question: Could *material* particles (those with mass, like electrons and protons) have a dual nature, too? Could particles ever act like waves?

It seemed unlikely. Indeed, de Broglie knew of no experiment that showed such behavior, and there was no wave nature in the actions of much larger objects like baseballs. But he persisted in searching for an answer to his questions. He wrote the mathematical expressions for the energy of a light wave and the energy of a particle moving at a constant speed (both as given by Einstein's theory of relativity, which we will discuss in Chapter 26) and compared them term for term. He found a connection between the momentum *mv* of a particle and a wavelength λ:

$$\text{momentum of particle} = \frac{\text{Planck's constant}}{\text{wavelength}}$$

In symbols,

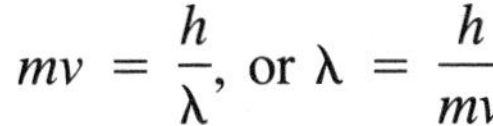

$$mv = \frac{h}{\lambda}, \text{ or } \lambda = \frac{h}{mv}$$

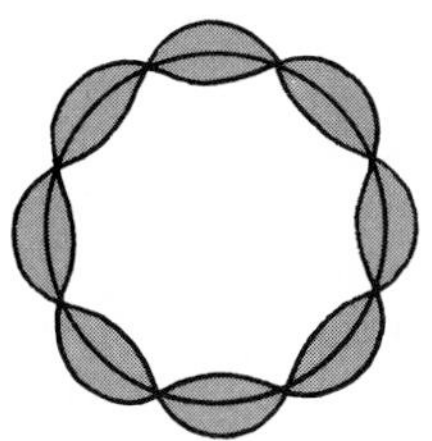

**FIGURE 25–1**
An example of a wave that has a whole number of wavelengths on a Bohr orbit. Consecutive passes of the wave around the orbit just retrace this pattern. Because the wavelength is related to the electron's energy, an electron can have only certain energies in the hydrogen atom, corresponding to a standing wave like this one.

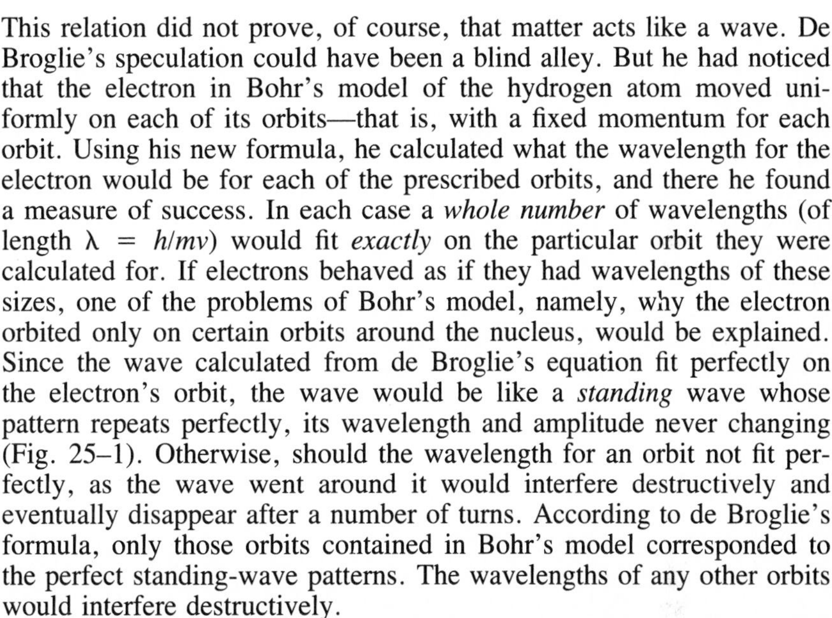

This relation did not prove, of course, that matter acts like a wave. De Broglie's speculation could have been a blind alley. But he had noticed that the electron in Bohr's model of the hydrogen atom moved uniformly on each of its orbits—that is, with a fixed momentum for each orbit. Using his new formula, he calculated what the wavelength for the electron would be for each of the prescribed orbits, and there he found a measure of success. In each case a *whole number* of wavelengths (of length $\lambda = h/mv$) would fit *exactly* on the particular orbit they were calculated for. If electrons behaved as if they had wavelengths of these sizes, one of the problems of Bohr's model, namely, why the electron orbited only on certain orbits around the nucleus, would be explained. Since the wave calculated from de Broglie's equation fit perfectly on the electron's orbit, the wave would be like a *standing* wave whose pattern repeats perfectly, its wavelength and amplitude never changing (Fig. 25–1). Otherwise, should the wavelength for an orbit not fit perfectly, as the wave went around it would interfere destructively and eventually disappear after a number of turns. According to de Broglie's formula, only those orbits contained in Bohr's model corresponded to the perfect standing-wave patterns. The wavelengths of any other orbits would interfere destructively.

Encouraged by this idea, de Broglie boldly presented them in his Ph.D. thesis. But there was no experimental evidence to support his ideas, and de Broglie's professors were skeptical. One of his professors decided to get an outside opinion; he sent de Broglie's thesis to Albert Einstein for evaluation. No one was happier with the reply than de Broglie. Though Einstein agreed that the ideas certainly looked crazy, he thought they were very important and, in Einstein's words, "sound." The prince received his Ph.D. degree in 1924 and in 1929 was awarded the Nobel Prize, to this day the only one ever awarded for a doctoral thesis in physics.

Before de Broglie's thesis was published in 1924, Clinton Davisson (working at the Bell Laboratories in the United States) was studying the behavior of vacuum tubes for radios. He aimed beams of electrons of known speeds (and hence known momenta) at metal targets and monitored the electrons as they scattered from the metal surface at various angles. In one such experiment with a metallic *crystal,* he found and reported a strange pattern in the scattered electrons. After de Broglie's work appeared in a journal, several European physicists made a remarkable observation: The American's pattern of electrons scattering from the crystal resembled an *interference* pattern, *as if the electron beam had the properties of a wave*. They convinced Davisson of the importance of his observations (he was unaware of de Broglie's hypothesis), and he did the experiment again, carefully measuring both $mv$ for the electron and the angles in the interference pattern. Soon he showed that the wavelength that would cause his interference pattern matched the wavelength de Broglie's formula predicted for the electrons in his beams. This experiment was the first to prove the existence of the wave nature of matter, and Davisson, too, was awarded a Nobel Prize. We'll return to his experiment after we've discussed the meaning of these waves.

## An Equation for Matter Waves

**FIGURE 25–2**
Mike, a physics major and a rock climber, going over the edge with Schrödinger's equation.

De Broglie's formula tells us little about the wave of a material particle. It gives a value for the wavelength without saying anything about the amplitude or, indeed, what the amplitude means, or *even what the wave really is*. (Are the particles waving as they go along?) Clearly the picture was incomplete.

The next step in understanding the behavior of subatomic particles was taken by Erwin Schrödinger, an Austrian physicist. Early in 1925 he began to search for an equation that would describe the amplitude of the wave of a particle. Even though he was not sure what significance the amplitude had, he assigned it a symbol, $\psi$ (Greek lowercase letter *psi*, pronounced "sigh"). Within a year he found an equation that predicted the amplitude of the waves of the electron in a hydrogen atom. These waves were standing waves in *three* dimensions, *not* confined to Bohr's planetlike orbits. However, each standing $\psi$ wave has a large amplitude at the position of its corresponding Bohr orbit in the atom. And the energies associated with the standing waves were precisely the same as the energy the electron had in its orbits. Schrödinger's equation was general; it would predict the waves for an electron (or a proton, and so on) moving under the influence of *any* force. The problem now concerned what $\psi$ actually meant in terms of the properties of the particle.

Schrödinger *thought* that these waves meant the electron was smeared out, as if a child pulled on the edges of a wad of bubble gum and stretched it. In his view, the amplitude of the wave would be big where most of the particle was and small at the particle's edges where the particle was thin. He thought the square of $\psi$, which would correspond to the intensity of a classical wave, might indicate the *density* (mass/volume) of the smeared-out electron. Figure 25–3 shows the wave clouds of $\psi$-squared predicted by Schrödinger for the hydrogen atom. Where the dots are thickest, the value of $\psi$-squared is largest, and so on. Although Schrödinger's wave clouds do represent the electron's behavior, his idea about the smeared-out electrons was *not* correct, as we'll see in the next section.

Here is a brief summary of what de Broglie and Schrödinger accomplished. De Broglie found an equation for the wavelength for a par-

**FIGURE 25–3**
Probability plots for the electron in some of its three-dimensional orbits, or states, in the hydrogen atom. Where the dots are closer together, the electron is more likely to be found if it is in that state. These patterns of probability are standing-wave patterns computed from the $\psi$ waves of the electron. The pattern at left is for the ground state, where at a given distance from the nucleus, the electron has equal probabilities for being found at any angle. The other patterns, moving from left to right are for increasingly excited states farther from the nucleus. For simplicity, all of these probability plots are shown with the same maximum diameter.

ticle with a constant momentum *mv*. Schrödinger found an equation to predict the wave's amplitude, even if the particle's momentum is changing, as when a force acts. In other words, de Broglie's formula for the waves of matter was something like Newton's first law, giving their property $\lambda$ if the particle had a constant momentum or speed. Schrödinger's equation, however, was something like Newton's second law for these waves, predicting what would happen to the wavelengths and amplitudes when a force acted on the particles. Schrödinger called his equation the **wave equation,** and the behavior of the waves is known as **wave mechanics** to distinguish it from *classical mechanics,* which describes the actions of larger particles using Newton's laws.

## What Matter Waves Really Mean

Max Born, another German physicist, quickly pointed out that Schrödinger's idea of an electron that was spread out along the wave did not agree with the facts. This was the problem: Electrons are found whole, not in parts, and at some *point,* not smeared out over an area or volume. Experiments never find a fraction of an electron. Born suggested that the density of Schrödinger's wave clouds around the proton represented only the *probability* for finding the electron at various places. In Born's view, the electron had a good chance of being found in a region only if the square of the wave amplitude $\psi$ was large there. Where that quantity was small, an experiment would have little chance of finding the electron. The electron, Born proposed, could be found at some *point* in the wave cloud, and *the density of the wave represents only the relative probability for the electron's presence.*

Strange as this idea sounds, it agrees with the experimental facts. Born's view of Schrödinger's wave clouds was the correct one, as many experiments since the 1920s have verified. The wave clouds in Fig. 25–2 show only the probability density for the electron. The particle is more likely to be found where the cloud is *dense* (large value of $\psi$-squared) than where the cloud is thin.

The electron, then, is not a wave, but it appears with probabilities that are calculated by a wave. The wave cloud has no physical reality of its own. A rain of photons cannot scatter from this imaginary cloud to give its image, as they do from a cloud of smoke over a forest fire, or from a billow of dust above a dirt road. These wave clouds don't have mass or energy of their own. However, the waves from Schrödinger's equation do predict the electron's behavior accurately, and this is the ultimate test—a valid physical theory must predict the observed facts.

Here, then, is the interpretation of the hydrogen atom according to wave mechanics. The clouds of Fig. 25–2 represent only the chance for finding the atom's electron, *which is as much as we can know about the electron's position*. Bohr's orbits are gone; a *path* for the electron is not to be found in wave mechanics, only probabilities for the electron's appearance. These clouds show us that if the electron has a certain probability to be in one area of the atom, it has an equal probability of being in the same area on the opposite side of the proton. (The clouds are symmetrical about the nucleus.) If at some instant the electron in a hydrogen atom absorbs a quantum of energy, its probability wave pattern changes to another standing wave pattern representing the higher energy and giving the electron a higher probability for being found farther from the nucleus. On the other hand, when the atom emits photons, the electron abruptly loses energy, to be represented by the pattern of a standing probability wave that is concentrated closer to the proton. The standing waves of probability that replaced Bohr's orbits are called **orbitals;** old ideas die hard.

The electron is a particle with wave aspects to its behavior, and there is nothing really like it on a scale that we can perceive with our senses. Schrödinger's equation gives us a representation of the electrons, but it doesn't tell us what they are really like. They compare to nothing in human experience.

## The Stuggle to Understand Wave Mechanics

Niels Bohr became convinced that Born's interpretation of the waves was correct, and his institute at Copenhagen, Denmark, soon emerged as the center of interpretation of the new wave mechanics. He and his colleagues came to the belief (sometimes called the **postulate** of wave mechanics) that the amplitude as calculated by Schrödinger's equation contains *all* the information we can obtain about the mechanical behavior of a subatomic particle, its position, momentum, energy, and so on.* That is, the $\psi$ wave describes the physical state of the particle as fully as possible.

At this point, Einstein, Schrödinger, Planck, and de Broglie all objected. They could not believe it was impossible to predict precisely where to find an electron in an atom, that probabilities for finding the positions of such small particles are all anyone could predict. They thought there must be another interpretation of the theory of wave mechanics, or perhaps more to be learned. Schrödinger, especially, was crushed. He had developed an equation that replaced Bohr's jumping electron with a progression of standing waves only to have others discover that his waves were only patterns of probability—that the electron was a pointlike particle that must still "jump" when its probability cloud pattern changed, emitting or absorbing a photon in the process.

---

*For very high speeds, the theory of relativity comes into play. Another equation, first found by the British physicist P.A.M. Dirac, must be used instead.

He said to Bohr, "Had I known that we were not going to get rid of the damned quantum jumping, I never would have involved myself in this business." As for the probabilities that the wave formula predicted, Einstein loudly (for him) protested, "God does not play dice . . ."*

But Born's interpretation of the wave cloud is consistent with the results of many experiments. We'll describe one such experiment next to illustrate how the theory of wave mechanics makes predictions about the behavior of particles.

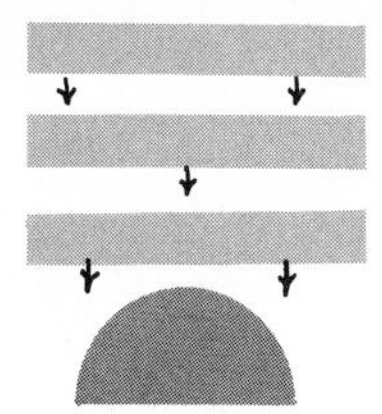

**FIGURE 25–4**
Parallel wavefronts of a ψ wave approaching an atom. Such regular wavelengths are associated with a particle of definite momentum that travels in a well-defined direction. This wave represents a low-energy electron that will have little interaction with the atom other than bouncing back when it strikes. (Higher energy electrons, having shorter wavelengths, could interact strongly with the atomic electrons, causing transitions.)

## An Experiment with Electrons

As we mentioned earlier, Clinton Davisson (assisted by Lester Germer) perfected the first experiment to demonstrate the wave behavior of electrons. Davisson and Germer directed a beam of low-speed electrons straight at one face of a crystal of nickel and detected the electrons as they bounced off the face. A large number rebounded straight back along the incoming beam's path, but the others bounced off at various angles to the sides.

The electrons coming back at angles to the incident beam were detected in large quantities at one certain angle and rarely appeared at another angle. This pattern, made by counting large numbers of electrons, had the look of an interference pattern where waves add constructively at one point and destructively at another. Wave mechanics explains this sort of interference pattern in the following way:† All of the electrons in the beam have the same momentum, so the $\psi$ wave for any electron also has the same wavelength, $\lambda$. Figure 25–4 pictures the wave of one such electron approaching one of the atoms on the surface of a crystal. When the electron arrives at the surface, it feels a repulsion from the electrons of the atoms there; the electron's wave, Schrödinger's equation says, will reflect from the edge of those atoms. These are low-energy electrons that cannot penetrate the atoms; each electron rebounds from the surface atoms with little or no loss of speed. (In terms of mass, it's like shooting a pea at a bowling ball.) Figure 25–5 shows what happens to the wave. The reflected wave goes off in all directions, like a wave from a pebble tossed into a pond.

**FIGURE 25–5**
Once the ψ wave in Fig. 25–4 strikes the atom, it reflects from the atom. The reflected wave is more or less spherical; this means the electron has a probability to be found traveling in almost any direction.

But a $\psi$ wave has breadth; it has a wavefront. It isn't a point as the electron is. It falls on and reflects from *many* atoms on the surface, not just one. Consider what happens as the electron's wave reflects from two *adjacent* atoms, as in Fig. 25–6. The waves traveling outward eventually *overlap,* and constructive or destructive interference results. Let's choose one direction to look at these waves as they move, for example, a direction where the waves add. The result is seen in Fig. 25–7.

This wave, remember, is not the electron but merely a predictor

*Bohr's answer was, "How do you know?"

†We'll examine the essential features only. The full analysis is more complicated, involving adjacent lines of atoms on the surface of adjacent planes of atoms if the electrons penetrate the surface and reflect from the atoms there.

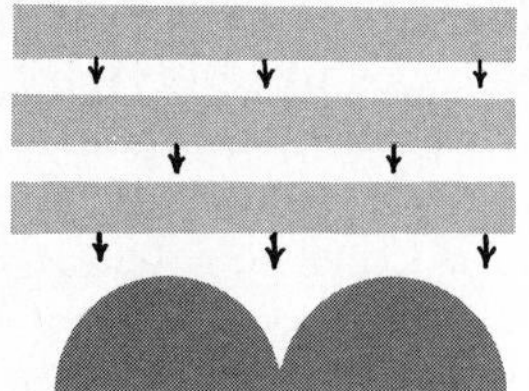

**FIGURE 25–6**
The $\psi$ wave of an electron approaching two atoms. Each atom will scatter the portion of the wave that strikes it, as the single atom does in Fig. 25–5.

that shows probabilities for where the electron will be found. Because portions of the wave interfere constructively at this particular angle, the wave's amplitude $\psi$ (hence $\psi$-squared) is large, and there is a large probability that reflected electrons will reach a detector placed at that angle above the target. At another angle where the wave portions from adjacent atoms interfere destructively, $\psi$-squared is very small, and few electrons will be counted because the probability is slight that electrons will travel in that direction.

The angles of high and low probability depend on the arrangement of atoms in the crystal and the wavelength (or momentum) of the electrons. When Davisson and Germer repeated the nickel crystal experiment, they found that de Broglie's formula gave a wavelength that accounted for their pattern. They quickly went on to experiments with other crystals, confirming de Broglie's hypothesis with a total of 19 different crystals. In 1930 other workers scattered whole hydrogen and helium *atoms* from crystal faces and found the same effect. Many more such experiments followed. Electrons, protons, neutrons, and atoms all show wave behavior, but particles of larger mass are represented by smaller wavelengths and interference effects become impossible to see.

**FIGURE 25–7**
*(a)* The spherical waves interfere. Where the interference is constructive, the net value of $\psi$ increases, leading to a greater probability for finding the electron at those places. *(b)* For clarity, only portions of the scattered waves are shown here. They are in step, crest to crest, and trough to trough. Because of this, when they converge at some distant point, they will interfere constructively. The electron, then, has a large probability for being scattered at the angle these portions of the waves are taking as they reflect.

In the Davisson and Germer experiment, the large numbers of electrons they counted give a distribution of electrons over all angles. The pattern they found thus revealed the values of the probability wave density. However, a single electron cannot by itself reveal this probability distribution because it goes to only *one* point, a point that cannot be predicted ahead of time. The probability clouds of wave mechanics do not predict precisely where any individual particle will go, and the postulate of wave mechanics says that such a prediction is impossible because $\psi$ tells us all we can know about the particle's position.

The strange behavior of these particles that we cannot see was only discovered when the means to investigate them experimentally came into being. The atom and its particles are still relatively new to our experience, and, as Einstein thought, there may be a better theory. But any theory that replaces wave mechanics will have to explain more than wave mechanics does. That will be a formidable act, because at the present time, the theory of wave mechanics seems as solid as a brick wall with perhaps no chinks to be filled. The picture it provides lets us interpret the results of nearly all the experiments at the atomic level and has led to predictions of enormous practicality, such as the technology of lasers and the solid-state electronics of computers.

## The Uncertainty Principle

In the experiment discussed in the previous section, the electron's probability wave was pictured coming into the face of a crystal and reflecting not from one atom but from many. Here we will find that it is, in fact, impossible in such experiments to aim an electron's wave at one single atom. Let's take a look at how wave mechanics describes a single electron traveling through a vacuum to see why this is so.

To simplify the description, we'll consider a $\psi$ wave moving in a single direction with an amplitude in one plane. (In reality, remember that the wave representing the electron is three-dimensional.) Suppose we know very closely where the electron starts and how fast it is moving. The electron's initial $\psi$ wave would be zero everywhere except in that region where it is known to be. For example, the electron may be known to be between two points along some direction, say the $x$ direction. The distance between these points is called $\Delta x$, and the length of $\Delta x$ tells how precisely we know the electron's position. That is to say, the particle's position is known only to within the value of $\Delta x$, so this quantity is called the **uncertainty** of the particle's position in this direction. The wave might be like the wave in Fig. 25–8*a*. In the case like this, where the particle (and its wave) is *localized* initially, a strange thing happens. The localized wave invariably *spreads out* as it moves along (see Fig. 25–8*b*).

When the wave spreads, there is a probability for finding the particle *ahead of* or *behind* that position where a definite known speed would take it. This means there is a probability that the particle will go faster or slower than we might expect. In short, there is an uncertainty in *speed* (and hence momentum) for a localized particle. The behavior of a localized wave, as calculated with Schrödinger's equation, follows this pattern. The localized wave generally spreads faster when $\Delta x$ is smaller, and vice versa. In other words, if the particle wave is very broad (large $\Delta x$) to begin with, the particle's position is less well known, and the spreading of the wave as it moves is generally slower. In practical terms, this means that if you know very well where a particle is at a certain time, you cannot know its speed very well.

Werner Heisenberg discovered (by other arguments, including many experimental results) that this relation is an inescapable part of wave mechanics. He concluded that there was by *necessity* a connection between the uncertainties in position and momentum, and he found

**uncertainty in position times uncertainty in momentum is greater than or equal to Planck's constant**

or



$$\Delta x \times \Delta mv \geq h$$

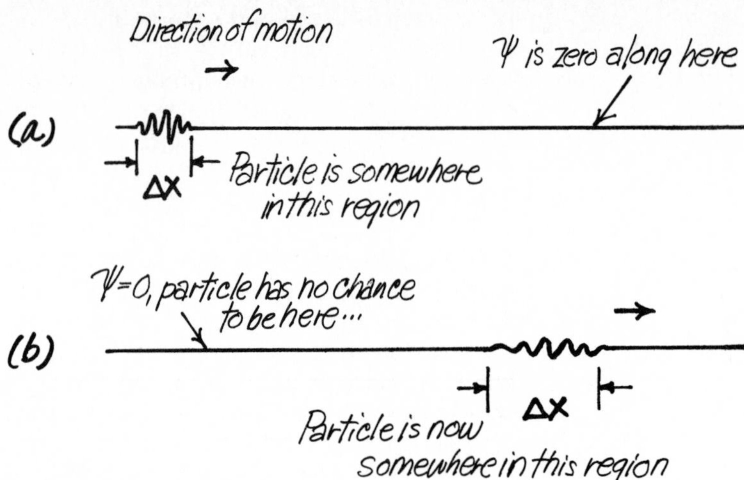


**FIGURE 25–8**
*(a)* The ψ wave for a particle that is known to be within a range of $\Delta x$ has a nonzero amplitude only in that region. *(b)* As the particle moves, the region where its ψ wave has an amplitude becomes larger. This behavior of ψ waves is predicted with Schrödinger's equation. It means that if you know the position of a tiny particle very well, to within a small $\Delta x$, you cannot know its speed (or momentum) very well. So as the particle moves, the region in which it may be found spreads out, corresponding to the range of speeds the particle might have.

This is known as the **uncertainty relation,** and it comes from the nature of the waves that must be used to predict the motion of matter's smallest particles.

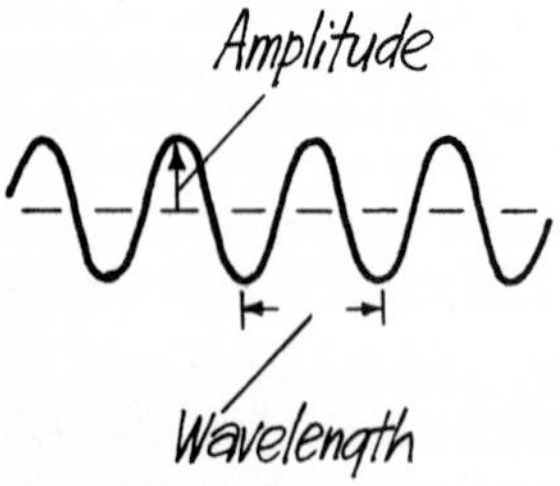

**FIGURE 25–9**
A representation of the ψ wave for an electron with a precisely known momentum (or speed). Its wavelength λ is equal to *h*/*mv*.

## "LOCATING" A PARTICLE WITH WAVES

An electron traveling through a vacuum can be accelerated to give it a firm value of kinetic energy or momentum, as in Davisson's experiment. Schrödinger's equation predicts the wave that exhibits an electron's probabilities if the electron moves with a constant momentum. The wave is simple: It has a constant wavelength (as de Broglie's formula implies) and a constant amplitude, wavelength after wavelength (Fig. 25–9).

Notice what this wave predicts about the electron's position. The mathematical square of the amplitude ψ gives the probability for finding the electron at any point. Because its wave train is continuous, the electron can be almost anywhere along that line of wavelengths. If the momentum is known with absolute certainty, then we lose all knowledge of where the particle is. But wait: An experiment can locate an electron and pin down its position to a small volume of space, so there must be a solution of Schrödinger's equation to represent an electron moving through a vacuum that has a significant amplitude in only a small region at every instant. There is, and to get a wave that has an amplitude in a small region requires adding together many waves that individually satisfy Schrödinger's equation. Two such waves of the same amplitude whose wavelengths differ by only 10 percent are labeled A and B in Fig. 25–10. If these waves are added point by point, wave A + B appears. This looks like a train of wave packets of a certain length. Next we add wave C, which has a wavelength between those of the first two waves and an amplitude 1½ times as large. Now the wave train has the pattern labeled A + B + C, reducing the two wave packets to the side while reinforcing the central packet. The addition of other such waves, close in wavelength and with varying amplitudes, builds up a wave packet such as the one in Fig. 25–8. The amplitudes and wavelengths of the combined waves determine the shape of the final packet.

What happens as a packet such as this moves along? The individual wavelengths that make up the packet represent somewhat different momenta, $\lambda = h/mv$. The narrower the packet is initially (small $\Delta x$), the wider the range of wavelengths needed for its construction (large $\Delta mv$). As the packet moves, it spreads out along the axis because the particle has a probability of going faster or slower than the average wavelength used to build the packet.

An electron localized in an experiment (its position well known) has a highly uncertain momentum, and as it moves, this uncertainty in its speed means we lose the certainty in its position as it travels.

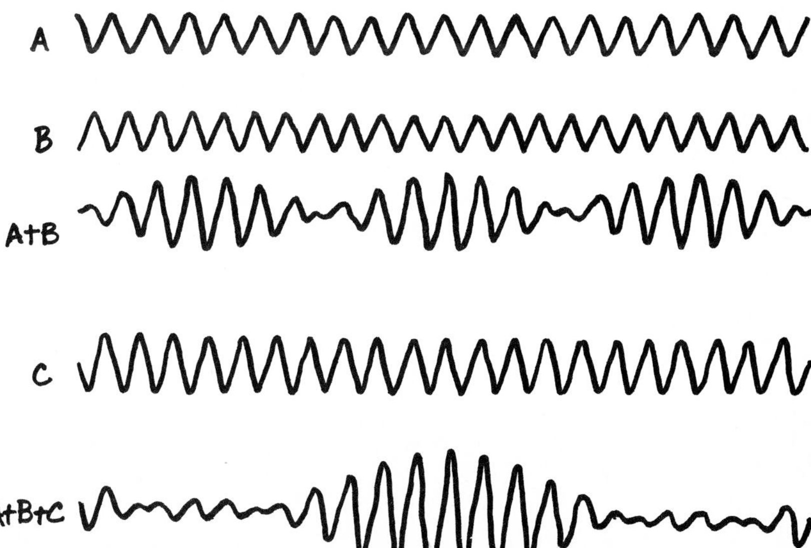


**FIGURE 25–10**
Waves A and B have wavelengths that differ by only 10 percent, and their amplitudes are the same. If they are added, the result is A + B. The wave then has regions where $\psi$ (and $\psi^2$) is large (the particle is likely to be found) and regions where $\psi$ (and $\psi^2$) is small (the particle is less likely to be found). Adding wave C to A + B gives the wave labeled A + B + C. Wave C has a wavelength $(\lambda_A + \lambda_B)/2$ and an amplitude 1½ times larger than A or B. The result with these three waves is one localized region where $\psi$ is large. By adding waves together like this, wave mechanics describes a particle known to be in a certain region of space at a certain time. If an infinite number of waves are used, the initial location of the particle can be very precise, with $\psi = 0$ everywhere but some small region of space, as in Fig. 25–8.

## Aiming Electrons

Suppose in an experiment such as Davisson's we want to aim the electrons in the beam at a *single atom* on the face of a crystal. If two plates, each with one small pinhole, block the path of the beam, very few electrons get past. (Fig. 25–11.) But those that do get past have traveled along a straight line between the pinholes that aims them toward a single atom on the crystal (if the pinholes are small enough). But even then many electrons that make it through both pinholes miss the target atom and hit another, as we can see by the following analysis.

The momentum that moves the electron along *in the line of the beam* can be very certain because the electron's position along the line is never known. There aren't any detectors along the way to pin down its location. But each electron must pass through the second pinhole to reach the crystal, and because it does, we know the electron's horizontal position to within $\Delta x$, the size of the pinhole (see Fig. 25–12). The uncertainty relation tells us what has to happen to the electron's wave.

*The small value of $\Delta x$ causes the momentum along that direction to have a large uncertainty*. So the electron's probability wave spreads out to the sides (or diffracts) as it travels toward the crystal face, and when the wave arrives, it reflects from all of the atoms in the vicinity of the target atom. The attempt to aim the electrons is foiled by the uncertainty principle. The reflecting probability waves interfere as before.

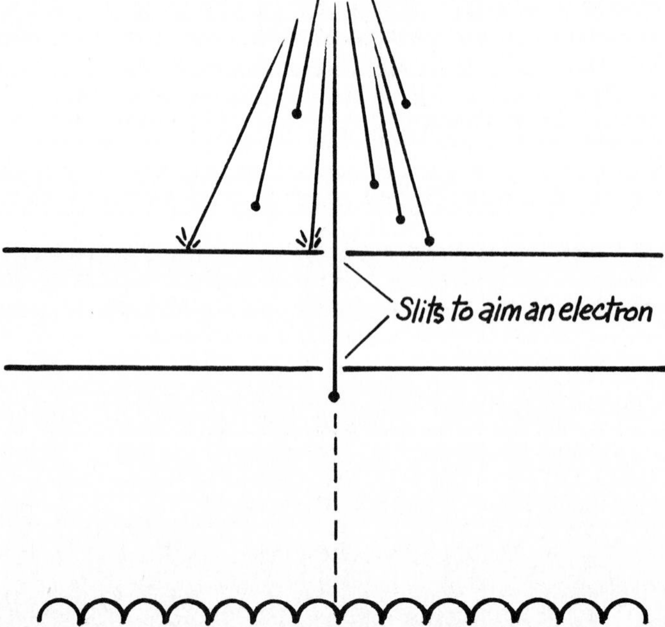


**FIGURE 25–11**
Pinholes in a pair of plates allow electrons to pass only if they are traveling in a single direction.

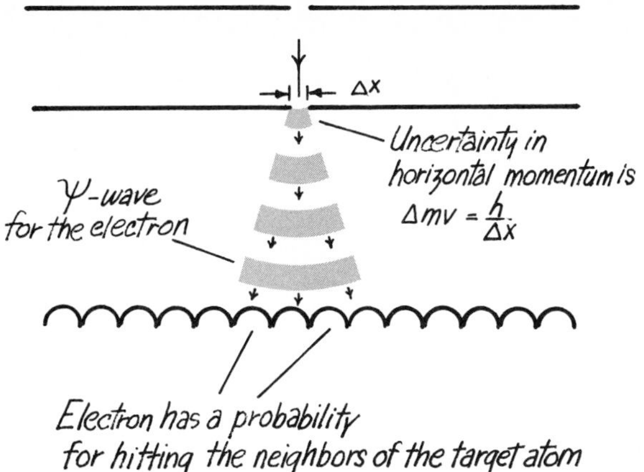


**FIGURE 25–12**
The uncertainty principle interferes with an attempt to aim an electron at an individual atom. Knowing the position of the electron well in the horizontal direction in this drawing means its momentum (hence speed) in that direction has a large uncertainty. That is, the $\psi$ wave diffracts at the opening, and the electron can be found to travel at a considerable angle to its original straight-line path.

The wave of a single moving particle, then, spreads as it moves along, reflects at any barrier, and spreads again after passing through any small openings. Though the $\psi$ wave is not observable or measurable, it diffracts and interferes much as water waves do. Wherever this wave appears, the particle has a chance to appear, and wherever the wave vanishes, the particle will not be found. Once an experiment finds an electron at some point, a new $\psi$ wave localized at that point can be followed with Schrödinger's equation to give the probabilities of the particle's behavior from that point on. And that probability wave, too, immediately begins to spread.

The sizes of the uncertainties in momentum or position are connected by Planck's constant, $h$. This is such a small number that for large collections of atoms like planets or people or even bacteria the uncertainties cannot be measured. But for individual electrons and protons and sometimes even whole atoms, uncertainty plays a large part in their behavior.

## Niels Bohr Brings Two Theories Together

After Niels Bohr published his 1913 theory of the hydrogen atom, he realized that the predictions of any new theory of nature at the subatomic level would have to agree with the observations of even larger particles. Otherwise, nature would be operating with two sets of laws, one for small particles and another for larger particles built from the smaller ones—and at some point in between, the physics would change abruptly.

After wave mechanics with its almost mysterious predictive powers came into being, Bohr was able to show that Schrödinger's equation predicts the same thing for everyday objects as Newton's laws do. In fact, when the mass and energy of an object are large (compared to the mass of a single atom, say), its predictions for that object boil down mathematically to *an equation that looks like Newton's second law.* Newtonian mechanics charts with great precision the motions and positions of moons and planets, pendulums, and falling apples. However, Newton's mechanics gives an upper limit to a theory that says we cannot know an object's speed and position with complete certainty. Niels Bohr's demand that wave mechanics and classical mechanics be logically consistent with each other was only a part of a general principle he promoted that is called the **correspondence principle.** Any two theories or hypotheses whose areas of prediction overlap should agree (or correspond) at those places; otherwise at least one of them is wrong.

## A Look Backward

Isaac Newton's laws of motion and his law of gravity were the very foundations of physics. Successful predictions followed whenever they were applied in the 1700s and 1800s, and the universe seemed to run like a clock at every level. For example, once the positions of the planets and their velocities were known at some instant, Newton's laws let physicists trace their histories both forward and backward in time.* Even historical eclipses could be predicted to check with old records. Physicists and philosophers alike naturally presumed atoms and molecules would follow the same rules.

The advent of wave mechanics, which contradicts this view, brought with it a reminder. Human beings produce the theories of physics. They use imagination to build abstractions that they pit against the cold, hard facts of nature found by other human beings. The theories are upheld only if they agree with the known facts and predict others. The abstract models of wave mechanics and the quantum theory of light passed tests that Newton's laws of mechanics did not. Historians of science have marveled that this revolution in understanding came to the most fundamental of the sciences in so short a time.

In the next chapter, we'll examine another revolution, one brought about largely by a physicist whose intellectual achievements are ranked alongside those of Isaac Newton. There, too, you'll see that our common ideas of our world don't serve to predict nature past our own familiar environment and sensations. The man was Albert Einstein, and the subject is space and time.

---

*The sole exception is a tiny precession of Mercury's orbit, which was explained by Einstein's general theory of relativity after 1915. Mercury's behavior was a chink in the brick wall of Newtonian mechanics.

## CALCULATIONS

EXAMPLE 1: **Find the wavelength λ associated with a nitrogen molecule in the air in the room you are in.** (The mass of a nitrogen molecule is about 28 amu, or about $28 \times 1.7 \times 10^{-27}$ kilograms, and its speed is about 500 meters/second.)

$$\lambda = \frac{h}{mv} = \frac{6.6 \times 10^{-34}\ \text{J·s}}{2.4 \times 10^{-23}\ \text{kg·m/s}} = \mathbf{2.7 \times 10^{-11}\ m}$$

This wavelength is only a little smaller than the diameter of the molecule, which is on the order of $10^{-10}$ meters. Recall from Fig. 16–16 that a wave diffracts more when passing through an opening that is about the size of its wavelength. If nitrogen molecules were passing through an opening about their own size, then you'd expect significant diffraction. Likewise, if a wave strikes an object, more scattering takes place if the object is about the same size as a wavelength. So the wave behavior of nitrogen molecules comes into play a little when two of them collide, for example.

EXAMPLE 2: Using the uncertainty relation and de Broglie's formula, **show that the uncertainty in momentum is large *only* if a particle can be located to within one de Broglie wavelength.**

$$\Delta x = \lambda = \frac{h}{mv}; \qquad \Delta x \times \Delta mv \geq h$$

If $\Delta x$ is approximately equal to λ, then

$$\frac{h}{mv} \times \Delta mv \geq h$$

$$h \times \Delta mv \geq h \cdot mv$$

$$\Delta mv \geq mv$$

When a golf ball is driven from a tee and soars quickly upward, its uncertainty in momentum is negligible so long as its associated wavelength is very small compared to the golf ball's size.

## REVIEW

**1.** Give de Broglie's equation for the wavelength associated with a particle of momentum $mv$.

**2.** Schrödinger's equation predicts the amplitude of matter waves, $\psi$, when the particles are moving under the influence of a force. True or false?

**3.** Schrödinger proposed that the density of matter waves represented the density of the particle. Born proposed that the density of matter waves represented the probability for finding the particle. Who was correct?

**4.** The Davisson–Germer experiment was the first to reveal the actions of matter waves by showing constructive and destructive interference in the reflection of electrons from planes of atoms in a crystal. True or false?

**5.** For objects like people and planets and even bacteria, does Schrödinger's equation predict the same motions as Newton's laws do?

**6.** As a consequence of matter's wave nature, a particle's momentum and position cannot both be known with absolute certainty. The uncertainty in these quantities is connected by $h$, a tiny constant, meaning cars and trees and pets are not significantly affected by these uncertainties. True or false?

**7.** What fundamental difference between wave mechanics and Newton's law led Einstein to say, "God does not play dice . . . "?

**8.** State the postulate of wave mechanics elaborated by Bohr.

# EXERCISES

**1.** Which of the following are true? (a) Bohr's model for the hydrogen atom predicted the hydrogen spectrum but failed to explain why the orbiting electron did not radiate. (b) Bohr couldn't explain why the electron should be found only on certain orbits. (c) De Broglie set out to find a better theory for the hydrogen atom. (d) De Broglie waves explained why the electron was found only on special orbits, or orbitals. (e) Electrons on orbitals in an atom don't radiate because the probabilities for finding them are given by standing-wave patterns that don't change unless the electrons make transitions between orbitals.

**2.** Name an experiment from the previous chapter that shows light behaves like particles when it interacts with electrons. Name an experiment from this chapter that shows electrons have wave properties.

**3.** Any molecule has a wavelength associated with it according to de Broglie's equation. To increase the wavelength of the molecule, must one (a) slow it down or (b) speed it up?

**4.** Does de Broglie's formula give any clue as to the amplitude of the wave associated with a particle?

**5.** Is an electron a wave? Explain.

**6.** Is it more correct to say that an electron is a particle found in atoms or that an electron is a wave cloud around an atom? Explain.

**7.** Sound waves move through the medium of matter. What medium does $\psi$ move through?

**8.** Give several facts that indicate a $\psi$ wave doesn't experience friction of any sort and die out as a wave on a clothesline does.

**9.** If a proton and an electron have identical speeds, how do their wavelengths compare?

***10.** What would be the uncertainty in the position of an electron that was known to have a speed of precisely 100 kilometers/second?

***11.** Would diffraction of particles play a larger role in our everyday lives if the value of $h$ were much larger? Explain with an example using $h = 10^{-3}$ rather than $6.6 \times 10^{-34}$.

**12.** An electron is confined to a shoebox. Can we know its speed with absolute certainty?

**13.** Explain why it is easier to predict future positions for a more massive particle than for a less massive particle.

**14.** Discuss what happens when atoms in a solid cool toward absolute zero and with less thermal motion their positions in the solid become less uncertain. (Atoms in a solid can never give up all their motion.)

**15.** It is a result of the uncertainty principle that electrons in the closest orbital to the nucleus cannot collapse onto the nucleus under its attractive electrical force. Explain why it would require extra energy to compress an electron in the ground state of the hydrogen atom into a smaller volume containing the nucleus. (So the uncertainty principle helps make the atom so huge compared to its constituents, the electrons and the nucleus.)

**16.** One of the effects the Schrödinger equation predicts about matter waves helps explain electrical conductivity. It shows that a free electron in a copper wire can often pass through a perfectly regular lattice of copper atoms without being reflected. Such an electron moves easily through the copper until it encounters a break in the regularity of the lattice, such as that made by an impurity atom, or a dislocation, or merely irregularities due to the thermal motions. In a copper wire at room temperature, an electron can travel past (or through) 100 atoms or more before it scatters; this means a copper wire has low resistance, or high electrical conductivity. Should this conductivity go up or down if the temperature of the wire decreases?

**17.** Because of diffraction, light won't reveal any details of objects that are about the size of the wavelength of light. Violet light has the shortest wavelength of visible light, about $3750 \times 10^{-10}$ meter, but electrons that pass through a potential difference of 75,000 volts have wavelengths of only $0.04 \times 10^{-10}$ meter, meaning they diffract far less than visible light. When used in electron microscopes, they reveal details of viruses and other objects far too small to form a clear image with visible light. Do you think an electron microscope could form an image of a single atom, with diameter about $1 \times 10^{-10}$ meter?

**18.** If $h$ were much larger, would the uncertainty principle play a larger role in your life? Explain with an example.

*19. Calculate the de Broglie wavelength of Lance the Cat (mass about 2 kilograms) running at 10 meters/second, and discuss the possibility of seeing wave behavior if you watched Lance streak past. (*Hint:* See Example 2 in "Calculations.")

*20. Calculate the de Broglie wavelength of an electron moving at $4.8 \times 10^5$ m/s, the result of moving through a potential difference of 54 V. (Davisson and Germer used electrons like these in their experiments. The planes of nickel atoms reflecting those electrons were about $0.9 \times 10^{-10}$ cm apart, and this explains why Davisson and Germer saw interference effects.)

# CHAPTER 26
# SPECIAL RELATIVITY

Back in class again. The lecture starts, and your ears and eyes catch messages carried by the sound and light coming from the front of the classroom. Luckily for you, the room is smaller than a football stadium where you might see a speaker's mouth move and then hear the sound later. In class you aren't distracted at all by the fact that light and sound travel at very different speeds. Light moves nearly a million times faster than sound, making it appear to travel instantaneously across earth's landscapes. Only when astronauts went to the moon did the finite speed of light introduce itself into the everyday world of humans. The electromagnetic waves used for communications (microwaves) took about 2.5 seconds for a round trip, causing a noticeable delay in the conversations between Mission Control in Houston and the moonwalkers.

Aside from its great speed, light's motion has another property that makes it fundamentally different from other types of wave motion. Albert Einstein derived a theory in 1905 based on that property, the special theory of relativity, that was as revolutionary as the wave theory of matter would be. Just as the waves of electrons and atoms are outside our everyday experience, so are Albert Einstein's remarkable discoveries. Only this time it is probably worse; relativity tells us that when relative speeds are very high, your common-sense notions about length and time and mass are all wrong. Fundamentally wrong. And when it comes to things as basic as your lifelong experience in the world around you, it's tough to be open-minded. Nevertheless, a short look at the world of relativity is worth your effort. Without going into derivations, the weird results of relativity can be appreciated through examples and arguments. ■

## What Does Light Move Through? The Ether

At the beginning of the 1800s, Young showed that light had wave properties. Beams of light could meet and vanish, much as sound and water waves do, when their crests and troughs cancel one another. In the

1860s Maxwell showed that waves of light are oscillations of electric and magnetic fields. But he was puzzled about the medium those oscillations traveled through. A wave travels along a stretched slinky because the spring is elastic; its tension tends to restore the spring from any displacement. Sound moves through air and water waves travel through water for the same reason—their media are elastic. Surely Maxwell's electromagnetic waves moved through some sort of elastic medium. Maxwell wondered about its properties in his papers. The elastic medium for light was given the name *luminiferous* (light-bearing) *ether,* but in the late 1800s physicists could only guess about its properties.

The ether's properties were bizarre. Since light arrived at earth from distant stars, this ether had to be almost everywhere, even between the stars where there is a very good vacuum. And it must be *thin,* their arguments went, or else its drag on the earth and other planets would cause them to slow and fall into the sun. (Surely any medium would resist something pushing through it.) Because light's speed is so great, they figured the medium was very *tense,* or stiff; if you pluck a guitar string or a clothesline, the wave you make travels fastest when the tension is greatest. The universe, it seemed, was filled with an extremely thin, astonishingly stiff, invisible stuff! The next step was to *detect* this ether. To see how the experimenters went about doing this, let's look for a moment at some properties of waves in another medium—water.

## Water Waves

If you've ever spent time sitting by a pool of quiet water in the woods, you've probably seen small waterbugs moving over the surface. Each kick of a bug's legs sends a wave outward in a growing circle. The water wave travels outward in every direction with the same speed—let's call it $s$.

Now if you find a slow-moving stream, deep and without surface ripples except those made by waterbugs, watch them from the bank. Sometimes a bug will kick just enough to stay in one place in the stream (Fig. 26–1), and you can see the effect of the slowly moving current on the circular wave. From your point of view (and that of the stationary waterbug), you'd see that the edge of the wave going *downstream* has a speed ($s + v_{\text{current}}$), while the edge going *upstream* has a different speed ($s - v_{\text{current}}$). The current—that is, the motion of the medium that the wave is traveling through—makes quite a difference to the relative speed of the water wave as seen by an observer. The same thing happens with sound waves in air.

Likewise, the argument went, light travels through the ether, and the motion of the ether past an observer should change the relative speed of the light. The light waves should either be swept "downstream" with the ether or travel more slowly if it moves "upstream." But the speed of the invisible ether could not be measured directly as the speed of a current of water can. However, there were instruments to detect how fast light travels, and in the last half of the 1800s, physicists tried to use them to find the motion of the ether. The differences in light's speed

**FIGURE 26–1**
Waves in water or sound waves in air or waves in solids move on or in a medium. If the medium has a motion of its own, the waves are carried with it. Here the speed of the current increases the speed of the wave in one direction and decreases it in another.

**FIGURE 26–2**
The idea behind an experiment to detect a medium that light travels through. Just as the speed of the water affects the relative speed of the wave in Fig. 26–1, if the medium for the waves of light (the so-called ether) was in motion, beams shining in opposite directions would have different speeds. Such experiments, however, have never found a difference in the speed of light traveling in different directions.

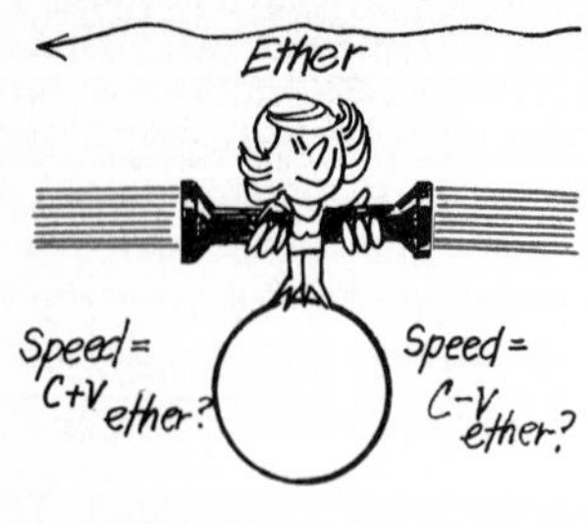


for different directions should show how fast the ether moved and in which direction, they argued.

## The Search for the Ether

The idea was simple: Shine a beam of light in one direction and measure its speed; then shine the beam in the opposite direction and find its speed again. If the beam moved *with* the ether, the speed would be $(c + v_{\text{ether}})$, where $c$ is the speed of the light waves though the ether. When the beam moved *against* the ether's motion, its speed would be $(c - v_{\text{ether}})$. If the experiment showed no difference in the speeds, the beams were probably traveling perpendicular to the ether's direction. Different directions would have to be tried until the beam chanced upon the ether's own direction of motion.

But doing this experiment proved too difficult. The problem had to do with light's great speed, more than 186,000 miles per second. If $(c - v_{\text{ether}})$ were only 100 miles per second less than $(c + v_{\text{ether}})$, the equipment of the day could not detect the difference between them. Finally, in 1887, two American physicists, Albert Michelson and E.W. Morley, found a way to measure to within a few miles per second any *difference* in the speeds light might have in various directions *without actually measuring the speeds themselves*. The first results were a shock; they found *no* difference in the speed of light in various directions. This brought up the question: Was the universal ether perfectly still at the earth's position in space when they did the experiment? This

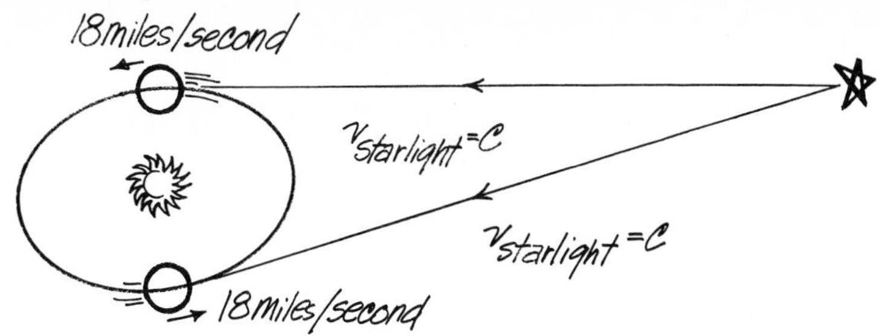


**FIGURE 26–3**
Another experiment to try to detect the effect of motion on light. The earth's motion carried the observers, Michelson and Morley, into the incoming starlight at one time and away from it at another. According to their instrument (which could detect very small changes in light's speed), the speed of the light from that star was unchanged regardless of the direction of earth's motion.

seemed improbable; but if it were so, all they had to do was wait and repeat the experiment. Six months later, they knew, the earth would be going in the opposite direction relative to when they first did the experiment. They did the experiment again, and again they found no difference in the speed of light in any direction. Since the earth's speed on its path around the sun is 18 miles per second, their instruments should have been able to detect such a difference in the speed of light.

They next asked if the earth could be dragging the ether along with it. In 1892 other investigators aimed light at the edges of a rapidly rotating mass to see if it would drag the ether closer to its spinning surface. They determined that the light was neither slowed nor speeded up by the rotating mass, ruling out drag as an explanation for the missing motion of the ether.

Meanwhile, Michelson and Morley used a distant star to see if its incoming light traveled at a different speed as the earth moved toward the star or away from it. They found no difference in the light's speed in any case. The behavior of light's speed that Michelson and Morley were observing is very different from anything in our experience, as the following thought experiment should help you see.

## What the Michelson–Morley Results Mean

Imagine you are in interplanetary space and you shine a flashlight in some direction. You could measure the speed of the beam as it leaves the flashlight, and you'd find it was traveling at $c$. Next, leaving the shining flashlight at rest, you could jump into a spaceship and accelerate alongside the beam until you are traveling at 30 percent of this speed, $0.3\ c$, with respect to the flashlight. Then while coasting at $0.3\ c$, you could use your measuring apparatus to measure once again the speed of that same light beam. Common sense says you should find the light traveling relative to you with 70 percent of the speed of light, or $0.7\ c$; all of the physics known to Galileo and Newton agrees with you. But the results of Michelson and Morley tell us that will not happen. The beam you are traveling alongside will still be moving with the speed $c$. You cannot reduce the relative speed of the beam by speeding along with it. Nor can you increase the relative speed of light by moving

**FIGURE 26–4**
Each observer in this thought experiment finds the light beam travels at the same speed, *c*, even though the observers have very different speeds.

toward the beam of the flashlight. *The relative speeds of the source and observer do not influence the speed of light.* Something seems terribly wrong with this the first time you hear it. The car doing 55 miles per hour that passes you while you are doing 50 miles per hour surely does not recede into the distance at 55 miles per hour as you look through your windshield—it moves ahead at only 5 miles per hour. And yet no matter how fast you chase a beam of light in space, it will still outrun you with a relative speed of *c*. This behavior isn't like anything anyone has ever experienced, but that is what the actual experiments with light (such as that by Michelson and Morley) show.

Albert Einstein took this invariance of the speed of light to mean something was wrong with the fundamental ideas of time and space, and he then made perhaps the boldest leap of imagination in physics. He showed how light's unchanging speed modified the old ideas, not just about space and time but about mass as well. His discoveries based on two postulates of relativity were a genuine revolution in scientific understanding.

## Frames of Reference and the Two Postulates of Relativity

Long ago, Galileo reasoned that people traveling in a ship with a perfectly constant velocity could not tell by on-board experiments if they were in motion. (There would have to be no waves at all for such smooth sailing.) If they rolled a marble, for example, it should go in a straight line with a constant speed so long as there was no net force on it; it would behave exactly as if it were on land. From the passengers' point of view, the ship could be at rest with the ocean moving past it.

Suppose the ship passed an island while the on-board experiment is taking place. The islanders could watch the marble and agree that it follows Newton's first law. The only difference in what the islanders and the shipmates see would be the addition (or subtraction) of the relative speed between the island and the ship. The passengers and islanders see (or measure) what is happening from different perspectives, that is, from different frames of reference. But the first law of motion holds (in the horizontal direction) for each of them because these are nonaccelerated frames of reference. Whenever Newton's first law holds—whenever the observer doesn't accelerate or rotate—we say the observer is in an **inertial frame of reference.** Galileo's observation can be put into simple words: *All inertial frames of reference are equivalent; that is, Newton's laws predict the motion of things equally well in any inertial reference frame.*

That certainly agrees with daily experience. If two buses traveling in opposite directions pass on a (smooth) highway, the passengers inside can see each other pouring tea, playing darts, or walking about as if they were all on the ground. Birds flitting about in cages on those buses might pass at relative speeds of 110 miles per hour, feeling quite at ease.

Einstein assumed this behavior extends to all the laws of physics, to everything we observe. The following statement is the first postulate of relativity, known to Galileo and to Isaac Newton, and restated by Albert Einstein in 1905: **The laws of physics are identical in *all* inertial frames of reference.**

Einstein's second postulate did away with the idea of an ether: **The speed of light through space is independent of any motion of the source or the observer.** In other words, a light beam travels at the same speed when measured by observers in any inertial frame of reference. Unlike other waves, light travels through space without a supporting medium of any sort. Apparently there is no medium hidden away in the vacuum of space. This postulate, Einstein showed, meant the old ideas of space and time and mass were wrong. A few thought experiments show the results of his special theory of relativity, which comes about logically from his two postulates.

## Strange Results

Two experimenters, a boy and a girl, take part in this thought experiment. The girl, positioned at the center of a long spaceship, glides low over level ground on an airless planet (with no air we won't have to worry about its effect on the speed of light or its drag on the spaceship). At the ends of the craft, fore and aft of her at precisely equal distances, are two light detectors connected to accurate atomic clocks. The girl's ship must fly with a uniform speed that is an appreciable percentage of the speed of light, say 50 percent or more, or else the relative differences we'll talk about would be very small.

At some point, the girl fires a flashbulb that sends a burst of light to be seen far and near in every direction (Fig. 26-5). Each detector records the time when the light from the flashbulb strikes it. Later, she checks the records of the detectors. By their clocks, each received the light at the same instant. The light covered the equal distances in equal times at the speed of $c$.

Suppose at the same time there is a row of similar light detectors on the ground directly beneath the flight path. They, as well, would

**FIGURE 26–5**
From her craft, the girl sees the landscape below her moving beneath the craft at a constant speed. A burst of light travels outward from the flashbulb at the center of her craft and reaches the light detectors on the ends at exactly the same time.

receive the burst of light and could record the time. Imagine a flagpole in the middle of the row of detectors. As the girl flies over, she could cause the bulb to flash directly above the tip of the flagpole (Fig. 26-6). The light spreads in all directions, once again, at the speed of $c$, and illuminates the two detectors on the ship. The closest ground detectors could note the exact time the light arrived to brighten the detectors on the ship. The boy could walk around and check the records of the ground detectors. He'd discover something different from the girl's findings. According to the detectors on the ground, the light did *not* arrive at the airborn detectors at the same instant. The rear detector on the ship brightened before the forward detector did, according to the ground detectors' clocks.

How did this happen? When the flash goes off at the flagpole, even though it is given off by a bulb moving along at a speed $v$ with the

**FIGURE 26–6**
Observers on the ground can make measurements of that same burst of light on the craft in Fig. 26–5. The flash of light is emitted over the flagpole as the craft moves by. That light travels outward at speed $c$ in every direction as measured by the detectors on the ground.

**FIGURE 26–7**
After a fraction of a second, the craft has moved to the right, and the burst of light has spread outward in all directions. As detected from the ground, however, the center of the burst is still over the flagpole. (Note that the spreading wavefronts have reached detectors that are at equal distances from the pole.) The detectors also observe the trailing end of the craft receiving the wavefronts before the leading end of the craft does. Compare this drawing to Fig. 26–5. Observers in different frames of reference (the girl in the ship and the boy on the ground) see the light reach the end points of the craft at different times because light's speed is independent of the speed of the light source or the speed of an observer.

girl's craft, the burst of light passes the ground detectors at the speed of *c* in each direction. As seen from the ground, the rear detector aboard the craft rushes toward this uniformly expanding shell of light, and the forward detector moves away from it (Fig. 26-7). The light hits the trailing detector first because the light goes a shorter distance than it does to catch the leading one. Hence, the ground detector records the trailing craft lighting up first. Yet the two on-board detectors register the light simultaneously as seen on their clocks; this lack of agreement is due to the second postulate of relativity. Light travels at the same speed over the ground as it does over the ship.

At this point, you may say, "The clocks must be wrong," or "They weren't set to the same time to begin with." So we'll consider just what it means to tell the time at different places and to set clocks to read the same.

## What It Means to Keep the Same Time

If you get together with some friends and put your watches side by side, and if the watches read the same time at the same instant, and if they stay precisely together as they tick off the seconds, we say the watches are keeping the same time.

Suppose some friends of yours had a watch that checked out at the get-together. Then they went home, some 2 miles across town. How could you be sure that their watch still agreed with yours? As they walked home, they moved away at some speed, meaning they changed inertial frames of reference. When they got there, their relative motion stopped, whereupon they rejoined your own inertial frame. But from what we have just seen about moving clocks, we might wonder if their watch would still agree with yours exactly.

**FIGURE 26–8**
Before the 1800s, each village and community across the country kept its own local time. Timepieces were set to 12:00 noon when the sun was observed to cross the local meridian, a north-south plane vertical to the ground. Time zones to standardize time by geographic region were encouraged by the railroads and enacted by Congress. (Cross-continental railway travel made the constantly changing local times a nuisance for the passengers and crews during east-west travel.) Portrayed here are a group of gentry simultaneously setting their watches to a gun fired at noon. (Or are they? See Fig. 26–9.)

Here is Einstein's prescription for making sure two separated clocks keep the same time. (He took great care in defining this notion when he published his theory in 1905.) Suppose you use a telescope to look at your friends' watch across town. You'd find it is ticking off seconds identically to your own; but if their watch is really keeping the same time, it will appear to show an earlier time than yours does at any given instant. This is because the light you see from their watch must travel the 2 miles to the telescope, and this takes time. If you want to make certain the watches really agree, you must know the distance and correct for the time the light needs to go that far. The distance covered is equal to the speed multipled by the time of flight, or $d = ct$. So if light travels a distance $d$, it takes an amount of time equal to $t = d/c$. This means that, in order for the watches to keep the same time, a watch a distance $d$ away should be set so that it appears through a telescope to read *d/c time units* behind the other watch. According to Einstein, they will then be keeping the same time.

If it bothers you to use light to help decide if the clocks read the same, there are other ways to do it. You could string a thin steel cord between the houses and agree to pull that cord at precisely 2:00 P.M. When your friends see the cord move at their end, they could check their watch to see if it is exactly 2:00 P.M. What's wrong with this experiment? A pull or a push at one end of anything, a steel cord, a long quartz crystal, or anything you can think of, does not start the other end moving instantaneously. Molecules at one end influence their neighbors,

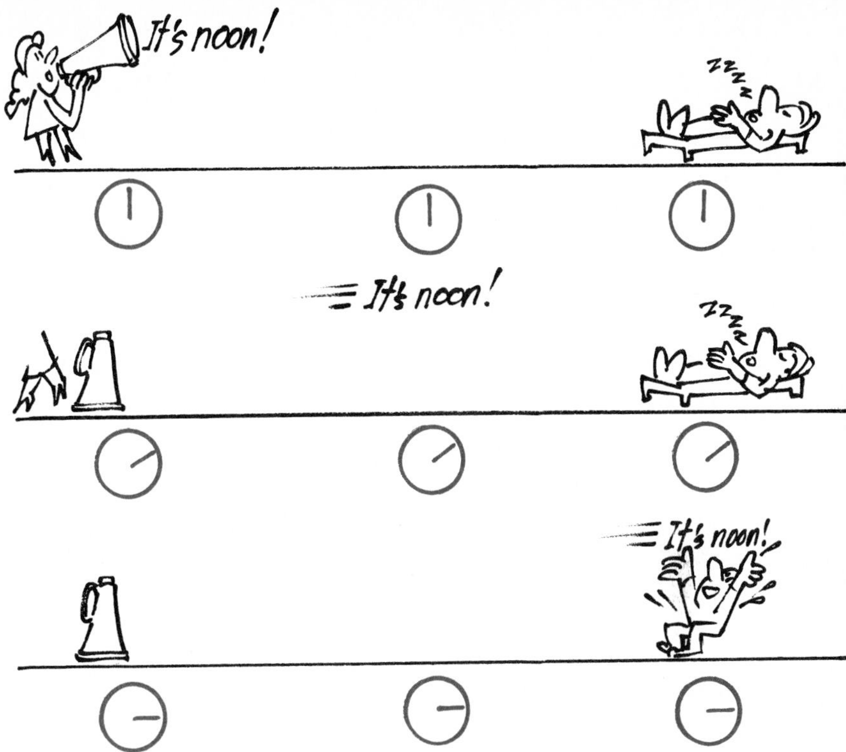


**FIGURE 26–9**
To make certain that clocks at different locations are set to the same time, so that they simultaneously read "12 noon," for example, the delay in communications must be taken into account. There is no method of communication that requires *no* time for travel between two locations.

and these in turn pass the push along, and so on. But just as in a long freight train, the car at the end doesn't move until some time after the engine starts to pull at the beginning. The important thing to understand is this: *To know the time at two different locations at once requires exchanging information in one way or another.* It doesn't matter that we "feel" time passes the same and "is" the same from one point to another in this world; we have to be able to measure it, to prove it.

## The Dilemma

Let's return to the flashbulb experiment. The boy could use Einstein's method to set all the clocks on the ground to read the same time, and the girl could do the same for the clocks in her spacecraft. But the results of the experiment would remain the same. The two could discuss this over their radios, and they would have to come to the following conclusions.

Let's remember, first of all, what the clocks moving over the ground do. The ground detectors see the rear clock on the ship receive the light first, and then a little later the rays catch the forward clock. Yet if the ground detectors record the times on the faces of the passing clocks as they are lit, the clocks will read the same time on each face—they *have* to, since the two flying detectors agree that they receive the light simultaneously. This tells us something immediately. *Detectors on the ground "see" the two clocks passing overhead as being out of sync, with the rear clock set ahead of the forward clock.* That is the only way these separate observations can agree.

But the detectors aboard the spaceship can observe the same sort of thing. At the instant the light strikes these detectors, they could record the time displayed on the face of the closest ground clocks. The girl could then say, "Aha, no wonder you have trouble. At exactly the same time in our frame, *your* clocks were out of sync. The clock on the ground that was next to my forward clock was, from my point of view, *ahead of* (in time) the clock on the ground that was next to my rear clock. That explains why you think our forward clock was struck at a later time." Observers in each frame of reference find that the other's clocks are out of sync, and the rear clock (or clocks) always leads the forward clock (or clocks) in time.

It is the constant speed of light that fouls up the notion that time should be the same for everyone else, even in other frames of reference. Both girl and boy know the clocks were running together in their respective frames of reference. How, then, can they both measure the same speed $c$ for the same beam of light when they are moving past each other? Einstein saw what this meant. Either that beam of light must be in two places at once or we have to accept new ideas about time and distance. He showed in 1905 exactly how time and distance must change when seen from different inertial frames of reference. The predictions he made based only on the two postulates of relativity have since that time been proved accurate by uncounted experiments. Now let's see what Einstein showed the girl and boy would discover.

## The Changing of Time and Space

The experimenters would find this next step easy. One of the flying clocks could be monitored by the detectors on the ground for a period of, say, 1 second. The ground detectors could record by their own clocks precisely when the flying clock's second started (tick) and when its second ended (tock). Of course, the ground detectors next to the flying clock at the end of the second would be far from the ones that saw its beginning. But all the detectors on the ground keep the same time, as checked by anyone standing still on the ground; so once again the boy could look at the ground clock's records. He would find that the moving clock ticks more slowly than his own; that is, *more than one second passes on the ground, in the inertial frame of reference where the observer is at rest, for every second that passes on the moving clock.* It is strange, but it is true. In the "Calculations" section we

TABLE 26–1

## THE RELATIVITY FACTOR

When a clock moves with a speed of $v$ past an observer, the observer would see (or detect) that clock moving more slowly compared to the observer's own clock. The dilation factor is $1/\sqrt{1 - v^2/c^2}$, which is listed in the right-hand column of this table. As an example, if a clock moves past at a speed of 0.5 $c$, or half the speed of light, then 1.15 seconds would elapse for the observer for each second that elapsed on the moving clock. Likewise, for every hour passing on the moving clock, 1.15 hours would pass for the observer who is at rest.

| $v$ | $\frac{1}{\sqrt{1 - v^2/c^2}}$ |
|---|---|
| Speed of space shuttle in orbit (about 29,000 km/hr or 18,000 mi/hr) | 1.0000000000036 |
| Speed of earth along its orbit (about 29 km/s or 18 mi/s) | 1.0000000047 |
| 0.1$c$ | 1.005 |
| 0.5$c$ | 1.15 |
| 0.8$c$ | 1.67 |
| 0.9$c$ | 2.29 |
| 0.99$c$ | 7.09 |
| 0.999$c$ | 22.37 |
| 0.9999$c$ | 70.7 |
| 0.99999$c$ | 223.6 |

discuss Einstein's formula for the precise difference in the rates of time, and some examples are given in Table 26–1. Notice that the faster the moving clock travels, the greater the stretching (or **dilation**) of its second as seen from the frame of reference of the observer. If the clock's speed is small compared to the speed of light, the effect is tiny. Otherwise, an ordinary car trip across the continent would cause watches to disagree noticeably (apart from any time zone changes). The results of the theory of relativity seem strange to us because they are completely outside our everday experience; only when relative speeds approach the speed of light do the effects of relativity become large.

The boy and girl can next compare the lengths of things as they move past. Suppose the girl stretches a chain between the rear detector and herself and flies over the ground. The detectors on the ground could mark where each end of that chain is at a predetermined time, say 10:00 A.M. sharp. Then the girl could stop her craft and bring the chain back and lay it down beside those marks on the ground. The marks would be closer together than the ends of the chain! The chain is shorter when it is moving than when it is at rest. *Whenever the length of a moving object is measured, it is shorter than it would be at rest.* We can see why if we watch this process from the girl's viewpoint. She would see a ground detector make a mark next to the forward end of the chain and another make a mark by the trailing end. But she would notice something strange going on. The detector that marked the front end of the

**FIGURE 26–10**
The length of a moving object contracts along its direction of motion when measured by an observer at rest, but the effect is extremely small unless the relative speed is close to the speed of light. This change in length is related to the fact that the observer's measurements of the end points of an object cannot take place at the same time as they do in the object's frame of reference. The length contraction shown here in an exaggerated way could be measured but not actually observed. (What one might "see" or photograph would be an image even more distorted since light reaches the eye or camera from different points along the moving object at different times, not simultaneously.)

chain would make its mark *first,* then the chain would move along for a short interval before the other detector marked the trailing end. The girl, then, would agree that marks on the ground should be closer together than the length of the chain at rest. She could also call "Foul." From her point of view, she is at rest and the line of detectors on the ground are rushing past her, her ship, and the chain. As with any line of synchronized clocks, when they are moving past an observer, the trailing clocks are ahead (in time) of the leading clocks. So she sees the front end of the chain being marked by a (trailing) detector when its clock reaches 10 A.M. *before* the (leading) detectors near the rear of the chain reach 10 A.M. and the position of that end of the chain is marked.

The width or height of a moving object, however, does not change. The contraction of a moving object occurs only along its direction of motion. For example, imagine a rod moving at a right angle to its length, with two experimenters riding on the ends. If a flashbulb goes off at the center of the rod, its light strikes both riders at the same time, according to their clocks. Observers on the ground also see the light travel an equal distance to each end of the rod (though the light travels at an angle to reach the ends; see Fig. 26–11). So this time even the ground observers agree the experimenters on the rod receive the light simultaneously. If the riding experimenters left paint stripes on the

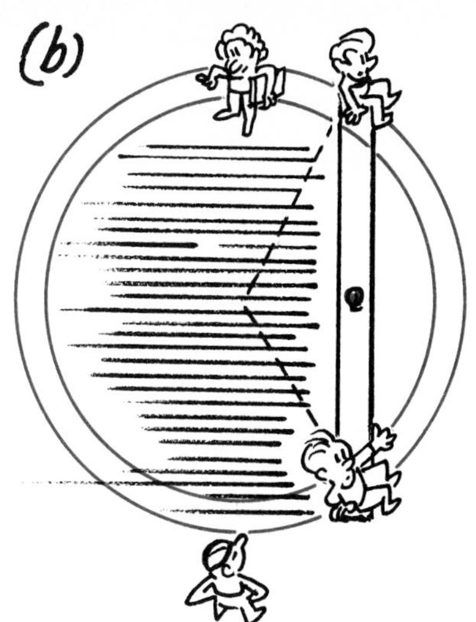


**FIGURE 26–11**
*(a)* A burst of light occurs at the center of a speeding craft moving sideways (that is, perpendicular to its length). There is a passenger at each end of the craft, and the burst occurs just as the craft passes two ground observers. *(b)* Observers on the ground agree with the observers on board that both ends of the craft receive the light at the same time. (Compare this with Figs. 26–5, 26–6, and 26–7.)

ground as they went along, then observers in *both* frames of reference would agree the actions happened at the same time. Because of this, a measurement would show the separation of the paint stripes on the ground to be the same as the length of the rod at rest.

## Mass and Energy in Relativity

Three centuries ago Isaac Newton defined inertial mass as the resistance of matter to a change in velocity with his second law, $\vec{F} = m\vec{a}$. The acceleration $\vec{a}$ is a change in speed per unit of time, and speed is distance over time. Because measurements of distance and time are relative to an observer, that changes Newton's second law. And since the definition of work is force times distance, energy too is affected by motion.

Einstein worked out the details for these changes and published his findings in 1906. He found that the mass of anything increases if it moves past you, according to the formula

$$m = \frac{m_0}{\sqrt{1 - v^2/c^2}}$$

where $m_0$ is its mass when it is at rest, that is, when its speed is zero, and $m$ is its mass as it moves past you with relative speed $v$. Table 26–1 shows what this mass formula of relativity means.* A bullet or rocket moving slowly past shows no noticeable change in mass; but at an appreciable fraction of the speed of light, a moving object becomes very massive. This is, remember, the mass you would measure as that

*Because the mass increases with speed by the same factor as the time dilates, the same table can be used for both. In other words, as you can see from Table 26–1, a 1-kilogram object moving past at $0.9c$ would have a mass in the reference frame of the observer (who is at rest) of 2.29 kilograms; likewise, 100 kilograms would increase to 229 kilograms, and so on.

rocket speeded past—not if it were stopped and then weighed. When it is at rest again in your own inertial frame, its mass is then $m_0$.

There is something else this formula tells us. What if a rocket ship or a cosmic ray or anything else passed you at exactly the speed of light? Its mass $m$ in Einstein's formula would become infinite because $m_0$ is divided by $\sqrt{1 - c^2/c^2}$, which is zero. Any number divided by zero gives *infinity* for an answer, an unattainable quantity. Let's see what this means.

If a rocket ship leisurely accelerates until it moves past you at $0.999c$ and then turns its engines on full blast, what would happen according to the mass formula of relativity? It wouldn't burst past the speed $c$ at all. As force was applied, the speed would go up very little despite all the work done by the rocket engine. Einstein found that the work done by the rocket goes into *increasing the mass*. No matter how much work is done on that rocket, it can never reach the speed of $c$. To push the speed of the rocket up to $c$ would take an *infinite* amount of work, and, of course, that much energy is not available anywhere in the universe.

So work goes into mass. In Chapter 6, we saw work change an object's kinetic energy, or go into frictional energy, which heats things up, or into the potential energy of the object. Does this mean there is some sort of equivalence between energy and mass? Yes, it does, and Einstein discovered the relation:

**energy equals mass times the speed of light squared**

or

$$E = mc^2$$

No doubt this is the most famous equation in all physics today—better known even than $F = ma$. If an amount of mass $m$ disappears in some reaction, it yields energy in an amount equal to $m$ times the speed of light squared. Since $c^2$ is such a large number, even a tiny mass represents a huge amount of energy. Mass is the source of energy in nuclear reactions, A-bombs, and H-bombs. And in those processes, only a tiny fraction of the mass present turns into energy. (In Chapter 28 we'll have a closer look at how this is done.)

## It Happens All the Time

The consequences of special relativity seem far removed from everyday experiences because everyone and everything we know share almost the same reference frame—the surface of the earth. Relative speeds here are usually a few miles an hour, which don't reveal the effects of relativity. The results of relativity seem contrary to experience, since we've never personally encountered relative motion near the speed of light.

Nevertheless, there are everyday examples of time dilation and length contraction. For example, cosmic rays strike the atmosphere of the earth from all directions day and night, traveling at speeds close to $c$. As they strike the nuclei of gas molecules in the upper atmosphere, the nuclei burst into small fragments that spray downward at high speeds. The particles that emerge from the shattered nuclei are not just protons and neutrons. Some are unstable particles that explode (or decay) after a short existence, turning into other particles. The *pi meson* is one such particle created in abundant numbers, and it quickly decays into less massive particles. Among these is the *muon,* another unstable particle that lives on the average hundreds of times longer than its parent meson. As these muons travel downward, we can measure the effects of relativity on them.

Muons created in physics experiments on earth last only about $2 \times 10^{-6}$ seconds in the frame of reference where they are at rest. But most of the muons created high in the atmosphere by cosmic rays travel downward at nearly the speed of light. If they decay on the average after $2 \times 10^{-6}$ second, they travel only about $d = ct = (3 \times 10^8 \text{ m/s}) \times (2 \times 10^{-6} \text{ s}) = 600$ meters before exploding. Detailed calculations show that for each sample of a million muons at an elevation of 9000 meters, none would be expected to reach sea level. But experiments show that about one-third of a million muons actually survive that journey. Relativity tells us why.

Since the muons travel at nearly *c,* their internal clocks run slower as seen from earth's atmosphere or the ground, and, from our point of view, they live (depending on their velocity) perhaps 15 times as long as they would at rest. That's why they can travel about 9000 meters before decaying.

But let's look at this another way, from the rest frame of the down-rushing muon. In their own frame, their clocks run normally, and they decay after $2 \times 10^{-6}$ second; but the thickness of the atmosphere, which they see rushing past with their own speed, is *contracted*. In fact, that 9000 meters of atmosphere, as we on earth measure it, is only about 600 meters long as seen from the muon's frame of reference. So the muons live to reach the earth's surface from *their* point of view too. It happens all the time.

## CALCULATIONS

Einstein's formula for length contraction of a moving object along its direction of motion is

$$l = l_0\sqrt{1 - \frac{v^2}{c^2}}$$

where $l$ is the observed length if the object is passing an observer at a speed of $v$ and $l_0$ is the length the observer would measure if the object were at rest.

Einstein's formula for time dilation of a moving object is

$$t = \frac{t_0}{\sqrt{1 - v^2/c^2}}$$

where $t$ is the time needed in the observer's frame of reference for a time interval $t_0$ to pass in a frame of reference that's moving past the observer with a speed of $v$. In other words, if $t_0$ is, say, 1 hour that passes on a clock in a spaceship moving past earth at speed $v$, then $t$ is the number of hours that elapse on earth during that time.

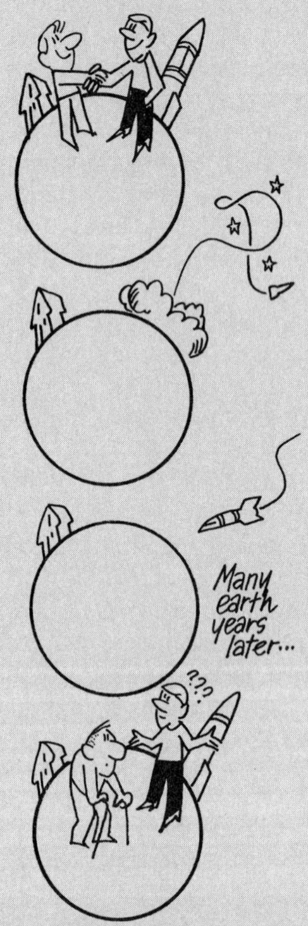


FIGURE 26–12

---

EXAMPLE 1: If a 1000-meter-long spaceship (as seen by the crew) passed over a colony on the moon at a speed of $0.95c$, **what length would careful measurements by the colonists reveal?**

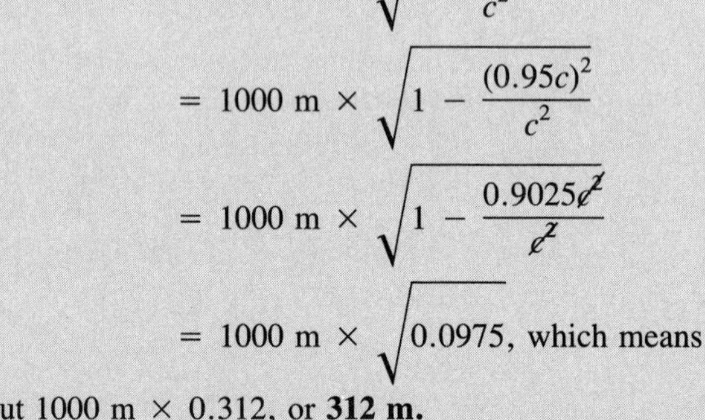

$$l = 1000 \text{ m} \times \sqrt{1 - \frac{v^2}{c^2}}$$

$$= 1000 \text{ m} \times \sqrt{1 - \frac{(0.95c)^2}{c^2}}$$

$$= 1000 \text{ m} \times \sqrt{1 - \frac{0.9025\cancel{c^2}}{\cancel{c^2}}}$$

$$= 1000 \text{ m} \times \sqrt{0.0975}\text{, which means}$$

$l$ is about 1000 m × 0.312, or **312 m.**

---

EXAMPLE 2: *The twin paradox*. Identical twins Peter and Paul were 20 years old when Paul left in a spacecruiser. After cruising at $0.99c$ for a year (by its clock), the ship turned around and came back to earth at $0.99c$. When Paul stepped out after landing, he was 22 years old. **How old was Peter?**

$$t_{\text{Peter}} = \frac{t_{\text{Paul}}}{\sqrt{1 - v^2/c^2}}$$

$$= \frac{2 \text{ years}}{\sqrt{1 - (0.99\ c)^2/c^2}}.$$

Referring to Table 26–1, $t_{\text{Peter}} = 2 \text{ years} \times 7.09 = 14.18 \text{ years}$

So Peter was 20y + 14.18y = **34.18 years old!**

*Paradox:* From Paul's point of view, *Peter* did the moving, so Peter's clock ran slower than his while the relative speed was $0.99c$. That is true. So why does Peter age more while Paul is gone? That's called the twin paradox.

Relativity gives the answer. Because the cruiser and Paul accelerated at takeoff, decelerated to turn around, accelerated to come back, and decelerated to stop and land, their frame of reference changed, while Peter and earth kept essentially the same frame of reference for the whole time. Their experiences aren't reversible, because each experienced something different. Paul felt the accelerations and changed frames of reference, but Peter didn't. *An analysis of a line of clocks placed along Paul's path would show that Peter spent extra time during Paul's accelerations and decelerations,* and the difference in elapsed time isn't entirely defined until they get back together again and compare clocks. We won't do the analysis here, but it is worth saying again that abundant experiments in atomic physics have proved that the predictions of relativity such as these are accurate.

## REVIEW

**1.** State the two postulates of relativity.

**2.** Relativity shows that the quantities of mass, length, and time are altered for any object that has speed relative to the observer. True or false?

**3.** Michelson and Morley found the ether's speed. True or false?

**4.** The length of a moving object is (a) shorter than, (b) longer than, (c) the same as its length would be at rest.

**5.** The mass of a moving object is (a) less than, (b) greater than, (c) the same as its mass would be at rest.

**6.** Whenever a second of time passes on the clock of a moving object, a time (a) less than, (b) more than, (c) equal to one second passes for the observer at rest.

**7.** Give the formula Einstein found for the relation between energy, mass, and the speed of light.

**8.** According to relativity, no material object can move as fast as the speed of light as seen from another object's frame of rest. True or false?

## EXERCISES

**1.** Which postulate of relativity is the most difficult for you to believe? Which postulate of relativity have you had personal experience with?

2. All starlight passes the earth at speed *c* regardless of the star's motions or the earth's. True or false?

**3.** Did Newtonian mechanics place a limit on how fast things could travel? Does relativity?

**4.** Can any everyday object move as fast as the speed of light past some other everyday object? Give a reason for your answer.

**5.** As a train approaches an intersection with a road, the light traveling down the track from its headlight travels at (a) $(c + v_{\text{train}})$, (b) $(c - v_{\text{train}})$, or (c) about $c$.

6. If observers on earth could watch an atomic clock ticking away on the moon, would they find the moon's clock ticking (a) faster than, (b) slower than, or (c) at an identical rate to an atomic clock on earth?

**7.** If the laws of physics weren't the same in all inertial frames of reference, discuss how a car trip might be very different from one you might ordinarily take.

**8.** The effects of relativity are not apparent in our world because the relative speeds we encounter are so much less than the speed of light. Discuss how things would be different if you could find a place where the speed of light in a vacuum was 10 meters/second, about the speed of a competitive sprinter.

9. Professor Phate decided to increase the value of his gold supply and launched it into space at a very great speed to increase its mass. Did his plan work? Explain.

10. As monitored by someone on the ground, is the time between heartbeats of people on a moving train (a) less than, (b) greater than, (c) the same as the the time between heartbeats of the same people when the train is stopped at the station?

11. Two ruffians sit 5 meters apart by the side of some railroad tracks. After synchronizing their watches, they simultaneously fire peas through peashooters, aiming straight at the side of a moving boxcar. Later, when the car is at rest, they measure the distance between the marks made by the peas. If they could make extremely accurate measurements, what would they discover?

12. A spacecraft moving between two stars a fixed distance apart travels at $0.999c$. Which statements are true? (a) Observers located at rest near the stars see the travelers age more slowly than they do. (b) The travelers feel they are aging more slowly than the observers. (*Hint:* See Example 2.) (c) The travelers see the distance between the stars as shorter than the observers do. (d) The observers at rest see the distance between the stars as shorter than the travelers do. (e) The observers at rest see the travelers moving about in slow motion compared to themselves.

**13.** Evaluate this statement: The distance from New York to San Francisco depends on how fast you travel.

**14.** Is the earth's mass increased because it is spinning?

15. According to relativity, a molecule that's moving has a greater mass than it has at rest. If you boil an egg, its mass is (a) slightly more than, (b) slightly less than, (c) the same as when the egg was taken from the refrigerator.

*16. Suppose you could look through a telescope and see

the watch of a friend who is 5 kilometers (3.1 miles) away. If they are keeping time simultaneously as defined by Einstein, by how many seconds should your watch be set behind your friend's watch?

*17. Find the energy in joules stored in the mass of 1 gram of water.

*18. When uranium nuclei split in a process called fission, about 0.001 of the rest mass, $m_0$, of the nucleus becomes energy that is carried away by the fragments. About how many joules of energy appear if one kilogram of uranium undergoes fission?

*19. Sirius A is a star that is 8.8 light-years (LY) from the sun. That is, it takes light traveling at $c$ some 8.8 years to make a one-way trip between Sirius and earth. Find how far in light years that distance would seem to someone in a spaceship traveling at $0.9999c$. (*Hint:* Use Table 26–1 and invert the value of $1/\sqrt{1 - v^2/c^2}$.)

## MORE DISCOVERIES AT THE TURN OF THE CENTURY

In the physics laboratories of the 1890s, some physicists had bad results with photographic plates. When developed, the plates sometimes turned out to be "fogged," just as if light had reached the plates from some source other than through the shutter of the camera. Searching for an explanation, Wilhelm Roentgen, a German physicist, wondered if the electrical discharge tube in his lab could be causing the accidental exposure of the film. (Discharge tubes were in common use in the physics labs of that time; see "New Beginnings," following Chapter 23.) He began to investigate what happened when cathode rays in a discharge tube were directed at different chemical compounds. He found that the rays caused many minerals to fluoresce.

One day he darkened the room so that the faint glow of the cathode rays was easier to see. When he threw the switch to activate the tube, something else in the room caught his eye. A piece of paper coated with fluorescent material and lying some distance away was glowing brightly in the dark. He carried the paper into the next room, and the glow continued—so long as the discharge tube was on. Roentgen had discovered a kind of radiation that could pass through a wall! He soon learned that heavier materials (heavy atoms), such as metals, stopped the radiation; yet it could easily pass through lighter things (that contained only light atoms), such as glass or water. He even directed these unidentified, or "X" rays, through his hand onto a photographic plate; and when he developed the plate, he saw images of the bones of his hand. Skeletal bones such as these are about 20 percent calcium, which is massive compared to carbon, hydrogen, and oxygen, the elements flesh is made from. Somehow the cathode rays, when they slammed into the matter at the end of a discharge tube, produced the X rays that traveled outward through the room. News of Roentgen's discovery, along with the photographs of his hand, was released in December 1895.

On January 20, 1896, the discovery of X rays was announced at a meeting of the French Academy of Sciences. Henri Becquerel, a French physicist, was in the audience. Becquerel was investigating the fluorescence of certain minerals when they were exposed to visible light. He wondered if Roentgen's X rays might be produced in the glow that visible light induced in his minerals. For a month he placed all kinds of naturally fluorescent crystals and ores in the sunshine to cause them to glow. Beneath them he put photographic plates wrapped in black paper to keep out the sunlight. When he developed the plates he found no fogging—until he happened to use a mineral that contained uranium. Placing coins under the samples containing uranium, Becquerel quickly repeated the experiments and saw outlines of the coins on the developed plates. The metal coins had stopped the radiation being given off by the uranium crystal. Becquerel, believing the sunlight had caused the radiation and that the radiation was probably the same as Roentgen's X rays, excitedly announced his results to the French Academy in February

1896. Both of Becquerel's assumptions were in error, but a real discovery was soon to come.

For several days after his report, the Paris skies were cloudy, and Becquerel could not do his experiments without sunlight. He tossed the minerals and paper-covered plates into a dark drawer to wait for a sunny day. Four days later he returned to his laboratory and decided to check some of his plates before resuming the experiments. When he developed them, he was amazed to find, not fogged spots, but black areas where the crystals had been touching the plates in the drawer. The uranium mineral was sending out rays without the aid of sunlight. Becquerel had accidentally discovered natural radiation, or **natural radioactivity.**

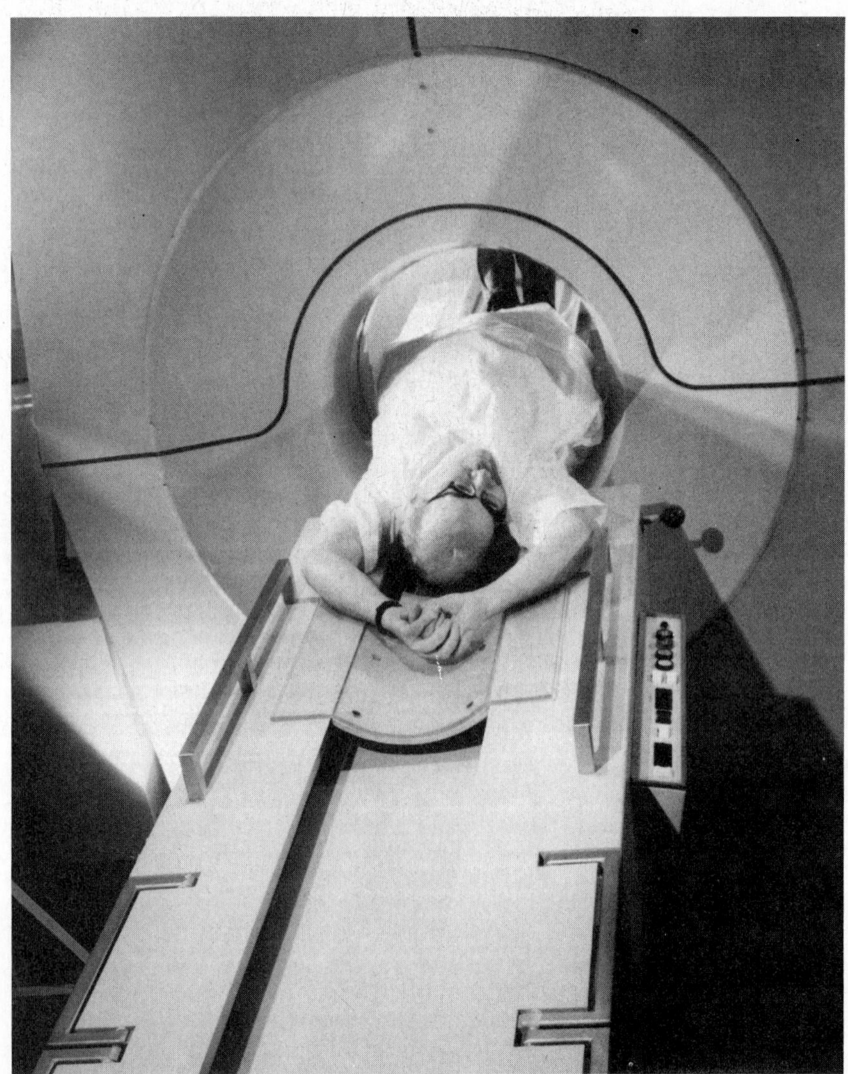

One of the legacies of Roentgen's discovery of X rays is this highly advanced computerized tomography (CAT) scanner, which takes detailed cross-section X-ray pictures of the human body in less than 5 seconds.

The natural radioactivity of uranium proved to be too weak to make pictures of bones, as Roentgen's X rays could do, and so most of the investigators of radiation preferred to work with X rays. But Becquerel pressed on with his research. When he found that a uranium ore called pitchblende affected photographic plates far more than it should have for the percentage of uranium it contained, he suspected that another, unknown radioactive element was present. He passed this idea on to a young Polish graduate student, Marie Curie, who was in search of a project for her doctorate. By early 1898 she and her husband, Pierre, managed to separate the element polonium (named in honor of her native country) from the pitchblende ore. Later in 1898 they also separated radium. The Curies shared a Nobel Prize in physics with Becquerel.

The discovery of polonium and radium, elements that are far more active than uranium, attracted the attention of Ernest Rutherford. Rutherford was born in 1871 in the then new settlement of New Zealand. The fourth of twelve children, Ernest did extremely well in school and upon graduation he applied for a new scholarship offered to students in England's colonial settlements by Cambridge University. His application came in second, but luck played a part. The student who placed first had to refuse the scholarship for family reasons, and Rutherford got the award. When the news arrived at his rural home, his mother went out to the garden where Ernest was working. Told of his good fortune, he threw down the hoe he was using and said, "That's the last potato I'll ever dig!" Much later Ernest Rutherford would be called the "Isaac Newton of the atom."

So it was that young Rutherford, an outspoken student from the newest of England's colonies, went to work for J. J. Thomson. He became a favorite student of Thomson's when he built a wireless apparatus to detect radio waves. He put the wireless set aside, however, to join Thomson in a study of how X rays ionize air and other gases, even though Rutherford believed the wireless invention would someday be valuable. It was, of course, and Guglielmo Marconi, an Italian electrical engineer, developed the wireless instrument that became the radio and made the money. Rutherford's experience with X rays paid off, however, when he began to investigate natural radioactivity. He found two types of radiation in the rays given off by polonium. One type ionized the air very strongly; the other ionized weakly by comparison. He named them alpha and beta rays, respectively. Gamma rays were discovered the next year, and so three types of natural radioactivity had been identified. In 1900 the Curies and Becquerel showed that beta rays have negative charge and the same mass as cathode rays. Both cathode rays and beta rays were electrons.

Working at McGill University in Canada, Rutherford (and Frederick Soddy, his co-worker there) discovered the details of the rate of decay of radioactive atoms and proposed that radioactivity caused an internal change in the atom itself. Rutherford and a chemist from Yale University, Bertram Boltwood, identified three chains, or families, of radioactive elements. He won the Nobel Prize in 1908 for these investigations. To Rutherford's surprise, the Nobel committee judged his work to be in chemistry rather than in physics, because it concerned the chemical elements. In his acceptance speech he related that in all of the transformations he had studied, he had never seen a faster one than his own transformation from a physicist to a chemist. In 1909 he showed alpha particles to be helium atoms that were missing their two electrons.

By this time Rutherford was a professor of physics at the University of Manchester in England, where he flourished. Called "Papa" by his students and co-workers, all of whom he called his "boys," he served tea and biscuits in his lab each day, and when things were going well he would burst into a stanza of "Onward Christian Solders." Research funds for his lab were often hard to come by, so he advised, "We've got no money, so we have to think." He and his colleagues established the area of physics called nuclear physics. In 1920 he predicted that a neutral particle would be found in the nucleus with a mass about equal to that of the proton, helping to account for the atomic weight of an individual atom. That particle, called the *neutron,* was discovered in 1932. But Rutherford's discovery of the nucleus of the atom was probably his greatest accomplishment, and that story is related in the opening section of Chapter 27. ■

# CHAPTER 27
# THE NUCLEUS

In 1908, the year Ernest Rutherford won the Nobel Prize, he and Hans Geiger were looking for a device to detect alpha rays.* They discovered that an alpha particle from a radioactive element would make a faint flash (or *scintillation*) when it struck a screen coated with zinc sulfide. Every scintillation showed precisely where an alpha particle landed, and Rutherford realized he could use such a screen to investigate how the alpha particles interacted with atoms. He sent a beam of them straight into a thin gold foil, and like bullets, they penetrated the foil to exit on the opposite side. He placed a zinc sulfide screen there to intercept them, and by observing how the alpha particles scattered as they passed through the gold atoms, Rutherford hoped to find clues to the structure of the atom. In 1909 the experiment was under way.

Geiger handled the counting because Rutherford was too impatient to sit still and record the tiny flashes. Geiger observed several thousand flashes. Almost all the alpha particles cut through the gold atoms in the foil like a hot knife through butter, deflecting only a degree or less. Wider deflections of, say, 5° were rare.

Geiger's results weren't unexpected. Only the year before, Rutherford had shown that alpha particles were totally ionized helium atoms, that is, helium atoms minus electrons. These alpha particles outweighed an electron by about 8000 to 1. At that time, electrons were the only identified constituents of atoms, and an energetic alpha particle certainly wouldn't be bothered by any electrons it hit. But Rutherford didn't drop the experiment; in the spirit of ''trying any damn fool thing,'' he asked a 19-year-old undergraduate student, Ernst Marsden, to spend some time looking for alpha particles deflected by as much as 45° (see Fig. 27–1). Geiger's apparatus was modified, and Marsden began his observations.

---

*Hans Geiger became famous for an electronic radiation detector, the Geiger counter, which he developed with W. Müller. Familiar to science fiction film buffs as the ticking machine that detects radiation, it is being replaced today by more sophisticated devices. But the ticking mechanism remains for portable, hand-held radiation detectors.

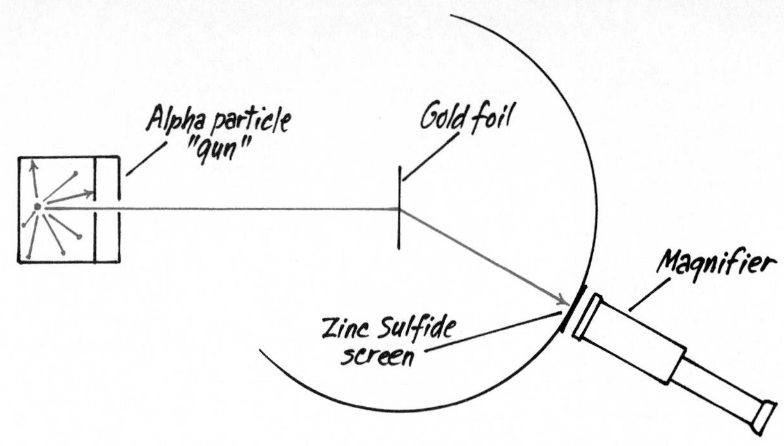


**FIGURE 27–1**
An arrangement to detect the scattering of alpha particles from the atoms in a thin gold foil. A movable screen that can swing around to any desired angle is used to intercept the alpha particles after they scatter. When they strike the screen, the alpha particles cause flashes, or scintillations, that an observer can see through the magnifier. By counting the flashes on the screen at various angles for equal amounts of time, Geiger and Marsden gathered statistics that led to a theory of how the particles must have interacted with the atoms.

By all accounts, the job was tedious. The equipment was in the basement of a building because the faint scintillations were best seen in the dark. The basement was cold and wet, with plumbing pipes both on the floor and overhead. Marsden had to sit in the dark for half an hour before his eyes were fully adjusted to the dark. Then he could peer through a telescope and see hundreds of flashes on the screen directly behind the gold foil. Swinging the screen and telescope away some 45°, Marsden saw a few pinpoints of light in the course of these sessions. But then Marsden did something else. He swung the screen and telescope around past 90° and found a few alpha particles were coming straight back from the *front* side of the gold foil. When Rutherford heard about this, he was astonished. He later said, "It was quite the most incredible event that has ever happened to me in my life. It was almost as if you had fired a 15-inch shell at a piece of tissue paper and it came back to hit you."

Then the serious work began. Geiger and Marsden counted over a million scintillations in order to record enough of the wide-angle events to form a statistically correct pattern of the scattering, and Rutherford sought a model of the atom that would predict what they were observing. He needed to explain the deflections as recorded by Geiger and Marsden. Knowing that the alpha particles would pass through several hundred gold atoms on their way through the foil, Rutherford knew the scattering was a statistical process. Though he was a brilliant experimentalist, he was not an accomplished mathematician, so he went to a statistics class. The students in that class in Manchester, England, were startled to find a Nobel laureate sitting with them and taking notes. Early in 1911, Rutherford proclaimed to Geiger, "Now I know what the atom looks like!" ■

**FIGURE 27–2**
The atom itself is immense when compared to the tiny nucleus at its center. Think of what this means. When you look at someone, you see them only because of reflected light from their atoms. And the electrons both reflect that light and account for the volume of atoms and molecules and hence the volume of that person. But it is the tiny nuclei that give that person inertia, that account for that person's mass. And it is the attraction of the protons in those nuclei for the atoms' electrons that actually keeps the atoms, and hence the person, together.

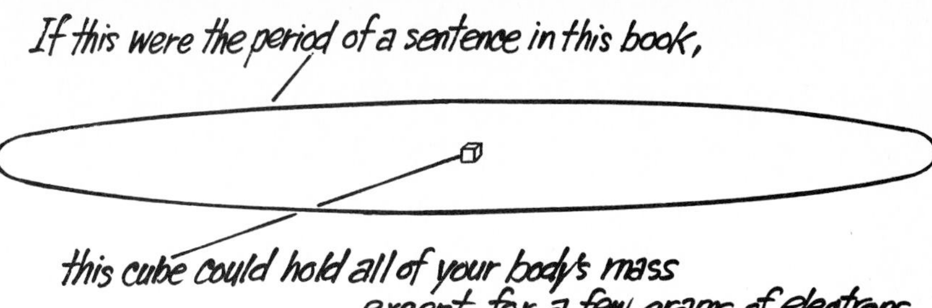

If this were the period of a sentence in this book,
this cube could hold all of your body's mass
except for a few grams of electrons.


## Properties of the Nucleus

Rutherford's findings were astounding. The atom had been pictured as a tiny, hard, billiard-ball affair, but he concluded that more than 99.9 percent of an atom's mass is in a mere speck at its center, the **nucleus,** occupying only about one part in $10^{14}$ of the atom's volume. The electrons orbit the nucleus in the remainder of the volume, leaving the atom with more empty space than anything else.

Because essentially all of an atom's mass is in the tiny nucleus, that structure is incredibly dense—about $10^{14}$ grams per cubic centimeter. In terms of weight, a cubic inch of nuclear matter weighs about a billion tons! Consider this: if the atoms in your body could be stripped of their electrons and just the nuclei packed into a box, that box would need to be only 0.0004 centimeter across (Fig. 27–2).* Your entire bodily mass (except for 15 grams or so of electrons) would fit into a cube that could not be seen with the naked eye.

At first impression, Rutherford's model of the atom did not seem to be in keeping with its physical characteristics, such as strength and incompressibility. However, the featherweight atomic electrons are firmly bound to the nucleus by the electrical attraction of its protons, and the exclusion principle and the uncertainty relation (Chapter 25) help keep them apart. So the atom has rigidity; any external push on the electrons is passed on to the heavy nucleus, which responds along with the electrons whenever there is a net push on the atom.

The small size of the nucleus was why Marsden found so few alpha particles at wide angles (only 1 in 20,000 was deflected by as much as 90°). But the positive charge of the massive gold nucleus could send any alpha particles that came directly at it straight back. Amazingly, Rutherford's conclusions were almost ignored by other physicists. Not until Niels Bohr's explanation of the hydrogen spectrum using the nuclear model in 1913 was serious attention given to Rutherford's results.

*The nuclei of your body could never be stacked this way, of course, because their positive charges cause a great repulsion between them.

## A Review of the Details of the Nucleus*

**FIGURE 27–3**
A representation of a copper ($^{63}_{29}Cu$) nucleus; the dark spheres represent protons, and the light-colored spheres, neutrons. This collection of spheres is a poor model for the nucleus, however. The nucleons are in rapid motion, and there exists no means to form an image of a nucleus.

The nucleus of the simplest atom, hydrogen, is a single proton, but the nucleus of any other element is a combination of protons and neutrons. The number of protons in a nucleus is its *atomic number,* and the number of its protons and neutrons together is its mass in atomic mass units, or its *atomic mass*.

Most atoms on earth are neutral, having one electron with its negative unit of charge to balance the positive unit of charge on each proton in the nucleus. Thus, in a neutral copper atom, a nucleus with 29 protons is surrounded by 29 electrons. These electrons, particularly those in the outer shell, govern the chemical and most of the physical properties of the copper atom. A neutral copper atom that loses an electron or two becomes a copper ion. This positive ion can easily pick up negatively charged electrons to become neutral again. But a nucleus with more or less than 29 protons *cannot* be a copper nucleus. Take away a proton, and its becomes a nickel nucleus (Ni), with atomic number 28. Add one proton, and the copper nucleus becomes a zinc nucleus (Zn), with atomic number 30.

On the other hand, the copper nuclei in a copper tea kettle or in copper home wiring have different numbers of neutrons. On the average, 7 of every 10 nuclei have 34 neutrons, so that the atomic mass of each is $29 + 34 = 63$. But the other three copper nuclei have two additional neutrons and an atomic mass of 65. These variations are called *isotopes* of copper. The symbols for the two isotopes of copper are $^{63}_{29}Cu$ and $^{65}_{29}Cu$. However, the atomic number is usually omitted from the symbol; if it is copper, it *must* have 29 protons.

Every chemical element has isotopes. The nucleus of the element hydrogen contains only one proton, but hydrogen has three isotopes, ordinary hydrogen ($^{1}H$), deuterium ($^{2}H$), and tritium ($^{3}H$). Other elements have even more isotopes: For instance, the tin (Sn) in ordinary tin cans has ten isotopes.

**The Attractions of Gravity, Magnetism, and Electricity, reach to very sensible distances, and so have been observed by vulgar Eyes, and there may be others which reach to so small distances as hitherto escape Observation. . .**

Isaac Newton

Except for the isotopes of hydrogen, every nucleus has more than one proton. Since like charges repel and the protons all have positive charge, the electric force acts in a way that would cause the protons to leave the nucleus explosively if there were not another, stronger force that acts to keep them there. This force within the nucleus, called the nuclear force, or the **strong interaction,** is an attraction between all *nucleons* (a nucleon is a neutron or a proton), but it is a force with very short range.† Experiments show that it is strong only when nucleons touch in the classical sense, that is, when their centers are about $2 \times 10^{-15}$ meters apart. In a uranium-238 nucleus ($^{238}_{92}U$), for example, a neutron or proton is in effect attracted only to the nucleons directly around it, its nearest neighbors. Experiments don't reveal a mathematical formula for the nuclear force; instead, they measure how energetically the nucleons in the various nuclei hold onto one another. In other words, we know how much energy it would take to pull the nucleons apart. We'll discuss this binding energy in the next chapter.

*This section reviews some of the material of Chapter 9.
†Atomic electrons don't feel the strong interaction at all.

## Unstable Nuclei

Most of the atoms of earth are **stable** (or *unchanging*) in their natural state. Of the 90 or so elements, there are about 190 stable nuclear isotopes, all of which have atomic numbers between 1 (hydrogen) and 83 (bismuth). The atoms of these stable nuclei give the matter in our world its permanence.

However, a tiny fraction of the atoms on earth are **unstable;** that is, they change, or **decay,** into different nuclei by emitting a particle of some kind. These nuclei are called **radioactive.** A nucleus will decay only if it can lose energy by doing so, and usually the particle carries away a great deal of energy from its parent nucleus. The heat in the earth's interior today comes from the decay of radioactive nuclei there; decaying atoms are small in number, but the energy released per decay is significant.

As you saw in ''More Discoveries at the Turn of the Century,'' between Chapters 26 and 27, the first three types of decay products to be identified from radioactive nuclei were alpha rays (helium nuclei), beta rays (electrons), and gamma rays (high-energy photons). Later some of the beta rays were found to be **positrons,** or *positive* electrons. The positron is called the *antiparticle* of the electron, because if a positron meets an electron, they are annihilated: Both disappear and two gamma rays are created which carry off their energy, including the energy of their mass, $E = mc^2$.*

We'll explore the reasons why some nuclei are unstable and why the unstable nuclei emit these four types of particles. But first we need to discuss the structure of the nucleus.

## The Shell Model

In Chapter 25 we saw that electrons and the actions the electrons take in atoms are described by wave mechanics (or quantum mechanics). The electrons' states are described by standing waves, which are sometimes called *stationary states*. In an atom these states correspond to shells around the nucleus. Electrons in the same shell have about the same energy, full shells don't react chemically, and an atom absorbs or emits photons by transitions of electrons between the shells.

The protons and neutrons in a nucleus also follow the rules of quantum mechanics. In 1949, using electron shells as an analogy, Maria Goeppert Mayer discovered a model that explained a great deal about the nucleus. She predicted shells of nucleons in the nucleus, something like the shells of the electrons in atoms. Each shell accommodates a certain number of nucleons, and these numbers are called the *magic*

*Antimatter was postulated by the British physicist P. A. M. Dirac in 1928 and discovered in experiments in 1932. Antiprotons, antineutrons, and even antiatoms have been created in experiments but do not seem to exist naturally. Physicists are still uncertain why the universe seems to be made of only normal matter, because the equations of physics don't seem to prefer matter over antimatter. In 1932, at the age of 30, Dirac was appointed to the chair once held by Isaac Newton at Cambridge.

*numbers*. Experiments have shown that nuclei with full outer shells are particularly stable, just as a full outer shell of electrons makes an atom nonreactive chemically. Similarly, all nucleons in a given shell have nearly the same energy. In 1963 she and J. Hans Jensen, who had independently come to the same conclusions about nuclear structure, shared a Nobel Prize.

## Gamma Emission

The shell model helps to explain how a nucleus loses energy by emitting high-energy photons, or gamma rays. Any nucleus except hydrogen's can be made to emit gamma-ray photons. For example, suppose a particle (perhaps from a cosmic ray shower) strikes a nucleus, giving the nucleus extra energy. As the nucleus absorbs the energy, a nucleon (or even more than one) may go into an excited state. As with electrons in excited states, such an excited nucleon returns to its ground state spontaneously. When the nucleon falls back into a lower energy state, a gamma ray takes away the energy the nucleon loses. Whereas an electron transition to a lower-energy state usually gives up a photon of visible light, a nucleon transition to a lower-energy state gives up a gamma ray. The difference in energy of these two corresponding actions is enormous because the nuclear force is so strong. A gamma-ray photon may carry a million times more energy than a visible photon does.

The shell model of the nucleus doesn't explain everything about gamma-ray emission. Whereas gamma rays are released when a proton in a nucleus has a transition to a lower-energy state, a gamma ray can also be emitted from a nucleus if a *neutron* jumps to a lower-energy state. Emission of electromagnetic waves is understood in terms of the acceleration of charged particles, but emission by neutral particles is a new wrinkle, not yet fully understood.

## Beta Decay

Beta decay of a nucleus, of either the electron or positron variety, stems from an entirely different process than gamma radiation. Furthermore, when beta decay occurs, the nucleus changes in character: It becomes a different nuclear species at the same instant the beta particle appears. To understand beta decay better, we first note an important fact about neutrons.

When a neutron finds itself *outside* any nucleus (as can happen when a cosmic ray hits a nucleus, for instance), *the neutron is unstable*. A solitary neutron survives on the average about 11 minutes before it disintegrates into a proton and an electron and another particle called an antineutrino.* We will say more about this particle in the section "The Little Neutral One." We can think of a neutron as a sort of metastable

*The neutron decays because of a manifestation of the electric force called the **weak interaction,** a force weaker than the electric force but stronger than gravity.

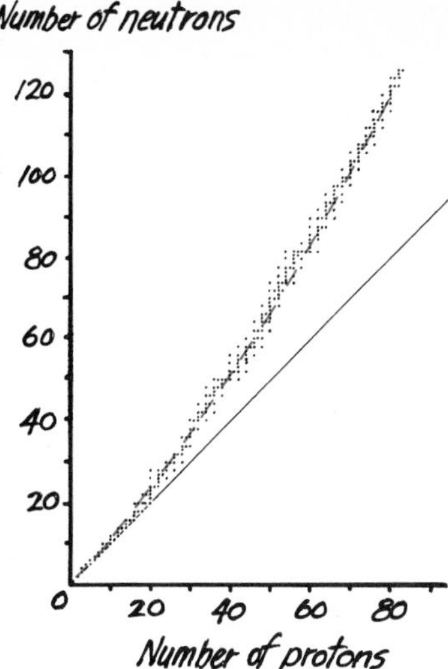


**FIGURE 27–4**
A plot of the number of neutrons versus the number of protons for the stable isotopes. The solid line is where a nucleus with equal numbers of protons and neutrons lies. All heavy and middleweight nuclei that are stable have more neutrons than protons. Such nuclei lie along an approximate line of stability indicated by the dashed line. Unstable nuclei to the left of that line tend to undergo beta decay; the resulting loss of a neutron and gain of a proton moves the daughter nucleus closer to the approximate line of stability. Elements to the right of this line tend to emit alpha particles, in the case of heavy elements, or positrons, and the daughter nucleus is more toward the dashed line of stability than its parent nucleus.

proton–electron combination that *can* be very stable when it is in a nucleus. But not just *any* nucleus. The stable nuclei in nature follow a distinct pattern. For low atomic mass, stable nuclei have almost equal numbers of protons and neutrons; high-mass nuclei need more neutrons than protons to be stable. A graph of the number of neutrons versus the number of protons reveals a *line of stability* for nuclei, shown by the dashed line in Fig. 27–4. Very often an unstable nucleus emits a negative or a positive electron to become a new nucleus that is closer to the line of stability. In either of these cases, a neutron is involved.

As an example, carbon-14 is a nucleus that undergoes beta decay because its neutron-to-proton ratio is too high. Such a nucleus is said to be *neutron-rich*. In essence, a neutron inside that nucleus decays into a proton and an electron and an antineutrino. The antineutrino and the electron escape from the nucleus immediately; neither is affected by the attractive nuclear force. But the proton remains, bound by the surrounding nucleons. The new nucleus has almost the same mass because the electron is light and the neutrino is (apparently) massless; but the atomic number of the new nucleus is one higher than the atomic number of the carbon nucleus because of the extra proton. We can describe its decay in equation form:

$$^{14}_{6}C \rightarrow {}^{14}_{7}N + e^{-} + \text{antineutrino}$$

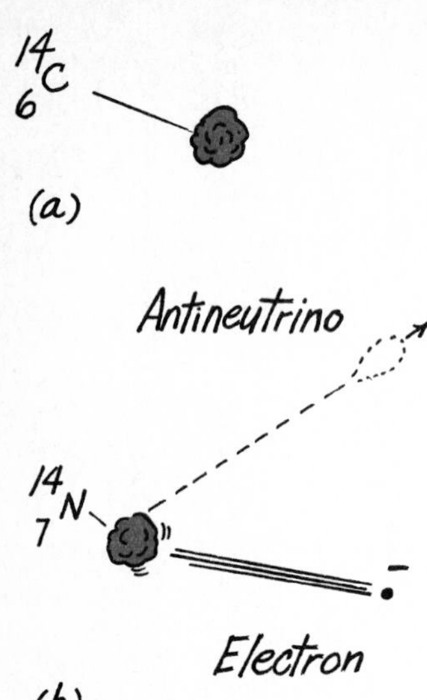


**FIGURE 27–5**
*(a)* A carbon-14 nucleus is unstable. When it decays *(b),* an electron, an antineutrino, and a nitrogen-14 nucleus appear.

where $e^-$ is the symbol for an electron. (See Fig. 27–5.) The carbon nucleus is called the **parent nucleus,** and the nitrogen nucleus is the **daughter nucleus.** The nitrogen nucleus has a lower neutron-to-proton ratio, which brings it toward the line of stability; it is, in fact, stable.

As another example, carbon-10 is a nucleus whose neutron-to-proton ratio is too low; such a nucleus is said to be *neutron-deficient.* Carbon-10 decays by emitting a positron and an ordinary neutrino from its nucleus. We can write its decay in equation form:

$$^{10}_{6}\text{C} \rightarrow {}^{10}_{5}\text{B} + e^+ + \text{neutrino}$$

where $e^+$ is the symbol for a positron and B is the symbol for the chemical element boron. This time the reaction involving the neutron is backward from before. A proton in the carbon nucleus, in essence, decays to become a neutron and a positron and an ordinary neutrino. The neutron stays in the nucleus, while the positron and neutrino escape. (A neutron is more massive than a proton; this means that as the proton disappears, extra energy must disappear with the proton to form the

## THE LITTLE NEUTRAL ONE

When a nucleus undergoes radioactive disintegration, a set amount of energy is lost in the process. This energy is shared by the decay products and the daughter nucleus (which recoils as the decay products leave). Furthermore, the laws of conservation of energy and momentum predict that if a nucleus decays by emitting only one particle, the decay particle and the daughter nucleus will share the energy and the momentum exactly the same way each time. The energies of the electrons emitted from a radioactive isotope undergoing beta decay, however, are not equal; the electrons have a range of energies as they emerge. When this was discovered, it seemed as if conservation of energy might not hold true for beta decay. Wolfgang Pauli, a German physicist, decided that was unthinkable. In 1930 he hypothesized that during beta decay an additional particle must be emitted along with the electron. Its properties, it turned out, make it one of the most unique particles ever imagined. The neutrino (''little neutral one'') has no mass and no charge and thus only very rarely reacts with matter. Having no mass, it travels, like photons, at the speed of light. Because it can leave the decaying nucleus in different directions and share the energy of decay with the electron and the nucleus in a variety of ways, the wide spectrum of energies for beta decay was explained with the neutrino. Neutrinos proved to be very hard to detect and were not found by experiments until 1956, even though they are emitted from the sun and from nuclear reactors in staggering numbers. These particles come close to being unstoppable; they can easily pass directly through the earth without deflection—in fact, if $10^9$ neutrinos were aimed at the earth's center, all but one would emerge unchanged on the other side. They also travel through the sun with impunity.

neutron. That energy comes from the energy stored in the nuclear bonds between the nucleons, from the binding energy of the nucleus. For this reason a proton outside the nucleus doesn't decay as a neutron does.) Boron-10 has a neutron-to-proton ratio on the line of stability; it is stable.

## Alpha Emission

In massive nuclei there are two competing forces at work. Nucleons that touch tend to stick together because of the nuclear force, but since the force is short ranged, the nucleons are attracted only to their nearest neighbors. On the other hand, the electric force is long ranged, so all the protons in the nucleus repel one another no matter how large the nucleus is. The electric force keeps a massive nucleus, one with many protons, under tension; if the electric repulsion is too great, the nucleus loses some of its positive charge. Uranium-238 provides an example. A

**FIGURE 27–6**
The three varieties of natural radioactivity. (Beta emissions can be either ordinary electrons or positrons.) A gamma ray that is emitted by a nucleus takes energy from the nucleus but no mass or charge. Therefore, the nucleus retains its identity after emission of a gamma ray. The other nuclear radiations, however, leave different (or daughter) nuclei after emissions. In alpha emission, the daughter has two fewer neutrons and two fewer protons than the parent nucleus had. In beta emission, the daughter has one fewer neutron and one more proton than the parent. In positron emission the daughter has one fewer proton and one more neutron than the parent nucleus. (Not shown are the antineutrino and neutrino.)

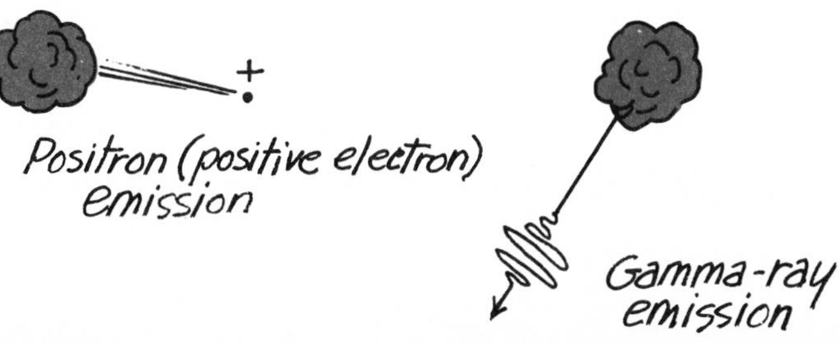

$^{238}_{92}U$ nucleus has 92 protons that repel one another. Eventually all uranium-238 will decay by losing positive charge in the process known as alpha decay.

To get rid of positive charge, the uranium nucleus doesn't simply shed one or more protons. The nucleus loses more energy in another way, by emitting an alpha particle, two protons bound to two neutrons. These tightly bound packages of two protons and two neutrons apparently exist within every large nucleus. More energy is released because the alpha particle has twice the charge of a proton, and so it is repelled by the emitting nucleus with essentially twice as much force. In addition and just as important, a more massive nucleus that loses four nucleons at once gets another bonus. Its remaining nucleons are bound tighter than they would be with the loss of fewer nucleons; the reason for this is subtle.

Once a massive nucleus loses an alpha particle, the nucleus that remains has a larger ratio of neutrons to protons (see Exercise 24). The protons in the daughter nucleus, then, are farther apart on the average than they were in the parent nucleus and being farther apart, experience less electrical repulsion. This extra padding (more neutrons per proton) tends to increase the stability of the nucleus against further alpha decay. All nuclei with atomic number greater than 83 (bismuth) are unstable because of the great electrical repulsion from within.

## The Lifetime of Radioactive Nuclei

When the nucleus of a radium atom decays, the alpha particle that escapes tears through the neighboring atoms, ionizing some atomic electrons and exciting others. As the electrons jump back down into lower-energy states, they emit photons of visible light, ultraviolet light, or even X rays. As a result, a specimen of radium literally glows in the dark. The glow is steady because at any time there are radium nuclei decaying. When Marie Curie first isolated radium from other elements, its steady rate of energy emission presented a puzzle. It appeared that the source of the radiation was constant and perpetual. But if that were true, the process would overthrow the principle of conservation of energy. The solution to this mystery came only after other radioactive elements were studied.

Ernest Rutherford and others experimentally isolated several radioactive elements that lost their radioactivity in days or even hours. Studying these, Rutherford (and Soddy, a coworker) became the first to understand what was happening in radioactive decay. Just as it is impossible to know precisely when a single excited atom will emit a photon, the time of decay of any given radioactive nucleus is unpredictable. Because of this, nuclear decays follow a statistical process: *A large collection of radioactive nuclei always loses the same fraction of its nuclei in the same amount of time.*

As an illustration, suppose it takes a year for 10,000 nuclei of some radioactive isotope to lose 10 percent of its nuclei. Years later, when there are only 1000 of the radioactive nuclei left, the specimen

**FIGURE 27–7**
*(a)* A rock containing a uranium-bearing material. *(b)* A photograph made in the dark by placing the film directly against the rock. The radioactive areas of the rock exposed the film.

*(a)*

*(b)*

will lose about 10 percent of that number during the next year. The more nuclei there are, the more closely the sample follows this rule.

A convenient benchmark in the life of a sample of radioactive material is the time it takes for half of the nuclei to decay. This time is called its **half-life.** If the half-life of an unstable isotope is 2 days, a sample of a million such nuclei would lose about 500,000 through decays in the first 2-day period. Over the next 2 days, it would lose about 250,000 of the 500,000 (half of the remaining nuclei), and so on. Figure 27–8 shows a plot of how samples of three unstable nuclei with different half-lives would decay as time goes by.

Rutherford then solved the mystery of why radium seemed to give off radiation at a constant rate. He found the half-life of radium ($^{226}Ra$) to be 1620 years; a specimen of radium has such a slow rate of decay that the change in the energy emitted per second was difficult to detect. Radium glows not because the number of decays per second is large, but because each nuclear decay emits so much energy. An alpha particle decays with as much energy as a million or more visible photons. Even when the radium sample's energy is spread over thousands of years, there is still enough energy released to make the radium glow warmly in the dark.

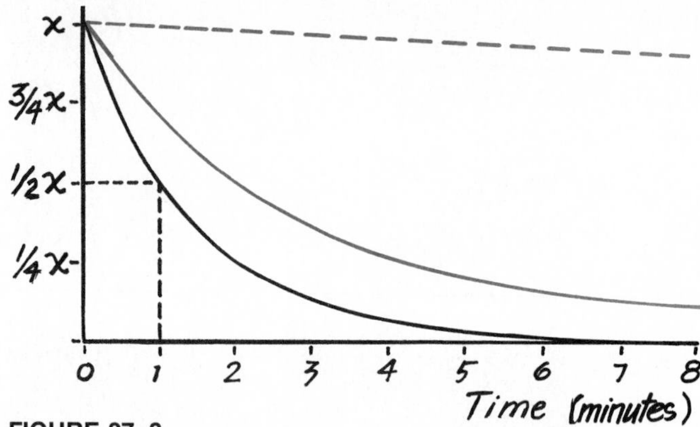


**FIGURE 27–8**
A plot to compare how three radioactive substances with different half-lives would diminish with time. (Solid black curve = 1-min half-life, solid color curve = 2-min half-life, dashed color curve = 50-min half-life.) Though the curve for each specimen differs, the quantity of nuclei in any given specimen tapers off by the same fraction over equal intervals of time. For example, the substance with a 1-minute half-life loses half of its initial number of nuclei in the first minute (see the black dashed lines that are parallel to the axes). During the second minute, it loses half of the number that remained after that first minute. Likewise, after 3 minutes, only half of the radioactive nuclei that were present at the end of 2 minutes remain, and so on. The longer the half-life, the slower the substance loses its nuclei, and the more level the curve that represents its decay.

## Secular (Long-Term) Equilibrium of an Isotope

Imagine what would happen if a machine could turn out nuclei of a radioactive isotope at a constant rate, say 100 per second. After 10 seconds, there would be 1000 nuclei, minus those that decayed during that time. Of course, the number that decay in 10 seconds depends on the half-life, but *it also depends on how many such nuclei are in the sample,* because a given fraction will always decay in a given time. As time passed and the total number of these unstable nuclei increased, eventually there would be enough nuclei (no matter what percentage of the nuclei decay per unit of time) so that 100 would decay every second. At this point, the rate at which nuclei were added would be equal to the rate at which they would disappear, and the *net amount of that radioactive substance would be constant*. Sometimes in nature a radioactive isotope is created at a constant rate, and such an isotope is present in a constant amount. Such an isotope is said to be in **secular equilibrium.**

Radium-226 decays by emitting an alpha particle to become radon,

**FIGURE 27–9**
To understand how the secular equilibrium of a daughter nucleus comes about, think of water being poured into a bucket with a hole near the bottom. The amount of water in the bucket increases until the pressure at the botton forces just as much water out per second as comes in at the top. The water level (and the amount of water in the bucket) then remains constant. In this illustration, a specimen of pure radium has a nearly constant number of decays per second, creating radon at a constant rate, analogous to the stream of water. The size of the hole in the bucket corresponds to the half-life of the radon. Thus, radon begins to accumulate, and the amount of radon present grows until just as many radon nuclei decay each second as are created by new radium decays. At that time, the net amount of radon present (mostly in the specimen of radium) is constant in time, or in secular equilibrium.

$^{222}$Rd, and because of radium's long half-life the number of decays per second from a specimen of radium is essentially constant over a human lifetime. That means radon, with a half-life of a mere 3.8 days, is created at a constant rate, and it is in secular equilibrium with its parent radium. The total amount of radon present because of radium decay is constant. (See Fig. 27-9.)

Another example of a radioactive isotope that exists in relatively constant numbers is found in the air. Cosmic rays crash into our atmosphere from space 24 hours a day. Hydrogen nuclei (about 90 percent), helium nuclei (about 9 percent), and heavier nuclei (about 1 percent) rain down at a more or less constant rate, and a certain percentage hit the nuclei of atmospheric molecules and blast them apart. Among the debris from the shattered atmospheric nuclei are neutrons. These neutrons often strike the nuclei of nitrogen atoms, which are plentiful in our atmosphere as molecular nitrogen, $N_2$. When this happens, the chances are good that a neutron will knock a proton out of the nitrogen nucleus ($^{14}_{7}N$) and the neutron itself will *be absorbed in the process,* thereby creating a carbon-14 nucleus ($^{14}_{6}C$). Carbon-14 is radioactive with a half-life of 5730 years. Carbon-14 is created at a reasonably steady rate and is, therefore, in approximate secular equilibrium; the amount of carbon-14 in the atmosphere remains relatively constant. The newborn carbon atoms combine with oxygen molecules to form molecules of carbon dioxide, $CO_2$. In the carbon dioxide in the air we breathe, there is roughly one carbon-14 atom for every trillion ($10^{12}$) carbon-12 atoms.

The steady percentage of carbon-14 in carbon dioxide in the air makes possible the technique known as carbon-14 dating of organic materials. Plants use carbon dioxide from the atmosphere (and emit oxygen) in their life processes, so any living plant has the same ratio of carbon-14 to carbon-12 atoms as the atmosphere does at that time.* Because all animals (including humans) depend on these plants through the food chain, they, too, have carbon-14 and carbon-12 in this same ratio. When a plant or animal dies, the exchange of carbon with food or atmosphere ceases, but the carbon-14 in the organic matter continues to decay. Gradually the radioactivity of the organic matter diminishes. The carbon-14 activity from a sample of old organic matter compared to the carbon-14 activity in the atmosphere at present indicates when the exchange of carbon dioxide stopped. Thus, the age of the wood from a sunken Spanish galleon, a scrap of bone from an ancient campfire, seeds from an Egyptian tomb, or the material from a Dead Sea scroll can be dated with fair precision—up to an age of about 50,000 years, after which time the carbon-14 activity becomes too small to measure reliably with today's instruments.

## DATING ROCKS

The half-life of uranium-238 is about 4.5 billion years. Uranium that is frozen in the solid structure of rocks decays, but the decay products are trapped at the same location. So the fraction of uranium in a rock compared to the fraction of its decay products gives an accurate dating of the age of the rock (the time since it became solid). With this technique geologists have learned that those rocks containing the oldest fossil shells of sea animals are about 500 million years old, and the rock strata surrounding the bones of our earliest ancestors date back about 4 million years. The first prize for the most ancient mountain formations goes to the Appalachians of the southeastern United States. The rock formations of Grandfather Mountain in North Carolina are 1 billion years old. But the oldest rocks on earth—some 4 billion years old—lie in Greenland. Moon rocks collected by the Apollo astronauts date back 4.6 billion years, only slightly preceding meteorite material that is 4.55 billion years old.

*Unless the organism is very long lived. Trees that live thousands of years have much less carbon-14 in their innermost rings, which are dead and so exchange no carbon with the atmosphere. The annual growth of rings of such trees provide corrections for carbon-14 levels over thousands of years and have brought dated artifacts from Egypt into excellent agreement with the calendars of the pharoahs.

## The Radioactive Series

All isotopes of the elements with atomic numbers greater than 83 are radioactive, emitting (almost exclusively) alpha and beta particles. These atomic species are disappearing. The heavy radioactive elements present on earth today have half-lives that are not much shorter than the age of the earth, or else they are the products of longer-lived isotopes. For example, radium nuclei that were present when the earth grew from materials orbiting the young sun some 4.5 billion years ago are long gone; they have all disintegrated. Radium's half-life of 1620 years guarantees that. Yet there is a constant amount of radium in uranium ore today, in secular equilibrium with its long-lived parent uranium.

The heavy radioactive isotopes found in minerals today occur in patterns, or series. The only way these isotopes lose significant mass is by alpha decay, when they lose four nucleons at once. Very often the large daughter nucleus that remains is also unstable and emits an alpha particle or a beta particle or a positron. If the second daughter is unstable as well, the process continues until what remains of the original nucleus is finally stable. For example, a nucleus of uranium-238 decays through a sequence of atomic masses. Without indicating the species of successive elements, the nuclear masses change as follows:

$$238 \rightarrow 234 \rightarrow 230 \rightarrow 226 \rightarrow 222 \rightarrow 218 \rightarrow 214 \rightarrow 210 \rightarrow 206$$

This is the way uranium-238 nuclei evolve into stable isotopes of lead ($^{206}_{82}Pb$).

Notice that each decay in the uranium-238 series skips three possible atomic masses; that is, between 238 and 234 there are possible

**TABLE 27-1** CONSECUTIVE NUCLEAR ISOTOPES IN THE URANIUM-238 SERIES OF RADIOACTIVE ELEMENTS

| ELEMENT | ISOTOPE | HALF-LIFE | TYPES OF RADIATION |
|---|---|---|---|
| Uranium | $^{238}_{92}U$ | $4.55 \times 10^9$ years | Alpha |
| Thorium | $^{234}_{90}Th$ | 24.1 days | Beta, gamma |
| Protactinium | $^{234}_{91}Pa$ | 6.6 hours | Beta, gamma |
| Uranium | $^{234}_{92}U$ | $2.48 \times 10^5$ years | Alpha |
| Thorium | $^{230}_{90}Th$ | $7.6 \times 10^4$ years | Alpha, gamma |
| Radium | $^{226}_{88}Ra$ | 1620 years | Alpha, gamma |
| Radon | $^{222}_{86}Rn$ | 3.8 days | Alpha |
| Polonium | $^{218}_{84}Po$ | 3.05 minutes | Alpha |
| Lead | $^{214}_{82}Pb$ | 26.8 minutes | Beta, gamma |
| Bismuth | $^{214}_{83}Bi$ | 19.7 minutes | Alpha, beta, gamma |
| Polonium | $^{214}_{84}Po$ | $1.6 \times 10^{-4}$ seconds | Alpha |
| Lead | $^{210}_{82}Pb$ | 22 years | Beta, gamma |
| Bismuth | $^{210}_{83}Bi$ | 5.1 days | Beta |
| Polonium | $^{210}_{84}Po$ | 138 days | Alpha, gamma |
| Lead | $^{206}_{82}Pb$ | Stable | — |

nuclear isotopes with masses of 237, 236, and 235. This means three *other* radioactive series can exist. In other words, if the atomic mass of each of these nuclei were reduced by four through an alpha emission, then four again, and so on, each could be part of another series that doesn't overlap with the others.

Two of these chains are found in the elements of earth. Uranium 235 decays through the sequence 235→231→227→223 . . . . and becomes stable at the isotope of lead ($^{207}Pb$). A natural isotope of uranium with a mass of 236 is nonexistent today, but thorium ($^{232}Th$) does occur and passes along the sequence 232→228→224→220 and so on to end with yet another stable isotope of lead ($^{208}Pb$). But the radioactive chain that would include 237, an isotope of neptunium, is not found on earth (except for its end product, the stable nucleus bismuth-209). The reason is that the longest-lived isotope in that chain, neptunium-237, has a half-life of only 2 million years. Since the earth is about 4.5 billion years old, whatever isotopes the earth may have had from that chain are gone, having passed through too many half-lives to survive.

## Nuclear Reactions

Both in the spring and at the "fall of the leaf," Issac Newton's secretary wrote, the great genius spent about 6 weeks in uninterrupted, intense investigations into the chemical changes of matter. He, like other alchemists of his day, probably tried to *transmute* chemical elements, that is, to turn one into another. Unlike Newton's work in physics, however, these sessions never paid off. The changes sought by these alchemists do not come from chemical reactions but from nuclear reactions.

Ernest Rutherford ushered in a new and vast field of physics when he first produced a nuclear reaction in 1918 (Fig. 27–10). He aimed alpha particles at a gas of nitrogen, and a few of the alpha particles struck the nitrogen nuclei and were absorbed. But as soon as this took place, the new nuclei ejected protons. Once again, Rutherford was able to grasp what these results meant. Given the proper subatomic particles for projectiles, he could probe the nuclei for more knowledge and perhaps transmute the isotopes of the elements throughout the periodic ta-

**FIGURE 27–10**
The first type of nuclear reaction produced by experiment. *(a)* An alpha particle collides with a nitrogen nucleus. *(b)* The nitrogen nucleus absorbs the alpha particle and emits a proton, transmuting the nucleus, which becomes an isotope of oxygen. In symbols, $^{4}_{2}He + ^{14}_{7}N \rightarrow ^{17}_{8}O + ^{1}_{1}H$. (The emitted proton is a hydrogen nucleus.)

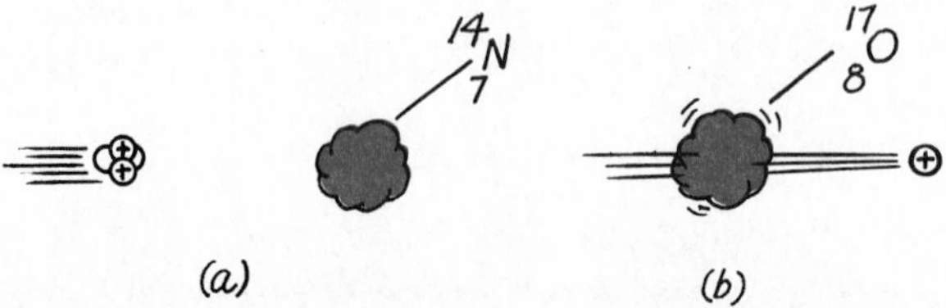

(a)

(b)

(c)

**FIGURE 27–11**
*(a)* An aerial view of Fermilab, at Batavia, Illinois. The circular ring is the main accelerator, with a radius of 1 kilometer. *(b)* The tunnel of the main accelerator at Fermilab. *(c)* The linear accelerator at Stanford University. The accelerator is 3 kilometers long.

ble. He had only the radiations from natural radioactive isotopes to work with for this purpose, but he began to think of a machine so designed that it could accelerate charged particles such as protons to such speeds that they could overcome their electric repulsion from the protons in a nucleus and penetrate it. He dreamed aloud of "a million volts in a suitcase" to accelerate such subatomic bullets to near the speed of light.

In 1932 two of "papa" Rutherford's "boys," John Cockroft and Ernest Walton, managed to build a device to accelerate protons, though it produced only 100,000 volts or so. When aimed at a lithium nucleus, the protons from their machine caused a transmutation to an isotope of beryllium not found in nature. With these experiments began the modern investigations of the properties of the nucleus. The first crude accelerators led to larger, more powerful machines and projectiles that can penetrate or even shatter a nucleus. Today these giant particle accelerators are the biggest machines on earth.

Over 100 *stable* nuclear isotopes have been created with accelerators, bringing the total number of stable isotopes to about 300. But even more unstable nuclei have been created with particle accelerators. Over 1300 artificial radioactive isotopes have been studied, and thousands of others are known to exist.

## Radiation and You

Your body is built from the food you eat and drink and the air you breathe, and because you come from the earth, you share its radioactivity. Everywhere in the rocks and topsoil of our planet are traces of the series of radioactive elements. Lighter radioactive elements such as hydrogen-3, carbon-14, and potassium-40 are created when incoming cosmic rays shred the nuclei of air molecules. Some cosmic rays (and their debris) penetrate the atmosphere to strike the ground, producing nuclear reactions that release even more radiation. But you should not be alarmed about all of this natural radioactivity, called **background radiation.** It is as much a part of your environment as sunshine and rain, and if our bodies couldn't tolerate it, we wouldn't be here.

Yet there is no question that radiation causes damage to molecules. When alpha, beta, and gamma rays penetrate matter, they ionize the molecules along their paths, an action that changes chemical properties. If a cosmic ray causes a transmutation in a nucleus in a molecule, the atom suddenly becomes an atom of a different element, which can play havoc with whatever molecule it is a part of. So we should be concerned with any additional radiation we receive.

Modern radiation detectors can monitor background radiation very accurately. For most of us, about 85 percent of the radiation we intercept each year comes from the natural radioactivity of the ground, air, and cosmic rays. Most of the other 15 percent comes from diagnostic X rays, color television, and extra cosmic radiation received during jet airplane travel—though 2 percent or so comes from fallout from atmospheric tests of nuclear weapons decades ago. (Since 1963 these have been banned by treaty by most of the major powers.) Nuclear power plants contribute only a minuscule amount to the 15 percent of nonnatural radiation we receive.

Here is a typical breakdown of the natural radiation. About 20 percent comes from the food and air you use, and at sea level about 30 percent comes from cosmic rays. About 35 percent on the average comes from the ground and the buildings you live and work in because construction materials come from the earth and contain traces of radioactive elements. When you are inside a building, these materials surround you, and you are exposed to this radiation from *all* directions, not just from the ground. For this reason, in a tent you get less than half of the radiation you get while inside a wooden building. Buildings made of concrete block or brick are a bit more radioactive than wooden buildings, and buildings of stone are the most radioactive of all.

At sea level the atmosphere is a good protective blanket against cosmic rays, but if you live high above sea level, your radiation profile changes. A person living in Denver, Colorado, is above much of the shielding atmosphere and gets more than twice as much cosmic radiation as someone who lives near sea level. Traveling by jet at an altitude of 36,000 feet puts you above 80 percent of the atmospheric shield and into the thick of the cosmic radiation. A round-trip flight between New York City and Los Angeles exposes you to the equivalent of half the dose of a chest X ray. Airline personnel are limited to a certain number of hours of air time a year because of this extra radiation.

In certain geographic locations there are concentrations of radioactive materials in the earth that raise the background radiation levels far above the average. Several of the locations are listed in Table 27–2. (See ''Calculations'' for an explanation of the units for radiation exposure.) Researchers have studied the population in Kerala, India, the hottest of these regions and compared them to a similar population group in another area in India with a normal background reading. Preliminary results show no disease or longevity difference between the people of the two regions, a good indication that such natural radioactivity levels don't statistically influence our health and well-being.

## CALCULATIONS

There are many units of measurement for radioactivity. Some deal only with the absolute properties of the radiation. The *curie* (Ci) is equal to $3.7 \times 10^{10}$ nuclear disintegrations per second. The *roentgen* (R) is the amount of radiation that will produce $2.58 \times 10^{-4}$ coulombs of electric charge per kilogram of air. The (absolute) unit that has been recommended as the SI unit of radiation is the *becquerel* (Bq), which is simply 1 disintegration per second.

Of more interest in medical therapy and radiation monitoring, however, is how much radiation energy is absorbed by tissue and bone. Until 1975 the unit used by almost everyone was the *rad* (*r*adiation *a*bsorbed *d*ose). One rad is 0.01 joule per kilogram. The recommended SI unit for absorption of radiation by people is the *gray* (Gy), which is equal to 100 rads. There is another unit that takes into account how damaging a given type of radiation is to living cells. This unit, the *rem* (*r*ad *e*quivalent *m*an), is equal to the rads absorbed times a *quality factor* that takes into account the effects of a certain type of radiation on living tissue. For example, a dose of 1 rad from X rays is not nearly as damaging as a dose of 1 rad from alpha particles: 1 rad from X rays is about 1 rem, while 1 rad from alpha particles is about 20 rems.

A safe radiation dose to someone in the general population is considered to be 0.5 rem per year, but federal laws restrict the average exposure to the general public to be 0.17 rem per year. However, people whose work involves

TABLE 27–2

**VALUES OF BACKGROUND RADIATION**

| LOCATION | MILLIREMS/YEAR |
|---|---|
| Average regions | 125 |
| France (granite regions) | 190 |
| Brazil (monazite regions) | 315 |
| Kerala, India | 830 |

exposure to sources of radiation are allowed by law to accumulate 5 rems per year.

Total *average* background radiation (cosmic rays, natural terrestrial radiation) for an average person amounts to 0.1 rem per year, or 100 mrem per year (1 mrem = 1 millirem = $10^{-3}$ rem). Such radiation is less at sea level and greater at high elevations. Approximately another 25 mrem comes from potassium-40, carbon-14, and other radioactive elements within our own bodies. So the populace at large in an average area is subjected to 125 mrem per year in natural radioactivity. Compare this to data given by the World Health Organization for hot spots on the earth in Table 27–2.

Total *average* radiation from other sources is as follows: Fallout from nuclear testing, 4 mrem per year; nuclear power plants, 0.003 mrem per year; diagnostic X rays (over half of us get one or more annually), 75 mrem per year; occupational, about 1 mrem per year. (Chest X rays are, or should be, on the lower end of the diagnostic scale, whereas some dental X rays are at the upper end. See Example 1.)

Incidentally, it takes some 5 mrem of radiation to fog a cannister of ordinary photographic film. X-ray devices for baggage inspection at airports in the United States supposedly subject bags to 1 mrem maximum. If you have a flight with a lot of stopovers, you might consider limiting your trips through the inspection line with a camera to four at the most.

EXAMPLE 1: A dental bitewing X ray gives a patient about 600 mrem of radiation. **About how many such X rays does it take to equal an average annual dose of natural background radiation?** Natural background radiation amounts to 125 mrem per year, so only **one-fourth** of such a dental X ray gives the patient the annual background dose. Remember, however, that people who work with radiation can legally have 5 rem (5000 mrem) a year. Compare that dental X ray to a chest X ray, which gives about 25 mrem to the patient and a single-tooth X ray that gives a dose of about 25 mrem.

EXAMPLE 2: If a powerful source of cobalt-60 gives 500 mrem per hour to a person who is 1 meter away, **how far away should that person stand to reduce the dosage to 10 mrem per hour?** The radiation travels out in all directions, like light, so its intensity diminishes as the inverse of the distance squared.

$$\frac{500 \text{ mrem/hr}}{10 \text{ mrem/hr}} = \frac{d^2}{(1 \text{ m})^2}$$

$$d^2 = 50 \text{ m}^2$$

$$d = 7.1 \text{ m}$$

## REVIEW

**1.** The nucleus contains the atom's protons and neutrons, also known as the nucleons. True or false?

**2.** Define atomic number and atomic mass number.

**3.** Individual neutral atoms of an element always contain the same number of **(a)** protons, **(b)** neutrons, **(c)** electrons, **(d)** nucleons.

**4.** Isotopes of an element have different **(a)** atomic numbers, **(b)** atomic masses, **(c)** numbers of protons, **(d)** numbers of neutrons.

**5.** Give the name of the short-range attractive force that holds the nucleus together.

**6.** Because the protons in the nucleus have like charge, they also experience another type of force. Which is it?

**7.** The more massive radioactive elements found naturally on earth emit radiation in the form of alpha particles. True or false?

**8.** In a specimen of radioactive nuclei, a precise number of nuclei always decay each second. True or false?

**9.** What is the name of the process whereby the number of unstable nuclei that are decaying is equal to the number being added by another radioactive substance or another nuclear reaction?

**10.** Natural background radiation from the earth and the atmosphere accounts for **(a)** a trace, **(b)** about ¼, **(c)** about ½, **(d)** more than ¾ of the radiation in an average person's environment.

## EXERCISES

**1.** When you look into a mirror, does anything of what you see come from the nuclei of atoms? When you push off to take a step, does that action have anything to do with the properties of the nuclei in your molecules?

**2.** Which particle can more easily approach a nucleus, **(a)** a proton or **(b)** a neutron? Why?

**3.** Why do most alpha particles pass through small thicknesses of any kind of matter without a great deflection in their path? How would you expect electrons of the same energy to behave?

**4.** Would you say that the decay of a radioactive nucleus is subject to the laws of classical (Newtonian) physics or modern physics? Why?

**5.** Why is the nucleus the seat of the most energy in an atom?

**6.** Early experimenters put samples of radioactive isotopes under high pressure to see if that would change the rate of decay. It didn't. Discuss why you should expect that result.

**7.** If a proton and a neutron have exactly the same speed and they are aimed at the same target, which will travel farther into the matter when it strikes it?

**8.** Which facts about atoms prevent you from causing nuclear reactions on your kitchen stove?

**9.** Tritium, ${}^{3}_{1}\mathrm{H}$, is radioactive. It decays by emitting an electron. What is the daughter nucleus?

**10.** If the nucleus is made of protons and neutrons, how can electrons be emitted from the nucleus?

**11.** Gamma rays can be used to sterilize food, which then keeps fresh at room temperature for times exceeding its safe storage time in refrigerators. Discuss how an intense gamma-ray source might be put to use in an underdeveloped country with little or no electricity available and no refrigerated trucks.

**12.** The only stable isotope of sodium is ${}^{23}\mathrm{Na}$. What sort of decay would you expect the neutron-deficient ${}^{22}\mathrm{Na}$ to undergo? What sort of decay should the neutron-rich nucleus ${}^{24}\mathrm{Na}$ undergo?

**13.** Fill in the decay products. See the periodic table in Chapter 9 to identify any missing elements.

$${}^{254}_{100}\mathrm{Fm} \rightarrow {}^{250}_{98}\mathrm{Cf} + \_\_\_\_\_$$

$${}^{212}_{83}\mathrm{Bi} \rightarrow \_\_\_\_\_ + {}^{4}_{2}\mathrm{He}$$

$${}^{39}_{17}\mathrm{Cl} \rightarrow \_\_\_\_\_ + e^{-} + \text{antineutrino}$$

$${}^{13}_{7}\mathrm{N} \rightarrow \_\_\_\_\_ + e^{+} + \text{neutrino}$$

**14.** Why can we still find short-lived isotopes on earth today?

**15.** Nuclear isotopes can be produced in any substance if it is exposed to radiation from a source such as a particle accelerator (or a nuclear reactor). Often the isotopes have short half-lives, making them valuable as *tracers*. Explain how such tracers could be used to locate leaks in an underground plumbing system or to monitor the wear of the rings of pistons in a car at a testing facility.

**16.** Sodium-24 is radioactive, with a half-life of 15 hours. Because of its short half-life, it is useful as a tracer for some purposes in the human body. Once in the bloodstream, its absence (detected by *lack* of decays) in some area will reveal blockages of circulation. Because of this short half-life, however, ${}^{24}\mathrm{Na}$ is not readily kept in stock. Suggest a way to make ${}^{24}\mathrm{Na}$ from ${}^{23}\mathrm{Na}$ at a hospital.

**17.** The gamma rays emitted from an excited nucleus

have definite frequencies, just like the photons emitted by atoms. That is, there is a nuclear emission and absorption spectrum just as the electron cloud has an emission and absorption spectrum. What similarity in structure does this come from?

18. One of the best techniques to identify atomic species in a sample of matter is called *neutron activation analysis*. The sample is irradiated with neutrons of medium energy, which merely bump into the nuclei and excite them. When a nucleus spontaneously loses its extra energy, it emits gamma rays of frequencies characteristic only to its spectrum. These nuclear fingerprints allow samples of pollutants in the air to be traced to their source, identify flecks of paint from hit-and-run accidents, and so on. Think of a way any product could be secretly protected against counterfeiting by a manufacturer using this technique.

19. Some nuclei will absorb a gamma ray and become so excited (or hot) that some of the neutrons are "evaporated" from them. Is such a gamma ray found in the emission or absorption spectrum of that nucleus? Discuss.

20. A nuclear physicist at the fair saw a child holding on to a buoyant helium balloon and remarked, "thanks to the uranium in the earth, that balloon is rising." What could she have meant?

*21. Show that if a nucleus is $10^{-5}$ atomic diameters across, the nucleus occupies $1/10^{15}$ of the atom's volume.

*22. Ashes from a campfire deep in a cave show a carbon-14 activity of only one-eighth the activity of fresh wood. How long ago was that campfire made?

*23. Uranium-238 has a half-life of $4.5 \times 10^9$ years. It's a coincidence that the earth appears to be $4.5 \times 10^9$ years old. How much uranium-238 exists in the earth today compared to the amount present when the earth was formed?

*24. $^{238}_{92}U$ decays by emitting an alpha particle, $^{4}_{2}He$. Form the ratio of neutrons to protons for $^{238}_{92}U$ and a similar ratio for $^{234}_{90}Th$ to show that the thorium nucleus will have a larger ratio of neutrons to protons than the uranium nucleus does.

25. Even though someone living in Denver gets about twice as much cosmic radiation as someone in New York, the total background radiation in Denver is only about 20 percent higher than the average at sea level. Explain the difference.

26. If you move to an elevation 100 feet higher than your present location, your background radiation goes up by about 1 millirem per year. Explain why.

*27. You add 1 millirem per year to your background level of radiation if you increase your diet by 4 percent. About how much of a radiation dose do you get per year just by eating normally?

# CHAPTER 28
# FISSION, FUSION, AND OUR ENERGY FUTURE

Those who first learned to start a fire—their names and places and dates—are lost in prehistory. But the consequences are clear. That chemical reaction brought a new world to our distant ancestors: In time they learned to use fire to smelt ores, bringing about the Bronze Age and then the Iron Age. Thousands of years later, the heat from their fires would drive metal steam engines and even later, with coal and then oil for fuel, their fires powered huge plants to generate electricity. Mastery of fire had its dark side, though. Those who fashioned plowshares could also forge swords, and the technology that built steam engines also built guns and tanks.

Then we learned to start another kind of fire. Its heat comes from changes in the nuclei of atoms rather than from rearrangements of their electrons, and it releases far more energy per unit of mass of fuel. Like the fire of the first type, it, too, has been used in war and in peace. While our mastery of the nuclear fire is not yet complete, the prospects for eventually harnessing the energy of the nucleus to satisfy the world's energy needs are excellent. However, we *do* know how and when and by whom this fire was discovered. ■

## The Discovery of Nuclear Fission

In 1938 Lise Meitner was forced to flee from the Nazis in Germany. As a physicist, she had worked in Berlin with a chemist, Otto Hahn, for over 30 years. She and Hahn were famous for their studies of radioactivity and nuclear transformations, and at the time they were studying the particles formed when the heaviest known natural element, uranium, was bombarded with neutrons.

Their interest in this came from previous work by Enrico Fermi, an Italian physicist who earlier fled Fascism to come to the United States. In Italy Fermi wanted to direct neutrons at the nuclei of each and every element on the chemists' shelves, because it was an easy way to bring about nuclear transformations. Unlike alpha particles and protons, neutrons are not repelled by the positive charge of the nucleus. Even slow-moving neutrons can enter a nucleus to create a new isotope that may or may not be stable. Before Fermi left Italy, he had bombarded uranium with neutrons, and he thought some of the particles emerging were unknown nuclei heavier than uranium. So Lise Meitner and Otto Hahn were trying to find such *transuranium* elements in their laboratory when she left Germany in the autumn of 1938. She went to the Nobel Institute in Stockholm, Sweden, and that winter her nephew Otto Frisch came to spend the holidays with her in the small Swedish town of Kungälv. Frisch was a young physicist in Bohr's Institute in Copenhagen, and he looked forward to telling her of his research—but he never got the chance. He later recalled his holidays as "the most momentous visit of my life."

On the first morning of his visit, Frisch walked out of his room to find his aunt poring over a letter from Hahn. After he read the letter, they took a walk in the snow to discuss what Hahn had found. Apparently uranium, upon absorbing a neutron, emits (at least part of the time) the element barium. This was totally unexpected. The uranium nucleus had split almost in half! The only nuclear transformations known until then involved emissions of only protons, electrons, alpha particles, or neutrons—small particles, in other words. Why should a small nudge from a single neutron cause the uranium nucleus of over 200 nucleons to split in two this way? (Later even Albert Einstein was to say, " . . . this was not something I could have predicted.")

Lise Meitner and her nephew paused in the middle of their walk and sat down to calculate what would happen. Two large nuclei, cleaved from a uranium nucleus, would push apart furiously because of their large positive charges, giving them *enormous* kinetic energy. When a carbon atom burns, it chemically unites with an oxygen molecule to make carbon dioxide. The rearrangement of the electrons to make the bonds releases 4.1 electron volts of energy, which appear as heat.* But the splitting (or **fission**) of one uranium atom's nucleus would release some 200 *million* electron volts of energy. They were astonished! Where could such energy come from? Lise Meitner knew the way to calculate the mass of the fragments, and a few minutes of scratching on scraps of paper from their pockets gave her the answer. The fragments from the ruptured uranium nucleus had less mass than the original nucleus. Almost 0.1 percent of the mass of the nucleus disappeared, which represents about 20 percent of the mass of a single nucleon. Here, then, was a very large-scale conversion of mass into energy. It was as if someone used a peashooter to fire a pea (the neutron) at a haystack and the pea set off a mighty explosion of dynamite (the uranium nucleus).

Otto Frisch went to Copenhagen to give Niels Bohr the news. He arrived only a few minutes before Bohr had to depart for a visit to the United States. Bohr listened while Frisch hurriedly told what they thought had happened in Hahn's experiment, and when he finished, Bohr exclaimed, "Oh, what idiots we have all been!" It was apparent

*The electron volt is the unit of energy most often used for atomic-sized processes. One electron volt is the energy given to an electron if it moves through a potential difference of 1 volt, and it is equal to $1.6 \times 10^{-19}$ joule.

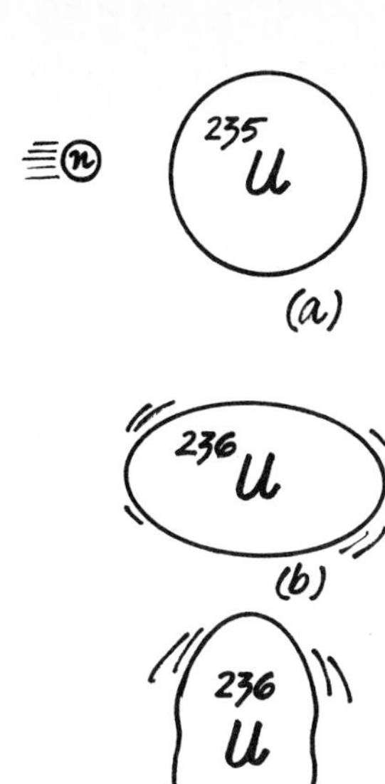

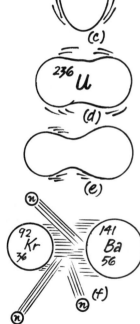

**FIGURE 28–1**
Steps in the fission of a uranium-235 nucleus. *(a)* The uranium-235 nucleus absorbs a slow neutron. *(b)* The neutron elongates the nucleus, even as it turns the nucleus into a uranium-236 isotope. In (c) and (d), the deformed nucleus oscillates as its surface tension tries to pull it back into a spherical shape. *(e)* But about 85 percent of the time the oscillations produce two large lobes on the nucleus that are connected by a smaller neck. *(f)* The mutual electrical repulsion of the protons in the two lobes then causes the nucleus to fission.

to Bohr at once how this reaction could take place because of an earlier model of the nucleus that he had put forth, as we'll see in the next section. Then Bohr sailed to America, and Frisch went to the lab and proved that this indeed was what had happened.

This is how the news of nuclear fission, the cracking apart of the uranium nucleus, came to America. Within a week after hearing of fission, American physicists had duplicated the experiments, and within a year almost 100 papers on this exciting and important discovery appeared in the physics journals of the United States, England, and Europe.

## Nuclear Fission

Niels Bohr's model of the nucleus drew on an analogy with the properties of a liquid. The nuclear force between the nuclear particles is effective over a very short range; so is the chemical attraction between the molecules of a liquid. In a small drop of water this attraction pulls the surface molecules inward, creating a surface tension that tends to shape the drop into a sphere. Bohr reasoned that the outermost nucleons in a nucleus would follow the same tendencies. His model had the nucleus acting like a liquid drop. If a neutron entered this nuclear drop, the surface might elongate and oscillate back and forth. If conditions were right, the quivering nucleus might stretch thin at its middle, whereupon the two positively charged lobes could repel each other enough to overcome the surface tension and tear apart.

Once in the United States, Bohr and the American physicist John Wheeler worked out the details with this model. They found that uranium 238, the most common isotope of uranium, could absorb a neutron without splitting, but that the rarer isotope uranium-235 had an 85-percent chance of splitting if it absorbed a *slow* neutron. (A fast neutron would not always cause the necessary oscillations; tap a bowl of Jello and it wiggles, but shoot a bullet through it and it hardly moves.)

## Another Surprise, with Sinister Implications

Lise Meitner's calculations had been correct. The fission of 4.5 grams of uranium-235 in a reactor today gives the amount of energy used by an average person in the United States in an entire year. The total energy used in the world in one year could be supplied by about 3500 tons of uranium-235. For comparison, the United States alone uses more than 600,000,000 tons of coal annually.

The large amount of energy released from the fission of uranium-235 nuclei was exciting. Uranium ores could be mined to obtain this energy, just as coal is mined. The uranium ore, once refined, would release energy if only some slow neutrons were aimed its way. The heat from the metal could turn water into steam that would drive steam turbines, which in turn would drive electrical generators—once again, just as with coal. No matter that those ores contain 99.3 percent nonfissioning uranium-238 and only 0.7 percent of the fissionable uranium-235; the energy release from uranium-235 fission was so great that its scarcity in uranium metal wouldn't prevent the development of such a fission energy plant.

But several weeks after the discovery of fission, another possibility dawned on some of the physicists who were working with this new nuclear process. When a uranium nucleus splits, other particles will splinter off; in particular, neutrons are certain to be emitted. Suppose a uranium-235 nucleus splits and its large fragments are a barium nucleus and a krypton nucleus. The massive uranium nucleus has more neutrons than the total number of neutrons two medium-sized nuclei ordinarily have. (Recall Fig. 27–4, which shows the higher neutron-to-proton ratio for heavier elements.) There are two possible ways neutrons could appear in fission: as splinters in the explosion of the uranium nucleus, or from unstable daughter nuclei that would shed neutrons quickly to become more stable. Both actually happen.

What this release of neutrons meant jumped out at the physicists. When a slow neutron causes a uranium-235 nucleus to fission, more neutrons emerge in the process. Under the right conditions, these fission neutrons could go on to cause nearby uranium-235 nuclei to fission. Those nuclei would release more neutrons to cause more fissions, and uranium could set itself off in a chain reaction!

So why doesn't a chain reaction proceed in uranium ore in surface deposits over the earth?* Slow neutrons are present at the earth's surface all the time, because of cosmic rays. Yet when one of those neutrons strikes a uranium-235 nucleus, it doesn't start a chain reaction. First of all, a uranium-235 nucleus fissions only if it captures a *slow* neutron, but the neutrons it emits are *fast* neutrons. These fast neutrons, having no charge, don't interact with the electrons, which aren't at all influenced by the neutron's nuclear force. But they do collide with other nuclei, which slow them down, especially if the nuclei struck are small. (Shoot a marble at a bowling ball, and it bounces off almost as fast as it came in. Throw that same marble at another marble instead, and it can lose much of its energy during the collision, giving it to the target marble.) However, there are many species of nuclei in uranium ore that may absorb neutrons, and in each ton of uranium ore there are only a few ounces of uranium-235. For a chain reaction to proceed, some of the fast neutrons from fissions must lose their speed by bouncing

*It may have once! See *Scientific American* for July 1976.

around, a process that takes about 1/1000 second, and then most of them must strike a nearby uranium-235 nucleus before other, nonfissionable nuclei catch them.

On the average, uranium-235 nuclei emit 2.5 neutrons per fission, so that a chain reaction is possible in a refined (concentrated) sample of uranium. If only one neutron released during each fission eventually causes another fission, a chain reaction takes place, and it will be self-sustaining. If less than one neutron per fission causes another fission, the chain reaction dies out. But think of what happens if more than one neutron per fission goes on to cause another fission. The number of uranium nuclei exploding each second grows, leading to a *runaway* chain reaction. So much energy is released so fast an explosion takes place, and that reaction was the basis for the first nuclear weapons.

In 1939 the construction of a nuclear bomb from the fissionable uranium-235 seemed possible. To achieve a runaway chain reaction, a sample of uranium must meet two criteria. First, the sample must contain enough uranium-235 per unit of volume for the slowed-down fission neutrons to cause more fissions rather than to be absorbed by other nuclei. This means the uranium sample should be nearly pure uranium-235 nuclei, because, as it turns out, uranium-238 nuclei are good absorbers of fast neutrons, which change them to uranium-239 nuclei. Finding a method to concentrate uranium-235 was a tough technical challenge. Chemically, uranium-238 and uranium-235 behave in exactly the same way, so other, nonchemical means had to be used to separate these isotopes. This was carried out in the Manhattan Project during World War II, the most expensive technological undertaking in history.

The second criterion had to do with the quantity of uranium-235 needed to initiate a runaway chain reaction. If the specimen is too small, many fission neutrons will escape from its surface and less than one neutron per fission would cause another fission—hence no chain reaction. Yet the instant a certain critical mass of uranium-235 is assembled (and the first fission occurs), the reaction runs away almost instantly, causing an explosion.

## Commercial Nuclear Reactors

Nuclear fission has long been harnessed (since 1957) in nuclear reactors to provide thermal energy to generate electrical power. Nevertheless, nuclear energy and nuclear reactors are often misunderstood by the public. It is impossible for a commercial nuclear reactor to undergo a nuclear explosion, and today they are designed to contain the worst internal accident. Only if the double containment buildings are breached—presumably by a major catastrophe of nature or possibly an act of sabotage—would a significant quantity of radioactive particles be released from a reactor into the nearby environment. That possibility, of course, is still a source of concern.

A nuclear reactor is a very simple machine (Fig. 28–2). Enough uranium metal is brought together to initiate a chain reaction, but the chain reaction is controlled. The object is to get an average of one neu-

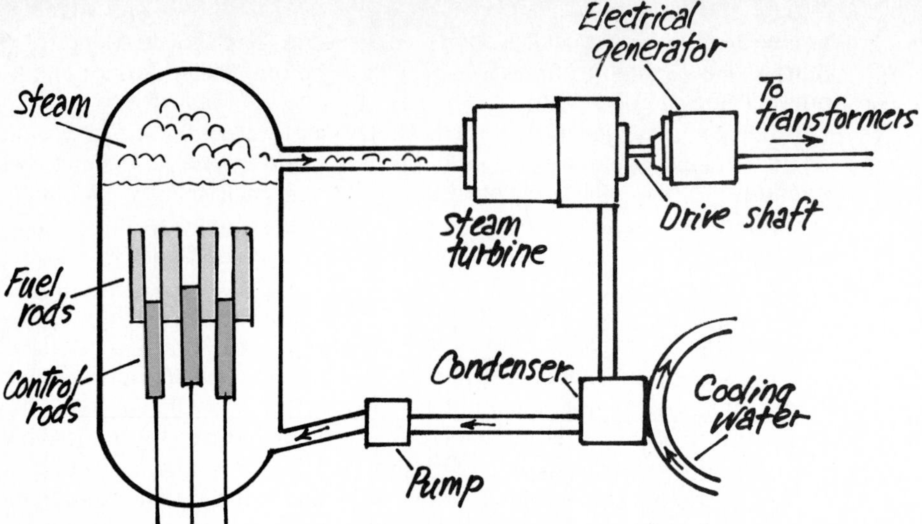


**FIGURE 28–2**
The position of the control rods determines how fast the fission proceeds in a nuclear reactor and thereby governs the rate at which energy is produced. Only when the fuel rods are exposed to each others neutrons does fission proceed at a high rate. Heat from the fission is carried away by steam (as shown here) or water under high pressure. Besides serving as the medium to transfer heat, water is the moderator that slows the fast neutrons emitted during fission. The steam goes to drive a turbine that drives an electrical generator. (In a high-pressure water reactor, the hot water makes steam in a steam generator.) Condensed with the help of cooling water, the reactor's water returns to be heated again.

tron per fission to cause another fission. Then the chain reaction proceeds and gives a steady release of energy. When a uranium-235 nucleus fissions, the fragments carry the energy away as kinetic energy; in other words, there is a tiny explosion. Those fragments quickly lose their energy as they rip through the uranium metal compound, running into many electrons and occasionally another nucleus. So their energy spreads as thermal energy, or heat.

The uranium metal in a typical reactor is assembled in long rods, with movable *control rods* between them (Fig. 28–3). The control rods contain an element (such as cadmium or boron) whose nuclei absorb the neutrons that strike them. As long as the control rods are in place, the uranium rods, with their large surface areas, lose too many neutrons to the control rods for a chain reaction to take place in the reactor. But when the control rods are slowly withdrawn and the uranium rods are exposed to each other's neutrons, the reactor can go critical and support a self-sustaining chain reaction.

Besides the control rods, there must be a neutron *moderator* to slow the fast neutrons emitted by fission so they can cause other fissions. This should be a material whose nuclei do not absorb neutrons, and unless those nuclei are relatively light they won't absorb much energy when a neutron strikes them. The first experimental nuclear reactors used bricks of solid graphite, a form of carbon, because carbon-12 is relatively light and doesn't absorb neutrons readily. (Graphite has one drawback. If heated in the presence of oxygen, it can catch fire. This happened in a reactor in England before the days of double containment buildings, and a very serious release of radioactive particles occurred.)

A reactor is more simply made if the neutron moderator can serve as the *coolant* at the same time. The coolant must be liquid or gas to circulate past the uranium rods and take the heat energy away to the steam generator. Water, $H_2O$, would seem to be ideal. When pressurized, it can be heated to high temperatures without boiling, and its two hydrogen nuclei, each with the same mass as a neutron, should slow the fast neutrons quickly. However, those hydrogen nuclei ($^1_1H$) absorb neu-

**FIGURE 28–3**
The fuel rod assembly in a modern fission reactor.

trons (becoming deuterons—$^2_1H$, or D), which removes the neutrons from the fission process. Nevertheless, water is an inexpensive, easily handled coolant, and it is used in U.S. commercial reactors. A price is paid for those lost neutrons, however; the uranium used in U.S. reactors must be enriched, containing from 1 percent to almost 4 percent uranium-235 rather than the naturally occurring ratio of 0.7 percent uranium-235 to 99.3 percent uranium-238. Canadian reactors avoid this—they run on natural uranium metal—by using pure *heavy water* ($D_2O$) as a coolant. The deuterium nuclei don't absorb neutrons, and their small mass serves well to slow the fast fission neutrons in the reactor. Although $D_2O$ is expensive, it eliminates the need for costly uranium isotope enrichment facilities.

The uranium metal rods, the control rods, and the moderator and coolant make up the reactor *core*. A typical pressurized water reactor in the United States uses about 200 tons of enriched uranium per year. In 1982 there were about 200 nuclear power reactors at about 75 sites in the United States, operated by about 40 different utility companies. In more than 30 other countries there were some 300 other nuclear reactors. When a reactor's fuel rods are depleted of uranium-235, the rods and their metal casings (which help them to withstand high temperatures) are treated as waste. The products of fission are highly radioactive, so the nuclear reactor waste must be carefully stored to prevent leakage into the environment. Nevertheless, nuclear fission today pollutes far less than coal, oil, or gas-fueled power plants. These methods of producing electricity are compared in the last section of this chapter.

## The Dwindling Supply of Uranium: Breeder Reactors

The known reserves of uranium ore in the world are relatively small. In 1980 about 300 uranium mines in the United States produced some 20 million tons of ore, but, as we've noted, in each ton of ore there are only several ounces of uranium-235. The cost of uranium is sure to climb dramatically as the richest ore deposits are depleted. Most experts think enriched uranium-235 reactors will be obsolete in a matter of a few decades because of lack of nuclear fuel.

One solution to the inevitable loss of the present nuclear power plants is to develop a reactor that, as it generates power, uses another nuclear reaction to produce, or *breed,* fuel. For example, plutonium can be made from the plentiful (but nonfissionable) uranium-238. Plutonium, like uranium-235, is fissionable, and it releases some 2.7 fast neutrons per fission on the average, so it too can fission by chain reactions.*

The first step in making plutonium is to let uranium-238 absorb a fast neutron to become uranium-239. This isotope of uranium, like all

*Another reaction that can breed nuclear fuel changes the element thorium-232 to the fissionable isotope uranium-233 by neutron bombardment. There is about four times as much thorium present in ores as there is uranium.

others, is unstable. With a half-life of about 20 minutes, it emits a beta particle to become $^{239}_{93}Np$, neptunium. This nucleus, too, is unstable, with a half-life of 2.3 days, and it emits a beta particle to become $^{239}_{94}Pu$, plutonium. Plutonium is also unstable, but it has a half-life of 24,360 years, so the plutonium in a reactor may be collected and used for fuel with almost no attrition through decay. Reactors that take advantage of these side reactions can produce more fuel than they use, and so they are called **breeder reactors.**

Because there is about 100 times more uranium-238 that could be converted to the nuclear fuel plutonium than there is fissionable uranium-235, plutonium breeders could increase the fuel supply for reactors by the same factor. A single breeder reactor cannot only supply its own fuel but make enough fuel to supply new reactors as well.

## The Problems with Breeder Reactors

Plutonium breeder reactors have more problems in a technological sense than do ordinary uranium-235 fission reactors. To breed fuel quickly, fast neutrons, which convert uranium-238 to uranium-239, must be abundant in the reactor. Though a uranium-235 nucleus fissions most easily by absorbing a slow neutron, it can also fission (with a lower probability) as a result of a strike by a fast neutron; if it does, it releases more than 2.5 neutrons on the average. So in a fast breeder reactor the moderator is eliminated, and the core is kept small to increase the number of fast neutrons per unit of volume. This not only produces plutonium faster but also helps to keep the chain reaction going, since fission of uranium-235 is less likely with fast neutrons.

The principal drawback of a small core and a high density of fast neutrons is the rate of heat production—about twice that of an ordinary nuclear reactor. And since the core itself must be small, it is more difficult to cool by a circulating fluid. Water isn't used for a coolant because it slows or absorbs neutrons. Liquid metal sodium, circulated through the core at high rates, is the current choice. It boils at 895°C, so it doesn't have to be under high pressure as water does. On the positive side, because of the higher temperature, we can get electrical energy from the thermal energy in a breeder reactor more efficiently, at a rate of about 40 percent rather than about 30 percent for ordinary nuclear reactors. You'll probably read and hear a lot more about *liquid metal fast breeder reactors* (LMFBR) in the future.

There are other tough problems with such breeder reactors. Liquid sodium burns in air and explodes in water, so it must be carefully contained in the plumbing of the reactor. The liquid sodium also becomes very radioactive because its nuclei can capture neutrons, though it is much less efficient than water at doing this. Moreover, it is dangerous to work with plutonium, because a very small amount is chemically poisonous. But the danger really comes from its radiation. The radioactivity from a dose of plutonium that is too small to poison a worker chemically would almost certainly cause cancer. Consequently, even spent fuel must be handled with extreme caution. Plutonium can also be used to build nuclear weapons, so security is a concern.

As soon as it produced power in 1966, the first experimental breeder reactor in the United States was shut down because of an accident. A piece of sheet metal that wasn't in the plans blocked some coolant openings into the core. The core promptly overheated, and some of the fuel units partially melted.

Great Britain, France, Japan, the Soviet Union, and West Germany already have breeder reactors in operation, producing electrical energy and nuclear fuel at the same time. (Because of the controversy breeder reactors have caused, there are none at present in the United States.) If there is to be a long-range future for nuclear fission as a major energy source, these reactors seem to be an answer. Breeder reactors could produce all of the world's present energy needs for several centuries.

## ENERGY FROM THE NUCLEUS

Today very fast-moving protons or electrons, whose energy comes from the machines called particle accelerators, are used as bullets to probe the nucleus. Careful analysis of what happens when these particles penetrate the nucleus tells much about how various nuclei differ. An important feature was discovered from such experiments: The average energy by which a nucleon is bound within a nucleus, called the binding energy per nucleon, varies from one nuclear species to the next and even from one isotope to the next of a single element. This binding energy per nucleon tells how much energy would be needed to separate the nucleons from the nucleus.

Because the binding energy per nucleon is different among the various nuclei, rearrangements of the nucleus by either breaking the nucleus apart or by merging nuclei together always releases or absorbs energy. A look at the binding energy per nucleon for various species lets us see which reactions give up energy. Figure 28–4 shows the *binding energy curve* for nuclei. If either a fission or fusion reaction leaves the average nucleon more tightly bound, the reaction gives up energy, just as in a chemical reaction. From the binding energy curve, we see that if the nucleus of a large element splits to form several smaller nuclei, energy is released, so long as the nuclei that emerge aren't smaller than iron, Fe. We also see that energy is almost certain to be given off in fusion so long as the final nucleus is no heavier than iron.

Einstein's formula relating energy to mass, $E = mc^2$, states these facts another way. The nucleus or nuclei before a nuclear reaction have a certain mass, and afterward the product(s) have a certain mass. If the mass is less after the reaction takes place, energy is released. That energy appears in the form of kinetic energy of the products and as radiation. The mass lost, $m$, is changed to energy, $E$, in the amount $E = mc^2$.

**FIGURE 28–4**
A graph comparing the binding energy per nucleon for nuclei of various masses. The iron ($^{56}Fe$) nucleus (at the highest point on the curve) is the most stable because its nucleons are on the average more tightly bound than the nucleons in other nuclei. Energy is released only by those nuclear reactions whose product nuclei are more tightly bound than the initial nuclei.

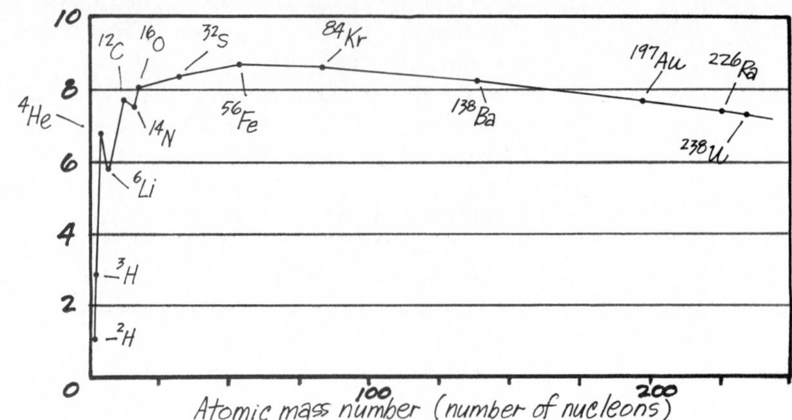


## Fusion

When a very large nucleus splits nearly in half, energy is released in the process called fission. Likewise, when two very small nuclei merge to form a single nucleus, energy is released. We call this process **fusion.** (See "Energy from the Nucleus.") Both fission and fusion are used in nuclear weapons. The so-called A-bomb, or atom bomb, is a detonation from uranium or plutonium fission. (It is clearly misnamed and should be called a nuclear bomb.) The H-bomb, or hydrogen bomb, uses a fusion reaction in which isotopes of hydrogen merge to form helium.

Although fission reactions are controlled today in nuclear power plants, fusion reactions have yet to be tamed. The fusion of deuterium, $^2H$, and tritium, $^3H$, to form helium releases about three times as much energy per gram as the fission of uranium-235 does. The supply of deuterium in the oceans would meet the energy requirements projected for the world for millions of years. (Of the hydrogen nuclei in water molecules, about 1 in 6500 is a deuterium nucleus.) Controlled fusion offers the brightest promise on the nuclear energy horizon, *if* it becomes a reality.

The difficulty in producing and controlling fusion comes from the electric force. For deuterium and tritium to meet and fuse, they must come together with a high relative speed, or else their positive charges turn them back before they can merge. So without great speed (very high temperatures) they have no chance to unite. In a hydrogen bomb, an ordinary fission bomb inside goes off first, raising the temperature of the enclosed deuterium and tritium. Some of those speeding nuclei overcome their electrical repulsion to merge during their random collisions, releasing an incredible amount of energy very quickly.

Because of the high temperatures required for fusion (about $10^8$ K—one hundred million degrees absolute), no ordinary container can hold fusion material without cooling the plasma or being vaporized. To be useful, the energy from fusion must be released in smaller amounts than in a bomb and over longer periods of time so that steam can be made from fusion's thermal energy to drive conventional steam turbines.

Two schemes to harness fusion are under intense investigation today. Since the 1950s, physicists have worked on a magnetic bottling technique whereby a plasma of deuterium and tritium can be simultaneously compressed and kept from the walls of the container by a large magnetic field. Although experiments with such devices are inching closer to the proper combination of plasma density and temperature, none has yet broken even in energy production. In other words, more energy is put into heating and pinching the plasma than is released by fusion.

The second idea is to use the tremendous power available in laser light to cause fusion of tiny amounts of deuterium and tritium. Tiny pellets of these isotopes, encased in a solid material, are centered at the focus of simultaneous laser pulses. That is, they are blasted from all sides at once. The material on the outside of the pellet vaporizes almost instantly and expands in all directions. This expanding gas compresses the deuterium and tritium at the center, dramatically increasing its tem-

**FIGURE 28–5**
These tiny pellets, almost hidden on this penny, contain isotopes of hydrogen (deuterium and tritium). When intense laser beams are simultaneously directed at such pellets from all sides, the surface material vaporizes and expands almost instantly, causing a shock wave to move toward their center. The isotopes are compressed tremendously, and both their densities and temperatures rise. Under such extreme conditions, the nuclei of the isotopes can come close enough to fuse and form helium, releasing energy in the process. (Note in Fig. 28–4 that the helium nucleons are much more tightly bound than the deuterium, $^2$H, and tritium, $^3$H, nucleons.)

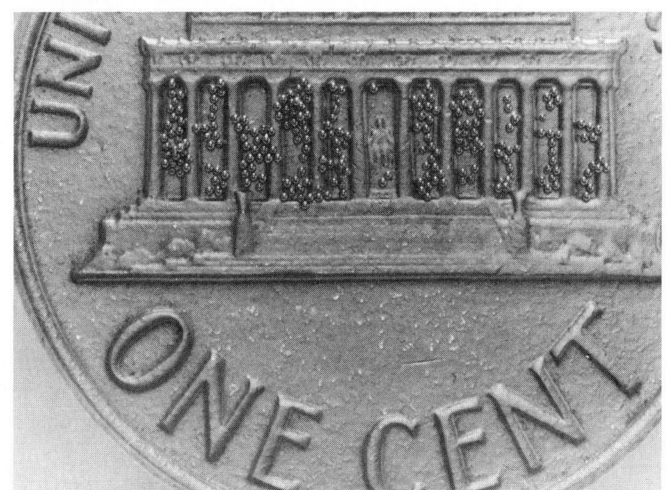

perature and making fusion possible. Though the prototypes of these laser fusion devices have caused fusion to happen, as yet the devices haven't been powerful enough to break even with respect to energy for more than a tiny fraction of a second.

## Problems with the Generation of Energy

Wood was the fuel used by our ancestors until well after steam engines arrived on the scene in the early 1700s. Coal, discovered in England in the twelfth century and perhaps used by others before then, also became a major fuel. The first producing oil well was sunk in Pennsylvania soil in 1859, and the energy from both coal and oil was used to industrialize civilization. The human population has grown so much since the Middle Ages that the energy needed to support the earth's peoples today could never again come from wood; trees don't grow fast enough. Besides this fact, wood fires share a problem with the burning of coal and oil for energy: They all produce carbon dioxide, $CO_2$, the product of combustion.

**FIGURE 28–6**
A Texas oil field in the early 1900s.

Combustion, whether used to generate electricity or to power cars, pours $CO_2$ into our atmosphere. Up to half of this carbon dioxide may dissolve into the ocean waters, but the rest remains in the air. Because $CO_2$ molecules in the atmosphere absorb heat so well, the earth's atmosphere might warm from this greenhouse effect. Environmental scientists monitor the annual rise in the level of carbon dioxide, and if the recent trends continue because of our heavy use of fossil fuels, by the year 2025 the carbon dioxide level in our atmosphere will be double the 1980 level. No one is certain what this doubling would mean, but scientists can build theoretical models of the atmosphere with computers. Recently these models suggest such an increase would cause the earth to warm an average of 1.5°C to 3.0°C, with polar regions having increases of 8° to 12°C. This warming would lead to a partial melting of

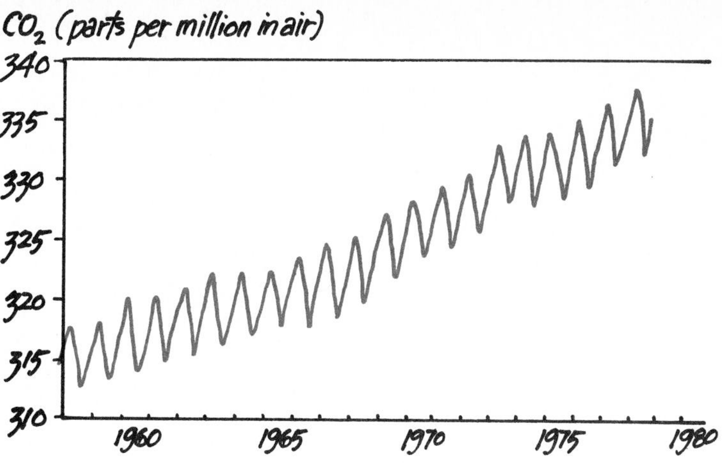


**FIGURE 28–7**
A graph of the carbon dioxide content in the atmosphere in the Northern Hemisphere. The concentration decreases in summer when plant photosynthesis proceeds at a high rate and carbon dioxide is absorbed from the air. In winter, however, much plant life decays and releases carbon dioxide and other gases into the air. (Data taken in the southern hemisphere reveals similar ups and downs in content except they are reversed since the seasons are opposite.) Notice the continual and increasing annual rise in the curve. That rise is attributed to the release of carbon dioxide into the air by human industrial and transportation activities.

the polar ice sheets, raising the global ocean levels, flooding coastal cities, and changing many climate patterns worldwide. In the future we can expect advances in atmospheric science to make such models more accurate.

Oil and coal (especially) pollute the air with other damaging impurities besides carbon dioxide. The list includes carbon monoxide (CO), sulfur dioxide ($SO_2$), various oxides of nitrogen, mercury, hydrocarbons, and even, in the case of coal, radioactivity from minerals. According to some estimates, a large coal-fired plant sends more radioactivity into the atmosphere than a comparable nuclear plant.

Sulfur dioxide, however, is probably the number one pollutant in the public eye. Released through smokestacks at fossil-fueled electrical power plants, it combines with water vapor to make sulfuric acid. When it rains, the water that falls is acidic. Downwind from huge industrial plants in Canada, land has been completely denuded, an effect of these acid rains.

Acidic rainwater runs off into lakes and slowly accumulates until the lakewater is unable to support life. Acid rains, largely from coal-burning plants in the American midwest, have killed many lakes in the northeast and Canada. The problem is everywhere. Ten thousand of Sweden's 100,000 lakes are completely dead as a result of sulfur dioxide being carried by the prevailing winds from coal-burning power plants in Germany, France, and Great Britain. The surfaces of both the Lincoln Memorial in Washington, D.C., and the Acropolis in Greece

**FIGURE 28–8**
Seventy-two miles north of San Francisco, geothermal steam from wells deep in the earth turns steam turbines that turn electrical generators. Known as The Geysers, the facility has grown from one commercial generating unit in 1960 to an expected 20 units in 1988 that will produce a total of 1.5 million kilowatts of electrical power. Other commercial geothermal units began operation in 1983 in Imperial Valley, California, an area that experts think could surpass The Geysers in electrical generating capabilities. Mexico, Iceland, Japan, New Zealand, the Philipines, Italy, and Russia also use geothermal power for electricity.

are being dissolved by acid rains. And the damage doesn't stop with acidity: Acid rainwater reacts with minerals in the ground to release heavy metals, some of them poisonous, into the water table.

Burning oil to produce energy pollutes less than coal, and natural gas pollutes even less. But these fuels are already in short supply compared to coal; only 2 percent of the known coal reserves have been used. The threat coal poses to the air and water demands we look elsewhere for power in the near future.

Elsewhere? Solar energy would be a perfect solution, with no pollution, no threat of contamination, and no waste products. However, up to now there is no way to convert sunlight to electricity in amounts needed for industrial activity. Hydroelectric power is close to pollution free. In the United States it provides about 12 percent of the energy generated by utility companies, and the potential is about four times as great. Geothermal energy is in use in a few places, such as Iceland and San Francisco, but unfortunately it contributes little to supplying the total energy needs of the world at present.

With regard to pollution, nuclear power is cleaner than any of the fossil fuels, emitting no carbon dioxide and no more radiation than emerges from the smokestacks of coal-fueled power plants. Their use requires no large-scale transportation; a nuclear power plant might use only two pounds of uranium while its coal-fired counterpart would burn 6,000,000 lb of coal. But safe storage of nuclear wastes is a problem, and even worse in the public eye is the chance, however small, of an accident that would cause considerable local radioactive contamination. No one can argue that nuclear energy is perfectly safe. But its cleanliness (compared to fossil fuels) is a fact worth considering. Each ton of coal that is burned releases over 80 lbs of sulfur dioxide into the air. Besides its effect on trees and lakes, the acid that results is harmful to humans. An estimated 8000 people die each year because of its presence, and another 2000 deaths are attributed to the tars in coal emissions.

Today New York City gets more than half of its electricity from nuclear energy, and Chicago gets 40 percent of its electricity from nuclear power plants. Overall, however, the United States derives only about 12 percent of its electricity from nuclear power, and construction of new nuclear plants has dropped off dramatically. Other countries are increasing their nuclear power capability. France has the most ambitious energy program toward ending fossil fuel use. Its goal is to produce 40 percent of its power by methods not employing fossil fuels by 1990.

If nuclear fusion can be harnessed for power, it could provide for the world's energy needs with minimal pollution for the foreseeable future. The use of nuclear fusion would check the present erosion of our atmosphere by combustion of fossil fuels. The fuel for fusion reactors presents no security problem, because to create a fusion bomb, one needs a fission bomb to set it off. Unfortunately, the technological problems with small-scale fusion have proved to be tough ones, and progress is slow. Only the most optimistic of the experts expect commercial fusion reactors by the year 2000, and most think 2020 would be a better guess.

## CALCULATIONS

In 1932 J.D. Cockcroft and E.T.S. Walton bombarded lithium-7 with protons to produce the first nuclear reaction with a particle accelerator. They observed high-speed alpha particles emerging with almost 70 times the kinetic energy of the incoming protons. A proton absorbed by a lithium nucleus causes the following reaction:

$$p + {}^{7}_{3}\text{Li} \rightarrow {}^{8}_{4}\text{Be}$$

But ${}^{8}_{4}\text{Be}$ (beryllium-8) disintegrates after a mere $10^{-16}$ seconds into two alpha particles. So we can also write the reaction as

$$p + {}^{7}_{3}\text{Li} \rightarrow {}^{4}_{2}\text{He} + {}^{4}_{2}\text{He}$$

These two helium nuclei repel one another strongly and go off at high speeds. The energy that is released was mostly in the form of potential energy, or binding energy, in the initial lithium nucleus. The amount of energy released in such a reaction can be found from $E = mc^2$, where $m$ is the actual mass that disappears in the reaction.

---

EXAMPLE 1: **Use the following masses to calculate the energy released in the reaction $p + {}^{7}_{3}\text{Li} \rightarrow {}^{4}_{2}\text{He} + {}^{4}_{2}\text{He}$.**

$$m_{\text{proton}} = 1.0073 \text{ amu}$$

$$m_{{}^{4}_{2}\text{He}} = 4.0016 \text{ amu}$$

$$m_{{}^{7}_{3}\text{Li}} = 7.016 \text{ amu}$$

where 1 amu = $1.66 \times 10^{-27}$ kg.

The mass of the proton plus the lithium nucleus is 8.023 amu, and the mass of two helium nuclei is 8.003 amu. The mass lost in the reaction is 0.020 amu, so

$$\begin{aligned} E &= mc^2 \\ &= 0.020 \text{ amu} \times (1.66 \times 10^{-27} \text{ kg/amu}) \times (3 \times 10^8 \text{ m/s})^2 \\ &= \mathbf{2.99 \times 10^{-12}\ J} \end{aligned}$$

and since 1 eV (electron volt) = $1.6 \times 10^{-19}$ J,

$$E = \mathbf{18.7\ MeV}$$

---

EXAMPLE 2: The tiniest speck of chalk dust from an eraser at a blackboard that your eyes can clearly resolve is probably about $10^{-5}$ meter (or $10^{-3}$ centimeter) in diameter. **If a single uranium-235 nucleus were to fission beneath that speck, and all of its energy went to raise the speck into the air, could you see it move?**

Presume the speck is spherical: its mass would be $\frac{4}{3}\pi r^3 \times \rho_{\text{chalk}}$. The density of chalk is about 2 g/cm$^3$, and the multiplication gives about $10^{-6}$ kilogram for the speck's mass. The energy from the fissioning nucleus is about 200 MeV, or $2 \times 10^8$eV. Converting this to joules,

$$E = 2 \times 10^8\text{eV} \times 1.6 \times 10^{-19} \text{ J/eV} = 3.2 \times 10^{-11}\text{J}$$

Converting this to potential energy for the speck of chalk,

$$mgh = E \quad \text{or,}$$

$$h = \frac{E}{mg} = \frac{3.2 \times 10^{-11}\ \text{J}}{10^{-6}\ \text{kg} \times 9.8\ \text{m/s}^2} = 3.3 \times 10^{-6}\ \text{m}$$

In other words, **the speck will rise about one-third of its diameter.** Such a motion just might be visible with the naked eye, revealing the power of nuclear fission since a speck of uranium-235 of the same size as that mote of chalk dust contains more than $10^{13}$ fissionable nuclei.

## REVIEW

**1.** In the process of fission, the nucleus splits apart, and the fragments are given enormous kinetic energy as some of their mass is converted into energy. True or false?

**2.** Explain how a chain reaction could occur in the fission of uranium-235.

**3.** What are the two criteria that determine if a sample of uranium-235 can undergo a runaway chain reaction?

**4.** A nuclear reactor produces a controlled chain reaction that results in a steady release of energy. True or false?

**5.** Give the function for each of the following components of a nuclear reactor: **(a)** the control rods, **(b)** the moderator, and **(c)** the coolant.

**6.** How does a breeder reactor differ from an ordinary nuclear reactor?

**7.** How does the fusion process differ from fission?

**8.** Which factor presents the greatest problem in controlling the fusion process?

**9.** How can lasers be useful in the fusion process?

**10.** What worldwide trend are atmospheric scientists watching for as a result of pollution from combustion of fossil fuels?

**11.** What is acid rain? In what ways is it harmful to our environment?

## EXERCISES

**1.** When you start a campfire, are you starting a chain reaction? Explain.

**2.** Discuss this statement: A slow neutron that happens to touch a nucleus can fall right on into it.

**3.** Which force in effect limits the maximum size of stable nuclei?

**4.** Why does burning wood not alter the percentage of carbon-14 in the carbon dioxide of the atmosphere?

**5.** Explain how the electrical force that resists fusion actually assists in the process of fission.

**6.** What single fact about the fission of uranium-235 makes a chain reaction possible?

**7.** Draw an analogy between the instability of a large liquid drop that tends to break apart easily and the instability of a large nucleus.

**8.** When small volumes of uranium-235 are brought together in a bomb to make the critical mass, has the surface area of the uranium per unit of mass increased or decreased? Does this affect what takes place next?

**9.** Explain why a fission reactor, though much larger than a fission bomb, cannot explode.

**10.** At the turn of this century, a certain Henry Adams talked about "attaching a motor to radium to take advantage of its emission of energy." How is a present-day nuclear power plant similar to his idea? How is it not similar?

**11.** When a rock solidifies, any uranium-238 minerals are frozen into place along with the rest of the rock. As time goes on, the uranium slowly decays, but each fission leaves a short track through the rock where the alpha particle escaped. Explain how such fission tracks could be used to determine the age of the rock.

**12.** Why must an atomic (fission) bomb be used to set off a hydrogen (fusion) bomb?

**13.** The efficiency with which heat can be transformed into energy is at best equal to $(T_{high} - T_{low})/T_{high}$, where the temperature is in degrees Kelvin, as discussed in Chapter 14. On this basis, which type of nuclear power plant can be more efficient, an ordinary fission reactor or a breeder reactor?

**14.** Carbon particles from coal-fired electric plants are found in the haze above the Arctic Circle. These particles absorb sunlight that might otherwise be reflected by the ice below. How might this affect the amount of ice in that region?

**15.** Pollution particles that make it into the cold air over the Arctic tend to stay there. Environmental scientists call the Arctic atmosphere a cold trap. Discuss what they must mean.

**16.** When atmospheric testing of nuclear weapons began in 1945, these devices (via neutron emission) created large amounts of carbon-14 in the atmosphere. By the time of the test ban treaty in the 1960s, the carbon-14 activity of our atmosphere had doubled. What effect will this have on archaeological dating in the distant future?

**17.** When humans began using coal and oil as their major source of fuel, they began to decrease the percentage of carbon-14 in the air. Coal and oil are so old compared to the half-life of carbon-14 (5730 years) that those fuels contain no carbon-14 nuclei to speak of. The carbon dioxide from the burning of these fossil fuels adds only carbon-12 atoms to the air and so dilutes the carbon-14 in the atmosphere. Since about 1850 the percentage content of radioactive carbon dioxide has decreased by 2 percent for that reason. How would this affect carbon-14 dating, if other carbon-14 pollution didn't interfere?

***18.** Calculate the energy your own mass represents (in joules).

***19.** Show that when $^2$H (2.0141 amu) and $^3$H (3.0160 amu) fuse to form $^4$He (4.0016 amu) and one neutron (1.0087 amu), approximately 18 million electron volts is released.

***20.** One ton of TNT releases about $4 \times 10^9$ J. The fission of 1 kilogram of uranium-235 releases about 23 million kWh. How many tons of TNT is represented by the energy available in 1 kg (which weighs about 2.2 pounds) of uranium-235?

***21.** The mass of a helium nucleus (2 protons and 2 neutrons) is 4.0016 amu. The mass of an individual proton is 1.0073 amu, and the mass of an individual neutron is 1.0087 amu. Find the binding energy per particle for the helium nucleus.

# APPENDIX 1

# UNITS OF MEASURE AND PHYSICAL CONSTANTS

(See Appendix 2 for metric prefixes, and page 4 for abbreviations of the units of length and time. See the Calculation section of Chapter 27 for the units of radiation dosage.)

## Units of Measure

### Length

1 m = 39.37 in. = 1.094 yd
1 cm = 0.394 in.
1 in. = 2.54 cm
1 km = 0.6215 mi
1 mi = 1.609 km = 5280 ft
1 LY (light-year) = $9.46 \times 10^{12}$ km

### Energy

1 J = 1 N·m = 0.738 ft · lb
1 ft · lb = 1.355 J
1 cal = 4.184 J
1 kcal = $4.18 \times 10^3$ J
1 Cal = 1000 cal
1 kWh = $3.60 \times 10^6$ J = 3.6 MJ

### Time

1 d (day) = $8.64 \times 10^4$ s
1 yr = $3.15 \times 10^7$ s

### Mass

1 kg = 0.0685 slug
1 slug = 14.6 kg
1 amu = $1.661 \times 10^{-27}$ kg

### Power

1 W (watt) = 1 J/s = 0.738 ft · lb/s
1 hp (horsepower) = 550 ft · lb/s = 746 W

### Force

1 N = 0.225 lb
1 lb = 4.45 N

### Pressure

1 atm = 760 torr = 14.7 lb/in.$^2$
1 Pa (pascal) = 1 N/m$^2$ = $1.45 \times 10^{-4}$ lb/in.$^2$

### Surface Area

1 m$^2$ = $10^4$ cm$^2$
1 yd$^2$ = 9 ft$^2$
1 ft$^2$ = 144 in.$^2$

### Volume

1 yd$^3$ = 27 ft$^3$
1 liter (L) = 1000 cm$^3$
1 ft$^3$ = 1728 in.$^3$
1 gallon (fluid) = 4 quarts = 8 pints = 231 in.$^3$ = 3.785 L
1 pint = 16 fluid ounces

## Mixed Units

454 grams of mass *weighs* 1 pound
1 kilogram of mass *weighs* 2.205 pounds
1 slug of mass *weighs* 32.2 pounds

## Other Units

The units of measurement for **momentum** are kg · m/s or slug · ft/s
The units of measurement for **impulse** are N · s or lb · s
The unit of measurement for the **coefficient of thermal conductivity** is W/(m · °C)
The unit of measurement for the **electrical resistivity** is Ω · m (or ohm · meters)

## Constants

Speed of light = $c$ = $2.998 \times 10^8$ m/s = $1.863 \times 10^5$ mi/s
Gravitational constant = $G$ = $6.672 \times 10^{-11}$ N · $m^2/kg^2$
Planck's constant = $h$ = $6.626 \times 10^{-34}$ J · s
Charge of electron = $e$ = $1.602 \times 10^{-19}$ C
mass of electron = $m_e$ = $9.11 \times 10^{-31}$ kg
Mass of proton = $m_p$ = $1.673 \times 10^{-27}$ kg
Mass of neutron = $m_n$ = $1.675 \times 10^{-27}$ kg
Boltzmann's constant = $k$ = $1.381 \times 10^{-23}$ J/k
The constant in Coulomb's law = $K$ = $8.988 \times 10^9$ N · $m^2/c^2$

# APPENDIX 2

# POWERS OF 10

When numbers are uncommonly large or small, a shorthand method makes them easier to write and handle. Consider the number *one thousand*. Because 1000 is the same as $10 \times 10 \times 10$, or 3 *factors* of 10, we can write 1000 more compactly as $10^3$. (Read $10^3$ as "ten to the three.") The number 3 is called the **exponent** of 10. It tells how many factors (or **powers**) of 10 are multiplied together to equal the large number. (Each factor of 10 is sometimes called an **order of magnitude.**)

Now consider the number *one hundred thousand* (100,000). This number is literally $100 \times 1000$. Or, factoring 100 and 1000 into powers of ten, we have $(10 \times 10) \times (10 \times 10 \times 10) = 10^5$. But returning to the form $100 \times 1000$, we can write $100 = 10^2$ and $1000 = 10^3$. Thus $10^2 \times 10^3 = 10^5$. *When powers of 10 are multiplied together, their exponents add.*

When powers of 10 are divided, the exponents subtract rather than add as they do in multiplication. For example,

$$\frac{1000}{100} = \frac{10^3}{10^2} = \frac{10 \times 10 \times 10}{10 \times 10} = 10^1$$

Each power of 10 in the denominator eliminates a power of 10 in the numerator; *thus the exponent of 10 in a denominator subtracts from the exponent of 10 in the numerator*. So we can write

$$\frac{10^3}{10^2} = 10^{3-2} = 10^1$$

Notice that the effect of dividing by $10^2$ is the same as multiplying by $10^{-2}$. So to take a power of 10 to the numerator from the denominator (or vice versa), just change the sign of its exponent:

$$10^4 = \frac{1}{10^{-4}} \quad \text{and} \quad 10^{-5} = \frac{1}{10^5}$$

As an example of how to express a number in terms of powers of 10, let's rewrite the approximate number of seconds in a year. In 365 days there are 31,536,000 seconds. If we factor, say, 3 powers of 10 from this number, we get $31{,}536 \times 10^3$. Notice that *each* factor of 10 removed from a number shifts the decimal point one place to the left. Continuing to factor powers of 10, we find

$3.1536 \times 10^7$ s = 31,536,000 s = the number of seconds in 365 days.

To help you recognize powers-of-10 equivalents, inspect Table A–1. Prefixes for some of the factors of 10 are given in Table A–2. For example, the radius of the earth is 6,370,000 meters. This distance can be expressed as 6370 km or 6.37 Mm.

**TABLE A–1**

| | |
|---|---|
| $10^1 = 10$ | $10^{-1} = 1/10$ or 0.1 |
| $10^2 = 100$ | $10^{-2} = 1/100$ or 0.01 |
| $10^3 = 1000$ | $10^{-3} = 1/1000$ or 0.001 |
| $10^4 = 10{,}000$ | $10^{-4} = 1/10{,}000$ or 0.0001 |
| $10^5 = 100{,}000$ | $10^{-5} = 1/100{,}000$ or 0.00001 |
| $10^6 = 1{,}000{,}000$ | $10^{-6} = 1/1{,}000{,}000$ or 0.000001 |

Also note $\frac{10^3}{10^3} = 10^{3-3} = 10^0 = 1$

**TABLE A–2**

| POWER OF 10 | PREFIX | ABBREVIATION |
|---|---|---|
| $10^{12}$ | tera | T |
| $10^9$ | giga | G |
| $10^6$ | mega | M |
| $10^3$ | kilo | k |
| $10^{-2}$ | centi | c |
| $10^{-3}$ | milli | m |
| $10^{-6}$ | micro | μ |
| $10^{-9}$ | nano | n |
| $10^{-12}$ | pico | p |

# ANSWERS TO SELECTED EXERCISES

## Chapter 1

**2.** One mile per hour is a slow pace for most walkers. (Speeds from 2 mph to 4 mph are normal.) The simplest way to ensure an *average* speed of 1 mph is to walk to the 1-mile mark, arriving early, watch the time, and take the last step to the line just as the hour ends.

**4.** (a) No; like runners, swimmers tire and slow and have other fluctuations in their speeds. (b) Yes; the hands should move at a uniform rate. (c) Yes. (d) No; gravity accelerates the rock unless it falls at terminal speed. (e) No; there is air resistance, and irregularities in the lane influence the ball's path.

**5.** There are *two* quantities used to derive or compute a speed, distance, and time. Your speed is distance of stride/time of stride, and a change in *either* will affect the speed. To increase your speed, you can increase the distance of your stride while taking the same time for each stride; or you can *decrease* the time of your stride (take more steps per minute) while keeping the same distance between your footfalls.

**7.** (a) In *C;* she traveled 2 miles in one hour, so her speed was 2 mph. (b) *B;* perhaps she was visiting a friend—she didn't change her distance from home for 4 hours. (c) 2 miles. (d) Average speed = distance/time. How far did she walk? 2 miles out plus 2 miles back, so she traveled 4 miles in 8 hours. Her average speed along the straight-line path was ½ mph.

**8.** If the motorist's *average* speed (which is all the authorities can compute) is greater than the speed limit, the jig is up. If the average speed is equal to the speed limit, however, there is no way to tell if the driver broke the limit. The car might have been speeding at one point and going slower at some other point. (Actually, the average speed should be slightly *less* than the speed limit because of the acceleration and deceleration periods at the toll booths.)

**9.** Speed and acceleration *describe* motion. They certainly don't *explain* motion; for example, they don't explain why gravity accelerates things at the rate of $g$ or why a typical terminal speed for a skydiver is about 120 mph. Although these two rates *aid* in predicting motion, they don't do any predicting. That is, if you know the speeds and accelerations, you can tell where something will be at a given time. But the only way to predict motion is to know these two quantities beforehand; it is these quantities that must be found for you to predict the motion. In Chapters 3 and 4 we'll introduce laws of physics that let us *predict* accelerations in some cases and, hence, predict motions.

**11.** Increased air resistance causes a slower descent rate, which gives more time in the air per trip. Also, a skydiver's speed slows abruptly to a new terminal speed when the parachute is opened. The less change in speed at this time, the smaller the jerk the diver receives.

**13.** *Between* the sweeps of the windshield wiper, visibility is limited; therefore, you lose information, just as you do when you view something with a strobe light. However, the faster the wipers go (or the strobe), the more information you are able to receive per unit of time. Naturally, you should decrease speed in a driving rain so that the distance you cover during the times when the view is obscured will not be so large.

**15.** Yes; *any* length/time/time is a unit of acceleration no matter which combinations you use.

**18.** (a) B and D. (b) A. (c) C and E. (d) B. (e) C.

**20.** The average acceleration is change in speed/time, so you need to estimate the change in speed between two times in the motion to get the average acceleration over that time interval. Take the distance between the first two positions and divide by ⅕ second to find the average speed over that interval; do the same for the ball's last two positions. Subtract these speeds to approximate the *change in speed* between those intervals. Then find the elapsed time between the midpoints of those intervals and divide that into the change in speed.

**21.** The least-blurred photos would be those exposed when the bee's wings were moving at their *slowest* speeds. Just like keys thrown into the air, the bee's wings slow near the bee's turning points, when their speed must pass through the value of zero.

**23.** The rate at which acceleration changes is change in acceleration/time it takes. Now the quantity "change in acceleration," or final acceleration minus initial acceleration, has the units length/time/time. When this is divided by another time, the units become length/time/time/time; if all of the time units are of the same kind, this is written as length/time$^3$.

**25.** When a cyclist travels *with* the flow of traffic, the relative speed between the bicycle and the cars is *smaller* than if the cyclist rides against the flow. This gives both rider and drivers more time to react and to brake (or accelerate, or turn) if necessary.

**27.** A fixed point near the outer edge of the record goes around a larger circle than a fixed point nearer to the center, yet it takes the same amount of time. So the outer portion of the record moves faster than the inner portion, and when the needle is close to the edge of the record, the groove beneath it moves faster than when the needle is close to the record's center.

**28.** Figure A-1*a* shows a fully inflated tire. One rotation of the axle carries the car about one tire circumference down the road. Figure A-1*b* shows a partially deflated tire. One rotation of the axle will take the car about the length of the dotted line. If the tire were fully deflated, the car would be rolling on the wheel's rim, and it would advance about one rim circumference for each rotation. A partially deflated tire stretches and deforms more than it is meant to during each revolution, which causes extra wear.

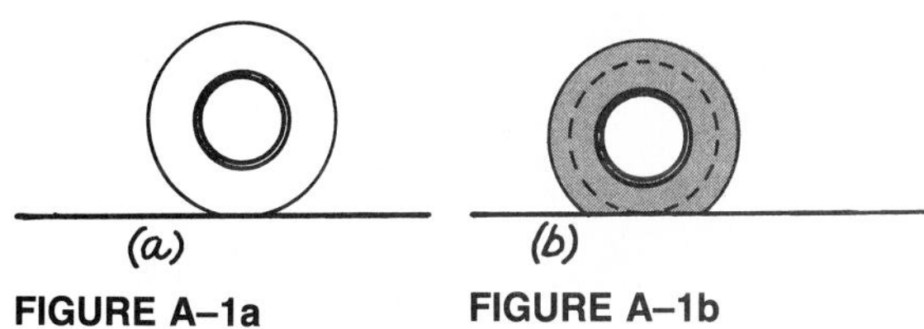


**FIGURE A–1a** **FIGURE A–1b**

**29.** In the winter a tire sags because the pressure from the air inside the tire drops with the decrease in temperature of that air. Professor Dolittle was probably too lazy to add air! (See also the answer to Exercise 28.) You may have heard that rubber (unlike most materials) shrinks when heated and expands when cooled. Although this is true, those effects on the tire are minor compared to the effect of changing the air pressure within.

**31.** 4 minutes/mile? Notice that this is not a speed, but it is the units of speed written upside down. If you turn this ratio over (that's called *inversion*), you will have a speed

$$\frac{1 \text{ mile}}{4 \text{ minutes}} = \frac{1}{4} \text{ miles per minute} = \frac{1}{4} \frac{\text{mile}}{(\frac{1}{60} \text{ hour})}$$

$$= \frac{60 \text{ miles}}{4 \text{ hour}} = 15 \text{ mph}$$

**32.** He averaged 94.24 miles per day. If he ran 24 hours each day, $\bar{v}$ = 3.9 mph. If he ran 12 hours each day, $\bar{v}$ = 7.85 mph, a little faster than 8-minute miles.

**33.** Without air resistance, the acceleration is $g$, or 22 mph/s. (See Example 3 of Chapter 1.) $v_f - v_i = at$, or 120 mph − 0 mph = 22 mph/s × $t$, so $t$ is between 5 and 6 seconds.

**34.** Use $v_{hair}$ = ½ inch/month in the formula $t = d/v$.

$$t = \frac{12 \text{ inches}}{\frac{1 \text{ inch}}{2 \text{ month}}} = 24 \text{ months, or 2 years}$$

To grow waist-length hair that is 2 feet long, for instance, then

$$t = \frac{24 \text{ inches}}{\frac{1 \text{ inch}}{2 \text{ month}}} = 4 \text{ years}$$

**35.** 1985 − 1492 = 493 years. If $\bar{v}$ = 2 centimeters/year, then $d = \bar{v}t$ gives 986 centimeters as the distance. Since 1 centimeter = 1/2.54 inches, 986 centimeters = 986(1/2.54 inches) = 388 inches = 388(1/12 foot) = about 32.3 feet.

**36.** In 8 hours, 4400 miles; in 8 minutes, about 73 miles. It would take about 44 minutes to go 400 miles, about 2 days (45 hours) to go around the world.

**37.** About 1.2 centimeters; it falls roughly that far before you can tell it's falling. (*Hint:* Use $d = \frac{1}{2}gt^2$ where $g$ = 980 cm/s$^2$.)

**38.** The average acceleration for the first minute was

$$\bar{a} = \frac{v_f - v_i}{t} = \frac{(\text{about}) \ 30 \text{ mph} - 0 \text{ mph}}{1 \text{ min}}$$

$$= (\text{about}) \ 30 \text{ mph/min}$$

At the end of 1 min the speed was 30 mph. At the end of 4 min the speed was 12 mph. The time interval, the time used to change the speed, was (4 min − 1 min) = 3 min. So

$$\bar{a} = \frac{12 \text{ mph} - 30 \text{ mph}}{3 \text{ minutes}} = -6 \text{ mph/min},$$

so on the average it was decelerating, losing speed.

**39.** 20 mph.

**40.** Because they are traveling in opposite directions, their relative speed is the sum of their speeds over the ground,

75 mph. It doesn't matter which has the 40 mph speed, and after they pass, their relative speed remains 75 mph. (The distance between them then increases rather than decreases.)

**41.** As the plane travels westward, the moving air decreases the speed of the plane *over the ground* by 100 mph. (Relative to the air, the plane still travels at 500 mph.) So $t = d/v = 2800$ miles/400 mph = 7 hours. The air's speed adds 100 mph to the plane's speed over the ground when it is moving eastward.

**42.** They are traveling in the same direction, so the difference in their speeds gives the relative speed. 8 mph − 5 mph = 3 mph. After he speeds up, their new relative speed is 12 mph − 8 mph = 4 mph. That's how fast the distance between them decreases. So that 100-foot distance is covered at a speed of 4 mph, and $d = \bar{v}t$, or $t = 100$ feet/4 mph = 100 feet/ (4 × 1.5 ft/s) = 16 ⅔ seconds.

**43.** About 3 $g$. (The easy way to get this answer is to use $g = 22$ mph/s.)

**44.** (c).

## Chapter 2

**2.** In essence the performers use only two dimensions; they swing only in a vertical plane between the two platforms. Gravity affects only their vertical motions on that plane as they swing.

**4.** To locate someone along a trail, you would need to know only how far from the starting point the hiker had walked, that is, one number. By staying *on the trail* the hiker moves essentially in one dimension. Of course, any trail, especially in the mountains, is a path through three dimensions. (There will be a distance out, curves to the right and left, and changes in elevation.) But the fact that the trail lies on the earth's surface means that you could pinpoint any position along the trail with two numbers, a latitude and a longitude on the surface. The trail, however, is a *line,* curved though it may be, and on that line only one number is needed to pinpoint a location.

**6.** By asking what happens one component at a time, we can understand this situation. If the plane flies at a steady altitude over level ground, the shadow of the plane moves across the ground with the speed of the plane. (See Fig. 2-27.) But suppose the plane has a vertical component of velocity, as when it gains or loses elevation. A helicopter rising, for example, has *only* a vertical velocity, and we see from Fig. 2-27 that such a velocity component can give a speed to the shadow on the ground that is not at all equal to the aircraft's speed; it depends on the angle of the sunlight. (If the sun is straight overhead, the vertical velocity will not affect the shadow's speed.) In general, an aircraft will have both horizontal and vertical components of velocity, so the shadow's speed most of the time is *not* exactly equal to the aircraft's speed.

**8.** The ball is tossed up when it is served and generally struck near the high point in its path. Reason 1: The higher the ball, the larger the angle into which it may be hit to pass over the net and yet still be in the service court. (Draw a sketch to convince yourself.) Reason 2: The ball travels slower near its highest point, making an easier target for the server.

**11.** (b) Temperature and (e) speed are not vector quantities.

**13.** No. A plane's velocity vector can be constant in speed and direction while it climbs or descends, meaning there is *no* acceleration. Even though a velocity vector has a component in the vertical direction, the velocity vector can still be constant.

**15.** Following the circle, you move tangent to the circle at each point. Your tangent motion does not move you in the perpendicular direction toward the center. Since you don't get closer, the line from you to the circle's center, called the *radius* of the circle, must be perpendicular to the tangent line that you are on. Motions in perpendicular directions are independent.

**16.** See Fig. A-2.

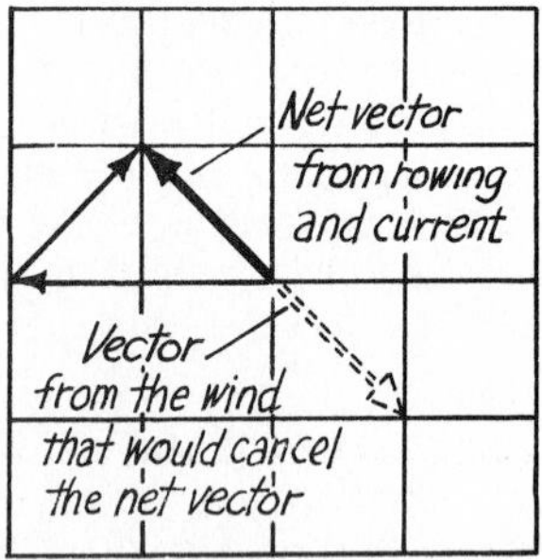


**FIGURE A–2**

**20.** The newspaper will have the car's speed as it is released and will keep pace *with the car's forward motion* as it falls to the ground. (a) The carrier, therefore, will see the newspaper move straight outward from her window and fall to the ground just as if she were not in a moving car when she threw the paper. (b) This person would see the newspaper curve downward in flight just as if the carrier had been standing still when she threw it onto the lawn. The observer would not see the component of motion toward (or away from) his or her position that the paper has because of the car's forward motion. (c) From the nearby tree an observer could see *all* of the components of motion over the ground, the paper arcing downward as it traveled and its forward motion from being launched from the moving car.

**21.** When you walk through the rain, *relative to you* the rain adds a component of velocity that is the opposite of your own velocity. So you must tilt the umbrella somewhat (depending on your speed) in the direction of your motion.

22. (a) You can't really tell. The road could be winding, and much of their motion would not increase their distance from home. (b) Once again, you can't tell. For example, the flat portion of the curve could mean they were at rest; or they might have been on a circular portion of the curve centered on their home, yet traveling at 100 mph. (c) Region II.

24. The second day at 9 A.M. you'll start at the top of the axis representing distance. (See the dashed line.) As you walk back, you'll get closer and closer to where you began, so the line that describes your distance at any time will drop down to zero distance at the 5 P.M. mark. Somewhere it will cross (at least once) your "line" from the previous day.

26. (d) All of these answers. Because the stars are so far away, we can't estimate their distances with the method called parallax, which works to a distance of about 1000 feet. However, astronomers can find the parallax of close stars by using telescopes to magnify starscapes and taking photographs at six-month intervals when the earth is at opposite extremes of its orbit. In effect, those photographic scenes are from "eyes" that are almost 200 million miles apart. The tiny shifts in the star's positions in the photographs let astronomers estimate the distances to about 1000 of the closest stars to within ±10 percent of their true distances. As for answer (b), if the earth's orbit were tremendously larger, parallax could be viewed with the naked eye over the course of a "year".

29. See Fig. A–3.

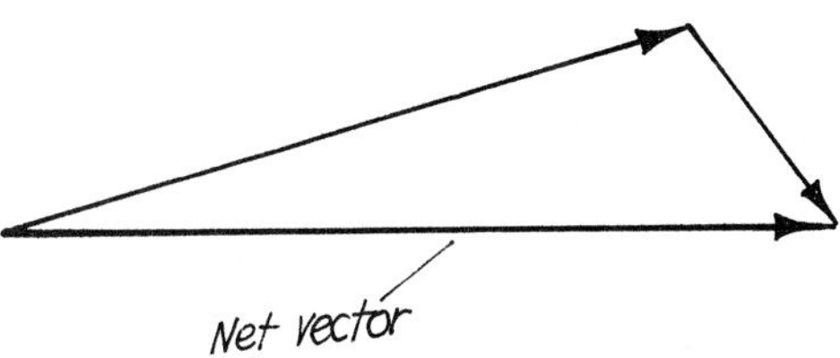


**FIGURE A–3**

30. See Fig. A–4.

**FIGURE A–4**

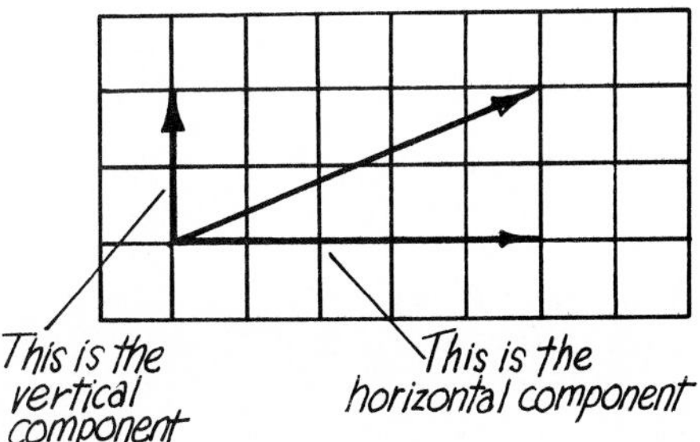


31. The net vector of any vector sum goes from the tail of the first vector to the tip of the last. Should those points be one and the same, the length of the net vector is zero. (See Fig. A–5).

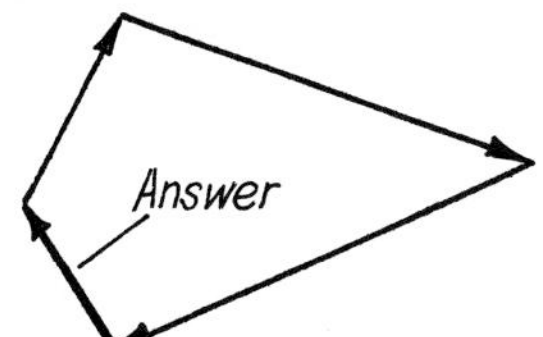


**FIGURE A–5**

32. You could determine this with a graph by finding the net velocity vector, which shows the direction the girl takes across the river, and measuring her displacement along the far bank. But you can also calculate that distance. The river is a mile wide, and she swims across at 2 mph, so she swims for 30 minutes (or ½ hour). During that time the 3-mph current will sweep her 1½ miles downstream. ($d = \bar{v}t$ = 3 mph × ½ hr = 1½ mi.)

33. (a) See Fig. A–6; (b) greater than; (c) less than.

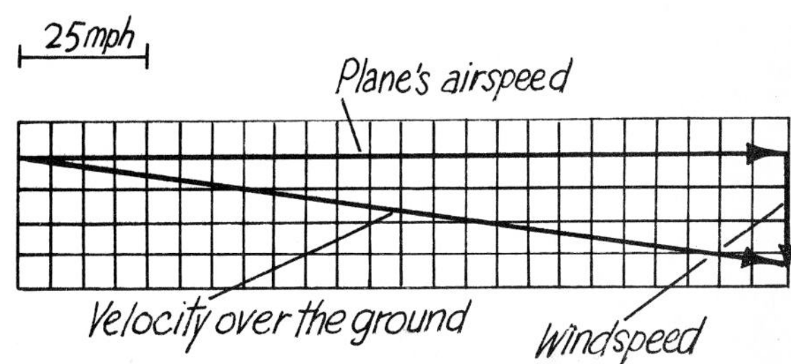


**FIGURE A–6**

34. The cross-field component of velocity takes the band across the field at a speed of about 42 percent of its actual speed. (Just measure the length of the component perpendicular to the side of the field and divide it by the length of the velocity vector. The answer is about 0.42.) The down-field component of velocity takes the band toward the goal line at about 91 percent of the band's actual speed. Your measurements should give about 40 percent and 90 percent for these values.

35. Just visualize what you would see from your car if you drive alongside another car at 55 mph. You could look at a wheel on the other car and see its center stay perfectly still; its relative speed is zero. The rest of the wheel and tire, however, rotate around this center, and the outer edge of that tire has a speed of 55 mph. Fig. A–7 shows the vector velocities of the points in question. To determine what velocities these same points would have relative *to the ground,* you need only to add the velocity of the wheel over the ground to these points. (This is the same as the

opposite of the ground's velocity as seen from the wheel, just as with the falling rain example in this chapter.) This is shown in View 2. To get the *net* velocity at each point, just add the two vectors at each point. Notice that where the wheel touches the ground its velocity is *zero;* see Demonstration 3 for a discussion of this. If the tire were spinning rather than just rolling, this would not be the case.

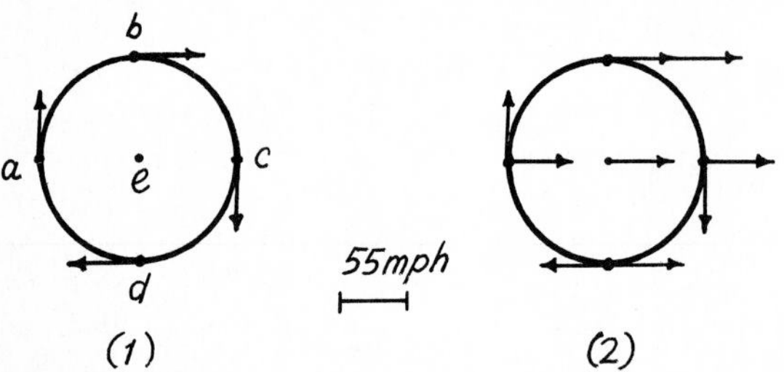


**FIGURE A–7**

## Chapter 3

**2.** Yes, you can feel a book's inertia. You judge inertia by sensing the force you apply compared to the acceleration the object gets in return. (The ratio of force to acceleration, $F/a$, is the measure of its inertia, $m$.) Thus, even when you don't know the numerical value for force and acceleration, you do know that if you push hard and the book accelerates little, its inertia is large. If you give it a shove and it accelerates a lot, it has little inertia.

**3.** The bully was wrong! *In every push action equals reaction.* Perhaps the victorious ruffian had the larger mass, in which case he'd respond to the same force with less acceleration than his opponent. Or perhaps he had a firmer foothold and mustered a greater frictional force to help balance the shove from the loser. Or maybe his push was directed somewhat upward. That would decrease his opponent's normal force from the ground and, thus, the friction on his opponent's feet. At the same time, the reaction force pushes the winner somewhat downward, increases his own normal force, and provides more friction to oppose that push.

**5.** If you are suddenly pushed on your shoulders from behind, your head does not actually snap backward. Your head just tends to stay at rest while your body accelerates forward. For your head to accelerate, there has to be a force, and in this case it must come from your neck. Your neck isn't absolutely stiff (or rigid), so your head is left somewhat behind as the rest of you accelerates. If your car is struck from the rear, a headrest will give your head a push along with the rest of your back, thus eliminating the whiplash action.

**6.** You can guess the answer right away: The paddle has so much more mass than the ball. But let's explain this with Newton's laws: In each case the second law applies, so we have $F_{\text{paddle}} = m_{\text{paddle}}a_{\text{paddle}}$ and $F_{\text{ball}} = m_{\text{ball}}a_{\text{ball}}$. However, action equals reaction, so the force the ball exerts on the paddle is the same in size that the paddle exerts on the ball. There's our answer. $m_{\text{paddle}}a_{\text{paddle}}$ must equal $m_{\text{ball}}a_{\text{ball}}$, and we know the mass of the paddle is many times the mass of the ball. $a_{\text{paddle}} = (m_{\text{ball}}/m_{\text{paddle}}) \times a_{\text{ball}}$, so $a_{\text{paddle}}$ is small compared to $a_{\text{ball}}$.

**8.** The forces the cars exert are equal (and in opposite directions). So $F_{\text{car 1}} = F_{\text{car 2}}$. In other words, $m_{\text{car 1}}a_{\text{car 1}} = m_{\text{car 2}}a_{\text{car 2}}$. This equation tells us that if car 1 is much more massive than car 2, then the acceleration of car 2 will be much greater than the acceleration of car 1; otherwise, the products of $m$ and $a$ could not be equal. The passengers in the more massive car will accelerate less than those in the less massive car. So if each car has a 65-kg passenger, the one in the less massive car will receive the greatest force during the collision.

**10.** You can usually *feel* a push or a pull. When a part of your body accelerates under a push or a pull, nerves detect it. Suppose someone pushes on your arm. Nerves react at the point of contact, and because the arm is attached to your body, some force is transmitted to move that part of you as well, and other nerves detect the shifting of parts. If you're sitting as you read this, you feel the push of the chair on your legs and buttocks, though we are so accustomed to gravity's force and the necessary reaction forces that we put them into our subconscious. But what about astronauts who fall freely in orbit around the earth? They *don't* feel the pull of gravity on their bodies even though it is a substantial force. They feel a little different because their internal organs do not need to be supported against gravity as they do on earth—livers, kidneys, and stomachs float with everything else in weightlessness. In that case, they do not sense the force directly.

**11.** When the wagon accelerates, there is no doubt that the ball rolls to the back of the wagon. But someone sitting on the ground beside the wagon could tell you that the ball does not move backward over the ground, that the wagon merely accelerates, leaving the ball behind. (*Almost.* There is a little friction from the wagon's surface that moves the ball forward a bit as seen from the ground.) When the wagon suddenly stops, the ball moves forward. But the motionless observer once again can tell you the wagon decelerated while the ball continued in a straight line at almost constant speed. The friction with the floor of the wagon slows the ball only a little before it hits the wagon's front edge.

**14.** This question is important to understand, since it applies to so many situations. Let's assume the car is on a level road. To push anything at rest and get it moving, you have to overcome both the object's inertia (its resistance to a change in motion) *and* any friction or other forces on it that might act to retard motion. But once the car (or other object) is moving at the speed you want, you no longer have to overcome its inertia, only whatever kinetic friction there is.

**17.** No matter where it is, earth or moon, the only quantity a bathroom scale measures is a *force*. The scale balances your weight with a force from a spring. On the moon the scale would balance your weight there and give an accurate reading of the pull of the moon's gravity on your body. The kilogram reading of a bathroom scale is not measured but *derived*. The spring measures $mg_{\text{earth}}$, your weight on earth. That kilogram reading on the scale is just the measured value of $mg_{\text{earth}}$ divided by $g_{\text{earth}}$. So on the moon that reading of mass would be in error. Why? $mg_{\text{moon}}$ divided by $g_{\text{earth}}$ won't be equal to your mass in kilograms. The scale would show ⅙ of the astronaut's earth weight, which is true, but it would also say the astronaut's mass was ⅙ of his mass on earth, which is not true. Few astronauts would have gone to the moon if they had thought they would lose ⅚ of their mass in the process!

**19.** Just grab one brick in each hand and shake them back and forth. The one that resists your push more (it is harder to accelerate) is the more massive lead brick.

**20.** Don't be tripped up on this question! *You could accelerate much faster on earth than on the moon.* To accelerate, you need a push in the horizontal direction from friction. The harder your feet press against the surface, the greater the friction. Therefore, the extra weight you have on earth is in your favor. If you pushed off against the moon's surface with your accustomed intensity, your feet would slip because of the reduced normal force on them. Any up-and-down motion would be easier on the moon than on the earth, however, because you would weigh less, and once you got up to your desired speed, it would be easier to maintain it on the moon.

**23.** No; you can only find the net acceleration due to the *net* force, and the net force by itself tells nothing about the individual forces that add to make it.

**25.** In $F = ma$, if $F$ definitely has a constant value, that does not mean that $m$ and $a$ must be constant as well. Only their product, $ma$, has to be constant. The mass, for example, could decrease and the acceleration rise by the same factor. A constant push on a leaking sandbag is an example. As more mass escaped, the same force would accelerate the sandbag more. (Also see Exercise 26 of Chapter 3.)

**26.** When a rocket first lifts off, its thrust is rarely much greater than $mg$, the weight of the rocket. But as fuel is used, the mass $m$ decreases and so does its weight. The thrust will remain about constant, so then it contributes most to the net force on the rocket. Because $F_{\text{net}} = ma$ and the mass $m$ decreases all the time, the constant thrust from the engine means the acceleration $a$ increases. The acceleration will be at a maximum value just before the engines shut off for lack of fuel.

**28.** The bonding forces between the particles make them stay together.

**31.** A normal force from the table is a reaction force that arises only because the book (or anything else) presses against it. And action is equal to reaction; the normal force equals the force that brings it about.

**32.** Three components of force: two to force the pen or pencil back and forth and up and down over the paper, and a third to press down on the pen or pencil to make the marks.

**33.** No; things don't always move in the direction of the net force, even though they accelerate in that direction. Recall how a marble on a table will move if you hit it with a ruler. When it leaves the tabletop, the net force on the marble is $mg$. Yet, though the marble gains speed toward the ground, it also keeps its horizontal speed as it moves along, never traveling straight along the direction of the net force.

**35.** The people dug a depression in the ground in front of the boulder and filled it with water. The morning after a hard freeze, they used levers to roll the boulder over onto the ice, where they could push and pull it for the length of their ice canal. Then another depression was dug, and filled with water, and they waited for another hard freeze. The line of wells was necessary to keep from traveling too far for the large volumes of water.

**37.** $F_{\text{weight}} = mg = 0.454 \text{ kg} \times 9.8 \text{ m/s}^2 = 4.45 \text{ N} = 1 \text{ lb}$. Only the mass would be correct on the moon; the weight would be about ⅙ lb.

**38.** 1 quart $= 1000 \text{ cm}^3/1.057$; ¼ quart $= 1000/(4 \times 1.057) \text{ cm}^3 = 236.52 \text{ cm}^3$, or 236.52 g of water. Since 454 g *weighs* 1 lb, $236.52/454 = 0.52$ lb, and a cup of water weighs about ½ lb.

**39.** 1800 lb/12 people = 150 lb per person.

**40.** $F_{\text{net}} = 128 \text{ lb} - 50 \text{ lb} = 78 \text{ lb} = ma$. Because $mg = 128$ lb, $m = 128 \text{ lb}/(32 \text{ ft/s}^2)$, so $a = 78 \text{ lb}/m = (78 \times 32)/128 \text{ ft/s}^2 = 19.5 \text{ ft/s}^2$.

**41.** $m = 50$ kg. From Appendix 1 of this text, 1 kg weighs about 2.2 lb, so 50 (2.2 lb) = 110 lb. To find your mass in kg, divide your weight in pounds by 2.2.

**42.** 1 lb = 4.45 N, so 1 N = 0.225 lb = 0.225(16 oz) = 3.6 oz.

**43.** $F = ma$, where the force and acceleration are both in the horizontal direction. $m = 900$ kg and $a = 4 \text{ m/s}^2$, so $F = (900 \text{ kg}) \times (4 \text{ m/s}^2) = 3600$ N (about 800 lb).

**44.** $F_{\text{horizontal}} = ma_{\text{horizontal}} = (65 \text{ kg}) \times (2 \text{ m/s}^2) = 130$ N. Most people would be aware of a force this size (about 29 lb), although it is applied fairly evenly on the area of the back.

**45.** The acceleration of the bullet is found from $(v_f - v_i) = \bar{a}t$. $380 \text{ m/s} - 0 = \bar{a}(0.002 \text{ s})$ or $\bar{a} = (380 \text{ m/s})/(0.002 \text{ s}) = 190{,}000 \text{ m/s}^2$. $F = ma$, so $F = 0.01 \text{ kg} \times (190{,}000 \text{ m/s}^2) = 1900$ N. But 4.45 N = 1 lb, so $F = 427$ lb, or more than 42 10-lb sacks of potatoes. The rifle causes an enormous force on such a small mass, hence the great acceleration.

46. Action equals reaction, so $F_{boy} = F_{girl}$. We know the girl's mass and acceleration, and we can calculate the force she received. $F_{girl} = (45 \text{ kg}) \times (1 \text{ m/s}^2) = 45 \text{ N}$. Since this is equal to the force on the boy, $m_{boy}a_{boy} = F_{boy} = 45 \text{ N}$. The acceleration of the boy is then $a_{boy} = 45 \text{ N}/60 \text{ kg} = 3/4 \text{ m/s}^2$.

47. He applies a force of 60 lb upwards, decreasing the contact force of the rock on the scale to only 40 lb. Because reaction equals action, the scale *he* stands on shows his weight plus the 60 lb of reaction to his upward push. The reading of the scale that holds both scales won't change. The weight subtracted from one scale is gained by the other.

48. (a) $F_{spring} = kd$. 10 lb $= k$ (1/6 ft); $k = 60$ lb/ft. (b) $F_{spring} = kd$. 1.8 lb $=$ (60 lb/ft) $\times$ $d$; $d = 0.03$ ft, or 0.36 in.

49. $F_{spring} = kd$, so 1.5 lb $= k \times$ (1/12 ft); $k = 18$ lb/ft.

50. (This exercise is rather tough.) We want to find the time it takes the car to stop on the wet road, so we need to know its maximum acceleration on the wet road. Then we can use: $t_{wet} = (v_f - v_i)/a_{wet} = (40 \text{ mph} - 0 \text{ mph})/a_{wet}$. We don't know $a_{wet}$ yet, but we can calculate $a_{dry}$: $a_{dry} = (v_f - v_i)/t_{dry} = 40 \text{ mph}/2.5 \text{ s} = 16 \text{ mph/s}$. How does this help us know $a_{wet}$? We know $F_{friction\ on\ wet\ road}/F_{friction\ on\ dry\ road} = (\mu_{wet}) \times F_n/(\mu_{dry}) \times F_n = 0.7/1 = 0.7$. But $F = ma$, so $a_{wet}/a_{dry} = 0.7$, and $a_{wet} = 11.2$ mph/s. That gives $t_{wet} = 40 \text{ mph}/(11.2 \text{ mph/s}) = 3.57 \text{ s}$. Next, use $d = \frac{1}{2} at^2$ in each case. Use 88 ft/s $=$ 60 mph, or 1.466 ft/s $=$ 1 mph to convert these accelerations. The answers are $d_{dry} = 73.3$ ft; $d_{wet} = 104.7$ ft.

51. $a_{slick}/a_{dry} = 0.003$, so $a_{slick} = 0.048$ mph/s or 0.07 ft/s$^2$. $t_{slick} = 40 \text{ mph}/(0.048 \text{ mph/s}) = 833.33 \text{ s}$. $d_{slick} = \frac{1}{2} a_{slick} (t_{slick})^2 = 24{,}305$ ft or about 4.6 mi!

52. $a = F/m = 178{,}000 \text{ lb}/(260{,}000/32) \text{ slugs} = 21.9 \text{ ft/s}^2$.

## Chapter 4

3. (a) You must exert an upward force equal to the weight of the groceries to oppose and cancel the downward force on them, which is the pull of gravity. (b) Since the groceries have a steady horizontal speed, the net horizontal force on the bag is zero. So once again your force must balance only the weight, as in (a). (c) You must exert an upward force somewhat greater than the bag's weight to overcome the pull of gravity and accelerate the groceries upward. (d) To lower the groceries, you'd support them initially with a force somewhat less than their weight and let them accelerate downward. Then you would apply a force slightly greater than their weight to decelerate the bag as it nears the trunk's floor.

5. The passengers going over a crest feel lighter. The car accelerates downward abruptly and suddenly supports the passengers less. The decrease in the normal force from the car's seats and floorboard means the passengers accelerate downward. They feel lighter, just as if they were in an elevator that starts down. When the car accelerates up a slope, the normal force from it on the passengers increases and becomes larger than their weight. Because of this, they feel heavier as the car pushes them up the slope.

6. Air resistance is the key. Whether empty or full, the can has the same shape and gets the same air drag at any given speed. But air drag will decelerate the *empty* can much more than the full one, since the empty can has much less mass. (Bullets are made of lead because lead is heavier per cubic centimeter than most cheap metals.)

8. When the approach is slower, it is easier to land safely, whether bird or plane. A tailwind tends to bring the bird or plane in faster, so a headwind that slows it is more desirable. There's another reason, especially for planes with fixed wings. When the wind's direction is the same as the plane's, the relative speed between air and wings is less, giving less lift and drag. Hence, the pilot has slightly less control over the descent rate. However, a headwind increases the air's speed against the wing and increases lift and drag. This is also an important consideration when an airplane takes off.

10. You must have a net force to accelerate and rise from a sitting position, so you must have an upward push from the floor greater than your weight. So (b) is the answer.

12. Even if the curb is sloped rather than vertical, you'd pull it up and over. A push increases the normal force against the curb, whereas an upward pull is in the direction the cart has to move.

16. On the moon there is no atmosphere and the rate of acceleration common to *every* falling object near its surface is $g/6$. Both feather and baseball will accelerate at the same rate toward the surface. If they are given the same vertical speed, they'll rise and fall together. If both are given the same horizontal speed at the same time they get identical vertical speeds, they will travel in a curve up and across the moon's surface and come down to luna at the same time and at the same distance away from where they were launched. In other words, if thrown in a way to give them identical initial velocities, their motions will be identical. Could you throw a feather farther? The feather has the least mass, and your arm could in all probability push it forward *faster* than it could heave the more massive baseball. You might well be able to give the feather the largest initial velocity, and the feather would go farther than the baseball would.

18. Two descending parachutists who do not weigh the same but who have identical chutes, both open, will have

different terminal speeds. The terminal speed, even with the chute open, is the speed at which the air drag balances the parachutist's weight. Air drag, remember, increases with speed. A heavier parachutist at terminal speed falls faster than a lighter one because it takes a greater air drag on that parachute to balance the greater weight. So the heaviest parachutist in the plane will have the greatest terminal speed and will be traveling faster when approaching the ground. (Does the air resistance on the person's body matter? Not when compared to the huge area of the chute. Terminal speeds are about 10 mph to 12 mph, and at that slow speed a person's air drag isn't much. What about the parachutist's terminal speed *before* the chute opens? Different-size people have different surface areas, of course. But we'll treat this question in detail in Chapter 10.)

**21.** A long rope would mean the angle of pull was more nearly horizontal, and the component of force in that direction would give a larger acceleration to the box.

**23.** The more tire area in contact with the sand, the less pressure there is on the sand. And as you can see from footprints made while jogging or walking, the less pressure there is, the less the sand is depressed or shifted when the foot pushes off. Less pressure means the car is less likely to dig a hole in the sand. (Vehicles that operate in the mud need larger tires for that reason, among others.)

**25.** Going uphill, less of the car's weight presses directly against the surface, so there's less normal force, which means there can't be as much frictional force (or traction). The tires slip (or spin off) more easily, especially since a component of gravity is pulling the car down the hill. The same thing applies when going downhill, but since a component of gravity pulls the car forward, less push from the wheels needs to be used to accelerate at a given speed. Nevertheless, when starting from rest going down a hill, it is easier for tires to slip than if the car was on level ground.

**27.** The flat sides of the bones have more surface area to support things; the pressure on them is somewhat less than if those bones were rounded. Besides being more comfortable, less pressure places less stress on bones and flesh.

**28.** Take the *outside* lane. The curve of the outside lane has a larger radius, reducing the centripetal force necessary to keep the car moving along the curve.

**29.** 25 lb toward the shoulders, and 5 lb toward the ground.

**30.** 300; 1/300.

**31.** The spring exerts equal but opposite pushes, then pulls, on each mass, causing them to pull apart, then come back together, and continue to repeat these motions. (The spring only pushes, not knowing what is on either end. A Cadillac would receive the same push as a Nissan.) The mass $m$ would have 3 times as great an acceleration as the mass $3m$ as both moved back and forth. Likewise, the lighter mass would travel farther—in fact, 3 times as far, as you can see in Exercise 42 in Chapter 5.

**33.** In this case the thrust is not horizontal but at an upward angle, along the plane's direction. One component of thrust is vertical, opposing gravity, while another is horizontal. Drag points backward along the plane's path, and lift points perpendicular to the airplane's path. (Draw a sketch to see this.) Therefore, drag plus one component of gravity balances the thrust, making the net force along the plane's path equal to zero. The lift is balanced by the other, perpendicular component of gravity, making the net force perpendicular to the plane's path equal to zero.

**35.** The pebble leaves *tangent* to the surface of the tire, in the direction of the pebble's velocity at the instant the centripetal force from the tire vanishes. The pebble is more likely to leave if the tire is turning fast, when the pebble needs a larger centripetal force to keep it going around with the tire.

**37.** True. If your car moves at a constant speed, the *sum of the external forces must be zero*. On a level surface, there are only two forces in the horizontal direction, air drag and the push of the road on the tires. In this case, they cancel.

**40.** Be sure to try this.

**42.** By turning the rope around once, any tension applied to the rope will tighten the loop and squeeze the rail, causing a large normal force and a correspondingly large frictional force to aid in supporting the tension.

**45.** In your sketch, make certain the normal force from the ground on the car is perpendicular to the surface of the banked road. Then the normal force has a component that points vertically, toward the center of the curve. This component contributes to the centripetal force that turns the car and reduces the amount of frictional force needed from the road to do that turning.

**47.** Friction on the legs is less because of those narrow seats.

**49.** The falcon is not only streamlined, it is also lightweight. That is, it has more surface area per gram of mass than a baseball. Air resistance, therefore, affects it a great deal more than it does a baseball.

**51.** No! It would take you a longer distance to stop for two reasons. First, gravity has a component that acts to pull the car downhill, and the friction on the car's tires has to balance that force. Second, the normal force between the car and the ground is less when the car is on a downhill slope. (Less of the car's weight presses directly against the road.) That means the maximum value of the static friction between the tires and the road is smaller (by the same percentage, in fact), so the car's brakes couldn't be applied as hard before the tires would skid.

**52.** (a) $a = (v_f - v_i)/t = (12 \text{ mph} - 120 \text{ mph})/2 \text{ s} = -54 \text{ mph/s}$. But $-54$ mph is the same as $-79.2$ ft/s, so

$a = -79.2$ ft/s$^2$. (b) $g = 32$ ft/s$^2$, so $a/g = -79.2/32$ (the units cancel) and $a/g = -2.5$, or $a = -2.5g$.

**53.** The maximum force the car's tires can get from the road without slipping is $F_{max} = m_{car}a_{max}$. When the trailer is added, the total mass to be accelerated is $3m_{car}$. The maximum force the car's tires get from the road won't change, provided the trailer is balanced and doesn't exert a great weight on the trailer hitch. The new maximum acceleration is found from $F_{max} = 3m_{car}a_{new}$. Using $F_{max} = m_{car}a_{max}$, we get $3m_{car}a_{new} = m_{car}a_{max}$, so $a_{new} = \frac{1}{3}a_{max} = \frac{2}{3}$ m/s$^2$.

**54.** On the moon the car presses down with 1/6 the force it does on earth, so the friction it can use to stop is also 1/6 as great. The car has the same mass $m$ on either surface, so $F = ma$ tells us that its (negative) acceleration on the moon will be 1/6 as great as on earth. The initial and final speeds on the moon and earth are the same, and since $\bar{a} = (v_f - v_i)/t$, on the moon $t$ must be 6 times as great as on earth to give an acceleration 1/6 as large.

**55.** The skydiver's acceleration begins with the value of $g$ and ends with the value zero (at terminal speed). The average (or uniform) rate of change of acceleration is $(a_f - a_i)/t = -g/12$ s $= (-32$ ft/s$^2)/12$ s $= -2\frac{2}{3}$ ft/s$^3$.

**56.** The lift on the 747 must balance its weight when it is in level flight. Therefore, the lift is just 600,000 lb. With 7000 sq ft of area, the average pressure under its wings is $P = F/A = 600{,}000$ lb/7000 ft$^2$ = 85 lb/ft$^2$ or about 0.6 lb/in.$^2$, since 1 sq ft is equal to 144 sq in. (12 in. × 12 in.). Atmospheric pressure at sea level is about 14.7 lb/in.$^2$, so $P_{wings}/P_{atmosphere} = 0.6/14.7 = 0.04$. The pressure on the wings during level flight is only four-hundreths of the pressure that the atmosphere exerts on the earth's surface.

**57.** $P = F/A$, and $F = 180$ gallons × (8.3 lb/gal) = 1500 lb. The area is 5 ft × 6.5 ft = 32.5 ft$^2$, so $P =$ 1500 lb/32.5 ft$^2$ = 46 lb/ft$^2$, or about 0.32 lb/in.$^2$. Let's see what this pressure is in atmospheres to compare it with the pressure under your feet when you are standing. (See Demonstration 3.) $P_{waterbed}/P_{atmosphere} = 0.32/14.7 = 0.02$. So $P_{waterbed} = 0.02$ atm. Not much, considering that the pressure under your feet is about 1/3 atm. It is the total weight of the waterbed, about 3/4 ton, not the pressure, that worries apartment owners.

**58.** The force it needs is $mv^2/r$. Its speed is 55 mph, or 80.7 ft/s. On a dry concrete road its maximum frictional force is about equal to its weight, so $mg = mv^2/r$. Then $r = v^2/g = (80.7^2/32)$ ft $= 200$ ft.

## Chapter 5

**2.** Any change in momentum $mv$ is equal to a force $F$ times the time $t$ the force is applied. If the momentum doesn't change, the net force $F_{net}$ on the object must be zero.

**4.** Each gets the same momentum change, but in opposite directions. However, because Dumbo's mass is so great in comparison to the mass of a peanut, Dumbo's final speed would be near zero.

**6.** Most bicycles weigh much less than most riders do, so the bike probably has a smaller mass. Since both move at the same speed, the bike also has the smaller momentum. The total momentum of bike and person is $(m_{bike} + m_{person}) \times v$. If the bike hits the curb, it get a tremendous normal force from the curb, one that might very well bring its momentum to zero. The rider, however, depends on friction from the seat and a good grip on the handlebars to stay on. Because the rider has a greater momentum, the rider requires a much greater force to stop in the same amount of time the bike stops. Therefore, the greater momentum may well carry the rider on over the handlebars.

**8.** The answer is (c). During the collision the total momentum is constant, so the change in momentum of either the bat or the ball is equal but opposite in direction to the change in momentum of the other. (Of course, if the batter is swinging the bat at the time of impact, the bat's momentum may be changing because of a force from the batter's hands.)

**10.** Yes, the crumpled parts decelerated faster; the force on them was larger. As they crumpled, the other parts of the car stopped more slowly and weren't damaged. It's the same thing that happens if an apple falls from a tree. Only the spot that hits the ground is bruised, and the rest of the apple is unharmed.

**12.** Suppose that immediately after the collision the relative speed of the ball and bat is about 50 percent of their relative speed before the collision. If the ball is pitched toward the bat, the ball and bat approach one another faster (the relative speed of the ball and bat is higher), which means that the relative speed after the collision will be greater. The bat continues moving forward immediately after the collision (in either case), so the relative speed immediately afterward is the speed with which the ball races ahead of the bat. Thus, a hit from a fastball leaves the bat to travel over the ground much faster than a hit from a ball that's merely tossed up.

**14.** The racket is a great deal more massive than the tennis ball. Equal but opposite changes in (mass × speed) mean the racket's speed changes very little compared to the tennis ball.

**16.** While standing, a racer's knee can flex and bend when the cycle hits a bump and act as shock absorbers by increasing the time over which the body's vertical momentum changes. The longer the time, the smaller the average force on the body.

**17.** If their total momentum *after* the collision is zero, it must have been zero *before* the collision. Although $(m \times \vec{v})$ for each player was the same, their velocities must have been pointing in opposite directions, so that the vector $\overrightarrow{mv}$

of one could add to the vector $\overrightarrow{mv}$ of the other to give zero. Then the perfectly inelastic collision can bring them to rest in midair.

**20.** To change the stuck vehicle's momentum from zero to a certain speed, a certain impulse must be delivered. If a rope is used, it can stretch a bit, increasing $t$ and decreasing $F$ as compared to a nonstretchable towchain. Besides preventing damage to the vehicles from the larger, more abrupt force a towchain would deliver, the towing vehicle's traction with the ground is less likely to be broken if the reaction force on it is less.

**22.** Perfectly elastic means with no deformation at all. Therefore, cars wouldn't bend and crunch during a collision, and repair bills would be less. But the relative speed afterward would equal the relative speed before, and the change in speed (or velocity) during a perfectly elastic collision is greater than for an inelastic collision. The passengers, then, would have greater changes in momentum, which means the forces on them would be larger.

**24.** The shoes' soles compress, giving your body extra *time* to come to a halt. That lowers the average force on your feet. (Your change in momentum, hence your impulse, is the same whether or not you have shoes on.)

**26.** The first reason is to increase the time of impact and so lower the average force that stops the fall. The second reason comes from Chapter 4. The larger the area of the body that takes the impact, the smaller the pressure $(P = F/A)$, and thus the less chance of an injury.

**28.** The boat moves forward with each step you take to the rear. The boat's change in momentum with each step is equal but opposite to yours, to the extent that the water's resistance can be neglected. (If you walk slowly, the boat moves slowly and there's less resistance from the water.)

**30.** (b). The rocket's thrust times 1 second is the impulse delivered during that second, which is equal to $(mv_f - mv_i)$ or $m(v_f - v_i)$. Initially $m$ is very large, so the change in speed over that second is small. Later, when $m$ is much smaller, the change in speed from an equal impulse is large by comparison. Acceleration is change in speed/time, so the rocket's acceleration is greatest when its mass is least.

**31.** The jettison rockets are located on the *spent* stage. Empty of fuel, its small mass responds more quickly to an impulse than the yet full upper stage(s). (Another practical reason is so the jettison rockets won't change the orientation of the upper stage(s) before the next burn.)

**33.** You'll move faster if you catch the ball and throw it back. Here's how to tell. *The ball's total momentum change is equal but opposite to yours.* It changes by a certain amount if you just catch it and stop it, but if you throw it back, it changes even more. You'll move faster in option (b) and slowest in option (a). Option (c) will give you a speed between (a) and (b). If it drops straight to the ground, the ball's momentum changes more than if you catch it and it continues to go forward at your speed, yet less than if you threw it back.

**34.** $m_{ball}v_{ball} = 0.014$ slug $\times$ 58.7 ft/s $= 0.82$ slug·ft/s. $m_{firetruck}\,v_{firetruck} = 100$ slugs $\times$ 7.33 ft/s $= 7333$ slug·ft/s.

**35.** Their momentum changes are equal and opposite, and one's mass is twice as large as the other. Both start at rest, so $m_1v_1 = m_2v_2$. Let $m_2 = \frac{1}{2}m_1$, and you'll see that $v_2$, the speed of the 100-lb person, is twice the speed of the 200-lb person.

**36.** $mv = (0.149 \text{ kg}) \times (40 \text{ m/s}) = 6$ kg·m/s. Using impulse = change in momentum, we have $F \times (\frac{1}{10}\text{ s}) = (0 \text{ kg·m/s}) - (6 \text{ kg·m/s})$ or $F = -60$ N, about 13½ lb. No wonder you can hear the leather of the mitt pop.

**37.** The impulse–momentum relation is $\overline{F} \times t = m(v_f - v_i)$ where $v_f$, the final speed of the jumper, is zero. Using $\overline{F} = m\overline{a}$, we see $m\overline{a} \times 0.0045 \text{ s} = m(0 - v_i)$, so $\overline{a} = -v_i/0.0045$ s. To calculate $\overline{a}$, then, we must find $v_i$, the jumper's speed at the instant of contact with the floor. The jumper falls 3 ft *(d)* with an acceleration of $g$. Equations (1–2) and (1–3) let us calculate that speed. $d = \frac{1}{2}gt^2$ and $v = gt$ apply, where $v$ is the speed of the jumper if $t$ is the time it takes to fall 3 ft. Let's eliminate $t$ in these two formulas. Square the second to get $t^2 = v^2/g^2$, and substitute this value of $t^2$ into the first to find $d = \frac{1}{2}v^2/g$. Solving for $v^2$, we find $v^2 = 2gd = 2(32 \text{ ft/s}^2)(3 \text{ ft}) = 196 \text{ ft}^2/\text{s}^2$. From this, $v = 14$ ft/s. So $\overline{a} = v_i/0.0045$ s gives $\overline{a} = 3111 \text{ ft/s}^2$ or $\overline{a}/g = -97$, so $\overline{a} = -97$ g.

**38.** For the first second of liftoff, we have $(F_{thrust} - mg) \times (1 \text{ s}) = m_{initial} \times (\text{change in speed})_{initial\ second}$. The fuel burned in one second is a very small percentage of the initial mass. For the last second of burn, we make the assumption that the rocket is still traveling straight up and that the acceleration of gravity at its new elevation is still approximately $g$. $(F_{thrust} - mg) \times (1 \text{ s}) = m_{final} \times (\text{change in speed})_{final\ second}$. If $F_{thrust}$ is the same, these two impulses are the same, so we can set the right-hand sides of these equations equal to each other. $(\text{change in speed})_f = (\text{initial mass/final mass})\,(\text{change in speed})_i = 17.5\,(\text{change in speed})_i$. In the last second of the burn, the Saturn V's change in speed (and hence acceleration) was about 17.5 times as great as during the first second of liftoff.

**39.** $F = (0.05 \text{ kg} \times 1 \text{ m/s})/0.1 \text{ s} = 0.5$ N (or about 0.11 lb, or 1.8 oz).

**40.** Initial momentum = final momentum. $(3\ m \times 5 \text{ mph}) + (1\ m \times 0 \text{ mph}) = 4\ m \times v_{final}$. Solving, we have $v_{final} = 3.75$ mph.

**41.** The initial relative speed is 5 mph. The final relative speed is 25 percent of 5 mph, or 1.25 mph. Because the massive engine (E) bumps the lighter flatcar (F), we'll assume they travel in the same direction. So

$$v'_F - v'_E = 1.25 \text{ mph} \quad (1)$$

where the primes mean speeds *after* the collision. The *net initial momentum* is equal to the *net final momentum,* or $m_E v_E + m_F v_F = m_E v'_E + m_F v'_F$. Using $m_E = 3m_F$, $v_F = 0$, and $v_E = 5$ mph, and canceling $m_F$, which now appears in every term of this equation, we find

$$15 \text{ mph} = 3v'_E + v'_F \quad (2)$$

From Eq. (1) above, we see $v'_F = v'_E + 1.25$ mph, and substituting this into Eq. (2), we have 15 mph $= 4v'_E + 1.25$ mph. Therefore, $v'_E = 3.44$ mph. Use this in Eq. (1) above to find $v'_F = 4.69$ mph. (Our assumption that the engine followed the car was correct; otherwise, the value of $v'_E$ would be negative.)

42. These masses start at rest, and each moves some distance from where it begins before the attached spring stops them and reverses their directions. If there is no friction, the equal and opposite forces from the spring are the only forces acting. That means the total momentum of the masses is constant. As in Example 5 in "Calculations," $3m \times d = m \times D$, the $m$'s cancel, and the less massive block travels 3 times as far as the one with three times as much mass.

43. (a) The relative speed before the collision is 90 mph + 70 mph = 160 mph, since the bat and ball travel in opposite directions. (b) Immediately after the collision, the bat and ball travel in the same direction toward the pitcher, so their relative speed is the difference between their new speeds. The relative speed afterward has a value of 80 mph (50 percent of 160 mph). If we *neglect* the speed the bat loses (since the bat is so massive compared to the ball), we can say that $v_{ball} - v_{bat} = v_{ball} - 70$ mph $= 80$ mph. Thus, $v_{ball} = 80$ mph + 70 mph = 150 mph. The same ball that the pitcher threw at 90 mph returns to him at 150 mph. (For a more accurate calculation, you could follow the procedure in the answer to Exercise 41. You'd need the mass of the bat and ball to do that.)

44. Human passengers inside an orbiting spacecraft can hover briefly without touching anything. Eventually, however, they'll drift about and touch the spacecraft. Then any motion by them exerts a push or pull on the spacecraft. The craft gets a change in momentum equal and opposite to the passengers' change in momentum. Such a change would affect a telescope's direction during a lengthy exposure. Even if the passengers were held motionless in seatbelts and harnesses, heartbeats and breathing actions would jiggle the camera.

## Chapter 6

2. The *least* amount of work is done if the bed is about the height of your navel. Here's how to tell that's true. If you sleep on the floor, you'll do the *most* work in getting up. Your head must be raised to almost your full height, and the *average* distance each pound of you is raised is about halfway up. It's easier to get up from the edge of your bed than from the floor because you don't have to lift quite as much of your mass and you don't have to lift it that far. In fact, if your bed were navel high, *no net work* would be done to get you to a standing position. The lower half of your body would drop (positive work by gravity) about as far as the upper half would rise (negative work by gravity). On the other hand, such workless beds wouldn't be as easy to get onto, and they'd be harder to make up. It would also be dangerous to roll off in your sleep.

4. When a wagon or car rolls along on wheels over level ground, its load moves along the road without any work against gravity. Each wheel rolls without rising or falling, and the people or animals or engine doing the pulling have to overcome only the inertia to get the load moving and then friction (and in the case of a car, air resistance). The Incas used circular shapes in designs and for other purposes. In fact, the reason they didn't use wheels was probably that they didn't have metallurgy. The first widespread use of wheels in Europe and the Middle East apparently wasn't until the Bronze Age, when thick wooden wheels could have metal rims for strength and could then be shaved down in weight. Solid wooden wheels are horribly heavy, and easy to split and cause tremendous wear and heat as the wooden axles turn in wooden sleeves.

8. Gravity does the same work $W$ on each, so if there's no friction, they'll both have the same speed $v$ (because $\frac{1}{2} mv_f^2$ will be the same for each). But the guy on the left will have the greater vertical acceleration and will get to the water first.

11. About the same amount. When you lift a leg, you move only its mass, not the mass of what you carry. Of course, the other leg would be under more strain because of the extra weight, and there'd be some extra muscular energy used there, but not much. Your bones do the supporting if the leg is straight and vertical. On the contrary, if you bob up and down a great deal as you walk, you lift the mass of your body plus the luggage, and that takes more work.

13. False. An object's velocity can change direction without changing speed, in which case the kinetic energy ($\frac{1}{2} mv^2$) is not affected. Net work has been done only if the values for mass and/or speed change in such a way that $\frac{1}{2} mv^2$ changes.

15. Unlike the guys pushing on the pillar in Fig. 6–1, this fellow is doing lots of work. Now the boat doesn't move, so he's *not* doing work on the boat. The water moves as he paddles, however, so he does work on the water.

16. As long as the chain is perpendicular to the swing's direction of motion, it can do no work. (If a child leans back and pumps, the chains may bend where the hands

pull on them, and then they do some work, but the net work comes from the child's pumping legs.)

**19.** About half, if the pile of dirt is initially level with the top of the wall. At the top the shovel won't have to do work against gravity; its motion could be horizontal. At the halfway mark, the shovel does work on each shovelfull to take it half as high as it will at the bottom. The *average* work would be about half as much if he started at the top.

**21.** With regard to *energy usage,* the same amount of work would be done against gravity in a quick elevator as a slow one. The power company charges not for kilowatts (power) but for *kilowatt hours* (energy), as you can see from your power bill. But the administrators would still have a valid reason to refuse Professor Quick. The electric motors needed for such great, sustained accelerations would be large both in size and cost. The expenses have to do with equipment, not energy. (Not to mention the discomfort of the passengers, of course.)

**23.** While "porpoising," less water resists a porpoise's forward motion. (Of course, the porpoise spends a little extra energy to arc out of the water, but it probably saves more energy by porpoising anyway, especially since it has to surface to breathe.)

**25.** It pulls itself up, doing work $\overline{F} \times d$ in steps up the tree. It gains a potential energy during each step equal to the work it does, because for each step $d$ the force of gravity does a negative amount of work $-mg \times d$. Gravity's work is equal to $\overline{F} \times d$, since the cat, hanging on with no net gain in speed, has no net work done on it.

**30.** It becomes heat. The brakes stop the car with a frictional force. The work of the frictional force of the touching brake pads and drums (or brake pads and disks) is negative, reducing the kinetic energy by converting it to heat.

**32.** If the bike frame bends as the rider pumps hard on the pedals, the rider does work in bending the frame through a certain distance. That work can't go to propel the bike along, so it's wasted effort. (The frame bends back, of course, but most of the energy that goes into stretching the metal turns into heat.) A perfectly rigid frame, one that doesn't bend, wastes none of the biker's energy in that manner.

**33.** The veins are somewhat elastic. When the heart pumps, they stretch, storing some of the work done to stretch them as potential energy. As the blood pressure drops, they squeeze down a bit as they collapse, helping to maintain a (reduced) pressure and hence flow of blood. This effect is greatest in the aorta, the huge artery leaving the heart. There large quantities of elastic tissue called *elastin* expand greatly with each heartbeat, store work as potential energy, and contract or shorten as soon as the beat is over. This helps to maintain the blood flow even between heartbeats.

**38.** When the crew raises or lowers a backdrop, no net work is done by gravity! As the backdrop moves up or down, the counterweight moves in the opposite direction for the same distance. The crew only needs to overcome a little friction and the inertia of the backdrop plus counterweight.

**40.** *The body is less efficient at converting energy into muscular force when the rate of using energy is high.* Training helps. (Untrained muscles have been shown to be only ⅐ as capable of using fatty acids as trained muscles, for example.) There are also the extra energy expenditures a runner has that don't go toward pushing the runner faster. An average male, for instance, inhales about 6 liters (a liter is a volume of about 1 quart) of air per minute while at rest. In a cross-country run the same man may average 150 liters per minute. To breathe that fast, he must force air in and out of his lungs, using perhaps 10 times as much energy per minute as he would for normal breathing. There's some air drag, and the extra extension of tendons and muscles in running uses up some energy. The heart works much harder to get the oxygen to the muscles and also to pump more blood to dilated vessels under the skin for cooling.

**43.** (a) ½ × (your weight in lb/2.2 lb/kg) × $(1\ \text{m/s})^2$ (b) $mgh$ = (your weight in lb) × (4.48 N/lb) × 10 $\text{m/s}^2$ × (½ × your height in meters) (c) The answer is the result of part (b) divided by 1 s.

**44.** For purposes of comparison, you needn't worry about units, so $\tfrac{1}{2}m(0 - 50^2) = \tfrac{1}{2}m(2500)$ and $\tfrac{1}{2}m(50^2 - 70^2) = -\tfrac{1}{2}m(2400)$. Thus the amounts of work are almost equal.

**45.** No; an empty roller coaster will fall as fast as a full one provided the air resistance isn't a factor. If we neglect air drag, the only force doing work is gravity, which gives the same acceleration to any mass. $F \times d_{\text{fall}} = (\tfrac{1}{2}mv_f^2 - \tfrac{1}{2}mv_i^2)$ and $F = mg$, so the $m$'s in each term of the equation can all be divided out. Thus the speed at the bottom is independent of mass.

**46.** Gravity does work $-mg \times d$ if you walk up a distance $d$ above the ground, so your positive work must equal $mg \times d$. 2500 Calories = $2.5 \times 10^6$ calories = $10.45 \times 10^6$ J. So ($10.45 \times 10^6$ J)/$mg$, your weight in newtons, equals $d$. If you weigh 125 lb (556 N), for example, one day's food supplies enough energy to lift you a distance $d$ equal to 18,795 m, over 2 times the height of Mt. Everest. Obviously your body cannot convert nearly all the food energy into physical work $W$.

**47.** $\tfrac{1}{2}mv^2 = \tfrac{1}{2}(0.005\ \text{kg}) \times (25\ \text{m/s})^2 = 1.56$ J.

**48.** The KE lost is $(\tfrac{1}{2}mv_f^2 - \tfrac{1}{2}mv_i^2)$. To see what percentage of the original KE is lost, divide by $(\tfrac{1}{2}mv_f^2)$ and multiply by 100. The mass $m$ cancels, and the percentage of KE lost is 19%.

**49.** You would use 90 Cal/hr × 24 hr/day = 2160 Cal/day. To estimate your body fat in lbs, multiply your weight by 0.22 (females) or by 0.15 (males). Multiply by

4200 Cal/lb and divide by 2160 Cal/day to see how long your fat supply could support you. For a 120-lb female, the estimate is 51 days.

50. About 52.5 days

51. The pound of tuna represents 480 calories, so the algae consumed (assume 1000 pounds) represents $4.8 \times 10^5$ Calories. The algae are only about 1 percent efficient at turning sunlight into molecular energy, so $4.8 \times 10^7$ Calories $= 4.8 \times 10^{10}$ calories $= 20 \times 10^{10}$ joules.

52. See Example 4. (a) $\bar{a} = -g\ (h/d) = -10g$; (b) $\bar{a} = -4g$.

## Chapter 7

2. $F_g = G(m_1m_2/d^2)$. If $m_2$, say, is doubled, the $F_g$ is doubled. So the sun's pull on earth would be twice as large, and *so would the earth's pull on the sun,* even though no mass was added to earth.

4. Yes. Suppose a spacecraft is directly between the earth and moon. Their pulls of gravity oppose one another; they point in opposite directions. If the amount of each pull is the same at any point between the surface of earth and the moon, the two forces will cancel. (For the earth and moon there is such a point; it lies about 24,000 miles from the moon, or about 216,000 miles from earth. Check this!)

7. Shrinking means $R_{\text{earth}}$ gets smaller. That makes $R^2_{\text{earth}}$ smaller, too, and dividing by a smaller number makes a ratio *larger*. As $R^2_{\text{earth}}$ decreases, the value of $g = G(M_{\text{earth}}/R^2_{\text{earth}})$ increases.

9. True. Your weight is the pull of gravity on your mass, and $F = ma$.

11. The bullet traveling in the barrel can't move vertically. But immediately upon leaving, it accelerates downward at the rate of $g$, regardless of how fast it's moving in any direction, and whether or not air resistance is slowing it in the horizontal direction.

12. The pull of the mountains to the side isn't balanced by the mountains on the other side. The mass of the Himalayas deflects a plumb bob (a lead weight hanging freely by a string) about 5⁄3600 of 1 degree from the vertical, for example, if you are in India near the border.

14. Only on the surface of a sphere is there no downhill. On a mass as large as a planet or a star, anything free to move on its surface tends to roll or move downhill, filling up the valleys and making the surface more spherical. When the planets and moons formed, they apparently accumulated by gravitational attraction of smaller bodies. During the collisions cracking, compressing, and heating took place and materials shifted about. As the mass grew larger, its attractions at its surface got larger, and the "rounding" proceeded.

17. The shuttle is freely falling and travels horizontally over the plane only instantaneously. Lift holds the plane and its contents in a horizontal path, so the passengers get a normal force from the floor for support.

19. Stopped in its orbit, it would continue to be attracted to the sun and would take a 93-million-mile free fall. (Incidentally, calculations show it would need about 204 days to reach the sun.)

22. They are at just the right distance to revolve around the center of the earth at the same rate as the surface of the earth revolves about earth's axis! If they were farther out, they'd travel slower and drift westward as seen from earth. Closer in, they'd orbit faster and drift eastward. (Because they orbit earth's *center,* they have to orbit in the plane of the equator to stay over one point. Otherwise, they'd swing from one hemisphere to the other during each revolution.)

26. (a) Four. (b) Elliptical; the apogee and perigee are apparent. (c) Its speed was increased, doing work to raise the spacecraft against the force of gravity. But in its new (and higher) orbit, its orbital speed is a bit slower.

28. No. Orbiting with the shuttle and the astronaut, a neglected monkey wrench, for example, would float along beside the shuttle in an orbit of its own. If the astronaut wanted it to drop to earth, he or she need only to throw it backward to reduce its speed. However, it would take a 17,000-mph heave to get it to drop straight down to earth.

29. He would drift up and slowly curve toward the west over his launch pad. (See Fig. 7–32.)

32. (a) 12 hr, 27.75 min (b) 12 hr, 25 min

34. (a) About 9:25 P.M.; (b) about 3:12 P.M.; (c) about 9:50 A.M.

37. *Larger,* because the moon's pull would have an even greater percentage change from one side of earth to the other.

38. The tidal forces do work on the atmosphere, but the effect is masked by the effect of the sun's rays. Each day the sun heats the air and the atmosphere swells, only to cool and shrink on the "night" side of earth.

40. The earth's circumference is about 25,000 miles, and a point on the equator travels that far around the earth's axis in 24 hours. The waters there move at about 1000 mph as the earth's surface turns beneath the moon (which takes about 27.3 days to orbit earth once). Should those waters follow the moon, those currents would exceed the speed of sound. Beachgoers wouldn't have time to duck!

42. Satellites are made of solid materials kept together with chemical and molecular bonds, not gravitational attraction. These bonds between these small particles are very strong compared to the attraction of gravity between them.

44. It couldn't be at rest. Action–reaction guarantees that the sun will move in response to the pulls of all its orbiting planets.

45. $v = d/t = 2\pi r/T = 2 \times 3.1416 \times 3.84 \times 10^5$ km/(27.3 × 24 × 3600s) = 1.02 km/s or 0.64 mi/s.

46. At the surface the force is equal to your weight, $mg$. That's equal to $F_g = G(m_{you}M_{earth}/R^2_{earth})$. (a) If you go 2 radii from earth's center, then $F_g = G(m_{you}M_{earth}/(2R_{earth})^2 = \frac{1}{4} G(m_{you}M_{earth}/R^2_{earth}) = \frac{1}{4}mg = \frac{1}{4}$ your weight at the surface. (b) If you were 3 earth radii out (2 radii from earth's surface), $F_g = \frac{1}{9}G(m_{you}M_{earth}/R^2_{earth}) = \frac{1}{9}$ your weight at the surface.

47. $a_{Mars}/g = 0.39$

48. $M_{earth} = (1.02 \times 10^3 \text{m/s})^2 \times (3.844 \times 10^8 \text{ m})/6.67 \times 10^{-11} \text{ N·m}^2/\text{kg}^2 = 5.99 \times 10^{24}$ kg. (The correct value to three significant figures is $5.98 \times 10^{24}$ kg.)

49. $M_{Jupiter} = 1.9 \times 10^{27}$ kg

50. $F_g = 2.835 \times 10^{-8} \text{ N} = 6.37 \times 10^{-9}$ lb

51. $F_g = G(m_{you}M_{sun}/d^2_{sun}) = m_{you} \times GM_{sun}/d^2_{sun} = m_{you} \times (6 \times 10^{-5} \text{ m/s}^2)$. If your mass is 60 kg, for example, $F_g = 3.6 \times 10^{-3}$ N (or $8 \times 10^{-4}$ lb).

52. $8g$

53. 29.8 km/s (or 17.6 mi/s)

54. $F_{sun}/F_{moon} = (Gm_eM_{sun}/d^2_{sun})/(Gm_eM_{moon}/d^2_{moon}) = 181$

55. $mg \times 2\text{ft} = m \times (.39g) \times h_{Mars}$; $h_{Mars} = 5.13$ ft

## Chapter 8

3. Because they have so much more rotational inertia. The torques provided by the air on the blades and friction at the fan's axle take longer to slow the larger fan blade's rate of rotation.

4. The farther mass is from the axis of rotation, the larger the rotational inertia and the greater the torque needed to rotate the upper body.

6. To get the door to move requires a certain minimum torque, so you'd have to push twice as hard because the lever arm at the center would be half as large.

8. The tennis racket, like the moon in orbit, was rotating as it revolved about her. For the moon to face earth, it rotates once for every revolution (see Exercise 5), and so does her tennis racket. When released, it continues to rotate as it moves off. (This is easy to see if you draw a sketch.)

12. When a jar's lid is on very tight, you must use a large torque to turn it. You can grip the lid firmly around the edge with your hand and give it a twist with your wrist, letting your hand swivel about your forearm. But if that is not enough, you can increase your torque this way. Tense your hand so it won't rotate about your forearm. Then your forearm and wrist act exactly like a wrench, with your forearm as the handle and your hand and wrist as the head that grips the jar's lid. When you use your upper arm to rotate this "wrench," it has a much longer lever arm, and you can exert more torque. This often works on stuck jar lids.

15. First of all, it is perfectly balanced and supported so that its center of mass doesn't move as the telescope and mount move. Second, only a very small torque is required to turn even such a huge object at the rate of one rotation per day.

17. No, the moment of inertia depends only on the mass and its distribution around the axis of rotation. However, if the distribution of mass changes because of a solid's rate of rotation, as when a ball of pizza dough is rapidly twirled, its moment of inertia does change.

19. Being large in diameter makes their rotational inertia very great unless the mass is really small. The greater the rotational inertia, the slower the angular spin will change. The second reason has to do directly with the bicycle's translational motion. Any increase in mass for the bike means that more force is required to accelerate it at a given rate.

21. The strength of the wood against twisting, that is, its ability not to crack while supporting itself at an angle.

24. If you stand vertically in a strong wind, your feet react to its push by summoning a frictional force from the ground. Unless the wind's speed is very great, your feet won't slip. Because they don't slip, however, the place where they touch the ground becomes a possible axis of rotation for your body. When you are vertical, the wind's push exerts a torque to rotate you over your feet. If the push is strong, you will fall unless you lean into the wind. Then the force of gravity, acting on your center of gravity, has a lever arm about your feet and exerts a torque that, unopposed, would rotate you in the other direction. If you lean at just the right angle, the two torques cancel, and you won't fall in either direction.

26. Yes! Your torque must equal (and slightly exceed) gravity's torque, but gravity has a smaller lever arm than you do. That means you can exert a smaller force than gravity. Draw a sketch to see this.

28. The frictional forces from the ground that accelerate a car point *forward*. This exerts a torque about the center of mass of the car that tends to rotate its rear end downward and its front end upward. Draw a sketch!

30. This increases the distance of his or her mass from the potential axis of rotation, where the front wheel touches the ground. The torque applied by gravity that counteracts the torque of friction from the road on the front wheel increases because of that increased lever arm.

32. A boot's strong, higher-than-ankle construction can prevent the wearer from twisting an ankle on a rough, uneven trail, especially if a hiker has a heavy backpack on.

34. They scoot to the right, courtesy of the centrifugal force, as seen by you on the seat.

36. Review the first page of this chapter. Launched eastward from Kennedy Space Center (near Orlando, Florida), a rocket already has a speed of about 915 mph around the

earth's axis, hence about earth's center, which the spacecraft will eventually orbit. Launched westward from California, a rocket must reach a speed of almost 900 mph before its speed becomes *zero* with respect to an *inertial* frame of reference.

**38.** Because the motorcycle is revolving about the center of a circle, it is in a rotating frame of reference and feels a centrifugal force field. The forces acting are the same ones that would act if the cycle were resting vertically at the edge of a rotating platform. The cyclist's helmet is pushed outward, the cyclist is pushed outward, and the cycle is pushed outward. Draw a sketch of this motorcycle and show the forces. The force that stops the cycle and cylist from moving outward in the rotating frame (or continuing in a straight line in the inertial frame) is the static friction as the tires are pressed against the ground. That force points inward, but more importantly, it defines an axis of rotation (so long as the cycle doesn't slip). The centrifugal force, then, exerts a torque that will rotate the vertical cycle about the tires, causing it to fall over. But if the cyclist leans the cycle inward, gravity acts at the center of mass and has a lever arm about the point where the tires touch the ground. The torque counteracts the torque of the centrifugal force and can prevent the rotation.

**40.** (a) The second from the left. (b) The one on the far right. Just observe how far their centers of gravity need to be lifted and moved to the sides if they rotate clockwise.

**42.** In the observer's inertial frame of reference, there is no centrifugal (center-fleeing) force, but there is centripetal (center-seeking) force. The observer sees earth rotate, and a mountain at the equator, for example, has to have a force $mv^2/r$ to keep it in its circular motion. The only force acting that can supply this force is the earth's gravitational attraction for the mountain. The observer would say that 0.35 percent of earth's gravity supplies the centripetal force; the rest presses that mountain against the surface. This explains why the equatorial matter is lighter than a similar amount of mass at the poles. This fact, in turn, explains the flattening.

**48.** There are pros and cons. He's right about the leverage. If the valve were tight, this could help. But the longer arm itself has a larger moment of inertia, although that shouldn't make much difference in this case. The extra torque, however, could wear out the washers (or other internal parts) much sooner, a real disadvantage.

**50.** As the amount of water at earth's equator increased, mass would have moved from near earth's axis of spin to farther away. That would increase the moment of inertia and decrease the angular speed.

**52.** Two reasons. First, as in Demonstration 3 (and Example 3 in "Calculations"), increasing the distance of the mass of your arms and shoes from your feet, a possible pivot point to rotate your center of mass over the log, makes your rotation about that point increase more slowly (for a given torque). This gives you more time to adjust your balance. Second, it increases your moment of inertia about a vertical axis through your body. Why does that help? Anytime you take a step, you move the mass of a leg about that axis, which twists your body in the other direction. The increase in your moment of inertia about that axis means your body twists around less with each step. This also makes it easier for you to keep your balance.

**53.** 33.3 rotations/min = 33.3 rotations/60 s = 0.555 rotations/s

**54.** $(F\perp \times d)_{\text{Melanie}} = (F\perp \times d)_{\text{bumper}} = 110 \text{ lb} \times 4.5 \text{ ft} = F\perp_{\text{bumper}} \times 1 \text{ ft}$, so $F\perp_{\text{bumper}} = 495 \text{ lb}$

**55.** 4.5 ft (see Exercise 54).

**56.** (33.3 rotations/min − 0 rotations/min)/3s = 11.1 rot./min/s

**57.** The earth would have to spin so fast that $mv^2/r$ would equal $mg$ at the equator. That is, $g = (v^2/r)$, where $g = 9.8 \text{ m/s}^2$ and $r = 6378 \times 10^3$ m. Then, $v^2 = 62.5 \times 10^6 \text{ m}^2/\text{s}^2$ and $v = 7.75 \times 10^3$ m/s (or 4.81 mi/s, or 17,340 mph, the orbital speed of something at earth's surface). How fast would earth turn around? $v = (2\pi r/T)$ or $T = (2\pi r)/v = 5171$ s or 1.44 hr, rather than the present rate of about 24 hr.

**58.** $(I\omega)_{\text{initial}} = (I\omega)_{\text{final}}$, so when $\omega_{\text{final}} = 3\ \omega_{\text{initial}}$, $I_{\text{final}} = \frac{1}{3} I_{\text{initial}}$.

**59.** The earth rotates about 360° in 24 hr, which is about ¼° per minute, or 1° in 4 minutes. So the sun will set about 2 minutes after it touches the horizon.

## Chapter 9

**2.** The number of materials around you is very much greater than 90, the total of the chemical elements. Differences in color, texture, density, flexibility, etc., are enormous. (Also see Exercise 14.)

**4.** According to some glaciologists, the runoff (melted ice and snow) at the foot of a summertime glacier provides a lubricant for the front edge of the glacier as it pushes forward. Those loose, almost incompressible molecules may be rolled on, apparently, speeding the glacier's progress.

**6.** Chemical bonds.

**8.** The *least* reactive, which either were found in the free state or were easiest to separate from compounds in the simplest chemical reactions. The ten were carbon, copper, gold, iron, lead, mercury, silver, sulfur, tin, and zinc.

**10.** Both the electric repulsion they have because of their like (negative) charge and the exclusion principle keep them apart.

**12.** The electrons occupy a huge volume around the nucleus, so they are more responsible for an atom's volume.

**14.** Carbon can form covalent bonds with up to four other atoms, including other carbon atoms. This provides a remarkable array of structural possibilities, including molecular rings, and molecular chains.

**16.** (a) Why mercury? Probably because their densities (mass/volume) are both high; that of mercury is 13.6 gm/$cm^3$ and that of gold 19 gm/$cm^3$. Mercury's appearance was silvery (they called it *quicksilver*), but mercury was closer to gold in density than silver is (10.5 gm/$cm^3$). (b) A look at the periodic table tells you that if they could have caused a *nuclear* reaction and removed 1 proton from the mercury nucleus, the mercury atom would become gold.

**18.** From the periodic table you can see that the atomic mass of gold (Au) is 197 amu. So the mass of a single gold atom is $197 \times (1.7 \times 10^{-24}$ g) $= 335 \times 10^{-24}$ g, or $3.35 \times 10^{-22}$ g.

**20.** The 18-amu oxygen atoms have two extra neutrons in their nuclei.

**22.** Yes, because it is in the same column of the periodic table, so its outer electron structure, and hence its bonds, should be similar. Metallic bonds provide the electrical conductivity of metals.

**24.** $Fe_2O_3$ = 160; $H_2O$ = 18; NaCl = 23 + 35 = 58 (see Example 1; the isotope of chlorine whose mass is 35 is the most abundant); $CO_2$ = 44; $N_2$ = 28; $O_2$ = 32.

**26.** The innermost electrons are pulled much closer to the 88 protons in the nucleus of the radium atom than the lithium's electrons are pulled to the three protons in its nucleus.

**27.** It is impossible to measure an individual atom's mass in the standard way; but macroscopic amounts of matter can be weighed to determine *relative* atomic or molecular weights.

**28.** They are much weaker than ionic or covalent bonds.

**29.** The electric force field of the protons in the nucleus.

**30.** About 0.4 lb. That is, since sodium's atomic mass is 23 and chlorine's is 35 (see Example 1), the molecular mass is 58. So 23/58 = 0.3966, or about 40 percent of the mass of sodium chloride is sodium.

**31.** Since 1 atom of carbon has a mass of 12 amu, or $12 \times (1.7 \times 10^{-24}$ g), then $10^{-6}$ g divided by $(20.4 \times 10^{-24}$ g/atom) $\simeq 5 \times 10^{16}$ atoms.

**32.** Yes for (a), (b), (c), and (d). Normally hydrogen gas, $H_2$, has a molecular mass of 1 + 1 = 2. But if one atom is the isotope deuterium, the molecular mass is 2 + 1 = 3. If one atom is tritium or if both atoms are deuterium, the mass is 3 + 1 = 4 or 2 + 2 = 4. If there's one deuterium and one tritium, the molecular mass is 3 + 2 = 5, and two tritiums give a mass of 3 + 3 = 6.

**33.** The molecular mass of $H_2O$ is 18 amu (16 + 1 + 1), so the hydrogen's mass is $2/18$ = 0.11111 or 11.111 percent (11 $1/9$ percent) of the total. Therefore, 1.7777 oz per pint is hydrogen. Have some hydrogen-it's light!

**34.** The volume of the cube is height × width × length, and each edge has about $10^8$ atoms along its length, so $10^8 \times 10^8 \times 10^8 = 10^{24}$ atoms occupy that cube.

**35.** One year is $365 \times 24 \times 60 \times 60 = 3.15 \times 10^7$ s. So $10^{24}/(80 \times 3.15 \times 10^7 \text{ s}) = 4 \times 10^{14}$ objects/s.

**36.** $10^{24}/10^9$ s $= 10^{24-9}$ s $= 10^{15}$ s. Now 1 yr $= 3.15 \times 10^7$ s, so $10^{15}$ s/$(3.15 \times 10^7$ s/yr) $= 0.317 \times 10^8$ yr = 32,000,000 yr.

**37.** $1 \times 10^{-8}$ centimeters $= 1 \times 10^{-10}$ meters, so $7 \times 10^{-6}$ meters/$1 \times 10^{-10}$ meters = 70,000 atom-widths across.

**38.** First find the number of revolutions the tire makes to use its tread. (40,000 mi × 5280 ft/mi × 12 in./ft)/(2 $\pi$ × 13 in.) = 31,000,000. Then divide the depth of the tread by this to see how much comes off during an average revolution of the wheel. 1 cm/$3.1 \times 10^7 = 0.32 \times 10^{-7} = 3 \times 10^{-8}$.

## Chapter 10

**2.** These huge stones depend on their weight and accurate fitting to keep them in place. Their shapes interlock somewhat, so there's no easy direction of yield. (Just as the Grand Coulee Dam is called a gravity dam because its sheer weight on its base holds back the water, these Inca walls could be called gravity walls.)

**3.** There are fewer carbon atoms per cubic centimeter in graphite than in diamond because of the different crystal structures. As can be seen in Fig. 10–8, the graphite structure has more space between the carbon atoms.

**6.** Molecular bonds. A molecule of water is very strong because the bonds between the oxygen and hydrogen atoms are strong. The molecular bonds that hold the $H_2O$ molecules together in the ice structure are weaker. When ice breaks (or just melts back into water), relatively few atomic bonds are broken.

**8.** Refer to Fig. 10–28, and consider the direction of the frictional forces from tires of cars. They will point along the road (in either direction). This tends to move the bricks against their broad sides rather than their narrow sides, reducing the pressure when one brick is pushed against the others as a car accelerates or decelerates. Also, the layers of bricks are staggered, so there is no easy plane of slippage as there would be if the layers were parallel to the direction of traffic rather than perpendicular to it. Finally, tires have a smaller lever arm to rotate the bricks than if they were pointed along the street.

**10.** When a large force is supported by a small area of contact, the pressure is large. The molecules of each surface that do touch press together with enough force to cause metallic bonding between the atoms in contact.

**11.** Ice under pressure melts. The loaded truck's tire patterns sink in a little, and as they do, they get a better grip on the ice. Tire chains do the same thing for small cars. By supporting the car on the smaller area of the chains, the pressure on the links is greater, and they sink into the ice a bit.

**13.** The whole beans have much less surface area per unit of volume than the ground beans will. They won't get stale (lose aroma and oils through their surface area to the air) as quickly as ground beans will.

**16.** The increased pressure eventually caused the glacier to move faster, since a glacier's motion is the result of its own weight.

**18.** The pressure of the needle is *huge!* The force isn't great, but the area is tiny, so $F/A$ is a large number. Needles of materials that aren't as hard as diamond wear out much more quickly. (By the way, the force on the needle is generally calibrated in grams, which isn't a force but a mass. What the audiophile means by saying the force is 1 gram, for instance, is that the force is the same as the *weight* of 1 gram of matter. That's sometimes called a *gram-weight.*)

**20.** There are two aspects to the answer, density and the relation between something's volume and its diameter. Remember from Chapter 9 that a carbon atom, for example, has 12 times the mass of a hydrogen atom and yet takes up about the same amount of space. The large molecule, then, could be made up of denser atoms to help account for its seemingly outlandish mass. But consider the relation between volume and diameter. For a cube, $V = l^3$. If the length $l$ of its edge (its diameter) doubles, the volume increases by a factor of 8; if its diameter increases by a factor of 10, its volume becomes 1000 times larger. A sphere's volume is $V = \frac{4}{3}\pi r^3$. If its radius is 10 times larger, its volume becomes 1000 times as great. So it is with any solid shape. Thus a sphere 1000 ($10^3$) times as large as another encloses 1 billion ($10^9$) times as much space. It is this volume/size factor, then, that explains why a molecule only 1000 times larger than the atom can weigh 10 million ($10^7$) times as much. (In fact, the density of the molecule in amu/cm$^3$ is less than the density of the hydrogen atom, so the molecule isn't tightly packed.)

**21.** Smaller particles have more surface area per unit of volume than larger ones. So the smallest particles of sand dust have a great surface area compared to their mass. For that reason, they are the easiest for the air to lift and blow around. For several days in 1982, and again in 1984, the skies over southern and central Florida were whitened with the glare of sunshine on such particles that had been carried by the wind across the Atlantic Ocean from the Sahara Desert.

**23.** The water comes out of a nozzle that helps to separate it into drops almost immediately. Those drops travel out at nearly the same angle and with essentially the same speed. *But the smaller drops slow faster because of air drag, so they travel a shorter distance before hitting the ground.* That's why the ground will get wet between the sprinkler and the outer circle where the largest drops fall. You can see this behavior under the stream of a garden hose as well. Why do the small drops slow faster? The smaller droplets certainly have less surface area than the large drops and have less total air drag. But they have more surface area per unit of mass, so *the air drag on a unit of area decelerates a smaller amount of mass in the smaller droplet.*

**25.** The cardboard box of wrapped eggs will have a huge surface area but very little mass, so its terminal speed will be very slow. The paper wrappings first yield to and then stop the eggs with a force over a large area on the side of the egg, minimizing the pressure.

**26.** The shape (and width) of the child and adult are roughly the same. Even though the larger adult has more total surface area than the child, the adult has *more volume per unit of surface area, and hence more mass per unit of surface area,* than the child. Thus the air drag, which is proportional to the surface area, decelerates the adult less. For each square centimeter that catches the air there is more mass to decelerate for the adult than for the child.

**27.** True, just as for Exercise 26.

**29.** True. The lollipop dissolves only at its surface. When there is more surface area, the rate of dissolving the candy will be greater in proportion.

**31.** True. The smaller the size of a given shape, the greater its surface area per unit of volume.

**33.** All of these factors.

**34.** If one cut gives two extra faces, you need three cuts to expose six extra faces, which have the same surface area as the original cube.

**35.** 100 percent gold

**36.** 33.2 cm$^3$

**37.** 58.3 million square miles

**38.** Mass 16.5 kg, weight 36.4 lb

**39.** 2.91 gm/cm$^3$

**40.** $6\frac{2}{3}$ cm$^3$

**41.** (9 ft × 170 ft)/(400 ft$^2$/gal) = 3.8 gallons

**42.** [(150 ft × 90 ft) − 1800 ft$^2$]/(2000 ft$^2$/bag) = 5.85 bags

**43.** 9 in. costs 7.4¢/in.$^2$; 13 in. costs 5.8¢/in$^2$; 15 in. costs 5.65¢/in.$^2$

**44.** Just find the pressure a column with a cross section of 1cm$^2$ (or any other cross section) exerts on its base and convert to the other units. The column is 1200 cm high, so 1 cm$^2$ at the base must support the weight $mg$ of the mass of (1.3 g/cm$^3$ × 1200 cm$^3$), or $mg$ = 1.56 kg × 9.8 m/s$^2$ = 15.29 N. So the pressure is 15.29 N/cm$^2$. Since 1 cm$^2$ = $10^{-4}$ m$^2$, $P$ = 152,900 N/m$^2$. Since 1 atm = 101,000 N/m$^2$, $P$ = 1.51 atm.

**45.** There's no need to do the entire calculation of Exercise 44 over again! $P = F/A$, $F = mg$, and $m = \rho v$, so $P_{al}/P_{wood} = \rho_{al}/\rho_{wood} = 2.7/1.3 = 2.08$. So $P_{al} = 2.08 \times P_{wood} = 3.14$ atm.

**46.** 3.24 cubic feet (1 cubic foot = 12 in. × 12 in. × 12 in. = 1728 in.$^3$)

**48.** $mg \simeq 2017$ lb

## Chapter 11

**2.** The lungs of someone who is snorkeling are about 1 ft below the water's surface. As the swimmer inhales, the diaphragm must work against the pressure of the water that is transmitted through the body to the lungs. Though the pressure at a depth of 1 ft isn't much (1/34 atm $\simeq$ 0.44 lb/in.$^2$), the snorkeler will tire from it after snorkeling for a long time.

**4.** (a) No; you didn't double the volume of water, just the depth. (b) No; the scale supports the weight, and the weight didn't double. (c) Yes! If the water is twice as deep, the pressure at the bottom will be twice as great. Here's why the water pressure on the bottom of the flask doesn't get through to the scale. The answer has to do with the shape of the flask. Remember that water pressure is transmitted in all directions, so wherever water presses against the flask, it pushes straight against it. The upper part of the flask slopes inward toward the neck, and the water there pushes partly to the sides and partly *upward*. The water on the sides gives an upward force that partially counteracts the downward force from water pressure at the bottom of the flask.

**7.** No. The water's pressure on the left face does give a force to the right, but the water's pressure on the angular surfaces on the right has components that point to the left and balance that push. Draw force vectors perpendicular to each surface to see this.

**9.** No; the ice floats at a level that displaces its weight in the water. When it melts, it still has the same mass, hence weight, and it still displaces its weight in the water. Its density is the same, so the space or volume it fits into is exactly equal to the volume displaced by the ice cube. The water level will neither rise nor fall.

**11.** When you lie down, the heart doesn't have to pump blood uphill from your toes to your scalp, so it gets a rest.

**16.** In the middle of the flow, farthest from the solid edges, which slow the flow to zero with friction.

**19.** Surface tension! Air-dried towels are stiff because the wet fibers of the towel on the line group together and stick in mats as a result of the surface tension of the water. As the water evaporates, they tangle, stiffening the fabric. In a clothesdryer the tossing motion and currents of warm air tend to keep the fibers moving and drying independently.

**22.** The high rate of flow through a small opening ensures turbulent flow, and eddies vibrate the valves and pipes and make noise.

**24.** The soap reduces the water's surface tension; but so does the extra temperature. At higher temperatures the water molecules move faster and are less able to hold onto one another.

**26.** Surface tension between the wet fibers of the thread hold them together in a point.

**28.** The buoyant force helps support them, keeping minimum stress on their bones and muscles.

**30.** Succulents, especially, wilt that way. The heat and wind take away water, which evaporates too quickly to be replenished. The internal water pressure helps keep many leaves and plant erect.

**32.** Though soapy water has less surface tension than pure water, that force still tends to minimize a bubble's surface area. A sphere has the smallest surface area per unit of volume. Look for bubbles of pure water in bathtubs or under waterfalls. They are always much smaller than soap bubbles because their surface tension is larger.

**36.** 1000 tons

**37.** Viscosity (oil's surface tension is less than water's).

**39.** Both the same! The pressure depends only on depth, not on how much water is behind the dam.

**40.** 63.96 lb/ft$^3$, or about 64 lb/ft$^3$

**42.** 2396 lb

**43.** $P = \rho gh$ and $\rho g = mg/V = 62.4$ lb/ft$^3$, so $P =$ 45,300.0 lb/ft$^2$ (or 314.5 lb/in.$^2$), or 21 atm.

**44.** 112,000 lb

**45.** $P_{H_2O}$ at 225 ft = 6.66 atm = 14,100 lb/ft$^2$. $F_{net} = \overline{P} \times A$ = 31 billion lb.

**46.** 4500 cm$^3$/80 cm$^3$ = 56, so 56 beats moves a volume of blood equal to his entire blood supply. Thus a red cell can make a cycle through a major artery–vein system in as little as one minute.

**47.** The buoyant force on completely submerged wood on Mars would be $\rho g_{Mars} V_{wood}$, where $\rho$ is the density of the wood, $V$ is equal to the volume of the water displaced, and $g_{Mars}$ is about $0.39g$ (from Table 7–3). The weight of the water displaced is $\rho_{water}\, g_{Mars}\, V_{wood}$, so if $\rho_{wood}$ is less than $\rho_{water}$, the buoyant force on Mars would cause the wood to float at the same level as it would on earth. The local $g$ affects the weight of the water displaced in the same way as it affects the wood's weight.

**48.** Think of the ocean as a rug you could wrap around the earth. It would have a surface area and a depth. The surface area would be about the same as earth's surface area, and the volume of the ocean would be its surface area times its depth. The surface area of a sphere is $4\pi r^2$, and $r$ for the earth is about $6.4 \times 10^6$ m. So depth = volume/surface area = $(1.4 \times 10^{18}\ \text{m}^3)/4\pi\,(6.4 \times 10^6\ \text{m})^2$ = 2720 m = 8900 ft (or 1.7 mi).

**49.** $F_{net} = mg - F_b = \rho_{object} g V_{object} - \rho_{water} g V_{object} = (\rho_{object} - \rho_{water}) g V$. If you could increase $g$, the acceleration due to gravity, you would increase the settling rate. You can't alter $g$, of course, but you can provide another acceleration to do the same thing. Devices called *centrifuges* in hospital laboratories separate the blood cells from the plasma (liquid portion of the blood) in a minute or two with a centrifugal acceleration that is approximately 2200$g$. Large protein molecules can be settled out in ultracentrifuges that produce accelerations equal to 300,000$g$.

## Chapter 12

**2.** Air, a fluid, has a sideways speed of zero at a solid boundary. This causes the air nearer the ground to move, generally speaking, much slower than the air does just a few meters up. (Review the section "Liquids in Motion" in Chapter 11 for a detailed explanation.)

**4.** The cabin's air is kept under pressure. Air from the compressors of the jet engines keeps the cabin pressure equivalent to the atmospheric pressure at an altitude of 5000 to 6000 feet.

**6.** Though it describes the sound, suction is a misnomer. A suction cup creates, by virtue of the rubber's elasticity, a partial vacuum beneath the cup. The air pressure on the *outside* of the cup presses the cup downward, holding it against the surface.

**8.** When a liquid escapes through a narrow opening under its surface and there's no inlet for air above its surface, a partial vacuum forms over the liquid as soon as some liquid escapes. The air pressure outside the container's opening quickly retards the flow until a bubble of air can pass into the can through the highest side of the opening, bringing the pressure above the liquid closer to atmospheric pressure. That's why a bottle of milk or soft drink gurgles or burps if you hold it at a large angle as you pour. If there is an air hole, as in a gasoline can, the flow can be smooth.

**10.** In Boston, most likely. If the hail is traveling at its terminal speed with no extra speed from wind, it will fall somewhat slower in Boston because earth's atmosphere is thickest closer to the ground. At Leadville, the highest incorporated town in the United States, the terminal speed for hail in the thinner air would be higher.

**12.** The $CO_2$ gas is kept in such brews only with artificial pressure. When that pressure is released the gas comes out of solution. The pressure such a drink is under before you open it is too small to burst the container ordinarily. (But the temperature matters too, as you may know from experience. Watch out for the ruffian with a warm can of pop, shaking it up and down. . .)

**14.** At the solid boundary of the fan blades, the air speed is zero even though the blades are pushing air along rapidly.

**16.** Unlike the liquid, a bubble of air can compress as the pressure on the brake fluid increases. The work done on the fluid goes partly to compress the bubble somewhat as well as to push on the brake shoes or brake pads. Therefore, the liquid beyond the bubble in the line won't move as far. (The bubble can't compress all the way and become incompressible liquid air; that would require about 1000 atmospheres of pressure.)

**18.** You may say you suck the liquid up the straw, but actually air pressure on the surface of the liquid outside the straw pushes the liquid into the straw. You merely create a modest partial vacuum (a centimeter or two of mercury below air pressure) in your mouth and in the top of the straw with your cheeks. Just don't try drinking from a straw that's over 34 feet long!

**20.** As the local air pressure goes up and down during the week, the density of the outside air changes. That means the box displaces different weights of air during the week. The buoyant force on the box changes, and it weighs different amounts on the scale. (Since the box is sealed, its volume may well change with different atmospheric pressures. Convince yourself that action would partially correct for the buoyancy lost or gained.)

**22.** Bernoulli's principle causes the partial collapse of the trachea. Then because the air pushed from your lungs goes through a smaller opening, it picks up speed, exerting a greater direct push on anything in its way. Both effects help to dislodge the particle.

**24.** You'll weigh the same. The air in your lungs is at 1 atmosphere of pressure, so it displaces the same volume in or out of your body and has the same density as the air on the outside. Only if there is a difference in density will one fluid "buoy" another one.

**26.** The 25 lb of pressure is the amount *over* atmospheric pressure. So there'd be fewer molecules in your tire in Denver to give the same pressure as in Miami at sea level, presuming the same temperature at both locations.

**28.** Water is incompressible, although fluid. So when you hit a bump, the tire could change shape but not volume. Filled with compressible air, the tire changes shape *and* volume when it hits a bump. This cushions the ride, but more energy is absorbed in working against friction. The heat given off by the tires is lost to the car's motion. The least elastic of all would be a solid rubber tire. These rigid tires would give the best gasoline mileage because less of the energy of the car's motion would be consumed by flexing the tires.

**30.** The birds would weigh less and be easier to support in the air. But the atmosphere would weigh less too. It would bulge outward, leaving the sea-level air much thinner. The birds would have to push much harder to get the same amount of force. The effects, therefore, would at least partially cancel.

**32.** Less. The air wouldn't be supporting the weight of as much air above it.

**34.** In space the exhaust gases don't have to push outward against the pressure of surrounding air—over a ton for every square foot of exhaust area at sea level.

**36.** Cooler air is heavier, making hot-air balloons more buoyant when the air is cooler. But more important, the atmosphere is quieter then, with less wind.

**38.** 3000 square inches, or about 21 square feet

**39.** About 77.5 lb

**40.** About 11.36 lb. You *can* tell the difference when you lift them.

**41.** $V_f = (71.2 \text{ ft}^3 \times 15 \text{ lb/in.}^2)/2265 \text{ lb/in.}^2 = 0.47 \text{ ft}^3$

**42.** Ordinarily the pressure is about 30 in. of mercury. So $\frac{1}{30} \times 14.7 \text{ lb/in.}^2 = 0.49 \text{ lb/in.}^2 = 0.49 \text{ lb/(ft}^2/144) = 70.56 \text{ lb/ft}^2$.

**43.** If you weigh 135 lb, you displace 2.16 $\text{ft}^3$ of air, which at sea level weighs 0.08071 $\text{lb/ft}^3$. The buoyant force on you is then = 0.174 lb. On Venus it would be 90 times that, or about 15⅔ lb.

**44.** A cubic foot of air at sea level weighs 0.08071 lb. A cubic foot of fresh water weighs 62.4 lb. To displace a cubic foot of either gives a buoyant force equal to the weight of the displaced fluid. So the buoyant force of air at sea level is 0.0013 times the buoyant force of water.

**45.** $P_1V_1 = P_2V_2$; $V_2 = \frac{1}{8} V_1$, so $V_2 = 2 \text{ in.}^3$. $P_2 = P_1V_1/V_2 = (1 \text{ atm} \times 16 \text{ in.}^3)/2 \text{ in.}^3 = 8 \text{ atm}$.

**46.** We can *estimate* the pressure 12,000 ft below by seeing how much extra pressure a column of sea-level air 12,000 ft high would produce at its base. $P_{extra} = \rho gh$, where $\rho g = 0.08071 \text{ lb/ft}^3$. $P_{extra} = 968.5 \text{ lb/ft}^2$. $P_{sea\ level} = 2117 \text{ lb/ft}^2$, so $P_{bottom} = P_{sea\ level} + P_{extra}$, or $P_{bottom} = 1.5$ atm. This underestimates the actual value, since the weight of the column of air compresses the air deep in the mine considerably, increasing its density.

**47.** (a) $V = \frac{4}{3}\pi r^3 = 9203 \text{ ft}^3$. The weight of that much air is the buoyant force $0.08071 \text{ lb/ft}^3 \times 9203 \text{ ft}^3 = 742.7$ lb. (b) The hydrogen gas weighs $0.07 \times 743 \text{ lb} = 51$ lb. So $743 \text{ lb} - 51 \text{ lb} = 692$ lb of lifting force.

## Chapter 13

**2.** The gas (usually water vapor) can't expand because of the rigid walls of the pressure cooker. Therefore, the faster-moving molecules pound harder on the container, increasing the pressure. Such a cooker has a movable weight covering a vent, or small hole, in the lid which fits tightly with a seal. When the vapor, or steam, inside comes to a certain pressure, the weight is lifted, and some steam escapes. More steam forms as extra heat comes into the pot, and the pressure remains steady. Pressure regulates the temperature at which the water in the pot boils. If a higher temperature is needed, a heavier weight over the steam vent will maintain higher pressure in the pot.

**4.** Expansion joints. Room to expand must be left in bridges whose great length means the change in length from a cold day to a hot one is significant. The temperature of the material in a bridge is influenced more by air temperature than the road is because air is under the bridge as well as over it. Roadways on the ground are closer in temperature to the ground's temperature.

**6.** As seen from Table 13–2, the coefficient of expansion of concrete is fairly close to the coefficient of expansion of steel. If the cement and aggregate are chosen properly, the coefficient of the resulting concrete is equal to the coefficient of expansion of steel.

**9.** The copper veins in the rock both expanded and contracted more than the surrounding rock. The heat made it expand, which broke up the rock, and the water cooled it. It could then be easily removed from the broken rock.

**11.** Phonograph records warp very quickly. Sunlight hits one side and causes uneven heating and deep wrinkles. A slow and even temperature change causes little harm, however, as when records are stored in a home.

**13.** To minimize the effects of uneven heating. Otherwise, the solid structure would have become very hot on the side exposed to sunlight and cold on the unlit side. Uneven expansion and contraction places stress on the connected materials.

**15.** Muscles in use give off much more heat than the normal resting metabolism does. They decouple their wings and vibrate the muscles very fast. (Also, the breeze caused by moving wings would cool rather than warm, as you will see in Chapter 14.)

**16.** Falling water gains kinetic energy, which is randomized when the water hits the pool at the bottom to become heat energy. In theory the water at the bottom should have been warmer than that at the top. However, Joule was disappointed because he found *no* change. You'll read about the reason in Chapter 14. (There's a story that Joule coincidentally met Kelvin on the path to the waterfall.)

**18.** The technique exposes maximum surface areas of food to the high temperature of the wok, heating and cooking it very fast.

**20.** Small regions on the surface of the droplet in contact with the pan burst into steam, which has a much greater volume than the liquid water it comes from. Those miniature explosions propel the drop of water one way and then another.

**22.** Increasing the surface area of the water in the puddle speeds evaporation, and the spot dries faster.

**24.** Look at Table 13–1. Silver contains much less heat energy than water does at a given temperature.

**25.** The flowing water takes a lot of heat energy from the engine. The water then sheds this heat in the radiator.

From there the cooler water is cycled back to the engine to repeat the process.

**26.** The first bubbles to form are tiny ones that cling to the pot where they pop up. You can see these in a glass teapot. As they pop into being, they push the water back, making a sound. As they grow and let go, they rise into cooler water (at lower pressure). There they lose heat and collapse into the liquid state before reaching the surface. When they collapse, the water around them vibrates.

**28.** The ends of the rails are heated to a very high temperature before they are welded in place. This technique simulates the expansion the length of rail would have on a summer's day. The rails aren't anchored so well that they can't move slightly to the sides. Finally, the small natural curves of the rails on turns and hills help them expand, as in Fig. 13–2, without great stress.

**29.** In such a thermometer, an increase in temperature would send the column of liquid into the bulb as the glass expanded faster than the liquid did. In that instance, the 0°C point would be located farther from the bulb, and the 100°C point would be closer to the bulb, just as Anders Celsius set it up originally. (See "Celsius, Fahrenheit, and Kelvin.")

**32.** No. Boiling proceeds when heat energy is added at such a great rate that the convection currents from the hot spots cannot get rid of the added heat fast enough. The water vaporizes only if its latent heat of vaporization is added before it can move away.

**33.** The temperature shown on the thermometer would stop rising or falling while the transformation from liquid to solid (or liquid to gas) was going on.

**35.** The kinetic energy of the bullet, $\frac{1}{2}mv^2$, turned mostly into heat energy as friction from the wood stopped the bullet.

**37.** 460,000 J

**38.** $56\frac{2}{3}$°C

**39.** 12°"Isaac"/37°C × 100°C = 32.4° "Isaac"

**40.** 4436 g, or about 4.4 liters

**41.** Each has the same temperature, so each type of molecule has the same average kinetic energy, $\frac{1}{2}mv^2$. But oxygen's mass is 16 times as great as hydrogen's, so hydrogen's value of $v^2$ must be 16 times as great as oxygen's $v^2$. Therefore $v_{hydrogen} = 4 \times v_{oxygen}$.

**42.** Yes. *mCT* is the measure of the heat energy, where *T* is the Kelvin temperature. If the masses are the same for objects one and two, then $C_1T_1 = C_2T_2$, if they are to have the same heat energy. The temperatures will be different if the specific heats are different.

**43.** A *change* in temperature of 20°C is equal to a *change* in temperature of 20 K. $L_f - L_i = \alpha_{steel}\, L_i\, (T_f - T_i) = 11 \times 10^{-6}\ K^{-1} \times 1200\ ft \times 20\ K = 0.264\ ft = 3.2$ in.

**44.** 0.302 meters, or about 30 centimeters

**45.** 7.2°C

**46.** First convert mph to meters per second. 220 mph = 98.35 m/s, and 35 mph = 15.65 m/s. Then, $\frac{1}{2}\,mv_f^2 - \frac{1}{2}\,mv_i^2 = \frac{1}{2} \times 1100\ kg \times (9672.7\ m^2/s^2 - 245\ m^2/s^2) = 5.18 \times 10^6$ J.

**47.** At room temperature atoms or small molecules typically move at speeds of 1000 mph (see Example 4, "Calculations"). A pitcher can deliver a fastball at about 100 mph, so the kinetic energy the pitcher gives the ball, $\frac{1}{2}mv^2$, is about *one-hundredth* of the kinetic energy the ball's molecules have as a result of their thermal motion. (The ball would have to be thrown 10 times faster than a fast ball.)

**48.** The heat *lost* by the copper is *gained* by the water, and eventually they come to the same temperature. (Styrofoam is a good insulator, so assume little of the heat energy is lost to it.) So $Q_{water} = -Q_{copper}$. $m_{water}C_{water}\,(T_f - T_i)_{water} = -m_{copper}C_{copper}\,(T_f - T_i)_{copper}$. (1000 g) × (1 cal/g/°C) × ($T_f$ − 0°C) = −(1000 g) × (0.092 cal/g/°C) × ($T_f$ − 100°C). Cancelling the units of the specific heat, multiplying, and rearranging give $1T_f + 0.092\ T_f$ = 9.2°C, or $T_f$ = 9.2 °C/1.092 = 8.4°C. (If 1 kg of water at 100°C had been added to the kg of water at 0°C, the final temperature would have been 50°C. Copper stores less heat energy at the same temperature than water, which is the property of matter specific heat measures.)

## Chapter 14

**2.** When the tip of your tongue sticks to the scoop, it's because the cold metal conducts heat away from your tongue so fast that it freezes the moisture on your tongue at the point of contact. A child's tongue can be seriously damaged if it's touched to an aluminum lightpole or other metal surface in subfreezing temperatures. A wooden spoon conducts heat poorly and won't cause that problem.

**5.** With no gravity no natural convection currents cause the gases of combustion to rise, bringing in fresh oxygen from the side. Placed under a ventilator that directs forced air, however, the candle could burn in the artificial convection current. Or if the candle was waved around while it burned, fresh oxygen from the air could get to the site of oxidation.

**7.** By exposing more surface area, the total heat lost by radiation per second is increased. (See the equations for heat lost by radiation in "Calculations.")

**9.** There can be little loss of heat from the water by convection currents, which ordinarily is an efficient way for a liquid to lose heat. Also, the solid surface impedes moving air, or winds, above it better than a liquid surface does, and still air is a good insulator.

**11.** The heat flow (whether by conduction or radiation) is proportional to the *temperature difference*. The heat flow is smaller if the inside and outside temperatures do not differ much.

**12.** First, the mud acts as a sun screen; light is absorbed by the mud rather than the elephant's back. It also acts as a crude air conditioner. The moisture in the mud absorbs its latent heat of vaporization as it evaporates, so the mud keeps cool by evaporation until it hardens. At that point, it's time for another mudding.

**14.** The fireplace bricks get hot and radiate a great deal of heat energy along with the fire.

**16.** The water heater is insulated; without the heater on, the water in the tank stays cool even when the daytime air around it becomes scorching hot. Meanwhile, the water moving through the uninsulated cold water pipes picks up heat from the ground and air and becomes hotter than the water stored in the insulated water heater.

**18.** When the person stands normally, the insides of the thighs radiate heat to each other and so reabsorb part of that heat radiation. Likewise, the insides of the underarms radiate to the body, and vice versa.

**19.** First, the beer has no ice. A lot of heat will be absorbed just to melt the ice in the tea. Second, thin aluminum walls will conduct heat faster than thick glass walls. But both beer and iced tea gain heat when water vapor condenses on the outside, giving up heat.

**21.** The answer is found in Table 14–1. Air is a very poor conductor of heat compared to water.

**23.** Bringing hot soup to the surface keeps the temperature difference between the soup and the air at the surface as large as possible, and cooling proceeds as rapidly as possible.

**25.** Heat energy must be absorbed through the surface, and a large turkey has *less* surface area per pound of meat. So a large turkey takes more time per pound to cook all the way through.

**28.** The spinning sets up artificial convection currents in the liquid that bring more of the liquid's heat energy to the glass to be absorbed by the ice. The bottle cools faster if packed in ice shavings with no water. Ice by itself is usually much colder than 0°C, whereas water can be no colder than 0°C. The greater the temperature difference, the greater the rate of heat flow.

**32.** When water boils, its volume increases dramatically. When a bubble of steam bursts forth underwater, its molecules occupy roughly a thousand times more volume than they did in the liquid state. No wonder rice pots overflow.

**34.** If the top layer of the permafrost thaws for some reason, the ice shrinks in volume about 10 percent as it melts into water; this causes the roadbed that rests on the permafrost to break up. Preventing this problem was expensive; 453 miles cost $100 million, and construction lasted from 1958 to 1980.

**36.** Just as the blades and runners of ice skates and bobsleds melt the ice under them with extra pressure, so can a heavy truck. This is why trucks that spread sand or salt don't usually need chains to grip an icy road. Increased pressure both lowers the freezing point and heats the roadbed. If the permafrost melts, refreezes, melts, and so on, the roadbed will break up.

**38.** When the leaves are curled into cylinders, less of their surface area is exposed, and there is less radiation. Also, the motionless air inside the cylinder conducts heat less rapidly than the air outside.

**40.** (a) The heat flow will be cut in half. The thickness of the material is in the denominator of the heat-flow equation for conduction. (b) The rate of heat flow is proportional to the temperature *difference* (see the heat-flow equation). So if the temperature difference is halved, the rate of flow of heat is also halved. $T_{new}$ = 6°C, halfway between 24°C and −12°C.

**41.** (a) Use $PV/T$ = constant, and let $V_i = V_f$. $P_f$ = 2333 lb/in.$^2$ (b) Yes, about 11 percent. (c) 2786 lb/in.$^2$

**42.** 11.99 lb/in.$^2$

**43.** (a) 15.78 lb/in.$^2$; (b) 777.6 lb

**44.** (Heat in − heat out)/heat in = 0.22 = work done/heat in. So work done = 0.22 × (32 × $10^6$ calories) = 7.04 × $10^6$ calories. The remainder, 24.96 × $10^6$ calories, is wasted heat.

**45.** Efficiency = 0.61 (or 61 percent).

**46.** Efficiency = (298 − 227)/298 = 0.07. The poor efficiency might be overcome because there's no fuel to buy and burn!

**47.** Heat lost per second = $e \times 5.7 \times 10^{-8}$ W/m$^2$/K$^4$ × surface area × $T^4$. (a) For the male: heat lost/s = (0.97) × 5.7 × $10^{-8}$ W/m$^2$/K$^4$ × (0.85 × 1.8 m$^2$) × (306 K)$^4$ = 742 J/s. For the female: heat lost/s = 536 J/s. (b) Male: 658; net loss per second = 742 − 658 = 84 J/s. Female: 474; net loss per second = 536 − 474 = 62 J/s. (c) Male: 1736 Calories; female: 1281 Calories.

## Chapter 15

**2.** Infrared radiation that is given off by the ground and objects on it rises and escapes through a clear sky at night. Because the air in an overcast, cloudy sky has a high content of water vapor, however, the radiation warms the air as it travels through it. Water vapor (along with carbon dioxide) absorbs infrared energy and converts much of it to thermal energy that is spread through collisions to other molecules in the air.

**4.** Those walls that take direct sunlight as the sun rises in the east and sets in the west are heated more than the walls with exposures to the north and south. Therefore, these walls should be the narrow ones with less surface area to absorb radiation from the sun.

**6.** As the temperature continued to fall, any water vapor in the air would condense as rain or freeze as snow. Cleared of water vapor, the air would let heat pass even faster, and soon carbon dioxide frost (dry ice) would form over the earth's surface. As the temperature continued to plunge, the atmospheric oxygen and nitrogen would descend to become liquid, then solid. In the solid state, the frozen atmosphere would be only some 30 feet thick over the ground. Fortunately for us, the sun always comes up to keep the molecules in gaseous form and provide a comfortable living zone approximately 10,000 feet deep.

**8.** A camel facing the sun presents less surface area to the sun's rays than its broader side would. Its head and neck can even shade its back to some extent. The layers of fat on top of its body insulate against heat radiation from the sunlight, while the absence of fat below allows body heat to pass through to the cooler sand it is lying in.

**10.** The long hours of daylight (up to 24 hours) in the area allow plankton to grow at much faster rates than in the middle latitudes.

**13.** Friction. The falling water drops push downward on the air in front of them and sweep it along.

**14.** The philosopher is correct. All of your body's chemical processes that produce heat depend on the energy you get from food, and the organic food chain begins with the sunlight absorbed by growing plants.

**15.** When your warm breath enters the cold air, it cools quickly. This raises its relative humidity to the point of supersaturation, and rapid condensation follows. But the cloud quickly disappears as the small droplets evaporate into the cold air around them which has a lower relative humidity.

**18.** Because hot air is less dense than cold air, it is buoyed up by any cold air around it. It rises in convection currents. But when a layer of hot air is above cold air, there is no tendency for the heavier cold air to rise. Since the colder low-lying air is still, pollutants from the ground layer are not mixed into the air and carried away as they normally are by convection currents.

**19.** The moving air speeds evaporation on the windward side of the finger, making it lose heat faster than the side away from the wind.

**20.** The rain falls into a region of warm air with a very low relative humidity, and the drops often completely evaporate on the way down. The gray streaks of saturated air beneath such rain clouds are called *virga*.

**22.** The percentage of the earth's surface covered by water in the Southern Hemisphere is much greater than for the Northern Hemisphere. Since water is slower to heat up and slower to lose heat, this fact tends to moderate the seasons, compensating for the effect of the Southern Hemisphere's being closer to the sun in summer and farther away in winter.

**24.** Drops of rainwater trickle down the leather strips, guided in this way away from the warm upper chest. Otherwise, the water would absorb heat from the body, causing a chill.

**27.** The upward breeze, called a *valley breeze,* rises as the sun heats the slopes, warming the air next to the slopes. Less dense, this air is buoyed up by the heavier air around it. It lasts from mid-morning to late afternoon. The downward breeze, called a *mountain breeze,* commences as the temperature of the slopes drops in the evening due to heat radiation. The air next to the slopes cools, becomes heavier than the air around it, and flows downward.

**28.** The vapor in the warm, saturated air condenses on the cool surface of the lens of the glasses. If the lips are pursed, the air expands and cools. This cooling lowers the relative humidity of the air before it strikes the cool lens, and the lens won't fog. (Blow on your hand with your mouth in each shape; feel the difference in temperature of the air.)

**29.** The daytime breeze most often comes from the direction of the ocean. Because land has a lower specific heat than water, its temperature rises rapidly under the sun. The air next to the warmer land heats and becomes less dense than the cooler air over the ocean. That cooler, heavier air moves inland, running under the warm air that it buoys upward. To the comfort of summer residents, the sea breeze may be felt as far as several miles from the beach. At night, the process reverses as the land cools off quickly while the ocean's temperature remains approximately the same.

**31.** Normally the beans that are on the arrow plant fall to the sand beneath, and the worms flip (by holding onto the silk at one end) to find a shady area. The heat from your hand affects the worms the same way. By flipping inside the pod, they cause it to move around. In the shade or out of your hand, the beans become still again. The beans can jump for 6 months, until the new moths emerge.

**33.** The rate of heat transfer between two points or surfaces is greater when the temperature difference is large. The transfer of heat between the penguin's blood vessels accomplishes two things. By cooling the warm blood on its way to the feet, the temperature difference between the feet and the surface of ice on the ground is less, so less heat is lost from the feet per unit of time. Simultaneously, the returning cold blood is warmed so that the body's interior doesn't lose so much heat.

**35.** The sunlight strikes the lunar soil less directly at that angle, and the surface temperature of the ground is less.

**36.** Where no convection currents occur, the gases that are lightest float on top of the heavier gases. Thus hydrogen and helium are found at the outermost distances of earth's atmosphere.

**38.** When warm moist air strikes a cold can of soda pop, that air loses heat rapidly, and its humidity rises. Because the metal can conducts heat well and because the drink (mostly water) inside requires a lot of heat to warm it, the water vapor cools to the point where it condenses on the surface.

**39.** The moving hot air surrounding the food keeps the temperature difference at the food's surface higher than it would be if the air were still. The resulting faster rate of heat flow to the food is why the external air temperature can be less.

**42.** The curves must eventually turn back to the left as they rise, since heating will occur due to friction with the wind. (For example, supersonic aircraft experience tremendous heating on the surfaces that strike even subzero air.)

**44.** At 90 degrees, the energy per hour is $2.75 \times 10^6$ calories; at 25 degrees, the energy absorbed is $1.16 \times 10^6$ calories. It takes 55 calories to raise the temperature of 1 gram of water from 15°C to 70°C, so at 90 degrees, the solar collector could heat $2.75/55 \times 10^6$ grams of water, or $5 \times 10^4$ grams. Since 1 gallon = 3785.4 cubic centimeters, the solar heater warms 13.2 gallons per hour. At the lower angle, it warms $1.16/2.75 \times 13.2$ gallons, or 5.6 gallons.

## Chapter 16

**3.** Only if raindrops landed at almost the same point at the same time could they cause waves that add constructively and travel together. But raindrops fall randomly on a lake's surface. Therefore, the waves they make spread from different points, losing amplitude as they do, and the interference of the various waves is just as often destructive as constructive.

**6.** Like that of any other wave, the earthquake wave's amplitude diminishes as the wave gets farther from its point of origin.

**9.** Such alligators would be close in size to the wavelengths of the wind-generated waves. As a result, they would be considerably disturbed by the passing waves on the water's surface. (See the section "Interacting with Waves.")

**11.** Waves travel faster through the solid string than through air. Also, when the wave travel along a one-dimensional medium, its energy doesn't spread out as it would in the air.

**12.** The passing subway trains set up compressional waves that are transmitted to the building. However, the cork and lead absorb the vibrations and act as a buffer between the subway and the hotel.

**16.** The compressional waves generated in an ice pack by the storm winds pushing it move much faster through the solid ice than the winds themselves can travel.

**17.** Within the container, a crest of the wave is at one side while a trough is at the other. A sketch will convince you that only one-half of a wavelength is present at any instant.

**18.** Ground vibrations during an earthquake sometimes cause low-frequency sound waves in the air if the earth moves up and down. However, most are of lower frequency than sound waves. Because air is very compressible compared to the solid earth, and because its density is so much less, the energy transmitted by the wave of the earthquake to the air is quite small and harmless.

**20.** The hairs on the mole can detect vibrations from compressional waves moving through the ground. In a similar way, the blind fish senses a disturbance in its vicinity by a wave traveling through the water.

**22.** The surface waves that disturb seismographs are caused by the winds that blow over the earth's surface (and the accompanying fluctuating air pressure). These waves resemble ocean waves in that the surface particles perform almost circular motions. However, the amplitudes are quite small and are noticeable only with sensitive seismographs and not at all by humans.

**24.** First refer to the answer for Exercise 17. The sloshing wave you can make in a glass of tea has a wavelength that is twice the diameter of the glass. In a nearly full glass of tea, the bottom will be a wavelength or more below the surface. A review of the section "Surface Waves on Water" will give you the answer. One wavelength below the surface, the tea is disturbed very little. But if you drank iced tea from a soup bowl or even a pie plate, the bottom of the tea would be much closer to the surface, and that wave would stir the sugar.

**27.** The ocean waves typically have wavelengths much less than 30 feet. Below the surface at a depth of one wavelength, the water is not much disturbed by a surface wave. (See "Surface Waves on Water.")

**28.** Marching in step, the troops exerted a large net force with a frequency that was close to a resonant frequency of the bridge. As they marched without interruption, the standing wave (normal mode) grew until the bridge collapsed. If the troops had broken formation while on the bridge, their footfalls would not have applied a large force with a definite frequency, and the resonant mode would not have been excited.

**30.** The impact of the solid object that caused the crater sent a wave outward. The pattern of rings consists of crests and troughs of the wave itself that remain in the crust of the moon, whose gravity was too weak to restore the displaced surface.

**33.** $v = d/t$, so $t = d/v$ = 5280 ft/(1090 ft/s) = 5 s.

**34.** 1/2s and $\lambda$ = 6 m, so $v = f \times \lambda$ = 3 m/s.

**35.** $v$ = length of rope/time for the wave to travel the length.

**36.** $d = vt$ = 1100 ft/s × 8 s = 8800 ft.

**37.** $I_J/I_e = r_e^2/r_J^2 = (1)^2/(5)^2$ = 1/25, or 4 percent

**38.** Intensity at minimum distance/intensity at maximum

distance $= r_{max}^2/r_{min}^2 = (1.017)^2/(0.983)^2 = 1.07$, or 7 percent more energy per second.

## CHAPTER 17

**3.** This is a complex question because the pressure in a gas depends on both its temperature and its density. If the temperature decreases, the molecules travel more slowly, and the pressure caused by their collisions decreases. The speed of sound, which depends on the speed of the molecules in the gas, also drops. But if the temperature of a gas remains the same while its *density* decreases, as when gas leaks from a container, the pressure also falls. The speed of sound in that gas is unaffected, however, since the speed of the remaining molecules is not changed.

**4.** The speed of sound in a gas is slower when the temperature (and hence the molecular speed) is lower. When the temperature is increased in a solid, the extra speed of its atoms or molecules causes the solid to expand. As the particles pull apart somewhat, the speed of sound drops as the solid's bonds become less effective as restoring forces in transmitting the wave.

**7.** Transmitted with the help of molecular bonds, sounds travel both faster and more efficiently through solids and liquids than through gases. Bathtubs are connected directly to the solid structure of the house, and their water pipes also serve to transmit sound from points far away.

**10.** Because of the Doppler effect, a bystander on the ground in front of the motorcycle would hear a higher frequency sound from the horn than the horn actually makes. That sound wave would travel on to reflect from the end of the canyon and return to the bystander, who will hear that higher frequency once again. However, the rider doesn't hear what the bystander hears in either instance. The rider initially hears the sound of the horn's own frequency, since the relative speed between them is zero. Later, the rider hears the reflected sound from the wall of the box canyon. Because the rider is moving into that reflected sound wave, the rider hears a frequency of sound that is even higher than the frequency the bystander hears.

**11.** Although all three instruments produce sound from vibrating strings, harps lack the thin, resonant sounding boards used in guitars and violins to produce a louder sound.

**14.** The law of conservation of energy tells us that the amount of energy the air receives from the plane moving through it can be no more than the amount of energy the plane loses. Because the air is thinner at high altitudes, the drag on the supersonic plane is less, so it loses less energy per second, and its shock wave carries away less energy per second.

**16.** Softer feathers make less noise when pushed through the air than hard feathers do; harder feathers create more and larger eddies in the air. Quieter flight helps the owl surprise its prey.

**17.** It's true that no sound can move through a vacuum, but the connected solids in the walls would carry sound anyway. More to the point, the enormous pressure of the air on each side of the wall would collapse it if it were made from ordinary building materials. Air pressure, remember, is more than a ton per square foot.

**19.** Much of the energy of the sound waves is reflected at the door's solid surface.

**21.** In Chapter 16 we saw that water waves are dispersed more by objects whose physical size is close to a wavelength. Sound waves behave the same way. The tiny pine needles scatter and reflect the high-frequency (short-wavelength) portion of a sound wave best, while the low-frequency portion passes by with less backward scattering.

**23.** Sound travels at over 300 m/s, so the deer can hear the twang of the bow string in only ⅓ s. At that instant, the arrow has traveled only 20 m. The deer has ample time to bolt before the arrow arrives.

**24.** Previously, the sound of the approaching boat reflected from the solid expanse of the rock. Experienced captains used these sounds to judge distance and position when visual conditions were not good.

**26.** Colder air is more dense and thus more viscous; the eddies that cause sound commence at lower speeds for cold air moving past a sharp corner than they would for warm air.

**29.** The space shuttle makes two distinctly separate sonic booms because of its large size. The nose and tail shock waves are created about 122 ft apart. Sound travels at approximately 1100 ft/s, so the booms are roughly 0.1 s apart.

**32.** The water that has been pushed aside by the propellers quickly collapses into the cavity that is formed. The sound comes from the impact of the inward-rushing water as the cavity fills.

**33.** (See Exercise 31 for inspiration.) It probably hears that frequency best. Sound at that frequency evidently excites a standing wave with large amplitude in the bat's ear cavity. This increases the response of the ear at that frequency and enables the bat to detect smaller intensities of sound at that frequency.

**35.** As the wave nears the ground, it picks up speed and refracts, bending forward. An observer on the ground hears the sound somewhat earlier than if the wave did not refract.

**36.** $f = v/\lambda$, so $f = (330\text{ m/s})/0.2\text{ m} = 1650\text{ Hz}$.

**37.** 10,500 revolutions per minute $= (10{,}500/60)$ Hz $=$ 175 Hz, a sound with a pitch between that of a bumblebee and that of a honeybee.

**38.** Sound waves are scattered best by objects about one wavelength in size, so $\lambda = v/f = (340\text{ m/s})/120{,}000\text{ Hz}$

$= 0.0028$ m $= 0.28$ cm. Insects about ¼ cm in size scatter (or reflect) such sound well.

**39.** $\lambda = v/f = (330 \text{ m/s})/20{,}000 \text{ Hz} = 1.65$ cm; $\lambda = (330 \text{ m/s})/20{,}000 \text{ Hz} = 16.5$ m.

**40.** Ten times farther away, the intensity drops by a factor of $(10)^2$, or 100. 90 dB corresponds to $10^{-3}$ W/m$^2$, and ten times farther away from the source, the sound intensity is $10^{-5}$ W/m$^2$, or 70 dB.

**41.** 100 W/($10^{-4}$ W/person) $= 10^6$ people; a million people talking emit sound energy per second equal to the heat and light energy given off by a 100-W light bulb each second.

**42.** A difference of 10 dB, as seen from Table 17–1, amounts to a factor of 10 in the intensity of the sound. (With respect to the *loudness* of the sound as perceived by the human ear, however, a 10-dB increase means the sound is only *twice* as loud. This should be sufficient.)

## Chapter 18

**2.** Dust has a huge amount of surface area per unit of volume or mass, and burning takes place at the surface. That's where the air's oxygen can react chemically with the material. An explosion occurs when the burning proceeds at a very great rate, which is due to the great surface area.

**4.** See Table 18–1. Vinyl and other hard plastics have large electron affinities. When the record disk (large surface area) is pulled from its plastic cover, there is a great deal of friction. As the needle tracks the groove of the record, there is more friction. In these ways, the disk becomes charged and attracts dust particles in the air. Wiping only increases the static charge, and the dust continues to cling to the surface.

**6.** As shown in Table 18–1, the charge is most likely negative. The chains on older gasoline trucks grounded the truck body, letting the charge that could cause a dangerous spark move off to the ground.

**8.** About 20 percent of lightning strokes occur between the positive area in a cloud and the earth. Positive charge isn't brought to earth, however; instead, electrons move from the earth to the positive region in the cloud. (The positive ions in a lightning bolt are thousands of times more massive than electrons and accelerate and move proportionally less than the electrons do.)

**10.** Normal matter always attracts other normal matter with the gravitational force. Only if another type of matter occurred that repelled normal matter could gravitational forces be balanced by others. Antimatter, discussed later in this text, may be repelled by normal matter. But experiments to create it and observe its attraction or repulsion for the earth have been inconclusive. The problem is that when particles of antimatter come into contact with particles of normal matter, they annihilate each other.

**14.** The electrons in the neutral atom shift toward the charge if it is positive or away from it if it is negative. This leads to an attraction by way of induction.

**16.** See the answer to Exercise 10. Normal matter is only attractive; there is no second type of matter that provides repulsion. With no second type of mass, there can be no mass dipole with one attractive "pole" and one repulsive "pole."

**19.** A beam of electrons in the tube excites the material on the screen and causes it to glow. The screen is negatively charged while in operation. The paper becomes charged by induction and sticks to the screen.

**22.** The electrons' strong electric attraction to the positive ions they would leave behind prevents the electrons from rushing to the surface because of the electrons' mutual repulsion.

**24.** It will be one-fourth as great. $1/(2d)^2 = 1/(4d^2)$.

**25.** The charge in shoes and wood would be $1/10^{10}$ as great. Since the charge appears twice and is multiplied in the formula for the force, the final acceleration of Phate is $1/10^{20}$ smaller than before, or $1.3 \times 10^5$ *g*.

**26.** The negative charge of an electron is about $1.6 \times 10^{-19}$ C. Dividing this into a coulomb gives about $6.2 \times 10^{18}$ electrons.

**27.** $E = q \times V = 0.001 \text{ C} \times 1.5 \text{ V} = 0.0015 \text{ J}$

**28.** $E = q \times V = (1.6 \times 10^{-19} \text{ C}) \times (3 \times 10^4 \text{ V}) = 4.8 \times 10^{-15}$ J. The energy $E$ will be in the form of kinetic energy, $\frac{1}{2}mv^2$. So $\frac{1}{2}mv^2 = 4.8 \times 10^{-15}$ J. $v^2 = 2E/m$, and $v$ is about $1 \times 10^8$ m/s, whereas $c = 3 \times 10^8$ m/s. (For speeds close to the speed of light, the formula $E = \frac{1}{2}mv^2$ is not precisely accurate. We'll see in Chapter 26 that the mass itself changes, but the effect in this example is very small.)

**29.** Energy $= V \times q =$ force $\times$ distance $= F \times d$. Or $V/d = F/q = \mathscr{E} = 10^8 \text{ V}/10^3 \text{ m} = 10^5 \text{ V/m} = 10^5$ N/C.

**30.** $\mathscr{E} = V/d = 1.5 \times 10^4 \text{ V}/0.01 \text{ m} = 1.5 \times 10^6 \text{ V/m} = 1.5 \times 10^6$ N/C.

**31.** $P = E/t = V \times q/t = 10^8 \text{ V} \times 25 \text{ C}/0.005 \text{ s} = 5 \times 10^{11} \text{ W} = 5 \times 10^8$ kW.

## Chapter 19

**3.** Yes; twice the cross-sectional area in the straw or the wire allows twice as much milkshake per second or twice as much charge per second to flow, as the case may be. Whereas the milkshake rises in the straw because of a difference in pressure, the charges flow in the wire because of a different electric potential, or a voltage difference. In either case, twice the cross-sectional area means that the flow is twice as great.

**4.** A short circuit would occur as a large current flowed through the metal of the screwdriver, which offers little resistance. This would cause rapid heating of the screwdriver. The battery has no circuit breaker, so the current would continue to flow until the circuit was broken or the battery was dead.

**6.** A current will flow only if there are charges that can move, as there are in a conductor. In an insulator, however, the charges of the atoms or molecules are too strongly held together for relatively small differences in electric potential (hence electric force) to cause them to move.

**7.** The wire would conduct a larger current if connected to the wall socket. $i = V/R$, and $V$ is 110 V for the home circuit but much less for the flashlight battery. The large current would produce a greater rate of heating ($P = i^2R$), and the wire would melt.

**12.** If the current is 60 hertz AC, the electrons oscillate or change directions 60 times per second.

**14.** The 5-ohm bulb on both counts. Less resistance means the bulb draws more current, heats more (becoming brighter), and drains the battery faster.

**17.** If a long and a short inclined plane are used to push an object up the same height, equal amounts of work are done, and in each case the object gains the same potential energy. Since $W = F \times d$ = change in PE for the object, the longer plane (larger $d$) requires a smaller force. When a long wire and a short wire are connected to points with the same change in electric potential energy, the same work is done on the charges in either wire as they are pushed between the terminals. The equation is: change in electrical energy $= W = F \times d$, or change in electrical potential energy/per unit of charge $= F/q \times d = \mathscr{E} \times d$ = voltage difference. The longer the wire $(d)$, the smaller the electric field ($\mathscr{E}$) on the charges in the wire.

**20.** The 60-Hz current passes through zero twice during each oscillation, once at each turning point. But those 120 instants of zero heating don't affect the intensity of light from an incandescent bulb, since the filament can't cool instantaneously. Indeed, the relatively slow cooling rate of the filament ensures that the emission of light is very steady through each second despite the rapidly changing current.

**21.** The electric potential of a battery of cells in series is the sum of the electric potential of the cells. Each flashlight cell gives a potential difference of 1.5 volts, so 1.5 volts/cell × 73 cells gives about 110 volts of potential difference. So you should use an ordinary incandescent bulb that you would use in a lamp connected to an ordinary household circuit.

**22.** 4; one watt is one joule per second.

**23.** Using $P = V^2/R$, we have $R$ = 240 ohms.

**24.** The higher voltage gives a much greater rate of heating for the same resistance, as seen through the formula $P = V^2/R$ (or even $P = i^2R$, since a higher voltage difference draws a greater current).

**25.** $P = iV$. The 1-W bulb draws a current of ⅓ A. The 40-W bulb draws a current of 40 W/115 V, or 0.35 A. The currents are approximately the same. Why, then, does the 40-W bulb emit so much more light and heat? The resistances are different. The resistance of the 1-W bulb is 9 Ω (see Example 1 in "Calculations"). The resistance of the 40-W bulb is 115 V/0.35 A = 329 Ω. And $P = i^2R$.

**26.** Current = charge/time = 2 C/10 s = 0.2 A.

**27.** $P = iV$, so 300 W = $i$ × 115 V, and $i$ = 2.6 A.

**28.** (a) $V = iR$, so $i$ = 6 V/30 Ω = 0.2 A; (b) $P = i^2R$ = 0.04 A × 30 Ω = 1.2 W; (c) $E = P \times t$ = 1.2 W × 60 s = 72 J.

**29.** The total power used in the circuit is (2 × 1100 W) + 1300 W + 600 W = 4100 W. $P = iV$, so $i$ = 4100 W/115 V = 36 A. The wiring could overheat, since a single wire in the circuit would carry the 36 A, whereas each appliance connected in parallel to that wire would draw only its normal current, much less than 36 A apiece. A 30-A circuit breaker would break the circuit under this load.

**30.** The two headlamps draw 6 A, so a fully charged 80-A-h battery would discharge in 80/6 h, or about 13 h.

**31.** $E = P \times t$ = 160 W × 24 h; 3840 Wh = 3.84 kWh, which is greater than 3.5 kWh.

**32.** If the power company charges 10 cents per kWh, then the cost of 3.5 kWh is 35 cents. Check the cost of, say, rice and beans that account for 3000 Calories. Chances are you will come up short. Today few students spend only 35¢ per day for food.

**33.** (a) $2.37 \times 10^{12}$ kWh/$2.32 \times 10^8$ persons = 10,216 kWh/person (for a year); (b) 3.5 kWh/day × 365 days = 1280 kWh (per person for a year); (c) 10,216 kWh/(1280 kWh/person) = 7.98 persons.

## Chapter 20

**2.** Along the line directly between the north and south poles, the magnetic lines of force point straight from the north pole to the south pole.

**3.** Find the two bars that behave like magnets, such that either end of one bar attracts one end of the other bar but repels the opposite end. The third bar is the nonmagnetic bar. Either of its ends would be attracted to any end of the magnetic bars by induction.

**8.** Angling the bar downward aligns it more nearly with the earth's magnetic lines of force. Its magnetic domains will, therefore, align all along the length of the bar and make it a stronger magnet.

**9.** It will be near the base. Remember that the earth's north magnetic pole is actually the south pole of a magnet.

**12.** We cannot notice or feel a change in the gravitational force because we cannot vary our distance $d$ to the center of the earth by a large percentage. We can, however, vary the distance $d$ between two small magnets.

14. Magnetic induction. The clapper is attracted to the energized electromagnet.

16. Near the magnetic equator, roughly the equatorial regions of the earth, where the magnetic field has little or no dip.

18. Greater; it's actually about 72°. (Also see Fig. 20–13.)

20. The temperature beneath the earth's surface increases with depth (see page 216). Only 100 miles deep, the temperature is 1500°C, well past the Curie point of iron that is not under pressure.

22. A varying electric field was sent through the coils, creating an oscillating magnetic field that could disalign the magnetic domains. The "permanent" magnetic field was eliminated by trial and error. This process was called "de-perming" the ship.

23. The earth-induced magnetic field can be canceled by looping a few copper wires around *inside* the ship's hull and passing a tiny current through these coils. Then when the ship turns to travel in the opposite direction, the direction of the current in the coils is reversed. The change in the direction of the magnetic field cancels the induced field once again.

25. A current in a solid always encounters resistance (with the exception of very low-temperature superconductors, unknown in Ampere's day). Such resistance to a current always causes heating. But current circulating entirely within atoms or molecules wouldn't necessarily have such resistance, Fresnel presumed. This was an interesting theory, long before the days of the discoveries of electrons, protons, and neutrons and any model of the atom that had charges circling other charges, which constitute current loops.

27. A horseshoe magnet is like a bar magnet that has been bent to bring the poles closer together. The closer the poles, the more intense the magnetic field between them will be. Also, the net magnetic field in the region of space between the closer poles is more uniform. The advantage of large magnets with close poles, then, is their almost uniform, intense magnetic field in the region between their poles.

29. If $D$ is about $2d$, the repulsive force on the south pole of $B$ is about *half* of the attractive force on the north pole of $B$. To see this, make a drawing and redo the calculation. The net force on $B$ is thus considerably less than the value for a pure $1/d^2$ force between the closest poles.

## Chapter 21

2. It made DC. The electrons in the copper disk crossed the magnetic field in only one direction; hence they were pushed in only one direction.

4. They would have had no effect. A sketch of such wires with their magnetic fields shows that the electrons in each wire travel parallel to the magnetic lines of force.

7. Nothing visible goes on because the copper atoms in the wire don't travel; about one electron per atom does the moving. Yet these electrons represent a lot of energy because of the strength of the electric force. Because those moving electrons oscillate back and forth rather than travel in only one direction, the net flow of electrons through an end of a wire is zero.

8. Yes, it could. A transformer only requires the magnetic field to change across the secondary coil. For instance, if AC flows through the primary coil, the magnetic field across the secondary one oscillates in direction and intensity with that current. On the other hand, if a steady DC flows, the magnetic field across the secondary coil is steady. To use a transformer with direct current, the DC can be regularly interrupted (with an oscillating off–on switch, for example). The DC then changes, and so does its magnetic field. A *changing* DC in the primary coil does generate a voltage across the secondary coil, and current will flow if there is a closed circuit.

9. The magnetic fields of the electromagnets exert only a turning force on the electrons, doing no work on them to change their speed. The electrons are deflected more if the intensity of the magnetic field is greater. By varying the strength of the magnetic field, the beam may be guided to the desired points on the screen.

11. "Transported" implies something material moves from one place to another, and that's not so in this case; the electrons only oscillate back and forth. However, energy is transmitted to points along a conducting wire from the source of the voltage difference across the wire.

16. Downward, toward earth. This isn't terribly easy to visualize; think of rotating Fig. 21–1 toward you by 90° so that the magnetic field lines point straight up. Then compare it to the left side of Fig. 21–4.

18. When the tape to be played passes through the magnetic head, the magnetic field of the tape induces in the head a current that can be amplified electronically to reproduce the original sound.

21. Mercury, Venus, and Mars do not have radiation belts because they have little or no planetary magnetic fields. Jupiter and Saturn have strong magnetic fields and intense radiation belts.

22. Read Exercise 8 and its answer. A switch interrupts the DC voltage, turning it off, then on, then off, and so on.

25. No, the path is not without interruption. Transformers are used at many points along the way, and their primary and secondary coils of wire are not connected to each other. They merely share the same magnetic field.

26. $N_p = 5$, $N_s = 3$, and $V_p = 110$ V, so $V_s = 66$ V.

27. $N_s/N_p = V_s/V_p = 15{,}000/12 = 1250$

28. Assuming the power out is equal to the power in, we have $i_pV_p = i_sV_s$ and $i_s = i_p(V_p/4V_p) = \frac{1}{4}i_p$.

## Chapter 22

2. Ultraviolet rays, X rays, green light, infrared waves, microwaves, AM radio waves.

4. The reason lies in the fact that an electron and a proton respond differently to equal forces. Because of its smaller mass, an electron has an acceleration nearly 2000 times as great as a proton that is experiencing the same force. The electric field of the electron, then, becomes much more distorted, and the resulting traveling kinks in the field lines represent the light that we see from the candle's flame.

5. TV signals, which are in the microwave region of the electromagnetic spectrum, pass through the ionosphere rather than reflect from it. Therefore, TV reception depends more on straight-line paths from the transmitter to the receiver than AM radio waves do.

9. Microwaves, like visible light, interact much less with the ionosphere, whereas ordinary radio waves reflect well from the ionosphere. Radio-wave reflections would cause interference in the space-to-earth (and earth-to-space) communications.

14. The beams on the moon would be invisible to an observer there unless they were aimed directly at the observer. There is no air or water vapor on the moon to scatter the light from a searchlight.

16. Cold air emits less infrared radiation than warm air. Cold air (higher pressure) seeps from the cave's entrance into the warmer air (lower pressure) outside, especially when a low-pressure front moves in. A photograph taken with a special camera that uses film sensitive to the infrared wavelengths will reveal a cave's entrance as a dark area because the cool air emits less infrared radiation than the surrounding air does.

17. The planet Venus is covered by clouds. Its surface gets almost no direct sunlight; the light it receives at the surface is thoroughly diffused by its dense atmosphere and comes in at all angles from the sky. Without direct beams of light, there are no sharp shadows on Venus, just as there are none on earth on a cloudy day.

18. The inner surface of a person's clothes may reach the surface temperature of the skin. The outer surface, however, will be closer to the air's temperature and thus will emit less infrared radiation per square centimeter than the skin beneath it does.

21. Microwaves (TV signals) and radio waves travel over 186,000 miles in the time it takes sound to travel about 1000 yards. There was ample time for microwaves and radio waves carrying the amplified signals from microphones to reach the United States before the sound waves traveling through air reached the local observers.

23. They scatter all wavelengths of light, so they diffuse the light to some degree, depending on their density in the air, thus giving the air a grayish-white appearance. The greater the concentration of the particles, the more opaque the air appears, as you know if you've seen smog.

25. (186,000 mi/s)/24,900 mi = 7.47 times per second.

26. The wavelength of that green light is $\lambda = c/f = (3 \times 10^8 \text{ m/s})/(5.5 \times 10^{14}/\text{s}) = 5.5 \times 10^{-7}$ m. That wavelength divided by the diameter of a hydrogen atom is about 5500. Therefore, some 5500 hydrogen atoms would fit across one wavelength of green light.

27. $\lambda = c/f = (3 \times 10^8 \text{ m/s})/(54 \times 10^6/\text{s}) = 0.056 \times 10^2 \text{ m} = 5.6$ m.

28. (a) $\lambda = c/f = 5 \times 10^6$ m, or 5 million meters. (b) Such long-wavelength waves carry very little energy.

29. The 108-MHz FM waves have the shortest wavelengths, 2.8 m, and the 88-MHz FM waves have the longest wavelengths, 3.4 m. A 550-kHz AM wave has a wavelength of 545 m, while a 1500-kHz AM wave has a wavelength of 200 m.

## Chapter 23

5. By coming from the direction of the sun, the kingfisher has two advantages. First, its prey will have poor visibility in the kingfisher's direction, since the direct sunlight is so intense and blinding. Second, this technique keeps the kingfisher from facing the glare of the sun on the water (except at high noon).

7. Backpackers on a mountain see the setting sun for a longer period of time than they would at sea level. (Draw a simple sketch to convince yourself; also recall the story of the balloonist on page 249 who saw two sunsets on the same day.) Seen from their elevation, the sun sets at a lower angle than horizontal, and the rays coming from the setting sun take a longer path through the densest part of the atmosphere than do the corresponding rays for observers at sea level. The setting sun thus appears both flatter and more crimson from the mountaintop than it does for observers at sea level.

12. It's simple! Keep one eye shut while the lights are on and open this eye as you turn them off.

13. Even with their night vision working full time, there is no light at all for the visitors to see, so the cave remains pitch-black.

15. The sun must be near sunrise or sunset.

18. The frosted bulb diffuses the light. Diffuse light is much more pleasant than light directly from the filament for two reasons. First, if you happen to look at the frosted bulb, the light is less intense than if you looked directly at the filament. Second, the shadows cast by a frosted bulb

are not as sharp as those from a clear bulb, making the light seem less harsh, or softer. That is, there is less stark contrast between lit and unlit areas. If the frosting is inside the bulb, it won't collect dust so easily, and the smooth outer surface is easier to clean than a frosted surface would be.

**19.** Movable electrons at the surface of the metal oscillate with the light's incoming electric field, absorbing its energy and reradiating it backward. Those same electrons act to cancel any electric field inside the metal, so the light can't penetrate the metal very far at all. Light penetrates farther into other materials with less mobile electrons and is absorbed to a much larger extent than with metals.

**20.** Visible light has wavelengths that are thousands of times longer than a molecule of silicon dioxide ($SiO_2$, glass). For example, see Exercise 26 in Chapter 22. The greatest scattering of light occurs when the irregularities are on the order of the size of a wavelength, as we saw in Chapter 16 for waves in general. Hence, the irregularities of about 20 atomic diameters hardly scatter the light striking it at all, and the mirror forms a very clear image.

**22.** Graphite is very soft, and maintaining the flatness of a single plane of atoms for areas large enough to make a macroscopic mirror would be difficult. A line of graphite a single centimeter long contains about $10^8$ atoms, as we saw in Chapter 9. Moreover, the graphite would have to be evenly coated with a metal in order to reflect light as well as possible (see Exercise 19).

**24.** The rays of light refract while they travel through the lens and do not come to a focus until they are well outside the lens. Only then is the light energy delivered per cubic centimeter per second great enough to weld and drill the metal parts.

**25.** Light from the moon's surface travels with a speed of $3 \times 10^5$ km/s $= d/t$. So $t = 384{,}000$ km/($3 \times 10^5$ km/s) $= 1.3$ s. The image of the moon is a little over a second old when you see it.

**27.** $\lambda_{diamond}/\lambda_{air} = 0.42$. However, a careful check of Fig. 23–30 reveals that $\lambda/\lambda_{air}$ is about 0.5. The transparent medium in the figure is not diamond because the wavelength of the transmitted light is not quite small enough.

## Chapter 24

**2.** (a). Photons of higher frequency light carry more energy than those of lower frequency light. As they pass through a gas, its electrons interact with the photons individually, not collectively. As a result, high-frequency light ionizes best, even at lower intensities (photons/second).

**4.** By performing a spectrographic analysis of the paint, a technician can tell its molecular (and atomic) composition. Molecules, as well as atoms, have definite frequencies (and even bands of frequencies) of light that they absorb and emit well. Identifying those frequencies with a spectrograph identifies the particular compounds used to make the paint and their relative proportions. The compounds and their proportions differ from manufacturer to manufacturer even for the same color of paint.

**6.** It would be a continuous spectrum where all frequencies are present.

**8.** The identical spectral lines come from the gas molecules in the earth's atmosphere. The light of every star we see passes through these gases on the way to us; therefore, these absorption lines appear in every spectrum of starlight. (They are called *telluric* absorption lines.)

**10.** The region above the surface of the sun, the corona, contains gases. As the bright sunlight passes through these gases, they absorb photons, causing the absorption lines in direct sunlight. But if during a solar eclipse the direct sunlight is missing and the corona is visible, the light gathered by a telescope produces the emission spectrum of the gases in the corona. The gases emit photons of the same frequency that they previously absorbed (and are still absorbing) from the sunlight passing through them.

**11.** Those glowing rocks emit visible light. However, an infrared lamp cannot excite their fluorescent atoms enough to cause them to glow, because visible photons contain more energy than infrared photons. Therefore, the lamp must be ultraviolet, and the minerals must fluoresce as the electrons return to the ground state in several stages, emitting photons of less energy than the ultraviolet photons they absorb. Those emitted photons have frequencies in the visible range.

**12.** A wave scatters best from imperfections that have dimensions of about one wavelength. Therefore, short-wave sound waves can reveal very small imperfections as they scatter from them.

**14.** Under these circumstances, sunlight would not contain all the frequencies of light as it now does (with the exception of the absorption lines). We would see only the spectral emission frequencies of the gases at the sun's surface, mostly those of hydrogen and helium.

**17.** The atoms in the filament of an incandescent bulb emit photons when they absorb enough energy from high-speed collisions with electrons. For every such high-energy collision, the electrons have undergone more lower-speed collisions with atoms. The collisions heat the filament tremendously, so that it loses a great deal of energy as heat. Fluorescent light occurs when the fluorescent atoms absorb photons of higher-than-visible energy, or ultraviolet photons, from the highly excited mercury vapor. They fluoresce by giving up their absorbed energy during several transitions, rather than by a single transition, emitting this absorbed energy as lower frequency photons. No random collision process is involved with the fluorescent atoms as it is with the atoms in the filament of an incandescent bulb, and much less energy is lost as heat.

**19.** The process is much the same as that which takes

place in a fluorescent lamp. Certain minerals in the material absorb the ultraviolet photons and emit their energy with two or more smaller energy (lower frequency) photons. At least one of the lower frequency photons is in the visible portion of the spectrum.

**21.** No; the photons of red light don't have enough energy to excite those atoms to the point where they make the higher energy transitions that are necessary to emit green light.

**22.** The low-pressure sodium vapor light generally emits light at two frequencies. Using spectrographs, astronomers can still observe starlight in the other portions of the spectrum. However, the high-pressure lamp emits a spread of frequencies due to collision broadening of the spectral lines. This causes more widespread interference with the spectrum.

**23.** Only 7 watts is carried away by visible photons; such a photon's energy is $hf = (6.6 \times 10^{-34}\ \text{J}\cdot\text{s}) \times (5 \times 10^{14}/\text{s}) = 3.3 \times 10^{-19}$ J. So in one second, $7\ \text{J}/(3.3 \times 10^{-19}\ \text{J}) = 2.1 \times 10^{19}$ photons are emitted by a 100-watt bulb.

**24.** $10\ \text{cm}/380{,}000\ \text{km} = 0.1\ \text{m}/(3.8 \times 10^8\ \text{m}) = 1/3.8 \times 10^9$ = about one part in $10^9$.

**25.** Down and back, the photon travels 20 m with a speed of $0.75c$. Its time in the water is $t = d/v = 20\ \text{m}/(0.75 \times 3 \times 10^8\ \text{m/s}) = 8.9 \times 10^{-8}$ s.

## Chapter 25

**1.** (a), (b), (d), (e).

**5.** No; it is a pointlike particle that has a wavelike nature in its behavior.

**7.** $\Psi$ is not a physical wave, and it has no medium through which it moves. It is a wave whose amplitude helps to predict probabilities for the actions of small particles.

**8.** The probability waves for atoms whose electrons are in ground states have constant frequencies and amplitudes, not decaying in time as waves on a clothesline would. If this were not true, all atoms would shrink in time. Also, and more fundamental, if the $\Psi$ waves for particles did shrink and die out, the probability for finding those particles in the universe would diminish as well. There is no evidence to suggest that particles are disappearing from the universe without a trace.

**10.** If its speed, and hence its momentum, was known precisely, its uncertainty in position would have to be infinite, that is, you could have no knowledge of its position. That is because $\Delta x \geq h/\Delta mv$, and when $\Delta mv = 0$, $\Delta x$ becomes infinite.

**11.** Here's one such example: Suppose a bullet of mass $10^{-2}$ kg was shot from a horizontally aimed gun whose bore (the size of the hole in the barrel) was 1 cm, or $10^{-2}$ m. When the bullet left the barrel, you'd know its *vertical* uncertainty to within $10^{-2}$ m. If $h$ had a value of $10^{-3}$ J·s (rather than $6.6 \times 10^{-34}$ J·s), the uncertainty in the bullet's *vertical* momentum would be $\Delta mv = h/\Delta x = 10^{-1}$ kg · m/s. Since $m = 10^{-2}$ kg, $\Delta v = 10$ m/s for the bullet. If the bullet traveled 300 m in ⅓ s, that uncertainty in its vertical speed would mean it could be 3.3 m (11 ft) higher or lower than when it was aimed, even after considering the effect of gravity, which would drop the bullet only 0.55 m (1.8 ft) over that distance. Diffraction of particles would indeed play a larger role in our environment.

**14.** As the thermal motion of an atom decreases, its position becomes more certain. But it can't become completely still, that is, with no $\Delta x$, no uncertainty in its position, because then $\Delta mv$ would become infinite. By the uncertainty principle, therefore, each atom, even at absolute zero, has motion. This is called *zero point* motion, and the atom cannot give that motion up as heat energy. $\frac{1}{2}mv^2 = \frac{3}{2}kT$, as discussed in Example 4 of Chapter 13, is not correct as the Kelvin temperature approaches zero.

**15.** If the electron distribution around the nucleus is compressed, the electrons' positions are known to within a smaller $\Delta x$. That means their uncertainty in momentum, $\Delta mv$, must increase. If the value of $v$ increases, the kinetic energy increases. This means the electrons are more likely to pull farther away from the nucleus, increasing rather than decreasing the volume. In this way the uncertainty principle ensures that the electrons in atoms cannot collapse into the nucleus, even though the nucleus attracts them.

**16.** The conductivity goes up (and the resistivity goes down) as the temperature decreases. That is because the ions in the copper wire vibrate less and their positions are more certain, making the lattice more regular for the passing electrons. (At several degrees Kelvin, the resistivity for some substances actually becomes zero. The explanation for the behavior of the electrons in these so-called *superconductors* is too complex to discuss here.)

**17.** With wavelengths equal to $\frac{1}{25}$ of an atomic diameter, such electrons can form a crude image of atoms in a device called an electron microscope.

**18.** See Exercise 11, and then calculate, for example, the uncertainty in the sideways speed of your car as it emerges from a narrow tunnel at 55 mph if $h = 1.0$ J · s.

**19.** $\lambda = h/mv = 6.6 \times 10^{-34}\ \text{J}\cdot\text{s}/(20\ \text{kg}\cdot\text{m/s}) = 3.3 \times 10^{-35}$ m. This wavelength is incredibly tiny, $10^{-17}$ times smaller than the diameter of a single atom. Diffraction of Lance, for example, would be impossible to observe because Lance's size is so great compared to his wavelength. Diffraction occurs best when the wavelength is on the order of the size of the opening a wave passes through. When Lance ran through a doorway, the probability for observing a diffraction of Lance is essentially zero, because Lance wouldn't pick up observable sideways momentum just by passing through the door.

**20.** $\lambda = 6.6 \times 10^{-34}$ J · s/(9.1 × $10^{-31}$ kg × 4.8 × $10^5$ m/s) = 1.5 × $10^{-9}$ m.

## Chapter 26

**2.** True.

**6.** (b). Because of the moon's motion around the earth, the clock there would be ticking slower than the observer's clock here on earth.

**9.** Yes, but only while the gold was in motion. If he stopped it to cash it in, the mass of the gold would become the same as it was before it was launched.

**10.** (a). As seen from the ground, a moving clock runs slower. Likewise, so would all physical processes. The heartbeats, as monitored by someone on the ground, would take place at a slower rate.

**11.** The distance between marks on the side of the boxcar would be less than 5 meters, which was the distance between the boys' positions.

**12.** (a), (c), (e).

**15.** (a). Because of its thermal energy, the egg's molecules are in more rapid motion.

**16.** $t = d/v$ = 5 km/(3 × $10^5$ km/s) = 1.67 × $10^{-5}$ s.

**17.** $E = mc^2$ = 0.001 kg × (3 × $10^8$ m/s)$^2$ = 9 × $10^{13}$ J.

**18.** $E = mc^2$ = 0.001 × 1 kg × (9 × $10^{16}$ m$^2$/s$^2$) = 9 × $10^{13}$ J.

**19.** $l = l_0 \times \sqrt{1 - v^2/c^2}$ = 8.8 LY × 0.014 = 0.12 LY.

## Chapter 27

**4.** Radioactive decay belongs to the arena of modern physics. With the laws of classical physics, motions and events were in principle predicted precisely; the time of decay of a radioactive nucleus is unpredictable.

**6.** The pressure that would compress an atom acts on its electrons, located far from the nucleus. Even if the electrons could be so compressed that they had a significantly larger probability of being found within the volume of the nucleus, they would not greatly affect the nuclear processes, since electrons are not subject to the nuclear force.

**7.** The neutron will travel farther. Because it has no charge, the neutron is not retarded by the electrons' attractions in matter as protons are.

**9.** One of tritium's two neutrons becomes a proton when the electron is emitted. The new nucleus contains 2 protons and 1 neutron. A glance at the periodic table (Table 9–1, page 193) confirms that the daughter nucleus is an isotope of helium.

**12.** Neutron-deficient $^{22}$Na undergoes positron decay, which raises its ratio of neutrons to protons and brings it closer to the line of stability for nuclei. Neutron-rich $^{24}$Na undergoes beta decay, which lowers its ratio of neutrons to protons and brings it closer to stability.

**13.** $^{254}_{100}\text{Fm} \rightarrow {}^{250}_{98}\text{Cf} + {}^{4}_{2}\text{He}$. $^{212}_{83}\text{Bi} \rightarrow {}^{208}_{81}\text{Tl} + {}^{4}_{2}\text{He}$. $^{39}_{17}\text{Cl} \rightarrow {}^{39}_{18}\text{Ar} + e^-$ + antineutrino. $^{13}_{7}\text{N} \rightarrow {}^{13}_{6}\text{C} + e^+$ + neutrino.

**18.** A manufacturer could include traces of a substance that would leave a clear fingerprint in the gamma-ray spectra of a product. The product could be effectively safeguarded from counterfeiting, since a check with neutron activation analysis could prove whether a specimen was genuine.

**19.** No; the energy would be too high. Such gamma rays alter the nucleus as ionizing radiation alters the electron structure of the atom, by removing one or more nucleons from the nucleus. Those gamma rays are too powerful to be found in the photons released by transitions between bound states in the nucleus.

**20.** Most of the helium that is in the earth's atmosphere today comes from the alpha decays of radioactive nuclei in the earth, and the most abundant parent nucleus of those elements is uranium.

**21.** The volume of the nucleus is $\frac{4}{3}\pi r^3_{\text{nucleus}}$, and the volume of the atom is $\frac{4}{3}\pi r^3_{\text{atom}}$, so the nucleus occupies a fraction of the atom equal to $V_{\text{nucleus}}/V_{\text{atom}} = r^3_{\text{nucleus}}/r^3_{\text{atom}}$. The radii are half of the diameters, so this is the same as $d^3_{\text{nucleus}}/d^3_{\text{atom}} = (d_{\text{nucleus}}/d_{\text{atom}})^3 = 10^{-15}$.

**22.** The half-life of $^{14}$C is about 5730 years. Three consecutive half-life periods will bring a sample of $^{14}$C to one-eighth of its original value (i.e., ½ × ½ × ½ = ⅛); so the ashes are approximately 17,200 years old.

**23.** Approximately one-half is present today.

**24.** (238 − 92)/92 = 1.587, and (235 − 90)/90 = 1.60. Thorium, the daughter nucleus, has a higher ratio of neutrons to protons than its parent.

**25.** Much of the background radiation we receive does not come from cosmic rays and hence isn't necessarily dependent on elevation. The natural radioactivity of the small traces of the heaviest elements of earth and those radioactive elements created by cosmic-ray bombardment account for most of the radiation dosage we receive on earth. Hence a doubling of cosmic-ray dosage doesn't double the background radiation.

**26.** The molecules of the atmosphere stop some of the cosmic radiation before it reaches earth's surface. At higher elevations, however, the atmosphere provides less protection. The one millirem of extra exposure brought by 100 feet of extra elevation is about 1 percent of an average person's yearly background radiation exposure.

**27.** If 4 percent extra adds one millirem, then your entire food intake (100 percent) must account for about 25 millirems/year. Since the average person's background total radiation exposure is only 125 millirems/year, we literally *eat* about one-fifth of the radiation we receive each year.

## Chapter 28

**1.** Yes, a campfire is a chain reaction. For burning (oxidation) to take place between carbon and oxygen, an activation energy must initially be added from an external source, such as a match, to start the reaction. Once the act of oxidation is underway, however, it gives up much more energy that provides the activation energy for nearby atoms and molecules. If enough fuel and air are present, as they are in a campfire, a chain reaction proceeds readily.

**4.** Live wood, or seasoned wood that has been cut for only a year or so, has the same percentage of $^{14}C$ to $^{12}C$ as the atmosphere, since its carbon came from the air during photosynthesis. Therefore, when it is burned, it releases $^{14}C$ and $^{12}C$ in the same ratio as found in the air, and the carbon isotope composition of the air isn't changed.

**8.** The larger the shape, the more volume per unit of surface area and the less surface area per unit of volume. The larger volume of uranium, then, has less surface area per unit of volume or mass, and it is less likely for a given fission neutron to be lost to the outside, increasing the possibility that it will cause another fission.

**9.** The $^{235}U$ fuel in the fuel rods of a reactor is too dilute, accounting for only a few percent of the nuclei present. For a fission chain reaction to proceed fast enough to cause an explosion, the sample of $^{235}U$ has to be very pure.

**15.** Cold air is heavy and does not rise as warm air does when it is surrounded by cooler air. Consequently there is little mixing of air in the atmosphere above the Arctic and Antarctic regions. Significant convection currents don't occur there, so pollution is trapped.

**17.** If this were the only effect (and it isn't), the technique of $^{14}C$ dating would become more complicated as time progressed. It would be more difficult to distinguish between an old sample whose $^{14}C$ level was reduced by time and a newer sample whose $^{14}C$ level was reduced to begin with by the smaller concentration of $^{14}C$ in the atmosphere.

**18.** $E = mc^2$. If you weigh 130 lb, your mass is about 59 kg. $E = 59 \text{ kg} \times (9 \times 10^{16} \text{ m}^2/\text{s}^2) = 5.4 \times 10^{18}$ J.

**19.** $E = mc^2$. The mass that disappears is (3.0160 + 2.0141) amu − (4.0016 + 1.0087) amu = 0.0198 amu = $0.0198 \times 1.7 \times 10^{-27}$ kg = $3.4 \times 10^{-29}$ kg. Converting this to energy gives $E = mc^2 = 3.4 \times 10^{-29} \text{ kg} \times 9 \times 10^{16} \text{ m}^2/\text{s}^2 = 3.1 \times 10^{-12}$ J. Dividing this by $1.6 \times 10^{-19}$ J/eV, we find $18.9 \times 10^6$ eV.

**20.** $23 \times 10^6 \text{ kWh} \times 3.6 \text{ MJ/kWh} = 8.3 \times 10^{13}$ J is released by the fission of 1 kg of $^{235}U$. Since one ton of coal releases $4 \times 10^9$ J when burned, one kg of $^{235}U$ represents the energy of $8.3 \times 10^{13}/(4 \times 10^9) = 2 \times 10^4$ = 20,000 tons of coal.

**21.** The binding energy per nucleon is equal to the missing energy per nucleon in a nucleus compared to the energy of those nucleons when they are far apart. The mass of 2 neutrons and 2 protons is 4.032 amu, whereas their combined mass when together in a helium nucleus is 4.0016 amu. The mass that has disappeared represents the binding energy of the nucleus. $E_{\text{nucleon}} = E/4 = mc^2/4$, which is $\frac{1}{4} \times 0.0304 \text{ amu} \times 9 \times 10^{16} \text{ m}^2/\text{s}^2 \times 1.7 \times 10^{-27} \text{ kg/amu} = 1.2 \times 10^{-12} \text{ J} = 7.3$ MeV.

# GLOSSARY

**absolute zero** (271): The zero point on the Kelvin scale of temperature; the temperature at which the molecules of a substance have no thermal energy that can be taken from them.

**absorption spectrum** (481): A continuous spectrum with dark lines in it; a spectrum missing certain definite frequencies (or wavelengths) of light.

**acceleration** (8): The rate of change of the velocity vector for a moving body. The acceleration may be either a change in direction or a change in speed, or both.

**adhesion** (234): The molecular attraction between molecules of different kinds.

**air resistance** (69): The frictional force received when a body moves relative to still air.

**alloy** (205): A mixture of two or more metals that have been melted and combined in order to form a new material with a special property such as increased strength or elasticity or durability.

**alpha emission** (543): The emission of an alpha particle from an unstable atom.

**alpha particle** (539): A helium nucleus, composed of two protons and two neutrons.

**alternating current, AC** (387): An electric current whose direction of flow oscillates.

**amorphous solid** (203): Solids such as plastics, glass, or cement in which the arrangement of the molecules or atoms is irregular.

**ampere** (381): A unit of measurement for an electric current equal to a flow of 1 coulomb of charge per second.

**amplitude** (319): The maximum distance that a particle in a medium is displaced from its undisturbed position by a wave moving through the medium.

**angular momentum** (167): The product of the quantities of the moment of inertia and the angular speed of a body. According to the type of rotational motion, the body may have either spin angular momentum or orbital angular momentum.

**angular speed** (160): The angle swept out per unit of time during a rotation or a revolution of a body.

**antinode** (325): A point or region with maximum amplitude in a standing wave.

**aphelion** (139): The farthest point from the sun that an orbiting body reaches on its elliptical path.

**apogee** (139): The farthest point that an orbiting body reaches on its elliptical path about the earth.

**Archimedes' principle** (227): The buoyant force on a floating object is equal to the weight of the water it displaces.

**atomic mass number** (191): The total number of protons and neutrons in a nucleus.

**atomic number** (192): The number of protons in the nucleus of an atom.

**average speed** (2): The distance an object moves in a specific amount of time divided by that time.

**axis of rotation** (159): An axis about which a body rotates or turns.

**background radiation** (552): The natural radiation from radioactive elements in our environment and from cosmic radiation.

**battery** (381): A combination of two or more cells joined to produce an electrical potential difference.

**Bernoulli effect** (248): The reduction of fluid pressure at the side of a stream or current of fluid.

**beta decay** (540): The process in which either an electron or a positron is emitted from a radioactive nucleus.

**beta particle or ray** (540): An electron or a positron.

**binding energy per nucleon** (565): The average energy by which a nucleon is bound within a nucleus.

**blackbody** (482): An object or surface that absorbs all the light that strikes it and reflects none, emitting radiant energy in a way that depends only on temperature.

**Boltzmann's constant** (274): A constant of proportionality that relates matter's temperature to the average kinetic energy of its particles.

**Boyle's law** (248): If the temperature of a confined gas does not change, the product of the volume and the pressure of the gas is constant. $P_1V_1 = P_2V_2$.

**breeder reactor** (564): A nuclear reactor with the ability to produce nuclear fuel as a by-product of the fission process.

**Brownian motion** (222): The constant, random movement of particles in a fluid, caused by collisions with the fluid's molecules in all directions.

**buoyant force** (227): The net force exerted by a fluid on an object immersed in it.

**capillary action** (235): The rising or sinking of a liquid into a small vertical space due to strong adhesion between the materials and surface tension of the liquid.

**capillary wave** (328): A wave on a liquid's surface that travels with surface tension acting as a restoring force.

**center of gravity** (135): The point at which the net gravitational force seems to act on a massive object. For the earth, the center of gravity is at its center.

**center of mass** (164): The balance point for an object; the point in (or outside of) an object that responds to an external force as if all the object's mass were concentrated there.

**centrifugal force** (173): A force observed from a rotating frame of reference that acts directly outward from the axis of rotation.

**centripetal force** (173): The turning force on a moving object. It is perpendicular to the object's path and points in the direction of the center of the circle that matches the object's curving path at any instant. $F_c = mv^2/r$.

**chain reaction** (561): The automatic progression that can occur when a nuclear reaction emits radiation or particles that can trigger similar reactions in neighboring nuclei.

**charge** (361): The property of matter responsible for the electric force between electrons and protons; like charges repel while unlikes attract.

**Charles's law** (285): The relation between the volume and temperature of a gas when those quantities change without a change in pressure of the gas. $V_1/T_1 = V_2/T_2$.

**chemical electric cell** (381): An arrangement that uses chemical reactions between two substances to transfer electrons.

**chemical element** (186): The simplest form of matter; it cannot be broken down into simpler substances by chemical means.

**classical physics** (479): Usually refers to any physical laws or principles discovered before 1900. (Though a product of research from 1905 on, relativity is sometimes called classical physics because it doesn't deal with the quantum theory or the wave theory of matter.)

**coefficient of friction** (59): A measure of how much the frictional force can be when two surfaces are touching. It is the ratio of the frictional force to the contact force.

**coefficient of thermal conductivity** (279): A constant that indicates the relative rate at which heat is transferred through a substance.

**coherent radiation** (490): Radiation whose waves (or photons) are in perfect step; the crests and troughs coincide, leading to maximum constructive interference.

**cohesion** (234): The molecular attraction of like molecules for each other.

**component vector** (29): A vector that is part of a chain of vectors adding to give a net vector.

**compression** (318, 341): A region in a wave where the particles in the medium move closer together, in contrast to rarefaction, where the medium's particles spread out.

**condensation nucleus** (304): A small particle in the atmosphere on which water vapor can condense.

**cones** (470): The nerve cells in the retina of the eye that are sensitive to light's colors.

**conservation of angular momentum** (168): In the absence of external torques, the angular momentum of a system of particles or bodies is constant.

**conservation of energy** (116): Anytime matter interacts, the total energy of that matter is constant even though the energy may change form.

**conservation of momentum** (93): The net vector momentum of a system of interacting objects before the interaction equals the net vector momentum after the interaction.

**contact, or normal, force** (54): The force of repulsion that occurs when the molecules or atoms of matter are pressed together.

**continuous** (19): Without interruption, as opposed to discrete.

**continuous spectrum** (488): See *Electromagnetic spectrum*.

**convection current** (281): A current of gaseous or liquid matter that carries heat energy as the matter is buoyed up by denser fluid matter around it.

**convergence** (34): Bringing to a focus.

**converging (convex) lens** (466): A lens that is thicker at its center than at the edges; parallel light rays passing through a converging lens meet at a point on the other side of the lens.

**Coriolis force** (173): The force observed in a rotating frame of reference that tends to turn anything that is moving in that frame of reference in a direction opposite to the direction of rotation of the frame of reference itself.

**correspondence principle** (508): The statement that any two theories or hypotheses whose areas of prediction overlap should agree at those places; otherwise at least one of them is wrong.

**coulomb** (363): The metric unit of measure of electric charge. One coulomb is equal to the collective charge of $6 \times 10^{18}$ electrons.

**Coulomb's law** (363): The law that describes the attractive or repulsive electric force between two charged objects.

$$F = K \frac{q_1 q_2}{d^2}$$

**critical mass** (561): The amount of mass of a substance undergoing nuclear fission that is required for the chain reaction to become self-sustaining.

**crystal face** (204): Planes of symmetry in the crystalline structure of a solid.

**crystalline solid** (203): A solid whose molecules and atoms are arranged in a pattern that repeats throughout the solid.

**daughter nucleus** (542): A nucleus that is brought into being when a radioactive (parent) nucleus decays.

**deceleration** (9): A negative acceleration, meaning that an object's speed is decreasing.

**density** (212): The ratio of an amount of mass to the space it occupies; density = mass/volume.

**dew point** (304): The temperature at which a given parcel of air becomes saturated with water vapor.

**diffraction** (328, 345): The bending of waves that pass adjacent to solid boundaries.

**dipole** (370): A positive and a negative charge separated by a short distance; found in many molecules (called polar molecules) such as water.

**direct current, DC** (387): An electric current that flows in only one direction.

**discrete** (19): Referring to individual or distinct parts of something.

**dispersion** (464): The separation of various frequencies of light or other waves due to their different speeds in a material.

**displacement** (27): The carrying or the going from one point to another; a translation.

**diverging** (concave) **lens** (466): A lens that is thinner through the center than at the edges; parallel light rays passing through a diverging lens are spread out on the other side of the lens.

**Doppler broadening (of light)** (488): The process whereby atoms emitting light (or absorbing it) have different relative velocities and emit (or absorb) a range of frequencies rather than their discrete spectral frequencies.

**Doppler shift** (348): A change in frequency of a wave due to motion of the source of the wave or to motion of the observer of the wave.

**drag:** See *Air resistance*.

**elasticity** (54): The property of solid matter that enables it to recover its shape after being deformed. Elasticity results from the separate actions of the intermolecular bonds that pull on the atoms and molecules and the contact force that pushes them apart when they are pressed together.

**electrical circuit** (387): A conducting path for electrons between points of different electric potential.

**electrical conductor** (364): A medium through which an electric current can easily move.

**electrical insulator** (364): A material that won't ordinarily carry an electric current; a nonconductor.

**electrical resistance** (384): The measure of opposition to the flow of charge through matter; measured in ohms.

**electric current** (381): A flow of charges between two points when there is a difference in electric potential. $i = q/t$.

**electric discharge** (372): The removal of electric charge from a charged area or an electrical energy storage device, as from a lightning bolt or a battery.

**electric force** (361): The strong, long-range force arising from the property of matter called charge; exerted by charged particles on one another.

**electric force field** (369): The region of electric influence around a charged particle.

**electromagnetic radiation** (439): In classical theory, a traveling oscillation in the electric field of a charged particle; in quantum theory, this radiation is composed of photons.

**electromagnetic spectrum** (440): The complete spectrum of all electromagnetic waves, from the shortest to longest wavelengths.

**electron affinity** (362): The attraction of an atom or molecule for its outermost electrons.

**electroscope** (364): A device used to detect the presence of an electric charge.

**emission spectrum** (481): A spectrum of bright lines produced by the emission of certain definite wavelengths of light by excited atoms or molecules.

**energy** (102): A physical property of matter and radiation that when considered in all its forms is constant in quantity. Related to mass, speed, temperature, position, and to the frequency of radiation, the conservation of this quantity is considered to be one of the most fundamental properties in nature.

**entropy** (289): A quantity that is a measure of the state of disorder or randomness of a system.

**escape velocity** (140): The speed that a projectile must have in order to escape from a planet's surface without being stopped and returned by the planet's gravitational attraction.

**evaporation** (269): The process by which a substance changes from a liquid to a gas.

**fluid** (245): The state of matter in which a substance may flow; liquids and gases are fluids.

**fluorescence** (489): The light-producing process by which an excited atom loses its extra energy as it quickly moves to a lower-energy state.

**focal length** (466): The distance from the center of a lens to its focal point.

**focal point** (466): The point where parallel light rays passing through a converging lens meet; also, the point from which parallel light rays passing through a diverging lens seem to originate.

**force** (45): A push or a pull.

**frictional force** (55): A reaction force between two solids in contact, or between a solid and a fluid, that acts to retard relative motion.

**gamma emission** (540): The process in which a nucleon of a radioactive nucleus moves to a lower-energy state and emits a gamma ray.

**gamma ray** (540): A high-energy photon, released when a proton or a neutron moves to a lower-energy state in a nucleus.

**gravitational force** (133): The mutual attraction of all particles in the universe for each other according to their masses:

$$F = G\frac{m_1 m_2}{d^2}$$

**greenhouse effect** (297): The trapping of heat in a region due to the absorption of infrared radiation by certain molecules. In earth's atmosphere, carbon dioxide and water vapor are the most effective at absorbing infrared rays and holding heat in the air.

**grounding** (368): To use a metal wire or cable to the ground as a conducting path for electric charge.

**ground state** (540): The lowest energy level of an electron in an atom.

**half-life** (545): The time needed for one-half of the nuclei of a given specimen of radioactive substance to disintegrate.

**heat** (262): Thermal energy; the random energy of motion of the molecules of matter and any radiation they emit as the result of such motion.

**impulse** (86): The product of an applied force and the time during which it acts. It is equal to the change in momentum of the object on which the force acts.

**incoherent radiation** (490): Waves (or photons) whose crests and troughs come at random positions or times, leading to random interference of the light.

**induction** (366): The process of separating charge on an object by bringing it near another charged body.

**inertia** (47): The resistance of matter to any change in its velocity.

**inertial frame of reference** (517): A fixed system of coordinates from which motion can be viewed and measured; a system where Newton's first law is obeyed.

**inertial mass** (48): The property of matter that resists a change in velocity. The ratio of force to acceleration when a force acts on a body.

**infrared radiation, or rays** (440): The invisible rays of the electromagnetic spectrum that have wavelengths slightly longer than that of the visible color red but shorter than that of microwaves.

**inorganic** (203): Inanimate matter; a substance not composed of molecules made by plants or animals.

**instantaneous speed** (6): The rate of travel that matter has at a particular instant in time (or at a particular point in space).

**insulation** (279, 364): A material that resists conduction of heat or electricity or sound.

**interference (wave)** (323): The addition or subtraction of two or more waves that meet in space or in some medium.

**isotopes** (191): The various species of a chemical element that have different atomic masses due to a different number of neutrons in their nuclei.

**kilowatt-hour** (396): A unit of energy or work; equal to the amount of energy provided by 1 kilowatt of power for 1 hour.

**kinetic energy** (107): The energy of motion of a body. It is equal to one-half of the object's mass times the square of its speed. $KE = 1/2\ mv^2$. It is the measure of a moving object's ability to do work.

**kinetic friction** (55): The frictional force between touching surfaces that have relative motion.

**laser** (490): A device to produce coherent light. When light waves add coherently, their electric field amplitudes add, giving rise to larger (amplified) electric fields. *Laser* stands for *l*ight *a*mplification by *s*timulated *e*mission of *r*adiation.

**latent (hidden) heat of fusion (melting)** (266): Heat released upon freezing or absorbed upon melting that does not change the temperature of the matter.

**latent (hidden) heat of vaporization (evaporation)** (268): Heat released upon condensation or absorbed upon evaporation that does not change the temperature of the matter.

**lever arm** (161): The perpendicular distance from a force that is applied to a body and the axis about which that body might rotate.

**lift** (70): The upward force on a wing of an aircraft due to the relative motion between the wing and the air.

**light ray** (453): A thin beam of light moving on a straight-line path.

**liquid** (221): The fluid state of matter where the molecules or atoms are in contact, but where the bonding forces are too weak to hold them still with respect to one another.

**longitudinal wave** (317): A wave in which the particles of the medium move back and forth along the same direction as the wave travels.

**macroscopic** (210): Large enough to be observed by the naked eye.

**magma** (203): Molten rock in the earth's interior.

**magnetic declination** (412): The deviation of a compass needle from the true geographical direction of north.

**magnetic domains** (409): The groups of atoms that align to produce a net magnetic field in a magnetic material.

**magnetic force** (404): A repulsive or attractive force between the poles of magnets; also, the force a charged particle experiences perpendicular to its velocity as it passes the pole of a magnet.

**magnetic force field** (405): The region of magnetic influence around a magnetic pole or a moving charged particle.

**magnetic inclination or dip** (412): The angle a freely hanging compass needle suspended at its center makes with the horizon. It is the angle at which the magnetic field lines of earth dip into the earth's surface.

**magnetic poles** (404): The areas on a magnet where its magnetic force is strongest.

**mass** (48): The quantity of matter in a body. Mass is a measure of a body's inertia; that is, the mass determines how a body responds to a force that acts on it.

**metastable state** (489): An excited state in an atom where the electron doesn't promptly drop to a lower-energy level.

**microscopic** (210): Too small to be seen without a microscope.

**mineral** (203): A naturally occurring inorganic compound.

**modern physics** (479): Usually referring to the physical principles and laws discovered after 1900.

**moment of inertia** (162): The measure of a body's resistance to a change in its rate of rotation.

**momentum** (86): The product of a body's mass times its velocity.

**neap tide** (148): Twice monthly tides that are the lowest-amplitude tides, when the sun and moon are about 90 degrees apart as seen from earth.

**net or resultant vector** (27): The single vector that by itself describes the successive or simultaneous addition of two or more (component) vectors.

**neutrino** (542): A neutral subatomic particle released with the electron during beta decay of an unstable nucleus.

**Newton's law of cooling** (283): The rate of cooling of an object is proportional to the temperature difference between the object and its surroundings.

**node** (325): A point or region with no motion in a standing wave.

**nuclear fission** (558): The splitting of a nucleus into two large parts that repel one another, converting a large amount of potential energy into kinetic energy.

**nuclear force** (538): The strong, short-ranged attractive force within the nucleus that keeps the nucleons in close contact with one another.

**nuclear fusion** (566): The process in which two or more nuclei join to form a single nucleus that is more tightly bound than its components were, emitting much energy.

**nuclear reactor** (561): A device that uses nuclear fission to generate heat that may be used to generate electrical power.

**nucleus** (537): The central part of an atom having a positive charge and almost all of the atom's mass.

**nucleon** (538): A proton or a neutron; a nuclear particle.

**ohm** (385): The unit of measurement of electrical resistance; a conducting body has a resistance of 1 ohm if a potential difference of 1 volt across the body produces a current of 1 amp.

**orbital** (500): A given energy level that an electron can occupy in an atom, characterized by a distribution in space given by a standing wave pattern of probability.

**oscillate** (319, 387): To move in a regular, cyclical manner; motion from side to side or back and forth so as to complete a cycle is an oscillation.

**ozone** (243): A molecular form of oxygen containing three atoms of oxygen in each molecule instead of two: $0_3$.

***p*-wave** (332): A compressional earthquake wave. (The particles are *p*ushed and *p*ulled.)

**parallel circuit** (394): An electrical circuit that provides more than one path along which current can move.

**parent nucleus** (542): A nucleus that decays to give birth to a different nucleus, called the daughter nucleus.

**partial pressure of a gas** (244): The pressure exerted by a specific gas in a mixture of gases, contributing to the total pressure exerted by the mixture.

**Pascal's principle** (222): At any point in a fluid, pressure is transmitted equally in all directions.

**perfectly elastic collision** (89): The relative speed of the colliding objects is the same before and after the collision.

**perfectly inelastic collision** (89): The relative speed between the colliding objects is zero after the collision.

**perigee** (139): The nearest approach to earth of a body orbiting earth on an elliptical path.

**perihelion** (139): The closest point to the sun for a body on an elliptical orbit about the sun.

**phosphorescence** (489): A light-producing process identical to fluorescence except that the excited atom stays in the excited state much longer. A phosphorescent material releases its light over a longer period of time than does a fluorescent material.

**photoelectric effect** (484): The action whereby certain materials absorb light and emit free electrons at their surfaces.

**pitch** (341): The highness or lowness of a sound, depending on its frequency.

**plane** (23): A flat two-dimensional surface.

**plane polarized waves** (442): Waves whose oscillations are in a single plane.

**plasma** (202): A gas of electrically charged particles, or ions; sometimes referred to as a fourth state of matter.

**polarization** (368): The shifting action of charge in an object whereby its atoms or molecules become more negatively charged on one side and positively charged on the other side.

**polar molecule** (369): A molecule that, though neutral overall, has outer regions of positive and negative charge.

**positron** (539): A particle with the mass of an electron but with an opposite charge to the electron's charge. A positron is the electron's antiparticle.

**potential energy** (111): The energy of a body due to its position. It is a measure of the potential of a force to do work on the body.

**power** (106): The rate of doing work or expending energy. $P = W/t$, or $P = E/t$.

**pressure** (72): The force applied per unit of area. $P = F/A$.

**projectile** (138): An object that once launched proceeds in a path influenced only by gravity and air resistance.

**quantum** (483): A quantity of energy that a charged particle absorbs or emits as it radiates; symbol, $hf$.

**radiation** (281, 439): The transport of energy by wave motion, usually synonymous with electromagnetic waves.

**radioactive nuclei** (539): Unstable nuclei that emit either gamma rays or particles in the act of losing energy.

**rarefaction** (318, 341): A region in a sound wave or other compressional wave where the medium's particles are spread out, in contrast to a compression where the medium's particles move closer together.

**rate** (2): A rate is a quantity divided by a time. A rate tells how fast something happens or by how much something changes in a certain amount of time.

**real image** (455): The image produced by light rays that converge at an image's location.

**refraction** (329, 456): The change in direction of a wave caused by a change in its speed when it enters a different medium at an angle.

**relative speed** (13): The speed of an object with respect to something else.

**resonance** (325): The inducing of standing waves in matter by an incoming wave.

**revolution** (159): The motion of an object that turns about an axis external to its body.

**Roche's limit** (150): The distance from a planet within which orbiting matter cannot coalesce due to its own gravitational attraction.

**rods** (470): The nerve cells in the retina of the eye that are sensitive to light but not to light's color.

**rotation** (159): The motion of turning around an axis within an object's body.

**rotational inertia** (162): A body's resistance to a change in its rate of rotation.

***s*-wave** (332): A transverse earthquake wave. (The particles move from *s*ide to *s*ide.)

**saturated vapor pressure** (303): The vapor pressure above an evaporating liquid when equal numbers of molecules are leaving and entering the liquid's surface each second.

**secular equilibrium** (546): Whenever the amount of a radioactive isotope remains constant because nuclei are being added to the isotope at the same rate at which nuclei are disintegrating.

**series circuit** (391): An electrical circuit that provides only one path for an electrical current to follow.

**shock wave** (350): A conical wavefront that is formed behind an object that is traveling faster than wavefronts can travel in the medium.

**short circuit** (390): A path of too low resistance in an electric circuit, leading to a great increase in the amount of current the (shorted) circuit can carry.

**solid** (201): Any material whose atoms are bonded in place with respect to their nearby neighbors, giving the matter shape and strength.

**sonic boom** (350): The explosive sound produced by a shock wave from an aircraft or other object traveling faster than the speed of sound.

**sound wave** (340): A compressional wave with a frequency between 20 hertz and 20,000 hertz.

**special theory of relativity** (512): The theory developed by Albert Einstein that shows how time and length and mass differ when measured by two observers in relative motion.

**specific heat** (264): The quantity of heat that will raise the temperature of 1 gram of a substance by 1 degree Celsius.

**spectrograph** (480): A spectroscope with a camera arrangement to photograph the spectrum.

**spectroscope** (480): An instrument composed of a prism (or other dispersive device) and used to separate wavelengths of light for visual observation through a telescope.

**spectrum (of white light)** (436): The continuous rainbow spread of colors that compose the electromagnetic radiation we call white light; it contains the colors of red, orange, yellow, green, blue, and violet.

**speed** (2): The rate at which an object moves.

**spring constant** (59): The constant for a given spring that relates how great the spring's force will be per unit of stretch or compression.

**spring tide** (148): The two tides of greatest amplitude each month when the sun, moon, and earth are along a straight line.

**stable atom** (539): An atom whose nucleus is stable, that is, unchanging in time (unless interfered with by an external source).

**standing wave** (325): A wave whose pattern of oscillation is stationary rather than in motion.

**static charge** (361): The charge on a nonconducting material; charge not in motion.

**static friction** (55): The frictional force that resists motion between two touching surfaces at rest.

**stereo vision** (33): Three-dimensional perception due to using two eyes simultaneously.

**stroboscope** (11): A device that illuminates an area at very regular intervals.

**sublimation** (270): The change of matter from the solid state directly into the gaseous state.

**temperature** (262): The measure of thermal energy (molecular kinetic energy) of a substance.

**terminal speed** (10): The limit to an object's speed because of air resistance on the object. A constant speed where the pull of gravity is balanced by the air's drag.

**thermal energy** (113, 262): Commonly called heat energy; the random kinetic energy of particles of matter that give rise to the property we call temperature.

**thermal equilibrium** (284): The state where an object's temperature is constant; when an object loses just as much thermal energy each second as it gains from its surroundings.

**thermodynamics** (116, 288): The part of physics dealing with heat and matter.

**thermodynamics, first law of** (116): The total energy of a system, in all of its forms including heat, is a constant even though it may change forms and spread out, becoming even more dilute.

**thermodynamics, second law of** (117): When the energy of a system changes from other forms into heat, the energy becomes more dilute—the heat energy spreads outward, becoming less concentrated. The heat energy will never naturally become as concentrated; the quantity entropy, which measures the state of randomness of a system, always increases (or remains the same). Entropy never decreases.

**tidal bore, or bore waves** (148): Waves of water that progress over the water in a channel when high tides are funneled into the channel.

**tidal force** (145): The stretching and squeezing force that arises on a body in a gravitational field because of the variation of gravity's pulls and directions over the object's dimensions.

**torque** (161): The product of an applied force and a lever arm to an axis of rotation. A net torque on an object causes a change in the rate of rotation of the object.

**torr** (246): A unit of pressure equivalent to the amount of air pressure that supports a 1-millimeter column of mercury in a barometer.

**trajectory** (138): The path taken by a projectile.

**transmutation** (550): The changing of one chemical element into another by alterations to its nucleus.

**transverse wave** (318): A wave in which the particles of the medium move perpendicular to the direction of the wave's motion.

**turbulence** (232): Interference within a moving fluid when portions of the fluid move at different speeds and directions.

**ultrasonic wave** (341): A compressional wave with a frequency higher than 20,000 hertz, beyond the limit of human hearing capabilities.

**ultraviolet radiation, or rays** (440): Invisible rays of the electromagnetic spectrum that have wavelengths shorter than that of the visible color violet but longer than X rays and gamma rays.

**uncertainty principle** (503): The fact that the position and the momentum of a particle, especially one of small mass, cannot be known precisely at the same time due to the particle's wavelike behavior.

**unstable atom** (539): See *Radioactive nuclei.*

**vacuum** (246): A space containing no atoms or molecules.

**vapor pressure** (303): The amount of pressure the molecules of a vapor exert in a given region.

**vector** (27): An arrow used to represent a quantity with magnitude and direction.

**velocity** (29): The speed of a moving object together with its direction of motion.

**virtual image** (455): An image that is formed by light rays that don't converge at the location of the image; that is, the image formed by light reflecting from the surface of a mirror.

**viscosity** (231): The internal friction in a liquid caused by collisions of its molecules.

**volt** (371): The unit of electrical potential energy per unit of charge.

**voltage difference (also called electric potential difference)** (372): The difference in electrical potential energy per unit of charge between two points.

**watt** (388): The metric unit of electrical power; equal to 1 joule per second.

**wave** (317): A moving disturbance that carries energy through matter or space. In a material medium, a wave travels without carrying the particles of the medium with it.

**wave frequency** (320): The number of oscillations that a wave makes at a given point over a given length of time.

**wavefront** (326): The region of a two- or three-dimensional wave where the amplitude at any instant is the same everywhere, such as in an ocean wave approaching the shore.

**wave intensity** (356): The energy per second per unit area of wavefront carried by a wave.

**wavelength** (320): The distance between successive crests (or troughs) in a regular or periodic wave pattern.

**wave mechanics** (499): The theory describing the wave behavior of subatomic particles.

**wave speed** (320); The speed of crests (or troughs) of a wave pattern.

**weight** (51): The force of the earth's gravitational attraction for any object on, below, or above the surface of earth.

**weightlessness** (53): An object experiences weightlessness when it has no apparent weight relative to any other object.

**work** (102): The product of an applied force and the distance through which it is applied.

# INDEX

The following abbreviations are used in the index: *f* for figure, *n* for footnote, and *t* for table.